Methods in Cell Biology

VOLUME 42

Flow Cytometry

SECOND EDITION, PART B

ASCB

Series Editors

Leslie Wilson
Department of Biology
University of California, Santa Barbara
Santa Barbara, California

Paul Matsudaira
Whitehead Institute for Biomedical Research and
Department of Biology
Massachusetts Institute of Technology
Cambridge, Massachusetts

Methods in Cell Biology

Prepared under the Auspices of the American Society for Cell Biology

VOLUME 42

Flow Cytometry

SECOND EDITION, PART B

Edited by

Zbigniew Darzynkiewicz

Cancer Research Institute
New York Medical College
Valhalla, New York

J. Paul Robinson

Department of Physiology and Pharmacology
Purdue University
West Lafayette, Indiana

Harry A. Crissman

Biomedical Sciences Division
Los Alamos National Laboratories
Los Alamos, New Mexico

ACADEMIC PRESS

San Diego New York Boston London Sydney Tokyo Toronto

Front cover photograph (paperback edition only): FISH with cloned DNA. For more details, see Color Plate 4 (Chapter 6).

This book is printed on acid-free paper. ∞

Academic Press, Inc.
A Division of Harcourt Brace & Company
525 B Street, Suite 1900, San Diego, California 92101-4495

United Kingdom Edition published by
Academic Press Limited
24-28 Oval Road, London NW1 7DX

International Standard Serial Number: 0091-679X

International Standard Book Number: 0-12-564143-5 (Hardback)

International Standard Book Number: 0-12-203052-4 (Paperback)

PRINTED IN THE UNITED STATES OF AMERICA
94 95 96 97 98 99 EB 9 8 7 6 5 4 3 2 1

I dedicate these volumes to my friends, the late Zalmen A. Arlin and Joel Brander, who recently lost their battles with cancer. The courage they showed fighting this disease inspires us in the continuing efforts to eradicate it from the face of the earth, so that future generations may be spared the tragedies that affected them and their families. We are also inspired by the support offered by Ann Arlin, Julie Brander, and Michael Bolton, through the "This Close" Foundation for Cancer Research, Inc., established in memory of Zal and Joel.

This book could not have been prepared without the help of my dearest friend Irene Logsdon.

Zbigniew Darzynkiewicz

CONTENTS

Contents of Volume 41
Flow Cytometry, Second Edition, Part A

CONTRIBUTORS

Numbers in parentheses indicate the pages on which the authors' contributions begin.

Nicholas Ashbolt (489), Australian Water Technologies, Science and Environment, Sydney, New South Wales 2114, Australia

Kenneth A. Ault (275), Maine Medical Center Research Institute, South Portland, Maine 04106

Wolfgang Beisker (149), GSF-Forschungszentrum für Umwelt und Gesundheit, Institut für Biophysikalische Strahlenforschung, D-85758 Oberschleissheim, Germany

Francis Belloc (59), Laboratoire d'Hématologie, Hôpital Haut-Lévèque, 33600 Pessac Cedex, France

Nancy C. Bigelow (263), Department of Clinical Pathology, William Beaumont Hospital, Royal Oak, Michigan 48073

Erik Boye (477), Department of Biophysics, Institute for Cancer Research, Montebello, 0310 Oslo, Norway

Carsten Brandt (71), Institute of Human Genetics, University of Åarhus, 8000 Åarhus C, Denmark

Marianne Briffod (177), Department of Pathology, Centre René Huguenin, 92211 St. Cloud, France

Sandra A. Brockman (131), Department of Biological Sciences and Center for Light Microscope Imaging and Biotechnology, Carnegie Mellon University, Pittsburgh, Pennsylvania 15213

Wayne O. Carter (423), Department of Physiology and Pharmacology, Purdue University, West Lafayette, Indiana 47907

Sue Chow (31), Ontario Cancer Institute, Princess Margaret Hospital, Toronto, Ontario, Canada M4X 1K9

L. Scott Cram (319), Life Sciences Division and Center for Human Genome Studies, Los Alamos National Laboratory, University of California, Los Alamos, New Mexico 87545

Bruce H. Davis (263), Department of Clinical Pathology, William Beaumont Hospital, Royal Oak, Michigan 48073

Larry L. Deaven (319), Life Sciences Division and Center for Human Genome Studies, Los Alamos National Laboratory, University of California, Los Alamos, New Mexico 87545

Bruce S. Dien (457), Department of Chemical Engineering and Materials Science, University of Minnesota, Minneapolis, Minnesota 55455, *and* Institute for Advanced Studies in Biological Process Technology, St. Paul, Minnesota 55108

Frank Dolbeare (1), Biomedical Sciences Division, Lawrence Livermore National Laboratory, Livermore, California 94550

Ricardo E. Duque (231), Department of Pathology, Lakeland Regional Medical Center, Lakeland, Florida 33805

Ralph E. Durand (405, 597), Medical Biophysics Department, British Columbia Cancer Research Centre, Vancouver, British Columbia, Canada V5Z 1L3

Françoise Durrieu (59), Laboratoire d'Hématologie, Hôpital Haut-Lévèque, 33600 Pessac Cedex, France

Donald Evenson (159), Olson Biochemistry Laboratories, Department of Chemistry, South Dakota State University, Brookings, South Dakota 57007

John J. Fawcett (319), Life Sciences Division and Center for Human Genome Studies, Los Alamos National Laboratory, University of California, Los Alamos, New Mexico 87545

Emma Fernández-Repollet (605), Department of Pharmacology, School of Medicine, University of Puerto Rico, San Juan, Puerto Rico 00936

Hanne Fischer (71), Institute of Human Genetics, University of Åarhus, 8000 Åarhus C, Denmark

David W. Galbraith (539), Department of Plant Sciences, University of Arizona, Tucson, Arizona 85721

Janis V. Giorgi (359, 437), Laboratory of Cellular Immunology and Cytometry, Department of Medicine, Division of Clinical Immunology and Allergy, University of California, Los Angeles School of Medicine, Los Angeles, California 90024

Ad C. Groenewegen (371), Department of Molecular Pathology, TNO Medical Biological Laboratory, 2280 HV Rijswijk, The Netherlands

Seth P. Harlow (95), Department of Flow Cytometry and Surgical Oncology, Roswell Park Cancer Institute, Buffalo, New York 14263

Richard P. Haugland (641), Molecular Probes, Inc., Eugene, Oregon 97402

David Hedley (31), Ontario Cancer Institute, Princess Margaret Hospital, Toronto, Ontario, Canada M4X 1K9

Johnny Hindkjaer (71), Institute of Human Genetics, University of Åarhus, 8000 Åarhus C, Denmark

Chris J. Janse (295), Laboratory of Parasitology, University of Leiden, 2300 RC Leiden, The Netherlands

Mette W. Jernaes (477), Department of Biophysics, Institute for Cancer Research, Montebello, 0310 Oslo, Norway

Lorna Jost (159), Olson Biochemistry Laboratories, Department of Chemistry, South Dakota State University, Brookings, South Dakota 57007

Jan F. Keij (371), Laboratory of Stem Cells, New York Blood Center, New York, New York 10021

Jørn Koch (71), Institute of Human Genetics, University of Åarhus, 8000 Åarhus C, Denmark

Steen Kølvraa (71), Institute of Human Genetics, University of Åarhus, 8000 Åarhus C, Denmark

Johannes Kramer (149), GSF-Forschungszentrum für Umwelt und Gesundheit, Institut für Biophysikalische Strahlenforschung, D-85758 Oberschleissheim, Germany

Awtar Krishan (21), Department of Radiation Oncology, University of Miami School of Medicine, Miami, Florida 33136

Francis Lacombe (45), Laboratoire d'Hématologie, Hôpital Haut-Lévèque, 33600 Pessac Cedex, France

Alan Landay (437), Department of Immunology and Microbiology, Rush Presbyterian-St. Luke's Medical Center, Chicago, Illinois 60612

James F. Leary (331), Department of Pathology and Laboratory Medicine, University of Rochester, Rochester, New York 14642

Benjamin D. Li (95), Department of Flow Cytometry and Surgical Oncology, Roswell Park Cancer Institute, Buffalo, New York 14263

Jonathan L. Longmire (319), Life Sciences Division and Center for Human Genome Studies, Los Alamos National Laboratory, University of California, Los Alamos, New Mexico 87545

John C. Martin (319), Life Sciences Division and Center for Human Genome Studies, Los Alamos National Laboratory, University of California, Los Alamos, New Mexico 87545

Thomas M. McHugh (575), Department of Laboratory Medicine, University of California, San Francisco Medical Center, San Francisco, California 94143

Birgit Mechtold (387), Institut für Genetik, Universität zu Köln, 50931 Köln, Germany

Beate M. Miller (149), GSF-Forschungszentrum für Umwelt und Gesundheit, Institut für Biophysikalische Strahlenforschung, D-85758 Oberschleissheim, Germany

Stefan Miltenyi (387), Miltenyi Biotec GmbH, 51429 Bergisch Gladbach, Germany

Jane Mitchell (275), Maine Medical Center Research Institute, South Portland, Maine 04106

Robert F. Murphy (131), Department of Biological Sciences and Center for Light Microscope Imaging and Biotechnology, Carnegie Mellon University, Pittsburgh, Pennsylvania 15213

Joe Narai (489), Commonwealth Centre for Laser Applications, Macquarie University, Sydney, New South Wales 2109, Australia

Padma Kumar Narayanan (423), Department of Physiology and Pharmacology, Purdue University, West Lafayette, Indiana 47907

Michael Nüsse (149), GSF-Forschungszentrum für Umwelt und Gesundheit, Institut für Biophysikalische Strahlenforschung, D-85758 Oberschleissheim, Germany

Stefano Papa (193), Istituto di Scienze Morfologiche, Universitá di Urbino, 61029 Urbino, Italy

Søren Pedersen (71), Institute of Human Genetics, University of Åarhus, 8000 Åarhus C, Denmark

Marvin S. Peterson (457), Department of Chemical Engineering and Materials Science, University of Minnesota, Minneapolis, Minnesota 55455, *and* Institute for Advanced Studies in Biological Process Technology, St. Paul, Minnesota 55108

Eckhard Pflüger (387), Miltenyi Biotec GmbH, 51429 Bergisch Gladbach, Germany

Shirley A. Pomponi (523), Division of Biomedical Marine Research, Harbor Branch Oceanographic Institution, Inc., Fort Pierce, Florida 34946

Andreas Radbruch (387), Institut für Genetik, Universität zu Köln, 50931 Köln, Germany

Mary C. Riedy (95), Department of Flow Cytometry and Surgical Oncology, Roswell Park Cancer Institute, Buffalo, New York 14263

J. Paul Robinson (423), Department of Physiology and Pharmacology, Purdue University, West Lafayette, Indiana 47907

Georg A. Schreiber (149), GSF-Forschungszentrum für Umwelt und Gesundheit, Institut für Biophysikalische Strahlenforschung, D-85758 Oberschleissheim, Germany

A. Schwartz (605), Flow Cytometry Standards Corporation, San Juan, Puerto Rico 00919

Jules R. Selden (1), Department of Safety Assessment, Merck Research Laboratories, West Point, Pennsylvania 19486

T. Vincent Shankey (209), Departments of Urology and Pathology, Loyola University Medical Center, Maywood, Illinois 60153

Kirsten Skarstad (477), Department of Biophysics, Institute for Cancer Research, Montebello, 0310 Oslo, Norway

Maja A. Sommerfelt (563), National Centre for Research in Virology, University of Bergen, Bergen High Technology Centre, N-5020 Bergen, Norway

Eric J. Sorscher (563), Department of Physiology and Biophysics, University of Alabama at Birmingham, Birmingham, Alabama 35294

Frédérique Spyratos (177), Department of Biology, Centre René Huguenin, 92211 St. Cloud, France

Friedrich Srienc (457), Department of Chemical Engineering and Materials Science, University of Minnesota, Minneapolis, Minnesota 55455, *and* Institute for Advanced Studies in Biological Process Technology, St. Paul, Minnesota 55108

Harald B. Steen (477), Department of Biophysics, Institute for Cancer Research, Montebello, 0310 Oslo, Norway

John A. Steinkamp (627), Los Alamos National Laboratory, University of California, Los Alamos, New Mexico 87545

Carleton C. Stewart (95), Department of Flow Cytometry and Surgical Oncology, Roswell Park Cancer Institute, Buffalo, New York 14263

Andreas Thiel (387), Institut für Genetik, Universität zu Köln, 50931 Köln, Germany

Earl A. Timm, Jr. (95), Department of Flow Cytometry and Surgical Oncology, Roswell Park Cancer Institute, Buffalo, New York 14263

Massimo Valentini (193), Laboratorio Analisi, San Salvatore Hospital, 61100 Pesaro, Italy

Duncan Veal (489), School of Biological Sciences, Macquarie University, Sydney, New South Wales 2109, Australia

Graham Vesey (489), School of Biological Sciences, Macquarie University, Sydney, New South Wales 2109, Australia

Silvia Viaggi (149), GSF-Forschungszentrum für Umwelt und Gesundheit, Institut für Biophysikalische Strahlenforschung, D-85758 Oberschleissheim, Germany

Philip H. Van Vianen (295), Laboratory of Parasitology, University of Leiden, 2300 RC Leiden, The Netherlands

Jan W. M. Visser (243, 371), Laboratory of Stem Cells, New York Blood Center, New York, New York 10021

Peter de Vries (243), Immunex Corporation, Seattle, Washington 98101

Eva Maria Weller (149), GSF-Forschungszentrum für Umwelt und Gesundheit, Institut für Biophysikalische Strahlenforschung, D-85758 Oberschleissheim, Germany

Jurina M. Wessels (149), GSF-Forschungszentrum für Umwelt und Gesundheit, Institut für Biophysikalische Strahlenforschung, D-85758 Oberschleissheim, Germany

Keith Williams (489), School of Biological Sciences, Macquarie University, Sydney, New South Wales 2109, Australia

Clarice M. Yentsch (523), Bigelow Laboratory for Ocean Sciences, West Boothbay Harbor, Maine 04575 *and* Biology Department, Bowdoin College, Brunswick, Maine 04011

PREFACE TO THE SECOND EDITION

The first edition of this book appeared four years ago (*Methods in Cell Biology,* Vol. 33, *Flow Cytometry,* Z. Darzynkiewicz and H. A. Crissman, Eds., Academic Press, 1990). This was the first attempt to compile a wide variety of flow cytometric methods in the form of a manual designed to describe both the practical aspects and the theoretical foundations of the most widely used methods, as well as to introduce the reader to their basic applications. The book was an instant publishing success. It received laudatory reviews and has become widely used by researchers from various disciplines of biology and medicine. Judging by this success, there was a strong need for this type of publication. Indeed, flow cytometry has now become an indispensable tool for researchers working in the fields of virology, bacteriology, pharmacology, plant biology, biotechnology, toxicology, and environmental sciences. Most applications, however, are in the medical sciences, in particular immunology and oncology. It is now difficult to find a single issue of any biomedical journal without an article in which flow cytometry has been used as a principal methodology. This book on methods in flow cytometry is therefore addressed to a wide, multidisciplinary audience.

Flow cytometry continues to rapidly expand. Extensive progress in the development of new probes and methods, as well as new applications, has occurred during the past few years. Many of the old techniques have been modified, improved, and often adapted to new applications. Numerous new methods have been introduced and applied in a variety of fields. This dramatic progress in the methodology, which occurred recently, and the positive reception of the first edition, which became outdated so rapidly, were the stimuli that led us to undertake the task of preparing a second edition.

The second edition is double the size of the first one, consisting of two volumes. It has a combined total of 71 chapters, well over half of them new, describing techniques that had not been presented previously. Several different methods and strategies for analysis of the same cell component or function are often presented and compared in a single chapter. Also included in these volumes are selected chapters from the first edition. Their choice was based on the continuing popularity of the methods; chapters describing less frequently used techniques were removed. All these chapters are updated, many are extensively modified, and new applications are presented.

From the wide spectrum of chapters presented in these volumes it is difficult to choose those methods that should be highlighted because of their novelty, possible high demand, or wide applicabilities. Certainly those methods that offer new tools for molecular biology belong in this category; they are presented in chapters on fluorescence *in situ* hybridization (FISH), primed *in situ* labeling (PRINS), mRNA species detection, and molecular phenotyping. Detection of intracellular viruses and viral proteins and analysis of bacteria, yeasts, and plant cells are broadly described in greater detail than before in separate chapters. The chapter on cell viability presents and compares ten different methods for identifying dead cells and discriminating between apoptosis and necrosis, including a new method of DNA gel electrophoresis designed for the detection of degraded DNA in apoptotic cells. The chapter describing analysis of enzyme kinetics by flow cytometry is very complete. The subject of magnetic cell sorting is also described in great detail.

Numerous chapters that focus on the analysis of cell proliferation also should be underscored. The subjects of these chapters include univariate DNA content analysis (using a variety of techniques and fluorochromes applicable to cell cultures, fresh clinical samples, or paraffin blocks), the deconvolution of DNA content frequency histograms, multivariate (DNA vs protein or DNA vs RNA content) analysis, simple and complex assays of cell cycle kinetics utilizing BrdUrd and IdUdr incorporation, and studies of the cell cycle based on the expression of several proliferation-associated antigens, including the G_1- and G_2-cyclin proteins. Approaches to discriminating between cells having the same DNA content but at different positions in the cell cycle (e.g., noncycling G_0 vs cycling G_1, G_2 vs M, and G_2 of lower DNA ploidy vs G_1 of higher ploidy) are also presented.

Many of the methods described in these volumes will be used extensively in the fields of toxicology and pharmacology. Among these are the techniques designed for analysis of somatic mutants, formation of micronuclei, DNA repair replication, and cumulative DNA damage in sperm cells (DNA *in situ* denaturability). The latter is applicable as a biological dosimetry assay. A plethora of methods for analysis of different cell functions (functional assays) will also find application in toxicology and pharmacology.

The largest number of chapters is devoted to methods having clinical applications, either in medical research or in routine practice. Chapters dealing with lymphocyte phenotyping, reticulocyte and platelet analysis, analysis and sorting of hemopoietic stem cells, various aspects of drug resistance, DNA ploidy, and cell cycle measurements in tumors are very exhaustive. Diagnosis and disease progression assays in HIV-infected patients, as well as sorting of biohazardous specimens, new topics of current importance in the clinic, are also represented in this book.

Individual chapters are written by the researchers who developed the described methods, contributed to their modification, or found new applications and have extensive experience in their use. Thus, the authors represent a "Who's Who" directory in the field of flow cytometry. This ensures that the essential details of each methodology are included and that readers may easily learn these techniques by following the authors' protocols. We express our gratitude to all contributing authors for sharing their knowledge and experience.

The chapters are designed to be of practical value for anyone who intends to use them as a methods handbook. Yet, the theoretical bases of most of the techniques are presented in detail sufficient for teaching the principle underlying the described methodology. This may be of help to those researchers who want to modify the techniques, or to extend their applicability to other cell systems. Understanding the principles of the method is also essential for data evaluation and for recognition of artifacts. A separate section of most chapters is devoted to the applicability of the described method to different biological systems. Another section of most chapters covers the critical points of the procedure, possible pitfalls, and experience of the author(s) with different instruments. Appropriate controls, standards, instrument adjustments, and calibrations are the subjects of still another section of each chapter. Typical results, frequently illustrating different cell types, are presented and discussed in yet another section. The Materials and Methods section of each chapter is exhaustive, providing a detailed, step-by-step description of the procedure in a protocol or cookbooklike format. Such exhaustive treatment of the methodology is unique; there is no other publication on the subject of similar scope.

We hope that the second edition of *Flow Cytometry* will be even more successful than the first. The explosive growth of this methodology guarantees that soon there will be the need to compile new procedures for a third edition.

Zbigniew Darzynkiewicz
J. Paul Robinson
Harry A. Crissman

PREFACE TO THE FIRST EDITION

Progress in cell biology has been closely associated with the development of quantitative analytical methods applicable to individual cells or cell organelles. Three distinctive phases characterize this development. The first started with the introduction of microspectrophotometry, microfluorometry, and microinterferometry. These methods provided a means to quantitate various cell constituents such as DNA, RNA, or protein. Their application initiated the modern era in cell biology, based on quantitative—rather than qualitative, visual—cell analysis. The second phase began with the birth of autoradiography. Applications of autoradiography were widespread and this technology greatly contributed to better understanding of many functions of the cell. Especially rewarding were studies on cell reproduction; data obtained with the use of autoradiography were essential in establishing the concept of the cell cycle and generated a plethora of information about the proliferation of both normal and tumor cells.

The introduction of flow cytometry initiated the third phase of progress in methods development. The history of flow cytometry is short, with most advances occurring over the past 15 years. Flow cytometry (and, associated with it electronic cell sorting) offers several advantages over the two earlier methodologies. The first is the rapidity of the measurements. Several hundred, or even thousands of cells can be measured per second, with high accuracy and reproducibility. Thus, large numbers of cells from a given population can be analyzed and rare cells or subpopulations detected. A multitude of probes have been developed that make it possible to measure a variety of cell constituents. Because different constituents can be measured simultaneously and the data are recorded by the computer in list mode fashion, subsequent bi- or multivariate analysis can provide information about quantitative relationships among constituents either in particular cells or between cell subpopulations. Still another advantage of flow cytometry stems from the capability for selective physical sorting of individual cells, cell nuclei, or chromosomes, based on differences in the variables measured. Because some of the staining methods preserve cell viability and/or cell membrane integrity, the reproductive and immunogenic capacity of the sorted cells can be investigated. Sorting of individual chromosomes has already provided the basis for development of chromosomal DNA libraries, which are now indispensable in molecular biology and cytogenetics.

Flow cytometry is a new methodology and is still under intense development, improvement, and continuing change. Most flow cytometers are quite complex and not yet user friendly. Some instruments fit particular applications better than others, and many proposed analytical applications have not been extensively tested on different cell types. Several methods are not yet routine and a certain degree of artistry and creativity is often required in adapting them to new biological material, to new applications, or even to different instrument designs. The methods published earlier often undergo modifications or improvements. New probes are frequently introduced.

This volume represents the first attempt to compile and present selected flow cytometric methods in the form of a manual designed to be of help to anyone interested in their practical applications. Methods having a wide immediate or potential application were selected, and the chapters are written by the authors who pioneered their development, or who modified earlier techniques and have extensive experience in their application. This ensures that the essential details are included and that readers may easily master these techniques in their laboratories by following the described procedures.

The selection of chapters also reflects the peculiarity of the early phase of method development referred to previously. The most popular applications of flow cytometry are in the fields of immunology and DNA content–cell cycle analysis. While the immunological applications are now quite routine, many laboratories still face problems with the DNA measurements, as is evident from the poor quality of the raw data (DNA frequency histograms) presented in many publications. We hope that the descriptions of several DNA methods in this volume, some of them individually tailored to specific dyes, flow cytometers, and material (e.g., fixed or unfixed cells or isolated cell nuclei from solid tumors), may help readers to select those methods that would be optimal for their laboratory setting and material. Of great importance is the standardization of the data, which is stressed in all chapters and is a subject of a separate chapter.

Some applications of flow cytometry included in this volume are not yet widely recognized but are of potential importance and are expected to become widespread in the near future. Among these are methods that deal with fluorescent labeling of plasma membrane for cell tracking, flow microsphere immunoassay, the cell cycle of bacteria, the analysis and sorting of plant cells, and flow cytometric exploration of organisms living in oceans, rivers, and lakes.

Individual chapters are designed to provide the maximum practical information needed to reproduce the methods described. The theoretical bases of the methods are briefly presented in the introduction of most chapters. A separate section of each chapter is devoted to applicability of the described method to different biological systems, and when possible, references are provided to articles that review the applications. Also discussed under separate subheads are the critical points of procedure, including the experience of the authors with different instruments, and the appropriate controls and standards. Typical results, often illustrating different cell types, are presented and discussed in the "Results" section. The "Materials and Methods" section of each chapter is the most extensive, giving a detailed description of the method in a cookbook format.

Flow cytometry and electronic sorting have already made a significant impact on research in various fields of cell and molecular biology and medicine. We hope that this volume will be of help to the many researchers who need flow cytometry in their studies, stimulate applications of this methodology to new areas, and promote progress in many disciplines of science.

Zbigniew Darzynkiewicz
Harry A. Crissman

CHAPTER 1

A Flow Cytometric Technique for Detection of DNA Repair in Mammalian Cells

Jules R. Selden* and Frank Dolbeare[†]

* Department of Safety Assessment
Merck Research Laboratories
West Point, Pennsylvania 19486

† Biomedical Sciences Division
Lawrence Livermore National Laboratory
Livermore, California 94550

I. Introduction

Genotoxicity is a crucial biological marker in the complex process of carcinogenesis. Mutagenic chemicals routinely bind covalently to the bases of the

DNA macromolecule, although other constituents are not exempt from attack. Regardless of the biological system traumatized (prokaryotes versus eukaryotes), or the nature or number of DNA insults per cell, the vast majority of these lesions are excised using error-free excisional or recombinational repair processes (Boyce and Howard-Flanders, 1964; Setlow and Carrier, 1964; Brunk and Hanawalt, 1967). In some organisms, initial repair is limited to sites of actively transcribed, crucial genes. Damaged flanking noncoding regions, and/or transcribed regions of less critical genes may persist for considerably longer periods (Mayne, 1984; Bohr *et al.*, 1985; Okumuto and Bohr, 1987; Thomas *et al.*, 1989; Wassermann *et al.*, 1990). Indeed, these cellular surveillance mechanisms are so proficient that only an extremely small minority of all DNA damage persists as permanent mutations that may perturb actively transcribed regions, and only a fraction of these pose any risk for initiating carcinogenesis. Despite this statistically remote circumstance, risk assessment focuses upon the profound health ramifications (viz, neoplastic development) arising from significant mutations.

Thus, short-term genotoxicity assays serve an indispensable function in expediting the detection of putative chemical carcinogens among potential human therapeutic candidates. Included among the repertoire of tests routinely performed are a variety of DNA repair assays, some which detect repair within bacteria [e.g., the rec assay with *Bacillus subtilis* (Kada *et al.*, 1972) and the polA assay with *Escherichia coli* (Slater *et al.*, 1971)], whereas others identify repair within mammalian cells [e.g., *in vitro* rat hepatocyte unscheduled DNA synthesis (UDS) assay (Rasmussen and Painter, 1964; Trosko and Yager, 1974)]. We have been developing a flow cytometric-based assay for detecting chemically induced DNA repair in mammalian cells to complement our existing battery of genotoxicity tests. The procedure utilized is a modification of the immunochemical technique of Dolbeare *et al.* (1983).

The central premise of this flow cytometric technique assumes that DNA damage arising from most chemical mutagens induces excision repair. This repair mechanism displaces numerous nucleotides adjacent to the sites of these DNA lesions in the process, requiring their eventual resynthesis. This restoration of these excised domains is the specific facet of the DNA repair process exploited by this assay. Molecules of a thymidine analogue, 5-bromodeoxyuridine (BrdUrd), are first incorporated during reassemblage, then recognized by an anti-BrdUrd monoclonal antibody, and finally visualized with a green fluorescent tag, fluorescein isothiocyanate (FITC). To permit simultaneous measurement of BrdUrd–antibody interaction and total cellular DNA content, limited DNA denaturation is performed since the anti-BrdUrd monoclonal antibody successfully binds to BrdUrd incorporated into denatured, but not intact (double-stranded) DNA, whereas propidium iodide (PI), employed to estimate cellular DNA content, intercalates into native DNA.

Cultures of diploid human fibroblasts (IMR-90 cells), rat hepatocytes, and rat kidney tubules have been exposed *in vitro* to a variety of mutagenic agents and pulsed with BrdUrd. *In vivo* DNA repair studies have also been performed with nuclei extracted from a heterogeneous population of renal cortical cells. After fixation in ethanol, cellular DNA is partially denatured by a combination of mild acid treatment and brief heat treatment in a low ionic solution, followed by sequential immunochemical staining of incorporated BrdUrd and counterstaining remaining native DNA with PI. Flow cytometric analysis of these dual-stained samples includes the construction of linear bivariate histograms to simultaneously detect low-level green fluorescence (representing BrdUrd incorporated in DNA of repairing cells) and the more abundant red fluorescence (characterizing the DNA content of individual cells).

II. Application

Widely utilized genetic toxicology assays for screening suspect carcinogens include several DNA repair assays. Some of these assays detect DNA repair in bacteria, typified by the *E. coli* polA and *B. subtilis* rec assays. In reality, neither of these assays directly examines DNA repair in these microorganisms; instead, they compare the relative growth patterns of paired wild-type and repair-deficient bacteria (i.e., DNA *pol*ymerase I and *rec*ombination deficient *E. coli* and *B. subtilis* strains, respectively) identically treated with test compound. It is well established that repair-deficient genotypes are substantially more sensitive to DNA damaging agents than their wild-type counterparts. Thus, selective cytotoxicity with the repair-deficient strain constitutes a "positive" response in these assays, a hallmark that the particular test chemical produced DNA lesions in these microbes.

Another representative DNA repair test, which utilizes mammalian cells, is the UDS assay and its many variations. In contrast to the microbial assays described above, evidence of DNA repair is obtained with this assay. In the classical version of the UDS test, cells are incubated with test compound and radioactively labeled thymidine. Incorporation of this labeled nucleoside into DNA occurs during DNA repair and replicative synthesis. By accommodating for the latter situation (e.g., using nonreplicative cellular populations; inhibiting cell division by serum starvation or arginine depletion), significant accumulation of radioactive label within target cells signifies DNA repair. Localization of this probe within the nuclear milieu of individual cells may be visualized using autoradiography, or alternatively, intracellular incorporation of this radioactive probe may be measured en masse using liquid scintillation counting. More recently, immunochemical techniques have been introduced which avert the use of radioactivity, yet exhibit sensitivities comparable to conventional techniques employing radioactivity.

The examination of DNA repair using flow cytometry has received little attention to date. Using the BrdUrd-FITC/PI technique and flow cytometric analysis, Beisker and Hittelman (1988) demonstrated DNA repair with UV-irradiated human fibroblasts. More recently, Affentranger and Burkart (1992) indirectly studied DNA repair with X-ray-irradiated Chinese hamster ovary (CHO) and bone marrow (M3-1) cells. After treatment with X-rays, DNA-damaged regions in these cells were selectively denatured by mild alkalinization, and then cells were stained with acridine orange. Acridine orange is a metachromatic dye which precipitates with single-stranded nucleic acids (emitting a red fluorescence), but intercalates between the helices of double-stranded nucleic acids (emitting a green fluorescence). Thus, the stained X-ray-damaged DNA regions of these cells emit a red fluorescence while the remaining stained undamaged or repaired DNA emit a green fluorescence. By allowing these cells to repair for varying periods after insult, the intensities of these acridine orange fluorescent signals shifted with time (red decreased, green increased).

We have employed a modification of the thermal DNA denaturation method (Dolbeare *et al.,* 1985) to prepare a variety of cell types for labeling with BrdUrd-FITC and PI to examine for evidence of DNA repair. Although other DNA denaturing methods compatible for use with this immunochemical procedure have been described, including concentrated HCl (Dolbeare *et al.,* 1983) or restriction nucleases (Dolbeare and Gray, 1988), and may be quite suitable for use here, we have not performed a systematic comparison of these protocols. To date, we have studied DNA repair in the following cell types: diploid human fibroblasts (IMR-90 cells), rat hepatocytes, rat kidney tubules, and a heterogeneous population of rat kidney cells. Over 30 chemicals, plus UV irradiation, have been evaluated; the bulk of these studies utilized rat hepatocytes. DNA repair in rat liver and kidney was detected after either *in vitro* or *in vivo* administration of chemical mutagens. Long BrdUrd pulses (viz, 24 hr or more) plus stringent denaturing conditions favor recognition of repairing cell populations. Limits of detection (i.e., lowest effective doses) were characterized for all agents eliciting DNA repair. A direct comparison between flow cytometric analysis and conventional autoradiography using isolated rat hepatocytes exposed *in vitro* to a series of chemical mutagens concluded that these techniques produce similar results.

III. Materials

1. *Reagents*. From Grand Island Biological Co. (Grand Island, NY): Dulbecco's phosphate-buffered saline (DPBS), without calcium and magnesium (310-4190); gentamycin (600-5750); Hanks' balanced salt solution (HBSS, 310-4020); L-glutamine (320-5030); Earle's minimum essential medium (320-1095); nonessential amino acids (320-1140); trypsin, 0.05%, and 0.53 m*M* ethylene diaminetetraacetic acid (EDTA) (610-5300).

From Sigma Chemical Co. (St. Louis, MO): Bovine serum albumin (BSA), globulin-free (A-7638); BrdUrd (B-5002); FITC-conjugated goat anti-mouse IgG antibody (F-9006); 5-fluoro-2′-deoxyuridine (5-FdU, F-0503); PI (P-5264); RNase, type III-A (R-5125); Triton X-100; Tween 20 (P-1379).

From Hyclone Laboratories, Inc. (Logan, UT): fetal calf serum (FCS, A-1111-L).

2. *Antibody diluting buffer.* DPBS with 1% Tween 20 and 1% BSA. Prepared fresh for each study.

3. *Antibody wash buffer.* DPBS with 0.5% Tween 20. Prepared fresh for each study.

4. *Anti-BrdUrd monoclonal antibody.* A list of commercial suppliers of anti-BrdUrd monoclonal antibodies has been published (see Table I in Dolbeare *et al.,* 1990). In these investigations, we utilized the IU-4 monoclonal antibody (Caltag Laboratories, Inc., San Francisco, CA). The stock antibody is divided into 8-μl aliquots and frozen at −90°C. Working solution of antibody contains a 1:1000 dilution of antibody in antibody diluting buffer and is prepared as follows: Stock antibody is diluted to 1:500 in antibody diluting buffer and transferred to microfuge tubes. Tubes are centrifuged in microfuge for 5 min, 25°C. Supernatant from tubes is recovered and pooled and volume measured. Final dilution (1:1) with DPBS produces ultimate antibody dilution of 1:1000. This working solution is stable for 1 week at 4°C.

5. *Fluoresceinated goat anti-mouse antibody.* Final titer in antibody diluting buffer is 1:20 and is prepared as follows: Stock antibody is diluted 1:10 in antibody diluting buffer and transferred to microfuge tubes. Tubes are centrifuged in microfuge for 5 min, 25°C. Supernatant from tubes is recovered and pooled and volume measured. Final dilution (1:1) with PBS produces ultimate antibody dilution of 1:20. Working solution is prepared fresh for each study.

6. *Propidium iodide.* Wear gloves when handling stain or stained material. The stock solution, 1 mg/ml PI in 70% ethanol, is protected from light and stored at 4°C. This is stable for at least 1 year. The working solution contains 10 μg/ml PI in DPBS, pH 7.2. Prepared fresh for each study.

IV. Sample Preparation and Staining

A. Overview

DNA repair has been detected in a variety of tissues after *in vivo* or *in vitro* administration of test compound. After *in vivo* administration of both test compound and BrdUrd, one can rapidly extract large numbers of labeled, treated nuclei directly from the target tissue, using any nuclear isolation technique, and immediately proceed with staining this material. With *in vitro* assays, however, one must conserve a highly viable cell population during the overnight incubation period with compound and BrdUrd. It is desirable to constrain the size of the *in vitro* replicative (S phase) cell population before and during compound administration/BrdUrd pulsing since the intensity of the replicative signal swamps any repair signal (the intensity of the former signal is orders of magnitude larger than that of the latter signal). A log phase culture of mammalian cells may contain more than 50% S phase cells. Quiescent cultures of viable IMR-90 cells containing no more than 4% S phase cells were produced via a combination of cellular contact inhibition and serum deprivation. Cultures were initially grown to confluency in medium containing 10% serum, and then the serum content of the medium was dropped to 5% for 3 days and then finally reduced to 0.5% for a minimum of 24 hr before being tested.

Optimal results are obtained when staining is performed within 48 hr after ethanol fixation. Stained samples, protected from light and stored at 4°C, may be preserved for several months without noticeable loss of resolution. A generic staining procedure, suitable for all tissues described in this report, is presented below. Methods employed for deriving cultures of IMR-90 cells, rat hepatocytes, kidney tubules, or isolated nuclei from renal biopsies are reviewed in the next section.

B. Sample Preparation and Fixation

1. Technique for Deriving Isolated Hepatocytes from the Rat Liver

The procedure described below, involving *in situ* enzymatic perfusion of a rat liver to procure a population of viable hepatocytes, is adapted from the technique of Seglin (1973).

1. A male Sprague–Dawley rat, weighing approximately 200–300 g is anesthetized with sodium pentobarbital (0.5 mg/kg, ip). The portal vein and inferior vena cava are exposed, and an 18-gauge needle is inserted into the former vessel, while a hemostat clamps off the latter vessel.

2. Using a peristaltic pump delivering 10–20 ml solution/min, the first perfusing solution (containing 0.5 m*M* EGTA in Hepes-buffered HBSS, pH 7.4) is delivered slowly (around 10 ml/min) via the portal vein, gradually swelling and blanching the organ.

3. The thoracic inferior vena cava is cannulated with a catheter via the right atrium. The catheter empties into a waste container (a beaker) to collect all excess perfusate. A total of 100 ml of the first perfusing solution is administered.

4. A total of 300 ml of a second perfusing solution (containing 100 units/ml collagenase I in Hepes-buffered HBSS, pH 7.4) is introduced (flow rate at 20 ml/min).

5. Following the perfusion with the second solution, the liver is transferred to a petri dish containing HBSS and trimmed of excess connective tissue. The clean liver is next transferred to a dish containing 30 ml second perfusing solution.

6. After stripping away the capsule, hepatocytes are gently released into the liquid with a blunt metal comb.

7. The cell suspension is poured through gauze into a 50-ml tube, and HBSS added to produce a final volume of 50 ml. Tubes are centrifuged at 75*g* for 5 min at 25°C, then cells are resuspended in 5 ml HBSS.

8. The cell suspension is layered onto cold 38% Percoll in 0.14 *M* NaCl, and centrifuged at 20,000*g* for 10 min at 4°C. Hepatocytes, located in the bottom layer, are collected and transferred to a tube containing 10 ml HBSS.

9. After microscopic examination to determine relative hepatocyte yield and viability, cultures are established in 100-mm petri dishes containing approximately 4×10^6 hepatocytes/dish. Cells are cultured in 10 ml of L-15 medium containing 10% FCS. After a 3-hr period, permitting attachment of viable hepatocytes to the stratum of the vessel, fresh medium is substituted, and treatment/pulsing begins.

10. After overnight incubation (18–20 hr), plates are rinsed with 3 ml 0.02% EDTA for 2 min and then replaced with 2 ml 0.05% trypsin and 0.53 m*M* EDTA. Plates are returned to the 37°C incubator for approximately 2 min until cellular detachment results.

11. A total of 5 ml HBSS containing 10% FCS is added to each plate, and the cellular suspension is transferred to a 15-ml centrifuge tube. Tubes are centrifuged at 80*g* for 5 min at 4°C.

12. After removal of supernatant, 5 ml 70% ethanol is slowly added while contents are vortexed. Samples are capped and stored overnight at 4°C.

2. Technique for Deriving and Treating Growth-Arrested IMR-90 Cells

The procedure described below is from Selden *et al.* (1993).

1. Monolayers of IMR-90 cells, derived from early passage inoculates, are produced using Earle's minimum essential medium supplemented with 20% FCS. Final passage is into 100-mm petri dishes.

2. Quiescent cultures (containing an excess of 92% of all cells residing in G_0 or G_1) are produced by reducing the serum content in the medium of confluent dishes in a stepwise fashion. Dishes are first supplemented with medium containing 10% FCS for 3 days, then the FCS content is reduced to 5% for an additional 3 days before cultures are maintained in medium with 0.5% FCS (a minimum of 24 hr) until cultures are used for experimentation.

3. During chemical introduction or after irradiation, conditioned medium is supplemented with BrdUrd (usual concentration 100 μM) and 1 μM 5-FdU. Dishes are returned to a humidified 37°C incubator for as long as 48 hr. After 24 hr, fresh medium containing BrdUrd and 5-FdU is added to all 48-hr cultures.

4. Following the labeling period, dishes are rinsed with 5 ml DPBS, and cells detached from the stratum with 5 ml DPBS containing 0.05% trypsin and 0.53 m*M* EDTA.

5. Cell suspensions are transferred to 15-ml tubes containing 5 ml HBSS with 10% FCS. After centrifugation at 145*g* for 5 min at 4°C, the resuspended cell pellets are fixed by slow addition of 5 ml cold 70% ethanol during vortexing, then are capped and stored overnight at 4°C.

3. Technique for Deriving Isolated Rat Renal Cortical Tubule Cultures

The procedure described below, which employs the enzymatic digestion of kidney tissue to derive a population enriched with cortical tubular elements, is adapted from the techniques of Richardson *et al.* (1982) and Thornthwaite *et al.* (1980).

1. Cortical regions from excised, decapsulated kidneys are minced in sterile HBSS into 1–2 mm fragments.

2. Tissue is transferred to sterile T25 flasks containing 25 ml William's E medium with 50 units collagenase and 0.125% trypsin per ml. These flasks are placed in a 37°C rocking water bath (speed set to 150 rpm) for 45 min.

3. Digested tissue is filtered through four layers sterile gauze into sterile 50-ml centrifuge tubes. An additional 25 ml of HBSS with 5% FCS is added to each tube, followed by centrifugation at 55*g* for 3 min at 25°C.

4. After supernatant is decanted and pellets are resuspended, 50 ml fresh HBSS with 5% FCS is introduced. Tubes are centrifuged at 55*g* for 3 minutes at 25°C.

5. After supernatant is decanted and pellets are resuspended, 3 ml William's E medium containing 20% FCS is added. An aliquot of sample is inspected under a microscope to determine relative tubule yield. One digested kidney routinely produces sufficient yield of tubules to create 10 cultures.

6. Tubule samples are divided into 100-mm petri dishes, containing 10 ml William's E medium, 20% FCS, and 100 μM BrdUrd, and incubated with compound at 37°C in a humidified incubator containing 5% CO_2 for 18–20 hr.

7. After supernatant is removed and saved, dishes are rinsed with 3 ml 0.53 m*M* EDTA and then replaced with 2 ml 0.05% trypsin and 0.53 m*M* EDTA. Dishes are incubated at 37°C for 2 min, or until tubule detachment is observed. Old supernatant is reintroduced, and samples are transferred to 15-ml tubes and centrifuged at 145*g* for 5 min at 25°C.

8. After supernatant is decanted and pellets are resuspended, 1.5 ml nuclear isolation medium (DPBS with 0.6% NP-40 and 0.2% BSA) is introduced, and the sample is gently mixed and then syringed twice through a bent 25-gauge needle to effectuate release of intact nuclei.

9. A total of 6 ml of absolute (100%) ethanol is rapidly added, and samples are capped and stored at 4°C overnight.

C. Staining Procedure

1. Centrifuge 5×10^6 ethanol-fixed cells or nuclei at 145*g* for 4 min at 25°C. Aspirate supernatant, and then gently resuspend cells in residual liquid.

2. Add 1 ml of DPBS with 0.5 mg/ml RNase, pH 7.0. Incubate for 20 min at 25°C.

3. Centrifuge at 145*g* for 4 min at 25°C. Aspirate supernatant and resuspend cells.

4. Add 1 ml, ice-cold 0.1 *M* HCl with 0.5% Triton X-100. Place tubes on ice for 10 min. Stop reaction with 5 ml dH_2O.

5. Centrifuge at 275*g*, 25°C, for 10 min. Carefully aspirate supernatant (cells pellet poorly here and may require additional centrifugation) and resuspend cells.

6. Add 1.5 ml dH_2O. Heat tubes in 90–100°C water bath for 10–15 min. Start clock after bath temperature returns to critical temperature. Optimal temperature and time of heating must be determined for different cells to achieve sufficient DNA denaturation without excessive cell loss. Denaturing at 90°C for 15 min is generally successful with a variety of fixed cell types and nuclei.

7. Chill tubes in ice water for 5–10 min. Add 5 ml antibody wash buffer (formula provided above).

8. Centrifuge at 275*g*, 25°C, for 6 min. Aspirate supernatant and resuspend cells. Examine for cellular clumping. If noted, vortexing (high speed, up to 20 sec) may disperse clumps.

9. Add 200 μl, 1:1000 dilution of IU-4 anti-BrdUrd monoclonal antibody in antibody diluting buffer. Incubate for 20 min at 25°C.

10. Add 5 ml antibody wash buffer. Centrifuge at 80*g*, 25°C, for 10 min. Check carefully for cellular clumping, and vortex if necessary. Aspirate supernatant and resuspend cells.

11. Add 100 μl, 1:20 dilution of fluoresceinated goat anti-mouse IgG antibody in antibody diluting buffer. Incubate for 20 min at 25°C.

12. Add 5 ml antibody wash buffer. Centrifuge at 80*g*, 25°C, for 10 min. Aspirate supernatant, resuspend cells, and transfer to labeled microfuge tubes.

13. Add 0.5–1.0 ml PI working solution. Protect tubes from light, and store specimens on ice or at 4°C.

14. Check specimen quality using a fluorescent microscope, noting particularly the staining quality and presence of clumps. Vortex samples, if indicated.

15. Filter samples through 40-μm nylon mesh prior to flow cytometric analysis.

V. Critical Aspects of the Procedure

1. The repair fidelity of the cell population examined and the nature and incidence of all DNA lesions represent the most critical aspects of this DNA repair assay. Not all repair processes remove nucleotides adjacent to the site of the DNA lesion. For instance, dealkylation of purine residues by endogenous transferases and other endonucleases is accomplished without any demolition

2. A minimum of two negative and two positive controls are configured into the design of these DNA repair studies. The two negative controls include an appropriate solvent control (routinely DMSO or dH_2O) pulsed with BrdUrd and a second negative control not receiving BrdUrd. Should a concern arise regarding potential compound-related autofluorescence (particularly with high, noncytotoxic doses), a third negative control may be configured into the experimental design which investigates cells receiving a high dose of compound in the absence of BrdUrd. The two positive controls routinely employed in our experiments include cultures exposed to a direct acting mutagen (e.g., 250 μM methyl methanesulfonate) and a mutagen requiring metabolic activation (e.g., 1 μM 2-acetylaminofluorene). All controls are subjected to every aspect of the staining process (i.e., denaturation, anti-BrdUrd antibody, goat anti-mouse IgG, and PI steps).

VII. Instruments and Setup for Data Collection

Any commercial flow cytometer should be adequate for these investigations. Our analyses have been performed using an EPICS 753 flow cytometer (Coulter Electronics), equipped with Innova 90-6 argon ion lasers (Coherent, Palo Alto, CA) and a 76-μm jet-in-air flow cell tip. One laser, tuned to 488 nm and delivering approximately 500 mW excites both FITC and PI. Two photomultiplier tubes are used to collect the spatially separated green and red fluorescent emissions. The filter combination for this two-color fluorescence analysis includes a 457–502-nm laser blocking filter, a 550-nm short-pass dichroic filter, a 530-nm (green) short-pass interference filter, and a 590-nm (red) long-pass absorbance filter.

Since the quantity of BrdUrd incorporation with repairing cells is extremely small relative to the amount observed with DNA replication, integrated linear scales with either 64- or 256-channel resolution were selected to present the BrdUrd-related fluorescence intensities of stained cells in bivariate or univariate histograms, respectively. While differences in the green fluorescent intensities between repairing and nonrepairing populations of G_1 and/or G_2 & mitotic cells are obvious using these linear scales, they also eliminate all replicative cells from the analysis since their relative green fluorescence intensities are driven off-scale.

Information from 10^4 cells/sample is routinely collected. Data collection is gated on forward light scatter and DNA content to eliminate cellular aggregates, occasional intact cells, and debris. The instrument's high voltage (ranging between 1050 and 1200) is adjusted until the mean green fluorescent signal from a negative control G_1 population routinely resides around channel 30 on a 256-channel univariate histogram, and then all remaining samples from the series are analyzed without altering instrument settings. If one desires to ensure that all green fluorescent distributions from a series of samples are on-scale without altering instrument settings during data collection, the procedure is modified as follows: all highly fluorescent treated samples are briefly analyzed to confirm the most fluorescent one. Using this sample, the operator adjusts the instrument's high voltage until this sample's highest green signals reside at the top of the bivariate histogram without spilling off-scale. This setting is employed for collecting information from all remaining samples in this series. Further investigation of this technique has demonstrated that this procedure resolves the lowest doses of agents which elicit significant DNA repair in target cells or tissues. Attempts to enhance this assay's sensitivity by adjusting the flow cytometer's electronics (i.e., high voltage or gain settings) and/or laser output with minimally fluorescent material proved futile.

Data from these studies are principally displayed as bivariate contour histograms with 64 × 64-channel resolution and integrated linear scales. The integrated linear red fluorescence (DNA content) distributions are presented on the *x*-axis, and the integrated linear green fluorescence (BrdUrd content) distributions are presented on the *y*-axis of these histograms. Two contour levels are routinely displayed on these histograms—single-cell distributions (represented as lightly shaded dots or regions) and locations containing 20 or more cells per site (represented as blackened dots or regions). Univariate displays with 256-channel resolution are also collected and are principally used to analyze the green fluorescence distributions within repairing versus nonrepairing cellular populations.

VIII. Results and Discussion

A. Confirmation of DNA Repair with UV-Irradiated Human Fibroblasts

Our first goal was to verify whether this assay detects DNA repair in mammalian cells. Since a stoichiometric relationship exists between UV dose and the consequent amount of cellular DNA damage up to saturation (Regan *et al.*, 1968; Setlow *et al.*, 1969; Ahmed and Setlow, 1979; Freeman *et al.*, 1986), we initially investigated UV-irradiated cultures of IMR-90 cells (a human fibroblast cell line) to ascertain whether such a relationship between UV dose and resultant levels of DNA repair in these cells could be demonstrated using this flow cytometric technique (Selden *et al.*, 1993). Stoichiometry was attained when two conditions were applied: (1) cultures were permitted to repair for a 48-hr period after irradiation, and (2) cellular DNA was denatured for 15 min at 90°C. After being normalized for autofluorescence, relative BrdUrd fluorescence intensities of IMR-90 cells from cultures irradiated with 2, 6, or 18 J/m^2 UV light increased an average of 12.4, 36.4, and 107.9 channels (total resolution = 256 channels), respectively (see Fig. 1). The magnitude of these relative fluorescence intensities (1×, 3×, and 9×) is identical to the magnitude of these dose increments, verifying that this relationship is indeed stoichiometric. Further investigations with IMR-90 cultures irradiated with lower doses of UV confirmed that

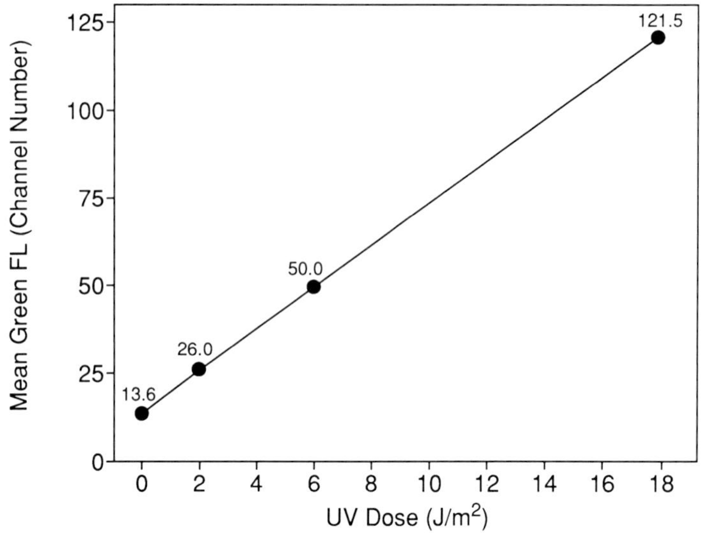

Fig. 1 Graphical representation of stoichiometric relationship between UV dose and magnitude of ensuing cellular DNA repair, the latter measured as relative fluorescence intensity emanating from FITC-labeled anti-BrdUrd antibody molecules (*y*-axis on graph). Cultures of IMR-90 cells received UV doses of 0, 2, 6, or 18 J/m^2 (*x*-axis on graph), followed by a 48-hr pulse with 100 μM BrdUrd, then cellular DNA was denatured for 15 min at 90°C before being stained. Mean green (BrdUrd) autofluorescent value of stained control culture is 13.6 channels; means of irradiated cultures are listed. Observed relative BrdUrd fluorescence intensity increases with cultures irradiated with 2, 6, or 18 J/m^2 are 12.4, 36.4, and 107.9 channels, respectively. The magnitude of these values (1×, 3×, 9×) is identical to the magnitude of these UV dose increments.

stoichiometry was conserved down to the lowest dose examined (0.025 J/m^2; see Selden *et al.*, 1993). The sensitivity of this assay compares quite favorably with reported limits of sensitivity with alkaline elution analysis of UV-irradiated human fibroblasts (0.1 J/m^2, Fornace *et al.*, 1976) and DNA supercoiling alterations within nucleoid bodies of irradiated human lymphocytes (0.01 J/m^2, Yew and Johnson, 1979).

B. Verification of DNA Repair in a Variety of Chemically Treated Cells and Tissues

In addition to IMR-90 cells, cultures of rat hepatocytes, rat kidney tubules, and an admixture of rat kidney cells have been exposed to a variety of genotoxic and nongenotoxic chemicals for evidence of DNA repair within these populations. In a head-to-head comparison between human fibroblasts and rat hepatocytes exposed to a series of eight prototypical mutagens (four direct acting agents, four agents requiring metabolic activation) and examined for evidence of DNA repair, lowest effective doses of all direct acting compounds (4-nitroquinoline oxide, ICR-170, methyl methanesulfonate, and ethyl methanesulfonate) with fibroblasts were consistently and substantially below minimally effective doses with rat hepatocytes (differences being as large as a factor of 20, see Table I). Figure 2 compares the DNA repair responses between cultures of IMR-90 cells and rat hepatocytes incubated with methyl methanesulfonate (MMS) and BrdUrd for a 24-hr period. Figure 2A presents the response of a culture of IMR-90 cells incubated with 5 μM MMS, representing an amount of compound near the lower limit of detection. Note that the green fluorescence intensity of the G_1 cellular population increases from a mean value at channel 9.0 with the control culture to a mean value at channel 16.3 with the treated culture (total resolution = 64 channels). Figure 2B presents the response of a culture of rat hepatocytes incubated with a 100 μM MMS (a 20-fold increase over the dose given to the IMR-90 culture in Fig. 2A), again representing an amount of compound near the lower limit of detection. Note that the resulting boost in mean green fluorescence intensity with treatment is modest (but statistically significant), increasing from a 9.6-channel mean for the G_1 cellular population within the control culture to a 11.8-channel mean for the same population in the treated culture (total resolution = 64 channels). However, results with human fibroblasts incubated with the four compounds requiring metabolic activation [2-acetylaminofluorene, benzo[a]pyrene, cyclophosphamide, and dimethylnitrosamine] were disappointing (see Table I), despite attempts to introduce a metabolic activation system (i.e., S-9) during the treatment period.

Table I
DNA Repair Results with Human Fibroblasts and Rat Hepatocytes Exposed to Chemical Mutagens

Compound	Dose range tested	IMR-90 results	Lowest positive dose	Hepatocyte results	Lowest positive dose
		I. Direct Acting Chemicals[a]			
4-NQO[c]	0.1–25 μM	Positive	0.25 μM	Positive	2.5 μM
ICR170	0.1–25 μM	Positive	0.10 μM[d]	Positive	0.25 μM
MMS	1–400 μM	Positive	5.0 μM	Positive	100 μM
EMS	0.25–7.5 mM	Positive	0.25 mM[d]	Positive	1.0 mM
		II. Metabolic Activation Necessary[b]			
AAF	5–100 μM	Negative	—	Positive	5.0 μM[d]
B[a]P	2.5–50 μM	Negative	—	Positive	2.5 μM[d]
Cyclophosphamide	1–25 μM	Negative	—	Positive	1.0 μM[d]
DMN	0.25–5 mM	Negative	—	Positive	0.25 mM[d]

[a] Cells incubated with compound, 100 μM BrdUrd, and 1 μM 5-FdU for 24 hr.

[b] Cells incubated with compound, 100 μM BrdUrd, and 1 μM 5-FdU for first 3 hr (S-9 added to fibroblast cultures). BrdUrd pulse continued for full 24 hr.

[c] Abbreviations used: 4-NQO, 4-nitroquinoline oxide; MMS, methyl methanesulfonate; EMS, ethyl methanesulfonate; AAF, 2-acetylaminofluorene; B[a]P, benzo[a]pyrene; DMN, dimethylnitrosamine.

[d] Lowest dose tested. Assay sensitivity below this dose.

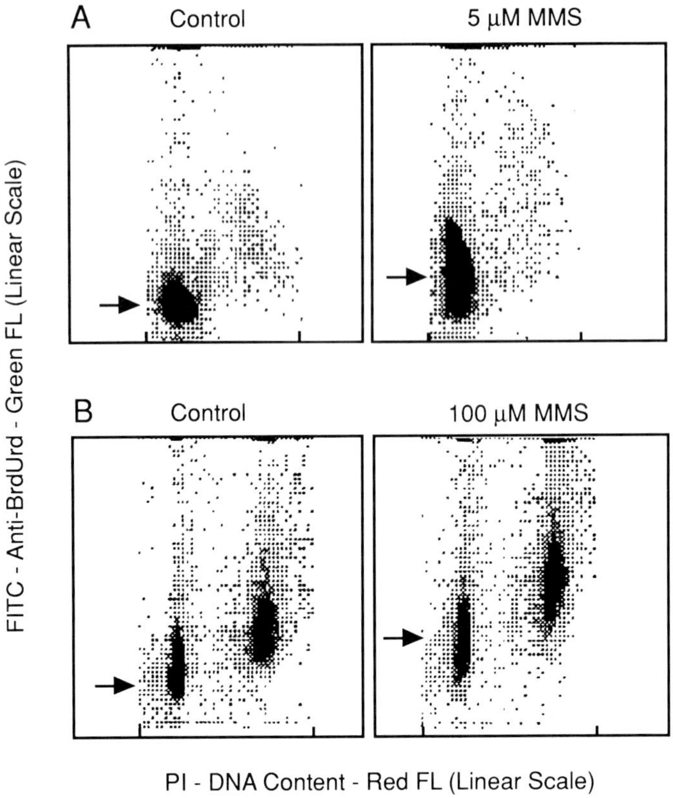

Fig. 2 Bivariate contour historgrams displaying DNA repair in two cellular populations exposed to doses of methyl methanesulfonate (MMS) near limit of detection. Each histogram is a composite of 10^4 cells fluorescently labeled to quantify both incorporated BrdUrd (*y*-axis) and total DNA content (*x*-axis, 64-channel resolution with both axes). Lightly shaded dots represent fluorescent emissions from single cells; black dots and blackened regions represent populations comprising 20 or more cells. Arrows identify the G_1 populations in these cultures. (A) DNA repair response with culture of IMR-90 cells incubated with 5 μM methyl methanesulfonate (MMS, right histogram) compared with concurrent negative (DMSO) control (left histogram). Mean green (BrdUrd-related) fluorescence of the G_1 cellular population significantly increases from a control value of 9.0 channels to 16.3 channels with the culture exposed to MMS. (B) DNA repair response with culture of rat hepatocytes incubated with 100 μM MMS (right histogram) compared with concurrent negative (DMSO) control (left histogram). Mean green (BrdUrd-related) fluorescence of the G_1 cellular population significantly increases from a control value of 9.6 channels to 11.8 channels with the culture exposed to MMS. Note the large population with 4C DNA content; this is typical of hepatocytes.

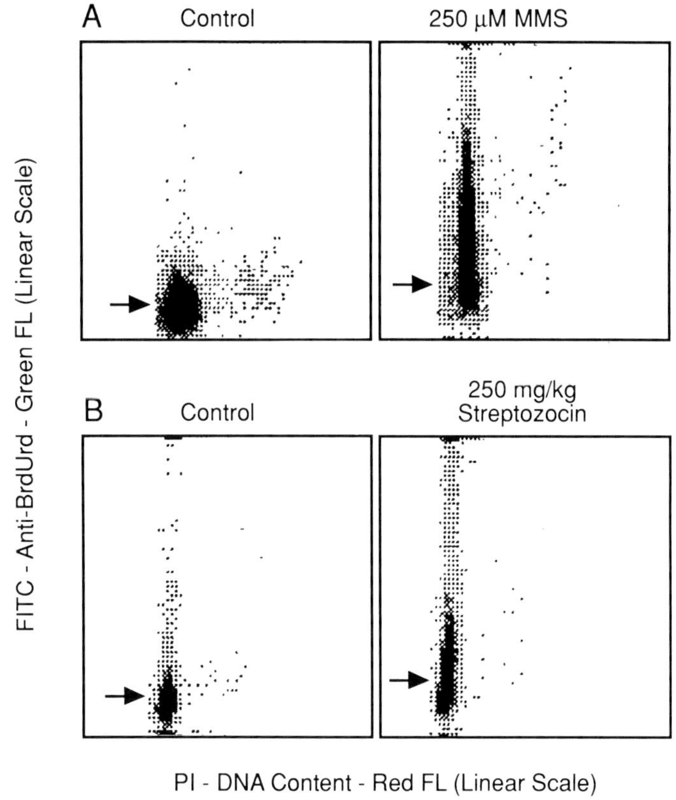

Fig. 3 Bivariate contour histograms displaying *in vitro* and *in vivo* DNA repair responses with rat kidney cells. Each histogram is a composite of 10^4 fluorescently labeled cells; contour levels are identical to those in Fig. 2. Arrows identify the G_1 cellular populations. (A) *In vitro* DNA repair response with culture of kidney tubules incubated with 250 μM MMS (right histogram) compared with concurrent negative (DMSO) control (left histogram). Mean green (BrdUrd-related) fluorescence of the G_1 cellular population significantly increases from a control value of 7.1 channels to 22. 7 channels with the culture exposed to MMS. (B) *In vivo* DNA repair response to streptozocin by mixed population of renal cortical cells (right histogram). Animal received bolus of streptozocin, 250 mg/kg, ip; three tablets of BrdUrd were implanted subcutaneously 3 hr earlier. Next day, renal cortical nuclei were harvested and stained. Mean green (BrdUrd-related) fluorescence of the G_1 cellular population significantly increases from a control value of 6.9 channels to 14.4 channels with streptozocin-treated population.

Further work is necessary to develop an *in vitro* DNA repair assay employing fibroblasts which converts compounds to their DNA-reactive moieties.

As a consequence of this preliminary investigation, our subsequent efforts have focused upon developing a flow cytometric DNA repair assay with rat hepatocytes. To date, we have studied cultures of rat hepatocytes from young male Sprague–Dawley rats dosed *in vitro* with over 30 different chemicals.

Samples were stained as described above after an 18- to 20-hour incubation period with compound plus BrdUrd. In addition, autoradiographic studies utilizing tritiated thymidine ([^{3}H]-thymidine) to detect DNA repair in these chemically treated rat hepatocyte cultures were performed in parallel with most of these experiments to directly compare the relative sensitivities of these two assays. We conclude that the presumptive classification of a compound being either genotoxic or nongenotoxic generally correlates with its induction of DNA repair in these rat hepatocytes (genotoxic compounds inducing repair). However, a poor correlation exists if one compares DNA repair in rat hepatocytes with a compound's presumptive carcinogenicity in the rat, even when the comparison is restricted to the induction of tumors within the liver of male rats. These investigations also reveal that this flow cytometric technique compares quite favorably with conventional autoradiography. No discrepancies arose between these assays, and their relative sensitivities were comparable. Thus, this flow cytometric assay represents a safe (i.e., nonradioactive), sensitive, rapid, and reliable means of identifying agents which induce DNA repair in mammalian cells (Selden *et al.,* 1994).

Representative DNA repair responses by rat kidney cells exposed to prototypical compounds *in vitro* or *in vivo* are presented (see Fig. 3). Figure 3A demonstrates DNA repair in cultures of isolated rat kidney tubules incubated *in vitro* with 250 μM methyl methanesulfonate and 100 μM BrdUrd for 20 hr before harvest and staining. Note that the mean green fluorescence intensity of the G_1 cellular population increases from 7.1 channels with the control culture to a mean value of 22.7 channels with the treated culture (total resolution = 64 channels). Figure 3B demonstrates DNA repair within the cortical region of a rat kidney after the animal received streptozocin (250 mg/kg, ip). Three hours before dosing, the rat was anesthesized, and three tablets containing BrdUrd (total dose = 150 mg,) were implanted subcutaneously near the scapulae. The kidney was recovered 24 hr later, and a wedge of tissue from the kidney cortical region was excised. Nuclei were harvested and stained as described above. In this study, the mean green fluorescence intensity of the G_1 cellular population from the control culture was 6.9 channels; the mean green fluorescence intensity of this same cellular population from the streptozocin-treated sample was 14.4 channels (total resolution = 64 channels).

References

Affentranger, M. I., and Burkart, W. (1992). *Cytometry* **13,** 31–38.

Ahmed, F. E., and Setlow, R. B. (1979). *Cancer Res.* **39,** 471–479.

Beisker, W., and Hittelman, W. N. (1988). *Exp. Cell Res.* **174,** 156–167.

Bohr, V. A., Smith, C. A., Okumuto, D. S., and Hanawalt, P. C. (1985). *Cell (Cambridge, Mass.)* **40,** 359–369.

Boyce, R., and Howard-Flanders, P. (1964). *Proc. Natl. Acad. Sci. U. S. A.* **51,** 293–300.

Brunk, C., and Hanawalt, P. C. (1967). *Science* **158,** 663–664.

Dolbeare, F., and Gray, J. W. (1988). *Cytometry* **9,** 631–635.

Dolbeare, F., Gratzner, H., Pallavicini, M., and Gray, J. W. (1983). *Proc. Natl. Acad. Sci. U. S. A.* **80,** 5573–5577.

Dolbeare, F., Beisker, W., Pallavicini, M., Vanderlaan, M., and Gray, J. W. (1985). *Cytometry* **6,** 521–530.

Dolbeare, F., Kuo, W.-L., Beisker, W., Vanderlaan, M., and Gray, J. W. (1990). *In* "Methods in Cell Biology" (Z. Darzynkiewicz and H. Crissman, eds), Vol. 33, pp. 207–216. Academic Press, San Diego.

Fornace, A. J., Kohn, K. W., and Kann, H. E. (1976). *Proc. Natl. Acad. Sci. U. S. A.* **73,** 39–43.

Freeman, S. E., Blackett, A. D., Monteleonc, D. C., Setlow, R. B., Sutherland, B. M., and Sutherland, J. C. (1986). *Anal. Biochem.* **158,** 119–129.

Kada, T., Tutikawa, K., and Sadaie, Y. (1972). *Mutat. Res.* **16,** 165–174.

Kirtikar, D. M., Cathcart, G. R., and Goldthwait, D. A. (1977). *In* "DNA Repair Processes (W. W. Nichols and D. G. Murphy, eds.), pp. 241–258. Symposia Specialists, Miami, FL.

Lieberman, M. W., Baney, R. N., Lee, R. E., Sell, S., and Farber, E. (1971). *Cancer Res.* **31,** 1297–1306.

Mayne, L. V. (1984). *Mutat. Res.* **131,** 187–191.

Okumuto, D. S., and Bohr, V. A. (1987). *Nucleic Acids Res.* **15,** 10021–10030.

Rasmussen, R. E., and Painter, R. B. (1964). *Nature (London)* **203,** 1360–1362.

Regan, J. D., Trosko, J. E., and Carrier, W. L. (1968). *Biophys. J.* **8,** 319–325.

Richardson, J. C. W., Waterson, P., and Simmons, N. L. (1982). *Q. J. Exp. Physiol. Cogn. Med. Sci.* **67,** 287–301.

Seglin, P. O. (1973). *Exp. Cell Res.* **82,** 391–398.

Selden, J. R., Dolbeare, F., Clair, J. H., Nichols, W. W., Miller, J. E., Kleemeyer, K. M., Hyland, R. J., and DeLuca, J. G. (1993). *Cytometry* **14,** 154–167.

Selden, J. R., Dolbeare, F., Clair, J. H., Miller, J. E., McGettigan, K. M., DiJohn, J. A., and DeLuca, J. G. (1994) *Mutat. Res.* In press.

Setlow, R. B., and Carrier, W. (1964). *Proc. Natl. Acad. Sci. U. S. A.* **51,** 226–231.

Setlow, R. B., Regan, J. D., German, J., and Carrier, W. L. (1969). *Proc. Natl. Acad. Sci. U. S. A.* **64,** 1035–1041.

Slater, E. E., Anderson, M. D., and Rosenkranz, H. S. (1971). *Cancer Res.* **31,** 970–973.

Thomas, D. C., Okumuto, D. S., Sancar, A., and Bohr, V. A. (1989). *J. Biol. Chem.* **264,** 18005–18010.

Thornthwaite, J. T., Sugarbaker, E. V., and Temple, W. J. (1980). *Cytometry* **1,** 229–237.

Trosko, J. E., and Yager, J. D. (1974). *Exp. Cell Res.* **88,** 47–55.

Wassermann, K., Kohn, K. W., and Bohr, V. A. (1990). *J. Biol. Chem.* **265,** 13906–13913.

Yew, F. F. H., and Johnson, R. T. (1979). *Biochim. Biophys. Acta* **562,** 240–251.

CHAPTER 2

Rapid Determination of Cellular Resistance-Related Drug Efflux in Tumor Cells

Awtar Krishan
Department of Radiation Oncology
University of Miami School of Medicine
Miami, Florida 33136

I. Introduction

Several studies suggest that cellular resistance to cancer chemotherapeutic agents, such as alkaloids and antibiotics (multiple drug resistance, MDR), is related to their rapid efflux from the intracellular environment (Kessel *et al.*, 1968; Dano, 1973; Skovsgaard, 1978). Analytical methods such as high-pressure liquid chromatography and spectrofluorometry can be used for monitoring of cellular drug retention and efflux but of necessity these methods are slow, need large samples, and cannot measure drug retention in single cells. Laser flow cytometry (FCM) offers a unique tool for monitoring of fluorescent antitumor drug retention and its modulation in tumor cells (Krishan and Ganapathi, 1979, 1980). Besides its rapidity, the laser FCM method can identify heterogeneity of drug retention (Krishan *et al.*, 1987) as well as allow for sorting of subpopulations for further biochemical or morphological characterization.

Anthracyclines such as doxorubicin and daunomycin are important cancer chemotherapeutic agents. Most of the anthracyclines are fluorescent and can be excited with the 488-nm laser line from an argon ion laser (Krishan, 1986). Thus, this method can be used for rapid monitoring of anthracycline transport and retention in drug-resistant and -sensitive tumor cells.

Cellular resistance to some of the clinically important anthracyclines has been suggested to be due to rapid drug efflux. Thus, drugs which block anthracycline efflux and thereby enhance retention can reduce cellular resistance to anthracyclines (Tsuruo *et al.*, 1981; Ganapathi and Grabowski, 1983; Krishan *et al.*, 1987). Several studies have shown that certain unrelated drugs, such as calcium channel blocker (e.g., verapamil) and phenothiazines (e.g., trifluoperazine), will inhibit anthracycline efflux from resistant cells and thereby render them drug sensitive. Similarly, reduced cellular retention of the vital DNA dye, Hoechst 33342 (Krishan, 1987), and the calcium indicator dye, Indo-AM (Krishan *et al.*, 1988), in certain refractory cells may also be related to rapid efflux. Drugs such as phenothiazines or calcium channel blockers can block Hoechst 33342 efflux and make it possible to generate DNA distribution histograms from living cells, which were heretofore difficult to stain with this vital dye (Krishan, 1987).

The use of efflux blockers has been advocated for chemotherapy of human malignancies (Ozols *et al.*, 1987; Miller *et al.*, 1988; Hait *et al.*, 1989; Sridhar *et al.*, 1993). It would be useful if before the administration of anthracyclines and agents that block their cellular efflux *in vivo,* tumor cells could be screened *in vitro* for their anthracycline retention and efflux characteristics. As shown in the following sections, we have used laser FCM to monitor anthracycline fluorescence in tumor cells and to monitor the effect of drug efflux blocking agents on drug retention. We have used this method to monitor anthracycline retention and its modulation by phenothiazines and amphotericin-B in P388 and doxorubicin-resistant cells (Krishan *et al.*, 1985a,b; 1986). Some of our data show that the effect of efflux blockers on doxorubicin retention is cell-cycle proliferation related (Krishan *et al.*, 1985a), and often one can identify subpopulations based on their differential response to efflux blockers. From these studies it follows that modulation of drug transport and thereby cellular resistance by phenothiazines or calcium channel blockers may not be uniform in a population but may be selective and confined to only certain types of cells and subpopulations.

II. Application

We initially used the laser excitation method for monitoring of anthracycline doxorubicin (adriamycin) transport and retention in tumor cells (Krishan and

Ganapathi, 1980). In subsequent studies, this method was used to monitor the effect of drugs which enhance influx (e.g., amphotericin-B; Krishan *et al.*, 1985b) and efflux blockers (such as verapamil and phenothiazine; Krishan *et al.*, 1985a). We used this method to determine heterogeneity of drug retention and response to efflux blockers in human tumor cells (Krishan *et al.*, 1987; Ramachandran *et al.*, 1993a,b; Sridhar *et al.*, 1993) and to monitor the effect of efflux blockers on retention of the DNA binding drug, Hoechst 33342, and the calcium indicator dye, Indo-AM (Krishan, 1987). Thus this rapid laser flow cytometric method can be used for monitoring of the following:

1. Cellular transport of fluorescent drugs (e.g., anthracyclines, rhodamine, Indo-Am, Hoechst 33342).
2. Effect of drugs which either increase drug influx (e.g., amphotericin-B) or enhance retention by blocking efflux (verapamil, phenothiazines, cyclosporins, tamoxifen, quinine).
3. Tumor cell heterogeneity in retention and response to efflux blockers.
4. Selection of ideal efflux blockers and protocols for possible clinical use.
5. Rapid identification and sorting of cells which have efflux as a major mechanism of drug resistance.

III. Materials

1. Suspension cultures of murine leukemic P388 and its Adriamycin-resistant cell line P388/R84 (Nair *et al.*, 1990) or P388/AdR (Johnson *et al.*, 1978).
2. Single-cell suspension of tumor cells from bone marrow aspirate, ascites, pleural fluid or after enzymatic digestion of solid tumor samples.
3. Doxorubicin (adriamycin, NSC-123127, Adria Labs, Columbus, OH), daunomycin (Cerubidine, NSC-821151, Ives Labs, NY), or Hoechst 33342 (Calbiochem, Inc., San Diego, CA).
4. Efflux blockers: prochlorperazine (Compazine, Smith, Kline and Beecham Labs, Carolina, Puerto Rico) and verapamil (Calan, Searle Pharmaceutical, Inc., Chicago, IL).

Suspension cultures of the P388 cell line and its Adriamycin-resistant subline (P388/R84) grown in RPMI 1640 medium supplemented with 10% heat-inactivated fetal bovine serum (FBS), penicillin, and streptomycin are used for calibration and as controls. In soft agar assays, the ID_{50} for the P388 and P388/R84 cells is 0.0875 and 8.4 μM of doxorubicin, respectively.

Human tumor cells are recovered by centrifugation of pleural fluid (lung cancer), bone marrow or peripheral blood (leukemia), ascites (ovarian, breast), or other solid tumor material after enzymatic digestion. Filtration through nylon mesh (40 μm) is used to remove clumps. The supernatant fluid is aspirated (after centrifugation) and the cell pellets resuspended and washed in Ca^{2+}- and

Mg^{2+}-free phosphate-buffered saline (PBS). After centrifugation, the pelleted cells are resuspended in fresh tissue culture medium supplemented with 10% heat-inactivated FBS. To remove erythrocytes from bone marrow aspirates and peripheral blood samples, specimens are diluted with Ca^{2+}- and Mg^{2+}-free PBS and centrifuged over a 70% preformed Percoll gradient (Pharmacia, Piscataway, NJ). Mononuclear cells are recovered and washed before incubation with the drugs at 37°C. Cytospin is used for analysis of morphological heterogeneity and for differential counting purposes.

Stock solutions of drugs are prepared in Ca^{2+}- and Mg^{2+}-free Hanks' balanced salt solution (HBSS). Fresh dilutions of the drugs are prepared in normal saline before each experiment.

IV. Staining Procedure

For generation of two-parameter dot plots based on cellular drug fluorescence and length of incubation (time), cell suspensions are directly mixed with the drug containing medium in the sampling cuvette of the flow cytometer (Coulter Epics 753 or Becton–Dickinson FACScan), maintained at 37°C. Final drug concentrations used are 1–3 μM doxorubicin or daunomycin.

For efflux blocking experiments, cells are incubated with or without the addition of prochlorperazine (25 μM or verapamil (10 μM). Samples can be run after 30–60 min of incubation.

V. Critical Aspects

Several parameters related to specimen preparation and instrumentation can cause artifacts and lead to generation of erroneous data. Special consideration should be given to the following factors:

1. Several anthracyclines quench their fluorescence on binding to DNA and other target molecules. It is important to keep this in mind and use nonquenching anthracyclines (e.g., AD-32, Krishan *et al.*, 1978) or Hoechst 33342 for critical experiments.

2. pH can have a major effect on drug fluorescence (Alabaster *et al.*, 1989) either by shifting the excitation maxima or by altering drug transport and retention.

3. Some efflux blockers may precipitate or bind to glass. Proper precautions should be taken to avoid these artifacts.

4. High anthracycline concentrations can overcome efflux and high levels of phenothiazines can damage cell membrane and thus increase cellular drug fluorescence. However, the cell membrane-damaged cells can be readily recog-

nized on the basis of their reduced light scatter signal and excluded from analysis.

5. Coated filters used in the flow cytometer can often, with age, develop pinholes which may result in light leaks. Similarly, certain commercially available filters are notorious for generating autofluorescence when excited with high laser excitation.

6. Dead cells (cells with damaged cell membrane) will rapidly stain with the dye and give erroneous results. Forward and/or right-angle light scatter should be used to isolate and identify these cells in multiparameter dot plots. Furthermore, dead cells can be selectively removed by treatment with DNase I.

VI. Controls and Standards

We regularly use adriamycin-sensitive and -resistant P388 cells coincubated with similar drug concentrations under identical conditions as controls for each experiment. Once the photomultiplier (PMT) high voltage, laser power, and amplifications are optimized and set, all samples are analyzed without altering any of these parameters. In general drug-sensitive P388 cells (ID_{50} 0.0875 μM), after incubation with 1.75–3.5 μM of doxorubicin for 30–60 min, are 20 times less fluorescent than human diploid nuclei stained with the propidium iodide (PI)–hypotonic citrate method (Krishan, 1975).

VII. Instruments

For FCM analysis, cell suspensions incubated *in vitro* with anthracyclines, with or without the addition of phenothiazines or verapamil, are analyzed for their total cellular fluorescence in a Becton–Dickinson FACScan cell analyzer interface with a HP Consort 32 data acquisition and analysis system. Fluorescence emission (>530 nm) and forward and 90° light scatter are collected in list mode. A minimum of 10,000 cells are analyzed for each sample and used for generation of gated histograms or dot plots.

For generation of two-parameter dot plots based on cellular drug fluorescence and length of incubation (time), cell suspensions are directly mixed with the drug containing medium in the sampling cuvette of the flow cytometer, maintained at 37°C. Time as a parameter is available in software from the Coulter MDADS and the Becton–Dickinson HP Consort 32-Lysis software.

For efflux blocking experiments, P388 and P388/R84 cells are incubated with doxorubicin (3.5 μM), for 120 min at 37°C, with or without the addition of prochlorperazine (25 μM) or verapamil (10 μM).

A Coulter Elite or a Becton–Dickinson FASCan/FACStar analyzer/cell sorter equipped with an argon laser is used to measure forward angle and

90° light scatter and fluorescence from cells excited with the 488-nm line for anthracyclines.

VIII. Results

Dot plots and single-parameter histograms (insets) are shown in Fig. 1. Figures 1A and 1B are of adriamycin-sensitive and -resistant P388/R84 cells, respectively, incubated with 3.5 μM of daunorubicin for 30 min. Forward angle light scatter (ordinate) and drug fluorescence (abscissa) in log scale are shown. Note that, by the fluorometric method, P388/R84 cells have four- to sixfold less drug content (due to rapid efflux) than the parental drug-sensitive cells.

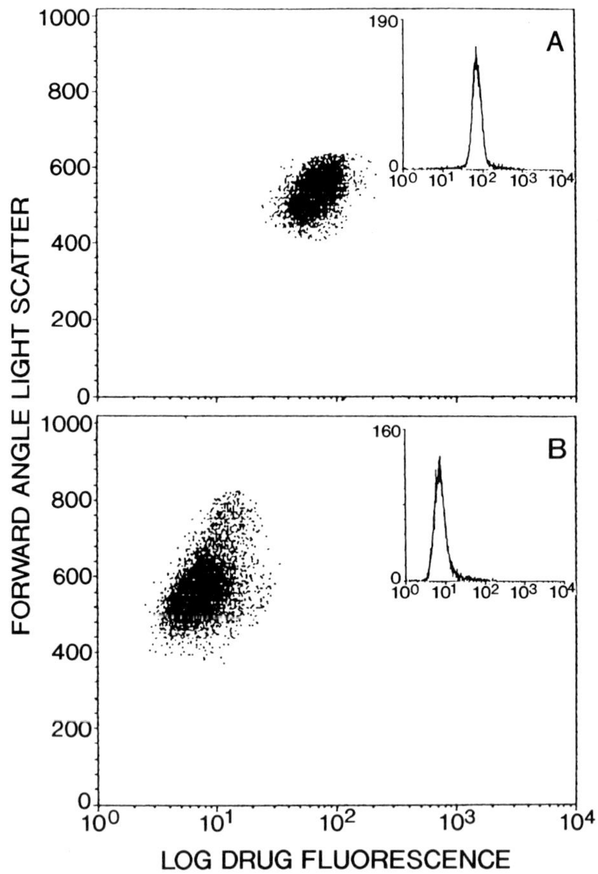

Fig. 1 Dot plots and single-parameter histograms (insets) of (A) adriamycin-sensitive and (B) -resistant P388 cells, respectively. Forward angle light scatter (ordinate) and drug fluorescence (abscissa) in log scale are shown.

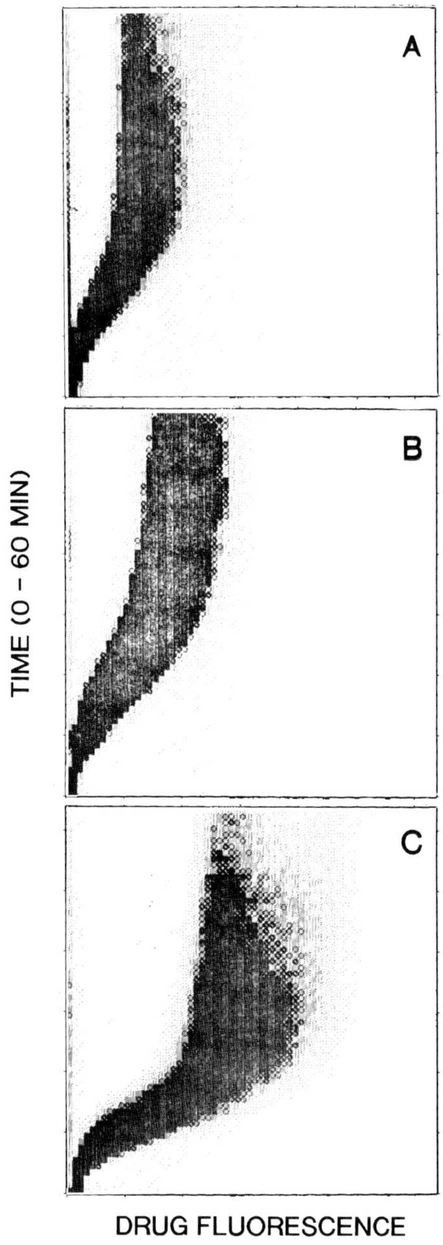

Fig. 2 Dot plots showing the effect of pH on doxorubicin fluorescence of P388 cells. (A) pH 6, (B) pH 7, (C) pH 8. Abscissa measures linear peak fluorescence whereas ordinate records time (60 min). Note that cells incubated at the alkaline pH of 8.0 (C) have much higher fluorescence than cells incubated in acidic medium (pH 6.0, A).

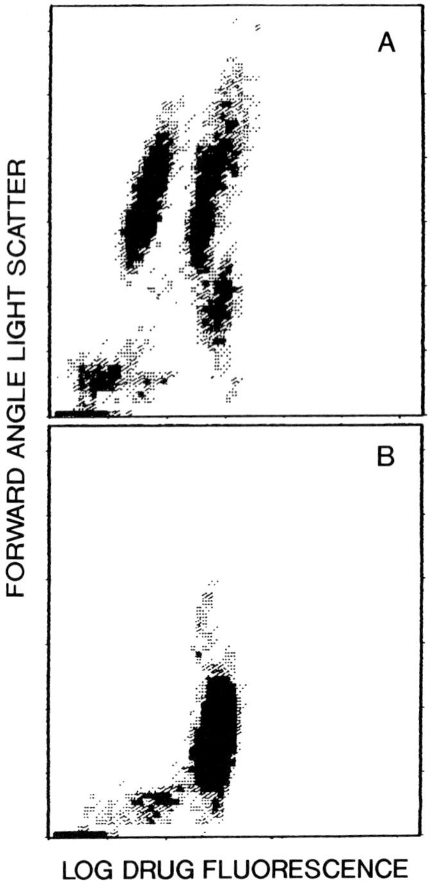

Fig. 3 Dot plots of cells from human solid ascites incubated with doxorubicin. (A) Note the appearance of four subsets. In cells coincubated with the efflux blocker chlorpromazine, a homogeneous population with high drug fluorescence emerges, suggesting that the subpopulations with the lower amount of drug fluorescence (in A) were possibly effluxing the drug.

Figures 2A, 2B, and 2C are dot plots showing the effect of pH 6, 7, and 8, respectively, on doxorubicin fluorescence of P388 cells. The abscissa measures linear peak fluorescence while the ordinate records time (60 min). Note cells incubated at alkaline pH of 8.0 (Fig. 2C) have much higher fluorescence than cells incubated in the acidic medium (pH 6.0, Fig. 2A).

The dot plots shown in Fig. 3 are of cells from a human tumor incubated with doxorubicin. Note the appearance of four subsets in Fig. 3A. In cells coincubated with the efflux blocker chlorpromazine, a homogeneous population with high drug fluorescence emerges, suggesting that the subpopulations with the lower amount of drug fluorescence (in Fig. 3A) were possibly effluxing the drug.

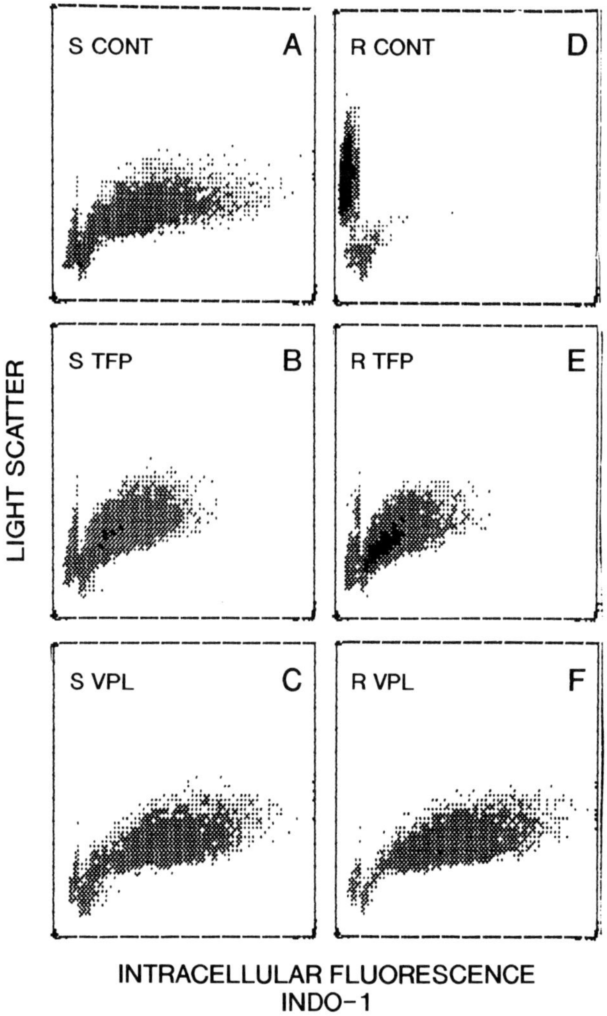

Fig. 4 Dot plots of P388 (A–C) and P388/R84 (D–F) cells incubated with the calcium indicator dye, Indo-AM. The efflux blockers trifluoperazine (B,E) and verapamil (C,F) were present as indicated. Note that the low fluorescence in P388/R84 cells (D) is significantly enhanced in the presence of these efflux blockers (E,F).

The dot plots shown in Fig. 4 are of P388 (Figs. 4A–4C) and P388/R84 (Figs. 4D–4F) cells incubated with the calcium indicator dye, Indo-AM. Note that low fluorescence in P388/R84 cells (Fig. 4D) is significantly enhanced in the presence of efflux blockers trifluoperazine (Fig. 4E) or verapamil (Fig. 4F)

References

Alabaster, O., Woods, T., Ortiz-Sanchez, V., and Jahangeer, S. (1989). *Cancer Res.* **49,** 5638–5643.
Dano, K. (1973). *Biochim. Biophys. Acta* **323,** 1466–1483.
Ganapathi, R., and Grabowski, D. (1983). *Cancer Res.* **43,** 3696–3699.
Hait, W., Morris, S., Lazo, J., Figlin, R. J., Durivage, H. J., White, K., and Schwartz, P. E. (1989). *Cancer Chemother. Pharmacol.* **23,** 358.
Johnson, E., Chitnis, M. P., Embrey, W. M., and Gregory, E. B. (1978). *Cancer Treat. Rep.* **62,** 1535–1547.
Kessel, D., Botterill, V., and Wodinsky, I. (1968). *Cancer Res.* **28,** 938–941.
Krishan, A. (1975). *J. Cell Biol.* **66,** 188–193.
Krishan, A. (1986). *In* "Techniques in Cell Cycle Analysis" (J. W. Gray and Z. Darzynkiewcz, eds.), pp. 337–366. Humana Press, Clifton, NJ.
Krishan, A. (1987). *Cytometry* **8,** 642–645.
Krishan, A., and Ganapathi, R. (1979). *J. Histochem. Cytochem.* **27,** 1655–1656.
Krishan, A., and Ganapathi, R. (1980). *Cancer Res.* **40,** 3895–3900.
Krishan, A., Ganapathi, R., and Israel, M. (1978). *Cancer Res.* **38,** 3656–3662.
Krishan, A., Sauerteig, A., and Wellham, L. (1985a). *Cancer Res.* **45,** 1046–1051.
Krishan, A., Sauerteig, A., and Gordon, K. (1985b). *Cancer Res.* **45,** 4097–4102.
Krishan, A., Sauerteig, A., Gordon, K., and Swinkin, C. (1986). *Cancer Res.* **46,** 1768–1773.
Krishan, A., Sridhar, K. S., Davila, E., Vogel, C., and Sternheim, W. (1987). *Cytometry* **8,** 306–314.
Krishan, A., Nair, S., Ganju, A., and Gordon, K. (1988). *Proc. Amer. Assoc. Cancer Res.* **29,** 305.
Miller, R., Bukowski, R., Budd, G., Purvis, J., Weick, J. K., Shepard, K., Midha, K. K., and Ganapathi, R. (1988). *J. Clin. Oncol.* **6,** 880.
Nair, S., Singh, S., Samy, T. S. A., and Krishan, A. (1990). *Biochem. Pharmacol.* **39,** 723–728.
Ozols, R., Cunnion, R., Klecker, R., Hamilton, T. C., Ostchega, Y., Perillo, J. E., and Young, R. C. (1987). *J. Clin. Oncol.* **5,** 641.
Ramachandran, C., Yuan, Z., Huang, X., and Krishan, A. (1993a). *Biochem. Pharmacol.* **45,** 743–751.
Ramachandran, C., Sauerteig, A., Sridhar, K., and Kishan, A. (1993b). *Cancer Chemother. Pharmacol.* **31,** 431–441.
Skovsgaard, T. (1978). *Cancer Res.* **38,** 1785–1791.
Sridhar, K., Krishan, A., Samy, T. S. A. Sauerteig, A., Wellham, L. L., McPhee, G., Duncan, R. C., Anac, S. Y., Ardalan, B., and Benedetto, P. W. (1993). *Cancer Chemother. Pharmacol.* **31,** 423–430.
Tsuruo, T., Lida, H., Tsukagoshi, S., and Sakurai, Y. (1981). *Cancer Res.* **41,** 1967–1972.

CHAPTER 3

Glutathione and Cellular Resistance to Anti-Cancer Drugs

David Hedley and Sue Chow

Ontario Cancer Institute
Princess Margaret Hospital
Toronto, Ontario M4X 1K9
Canada

I. Introduction

Glutathione is a tripeptide (*glu-cys-gly*) which is present at millimolar concentrations in probably all eukaryotic cells. It is a major intracellular reducing agent involved in numerous metabolic processes (Meister and Anderson, 1983), but GSH has particular importance in oncology because of its central role in the metabolism or detoxification of reactive species which can damage DNA (Farber *et al.*, 1990; Reed, 1990; Waxman, 1990; Boobis *et al.*, 1989). Broadly speaking, damage to purine or pyrimidine bases causes mutations and forms the basis of chemical or radiation carcinogenesis, while damage to the sugar-phosphate backbone causes DNA strand breaks and is the mechanism whereby many forms of treatment kill cancer cells. Glutathione therefore plays a key role in both carcinogenesis and resistance to cancer treatment.

Drug resistance is the central problem in cancer chemotherapy. Although clearly multifactorial, the emergence of drug resistance in cancer patients can to a large extent be explained by overactivity of a limited number of cell processes which are normally involved in detoxification or repair. Conceptually the simplest and best understood of these is the P-glycoprotein efflux pump (see Chapter 2 of this volume by Krishan), but alterations in GSH chemistry, enhanced activity of DNA repair enzymes, and alterations in topoisomerase II also play an important role. Of particular relevance to clinical oncology is the development of drug resistance modulating agents (Mitchell *et al.,* 1989; Coleman *et al.,* 1988), drugs which while not cytotoxic themselves are capable of increasing cell kill by reversing specific drug resistance mechanisms. Although these agents are effective in experimental systems, inappropriate use in cancer patients would be expected to increase normal tissue side effects because drug resistance mechanisms are not cancer specific. There is therefore an urgent need to develop methods for monitoring drug resistance at the clinical level, and flow cytometry is particularly attractive because it is inherently quantitative, gives an indication of cellular heterogeneity, and can be applied to small biopsy samples (Hedley, 1993).

A number of flow cytometric methods for measuring cellular GSH have been described, using probes which form fluorescent adducts with GSH (Poot *et al.,* 1986,1991; O'Connor *et al.,* 1988; Rice *et al.,* 1986; Treumer and Valet, 1986; Durand and Olive, 1983), but there are a number of practical difficulties with them, and none has entered routine clinical practice. An ideal staining method for GSH needs to be:

1. Specific

Protein sulfhydryls inside cells are usually maintained in the reduced form, i.e., they contain free -SH groups. Although these tend to be less reactive with GSH probes, or may be configured in forms which are unavailable for binding, protein thiols significantly outnumber GSH in most cells and can produce high background staining. Low-molecular-weight sulfhydryls such as cysteine may also be reactive with GSH stains and represent another source of background staining, although these are usually much less abundant than GSH.

2. Quantitative

A typical epithelial cell contains over 1×10^8 molecules of GSH. It is unlikely that all are available for staining, however, since some GSH is localized in subcellular compartments such as mitochondria, while a variable proportion is bound to protein thiols via mixed disulfide bonds. Furthermore, stains such as monochlorobimane which are bound to GSH by glutathione S-transferase are unlikely to saturate the entire pool of available GSH, because these enzymes are inhibited by high concentrations of glutathione adducts (Cook *et al.*, 1991). Nevertheless, in order to give a reliable estimate of cellular GSH content, staining methods need to be stoichiometric, so that fluorescence is proportional to GSH content, and the extent of GSH labeling should be similar for a wide range of cell types.

Although cells containing increased GSH show increased resistance to some forms of cancer treatment, this is unlikely to be seen simply as the emergence of distinct "GSH-positive cells" in clinical samples. Furthermore, to have clinical utility results should be predictive of treatment outcome, which will only become possible following prospective studies where large numbers of assays are performed using a standardized procedure, and the results compared to clinical response. This will require painstaking attention to laboratory standards and quality controls, while in order to compare results between different laboratories, standard units of measurement will be needed, such as number of GSH molecules per cell.

II. Flow Cytometric Glutathione Probes

A large number of fluorescent sulfhydryl-reactive stains are available, and several show sufficient specificity for GSH to have been published as potentially useable in flow cytometric assays (Fig. 1). This specificity can be due to selective binding to GSH under the action of glutathione S-transferase or to differences in the spontaneous chemical reaction with GSH compared to other sulfhydryls. We have recently conducted an extensive assessment of the available flow cytometric GSH probes (Hedley and Chow, 1994). None is entirely satisfactory for labeling human cells, and it is essential that a laboratory attempting to measure GSH using flow cytometry be aware of their limitations. These probes are summarized as follows:

A. Probes Conjugated by GSH S-Transferase

1. Monochlorobimane

This an analogue of monobromobimane (see below). It has greater specificity for GSH because of a low spontaneous reaction rate with protein sulfhydryls, while being efficiently conjugated to GSH under the action of glutathione S-transferase, yielding a highly fluorescent adduct (Rice *et al.*, 1986; Cook *et al.*, 1991; Shrieve *et al.*, 1988). Although monochlorobimane is the most specific of the available GSH probes, staining of rodent cells being closely correlated with biochemically determined GSH content, most human cells are deficient

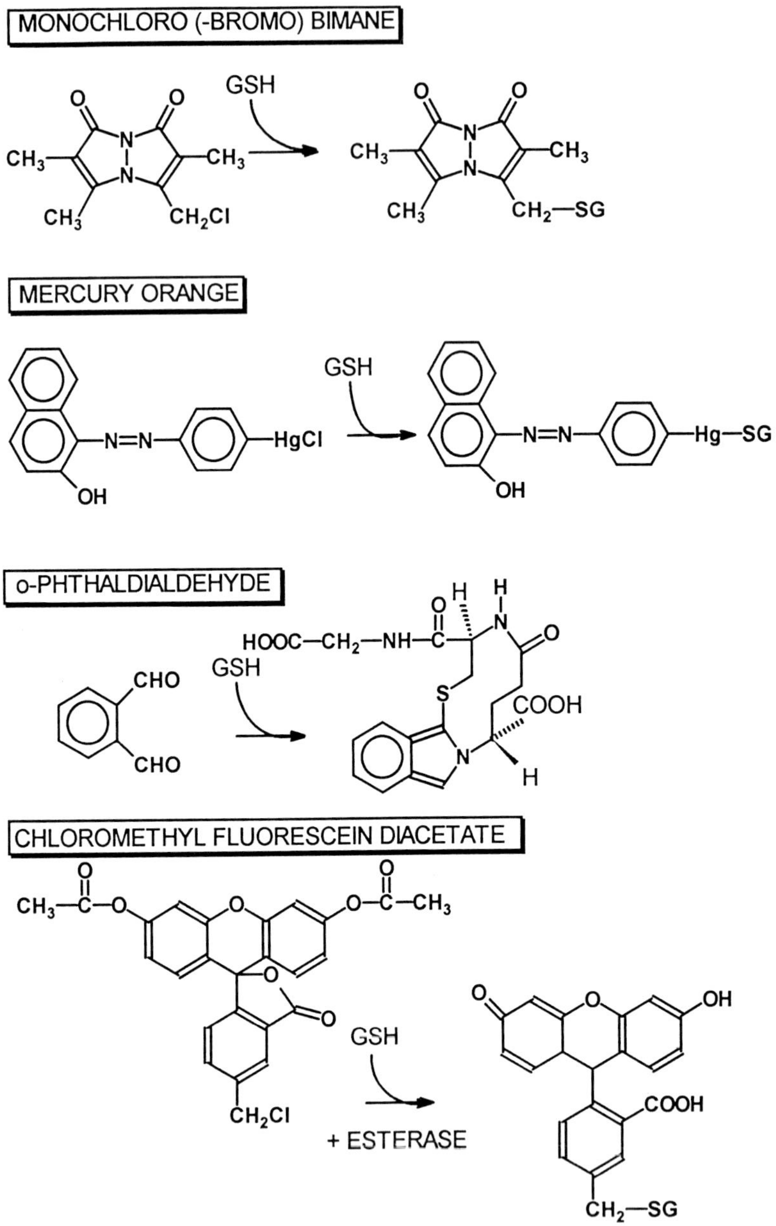

Fig. 1 Structure of flow cytometric glutathione probes, showing the chemical reaction with GSH. This involves binding of the sulfhydryl group of GSH to the probe, except for *o*-phthaldialdehyde, where binding also takes place through the terminal amine group.

in the relevant glutathione S-transferase isoenzyme, resulting in a serious underestimate of cellular GSH (Cook *et al.*, 1991; Ublacker *et al.*, 1991). Increased monochlorobimane concentrations give more complete labeling of GSH in human cell lines, but at the expense of greater background staining. Staining protocols for monochlorobimane are given under Methods.

2. Chloromethyl Fluorescein Diacetate

Chloromethyl fluorescein (CMF, Molecular Probes) is one of the few GSH probes capable of being excited using a 488-nm argon laser (Poot *et al.*, 1991). Unlike the others, it is loaded into cells as the membrane-permeant ester, chloromethyl fluorescein diacetate (CMF-DA), which is hydrolyzed by intracellular esterases to yield the fluorescent, membrane-impermeant CMF. The Molecular Probes catalog states that chloromethyl fluorescein reacts with GSH under the action of glutathione S-transferase to form a fluorescent adduct. Cells labeled with CMF are extremely bright using fluorescein filter sets, but in contrast to monochlorobimane there is essentially no change in the emission spectrum or quantum efficiency on binding to GSH, and cellular fluorescence is therefore the sum of free and GSH-conjugated CMF. Free CMF is retained in cells that have been depleted of GSH, as can be shown by fractionating cell extracts on Sephadex G-25 columns (Hedley and Chow, 1994), and because of this high background the stain is unsuitable for routine use.

B. Probes That Are Not Conjugated by Glutathione S-Transferase

1. Mercury Orange

Originally developed as a cytochemical stain for sulfhydryls (Bennett, 1951), mercury orange was later found to react much more rapidly with GSH than with protein thiols, so staining times of about 5 min (as compared to several hours) gave a fair degree of specificity for GSH (Asghar *et al.*, 1975). Mercury orange was the first GSH stain to be used in flow cytometry capable of excitation at 488 nm (O'Connor *et al.*, 1988), but in our hands it is difficult to obtain reproducible staining on a day-by-day basis, and even under ideal conditions about 40% of cellular fluorescence is non-GSH derived. These problems appear to be at least partly due to the fact that staining with mercury orange takes place in a high concentration of organic solvent such as acetone, so that cellular retention of fluorescent reaction product is dependent on this precipitating out, rather than on surface membrane integrity (Laurrauri *et al.*, 1987). Staining procedures for mercury orange are discussed in detail below.

2. Monobromobimane

The presence of a bromine atom renders this stain more chemically reactive than monochlorobimane (Rice *et al.*, 1986), both probes yielding the same reaction product with GSH (Fig. 1). Background staining of protein sulfhydryls is greater than with monochlorobimane, but as with mercury orange, binding to GSH is more rapid, and a degree of specificity is obtained when staining time and concentration are controlled. Because the reaction with GSH does not require glutathione S-transferase activity, monobromobimane is a more satisfactory stain for human cells, and staining protocols are given in detail below.

3. o-Phthaldialdehyde

This is an interesting thiol-reactive stain which forms chemically distinct conjugates with GSH and protein sulfhydryls (Treumer and Valet, 1986). Because the former emits at longer wavelengths, o-phthaldialdehyde (oPT) in principle might allow simultaneous measurement of GSH and protein thiols in flow cytometry. In practice, however, the spectral separation of the two forms is not large, while the cellular content of protein sulfhydryls generally far exceeds that of GSH, so that the GSH signal is swamped by spillover from proteins.

III. Materials

A. Stains for Glutathione

Monochloro- and monobromobimane: Both of the bimanes are obtainable from Calbiochem or Molecular Probes. Make up 4 mM solutions in 100% ethanol, and store refrigerated or at −20°C. Solutions are stable for several weeks.

Mercury orange: Obtainable from Sigma. Stock solution is 100 μM mercury orange in 100% acetone. This is stable for at least several weeks when stored at −20°C. Staining solution is 40 μM in acetone, with 5% distilled water added prior to use.

B. Glutathione Depleting Agents

Diethyl maleate: Diethyl maleate (DEM) is a liquid, obtainable from Sigma. Dilute into 100% ethanol to give a final concentration of 10 mM.

N-ethyl maleimide: N-ethyl maleimide (NEM) is obtainable from Sigma. Store at room temperature. Working solution is 10 mM in PBS. Note that NEM is a biohazardous material.

Buthionine sulfoximine: Buthionine sulfoximine is also obtainable from Sigma, and made up as a 100 mM solution in distilled water. This should be sterilized using a 0.22-μm millipore filter before being added to cells in culture.

C. Biochemical Assay of Cellular GSH Content

The following solutions should be prepared for spectrophotometric determination of cellular GSH.

Buffer A: This buffer is used for making up all the other solutions and consists of 0.1 *M* sodium phosphate buffer, pH 7.4, containing 0.5 m*M* disodium ethylenediaminetetraacetic acid (EDTA). The sodium phosphate buffer is made up by dissolving 1.2 g NaH_2PO_4 plus 5.68 g Na_2HPO_4 in 500 ml distilled H_2O.

DTNB (5,5'-dithiobis (2-nitrobenzoic) acid): DTNB is obtained from Sigma and stored at 4°C. Make up at 60 mg per 100 ml in buffer A. Store solution in dark at 4°C for up to 1 month.

Sulfosalicylic acid: Make up a 0.6% solution of sulfosalicylic acid (Sigma) in distilled water. Note that this is very acidic. It can be kept at 4 or −20°C.

NADPH: NADPH (β-nicotinamide adenine dinucleotide phosphate, reduced form, tetrasodium salt, type III) is obtainable from Sigma and stored at −20°C. It is light sensitive. Make up at 20 mg per 10 ml buffer A. Solution is stable in refrigerator for 1 week only.

Glutathione reductase: Glutathione reductase (GR) is obtainable from Sigma, and stored at 4°C. Make up fresh at 20 units per 1 μl buffer A to the required volume.

GSH (reduced glutathione): Obtainable from Sigma. Make up a stock solution of 10 mg in 10 ml buffer A, and dilute 10 μl of this solution in 10 ml buffer A containing 100 μl 1 *N* HCl (added to keep the GSH reduced). The undiluted stock solution of GSH can be aliquoted and stored at −20°C.

IV. Methods

A. Preparation of Controls and Standards

For standardization and quality control, it is strongly recommended that the results of flow cytometric GSH assays be checked against the biochemically determined mean GSH content per cell across a range of cellular GSH contents. This also provides a means for calibrating the flow cytometry assay so as to give results as femtomoles GSH per cell (see below). As an absolute minimum a duplicate sample of the cells should be depleted to low levels of GSH prior to staining, since background labeling of protein sulfhydryls is a major problem with flow cytometric determinations of cellular GSH.

1. Depletion of Cellular GSH

Glutathione can be depleted either by using sulfhydryl-reactive agents, such as DEM or NEM, or by inhibiting synthesis with 100 μM buthionine sulfoximine (BSO). The latter is more specific, but the rate of turnover of GSH in most cells is such that overnight incubation with BSO is usually required before depletion is sufficient to allow an estimate of background staining. Of the sulfydryl-reactive agents, DEM is more specific for GSH than NEM, since binding is catalyzed by glutathione S-transferase. A concentration of 100 m*M* is satisfactory for rodent cells, but DEM is a weaker substrate for the human isoenzymes and concentrations of 500 μM or greater are often required to deplete GSH in human cells. These concentrations are toxic and can also deplete protein sulfydryls, and we therefore prefer to use 100 μM NEM when staining human cells. Although this is a more reactive sulfydryl reagent, the comparatively low concentration used has little effect on total protein sulfydryl, while causing profound depletion of GSH. The usual incubation time for DEM is 30 min at 37°C, whereas NEM depletes GSH within seconds at room temperature.

2. Spectrophotometric Measurement of Cellular GSH

Although methods based on HPLC require fewer cells than spectroscopy, the latter can be done using 1×10^6 cells or less, is simpler, and uses readily available equipment. The principle of the assay is the reduction by GSH of the disulfide 5,5′-dinitro-2-nitrobenzoic acid (Ellman's reagent) yielding 2-nitro-5-thiobenzoic acid which absorbs at 412 nm (Eyer and Podhradsky, 1986; Tietze, 1969). In the process GSH is oxidized to GSSG, and the reaction is maintained by adding the enzyme glutathione reductase and the electron donor NADPH, so that the concentration of GSH is rate limiting. The assay is calibrated using dilutions of GSH and comparing rate of generation of color with the test sample.

3. Assay Procedure

a. Sample Preparation

1. Treat the cells with GSH depleting agent, as required.
2. Pellet 1×10^6 cells in Eppendorf tube.
3. Aspirate supernatant and disperse cell pellet.
4. Resuspend cells in 0.5 ml cold 0.6% sulfosalicylic acid.
5. Stand for 1 hr on ice, then spin down samples at 14,000 rpm for 15 min at 4°C.
6. Transfer supernatant to new Eppendorf tube and store at −20°C until assay.

b. Glutathione Assay

1. Use disposable cuvettes with 1 cm path length and 1.5 ml volume. Set spectrophotometer at 412 nm and chart speed 20 sec per cm. Zero spectrophotometer with 900 μl buffer A plus 100 μl NADPH.
2. Run sample. Add 50 μl of sample to 700 μl buffer A; add 100 μl DNTB, 100 μl NADPH, and 50 μl glutathione reductase.

Keep the samples and all solutions except buffer A on ice. To expedite the assay, leave buffer A at room temperature, and keep a 1:1 mixture of DNTB and NADPH solution on ice protected from light. Add everything to the cuvette except glutathione reductase, take an initial reading, then add the enzyme, and let the sample run for 2 to 2 1/2 min. While one sample is running, prepare the next sample. In order to optimize accuracy, try to be consistent in the order and amount of solutions added and time interval between runs.

c. Calibration

The standard curve is constructed by running samples with 0, 25, 50, 75, and 100 μl GSH in buffer A to make up 750 μl volume. The calibration curve is a plot of the slope (the rate of the reaction) against the concentration of GSH and should show a high and linear correlation. The background slope is obtained from the sample without any GSH added. From the calibration curve, read off the GSH value which corresponds to the slope of the test sample.

B. Flow Cytometric GSH Assays

1. Monochlorobimane

This is the method of choice for staining rodent cells, but it requires UV excitation. This is ideally achieved with a mercury arc lamp or the UV doublet of a water-cooled argon laser, but an air-cooled HeCd laser emitting at 325 nm can also be used, despite the fact that the bimane–GSH adduct has an absorption maximum of nearly 400 nm. Collect fluorescence using a band pass filter centered on 450 nm.

a. Staining Procedure

Resuspend cells at a concentration of 1×10^6/ml in a buffered medium (PBS is adequate). Add 10 μl of 4 m*M* stock solution of MClB, giving a final concentration of 40 μM, incubate at room temperature for 10 min, and either run immediately or store on ice for up to 1 hr. Samples do not require resuspending or washing before being run on the flow cytometer. For most purposes it is useful to record data on a linear scale.

b. Critical Aspects of the Method

This is a live cell assay, and because both GSH and the GSH–bimane adduct are water soluble, they are rapidly lost if the cell membrane is disrupted. Because the staining intensity of some cell types decreases when they are left at room temperature for prolonged periods, it is recommended that unstained cells be kept on ice. Incubation time, cell number, and stain concentration are all important when using monochlorobimane and should be kept consistent. The above staining conditions are close to optimum for most cells, with a moderate excess in the ratio of number of stain molecules to the total amount of GSH in the cells. For rodent cells which have been variably depleted of GSH using different concentrations of DEM, there is a close linear correlation between mean fluorescence and biochemically determined GSH content, and this line intercepts near the origin, suggesting that little background labeling of other sulfydryls takes place (Fig. 2).

Binding to GSH is dependent on the activity of a form of glutathione S-transferase whose expression varies between species. It is usually abundant in rodent cells, but expressed at low levels in human cells (Cook *et al.*, 1991; Ublacker *et al.*, 1991), while there appear to be no published data concerning other species. The staining conditions detailed above seriously underestimate the GSH content of human cells. Staining intensity increases with higher concentrations of monochlorobimane and longer incubation times, but in a range of human cell lines tested this failed to reach saturation at concentrations of up to 1 m*M* and incubation times of 1 hr. Furthermore these gave increased background fluorescence, probably due to binding of MClB to protein sulfhydryls.

2. Monobromobimane

a. Staining Procedure

Instrument setup and staining conditions are essentially the same as for monochlorobimane. Monobromobimane is a more reactive stain, and incubation time should be limited to 10 min since binding to GSH is then effectively complete, while longer periods increase the level of protein binding. It is essential to use a control where the GSH content has been depleted in order to obtain an estimate of this background staining, and our current practice is to add 100 μM *N*-ethyl maleimide immediately prior to the monobromobimane, as originally described by Poot *et al.* (1986).

b. Critical Aspects of the Method

In our experience monobromobimane is the most satisfactory stain currently available for labeling GSH in human cells (Hedley and Chow, 1994). Unlike monochlorobimane, the rate of staining follows a clearly biphasic pattern, with a rapid increase in fluorescence over the first 5–10 min, followed by a much slower rise. The time course of staining GSH-depleted cells fails to show the initial rapid build up, and parallels the later slow increase in fluorescence seen in nondepleted cells (Fig. 3). Note that even when using the optimum staining time, however, up to 20% of cellular fluorescence is due to background. In principle it should be possible to correct for this background by subtracting the mean fluorescence of a GSH-depleted control, although this approach remains to be validated in routine practice.

3. Mercury Orange

a. Instrument Setup

Mercury orange-stained cells are readily excited by the 488-nm line of an argon laser. Fluorescence emission maximum is around 580 nm, and the filter setup for phycoerythrin is therefore ideal.

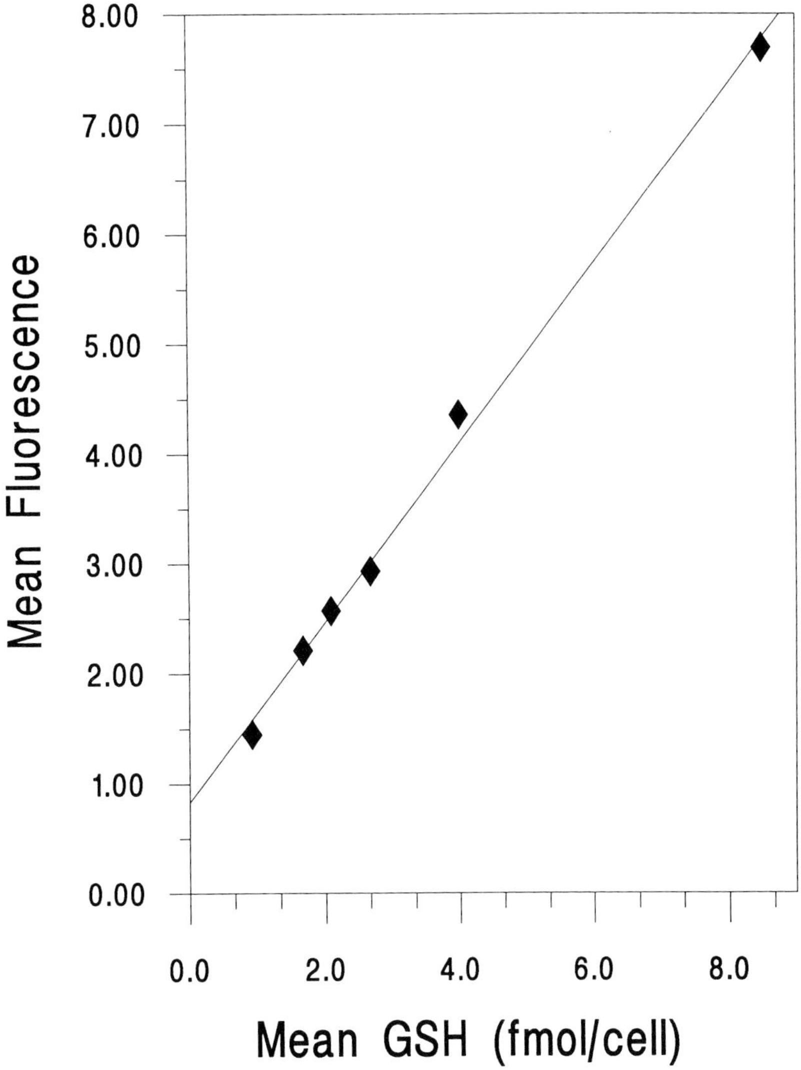

Fig. 2 Relationship between biochemically determined GSH content and mean fluorescence of murine EMT-6 cells stained with 40 μM monochlorobimane, standardized against fluorescent calibration beads. Note the close linear relationship between the two assays across a wide range of GSH values and the low value for the *y* intercept, which suggests that nonspecific binding is low compared to GSH binding.

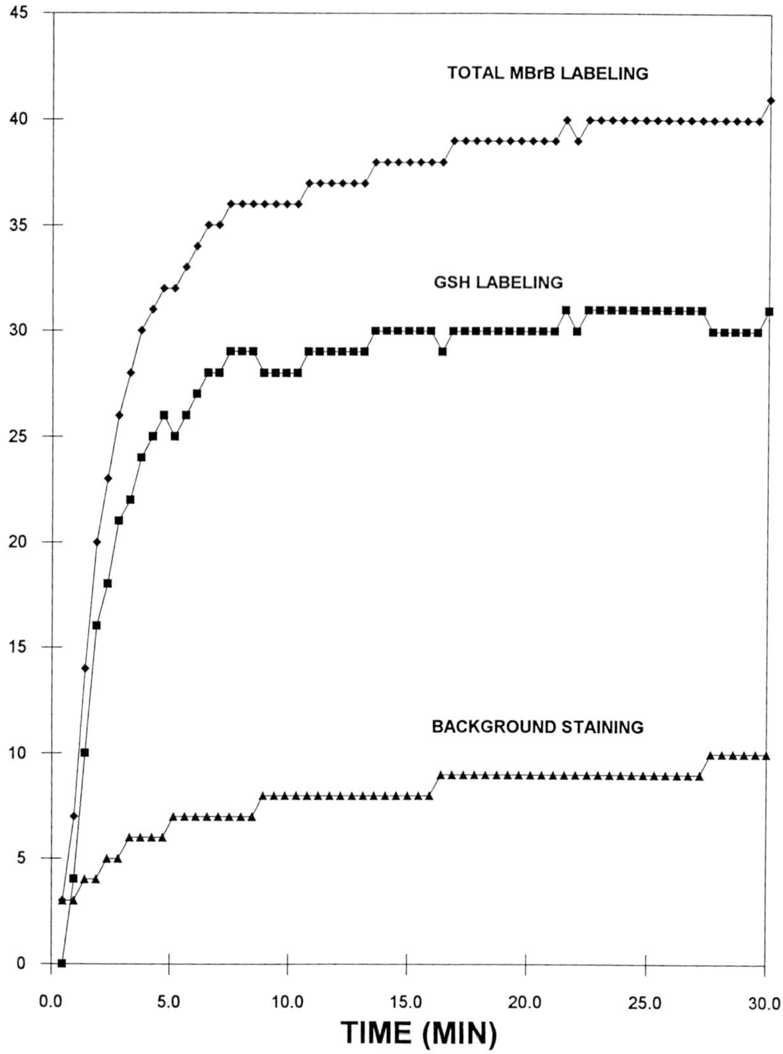

Fig. 3 Time course of staining using 40 μM monobromobimane. Stain was added to cells at time zero, and fluorescence continuously recorded. A replicate sample was first depleted of GSH by treatment with 100 μM NEM, to give a measure of the rate of nonspecific binding. From these two curves the rate of GSH binding was obtained by first converting the flow cytometry data into mean fluroescence versus time using Multi Time (Phoenix Flow Systems) and exporting this in ASCII format. The two curves were then subtracted using a spreadsheet. Note that while GSH binding is essentially complete by 10 min, the background shows a slow increase beyond this point, probably due to protein sulfhydryl binding.

b. Staining Procedure

Pellet 1×10^6 cells in a 5-ml polypropylene Falcon tube (No. 2063), aspirate as much supernatant as possible, and disperse the pellet by flicking the tube. Add 1 ml of mercury orange working solution dropwise, with constant vortex mixing, and then place on ice for 5 min. Spin down and resuspend in 0.5 ml ice-cold PBS. Washing the cells after they have been resuspended has no significant effect on background staining using this method. Samples can either be run immediately or stored at 4°C overnight.

c. Critical Aspect of the Method

Obtaining reproducible staining with mercury orange is more difficult than with the bimanes, and we have tried numerous variations of the above method. The high concentration of acetone used causes cell shrinkage, degrading the forward and orthogonal light scatter signals, and cell clumping, and the cells stick to the sample tube. Although a range of organic solvents will dissolve mercury orange, the choice is limited by the need for the mercury orange–GSH complex to precipitate inside the cells (Larrauri *et al.*, 1987). Addition of small amounts of water increases cellular fluorescence, probably because the reaction between GSH and mercury orange takes place in an aqueous phase (Larrauri *et al.*, 1987), although cell clumping is a serious problem at water concentrations above 10%. Fluorescence intensity is often highly dependent on cell concentration, but this is unpredictable and may either increase or decrease with increasing cell concentration, depending on cell type.

Time course and dose/response curves for mercury orange staining show only small increases in cell fluorescence at concentrations above 50 μM and beyond 5 min incubation. Fluorescence slowly increases over a period of hours, probably due to protein binding. Unfortunately, over a wide range of staining conditions mercury orange produces a high background fluorescence in cells which have been depleted to low levels of GSH, and controls for nonspecific binding are therefore essential. Technical aspects of staining biopsies from human cancer using mercury orange are discussed in detail in Coates and Tripp (1991).

C. Calibrating Flow Cytometric Assays

Human tumors are extremely heterogeneous with respect to cellular GSH content, but because all nucleated cells contain appreciable amounts of GSH it is not sufficient to express the results of clinical assays simply as "percent positive cells," as is common practice in clinical flow cytometry. The alternative of using "mean channel numbers" is equally meaningless without some form of external standard. It has been proposed that a standard biochemical assay for GSH be performed in parallel with the flow cytometric assay, and the latter then used simply to give an indication of cellular heterogeneity, but this may be impractical for routine clinical use.

It is in principle possible to calibrate flow cytometric GSH assays so that the results can be described in terms of number of GSH molecules per cell (Hedley *et al.*, 1990). This is done by comparing the fluorescence of cells with that of calibration beads. Cells can be variably depleted of GSH *in vitro*, and the actual GSH content then determined biochemically. An aliquot of cells is stained for flow cytometric GSH determination using a strictly standardized procedure, and mean fluorescence compared to that of the calibration beads run on identical instrument settings (Fig. 2). This allows the beads to be assigned a "GSH equivalent fluorescence." Test samples prepared under identical conditions can then be compared to these beads in order to determine mean cellular GSH content.

Acknowledgment

This work was supported by a grant from the National Cancer Institute of Canada.

References

Asghar, K., Reddy, B. G., and Krishna, G. (1975). *J. Histochem. Cytom.* **23,** 774–779.
Bennett, H. S., (1951). *Anat. Rec.* **110,** 231–248.
Boobis, A. R., Fawthrop, D. J., and Davies, D. S. (1989). *Trends Pharmacol. Sci.* **10**(7), 275–280.
Coates, A., and Tripp, E. H. (1991). *Melanoma Res.* **1,** 327–332.
Coleman, C. N., Bump, E. A., and Kramer, R. A. (1988). *J. Clin. Oncol.* **6**(4), 709–733.
Cook, J. A., Iype, S. N., and Mitchell, J. B. (1991). *Cancer Res.* **51,** 1696–1612.
Durand, R. E., and Olive, P. L. (1983). *Radiat. Res.* **95,** 456–470.
Eyer, P., and Podhradsky, D. (1986). *Anal. Biochem.* **153,** 57–66.
Farber, J. L., Kyle, N. E., and Coleman, J. B. (1990). *Lab. Invest.* **62**(6), 670–679.
Hedley, D. W. (1993). *Ann. N. Y. Acad. Sci.* **677,** 341–353.
Hedley, D. W., Hallahan, A. R., and Tripp, E. H. (1990). *Br. J. Cancer* **61,** 65–68.
Hedley, D. W., and Chow, S. (1994). *Cytometry* **15,** 349–358.
Larrauri, A., Lopez, P., Lechon-Gomez, M. J., and Castell, J. V. (1987). *J. Histochem. Cytochem.* **35,** 271–274.
Meister, A., and Anderson, N. E. (1983). *Annu. Rev. Biochem.* **52,** 711–760.
Mitchell, J. B., Cook, J. A., DeGraff, W., Glatstein, E., and Russo, A. (1989). *Cytometry* **9,** 529–532.
O'Connor, J. E., Kimler, B. F., Morgan, M. C., and Tempas, K. J. (1988). *Cytometry* **9,** 529–532.
Poot, M., Verkert, A., Koster, J. F., and Jongkind, J. F. (1986). *Biochim. Biophys. Acta* **883,** 580–584.
Poot, M., Kavanagh, T. J., Kang, H. C., Haugland, R. P., and Rabinovitch, P. S. (1991). *Cytometry* **12,** 184–187.
Reed, D. J. (1990). *Annu. Rev. Pharmacol. Toxicol.* **30,** 603–631.
Rice, G. C., Bump, E. A., Shrieve, D. C., Lee, W., and Kovacs, M. (1986). *Cancer Res.* **46,** 6105–6110.
Shrieve, D. C., Bump, E. A., and Rice, G. C. (1988). *J. Biol. Chem.* **263,** 14107–14114.
Tietze, F. (1969). *Anal. Bioch.* **27,** 502–522.
Treumer, J., and Valet, G. (1986). *Exp. Cell Res.* **163,** 518–524.
Ublacker, G. A., Johnson, J. A., Siegel, F. L., and Mulcahy, R. T. (1991). *Cancer Res.* **51,** 1783–1788.
Waxman, D. J. (1990). *Cancer Res.* **50,** 6449–6454.

CHAPTER 4

Assay of Cell Resistance to Ara-C

Francis Lacombe
Laboratoire d'Hématologie
Hôpital Haut-Lévèque
33600 Pessac
France

I. Introduction

Ara-C is one of the most effective agents for treatment of human acute leukemia. Ara-C is an S phase-specific agent and its active metabolite, ara-CTP, is a substrate for DNA polymerase and is incorporated into DNA. Ara-C incorporation correlates strongly with cytotoxicity (Kufe *et al.*, 1980; Major *et al.*, 1981, 1982; Rustum and Preisler, 1979) and results in the inhibition of DNA synthesis (Rustum, 1979). Correlations between the inhibition of DNA synthesis and response to treatment with ara-C were found by Harris and Grahame-Smith (1980) and Preisler *et al.* (1984).

In flow cytometry (FCM), the use of bromodeoxyuridine/DNA (BrdUrd/DNA) staining with a monoclonal antibody anti-BrdUrd and propidium iodide (Dolbeare *et al.*, 1983, 1985) allows for very good and simple determination of the cell-cycle phases and an evaluation of the rate of DNA synthesis. Alterations in the rate of DNA synthesis during S phase can be quickly and accurately evaluated by FCM from bivariate BrdUrd/DNA distributions, as the amount of BrdUrd incorporated over a short period is proportional to the amount of DNA synthesized during that period (Dean *et al.*, 1984).

Pallavicini *et al.* (1985) measured the cytokinetic properties of asynchronous and cytosine arabinoside perturbed murine tumors by simultaneous BrdUrd/DNA analysis using dedicated software. Cell resistance to ara-C has been evaluated from BrdUrd/DNA distributions in cell lines sensitive or resistant to ara-C by Waldman *et al.* (1985) and Ross *et al.* (1987). Katano *et al.* (1989) used this technique to detect cell resistance to ara-C in a small series of patients.

To have a reliable evaluation of the degree of resistance of leukemic cells to ara-C in clinical situations it is necessary to determine the proportion of S phase cells resistant to ara-C and the level of BrdUrd incorporation in ara-C-resistant cells. We have described a computerized method for the analysis of BrdUrd/

Table I
Brief Review of the Literature

Author	Comment
Harris and Grahame-Smith (1980)	Variation in sensitivity of DNA synthesis to ara-C in acute myeloid leukemia. (First paper to directly correlate ara-C resistance and inhibition of DNA synthesis using [^{3}H]dThd technique.)
Preisler *et al.* (1984)	Inhibition of DNA synthesis by cytosine arabinoside: Relation to response of acute nonlymphocytic leukemia to remission induction therapy and to stage of the disease. (Use of [^{3}H]dThd technique.)
Waldman *et al.* (1985)	Detection of ara-C-resistant cells at low frequency using the BrdUrd assay. (No quantitation of the degree of resistance of cells resistant to ara-C.)
Ross *et al.* (1987)	Study of ara-C metabolism in acute leukemia. (Detection of cells resistant to ara-C using a single staining with BrdUrd.)
Andreeff (1990)	(Evocation of the possibility of the use of the BrdUrd/DNA method to detect ara-C-resistant cells in clinical practice.)
Katano et al. (1989)	Sensitivity of S phase cells to cytosine arabinoside in childhood acute lymphoblastic leukemia using BrdU/PI technique. (An evaluation of the number of cells remaining in the S phase box after incubation with ara-C.)
Lacombe *et al.* (1992)	Quantitation of resistance of leukemic cells to cytosine arabinoside from BrdUrd/DNA bivariate histograms. (Study with HL-60 leukemic cells.)

DNA bivariate distributions of HL-60 cell lines that were sensitive or resistant to various doses of ara-C (Lacombe *et al.*, 1992). The degree of resistance was determined automatically using an index of resistance to ara-C (RI). We showed that it was possible to analyze accurately the BrdUrd/DNA distributions and hence determine small numbers of slightly resistant cells from the RI. We published results concerning the application of this method to the detection of ara-C resistance in patients with acute myeloid leukemia treated with ara-C (Lacombe *et al.*, 1994).

II. Application

This method of analysis of BrdUrd/DNA bivariate distributions can be applied to leukemic cell lines sensitive or resistant to ara-C and to leukemic cells of patients with acute myeloid or lymphoblastic leukemia. Moreover, it can be applied to all the drugs acting on DNA synthesis such as 5-F-ara-A and anthracyclines. Of course, concentrations used for *in vitro* tests must be compatible with patient plasma levels of the drugs.

It should also be possible to analyze the resistance of leukemic cells after *in vivo* infusion of BrdUrd and treatment with cytotoxic drugs. This method has also been applied to the quantitation of the inhibition of DNA synthesis by antisense oligonucleotides in chronic myelogenous leukemia (Mahon *et al.*, 1993). A brief review of the literature on the measurement of the inhibition of DNA synthesis by ara-C is presented in Table I.

Table II
Chemical Products

Product	Vendor	Comment
Bromodeoxyuridine (BrdUrd)	Calbiochem, San-Diego, CA	Stock solution, 30 m*M* stored at −20°C
RPMI 1640	Gibco BRL, Grand Island, NY	
Lymphoprep	Nyegaard, Norway	
5637 cells	ATCC, Rockville, MD	HTB9 cell line
Tween 20	Sigma Chemical Co., St. Louis, MO	
Anti-BrdUrd (IU-4)	Gift from Frank Dolbeare; available from TEBU, Le Perray en Yveline, France	
Anti-mouse IgG-FITC $F(ab')_2$	Sigma Chemical Co.	
Propidium iodide (PI)	Calbiochem	Stock Solution, 1 mg/ml
RNase	Boehringer-Mannheim, Germany	Stock solution, 2.5 mg/ml
ara-C	Upjohn (Lab), Kalamazoo, MI	Stock solution, 20 mg/ml stored at −20°C

III. Materials

The main chemical products necessary for this technique are listed in Table II.

IV. Cell Preparation and Staining

A. Sample Processing

Bone marrow aspirates were collected in sodium heparinate containing tubes. Bone marrow mononuclear cells (BM-MNC) were isolated by centrifuging one volume of bone marrow aspirate diluted with one equal volume of RPMI 1640 over one volume of lymphoprep for 40 min at 400*g*. BM-MNC were washed in RPMI 1640 and resuspended in RPMI 1640 supplemented with 2 m*M* L-glutamine, penicillin (100 μg/ml), streptomycin (100 μg/ml), 20% (v/v) heat-inactivated fetal calf serum (FCS), and 10% (v/v) 5637 supernatant of bladder cell line. The cell concentration was adjusted to 1×10^6 cells/ml and cells were cultured for 24–48 hr in a fully humidified 5% CO_2, 37°C incubator.

B. Cell Incubation with Ara-C and BrdUrd

Viability of the cultured cells was first assessed by trypan blue exclusion. Approximately 1 ml of the cultured cells (adjusted to 1×10^6 viable cells/ml) was added to four tubes. Two samples were incubated without ara-C, one sample was incubated with 0.1 μg/ml of ara-C, and the remaining sample with 1 μg/ml of ara-C for 3 hr. The samples were then washed in RPMI 1640 to remove the drug. Finally, the blast cells were pulse labeled with 10 μM of BrdUrd for 30 min at 37°C. A sample that had not been incubated with BrdUrd or with ara-C was included as a negative control. After washing and centrifugation, the cells were fixed with 50% ethanol in 50% PBS and stored at −20°C.

C. BrdUrd/DNA Staining

The cells were washed two times in PBS and pelleted. Cells were resuspended in 1 ml 2 *M* HCl for 30 min at room temperature to partially denature DNA. The cells were then carefully washed four times with PBT buffer (PBS containing 0.5% Tween 20). Cells were then pelleted and resuspended in 100 μl of PBT buffer containing the monoclonal antibody anti-BrdUrd diluted 1 to 1000. After 30 min incubation at room temperature, 5 ml PBT buffer was added, the cells were centrifuged, washed again in 5 ml PBT buffer, and resuspended in 100 μl of PBT buffer containing the anti-mouse IgG-FITC $F(ab')_2$ diluted 1 to 10. After

30 min incubation at room temperature, 5 ml PBT buffer was added. The cells were then centrifuged, washed once in 5 ml PBT buffer, and resuspended in 1 ml PBS containing 10 μg/ml PI and 25 μg/ml RNase. Cells were analyzed by FCM the same day.

V. Critical Aspects of the Procedure

Bone marrow samples have to be aspirated on sodium heparinate containing tubes and not lithium heparinate so as to maintain good viability of the blast cells. The 24- to 48-hr culture with the 5637 cell line supernatant is not necessary for the success of the experiment, but it allows a greater flexibility in the schedule of the BrdUrd and ara-C incubations. Nevertheless, the presence of a large number of dead cells after 48 hr of culture would seem to be a sign of bad prognostic for patients, a more detailed analysis of this phenomenon is now under investigation in our laboratory. The duration of incubation of the cells with ara-C (3 hr) was based on the data in the literature and the results of preliminary experiments. The 3-hr incubation period was designed for maximum incorporation of ara-C into DNA and maximum inhibition of DNA synthesis. The maximum decrease in BrdUrd incorporation was observed after 3 hr exposure to ara-C and thus this duration was retained for all subsequent experiments. We demonstrated that it was possible to perform the BrdUrd/DNA staining several weeks after fixation of the blast cells in ethanol at −20°C.

The most critical step in the BrdUrd/DNA staining procedure is the partial DNA denaturation by HCl. The length of the denaturation (30 min) and the concentration of HCl (2 *M*) must be scrupulously respected. It is impossible to correctly evaluate more than 10 tubes simultaneously if the experiment is to be reproducible. The concentration of the monoclonal anti-BrdUrd must be verified for each clone. To ensure reproducibility, it is essential to use excess antibody and to have the same ratio "concentration of anti-BrdUrd/number of cells" in each experiment.

VI. Instruments

A. FCM Assay

Cells were analyzed with an ATC 3000 cell sorter (Odam-Brucker, Wissembourg, France). An argon ion laser (2025 Spectra Physics, Les Ulis, France) was adjusted to emit 500 mW at 488 nm. Green fluorescence was measured through a band pass 530-nm filter, which gave a measure of the amount of bound anti-BrdUrd antibody. Red fluorescence was measured through a long-pass 600-nm filter, which determined the amount of bound PI. The bivariate BrdUrd/DNA (green/red) fluorescence distributions were displayed as dot plots

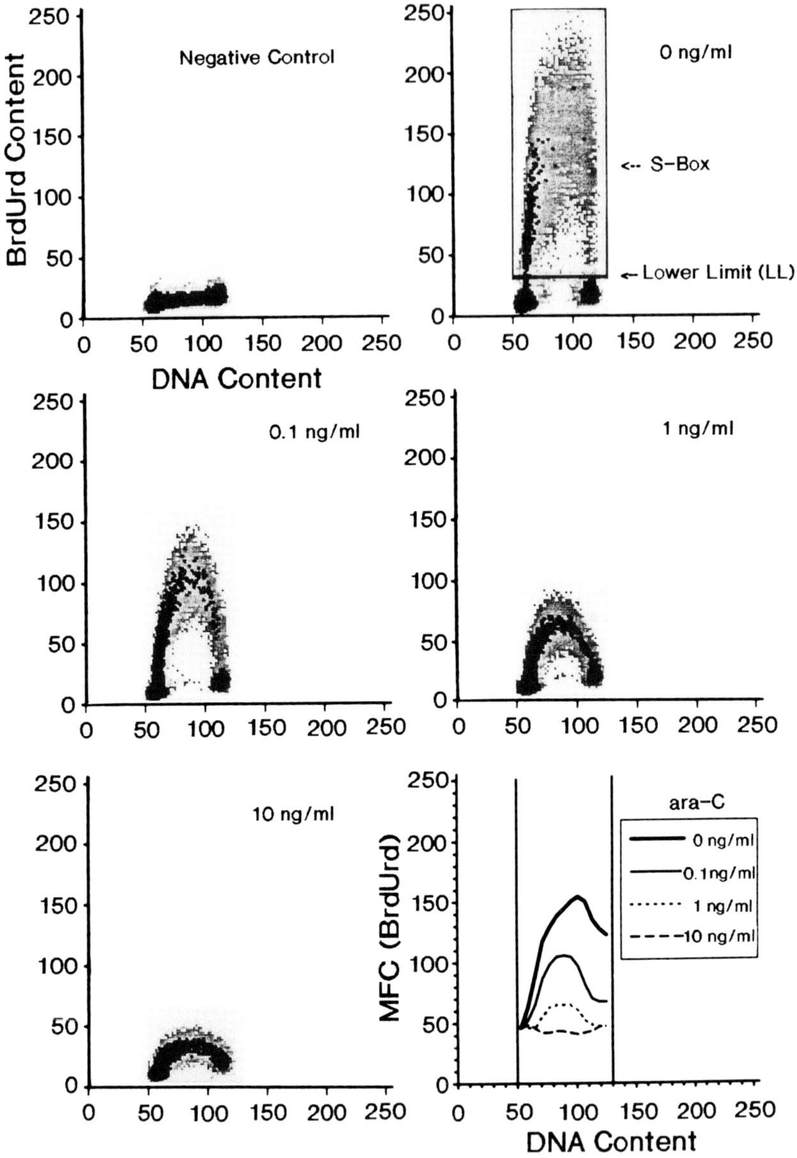

Fig. 1 BrdUrd/DNA bivariate distributions of HL-60 cells incubated without BrdUrd or ara-C (negative control, upper left), incubated for 3 hr without ara-C (0 ng/ml, upper right), incubated with 0.1, or 1, or 10 ng/ml of ara-C. The S-box and the lower limit (LL) of the S-box indicated in the upper right panel are the same for all the BrdUrd/DNA distributions. The plots of the rate of DNA synthesis corresponding to the S-box of the BrdUrd/DNA bivariate distributions (see text) are plotted in the lower right panel.

(256x256 channal arrays). Each analysis was carried out on a minimum of 50,000 cells. Doublets were eliminated by gating all data on the DNA fluorescence peak–area cytogram. The green signal was corrected for red contamination by electronic compensation until the biparametric FITC/PI histogram assumed an horizontal shape (mean green fluorescences of G_1 and G_2M phases approximatively equal).

B. Analysis of BrdUrd/DNA Bivariate Distributions

We developed a program using Pascal to analyze the BrdUrd/DNA bivariate distributions (cytograms) which runs on the ATC 3000 cell sorter and PC compatible computers using files recorded or converted into FCS 2.0 format as formulated by the Data File Standards Committee of the Society for Analytical Cytology (1990).

All the curves and calculations were carried out with data corresponding to a box enclosing the cell population in S phase (S-box). BrdUrd/DNA histograms of HL-60 cells sensitive to ara-C and incubated with and without various doses of ara-C are shown in Fig. 1.

C. DNA Synthesis Curves

As described by Dean *et al.* (1984) and Dolbeare *et al.* (1983,1985), the intensity of the green fluorescence is proportional to the amount of incorporated BrdUrd and the change in the rate of DNA synthesis across S phase can be estimated from the mean of the BrdUrd distribution for each channel of DNA content.

The mean rates of BrdUrd incorporation [mean fluorescence channel of BrdUrd-FITC; MFC(BrdUrd)] for each DNA content (channel by channel), i.e., the change in the mean rate of DNA synthesis for cells incubated with increasing concentrations of ara-C and enclosed in the S-box, are plotted in Fig. 1 (lower right). The peak of these curves corresponds to the maximum rate of DNA synthesis expressed in arbitrary units. DNA synthesis from the same experiment can thus be readily compared from these plots.

D. Calculation of the Resistance Index to Ara-C

An area within the BrdUrd/DNA bivariate distribution is defined enclosing BrdUrd staining cells with S phase DNA content (S-box). The S-box must also be plotted from the BrdUrd/DNA cytogram of the negative control sample and from that of the sample incubated without ara-C. A ratio RS is then calculated to quantify the percentages of cells incorporating BrdUrd in the presence of ara-C as follows:

$$\mathrm{RS} = \frac{\text{percentage of cells in S-box incubated with ara-C}}{\text{percentage of cells in S-box incubated without ara-C}} \times 100. \qquad (1)$$

The percentages are corrected taking into account the percentage of cells in the S-box of the negative control.

In the S-box, the mean fluorescence channel of the green BrdUrd fluorescence [MFC(BrdUrd)] corresponding to the rate of DNA synthesis but referred to the lower limit (LL) of the S-box on the *y*-axis is calculated. Inhibition of the rate of DNA synthesis by ara-C in the cells incorporating BrdUrd is evaluated from the ratio MS as follows:

$$\text{MS} = \frac{\text{MFC(BrdUrd) of cells in S-box incubated with ara-C}}{\text{MFC(BrdUrd) of cells in S-box incubated without ara-C}}. \qquad (2)$$

The program calculated MFC(BrdUrd) taking account of the value of MFC(BrdUrd) of the negative control.

The index of resistance is defined as

$$\text{RI} = \text{RS} \times \text{MS}. \qquad (3)$$

The program requires the parameters of the S-box, the data corresponding to the cells incubated without the drug, and the data corresponding to the negative control. For each sample incubated with various doses of ara-C, the program produces the RI and the curves of the rate of DNA synthesis.

VII. Results and Discussion

Using this technique with leukemic cell lines, we proved that it was possible to quantify the inhibition of DNA synthesis by ara-C (Fig. 1) Lacombe *et al.*, 1992). We applied this method to 121 patients with *de novo* AML that were treated with a protocol which included daunorubicin (50–60 mg/m^2/day, 3 days) and ara-C (100 mg/m^2/day, continuous infusion, 7–10 days). In order to properly

Table III
Clinical Characteristics of Patients

		Failure	CR
Gr 1	*N*	1	62
Gr 2	*N*	5	19
Gr 3	*N*	9	0
Total	*N*	15	81
Total	Blast cells (%)	79.3 ± 14.4	69.2 ± 22.4
	Age	53.6 ± 19.9	54.2 ± 15.9
	Leucocytes (/mm^3)	91,018 ± 24,875	36,035 ± 57,372
	S phase (%)	24.3 ± 12.1	18.0 ± 9.8
	RI(0.1)	17.8 ± 8.4	6.4 ± 6.3
	RI(1)	9.4 ± 7.0	2.3 ± 2.0

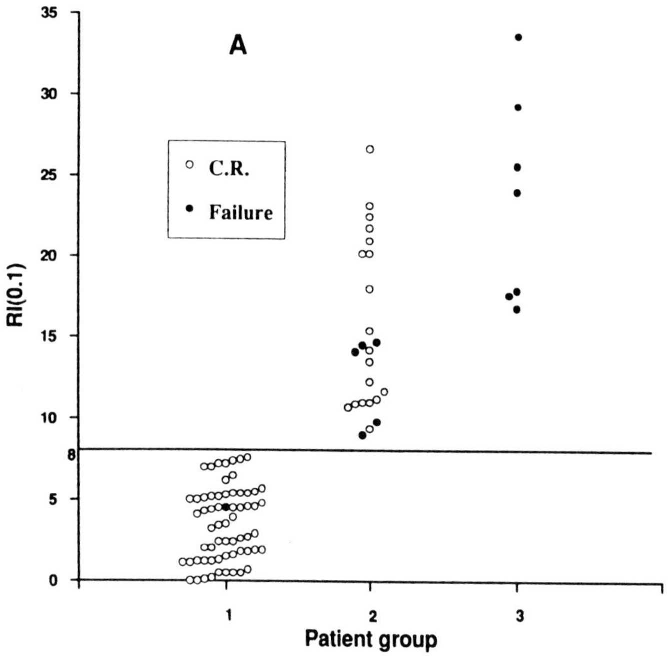

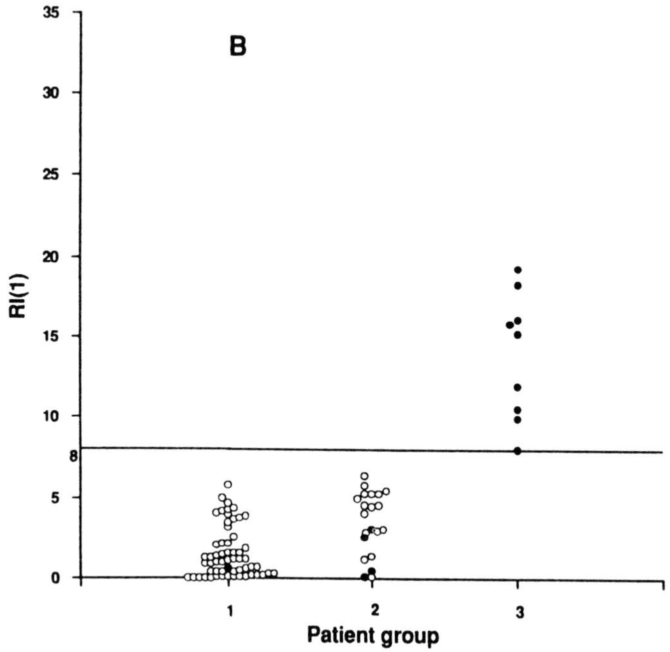

Fig. 2 Distribution of the 96 patients with AML in the three groups determined with the RI (0.1) and RI(1) values (see text). Group 1 corresponds to patients with values RI(0.1) and RI(1) < 8, group 2 to patients with RI(0.1) > 8 and RI(1) < 8 values, and group 3 to patients with RI(0.1) and RI(1) > 8 values. (A) RI(0.1) values for patients in complete remission or in failure distributed in the three groups. (B) RI(1) values for the same patients.

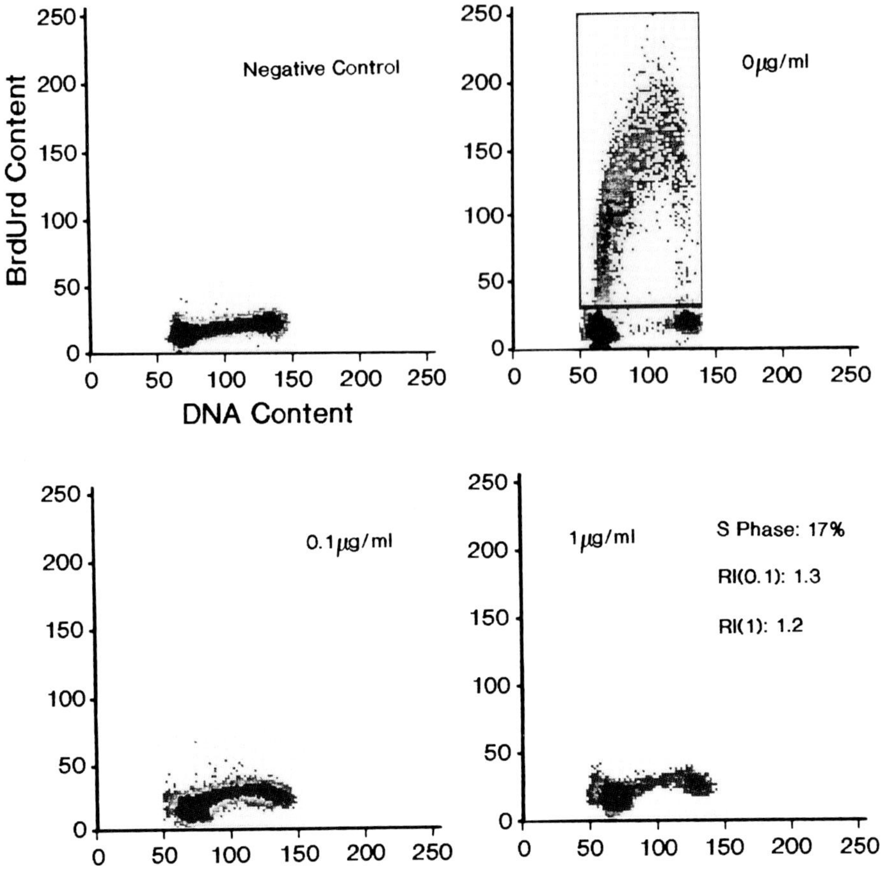

Fig. 3 BrdUrd/DNA bivariate distributions for a patient in CR. The S-box as indicated in the upper right figure corresponds to the sample which is not incubated with ara-C (0 μg/ml) and is the same for the four distributions; the percentage of cells having incorporated BrdUrd is indicated in the lower right figure. RI indicates the distributions for the leukemic cells incubated with 0.1 and 1 μg/ml ara-C.

evaluate the RI, a sufficient percentage of cells in S phase (3% in this study) after 48 hr culture was considered to be necessary. We found 96/121 patients in this series with more than 3% of cells in S phase. Among the 96 evaluable patients using our method, 81 (84.4%) achieved a complete remission (CR) and 15 (15.6%) did not. For each patient, we calculated the RI with two doses of ara-C encompassing the presumed plasma concentration of ara-C, i.e., 0.1 and 1 μg/ml. RI(0.1) and RI(1) corresponded to the two doses of ara-C used.

Three groups of patients were created:

- Group 1 with values of RI(0.1) and RI(1) < 8,
- Group 2 with a value of RI(0.1) > 8 and a value of RI(1) < 8,
- Group 3 with values of RI(0.1) and RI(1) > 8.

The numbers of patients (N), the mean and standard deviation of percentages of bone marrow blast cells at diagnosis, patient age, leucocytes (number/mm^3), percentages of cells in S phase, RI(0.1), and RI(1) are shown in the Table III for the cases in failure and in CR.

The distribution of the patients in CR and in failure among the three groups are presented in Fig. 2. In group 1, sixty-two patients achieved CR and 1 patient failed to achieve CR. In group 3 all the nine patients were in failure. In group 2, nineteen patients achieved CR and five were in failure; the distribution differs significantly from that seen in groups 1 and 3. In our series, the threshold, 8, is the optimal value to discriminate patients that are sensitive or resistant to ara-C. This value could be different in other series and in other hands and needs to be validated. The BrdUrd/DNA bivariate histograms obtained from a sample of a patient who achieved CR are shown in Fig. 3. The bivariate histograms of a patient who was resistant to treatment are shown in Fig. 4.

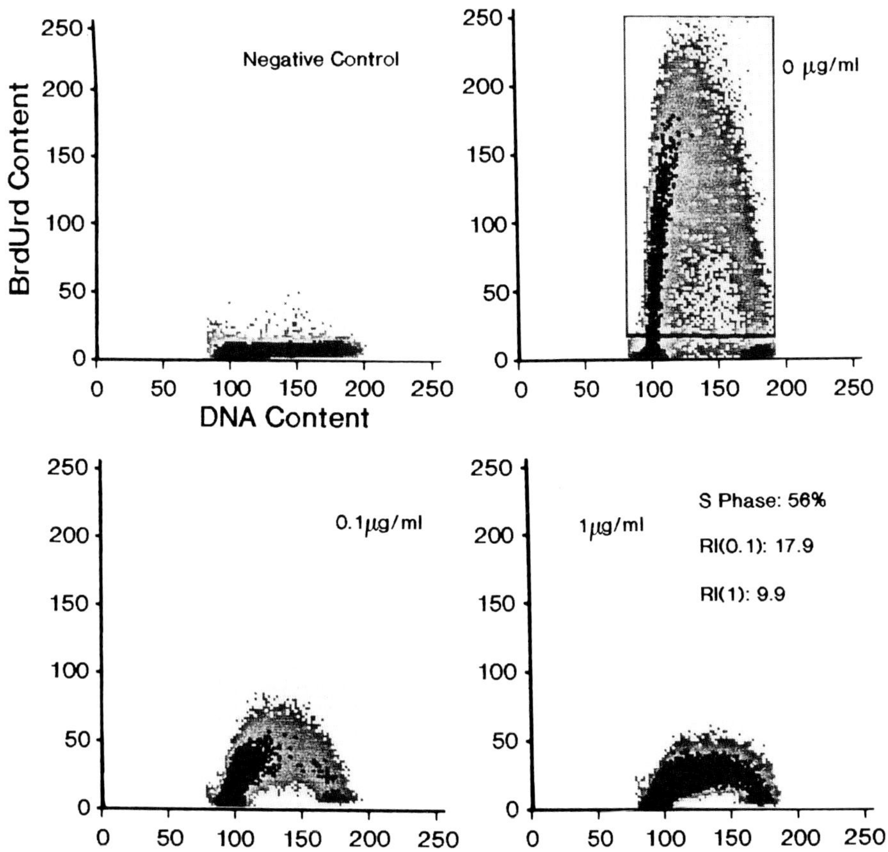

Fig. 4. BrdUrd/DNA distributions for a patient in failure. See Fig. 3 for the comments.

The S phase boxes used for the calculation of the RI are indicated in the histograms.

The RI is independent of the shape of the BrdUrd labeling distribution and can detect slightly resistant homogeneous populations of blast cells as well as resistant cells among ara-C-sensitive cells.

We believe it is essential to determine:

1. The percentage of cells remaining in the S-box after treatment by ara-C. These are the cells potentially resistant to a given dose of ara-C.
2. The extent of BrdUrd incorporation *in these cells alone*, which is the best way of assessing their degree of resistance to a given dose of ara-C.

Katano *et al.* (1989) assayed ara-C-resistant blast cells from the proportion of S phase-resistant cells in 20 children with acute lymphoblastic leukemia and correlated the results with their disease status. No significant results were found, perhaps due to the loss of information from the absence of BrdUrd incorporation data.

VIII. Comparison of the Methods

The measurement of the inhibition of DNA synthesis by ara-C is essentially assessed with two methods: the inhibition of [^{3}H]dThd incorporation and the inhibition of BrdUrd incorporation into DNA.

A. Inhibition of [^{3}H]dThd Incorporation

The inhibition of [^{3}H]dThd incorporation is a standard method for evaluation of the rate of DNA synthesis and has been used with both cell lines (Harris and Grahame-Smith, 1980; Ross *et al.*, 1987) and leukemic blast cells (Preisler *et al.*, 1984; Ross *et al.*, 1986). The limitations of this method are essentially associated with the use of radioactivity and the impossibility of cell-by-cell analysis to distinguish between ara-C-sensitive and ara-C-resistant cells and to quantify the degree of inhibition of DNA synthesis in the resistant cells alone.

B. Inhibition of BrdUrd Incorporation

The inhibition of BrdUrd incorporation for evaluation of the ara-C resistance of leukemic cells has been described in two publications.

Ross *et al.* (1987) reported the effect of ara-C on the BrdUrd incorporation into DNA of HL-60 cell lines. They measured the incorporation of BrdUrd (evaluated from MS) in ara-C-sensitive and -resistant HL-60 cells, but they did not measure DNA content and the rate of DNA synthesis could not be determined due to the long exposure of HL-60 cells to BrdUrd (16 hr). Moreover, the inhibition of BrdUrd incorporation by ara-C was only compared between sensitive and highly resistant cell lines (100 n*M*).

Waldman *et al.* (1985) studied the proportion of S phase-resistant cells remaining in a trapezoidal S-box (evaluated from RS) after incubation with 100 μM ara-C, without reference to the degree of inhibition of the rate of DNA synthesis by ara-C. The cells used in the experiments were completely resistant to 100 μM ara-C and so there was no inhibition of DNA synthesis. A small number of highly resistant cells was detected. However, levels of ara-C as high as this are not found in patients' plasma and complete resistance to therapeutic doses of ara-C is probably rare. Partial resistance of blast cells to the therapeutic doses of ara-C is more commonly encountered. The concomitant slight decrease in cell DNA synthesis is only detectable by measurement of the incorporation of BrdUrd.

Our assay combines the advantages of [^{3}H]dThd pulse labeling to quantitate the rate of DNA synthesis with those of FCM for the analysis of DNA content distributions.

References

Andreeff, M. (1990). *In* "Flow Cytometry and Sorting" (M. R. Melamed, T. Lindmo, and M. L. Mendelsohn, eds.), 2nd ed., pp. 697–724. Wiley-Liss, New York.

Data File Standards Committee of the Society for Analytical Cytology (1990). *Cytometry* **11,** 323–332.

Dean, P. N., Dolbeare, F., Gratzner, H., Rice, G. C., and Gray, J. W. (1984). *Cell Tissue Kinet.* **17,** 427–436.

Dolbeare, F., Gratzner, H., Pallavicini, M. G., and Gray, J. W. (1983). *Proc. Natl. Acad. Sci. U.S.A.* **80,** 5573–5577.

Dolbeare, F., Beisker, W., Pallavicini, M. G., Vanderlaan, M., and Gray, J. W. (1985). *Cytometry* **6,** 620–626.

Harris, A., and Grahame-Smith, D. (1980). *Br. J. Haematol.* **45,** 371–378.

Katano, N., Tsurusawa, M., Niwa, M., and Fujimoto, T. (1989). *Am. J. Pediatr. Hematol./Oncol.* **11,** 411–416.

Kufe, D., Major, P., Egan, E., and Beardsley, P. (1980). *J. Biol. Chem.* **255,** 8997–9000.

Lacombe, F., Belloc, F., Dumain, P., Puntous, M., Lopez, F., Bernard, P., Boisseau, M. R., and Reiffers, J. (1992). *Cytometry* **13,** 730–738.

Lacombe, F., Belloc, F., Dumain, P., Puntous, M., Cony-Makhoul, P., Sausc, M. C., Bernard, P., Boisseau, M. R., and Reiffers, J. (1994). *Blood.* In Press.

Mahon, F. X., Belloc, F., and Reiffers, J. (1993). *Lancet* **341,** 566.

Major, P. P., Egan, E. M., Beardsley, G. P., Minden, M. D., and Kufe, D. W. (1981). *Proc. Natl. Acad. Sci. U.S.A.* **78,** 3235–3239.

Major, P. P., Egan, E. M., Herrick, D. J., and Kufe, D. W. (1982). *Biochem. Pharmacol.* **31,** 2937–2941.

Pallavicini, M. G., Summers, L. J., Dolbeare, F. D., and Gray, J. W. (1985). *Cytometry* **6,** 602–610.

Preisler, H. D., Epstein, J., Raza, A., Azarnia, N., Browman, G., Booker, L., Goldberg, J., Gottlieb, A., Brennan, J., Grunwald, H., Rai, K., Vogler, R., Winton, L., Miller, K., and Larson, R. (1984). *Eur. J. Cancer Clin. Oncol.* **20,** 1061–1068.

Ross, D. D., Thompson, B. W., Joneckis, C. C., Akman, S. A., and Schiffer, C. A. (1986). *Blood* **68,** 76–82.

Ross, D. D., Thompson, B. W., Joneckis, C. C., Akman, S. A., and Schiffer, C. A. (1987). *Semin. Oncol.* **14,** Suppl. 1, 182–191.

Rustum, Y. M. (1979). *Cancer Res.* **38,** 543–549.

Rustum, Y. M., and Preisler, H. D. (1979). *Cancer Res.* **39,** 42–49.

Waldman, F., Dolbeare, F., and Gray, J. W. (1985). *Cytometry* **6,** 657–662.

CHAPTER 5

Detection of mRNA Species by Flow Cytometry

Francis Belloc and Françoise Durrieu

Laboratoire d'Hématologie
Hôpital Haut-Lévêque
33600 Pessac
France

I. Introduction

Flow cytometry is now widely used for determination of total or nuclear RNA after staining with nucleic acid binding fluorochromes such as acridine orange (Darzynkiewicz *et al.*, 1980), pyronine Y (Darzynkiewicz *et al.*, 1987), thioflavine (Sage *et al.*, 1983), and thiazole orange (Lee *et al.*, 1986). These methods have been employed successfully to study alterations in cellular RNA content during the cell cycle (Darzynkiewicz *et al.*, 1980; Staiano-Coico *et al.*, 1989; Campan *et al.*, 1992) or to discriminate reticulocytes from mature red blood cells (Sage *et al.*, 1983; Lee *et al.*, 1986). However, these staining methods have been little used to follow the effect of drugs on RNA metabolism as they essentially measure ribosomal and transfer RNA, which together make up 90% of the total RNA and are relatively stable species. It would be a great advantage to be able to analyze the effect of drugs on mRNA as the rapid turnover of these species makes them specially sensitive to drugs affecting RNA metabolism.

Usually, mRNA is quantified from cell lysates by affinity separation, which exploits the presence of poly(A)$^+$ sequences at the 3′ end of eukaryotic mature mRNA (Jacobson, 1987). Columns of polymers coupled to poly(U) or oligo(dT) are used for their ability to hybridize with the poly(A)$^+$ sequences of mRNA. The bound RNA are then eluted from the column and quantified by spectrophotometry. Unfortunately this method requires a large number of cells, has a variable preparative yield, and has the drawbacks of batch analysis applied to heterogeneous populations.

Various attempts to detect rRNA or mRNA by flow cytometry (FCM) have been described using either biotinylated probes and fluorescent *in situ* hybridization (FISH) (Bauman and Bentvelzen, 1988; Bauman *et al.*, 1990; Bayer and Bauman, 1990) or primed *in situ* labeling (PRINS) (Mogensen and Kolvroa, 1991; Bolund *et al.*, 1991). Such methods are derived from microscopic and morphologic methodologies. Analysis of FISH and FCM has been hampered by two main problems. First, although the photomultipliers used in FCM can detect low fluorescence intensities, they measure the entire fluorescence signal for each cell and cannot resolve the pattern of fluorescence within the cell. Target-specific fluorescence is not therefore readily discriminated from background fluorescence. The second problem is the "sponge-like" behavior of fixed permeabilized cells in suspension, which tend to trap macromolecules in their cytoplasm. This considerably increases the nonspecific fluorescence when fluorescent proteins such as avidin or antibodies are used to reveal the probes. In view of these drawbacks, a more specific labeling method is required for flow cytometric analysis of FISH, which takes both the characteristics of the detectors and the properties of the cell suspensions into account.

In this chapter we describe two methods for revealing poly(A)$^+$ RNA, which are particularly suited to flow cytometric analysis (Fig. 1).

- FISH using FITC-coupled olido(dT)$_{15}$ [FITC-o(dT)] as a probe: this small, directly fluorescent oligonucleotide enables one-step staining of the cells, thereby reducing cell handling and damage. Background fluorescence is also decreased as this probe is easily washed out of cells.
- PRINS using oligo(dT) as a primer and reverse transcriptase as a polymerase: the poly(A)$^+$ RNA is used as a template to incorporate FITC-dUTP into complementary strand. Incorporation of fluorescence into a macromolecule is conditioned by hybridization of the primer with the target and so confers high specificity for the fluorescent label.

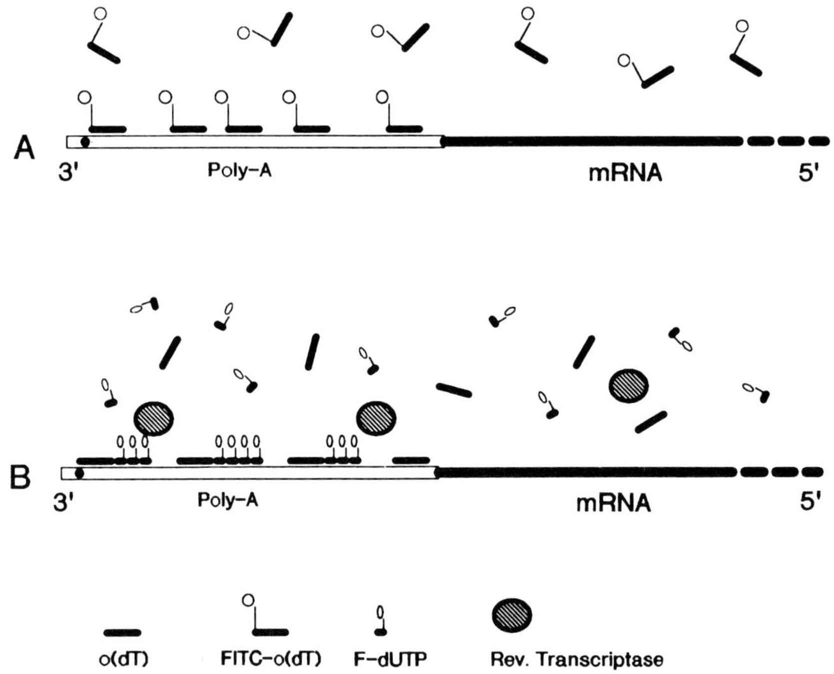

Fig. 1 Schematic representation of the two procedures described for discriminating poly(A)$^+$ RNA by flow cytometry. (A) The oligo(dT) was chemically coupled to a molecule of FITC and directly used as a probe for FISH. (B) The oligo(dT) was used as a primer for PRINS. Labeling was introduced from fluorescein-12-dUTP by elongation of the primer using reverse transcriptase *in situ*.

II. Applications

Multiparametric analysis can be employed to detect and quantify mRNA by FCM. The different populations from heterogeneous samples can be identified (on the basis of their scatter properties for example) and mRNA content in a given population can by analyzed. Cell membranes can also be immunolabeled prior to fixation of the cells in order to study alterations in mRNA content in a population expressing a particular antigen. The FCM methods developed in our laboratory were designed principally to study the effect of anti-neoplastic drugs on mRNA metabolism. These drugs usually induce a block in one of the phases of the cell cycle. Comparison of the mRNA content in bulk of a blocked cell population with that of a growing population provides little information, as the mRNA content of growing cells varies considerably during the cell cycle (Belloc *et al.*, 1993a). Moreover, some mRNA content of anthracycline-treated cells (blocked in G_2) with growing control cells (which are mostly G_1) is meaningless. However, the mRNA content of individual cells and, for example, gated G_2M populations, can be assessed by FCM after DNA and mRNA are double-stained.

III. Materials

Oligothymidylate [$o(dT)_{15}$] was purchased from Pharmacia (Saint Quentin en Yvelines, France); chemical reagents were from Aldrich (Strasbourg, France); and FITC isomer I adsorbed on Celite, poly(A), and poly(U) were from Sigma (St. Quentin Fallavier, France). Pancreatic ribonuclease A (RNase) was purchased from Boehringer-Mannheim (Meylan, France) and propidium iodide from Calbiochem (Meudon, France). Fluorescein-12-dUTP was from Boehringer-Mannheim, and M-MLV reverse transcriptase and the 5× RT buffer were from Gibco BRL (Cergy Pontoise-France).

The promyelocytic leukemia cell line HL-60 was routinely cultured in suspension in RPMI supplemented with 10% fetal calf serum, glutamine, Hepes, buffer, penicillin, and streptomycin. The cells were maintained in exponential growth by dilution to 2×10^5/ml every other day.

A. Synthesis of Fluorescent Oligo(dT)

All the solvents must be anhydrous. The 8% cetyltrimethylammonium bromide solution can be stored at 4°C, and must be warmed before use to facilitate dissolution. The FITC adsorbed on Celite and the dipyridyldisulfide (Aldrithiol) were stored at 4°C. The other reagents were stored desiccated at room temperature.

The procedure has been described elsewhere (Godovikova *et al.*, 1986). A total of 200 μg of 5′ phosphorylated o(dT)(15 mer) was dissolved in 30 μl of distilled water and precipitated as the cetyltrimethylammonium salt by addition of 3 μl of 8% hexadecyltrimethylammonium bromide (CTAB). The precipitate was pelleted 5 min at 10,000*g*. CTAB (1.5 μl) was added and the suspension was centrifuged again. This stepwise addition of CTAB was repeated until no more precipitate was formed. At this stage the oligonucleotide was in its CTAB salt form, which is soluble in DMSO. The pellet was resuspended in dry methanol and evaporated to dryness in a vacuum drier (Speed Vac). The CTAB salt of o(dT) was dissolved in 200 μl of water-free DMSO containing *N*-methylimidazole (0.6 *M*), dipyridyldisulfide (0.3 *M*), and triphenylphosphine (0.3 *M*). The mixture was incubated for 15 min at room temperature to activate the terminal phosphate, and ethylenediamine was added (1.2 *M* final concentration) and incubated for a further 30 min. At this stage, the diamine was coupled to the 5′ phosphate. The o(dT) derivative was precipitated by adding 1 ml of acetone containing 3% of $LiClO_4$ and the precipitate was washed twice with acetone by centrifugation. The pellet was dried by evaporating the acetone at 60°C and was then dissolved in 100 μl of 0.3 *M* triethylamine in water. The 5′-terminal amino group was coupled to FITC by adding 2 mg of Celite containing 10% FITC and incubating the mixture overnight with constant agitation. The FITC-5′o(dT) was separated from the Celite by centrifugation and precipitated by adding 0.1 volume of 3 *M* sodium acetate, pH 5, and 2 volumes of ethanol. After 1 hr at −20°C the precipitate was recovered by centrifugation for 15 min at 12,000*g*, washed with 70% ethanol, and dissolved in TBE buffer. The fluorescent oligonucleotide was purified from the uncoupled reaction products by electrophoresis on a 20% acrylamide sequencing gel. The brightly fluorescent band corresponding to the FITC-o$(dT)_{15}$ was excised, the acrylamide gel was crushed, and the FITC-o(dT) was eluted overnight with 0.5 ml of 0.1% Triton X-100, 0.3 *M* $LiClO_4$ in H_2O. The FITC-o(dT) was precipitated by addition of 3 ml of acetone containing 3% $LiClO_4$ (1 hr at −20°C), washed with acetone, dried, and dissolved in 200 μl of H_2O. The concentration can be measured by spectrofluorometry, using dilutions of FITC as standard.

B. Cell Fixation

- FISH: 4×10^6 HL-60 cells were washed in cold PBS and resuspended in 1 ml of 1% paraformaldehyde in PBS for 5 min at room temperature. After centrifugation (10 min, 200*g*, 4°C), the cells were resuspended in 1 ml of PBS, and 2.3 ml of absolute ethanol was added. After centrifugation for 3 min at 1000*g*, the cells were resuspended in 1 ml of 70% ethanol and conserved at −20°C for several weeks.
- PRINS: prefixation with PFA was found to significantly reduce elongation of the primer by the reverse transcriptase. The samples for PRINS labeling were the fixed at a concentration of 4×10^6 cells/ml with 70% ethanol.

C. RNase Treatment

In some experiments, 4×10^5 fixed cells were pelleted for 20 sec at 12,000*g*, resuspended in PBS containing 10 U/ml of RNase A, and incubated for 30 min at 37°C. The cells were then pelleted and conserved in 70% ethanol.

IV. Staining Procedures

All the following manipulations were carried out in autoclaved conical microtubes; the centrifugations were for 30 sec at 12,000*g* in a microfuge; all solutions were sterilized by filtration at 0.22 μm and prepared with DEPC-treated distilled water.

A. *In Situ* Hybridization with FITC-o(dT)

- Add 3 μl of a 10% solution of DEPC in ethanol to 50 μl (2 × 10^5) of the fixed cell suspension. Incubate for 5 min at room temperature to inhibit endogenous RNase.
- Pellet the cells and wash with 100 μl of PBS containing 0.5% Tween 20 (PBST). Add 100 μl 5× SSPE (1× SSPE: NaCl, 180 m*M;* EDTA, 1 m*M;* Na_2 HPO_4, 5 m*M*) and leave to equilibrate for 1 hr at room temperature. Pellet the cells.
- Resuspend the cells in 10 μl of the hybridization mixture (HM): 0.1% SDS, 0.1% Ficoll, 0.1% polyvinylpyrrolidone, 0.1% albumin, 0.5 mg/ml calf thymus DNA, and 0.5 μg/ml FITC-o(dT) in 5× SSPE. The mixture was incubated for 2 hr at 35°C under constant agitation.
- Add 40 μl of H_2O and incubate for 30 min at 35°C. Pellet the cells.
- Wash the cells in 100 μl of PBST; resuspend on 0.7 ml of PBST containing 1 μg/ml of propidium iodide and 1 U/ml of RNase. After 10 min at 20°C, the samples are ready for flow cytometry.

B. PRINS Labeling with o(dT) as Primer

- Pellet 100 μl of fixed cell suspension (4 × 10^5 cells) and resuspend in 10 μl H_2O. Add 1 μl of o(dT). Incubate for 10 min at 70°C, and cool on ice.
- Add 4 μl of 5× RT buffer; 2 μl of 0.1 *M* DTT; 2μl of a mixture of 1 m*M* of each dGTP, dATP, and dCTP; 0.7 μl of 1 m*M* fluorescein-12-dUTP; and 1 μl (200 U) of reverse transcriptase.
- Incubate for 60 min at 37°C. Add 500 μl of PBST. Pellet the cells.
- Resuspend the cells in 0.7 ml of PBST containing 1 μg/ml of propidium iodide and 1 U/ml of RNase. After 10 min at 20°C, the samples are ready for flow cytometry.

V. Critical Aspects of the Procedure and Controls

Avoidance of any contamination with RNase is crucial to the success of any work concerning mRNA. The operator must wear gloves throughout the experiments. All the conical microfuge tubes and pipette tips must be autoclaved and kept separate from the rest of the laboratory supply. All the solutions were prepared using DEPC-treated distilled water.

The duration of the hybridization step depends on the accessibility of the RNA to the probe and may vary from one experiment to another. It can be affected by the nature of the cells, the size of the oligonucleotide, and the fixation procedure, but 1 hr will usually be sufficient.

Table I
Verification of Specificity of Fish for Poly(A)$^+$ RNA

Sample	Ratio	%
Positive (0.5 μg/ml FITC-o(dT)	3.7	100
Negative (no probe)	1	0
FITC-o(dT) + poly(U)(3 mg/ml)	1.9	33
FITC-o(dT) + poly(A)(3 mg/ml)	1.1	4
Fixed cells treated with RNase	1.5	19
Cells incubated with 1 μg/ml Act-D for 6 hr.	2.1	39

Note. The ratio was calculated as the mean fluorescence channel of the sample divided by the mean fluorescence channel of the negative control.

On counterstaining DNA with propidium iodide (PI), very low concentrations of PI must be used (1 to 3 μg/ml are enough to obtain a "cell-cycle shape" on the histogram) to avoid an excessive contamination of the faint green fluorescence by PI fluorescence.

The specificity of the label must be assessed by running controls. For FISH (Table I), the specific labeling is sensitive to pretreatment of the cells with RNase; its intensity was significantly decreased by competitive unlabeled probes [o(dT) or poly(U)] and by competitive target [soluble poly(A)]. There was a significant decrease in labeling after incubating the cells with 1 μg/ml of D-actinomycin for 6 hr, showing the accuracy of the method for detecting the effect of drugs on mRNA metabolism. For PRINS (Table II), the labeling was sensitive to pretreatment with RNase and was dependent on the presence of both primer [o(dT)] and polymerase (reverse transcriptase). Excess TTP reduced the labeling, probably by competing with fluorescent dUTP for the transcriptase.

Table II
Verification of Specificity of Prins

Sample	Ratio	%
Complete reaction mix	4.1	100
Without reverse transcriptase	1	0
Without o(dT)	2	32
+70 μ*M* TTP	2.4	38
Fixed cells treated with RNase	1.8	25
Cells incubated with 1 μg/ml Act-D for 6 hr.	2.3	44

Note. The ratio was calculated as the mean fluorescence channel of the sample divided by the mean fluorescence channel of the negative control.

Incubation of the cells with D-actinomycin induced a decrease similar to that observed with the FISH method. For both methods, incubation of the cells with RNase after the labeling step will not significantly affect the fluorescent signal as the labeling (either hybridization or primer elongation) produces an RNase-resistant RNA/DNA hybrid.

VI. Instruments and FCM Analysis

For FCM analysis, we used an ATC 3000 cytometer (ODAM-Brucker; Wissembourg, France) equipped with a 2025 Spectraphysics argon ion laser. The beam was tuned to emit 500 mW at 488 nm. The emission was split into green and red fluorescences by a 600-nm short-pass filter (Melles-Griot, Arnhem, Holland). The mRNA-specific green fluorescence (FITC) was collected through a 530 ± 30 nm band pass filter (Oriel, Paris, France). The DNA-specific red fluorescence (PI) was collected through a 600-nm long-pass filter (Melles-Griot). For each experiment, a sample of RNase-treated control cells was labeled and analyzed and the green signal was corrected for red contamination by electronic compensation until the biparametric FITC/PI histogram assumed a horizontal shape (Fig. 2). All further analyses were carried out at this electronic setting.

The mRNA content of cells within specific regions of the cell cycle can be determined by gating the signal. For example, a gate can be set on the DNA histogram to provide the mRNA content of G_2M cells before and after treatment with anthracycline. The data are collected and stored in list-mode fashion, and

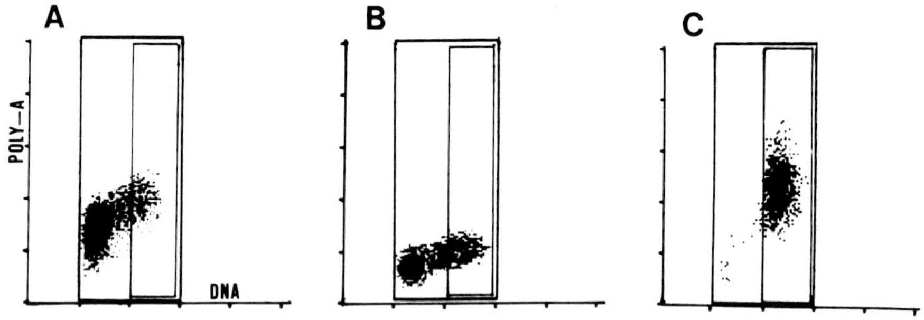

Fig. 2 FCM gated analysis of mRNA content on HL-60 cells. FISH using FITC-o(dT) was performed on control cells (A), on cells digested with RNase (B), and on cells treated with 10 n*M* idarubicin for 24 hr (C). The cells were counterstained with 1 μg/ml of PI and analyzed by FCM. The red fluorescence (DNA) is on the horizontal axis and the green fluorescence (mRNA) on the vertical axis. The green fluorescence in B was electronically compensated to obtain a horizontal distribution. The samples in A and C were then analyzed with the same electronic settings. The box encompassing the G_2M cells was used to gate the analysis of mRNA content in G_2M cells. The mean fluorescence value of control (in A) and Ida-treated cells (in C) was corrected for the value of RNase-treated cells (in B) and could then be compared.

the measurements can then be retrieved for cells in a given phase of the cell cycle. For accurate measurement of the effect of a drug on the mRNA content, the mean fluorescence channel of the sample was corrected for the value of the RNase-treated sample (RNase-sensitive fluorescence).

VII. Results and Discussion

Particular species of mRNA can be detected and quantified relatively by FISH using FITC-o(dT) as a probe followed by FCM (Table I and Belloc *et al.*, 1993a). The effect of drugs on mRNA metabolism can be evaluated by multiparametric analysis with gating on DNA content (Fig. 3 and Belloc *et al.*, 1993b). It can be seen in Fig. 2A that G_2M cells contain more mRNA than G_1 cells. Moreover, the mRNA in G_2M are probably different than those of the G_1 cells (Campan *et al.*, 1992). By gated acquisition, the mRNA content of G_2M cells from idarubicine-treated cells (Fig. 2C) can be readily compared with the mRNA content of G_2M control cells (Fig. 2A). The mRNA content was defined as the RNase-sensitive green fluorescence: the mean green fluorescence channel of the defined cell population minus the mean green fluorescence of the same population in the RNase-treated sample. Using this method, we were

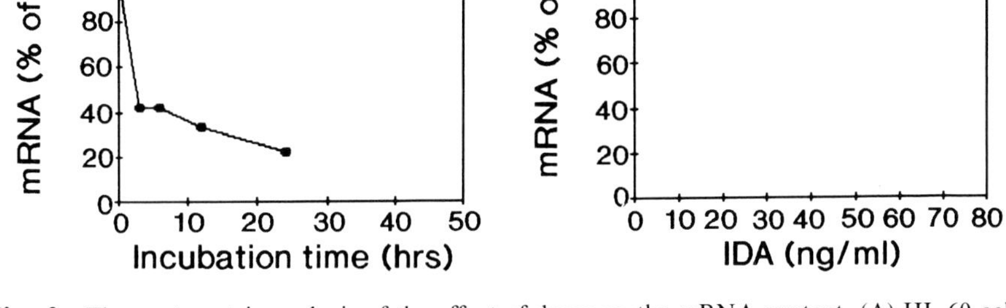

Fig. 3 Flow cytometric analysis of the effect of drugs on the mRNA content. (A) HL-60 cells were treated with 10 ng/ml of idarubicine (open symbols) or 500 ng/ml of actinomycine D (closed symbols) for different periods, fixed, and hybridized with FITC-o(dT). The RNase-sensitive mRNA content of G_2M cells was measured by flow cytometry as described in Fig. 2. The results are expressed as percent of initial mRNA content as a function of the duration of the incubation. (B) HL-60 cells were incubated for 8 hr with increasing concentrations of idarubicine. The RNase sensitive mRNA content of G_2M cells was plotted as a function of the drug concentration.

able to detect an accumulation of mRNA in Idarubicine-treated HL-60 cells and a decrease in mRNA in actinomycine-D-treated cells (Fig. 3 and Belloc *et al.,* 1993a,b). The tedious and time consuming synthesis of FITC-o(dT) is the main drawback of this method, although it is relatively inexpensive. Much time could be saved by replacing the chemical coupling method with an enzymatic tailing of the o(dT) using terminal deoxynucleotidyl transferase and commercially available fluorescent deoxynucleotide triphosphates. The main drawback of such an enzymatic procedure would be difficulty in controlling the reaction to obtain probes that are homogeneous in size and fluorescence. Dideoxy fluorescent nucleotides triphosphates can be tried for this purpose.

The PRINS labeling method using fluorescein-12dUTP as a label, o(dT) as a primer, and reverse transcriptase as a polymerase represents another approach, which in principle is an improvement on the preceding methods (Belund *et al.,* 1991). Hybridization and fluorescence incorporation are performed in a single step; the primer confers the specificity, while the sensitivity is defined by the enzyme efficiency. In addition, primer molecules retained nonspecifically in the cytoplasm do not give rise to fluorescence incorporation in the absence of template, and several fluorescent molecules can be incorporated for each o(dT) molecule hybridized. Together these advantages should produce a very high signal/noise ratio. So far we have only obtained a signal to noise ratio of the same order of magnitude as that of FITC. Table II shows that fluorescence incorporation was RNase sensitive, was influenced by treatment of the cells with D-actinomycin, and was dependent on the presence of both o(dT) and reverse transcriptase, which is indicative of good specificity for poly$(A)^+$ RNA. In practice, the PRINS method was found to be a rapid and sensitive way of detecting poly$(A)^+$ RNA by FCM. However, relative quantification is only accurate if one operates within the linear part of the enzyme kinetics. If this condition is met, the incorporated fluorescence should be proportional to the amount of hybridized o(dT).

References

Bauman, J. G. J., and Bentvelzen, P. (1988). *Cytometry* **9,** 517–524.

Bauman, J. G. J., Bayer, J. A., and van Dekken, H. (1990). *J. Microsc. (Oxford)* **157,** 73–81.

Bayer, J. A., and Bauman, J. G. J. (1990). *Cytometry* **11,** 132–143.

Belloc, F., Lacombe, F., Dumain, P., Mergny, J. L., Lopez, F., Bernard, P., Reiffers, J., and Boisseau, M. R. (1993a). *Cytometry* **14,** 339–343.

Belloc, F., Lacombe, F., Dumain, P., Lopez, F., Bernard, P., Reiffers, J., and Boisseau, M. R. (1993b). *Cytometry, Suppl.* **6,** 38.

Bolund, L., Hindkjaer, J., Junker, S., Koch, J., Kolvraa, S., Mogensen, J., Nygaard, M., and Pedersen, S. (1991). *Cytometry, Suppl.* **5,** 61.

Campan, M., Desgranges, C., Gadeau, A. P., Millet, D., and Belloc, F. (1992). *J. Cell. Physiol.* **150,** 493–500.

Darzynkiewicz, Z., Traganos, P., and Melamed, M. R. (1980). *Cytometry* **1,** 98–108.

Darzynkiewicz, Z., Kapuscinski, J., Traganos, F., and Crissman, H. (1987). *Cytometry* **8,** 138–145.

Godovikova, T. S., Zarytova, V. F., and Khalimskaya, L. M. (1986). *Bioorg. Khimi.* **12,** 475–481.

Jacobson, A. (1987). *In* "Methods in Enzymology" (S. L. Berger and A. R. Kimmel, eds.), Vol. 152, pp. 254–261. Academic Press, San Diego.
Lee, L. G., Chen, C. H., and Chiuu, L. A. (1986). *Cytometry* **7,** 508–517.
Morgensen, J., and Kolvroa, S. (1991). *Exp. Cell Res.* **196,** 92–98.
Sage, B. H., Jr., O'Connell, J. P., and Mercouno, T. J. (1983). *Cytometry* **4,** 222–227.
Staiano-Coico, L., Darzynkiewicz, Z., and McMahon, C. K. (1989). *Cell Tissue Kinet.* **22,** 235–243.

CHAPTER 6

Primed *in Situ* Labeling (PRINS) and Fluorescence *in Situ* Hybridization (FISH)

Hanne Fischer, Johnny Hindkjaer, Søren Pedersen, Jørn Koch, Carsten Brandt, and Steen Kølvraa

Institute of Human Genetics
University of Århus
8000 Århus C
Denmark

I. Introduction

PRimed *IN Situ* labeling (PRINS) and fluorescence *in situ* hybridization (FISH) are methods for visualizing specific DNA sequences directly on chromosome spreads, thereby demonstrating the presence of a specific sequence in a certain cell and at the same time localizing the sequence at a specific site on the chromosome.

The PRINS reaction (Koch *et al.*, 1989) uses unlabeled DNA as probes, which after hybridization serve as primers for an enzyme-catalyzed DNA synthesis *in situ* from labeled nucleotides using the chromosomal DNA as a template for the polymerization. After the chain elongation the labeled DNA synthesized is visualized using specific fluorochrome-conjugated biological reagents, unless fluorochrome-labeled dUTPs are used (Koch *et al.*, 1992). A multicolor PRINS using two probes and elongation with differently labeled dUTPs has been developed (Hindkjaer *et al.*, 1993). FISH (Rudkin and Stollar, 1977) uses labeled probes, which are hybridized to their specific target sequence and then visualized using the same biological reactions as described for the PRINS reaction. In this chapter we describe protocols for the fluorescence labeling of specific sequences in chromosomes fixed to a microscope slide using either PRINS or FISH. For the labeling we use either the biotin–avidin or the digoxigenin–antidigoxigenin system, conjugated with various fluorochromes.

II. Applications

PRINS and *in situ* hybridization are both methods originally developed for microscopy of standard metaphase chromosome spreads. The two techniques have been used in many situations, in gene mapping, in the deciphering of complex chromosome aberrations, and in the simple determination of aneuploidy.

Two applications are especially related to flow cytometry, namely:

1. The identification with chromosome-specific markers of sorted chromosomes, which have been positioned on and fixed to microscope slides. This is

a very useful application since, often, chromosomes cannot be identified by standard banding techniques after flow cytometry (see Section VII,A,2)

2. CISS hybridization with chromosome libraries produced by PCR of specific flow-sorted chromosomes can be used for painting of metaphase chromosomes, e.g., to identify translocations (see Sections IV,C,2 and VII,B,2)

In Section VII we illustrate the various applications with concrete examples.

III. Materials

A. Chemicals

1. Restriction enzymes and buffers (Boehringer-Mannheim). Store at −20°C.
2. Taq DNA polymerase (Boehringer-Mannheim or Perkin–Elmer Cetus). Store at −20°C.
3. DNase (Worthington Biochemicals), 1μg/μl in standard solution (50% glycerol, 150 m*M* NaCl, 10 m*M* NaH_2PO_4, pH 7.2, 0.1% bovine serum albumin, 1 m*M* dithiothreitol, DTT). Store at −20°C.
4. DNA polymerase I (Kornberg)(Boehringer-Mannheim). Store at −20°C.
5. Avidin alkaline phosphatase (AAP) solution: 1 μg AAP (Sigma)/μl in 1 vol developing buffer and 1 vol glycerol. Store at −20°C.
6. Cot1-DNA (Gibco BRL).
7. Sonicated human placenta DNA.
8. Primer 6 MW: 5′-CCG ACT CGA GNN NNN NAT GTG G-3′, 95 pmol/μl. Store at −20°C.
9. Biotin-11-dUTP, 1 m*M* (Sigma). Store at −20°C.
10. Digoxigenin-11-dUTP, 1 m*M* (Boehringer-Mannheim). Store according to manufacturer.
11. dVTP mixture: dATP, dCTP, dGTP, 10 m*M* each. Store at −20°C.
12. ddNTP mixture: ddATP, ddGTP, ddCTP, and ddTTP, 2.5 m*M* each. Store at −20°C.
13. dTTP, 10m*M*. Store at −20°C.
14. 10× Taq buffer: 500 m*M* KCl, 100 m*M* Tris–HCl, pH 8.3, 15 m*M* $MgCl_2$, 0.1% BSA.
15. 10× Taq buffer (without Mg^{2+}): 500 m*M* KCl, 100 m*M* Tris–HCl, 0.1% gelatine, pH 8.3. Store at −20°C.
16. Stop buffer: 50 m*M* NaCl, 50 m*M* EDTA, pH 8.0.

17. Wash buffer: 4× SSC (20× SSC: 3 *M* NaCl, 300 m*M* sodium citrate), pH 7.0, 0.05 Tween 20.
18. 10× nick buffer: 500 m*M* Tris–HCl, pH 7.5, 100 m*M* $MgCl_2$, 10 m*M* DTT, 0.05%(w/V) bovine serum albumin. Store at 4°C.
19. Basis buffer: 0.5 *M* NaCl, 0.1 *M* Tris–HCl, pH 9.0.
20. Developing buffer: 0.1 *M* NaCl, 0.1 *M* Tris–HCl, pH 9.6, 0.01 *M* $MgCl_2$.
21. Hybridization buffer: 2 g glycin, 20 g dextrane sulfate, 0.4 g polyvinylpyrrolidon 90, 0.4 g Ficoll 70, 0.4 g bovine serum albumin, 30 ml 20× SSC, 20 ml 0.5 *M* $NaPO_4$ buffer, pH 6.8, and 21 ml redistilled water are mixed and cooled to 4°C. Then add 8 ml herring sperm DNA (5 mg/ml), which has been boiled for 10 min and 100 ml deionized formamide. The buffer is kept at 4°C. Shake well before use.
22. Hepes buffer: 6.25 m*M* Hepes, 50 m*M* KCl. Adjust to pH 8.0 with KOH and store at 4°C for not more than 1 week.
23. TE buffer: 10 m*M* Tris–HCl, 1 m*M* EDTA, pH 7.5. Store at 4°C.
24. Blocking solution: 5% (w/V) skimmed milk powder in wash buffer (**17**). Store at 4°C. Make fresh each day.
25. FITC-avidin DCS (Vector Lab.). Store according to manufacturer.
26. Rhodamin-conjugated avidin (Vector). Store according to manufacturer.
27. Biotinylated anti-avidin D (Vector Lab.). Store according to manufacturer.
28. Anti-digoxigenin fluorescein, Fab fragments (Boehringer-Mannheim). Store according to manufacturer.
29. Anti-fluorescein, mouse monoclonal antibodies (Boehringer-Mannheim). Store according to manufacturer.
30. Fluorescein-conjugated rabbit anti-mouse antibodies (DAKO, Copenhagen). Store according to manufacturer.
31. Fluorescein-conjugated swine anti-rabbit antibodies (DAKO, Copenhagen). Store according to manufacturer.
32. BCIP: 16 mg 5-bromo-4-chloro-3-indolylphosphate(Sigma)/ml 100% dimethylformamide. Store at −20°C in a lightproof *glass* container.
33. NBT: 37.5 mg nitroblue tetrazolium (Sigma)/ml 70% dimethylformamide. Store at 4°C in a lightproof *glass* container.
34. Ethidium bromide 0.5 ng/ml in 1× TBE (10× TBE: 900 m*M* Tris base, 900 m*M* boric acid, 1 m*M* EDTA). Hoechst 33258 DNA stain, dissolve 0.5 μg/ml in PBS. Store at 4°C.
35. Antifade solution: 10 mg *p*-phenylenediamindihydrochloride/ml 80% glycerol (V/V), 0.1 *M* Tris–HCl, pH 9.0. Store at −20°C in a lightproof container.

36. Propidium iodide, 2 μg/μl in H_2O. Store at 4°C.
37. Hoechst 33258 DNA stain; dissolve 100 μg/ml distilled water. Store at 4°C.
38. Hoechst 33258 DNA stain; dissolve 0.5 μg/ml distilled water. Store at 4°C.
39. Chromomycin A3 DNA stain, 2 mg/ml absolute ethanol. Store at −20°C.
40. Colcemid stock solution, 10 μg/ml. Store at −20°C.
41. Ammonium acetate, 1 *M*.
42. Glycerol, 87%.
43. Paraformaldehyde, 3% in PBS.
44. $MgSO_4$, 100 m*M*. Store at −20°C.
45. Dithiothreitol (DTT), 120 m*M* in H_2O. Store at −20°C.
46. Sodium citrate, 100 m*M*. Store at 4°C.
47. Sodium sulfite, 250 m*M*. Store at 4°C.
48. Sodium azide, 2% in water. Store at 4°C.
49. Triton X-100 stock solution, 2.5%. Filter through a 0.22-μm sterile filter.
50. $MgCl_2$ 100 m*M* stock solution. Store at −20°C.
51. Sodium acetate, 3 *M*, pH 5.2. Store at 4°C.
52. Agarose.
53. Bovine serum albumin. Store at 4°C.
54. Paraffin oil.
55. Ethanol.
56. Cell fixative—methanol/acetic acid 3 : 1.

Unless noted otherwise, the chemicals are stored at room temperature.

B. Instruments

DNA gel electrophoresis apparatus.
Vacuum oven.
Heating cupboard.
Thermo-block.
Water bath with humidified chamber.
Hybaid thermal reactor.
Vacuum centrifuge.
Fluorescence microscope (Leitz).
Confocal laser scanning microscope (Wild Leitz)(optional)
FACStar Plus (Becton–Dickinson).

IV. Protocols

A. Metaphase Chromosome Preparation

Bold numbers refer to the materials list.

1. Standard Chromosome Spreads

1. Standard methanol/acetic acid (**56**) fixed cells are spread on a microscope slide, which has been cooled in redistilled water at 4°C for no less than 30 min and then drained (not dried!). The best spreads are seen when two to four drops of the cell suspension are dripped from a Pasteur pipette onto the slide from a distance of approx. 30–50 cm. The slide is cautiously drained for surplus water and cell suspension and air-dried.
2. For CISS hybridization slides are then prewarmed for 5 min at 37°C in a humidified chamber followed by incubation with 0.5 μg/ml proteinase K in 2 m*M* $CaCl_2$, 20 m*M* Tris–HCl, pH 7.4, at 37°C for 7 min (100 μl/slide). Slides are washed for 3 min in proteinase K buffer and for 3 min in PBS/50 m*M* $MgCl_2$.

2. Spreads of Sorted Chromosomes

1. Chromosomes (1000–4000) are positioned directly on the slide by selecting and sorting the relevant chromosome region of the flow karyogram (see Figs. 1 and 2).
2. The sorted chromosomes are allowed to sediment to the surface of the slide, and then standard chromosome fixative (**56**) is applied around the sorted chromosome spread and allowed to diffuse into the chromosome suspension. The water is drained and the slide is air-dried. Prior to use in the PRINS reaction described in Section IV,B,2, the slide is washed for 2 × 5 min in PBS.

3. Paraformaldehyde Fixation

1. For FISH of standard metaphase spreads the dried slides are incubated with 200 μl of 3% paraformaldehyde (**43**) for 2 min under a glass coverslip. The coverslip is removed by tipping the slide, which is then dehydrated through an ethanol series (70, 90 and 99%, 3 min each) and air-dried.
2. For CISS hybridization slides are postfixed in 1% paraformaldehyde in PBS (with 50 m*M* $MgCl_2$) for 5 min at room temperature (100 μl/slide), followed by washing 2 × 2 min in PBS (with 50 m*M* $MgCl_2$) and 3 min in 2× SSC.

The slides can be stored at room temperature for a couple of weeks, but better results are seen with freshly made spreads.

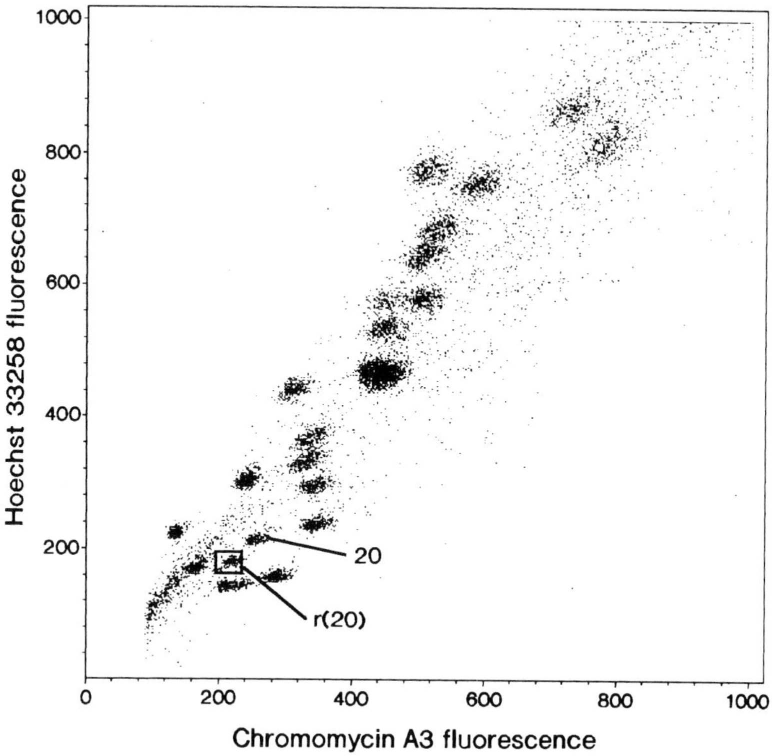

Fig. 1 PRINS of flow-sorted chromosomes. Dot plot of a bivariate distribution of 10^7 chromosomes from a patient with a ring chromosome 20. Chromosomes were stained in suspension with Chromomycin A3 and Hoechst 33258 and analyzed in a Facstar^{Plus} dual-laser cell sorter. The locations of the normal chromosome 20 and the smaller r(20) are indicated in the figure. These two chromosome peaks were individually gated and sorted directly onto slides, which were later used in a centromere-specific PRINS reaction. This verified the chromosome constitution of these peaks. Chromosomes (10^3) from the indicated peaks were also sorted directly into PCR tubes and used for the generation of painting material with the DOP-PCR method as described in Section IV,C,2,a and b.

4. Preparation of Cells in Suspension for Production of Chromosome-Specific Libraries

a. Culturing of Cells for Production of Chromosomes in Suspension

1. Lymphoblastoid cell lines, transformed with Epstein–Barr-virus (EBV), are cultured in RPMI 1640 medium containing 10% fetal calf serum, 2 m*M* L-glutamine, and antibiotics.[1]
2. The cells are arrested in metaphase by adding 100 μl colcemid solution (**40**)/10 ml cell culture. The cells are left for 17 to 20 hr and then the

[1] When culturing cell lines it is essential that most of the dead cells are removed from the suspension and a pool of rapidly proliferating cells are maintained.

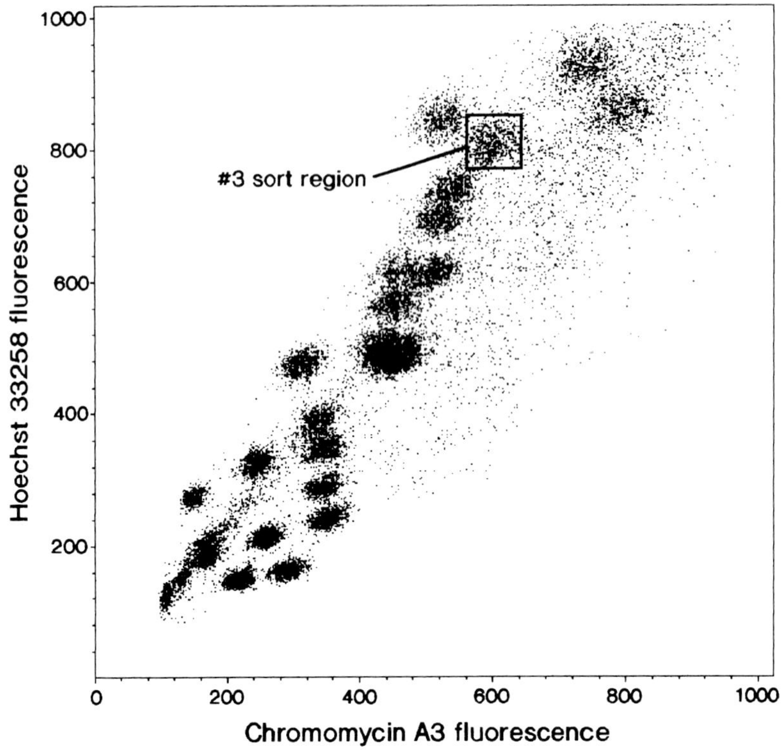

Fig. 2 Painting with flow-sorted chromosome libraries. Bivariate chromosome distribution from a normal male (46,XY). Normal chromosomes 3 (10^3) were sorted directly into PCR tubes from the indicated region and used for the generation of chromosome-specific painting material according to the DOP-PCR method described in Section IV,C,2,a.

chromosomes in suspension are isolated (method described by van den Engh *et al.*, 1988):

b. Isolation and Staining of Chromosomes in Suspension

1. A total of 10 ml of chromosome isolation buffer [1 ml 100 m*M* $MgSO_4$, 8 ml Hepes buffer (**22**), 0.25 ml DTT solution (**45**), 0.75 ml H_2O] is filtered through a sterile 0.22-μm filter.
2. The colcemide-arrested cells are transferred to a 15-ml centrifuge tube and centrifuged at 800 rpm for 10 min at room temperature.
3. The supernatant is removed completely (the last drops are absorbed with a Kleenex tissue) and the pellet is loosened by tapping.
4. Chromosome isolation buffer (5 ml) is added and the sample is incubated for 10 min at room temperature.
5. Triton X-100 solution (500 μl) (**49**) is added and the sample is incubated for another 10 min on ice.

6. The sample is mixed for 20 sec on a whirl mixer.
7. Both 125 μl Chromomycin A3 (**39**) and 125 μl Hoechst 33258 (**38**) are added and mixed with the sample.
8. A total of 50 μl sodium azide (**48**) is added, and the sample is incubated in the dark at 4°C overnight.
9. The following day, 800 μl of the chromosome suspension is transferred with a 1-ml syringe to a new tube containing 100 μl sodium citrate (**46**) and 100 μl sodium sulfite (**47**). The chromosome clumps are disintegrated by syringing three times through a 22-gauge needle.
10. After 15 min incubation the sample is analyzed in the FACS, and after the relevant chromosome peak is identified, this region is gated and 500 to 2000 chromosomes are sorted directly into a PCR tube.
11. The PCR tubes are centrifuged in an Eppendorf centrifuge for about 10 sec and the chromosomes are stored at −20°C until use.

B. Primed *in Situ* Labeling

1. Preparation of Primer DNA

PRINS can be performed with oligonucleotides as well as longer double-stranded DNA as primers, but whereas short oligonucleotides can be used directly, the cloned DNA and longer PCR products have to be cut into shorter fragments, not exceeding a few hundred base pairs, using one or more restriction enzymes, before being used in the PRINS reaction.[2]

1. The DNA is incubated with the appropriate restriction enzyme (1 U/μg DNA) in the appropriate buffer (see supplier's instructions) at 37°C for 2–3 hr. The final concentration should be no less than 1 μg DNA/1 μl.
2. The cleavage is analyzed by gel electrophoresis using at least 100 ng DNA in a 2% agarose gel which is afterward stained for at least 10 min in ethidium bromide (**34**) and evaluated at the UV transilluminator.

2. Single-Color PRINS Reaction

1. Reaction mixture is prepared in an Eppendorf tube: 0.5 μg of oligonucleotide DNA or 2 μg of cloned DNA, 0.5 μl dVTP mixture (**11**), 0.5 μl biotin-11-dUTP (**9**) or digoxigenin-11-dUTP (**10**), 2.5 μl glycerol (**42**), 5 μl 10× Taq buffer (**14**), and redistilled water added to a total volume of 50 μl. Then 1 U of Taq polymerase is added to the reaction mixture.
2. The slide is preheated on a thermo bloc at 94°C for 15 sec.
3. The reaction mixture is applied to the slide, spread with a coverslip, and left at 94°C for 4 min, during which time the DNA is denatured.

[2] We find that the restriction enzyme *Dde*I is useful with most probes, almost regardless of their nucleotide sequence.

4. The slide is transferred to a preheated humidified chamber in a water bath at stringent hybridization temperature (according to the probe used) and incubated for about 30 min. During this time the probe is annealed and the chain elongation takes place.
5. Then the slide is washed in stop buffer (**16**) for 1 min at hybridization temperature.
6. The slide is washed for 5 min in wash buffer (**17**) at room temperature and is ready for visualization (see Section IV,D).

3. Multicolor PRINS Using Two Different Probes

With a multicolor PRINS it is possible to label two different DNA sequences in two separate colors, using a two-step procedure with diversely labeled dUTPs. The method also makes it possible to use two probes with different melting temperatures. In this case it makes no difference which probe is used in the first step, and which in the second, since the chain elongation product achieved in the first step will not dissociate from the chromosomal DNA in the second PRINS reaction, even if this is performed at a higher temperature. In the following example an oligonucleotide and digoxigenin-labeled dUTP are used in the first step and cloned DNA and biotin-labeled dUTP in the second step.[3] Between the two steps the chain elongation is terminated with dideoxynucleotides.

a. First PRINS Reaction

1. A PRINS reaction is carried out according to steps 1–4 (Section IV,B,2), using an oligonucleotide as probe and digoxigenin-labeled dUTP for labeling.
2. The slide is washed for 2 × 30 sec at hybridization temperature in 100 ml preheated stop buffer (**16**).
3. The slide is dehydrated in a 4°C ethanol series (70, 90, 99%, 3 min each).

b. Blocking with Dideoxynucleotides

1. Fifty microliters of dideoxynucleotide reaction mixture is prepared: 2 μl ddNTP mixture (**12**), 2.5 μl glycerol (87%), 5 μl 10× Taq buffer (**14**). Water is added to a final volume of 50 μl. Taq polymerase (1 U) is added (**2**) and mixed gently.
2. The slide is drained, placed in the humidified chamber at the first hybridization temperature, and preheated for 1 min.
3. The preheated slide is incubated with the dideoxynucleotide reaction mixture at the first hybridization temperature for 15 min.

[3] Since digoxigenin-labeled dUTP gives the strongest symbols, we use this dUTP with the probe suspected to give the weakest symbol.

4. The slide is washed for 2 × 30 sec at hybridization temperature in 100 ml preheated stop buffer (**16**).
5. The slide is dehydrated in a 4°C ethanol series (70, 90, 100%, 3 min each).

c. Second PRINS Reaction

1. The slide is drained and placed in the humidified chamber at the second hybridization temperature and preheated for 1 min.
2. The reaction mixture with cloned DNA as primer and digoxigenin-11-dUTP for labeling is denatured at 100°C before the Taq-polymerase is added.
3. The reaction mixture is placed on the slide and spread with a coverslip before a second PRINS reaction is carried out from steps 4–6 (Section IV,B,2). Please note that the denaturation step (3) should be omitted.

d. Visualization of Multicolor PRINS

First perform the digoxigenin detection as described in Section IV,D,2 and then the biotin detection as described in Section IV,D,1, but with rhodamin-conjugated avidin (**26**) instead of fluorescein-conjugated avidin.

N.B. Counterstain with Hoechst:

1. The slide is incubated with Hoechst (**37**) for 1 min, then washed for 30 sec in PBS, and drained.
2. Finally the slide is mounted in one drop of pure antifade (**35**) and evaluated in a fluorescence microscope.

C. Fluorescence *in Situ* Hybridization

1. FISH with Cloned Probes

a. Probe Labeling (Nick Translation)

We use the two-step nick translation described by Koch *et al.* (1986), separating the DNase treatment and the polymerization.

DNase treatment:

1. The standard DNase solution (**3**) is serially diluted with redistilled water in Eppendorf tubes to concentrations of 10 ng, 1 ng, 100 pg, 10 pg, and 1 pg DNase/μl.
2. A mixture of 1 μg plasmid DNA [in TE (**23**)][4], 1 μl 10× nick buffer (**18**) and redistilled water to a total volume of 9 μl per DNase concentration is distributed in a number of Eppendorf tubes.

[4] The scale of the reaction can easily be increased if it is desirable to produce larger quantities of labeled DNA: simply add more DNA in the DNase step and increase the amount of nucleotides correspondingly. It is, however, rarely necessary to increase the amount of DNA polymerase.

DOP primer (primer 6 MW, **8**), 0.25 μl Taq DNA polymerase (**2**), and 65.25 μl H_2O] is added to each PCR tube of 1000 chromosomes in approx. 10 μl sheath liquid and mixed.

2. The mixture is overlaid with two drops of paraffin oil.
3. The tubes are placed in the PCR thermocycler and the chromosomes are denatured for 10 min at 93°C.
4. Then five identical cycles are performed with denaturation at 94°C for 1 min, annealing at 30°C for 1.5 min, increasing the temperature gradually to 72°C over 2 min, and then holding the temperature at 72°C for 3 min.
5. Then 35 cycles are performed with denaturation at 94°C for 1 min, annealing at 55°C for 1 min, and then extension at 72°C for 3 min, increasing the extension step by 1 sec for each of the first 34 cycles.
6. The final extension is performed for 10 min.
7. The paraffin oil is removed by placing the PCR mixture on a strip of parafilm and adsorbing the oil, and the PCR mixture is transferred to an Eppendorf tube.
8. Sodium acetate (10 μl) (**51**) is added to the sample and mixed, and then 220 μl ice-cold absolute ethanol is added and the sample incubated for 1 hr at −70°C to precipitate the DNA.
9. The precipitate is spun down for 15 min in a cooled Eppendorf centrifuge.
10. The ethanol is removed, 100 μl ice-cold 70% ethanol is added, and the centrifuging is repeated.
11. The ethanol is removed, and the pellet is dried in a vacuum centrifuge for approx. 10 min, after which the pellet is dissolved in 20 μl TE buffer (**23**) and stored at −20°C until use.

Labeling of the amplification product with biotin. The probe is labeled with biotin-11-dUTP during a second PCR procedure:

12. One microliter of the PCR product is added to 99 μl biotin DOP-PCR mixture [6 μl dVTP mixture (**11**), 1.4 μl dTTP (**13**), 6 μl biotin-11-dUTP (**9**), 10 μl 10× Taq buffer (without Mg^{2+}) (**15**), 6 μl $MgCl_2$ (**50**), 2 μl DOP primer (primer 6 MW, **8**), 0.25 μl Taq DNA polymerase (**2**), and 65.75 μl H_2O].
13. PCR cycles (35) are performed with denaturation at 94°C for 1 min, annealing at 55°C for 1 min, and then extension at 72°C for 3 min (the first 34 cycles).
14. The final extension is performed for 10 min.
15. The new PCR product is precipitated as described above (steps 7–10).

16. The biotin-labeled DNA is dissolved in 10 μl TE buffer (**23**) and stored at −20°C until further use.

b. CISS Hybridization

For CISS hybridization the chromosomal DNA is denatured before application of the hybridization mixture with the labeled DNA.

1. Chromosomes are incubated in 70% deionized formamide, 2× SSC for 2 min at 70°C and then dehydrated in a 4°C ethanol series (70, 90, 99%, 3 min each).

Labeled library DNA and carrier DNA are mixed, denatured, and allowed to partially reanneal before application on the denatured slides.

2. Labeled library DNA (50–200 ng) is mixed with 10- to 40-fold excess of a 1 : 1 mixture of human Cot1-DNA (**6**) and sonicated human placenta DNA (average length of 500 bp). Then 2.5 vol cold ethanol is added to precipitate the DNA and the sample is dried in a vacuum centrifuge.
3. The dried sample is dissolved in 12 μl hybridization mixture (**21**).
4. The mixture is now incubated in a water bath at 70°C for 5 min to denature the DNA and then immediately transferred to a 37°C incubator and preincubated for 30–60 min at 37°C.
5. Hybridization mix (12 μl) is applied to the slide, spread with a 18 × 18-mm coverslip, and hybridized overnight (16 hr) at 46°C.
6. The slide is washed 2 × 10 min in 50% formamide, 2× SSC at 46°C, 2 × 5 min in 2× SSC at 46°C, and then finally in wash buffer (**17**) for 15 min at room temperature with gentle agitation. The slide is now ready for visualization (see Section IV,D).

D. Visualization of Labeled DNA

For *in situ* hybridization and PRINS described above, the visualization process is identical. All reactions described below are performed at room temperature, and incubations are carried out under a lightproof lid to minimize any fluorochrome fading prior to microscopy.

1. Visualization of Biotin-Labeled Nucleotides

1. The slide is incubated with 100 μl blocking solution (**24**) for 10 min under a coverslip cut from a plastic folder (easy to remove after the incubation).
2. The coverslip is gently removed and 100 μl of blocking solution containing 1 μg FITC avidin (**25**)/ml blocking solution is added. The same coverslip is reused and the slide is incubated for 30 min.
3. The slide is washed for 3 × 5 min in wash buffer (**17**).

If the probe used represents highly repetitive sequences the signal might be strong enough to be seen in a fluorescence microscope, but for less-repeated sequences it might be necessary to enhance the signal using a layer of biotinylated anti-avidin and an additional layer of FITC-avidin. The enhancement procedure could be repeated if necessary [if the chromosomes have been counterstained for microscopy they were washed for 2 × 5 min in wash buffer (**17**) before enhancing signal and the glass coverslip was removed after the first 5 min].

4. The slide is incubated with 100 μl of blocking solution containing 1 μg biotinylated anti-avidin (**27**)/ml blocking solution under a new coverslip for 30 min.
5. The slide is washed for 3 × 5 min in wash buffer.
6. Then the slide is incubated with an additional layer of FITC-avidin as described above (step 2).
7. The slide is washed for 3 × 5 min in wash buffer.
8. Prior to fluorescence microscopy the slide is mounted with one drop of antifade solution (**35**) containing 1 μl propidium iodide (**36**)/2 ml of antifade solution under a glass coverslip.

2. Visualization of Digoxigenin-Labeled Nucleotides

1. The slide is incubated with 100 μl blocking solution (**24**) for 10 min under a coverslip cut from a plastic folder (easy to remove after the incubation).
2. The slide is incubated with 50 μl anti-digoxigenin-fluorescein, Fab fragments (**28**) in blocking solution (0.1 μg/50 μl) under the same coverslip for 30 min.
3. The slide is washed for 3 × 5 min in wash buffer (**17**).

If the probe used represents highly repetitive sequences the signal might be strong enough for visualization in a fluorescence microscope, but it might be necessary to enhance the signal using one or more layers of fluorescein-conjugated antibodies [if the chromosomes have been counterstained for microscopy they were washed for 2 × 5 min in wash buffer (**17**) before enhancing signal, and the glass cover slip was removed after the first 5 min].

4. The slide is incubated with 50 μl anti-fluorescein, monoclonal antibodies (**29**) in blocking solution (0.25 μg/50 μl) under a plastic coverslip for 30 min.
5. The slide is washed for 3 × 5 min in wash buffer.
6. The slide is then incubated with 100 μl fluorescein-conjugated rabbit anti-mouse antibodies (**30** diluted 100× in blocking solution) for 30 min.

If the signal is still not strong enough steps 5–6 of Section IV,D,2 are repeated this time using fluorescein-conjugated swine anti-rabbit antibodies (**31** diluted 100× in blocking solution). It is possible to repeat steps 3–6 of Section IV,D,2 for further signal amplification.

V. Critical Aspects of the Procedures

A. Metaphase Chromosome Preparation

The quality of the chromosomes is very important if subsequent Q banding has to be performed. Incubating the slides for 2 min with 200 μl 3% paraformaldehyde under a glass coverslip and dehydrating them in an ethanol series (70, 90, and 99% for 3 × 5 min) will conserve the chromosomal structure better after the denaturation.

For slides used for PRINS it is important that there are not an excessive number of intrinsic nicks in the chromosomal DNA, since these could act as primers and create unspecific signals. This is usually no problem with freshly prepared slides (0–2 weeks old), but may be a problem with older slides. It can be avoided by pretreating the slides with T4 DNA ligase (1 U per slide in 50 μl ligase buffer for 1 hr at room temperature), washing for 1 min in stop buffer (**16**) at room temperature, and dehydrating the slide in an ethanol series (as above) (Koch *et al.*, 1991).

B. Primed *in Situ* Labeling

1. Preparation of Primer DNA

When cloned DNA is used as a primer, it is important to inactivate the restriction enzyme totally before using it for the PRINS procedure, or it might produce nicks in the genomic DNA, thereby creating a banding pattern. To obtain this inactivation, the reaction mixture can be boiled for 5 min and centrifuged briefly to regain the vapor *prior* to the addition of the Taq polymerase. Also it is preferable not to let the total volume of primer DNA exceed 10% of the PRINS reaction mixture, since the restriction enzyme buffer is still in the DNA solution.

Lack of signal using cloned DNA may be due to insufficient cleaving by the restriction enzyme.

2. PRINS Reaction

New primers should be tested at hybridization temperatures other than the theoretical melting temperature[6] and with different concentrations, good starting points being 0.5 μg of oligonucleotide or 2.0 μg of cloned DNA.

[6] Shorter oligonucleotides often work well above the melting point, whereas longer oligonucleotides often work well below the melting point.

It is very important to keep humidity a high during the reaction, since the high temperatures may cause the slide to dry out. Addition of glycerol (up to 5%) to the reaction mixture helps prevent this.

C. Fluorescence *in Situ* Hybridization

1. Cloned Probes

a. Probe Labeling

For the *in situ* hybridization to succeed with as strong a signal and as little a background staining as possible, it is essential to get the optimal length of probes, and we have found that the optimal single-stranded length is between 100 and 1000 nucleotides (Fischer *et al.*, 1992). Since not all cloned DNA is equally sensitive to DNase, and the cloned DNA in different preparations is not equally sensitive to the DNase, it is not possible to state a standard concentration of DNase to be used for the nick translation. Therefore, it is wise to treat the DNA with a series of DNase concentrations with each new probe preparation and test them all by *in situ* hybridization.

b. Determination of Labeling

The purpose of the dot blot is to make certain that the probe has been labeled and to what extent. For the probe to be sufficiently labeled, you should ideally be able to see coloring of the 0.1-pg dot.

c. In Situ Hybridization

It is essential to shake the hybridization buffer thoroughly before use, since the contents will divide into two phases during storage.

The hybridization time (normally 16 hr) is not very strict, and can be both shortened (for highly repetitive target sequences 1 hr may be sufficient for routine testing) and prolonged (shorter, less repetitive target sequences). If longer hybridization times are used, you should, however, be especially careful about the humidity in the chamber, since the slides may dry out. Alternatively, sealing with rubber cement can be used.

2. Chromosome Libraries

a. DOP-PCR

For the painting reaction it is essential that the specificity of the complex probe is not reduced by enrichment for repetitive DNA in the PCR product. This can be tested by electrophoresis in a 2% agarose gel in TBE buffer, 5–10 μl of the PCR product, and appropriate markers. The gel is stained shortly with ethidium bromide and evaluated under UV light. The PCR product should produce a smear of fragments with lengths of 150–1000 bp and no discrete bands.

The biotin-11-dUTP can be replaced with other labeled nucleotides, e.g., digoxigenin-11-dUTP, fluorescein-12-dUTP, rhodamine-4-dUTP, or AMCA-4-dUTP, all of which we have utilized with success, using the same molar concentration as for biotin-11-dUTP. When using the fluorochrome-conjugated dUTPs the DOP-PCR products are directly visible in the microscope and in agarose gels evaluated under UV light.

b. CISS Hybridization

We use both Cot1 DNA and sonicated placenta DNA to avoid background from repeated sequences. Results can be obtained using only Cot1 DNA or only sonicated human placenta DNA, but in our hands the best result is obtained using a mixture of the two.

We use a rather high hybridization temperature of 46°C. Results are also obtained at 37°C, but less background signal is seen at the higher temperature. The optimal hybridization temperature is dependent on probe length and should therefore be determined empirically.

The proteinase K digestion clearly enhances the signal. Comparing painting on metaphases with or without proteinase K treatment, a clear signal is obtained after one layer of FITC-avidin with proteinase K treatment, whereas the same strength of signal on untreated metaphases is only obtained after two layers of FITC-avidin.

D. Visualization of Labeled DNA

The most important step of the visualization procedure is getting the skimmed milk powder properly dissolved in the wash buffer, since undissolved powder will produce significant background staining. Therefore it is a good idea to mix the blocking solution carefully and then centrifuge shortly in an Eppendorf centrifuge and use only the supernatant. The blocking solution should be made fresh each day. Incubation times and washing times are not strict, but the latter should not be minimized, whereas the former can be both shortened (weaker signal, but less background staining) and prolonged (stronger signal, but also heavier background staining).

VI. Instruments

A. Fluorescence Microscope

The standard fluorescence microscope was a Leitz Diaplan, and it was equipped with standard filter blocks for detection of fluorescein, rhodamine, and Hoechst 33258.

B. Confocal Laser Scanning Microscope (CLSM)

In some cases slides were evaluated with a Confocal laser scanning microscope. This is a Leica CLSM system and is composed of an integrated fluorescence microscope, laser scanner, and picture acquisition and analysis software. The system acquires gray-scale pictures with relevant filter sets and offers the possibility of modifying each picture, e.g., in the form of enhancing intensities, before combination of the pictures and application of artificial colors.

VII. Results and Discussion

A. Primed *in Situ* Labeling

1. PRINS of Standard Metaphase Spreads Using Oligonucleotides as Primers

On metaphase spreads made from a male infant with a structurally abnormal chromosome replacing one normal chromosome 18 we performed PRINS according to the description in Section IV,B,2, using digoxigenin-labeled nucleotides and a 43-mer oligonucleotide specific for the α-satellite family on chromosome 18. The slides were prepared and fixed as described in Section IV,A,1 and 3. After the PRINS reaction, slides were stained for digoxigenin according to the procedure in Section IV,D,2. As seen in Color Plate 1 three strong and specific signals were detected. One was at the centromere of the normal chromosome 18, and two were situated on the abnormal chromosome showing this to be a dicentric chromosome 18. The entire procedure took only about half an hour from start to finish.

The example chosen illustrates not only the speed and simplicity of the PRINS reaction but also the fact that within most repeat subfamilies it is possible to find short specific motifs that can be utilized in the design of primer. Since the labeling in the PRINS reaction is independent of primer length this means that the better penetration and hybridization kinetics of oligos can be utilized also in the detection of repeated sequences.

2. PRINS of Flow-Sorted Chromosomes

This rather special application of the PRINS reaction is illustrated by a case where the exact composition of a suspected ring chromosome 20 was to be determined by reverse painting. As a first step the ring chromosome had to be isolated by chromosome sorting. A bivariate dot display of the total chromosome complement of the patient is shown in Fig. 1. Based on visual evaluation of the size of the ring it was assumed to lie within the gating depicted as a square in Fig. 1. For definite verification of this, chromosomes within the square were sorted onto a slide using the procedure described in Section IV,A,2, and a PRINS reaction using as primer a *Dde*I-digested cloned α-satellite DNA specific for chromosome 20 (p20Z1) was performed after the procedure in Section IV,B,2.

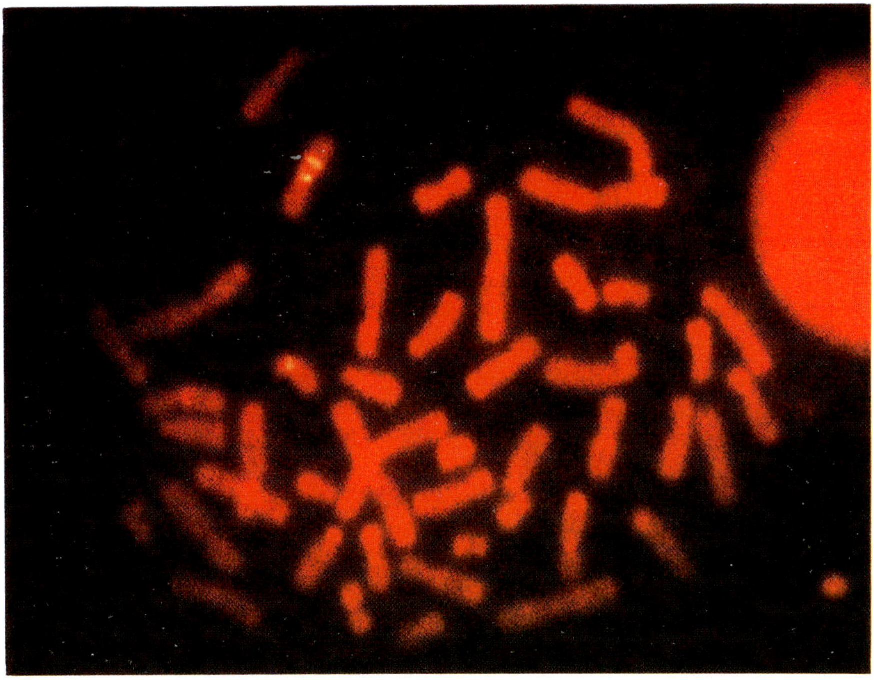

Color Plate 1 (Chapter 6) PRINS with oligonucleotides. Standard metaphase spreads were made from cultured skin cells from an individual with the karyotype 46,XY, psu dic (18). As primer we used a 43-mer oligonucleotide-specific for the α-satellite family at the centromere of chromosome 18. Single-color PRINS was performed at 60°C with 10-min reaction time. We used digoxigenin-labeled dUTP for labeling and the chain elongation product was visualized with only one layer of antidigoxigenin fluorescein (15-min incubation). Metaphases were photographed directly from the fluorescence microscope. One metaphase is shown and identifies the marker chromosome as an isodicentric chromosome with two chromosome 18 centromeres.

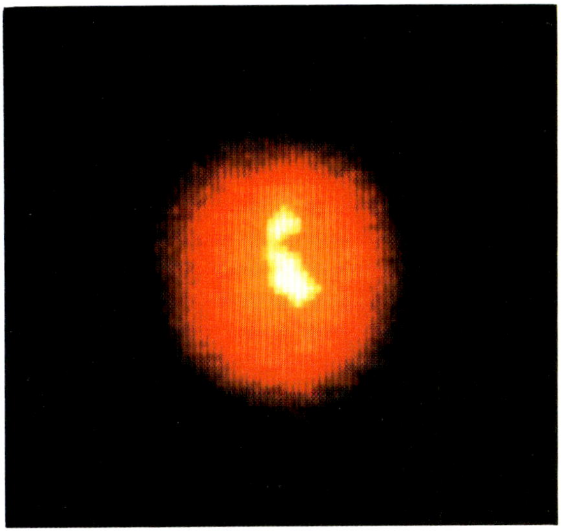

Color Plate 2 (Chapter 6) PRINS of flow-sorted chromosomes using cloned DNA as primer. Flow-sorted ring chromosomes were gated directly onto a microscope slide and fixed with 3% paraformaldehyde for 2 min. Single-color PRINS was performed with cloned DNA specific for chromosome 20 at 77°C for 30 min. We used digoxigenin-labeled dUTP for labeling and the chain elongation was visualized with one layer of fluorescein-conjugated antidigoxigenin. The chromosomes were scanned using the CLSM and pictures were photographed from the computer monitor. One characteristic ring chromosome is shown and the specific signal identifies the chromosome as a r(20).

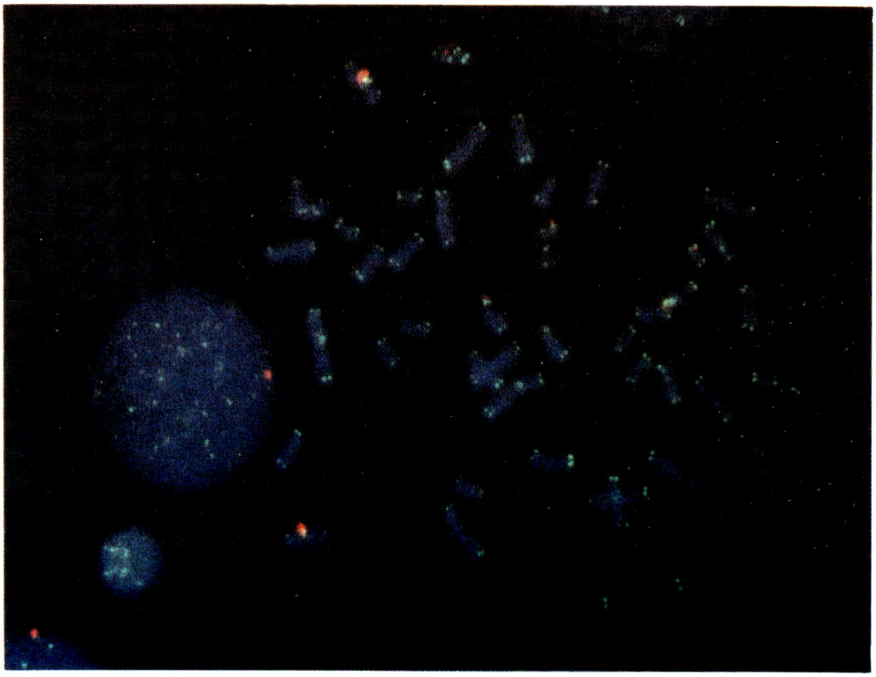

Color Plate 3 (Chapter 6) Multicolor PRINS. We performed PRINS of standard human lymphocyte metaphase spreads. First, we used a 42-mer oligonucleotide specific for telomeres and digoxigenin-labeled dUTP. Second, we cloned DNA specific for the α-satellite DNA of chromosome X and biotin-labeled dUTP. The oligonucleotide PRINS was carried out at 55°C for 30 min and the cloned DNA PRINS was carried out at 70°C for 30 min. The visualization was first with one layer of fluorescein-conjugated antidigoxigenin and second with two layers of rhodamin-conjugated avidin. Counterstaining was with Hoechst 33258. The metaphases were photographed directly form the microscope by double exposure.

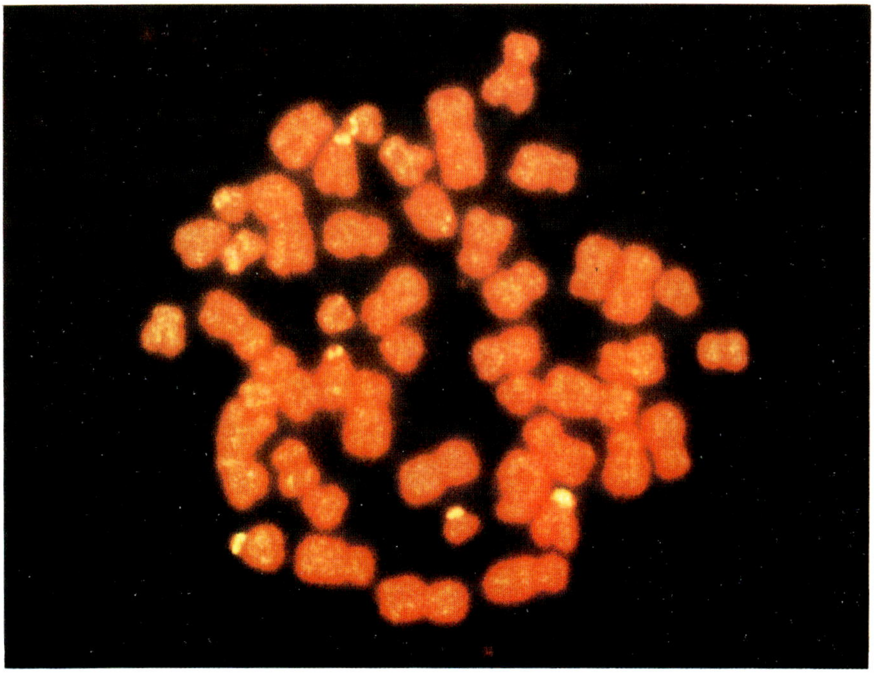

Color Plate 4 (Chapter 6) FISH with cloned DNA. FISH was performed on standard metaphase spreads from a female. We used cloned DNA representing a 5000-bp RNA spacer as probe, labeled with biotin-11-dUTP. The *in situ* hybridization was carried out at 48°C for 16 hr. The probe was visualized using FITC avidin and enhancing the signal with one layer of biotinylated anti-avidin and a second layer of FITC-avidin. The metaphases were photographed directly through the fluorescence microscope.

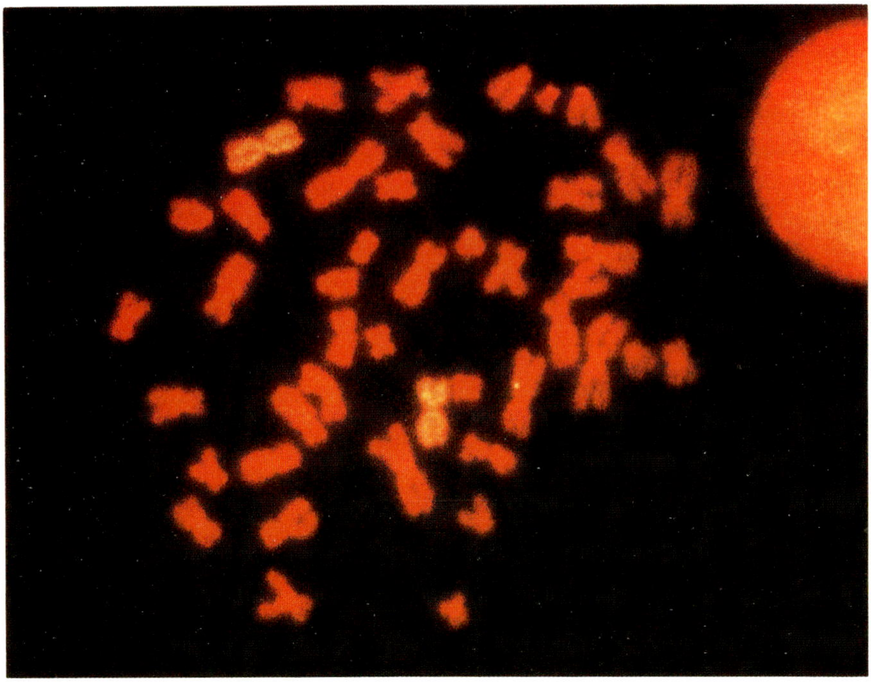

Color Plate 5 (Chapter 6) Painting with flow-sorted chromosome libraries. CISS hybridization was carried out on standard metaphase spreads using the biotin-labeled DOP-PCR DNA from chromosome 3 probe. The hybridization was carried out at 46°C for 16 hr and the probe was visualized with only one layer of FITC-avidin.

This verified that the majority of the chromosomes in the square carried a centromere 20 (one example is shown in Color Plate 2). Since the normal chromosome 20 is located outside the square (Fig. 1) this clearly verifies that it is the r(20) that is isolated with this gating. The r(20) library achieved by sorting with this gating was later used for CISS hybridization onto a reference metaphase after the procedure given in Sections IV,C,2,a and b. No deletion of chromosome 20 material was evident here, probably because of the small size of the deletion in both telomeres (data not shown).

3. Multicolor PRINS

This application of the PRINS is illustrated by simultaneous two-color detection of telomeres and the X centromere on a standard human metaphase. 42-mer oligonucleotides specific for the human telomeres were used for detection of telomeres and a *Dde*I digest of the cloned α-satellite DNA from chromosome X was used for centromere staining. The procedure used is that given in Section IV,B,3.

The results obtained is depicted in Color Plate 3. The example chosen illustrates, apart from the two colors, the fact that this two-color PRINS, contrary to two-color *in situ* hybridization, works even in situations where the melting temperature of the two probes is very different.

B. Fluorescence *in Situ* Hybridization

1. FISH of Standard Metaphase Spreads Using Cloned Probes

We performed *in situ* hybridization after the description in this chapter (Section IV,C,1) using a 5000-bp RNA spacer, which is situated on the short arm of the acrocentric chromosomes (13, 14, 15, 21, and 22) of human cells in a tandem array of approx. 40 copies on each chromosome (Sylvester *et al.,* 1986), thereby representing a target sequence of approx. 200 kbp on each chromosome. We used standard methanol/acetic acid-fixed metaphase preparations of human female lymphocyte culture, spread on slides, and pretreated with paraformaldhyde according to Section IV,A,3. *In situ* hybridization was performed according to the protocol in Section IV,C,1 and the labeled DNA was visualized with the biotin–avidin system (see Section IV,D,1).

We found a strong signal on the short arm of approx. 8 of the 10 acrocentric chromosomes (Color Plate 4). The chromosomes without signal were No. 22 (most metaphases) 13, or 14 (never one whole pair without signal).

2. Painting with Flow-Sorted Chromosomes

Generation of and painting with a chromosome-specific library produced by chromosome sorting is illustrated with chromosome 3.

Initially about 5000 chromosomes 3 were isolated by the method given in Section IV,B,3,a. A bivariate dot display of the total chromosome complement of a normal person is shown in Fig. 2, showing the gating used for isolating chromosome 3.

DNA from the isolated chromosomes was then amplified with degenerated primers (Telenius *et al.*, 1992) according to the procedure in Section IV,A,4,b, giving a mixture of fragments of the size 300–1000 nucleotides.

The amplified DNA was then labeled according to Section IV,C,2,a, and finally chromosome painting of a normal metaphase was performed with this library according to the procedure in Section IV,C,2,b.

The result obtained is shown in Color Plate 5. It is seen that a strong and homogeneous paint of the whole chromosome was obtained, after only one layer of FITC-avidin.

VIII. Comparison of Methods

For a number of applications PRINS and traditional *in situ* hybridization are equally effective. This applies to situations where cloned probes are used for detection of highly repeated sequences. Here the sensitivity is similar. The PRINS is, however, faster and probably somewhat gentler to chromatin structure, resulting in better Q-bands after PRINS compared to *in situ* hybridization. The incubation times given in the protocol section are ample and may be reduced if necessary, once a sufficient routine is obtained. This allows centromeres with digoxigenin- or biotin-labeled dUTP to be detected within half an hour, and with fluorescein-dUTP within a few minutes.

When trying to vizualize low-copy number or unique sequences we find also very similar sensitivities. The only situation where a difference may be present is when cDNAs are used as probes for the detection of genes with very short exons. In this situation, it has been suggested that a PRINS approach gives a higher sensitivity, probably due to labeling of intron sequences during the chain elongation. When performing chromosome painting one has to use *in situ* hybridization. This is due to the fact that we so far have not been able to develop a PRINS-based variant of CISS hybridization, simply because annealing of the carrier DNA contributes extensively with 3′ ends available for the chain elongation. The same problem arises when using cosmid clones.

The original intention with developing the PRINS method, namely, the *in situ* detection of minor sequence variations, still holds true. The problem is, however, that as long as the sensitivity obtained does not allow one to see single priming events *in situ,* the detection of sequence variation in coding sequences is not possible. We can, however, see sequence variation within subfamilies of repeats as illustrated in Color Plate 1, and we have also been able to detect sequence variations on mRNA's by PRINS of RNA, simply because of the presence of many copies of the mRNA in question within one cell.

References

Fischer, H., Koch, J., Hindkjaer, J., and Askholm, H. H. (1992). *App. Flouresc. Technol.* **4,** (2&3), 14–19.

Hindkjaer, J., Koch, J., Mogensen, J., Kølvraa, S., and Bolund, L. (1993). *Methods Mol. Biol.* (in press).

Koch, J. E., Kølvraa, S., and Bolund, L. (1986). *Nucleic Acids Res.* **14,** 7132.

Koch, J. E., Kølvraa, S., Petersen, K. B., Gregersen, N., and Bolund, L. (1989). *Chromosoma* **98,** 259–265.

Koch, J., Hindkjaer, J., Mogensen, J., Kølvraa, S., and Bolund, L. (1991). *Genet. Anal. Techn. Appl.,* **8,** 171–178.

Koch, J. E., Mogensen, J., Pedersen, S., Fischer, H., Handkjær, J., Kølvraa, S., and Bolund, L. (1992). *Cytogenet. Cell Genet.* **60,** 1–3.

Rudkin, G. T., and Stollar, B. D. (1977). *Nature (London)* **265,** 472–473.

Sylvester, J. E., Whiteman, D. A., Podolsky, R., Pozsgay, J. M., Respess, J., and Schmickel, R. D. (1986). *Hum. Genet.* **73,** 193–198.

Telenius, H., Carter, N. P., Bebb, C. E., Nordenskjöld, M., Ponder, B. A. J., and Tunnacliffe, A. (1992). *Genomics* **13,** 718–725.

van den Engh, G., Trask, B., Lansdorp, P., and Gray, J. (1988). *Cytometry* **9,** 266–270.

CHAPTER 7

Molecular Phenotyping by Flow Cytometry

Benjamin D. Li, Earl A. Timm, Jr., Mary C. Riedy, Seth P. Harlow, and Carleton C. Stewart

Departments of Flow Cytometry and Surgical Oncology
Roswell Park Cancer Institute
Buffalo, New York 14263

I. Introduction

Flow cytometry has been the method of choice for identifying and quantifying the binding of fluorochrome-labeled antibodies to specific protein antigens on the membrane or inside of cells. The methodology to do this is now well developed and reliable and has important implications in both the research and clinical environments.

The polymerase chain reaction (PCR) has become an indispensable research tool in the field of molecular biology. Starting with a minute amount of DNA or RNA, the PCR can amplify a specific nucleic acid sequence to the point where it can be easily detected on an ethidium bromide-stained agarose gel. This completely eliminates the need for radioactive probes for detection. Many variations of this procedure have been developed such as inverse PCR, anchored PCR, and asymmetric PCR to accommodate the needs of individual researchers (Innis *et al.,* 1990; Kawasaki *et al.,* 1988; Saiki *et al.,* 1988).

Gene amplification can be quantified using competitive PCR (Clementi *et al.,* 1993; Harlow *et al.,* 1993; Li *et al.,* 1993b). For evaluating gene expression, reverse transcription PCR (RT-PCR) is used when extracted RNA, whether total RNA or messenger RNA, is first transcribed into DNA and then amplified by the PCR. Gene expression can then be quantified using competitive RT-PCR (Clementi *et al.,* 1993; Harlow *et al.,* 1993). The major disadvantage, however, to all of these technologies is that once the RNA or DNA is extracted from the heterogeneous cell populations, it becomes extremely difficult to relate the data obtained by PCR to the specific cells from which the product was derived (Harlow *et al.,* 1993; Li *et al.,* 1993a; Singleton and Strickler, 1992).

One strategy is to use multiparameter flow cytometry (FCM) to identify and sort specific populations of cells out of a heterogeneous cell population for further analysis by PCR (Harlow *et al.,* 1993). The use of PCR to then amplify the DNA or RNA of interest is highly advantageous since very few cells are required for analysis. Thus by combining immunophenotyping with molecular biology, quantitative analysis of gene amplification and expression can be achieved in a specific cell population. Even though further development is necessary, it also seems reasonable that direct molecular phenotyping on a cell-by-cell basis utilizing *in situ* hybridization (FISHES) or *in situ* PCR (FLIP) can be accomplished with high sensitivity using fluorochrome-labeled probes (Timm and Stewart, 1992).

In this chapter, we describe methods to identify and sort for specific populations of cells and, using comopetitive PCR and RT-PCR, quantitatively determine gene copy number and mRNA copy number in as few as several hundred sorted cells. We then describe the current methodology for the direct evaluation of gene expression using *in situ* hybridization and *in situ* PCR in suspension by flow cytometric evaluation.

II. General Materials and Methods

Several common materials and methods are used throughout. To avoid repeating the same detailed description, they are defined here and referred to in each specific process.

A. Flow Cytometry

1. Cell Wash

1. Add indicated buffer, generally Dulbecco's phosphate-buffered saline (PBS) (Gibco BRL Life Technologies, Inc., Grand Island, NY).
2. Incubate 20 min and centrifuge cells at 2000*g* for 3 min.
3. Discard supernatant and resuspend cells in residual buffer.

2. Fixation and Permeabilization

1. Place cells in microcentrifuge tube and wash.
2. Add 0.5% ultrapure paraformaldehyde (Sigma Chemical Co., St. Louis, MO) in PBS for 5 min, and then add 0.5 ml of 0.1% Triton (Sigma) in PBS for 3 min. (Alternative fixation and permeabilization procedures may be required for some cell types).
3. Wash cells with PBS and resuspend in 1 ml 0.5% nuclease-free bovine serum albumin (Sigma) in PBS.

3. Staining Cells

Detailed methods for staining cells are found in other chapters of this volume. A brief description is described here:

1. Aliquot 100 μl of fixed and permeabilized cells into two microcentrifuge tubes.
2. Incubate cells with 10 μg of goat IgG (Sigma) for 10 min to reduce Fc receptor and nonspecific binding; this block is more effective than using whole serum.

3. Add correct amount of the primary monoclonal antibody to one sample and correct amount of isotype control IgG to the second.
4. Incubate 30 min at room temperature and wash.
5. Add a fluoresceinated (FITC) or a phycoerythrin (PE)-conjugated goat anti-mouse IgG titered for 10 μl and wash.
6. If a second directly labeled antibody(s) will be used, reblock with 10 μg MsIgG (Caltag Laboratories, San Francisco, CA) for 10 min.
7. Add directly conjugated antibody to one sample and an equivalent amount of directly conjugated isotype control IgG to the second, incubate for 30 min and wash and resuspend cells.
8. If a DNA stain is desired, approximately 1 hr prior to sorting cells, resuspend the cells in 0.8 ml of 1.5 μg/ml Hoechst 33342 (Calbiochem, Corp., La Jolla, CA).
9. Just prior to sorting cells, filter cells through a 37-μm monofilament nylon cloth.

4. Laser Setup

Samples are run on the FACStarPLUS cytometer (Becton–Dickinson, San Jose, CA). An argon laser operating at 488 nm is used for excitation of PE and FITC and one operating in the UV range of 350 nm is used for excitation of Hoechst. Fluorescence emissions from FITC and PE are detected by selectively collecting 530 ± 15 and 575 ± 13 nm emission, respectively, using log amplifiers. Hoechst emission is detected by selectively collecting 424 ± 22 nm emission through a linear amplifier.

5. FCM Analysis

For each sample, 10,000 or more events are collected to establish sort regions. It is recommended the aggregates be excluded by gating on single cells, utilizing area versus width of the DNA pulse. Data analysis is performed using LYSIS II software (Becton–Dickinson) on a HP340 desktop computer (Hewlett–Packard, Fort Collins, CO).

B. Polymerase Chain Reaction

1. Reagents and Solutions

a. Treatment with 0.5% Diethyl Procarbonate (DEPC)

This procedure is performed to help inhibit potential RNase contamination.

1. Add 500 μl of DEPC (Sigma) per 500 ml of aqueous solution.
2. Incubate at 37°C overnight.
3. Autoclave the solution for 45 min.

b. 50× TAE Buffer

Trizma base (Sigma)	242 g
Glacial acetic acid (Sigma)	57.1 ml
0.5 *M* EDTA (Sigma), pH 8.0	100 ml

c. 10× PCR Reaction Buffer (Boehringer-Mannheim Corp., Indianapolis, IN)

100 m*M* Tris–HCl

15 m*M* MgC12

500 m*M* KCl

pH 8.3

d. DNA Lysing Buffer (Higuchi, 1989a)

0.45% NP-40 (LKB-Produker-AB, Broma, Sweden)

0.45% Tween 20 (Sigma)

6 μg/100 μl proteinase K (Boehringer–Mannheim)

50 m*M* Tris–HCl (pH 8.3), 1.5 m*M* MgCl2, 50 m*M* KCl (1× PCR reaction buffer).

e. RNA Extraction Buffer

4 *M* guanidinium thiocyanate (Sigma)

25 m*M* sodium citrate (Sigma), pH 7.0

0.5% sarcosyl (Sigma)

0.1 *M* 2-mercaptoethanol (Sigma).

f. Reverse Transcription Mix

1. Substrate	1 m*M* dNTPs (Pharmacia LKB Biotech, Inc., Piscataway, NJ)
2. RT-PCR kit (Gibco BRL)	1 m*M* dithiothreitol
	200 u reverse transcriptase
	50 u RNase inhibitor (Ambion Inc., Austin, TX)
	50 m*M* Tris–HCl, pH 8.3
	75 m*M* KCl
	6 m*M* Mg Cl_2.

g. FISHES/FLIP Buffers

1. Prehybridization
   a. 20× SSC
      i. 3.0 *M* NaCl
      ii. 0.3 *M* Na citrate
      iii. pH to 7.0 with HCl
   b. 0.1 *M* phosphate buffer, pH 7.0
   c. Placental RNase inhibitor (Ambion).

2. Polymerase chain reaction
   a. 10× PCR Buffer (make in lab as many commercial products contain detergents that may lyse cells during the cycling)
      i. 100 m*M* Tris–HCl (pH 8.3)
      ii. 500 m*M* KCl
      iii. 15 m*M* $MgCl_2$ (the final concentration may need to be optimized for each PCR experiment).
   b. 1.0 m*M* digoxigenin dNTP mix (Boehringer–Mannheim).3. Formamide
   a. 5 ml AG 105-8× resin (Bio-Rad Laboratories, Hercules, CA)
   b. 100% formamide (Sigma) to 50 ml
   c. Store at room temperature.

h. Primers and Probes

The sequence for the desired primers is derived from a data base, e.g., Genbank, for the genomic regions of interest. To determine the oligomer primer sequence that is desired, we recommend using the Genetic Computer Group software (Devereaux *et al.*, 1984) to find the gene sequence and the OLIGO software (National Sciences, Hamel, MN) to detemine the optimal sequence for the primer. Primer oligodeoxynucleotides are generally 20 mer, synthesized on a DNA synthesizer (Applied Biosystem, Foster City, CA), and purified using the Poly-Pak (Glen Research Corp., Sterling, VA) cartridge purification technique.

2. Basic PCR

a. PCR Reaction Mix

The basic PCR reaction mix (total volume = 100 μl) consists of the following:

10× PCR reaction buffer (Boehringer-Mannheim)	10 μl
2.5 μ*M* dNTP (Pharmacia LKB)	12 μl
10 pmol/μl upstream (us) primer	1 μl
10 pmol/μl downstream (ds) primer	1 μl
5 units/μl Taq polymerase (Boehringer-Mannheim)	0.5 μl
DNA template	1–2 μl
CRS template	5–10 μl
Nuclease-free water	Up to reaction volume = 100 μl.

b. Thermocycling

Amplification of DNA is carried out in a thermal cycler (Perkins–Elmer, Noralk, CT). The following represent typical settings for the HER-2/gastrin competitive PCR; the settings of individual PCR reactions need to be adjusted according to the specific requirements for the PCR template used.

Standard Setting		
Temperature (°C)	Time (min).	
94	1	
60	1	
72	2	× 35–40 cycles
72	2	× 35–40 cycles for extention
72	7	

c. Gel Electrophoresis

Use gel electrophoresis for detection of PCR products ranging in size from 50 to 1500 bp.

1. Prepare a 2 or 3% agarose gel (Clonotech Laboratories, Inc., Palo Alto, CA) in 150 ml 1× TAE buffer.
2. Load samples in appropriate lanes and electrophorese at 90 V for 2 hr. Flanking lanes should contain sizing DNA standards (Bio Ventures, Inc., Murfreesboro, TN).
3. Stain gel with solution containing 0.5 μg/ml ethidium bromide (Sigma) in 1× TAE for 10 min.
4. Destain by incubating gel in water for 10 min.
5. Visualize DNA bands with UV transilluminator (International Biotechnologies, Inc., New Haven, CT) and photograph using Polaroid 667 film (Polaroid, Inc., Cambridge, MA).

d. Densitometry

Densitometry can be performed by scanning the Polaroid image of the gel using a flatbed scanner [e.g., Scan Maker 600GS (Microtek Lab., Inc., Torrance, CA)] and an acquisition software such as Adobe Photoshop (Adobe Systems Inc., Mountain View, CA). The Adobe software used in a Macintosh IIci computer (Apple Computer, Inc., Cupertino, CA) can invert this image so that it appears like an X-ray film image. Densitometry software such as Scan Analysis (Biosoft, Cambridge, England) can be used to quantify this inverted image. The data are plotted on statistical software such as Starview II (Abacus Concepts, Inc.).

III. Measurement of Gene Amplification Using Competitive PCR

A. Introduction

In heterogeneous populations of cells, undesired cell types contaminate and dilute the specific DNA of the desired cells. This can often lead to an underestimation of the degree of gene amplification. If the desired population can be resolved using specific stains such as DNA stains or immunophenotyping, the specific cell population can be sorted. This approach was not feasible using classical molecular techniques because too many cells were required. This is no longer a problem because very few cells are required when utilizing PCR.

B. Strategy

The PCR is a technique whereby a small amount of a genetic material of interest can be exponentially amplified from a few cells (e.g., 100 cells is a practical number). By introducing known concentrations of a mutant template, called the competitive reference standard (CRS), to an unknown amount of the genomic DNA, PCR of the two templates can be carried out in a competitive fashion, under identical conditions, in the same test tube. The mutation in the CRS allows this template to be distinguished from the genomic template by gel electrophoresis (see Fig. 6). Densitometric analysis of the genomic template PCR product can then be quantified relative to the CRS template PCR product. Since the ratio of the concentration of the final PCR products reflects the ratio of the concentrations of the initial templates, the concentration of the unknown genomic template can be calculated (see Fig. 7).

There are two approaches to synthesizing a CRS template. The first is to introduce a restriction site somewhere within the desired template by site-directed mutagenesis. The advantage of this approach is that the templates for PCR are virtually identical for both the CRS and the genomic or target sequence. The disadvantage is the amount of verification required to show that all CRS molecules are cut by the restriction enzyme. The second method is to produce the CRS by deleting a short segment of the desired genomic template so it can be resolved from the original sequence. This CRS is easier to produce but the size difference from the original template needs to be accounted for. The deletion segment should be kept as short as possible while maintaining the ability to differentiate the CRS from target template electrophoretically. Both methods are demonstrated in this chapter: point mutation in PCR and deletion in RT-PCR.

The basic scheme of incorporating a point mutation into a DNA template to introduce a restriction site is shown in Fig. 1. The general strategy of site-directed point mutation by PCR has been described by Higuchi (1989b). An upstream primer (us Primer A) and a downstream primer (ds Primer B) are prepared that span the desired region of the DNA to be amplified. Using the mutation introducing primers, muPrimer C and muPrimer D, a single mismatch base pair converts the nucleotide (X) to cytosine (C) and the nucleotide (Y) to a guanidine (G) (see Fig. 1A). The sequence-specific *Sma*I restriction endonuclease (New England Biolabs, Inc., Beverly, MA) recognizes the CCC–GGG sequence, resulting in the digestion of the CRS template into two smaller DNA fragments that can be separated from the genomic template by gel electrophore-

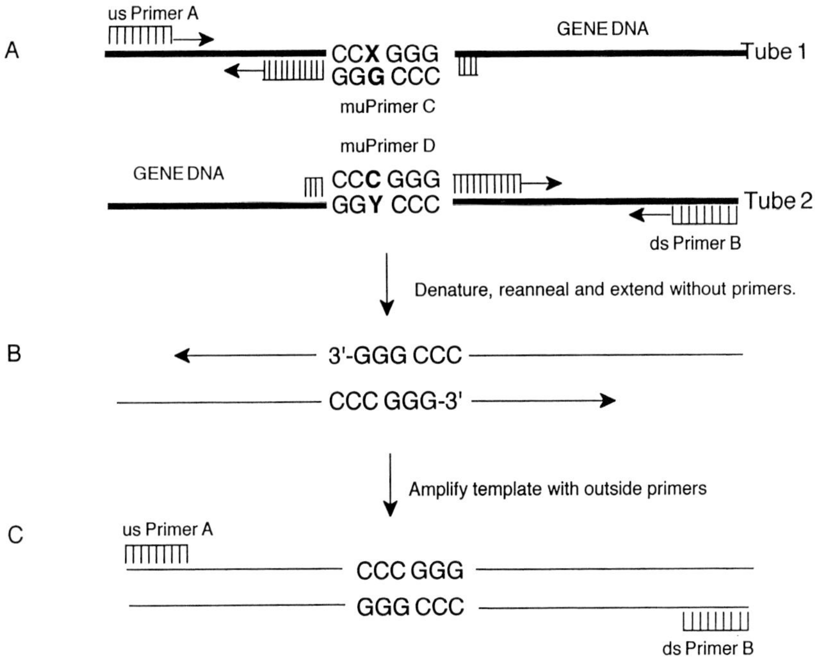

Fig. 1 Schematic representation for the synthesis of a DNA-CRS template containing the Sma-1 site (CCC-GGG). This diagram illustrates the primer configurations and the steps involved in introducing the necessary point mutation.

sis. The site of mutation is selected to be near but not at the center of the template for optimal separation of the two digested fragments.

In separate tubes, using us primer A with ds primer muPrimer C as one set and using us primer muPrimer D and ds primer B as the other set, PCR amplifications are carried out. The PCR products are then gel electrophoresed and the products purified with Magic PCR prep (Promega Corp., Madison, WI). The contents of the separate reactions are combined, denatured, and, in the absence of primers, allowed to reanneal. Selective extention of the 3′-heteroduplexes by Taq polymerase in the absence of primers results in the synthesis of the desired CRS with the point mutation (Fig. 1B). By adding back outside us primer A and ds primer B, the CRS is then amplified (Fig. 1C).

It is also advantageous to perform competitive PCR with a nonamplified gene that can serve as a reference gene for precise determination of gene amplification. Since two separate CRS are required (one for the desired genetic material, i.e., target gene, and one for the reference gene) and because a slight difference in concentrations between the two is exponentially amplified during the PCR causing significant quantitation error, a hybrid of the two CRS linked in tandem as a single template is prepared.

The basic scheme to link the target and reference templates together (Higuchi, 1989b) is shown in Fig. 2. The sequence for linker primer A consists of ds primer A and the complementary sequence of us primer B attached to its 5′ end. The sequence for linker primer B consists of us primer B and the complementary sequence of ds primer A attached to its 5′ end. By denaturing the templates and allowing the complimentary 3′ ends to anneal, extension using the Taq polymerase results in the synthesis of the hybrid template which can then be amplified using the outside primers, us Primer A and ds Primer B.

The tandem linkage of the target and reference CRS templates allows for the quantification of the target gene relative to the nonamplified reference gene. The reference gene should reside on the same chromosome as the target gene. Thus when duplication of the chromosome occurs (e.g., in G_2), erroneous

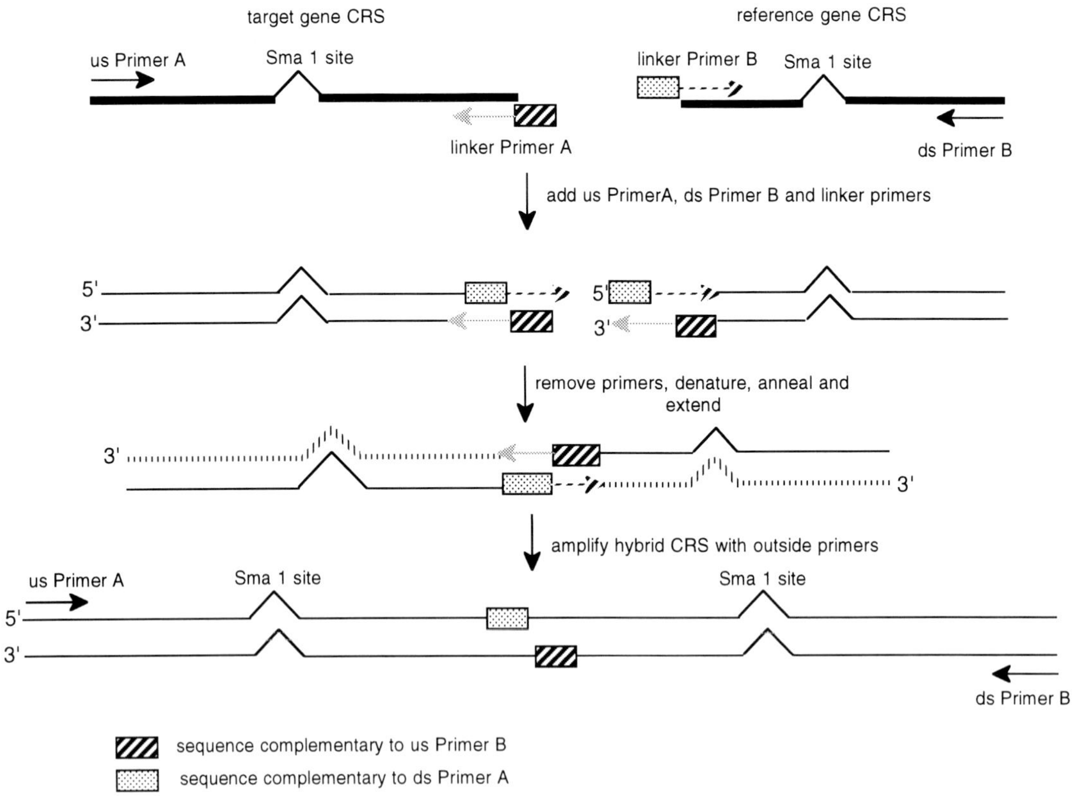

Fig. 2 Schematic representation for the linkage of a target gene CRS with a reference gene CRS. This diagram illustrates the primer configurations and the steps involved in linking the two CRS.

quantification of gene amplification is avoided because the target gene is quantified relative to the reference gene. Another reason for preparing a hybrid CRS is that PCR can amplify small errors introduced by minute variations in dilution or pipetting reagents. Linkage of the two templates in tandem and introducing them as a single CRS template to the reaction mixture minimizes this error.

In competitive PCR a fixed concentration of the extracted genomic DNA from sorted cells is added to serially diluted hybrid CRS templates. Using the appropriate primers, the desired templates are coamplified in the same tubes under identical conditions. The concentrations of the final products reflect the initial concentrations of the starting templates. When the PCR products are digested with the *Sma*I restriction endonuclease, the CRS template is susceptible to site-specific cleavage. The genomic template however does not contain the *Sma*I endonuclease site and therefore remains undigested. Using gel electrophoresis, the PCR products are resolved. By densitometric analysis, the relative concentration of the two starting templates is extrapolated. The degree of amplification is calculated by dividing the concentration of target gene by reference gene (Clementi *et al.*, 1993; Harlow *et al.*, 1993; Wang and Mark, 1990).

C. Step-by-Step Procedure

For the purpose of illustrating how this protocol can be applied to quantify gene amplification, HER-2 oncogene amplification specifically in breast ductal cells from breast tumor specimens are used. The HER-2/*neu* oncogene is used as a specific example of a target gene for studying oncogene amplification and the gastrin gene serves as the reference gene. Needless to say, this protocol could be applied to quantify other genes (e.g., c-myc) in other selective tissue types (e.g., bone marrow).

1. Identifying and Sorting for Breast Ductal Cells by FCM

Breast tumor specimens received from surgery are immediately frozen and stored in liquid nitrogen.

1. Mince finely breast tissue with surgical scalpel in RPMI 1640 (Gibco BRL) supplemented with 10% fetal bovine serum (Gibco BRL).
2. Filter serially through wire cloth, stainless steel mesh No. 60 and mesh No. 150 (Small Parts, Inc., Miami, FL).
3. Wash cells with PBS and fixed cells per Section II,A,2, Fixation and Permeabilization (cell concentration > 10,000 cells/ml).
4. Stain cells per Section II,A,3, Staining Cells, using the following fluorescence probes:

   a. Anti-HER-2 antibody

      i. Primary antibody — 0.067 μg/10 μl anti-HER-2 antibody (MsIgG) (Cambridge Research Biochemicals, Inc., Wilmington, DE)

4. Denature the two CRS templates in the same test tube and allow for "sticky ends" to reanneal and from heteroduplexes.
5. Allow for selective 3′ extention of the CRS 3′-heteroduplexes by adding Taq polymerase and the basic PCR mix in the absence of primers.
6. Amplify by PCR (30 cycles) the hybrid CRS using the outside primers neu 1c and gast 3a.
7. Purify the HER-2–gast hybrid CRS (see Fig. 4) as previously described using gel electrophoresis and Magic PCR prep.

6. Competitive PCR

1. For HER-2 PCR, label 10 sets of tubes 1A through 1J and similarly label for gastrin PCR 10 sets of tubes 2A through 2J.
2. Previously sorted cells are incubated with PCR DNA lysis buffer at the final concentration of 100 cells/μl at 60°C for 60 min, followed by inactivation at 95°C for 10 min. This DNA extract is ready for comparative PCR as the native template or storage at −20°C.
3. Add 10 μl of serially diluted hybrid CRS template or nuclease-free water to each labeled tube as shown in Table I.
4. Master mixes are always used to minimize any tube to tube variations. The following master mix is made:

a. 190 μl	10× reaction buffer
b. 9.5 μl	Taq polymerase
c. 1225.5 μl	H_2O
d. 19 μl	DNA final concentration is 100 cells equivalent/tube).

5. Add 684 μl from each master mix into two tubes (see Fig. 5).

a. To master mix 1 (HER-2 PCR)	9 μl neu 1c
	9 μl neu 3b
b. To master mix 2 (gastrin PCR)	9 μl gast 1
	9 μl gast 3a.

6. Aliquot 78 μl of the master mix 1 into tubes 1A–1H and 78 μl of the master mix 2 into tubes 2A–2H.

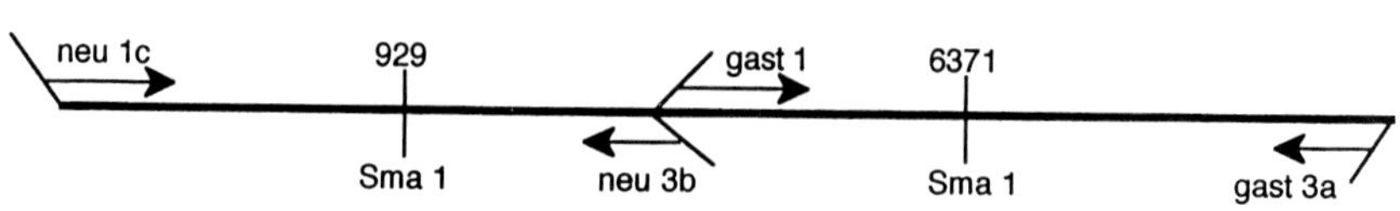

Fig. 4 HER-2-gast CRS Hybrid. The HER-2-gast CRS hybrid consist of a gene sequence for HER-2 spanning base 821 to 992 with the SMA-1 restriction site at base 929. Hybridized to this is the gastrin reference gene sequence spanning base 6290 to 6437 with the Sma-1 restriction site at 6371. The various primers for this hybrid are as shown.

Table I

Tubes	CRS conc (attomols/μl)	(μl) Amount in each tube	Sample designation
A	dH_2O	10	Native template only
B	0.0003	10	Native template and CRS
C	0.0006	10	Native template and CRS
D	0.0012	10	Native template and CRS
E	0.0024	10	Native template and CRS
F	0.0048	10	Native template and CRS
G	0.0096	10	Native template and CRS
H	0.0192	10	Native template and CRS
I	0.0003	10	CRS only
J	dH_2O	10	DNA contamination control

7. Prepare a control master mix containing

a. 10× reaction buffer	50 μl
b. Taq polymerase	2.5 μl
c. dH_2O	327.5 μl

8. To tubes 1I and 1J, add 1 μl neu 1c and neu 3b, and to tubes 2I and 2J, add 1 μl gast 1 and gast 3a.

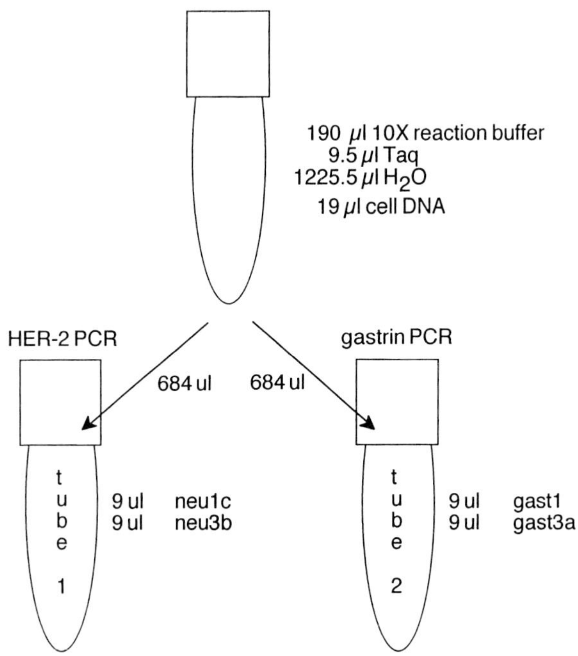

Fig. 5 Master mixes are used to minimize tube to tube variations. The master mix contains the reaction buffer, Taq polymerase, and DNA extract. This is then aliquoted into a second set of master mixes, tube 1 containing the neu primers and tube 2 containing the gast primers.

9. Aliquot 76 μl of the control master mix to tubes 1I, IJ, 2I, and 2J.
10. Add 75 μl of nuclease-free mineral oil to each tube to minimize evaporative losses during thermocycling.
11. Place tubes in thermocycler with settings as described in Section II,B,2,b, Thermocycling.
12. When the thermocycler reaches 94°C, "hot start" the PCR reaction by adding the 12 μl 2.5 μ*M* dNTP mix to each tube.
13. Perform PCR for 35–40 cycles.
14. Digest the PCR products with restriction endonuclease *Sma*I.
    a. Add 3 μl (1 unit/μl) *Sma*I to 45 μl of PCR products.
    b. Incubate at 25°C for 120 min.
15. Electrophorese on 3% agarose gel and stain with ethidium bromide (see Fig. 6).
16. Determine density of each band as described in Section II,B,2,d, Densitometry.

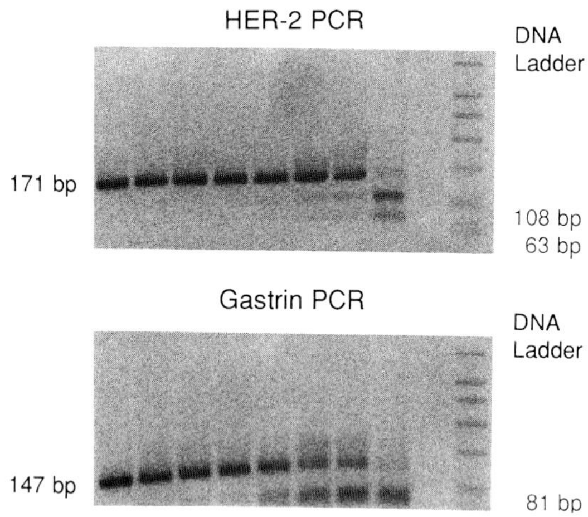

Fig. 6 The photographed image of an ethidium bromide stained gel illuminated by UV light demonstrating a concentration titration of HER-2 CRS versus genomic HER-2 DNA template (upper gel) and gastrin CRS versus genomic gastrin DNA template (lower gel). The PCR products had been previously incubated with Sma-1 at 25°C for 120 min. In the HER-2 gel, lane 1 contains the HER-2 PCR product of genomic DNA but no CRS and is therefore not susceptible to Sma-1 digestion. Thus, there is a single band of 171 bp as shown. In contrast, in Lane 8, the PCR products of the HER-2 CRS with no genomic DNA is shown. The HER-2 CRS is susceptible to the cleavage by SMA-1 into two bands of 108 bp and 63 bp respectively. Lanes 2 through 7 represent increasing concentration of CRS versus a fixed unknown concentration of genomic HER-2 DNA template (0.006, 0.012, 0.024, 0.048, 0.096, and 0.192 attomoles respectively). The density of each band is quantified by densitometry. An analogous set up is demonstrated by the lower gel using the gastrin reference gene.

17. Calculate gene amplification:
    a. Plot the ratio of the density of the HER-2 or gastrin CRS divided by the respective density of their genomic bands (*y*-axis) versus the CRS concentration (*x*-axis) (see Fig. 7).
    b. Where the ratio of genomic template to CRS template equals one, the two concentrations are equal. Therefore the concentration of the genomic template equals that of the CRS template.
    c. Determine gene amplification by dividing the concentration of the HER-2 oncogene by the concentration of the reference gastrin gene.

D. Critical Aspects

PCR perpetuates and exponentially amplifies even minute errors. Small variations of concentrations from tube to tube, introduced by pipetting or serial dilutions, can result in significant errors. Thus, it is critical that a master mix be used whenever possible to minimize this variation in concentrations.

We have also addressed the critical requirement to keep the ratio of target gene CRS to reference gene CRS constant since amplification is quantified relative to the reference gene. This is accomplished by linking the two CRS in tandem (see Fig. 4). This strategy allows us to serially dilute these two templates as a single template and introduce it into each reaction tube as a single aliquot,

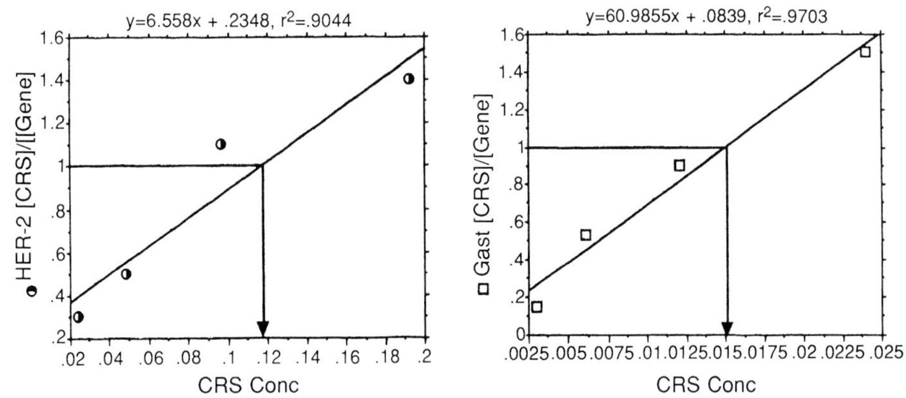

Fig. 7 HER-2 competitive PCR (left) and gastrin competitive PCR (right). Since both the genomic and the CRS templates compete for amplification, the ratio of the PCR products (CRS divided by genomic) reflects the ratio of the initial concentrations of the two templates. As this is a linear relationship, when the density of the cleaved CRS template equals the density of the uncleaved genomic template, their concentrations are equal. The relative amplification of the HER-2 oncogene can then be obtained by dividing the concentration of the HER-2 oncogene template (0.12 attomoles/μl) by the concentration of the gastrin gene template (0.05 attomoles/μl). Therefore, HER-2 is 8 times amplified relative to gastrin.

minimizing the inadvertent tube-to-tube variations associated with two disparate CRS.

When target gene is highly amplified relative to the reference gene, the equivalence point where the [CRS template] = [genomic template] may be missed. In this case, the hybrid CRS template will need to be further serially diluted and the number of cycles of amplification increased in order to bracket the equivalence point.

As with any experimental protocol, controls are critical. In each assay, we have a control where only the CRS template is used (tube I). This control is needed to demonstrate that the *Sma*I can recognize and completely digest the PCR-amplified CRS template. This is critical because our quantitative analysis is based on densitometric quantification of the genomic DNA template relative to a cleaved CRS. Incomplete digestion of the CRS will lead to overestimation of the genomic template concentration.

Incomplete CRS digestion may also result from a problem with the fidelity of the Taq polymerase, leading to PCR product that has lost the CCC–GGG sequence. To minimize nonspecific nucleotide-to-nucleotide binding, the "hot start" technique, i.e., adding the nucleotides into the reaction mixture after the DNA is denatured, is advocated. Furthermore, as few cycles of amplification as possible leading to detectable PCR products are performed to minimize amplification errors from the inadvertent misincorporation of bases by Taq, estimated to occur at a frequency of 2×10^{-4} per nucleotide per cycle (Saiki *et al.,* 1988).

The incomplete digestion of the CRS template can also result from the presence of DNA, other than the CRS, as a contaminant starting template. For this specific reason, each assay has a second control (tube J), containing all components of the reaction mixture, including upstream and downstream primers, but without any DNA template. If there is DNA contamination in the tubes or in any of the reagents, it will be amplified and detected. If this occurs, the experiment is to be discarded since this contaminant DNA will lead to erroneous quantification of the target gene.

E. Expected Results

Figure 8A is a representative histogram of a breast cancer stained with FITC-conjugated isotype control. By setting the gate at R1 and R2, 90% of the nonspecific FITC staining is excluded such that when the same cells are stained with FITC-conjugated anti-cytokeratin antibody, the R1 and R2 regions now gate for cytokeratin-positive cells (see Fig. 8B). These cells are then sorted by flow cytometry into cytokeratin-positive (R1 and R2), predominantly breast ductal cells and cytokeratin-negative (R7), predominantly stromal cells.

That successful sorting for cancerous ductal cells has been achieved with this strategy is demonstrated by the change from the predominantly diploid DNA profile of the cytokeratin-negative cells (see Fig. 8C) to the predominantly

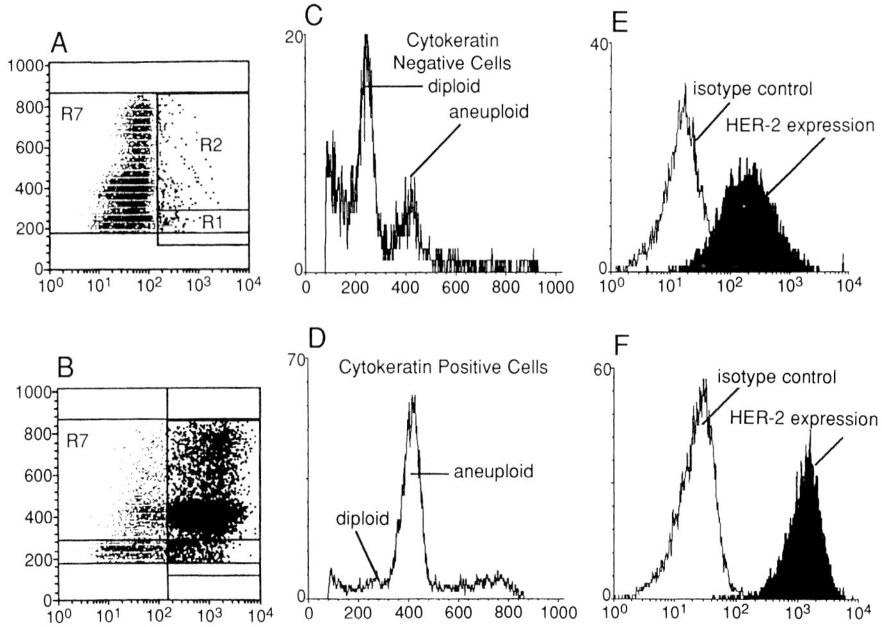

Fig. 8 Sorting cells for molecular phenotyping. Sort regions were established for CK+ (R1,2) and CK− (R7) cells based on the isotype control (A,B). The sorted cells were reanalyzed and the DNA histograms for CK+ (D) and CK− (C) cells were obtained to verify the purity of the sort. In this tumor, CK+ fraction contained predominantly aneuploid cells and the CK− fraction contained predominantly diploid cells. Histological examination of the CK+ specimen revealed that virtually all cells were CK+ intact cells. In contrast, the CK− fraction contained both CK− intact cells as well as cell nuclei. Thus, the CK− fraction was contaminated with aneuploid tumor cells that may have CK+ before cytoplasmic lysis. However, this fraction is small. Simultaneous determination by FCM of $p185^{erbB\text{-}2}$ expression in tumor cells is shown in E (baseline expression) and F (overexpression).

aneuploid DNA profile of the cytokeratin positive cells (see Fig. 8D). Some aneuploid cells contaminate the diploid fraction and most of these bear nuclei appearing as Cytk(−)/HER-2(−) cells. Clearly, Fig. 8D shows very low contamination of aneuploid cells by diploid cells. Competitive PCR to detect and quantify HER-2 amplification is then performed on these two subpopulations of cells.

To verify that amplification of HER-2 is in fact correlated with HER-2 oncoprotein overexpression, FCM analysis was performed simultaneously on these tumor cells, using an anti-HER-2 antibody, labeled with a second antibody conjugated to PE. Overexpression of HER-2 is defined as when HER-2 mean fluorescence intensity is greater than two standard deviations above the HER-2 mean fluorescence of the baseline HER-2 samples. Overexpression of the HER-2 oncoprotein ($p185^{erbB\text{-}2}$) in HER-2-amplified cytokeratin-positive cells is shown in Fig. 8F as compared to the p185^{erbB-2} expression in nonamplified cytokeratin-positive cells (see Fig. 8E).

Gel electrophoresis of the competitive PCR products, after *Sma*I digestion, is shown in Fig. 6. Since both the genomic and the CRS templates compete for amplification, the ratio of PCR products (native versus CRS) reflects the ratio of the initial concentration of the two templates. As this is a linear relationship, when the ratio of the density of the cleaved CRS template divided by the density of the uncleaved genomic template equals one, their concentrations are equal. For example, as shown in Fig. 7, when HER-2 [CRS]/[genomic] = 1, the initial [HER-2] = 0.12 attomol/μl. The relative amplification of the HER-2 oncogene can then be obtained by dividing the concentration of the HER-2 oncogene template by the concentration of the gastrin gene template (0.015 attomol/μl). In this example, HER-2 amplification is eight times that of gastrin.

Table II shows the results of cell lines tested to verify the accuracy of this technique. In the SK-BR-3 breast cancer cell line (ATCC, Rockville, MD), known to have HER-2 amplification, detectable amplification using this protocol is 10× reference gene. This compares favorably with Southern blot analysis demonstrating a 9× amplification. In MCF-7 and MDA-MB-231 breast cancer cell lines (ATCC), known to have normal HER-2 gene copy number, competitive PCR demonstrates no detectable amplification. These findings are in agreement with values reported in the literature (Lupu and Lippman, 1993; Maguire and Greene, 1989; Singleton and Strickler, 1992; Tandon *et al.,* 1989) and by one of the authors (Li *et al.,* 1993a).

In our hands, HER-2 amplification, defined as [HER-2] > 2× [gastrin], has been detected in >40% of breast cancers (Li *et al.,* 1993b). This is higher than the 20–30% reported in the literature (Lupu and Lippman, 1993; Singleton and Strickler, 1992; Tandon *et al.,* 1989). In all the samples tested to date, HER-2 amplification is higher in cytokeratin-positive cells when compared to cytokeratin-negative cells of the same specimen [7.2 + 2.8× versus 3.2 + 1.1× (mean ± SEM)]. As expected, all the cytokeratin-negative cells with detectable HER-2 amplification have corresponding HER-2 amplification detectable in the cytokeratin-positive subpopulation of cells. However, there are samples with

Table II
c-erbB-2 Amplification/Overexpression in Cell Lines

Cell lines	Amplification		p185 overexpression (by FCM)
	PCR	Southern	
SK-BR-3	10×	9×	Positive
MCF-7	1×	1×	Negative
MDA-MB-231	1×	—	Negative

HER-2 amplification detected in the cytokeratin-positive cells but not in the cytokeratin-negative cells. Therefore there is a potential for missing HER-2 amplification in these tumor samples if breast ductal cells were not identified and sorted by FCM prior to oncogene quantification. These results reinforce the basic tenet of this strategy, i.e., *by selecting specifically for cytokeratin-positive, breast ductal cells, stromal dilution is minimized, thus resulting in a higher sensitivity for the detection of HER-2 amplification.* Finally, in tumor samples that were HER-2 amplified, HER-2 oncoprotein overexpression is detected by FCM.

IV. Measurement of Gene Expression Using Competitive RT-PCR

A. Introduction

Classical Northern blot analysis requires approximately 10^6 cells for extracting and measuring mRNA. Yet, quantitation of gene expression (mRNA) in small subsets within a heterogeneous population of cells is often desired. A reverse transcription of the mRNA of interest in combination with the polymerase chain reaction provides a means for detecting gene expression on as few as 100–1000 cells. When quantitative RT-PCR is performed, the reliability of the data can be highly subjective due to the efficiency of both RT and PCR steps. This subjectivity can be eliminated by a technique for quantitating specific RNA molecules using an internal RNA-CRS, which is identical to the sequence of interest except for an internal deletion of 50–100 bases.

B. Strategy

The strategy previously described for measurement of gene amplification can be applied to the measurement of gene expression by preparing an RNA-CRS. A titration of the RNA-CRS and a constant amount of native mRNA are reverse transcribed together to produce cDNA of both the native mRNA and the RNA-CRS. The cDNA is amplified together by the PCR. The PCR products are separated by size in a 2% agarose gel electrophoresis, stained with ethidium bromide, photographed, and subjected to scanning densitometric analysis. Specific cell subset from a heterogenous population can be sorted out using flow cytometric techniques, the mRNA extracted, and the number of RNA molecules determined.

Instead of a point mutation that creates a *Sma*I site, the RNA-CRS is synthesized with a deletion of internal bases in the template DNA. This template also contains the sequences necessary for binding the T7 polymerase. A transcription reaction, including the T7 polymerase and the cDNA template, allows the DNA

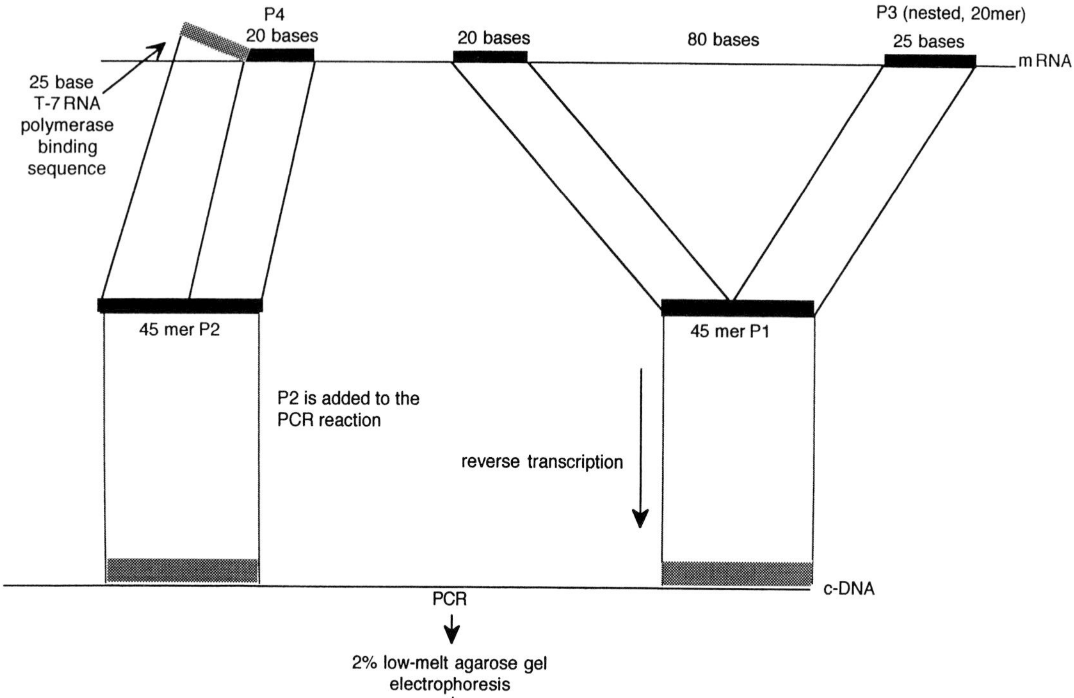

Fig. 9 Schematic representation of the synthesis of a RNA-CRS cDNA template (containing the binding site of the T7 polymerase). This diagram illustrates the primer configurations and the steps involved in making a RNA-CRS with an 80-base delection.

to be transcribed into RNA. A schematic outline for the procedure is shown in Fig. 9. The deletion is produced by preparing a single 45-mer downstream primer, Primer 1 (P1), in Fig. 9 that has a complementary base sequence to the 3′ desired target RNA but has the 80-mer deletion as shown. The native mRNA that will be used to illustrate this methodology is from the Src-family of nonreceptor tyrosine kinase protooncogenes. This protooncogene, c-fgr, is found exclusively in macrophage/monocyte lineages. The murine sequence will be used and the region chosen to use for quantitation has a sequence length of 230 bp (Willman and Yi, 1989). The primers were chosen from the published sequence obtained from the Genbank sequence program (Devereaux *et al.*, 1984).

The upstream primer is also a 45-mer oligonucleotide (P2, Fig. 9) that has a sequence analogous to the 5′ desired region of the mRNA (20 bases) linked to the 25-base sequence encoding the T7 RNA polymerase binding site [5′AATT-TAATACGACTCACTATAGGGA3′ (Guatelli *et al.*, 1990)].

C. Step-by-Step Procedure

1. RNA Extraction (Extracts Total RNA: rRNA to tRNA and mRNA)

It is advised (whenever possible) to extract RNA from nonfixed viable cells. The fixation process (i.e., formalin, formaldehyde, ethanol, etc.) causes a reduction in the amount of RNA extracted as compared to the same amount of viable cells (author's experience).

1. Wash cells in PBS (Gibco, Grand Island, NY) and add to the pellet 100 μl RNA extraction buffer per 10^6 cells as determined from a cell count.
2. Add sequentially 0.1 ml 2 *M* sodium acetate (Sigma), pH 4.0, 1 ml water-saturated phenol (Sigma) and 0.2 ml chloroform-iso amyl alcohol (49 : 1) (Sigma) per ml of the extraction buffer. Centrifuge 10,000*g* for 30 min at 4°C.
3. Transfer aqueous phase to new tube and add an equal amount of isopropanol (Sigma). Incubate at −20°C for 1 hr to precipitate RNA.
4. Centrifuge 10,000*g* for 30 min at 4°C, discard supernatant, and dissolve RNA in 0.3 ml RNA extraction buffer. Add equal volume of isopropanol at −20°C for 1 hr (or can be precipitated overnight).
5. Centrifuge at 14,000*g* at 4°C for 30 min, discard supernatant, and resuspend pellet in 1 ml of 75% ethanol.
6. Centrifuge at 14,000*g* at 4°C for 30 min, discard supernatant, and dry pellet.
7. Suspend dried total RNA in 100 μl DEPC-treated distilled water. Prepare a 1/100 dilution in water and determine RNA concentration by spectrophotometry at a wavelength of 260 nm.
8. 1 OD unit = 40 μg single-stranded RNA.

Note: Alternative methods for isolation of mRNA in the form of kits (i.e., Micro FastTrack, Invitrogen, San Diego, CA) are currently available and may be used instead of the above method. The directions supplied with the kit should be followed.

2. Preparation of Deletion RNA-CRS

1. Add 200 ng mRNA (2.5 μg total RNA) to make a final volume of 30 μl of RT reaction mixture containing 10 pmol primer 1 (P1, Fig. 9).
2. Incubate 45 min at 37°C. To the RT mixture, add 10 μl 10× PCR buffer (Boehringer-Mannheim), 2.5 U Taq (Boehringer-Mannheim), 10 pmol primer 2 (P2, Fig. 9) and enough DEPC-treated H_2O to make a final volume of 100 μl.

   One cycle: Melting: 94°C—1 min.

Annealing: 55°C—1 min. (The annealing temperature may vary from sequence to sequence based upon the nucleotide content.)
Extension: 72°C—2 min.
Amplify for 30 cycles.

3. Electrophorese 30 μl of the PCR product on a horizontal 2% low-melt agarose gel (CloneTech Laboratories) at 90–100 V for 1–½ hr. Low-molecular-weight standards (Bio Ventures, Inc.) are run in a separate lane to verify product size.
4. Visualize gel with a UV transilluminator and excise the band from the agarose gel containing the desired deleted sequence and purify using the Magic PCR preps DNA purification system (Promega).
5. Determine the cDNA template concentration on a spectrophometer at a wavelength of 260 nm. Make a 1 : 100 dilution of the cDNA in nuclease-free ddH_2O and read the optical density at 260 nm.

Note: Because the intermediate cDNA CRS template sequences are stable at −20°C and the amounts purified are adequate for many transcription reactions, this template can be stored for quite some time and then easily amplified to synthesize more when needed. Because of the instability of RNA and the ease of making new material, we do not recommend storing this product for more than 30 days at −70°C. Since the RNA-CRS concentration is known and since its sequence is nearly identical (except for the deletion) to the sequence of interest, the efficiency of all reactions, both reverse transcription and PCR, are controlled for more accurate quantitation.

3. Preparation and Verification of RNA-CRS

1. The Megascript T7 *in vitro* transcription kit by Ambion (Austin, TX) was used to produce and purify the RNA-CRS.
2. Prepare five serial 1/10 dilutions of the RNA-CRS (attomoles); each dilution is added to 30 μl of the reverse transcription mixture containing 10 pmol primer 3 (P3, Fig. 9).
3. Using the highest amount of RNA-CRS prepare a tube using transcription mixture without reverse transcriptase to check for any contaminating DNA template. Finally, prepare a tube without the RNA-CRS to control for RNA or DNA contamination in any of the reagents.
4. Follow steps 2 and 3 in the preceding section for the reverse transcription reaction and the PCR.
5. Run standard gel, inspect for proper band position, and verify that the control tubes show no bands. If there are bands in either or both of the control tubes, new reagents must be prepared and the reactions repeated or the RNA-CRS should be purified a second time to remove residual cDNA template.

4. Competitive RT-PCR

Follow the method described in Section III,C,6, Competitive PCR, but instead of adding extracted DNA, add extracted total or messenger RNA. Also, instead of adding the hybrid DNA competitive reference standard, add the RNA competitive reference standard in a serial dilution scheme. Figure 10 illustrates a schematic of the tube configuration and control tubes for the competitive RT-PCR. Perform the reverse transcription step using a nested primer (P3 from Fig. 9) as described above. In the PCR steps, use the nested primer P4 (Fig. 9) as the upstream primer. There is no need for *Sma*I digestion. The deletion in the CRS will cause this product to migrate faster in the agarose gel thereby resolving the native cDNA product from the RNA-CRS cDNA product. Densitometric analysis is performed as previously described (see Section II,B,2,d, Densitometry) and the results are plotted as concentration of the RNA-CRS (attomoles) vs the ratio of mutated : native band densities. The point of equivalence (ratio = 1) is the concentration of the native mRNA. This value can then be converted to molecules of RNA per cell (1 attomole = 602,252 molecules).

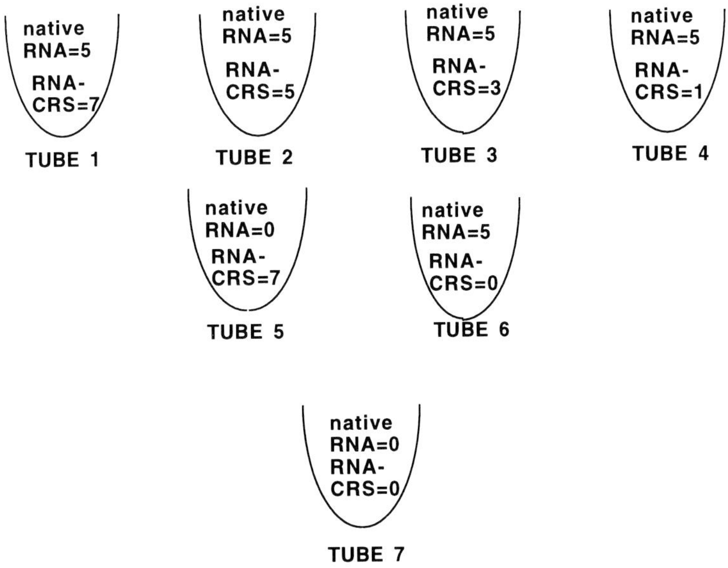

Fig. 10 The tube configuration for a quantitative PCR reaction. An illustration of the dilution scheme for the RNA-CRS along with the native RNA template, which is kept at a constant unknown concentration. Two tubes are established that contain the RNA-CRS or the native RNA templates only. This assures accurate sizing of each band individually as well as controls for any contaminates. One tube contains only the reagents of the reverse transcription and PCR reactions without RNA templates. This will control for any contaminates in these reagents.

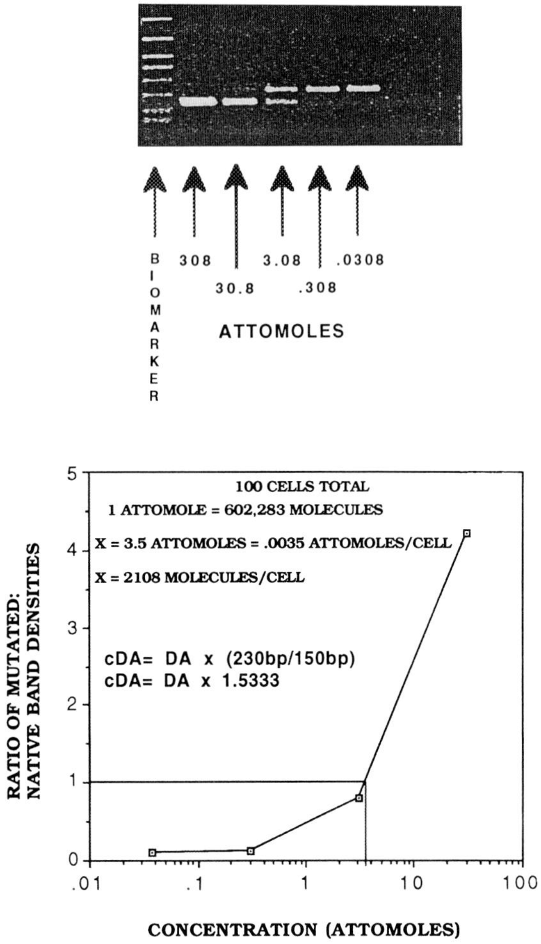

Fig. 11 G_1 phase of the cell cycle. c-Fgr mRNA quantitation in 32-hr postsynchronous murine bone marrrow macrophages. Photograph of a 2% agarose gel electrophoresis of c-fgr cDNA and the c-fgr RNA-CRS cDNA with an 80-base deletion. The photograph was subjected to scanning densitometry and the band densities are plotted as a ratio of the RNA-CRS (mutated) to the native against the known concentration of the RNA-CRS. The equivalence point (ratio = 1) is the concentration of native c-fgr mRNA added to the reverse transcription reaction.

D. Critical Aspects

The RNA extraction is based on the method of Chomczynski (Chomczynski and Sacchi, 1987). Because of the RNAs extreme lability, viable cells should be processed quickly. For example, if sorting cells, they should be sorted directly into the RNA extraction buffer. This procedure is sensitive enough to quantify gene expression in less than 1000 cells. RNA should be stored frozen at −70°C for less than 1 month.

E. Expected Results

Figure 11 outlines an example of a typical quantitative RT-PCR using a deletion RNA-CRS. From the 2% agarose gel, the products of the RT-PCR of the protooncogene c-fgr and its corresponding RNA-CRS are easily separated by size. A scanning densitometer (as described above) is used to analyze the band densities from the photograph of the gel. These results are plotted as the concentration of the RNA-CRS (*x*-axis) vs. the ratio of the RNA-CRS (product band density to the native cDNA product band density (*y*-axis). The point where the ratio is equal to 1 is the concentration of the native c-fgr mRNA present. From the graph, after 32 hr postsynchrony of murine bone marrow macrophages (synchronized in G_0), there are 2108 molecules of c-fgr per G_1 phase cell.

V. Fluorescence *in Situ* Hybridization En Suspension (FISHES)

A. Introduction

The quantitation of mRNA or DNA extracted from sorted cells provides one means for obtaining their molecular phenotype. A more direct approach would be the use of fluorochrome-labeled probes that are hybridized to the specific sequences *in situ*. *In situ* hybridization can be performed in slide mounts (Bakkus *et al.*, 1989; Nederlof *et al.*, 1990; Singer and Ward, 1982; Singer *et al.*, 1986; Taneja *et al.*, 1990) or in a more recent development, *in situ* hybridization can be performed using cells in suspension that are subsequently analyzed by a flow cytometer (Bauman and Bentvelzen, 1988; Bauman and van Dekken, 1989; Pennline *et al.*, 1992). Fluorescence *in situ* hybridization (FISH) has become a very powerful means for evaluating cytogenetic abnormalities by microscopy (Kallioniemi *et al.*, 1992). Flow cytometry for quantifying labeled cells has not been widely used because of its limited sensitivity for detection of specific hybridization (Timm and Stewart, 1992; Yu *et al.*, 1991).

When gel electrophoresis is used, product bands (signal) are separated so they can be properly identified by molecular weight. Background noise is represented by a smear uniformly distributed throughout the electrophoresis lane. When *in situ* hybridization is performed, the resolution of signal from noise can also be resolved microscopically but not by flow cytometry. Because of this background, the detection sensitivity is limited to a minimum of about 300 copies, depending upon the length of the sequence being studied and the number of fluorochrome molecules attached to the probe.

B. Strategy

The requirements for a successful *in situ* hybridization begin with an appropriate fixation protocol. The cells must be fixed in a manner that preserves the basic cell morphology yet at the same time makes large enough openings in the cell membrane so that the reagents for the *in situ* work have easy access to the internal nucleic acid sequences. The strength and timing of the fixative are important considerations when preparing the cells. Too strong a fixative or excessive time in the fixative may crosslink the cellular proteins to the point that the target sequences become inaccessible to the reagents and probes (Singer *et al.,* 1986). Although Ultrapure formaldehyde is used in this protocol, other fixatives may work equally well or better.

After fixation and permeabilization, cells are reconstituted in buffer and then incubated with the appropriate probe under low-stringency hybridization conditions to optimize binding. The cells are then washed under increasing stringency conditions to remove any probe bound with low specificity. An incubation period in the proper wash buffer between each centrifugation is provided to ensure diffusion of unbound material from the cell. The bound probe is then detected directly if a fluorochrome has been incorporated in the probe or it is detected using a secondary reagent such as a labeled avidin for a biotinylated probe.

C. Step-by-Step Procedure

1. Harvest cells, and obtain a cell count. A wash step in a nuclease-free buffer can be done at this point before adding the fixative. Add 10% Ultrapure formaldehyde (Polysciences, Inc. Warrington, PA) to the cells so that the final concentration is 1% in the cell suspension. Incubate at room temperature for 25 min.

2. Centrifuge cells at 1500*g* in a swing bucket centrifuge for 3 min and resuspend at a concentration of 2×10^6 cells/ml in 70% ethanol while slowly vortexing cells; incubate 60 min or store the cells at −20°C until ready to use.

3. Centrifuge cells at 1500*g* and resuspend in 500 μl 1× SSC with 500 μg/ml of nuclease-free BSA (Gibco BRL Gaithersburg, MD) for 20 min using a 1.7-ml siliconized microcentrifuge tube.

4. Centrifuge cells as before. Resuspend in 100 μl same. Again centrifuge cells, but this time use a microcentrifuge (Eppendorf Centrifuge 5415) at the same speed. Aspirate the supernatant using a gel tip (the type used for loading sequencing gels). Resuspend the cells in 20 μl hybridization mix containing 10 ng −100 ng of the probe.

5. If using a double-stranded DNA probe prepared by PCR, heat sample to correct melting temperature (T_m) to dissociate the probe strands in the mix before adding to the cells. The correct temperature for melting will depend upon the base sequence and time of incubation, e.g., a 200-base pair DNA

probe in 4× SSC containing 50% formamide at or above 72°C (melt for 10 minutes then put this mix in an ice bath for 2 min before distributing to the samples). The optimal conditions need to be determined for each probe and this step can be omitted for single-stranded probes. In general as incubation time increases so does nonspecific binding so the hybridization time is limited to 3 hr. Stringency conditions for hybridization are low for this step to optimize binding of the probe to the target.

6. Three washes of increasing stringency are performed to remove mismatches and reduce noise as follows: Add 3 ml 2× SSC with 500 μg/ml BSA + 50% formamide at 42°C for 30 min; centrifuge (1500*g*). Remove supernatant and resuspend cells in 3 ml 1× SSC with 500 μg/ml BSA + 50% formamide at 42°C for 50% formamide at room temperature for 15 min; centrifuge. Last, incubate the cells in 3 ml 1× SSC with 500 μg/ml BSA at room temperature for 15 min.

7. This step is not necessary if a directly conjugated probe is to be used. Transfer cells to 3.0 ml PBS + 0.1% azide + 5% BSA (PAB buffer) and centrifuge 1500*g* for 3 min. Decant and blot the tubes on paper towels. Add 5 μl of the second reagent to the cells in the residual buffer. Incubate cells 4°C for 30 min, and mix occasionally. Add 3 ml PAB, repeat wash, and incubate cells in 3 ml PAB for 30 min before centrifuging at 1500*g*.

8. Resuspend cells in 0.5 ml PAB and analyze by flow cytometry. Wet or Cytospin preparations can also be prepared for microscopic evaluation.

D. Critical Aspects

An important part of the overall protocol is the proper controls. For the FISHES protocol some of these controls should include

1. Cells treated with RNase to eliminate the mRNA.
2. Sense and antisense probes whenever possible.
3. Cells that have gone through the entire procedure without the probe and/or the secondary detecting reagent (the procedure causes an increase in cellular autofluorescence).
4. A nonsense probe if possible.

The selection of a proper probe to perform the FISHES is necessary. The menu of nucleic acid probes includes double-stranded DNA, 20- to 50-mer oligonucleotides, or single-stranded RNA probes of a few hundred bases. Each type of probe has its advantages and its disadvantages. The double-stranded DNA probes are more resistant to degradation; easy to label using a tagged deoxynucleotide during a PCR, nick translation, or random priming protocol with a moiety for fluorescence detection such as digoxigenin, biotin, fluorescein, or CY5; and can be made to any size that is appropriate for hybridization *in situ*. The PCR-generated probe offers the best control over probe size and it is easily evaluated for tagged nucleotide incorporation on an agarose gel.

When performing FISHES with DNA probes, the optimal probe size appears to be between 150 and 300 bp. A disadvantage for using this probe is the competition between the target DNA for the probe strands and the complementary probe strands reannealing to themselves before reaching the target. To overcome this, a larger amount of probe is required in order to saturate the target therefore leading to increased nonspecific binding of the probe. Large probes may also hybridize with some degree of base mismatching causing reduced specificity unless high-stringency washes are employed. Finally, if too large, the probe may enter the cells poorly causing a loss of signal.

In order to overcome the problem of strand competition as seen with the double-stranded DNA probes, short single-stranded probes such as DNA oligonucleotides synthesized *in vitro* can be utilized. The size of an oligo probe is generally between 20 to 50 bases. The fluorochrome can be conjugated to either end during the synthesis of the oligonucleotide or enzymatically after the synthesis. Furthermore, oligonucleotide probes can easily be site directed to very specific targets on the desired product inside the cell and the problem of self-competition is reduced. Because of the limited amount of label that can be incorporated into an oligonucleotide probe, however, it becomes necessary to construct several of them in order to "paint" the sequence. "Painting" a sequence will allow for more signal per specific target. When using short probes to paint the desired product the above problems are minimized but the likelihood of homology with undesired sequences within the genomic message pool or ribosomal RNA is much higher. This will lead to a possible misinterpretation of the data because of biological noise in the readout system. On a gel this noise appears as bands outside the expected molecular weight region or as part of the heterogeneous material found in the lane. In cells, this same material is part of the "nonspecific" background fluorescence.

Another option for making single-stranded probes is to build a RNA probe by applying the method described above for making the RNA-CRS–use the T7 promoter but without any deletion. The single-stranded RNA can be tagged with a digoxigenin, biotin, fluorescein, or CY5 ribonucleotide during the synthesis. The RNA can be made to any length and there is no strand competition. A disadvantage is that single-stranded RNA is highly susceptible to ribonucleases and precautions should be taken to inhibit those nucleases and prevent probe degradation.

When using secondary reagents, a titration of the second reagent is necessary it order to minimize the noise caused by its nonspecific binding to cellular products. Probes directly tagged with a fluorochrome offer the advantage of reduced noise levels but may also have reduced signal levels as well. In addition, fluorochromes that emit in the red region of the spectrum (i.e., CY5) give the best signal to noise ratios (Yu *et al.*, 1991) due to lower cellular autofluorescence emissions.

E. Expected Results

FACS acquisition and analysis for these examples were done using a FACScan flow cytometer (Becton–Dickinson). Although a fluoresceinated antibody was used for these experiments, other fluorochromes can be used so that multicolor analysis of different DNA or RNA species can be accompanied. The double-stranded DNA probe was produced using the PCR. As seen in Fig. 12, distribution 1 is the baseline autofluorescence from the cells and distribution 2 is the background noise caused by the FISHES procedure plus the addition of the detecting Fl-antiDig antibody. Both the Dig-11-dUTP nucleotide and the detecting antibody are obtained from Boehringer-Mannheim (Indianapolis, IN). The best way to reduce this noise to a minimum is to titrate the concentration of the antibody that is used. Too much antibody will result in high levels of noise and too little antibody will prevent the detection of bound probe. Figure 12, distribution 3 shows an example of a direct FISHES as detected by a 191-base digoxigenin-tagged probe used for the detection of the cytoskeletal actin mRNA in the murine fibroblast line L929. Because of the abundance of this

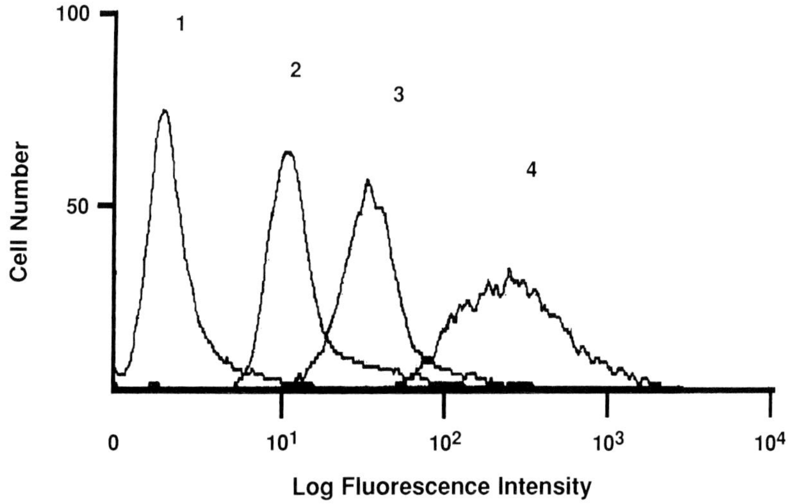

Fig. 12 Actin mRNA expression in L929 fibroblasts. Fluorescence *in Situ* Hybridization En Suspension (FISHES) of a digoxigenin-tagged DNA probe directed against *g*-actin mRNA is shown. A FITC-conjugated antiDig antibody was used to detect the probe. Cell distribution 1 represents cellular autofluorescence before the FISHES. Cell distribution 2 represents the background fluorescence due to the hybridization procedure and the nonspecific binding of the antibody in the absence of the probe. Cell distribution 3 is the complete FISHES procedure with the probe. The mean channel fluorescence (MCF) of these cells is 4 times that of the background noise (cell distribution 2). Fluorescence *in situ* PCR (FLIP) for the same cell type is shown in cell distribution 4. A reverse transcription of the mRNA was done to produce a cDNA and the cDNA was amplified using the PCR. The original template was digested with RNases after the PCR and the amplified product was detected using the FISHES procedure. The MCF of these cells is 25 times that of the background.

mRNA (2000–3000 copies per cell), direct detection of the mRNA can easily be seen on the cytometer as demonstrated here.

VI. Fluorescence *in Situ* PCR (FLIP)

A. Introduction

The major problem with FISHES is low sensitivity caused by high background. The number of copies per cell must be high enough to exceed the background. The sensitivity of FISHES is about copies 300 copies of mRNA sequences. Unlike protein molecules that are usually several thousand per cell, most mRNAs are less than 1000 copies/cell and only 2 copies of DNA are present. To improve sensitivity, some researchers have performed the PCR *in situ* using cells attached to slides to amplify viral DNA sequences before detection by FISH (Haase *et al.*, 1990; Kuo-Ping *et al.*, 1992; Nuovo *et al.*, 1991). The *in situ* PCR in suspension is being developed to help increase the number of target molecules of DNA or mRNA converted to cDNA so that these can readily be detected on a flow cytometer.

B. Strategy

The *in situ* PCR cells are fixed in formaldehyde so they remain intact during the rigorous temperature cycling and subsequent *in situ* hybridization. The fixed cells are permeabilized by an alcohol treatment step to produce openings in the cell membrane large enough for the PCR and FISHES reagents to enter. A major drawback of the procedure is that openings that allow easy access for the reagents also allow easy egress of the amplified DNA. While digoxigenin-tagged nucleotides incorporated into the PCR product have been shown to reduce this leakage (Long *et al.*, 1993; Nuovo *et al.*, 1991), the amplification efficiency of Dig-substituted product is very poor and the detection of specific product using the FISHES is problematic due to low-efficiency hybridization of a hapten or fluorochrome-tagged probe to a digoxigenin-11-dUTP-substituted PCR product. Thus, end-labeled oligoprobes or 30- to 50-mer directly conjugated probes that can interdigitate between the Dig-11-dUTP regions are more effective. The best method for the *in situ* PCR would be one which uses untagged nucleotides and still retains the amplified product. Alternative methods for retaining the product within the cell are needed because, even with Dig-tagged nucleotides, there is still some degree of product leakage as demonstrated by Long *et al.* (1993). To confirm that the desired PCR product was actually amplified inside the cells or to determine the mRNA copy number in the cells, the procedures including mRNA and DNA extractions described in Sections III and IV can be performed.

Cell loss has been a common finding during the PCR and this has been often interpreted to indicate cell degradation. This interpretation is not correct because most cell loss is caused by adherence of cells to the reaction tube surface during the initial PCR cycles. It is critical that a nuclease-free (nf) albumin be used to coat the tubes prior to the PCR to prevent this from occurring and this is done by using nfBSA in the washes prior to and after the PCR.

The application of the FISHES and *in situ* PCR is accomplished using cells in suspension so that they can be analyzed by a flow cytometer and observed under a fluorescence microscope. The procedure for amplifying mRNA targets requires that the mRNA sequence be converted to a complementary DNA (cDNA) first before the PCR can take place. A reverse transcription step is necessary to construct the cDNA. The reverse transcription procedure presented uses a prehybridization step to anneal a starting oligo primer (complementary to a 20- to 40-base stretch of the mRNA at the 3′ end) to the target. This prehybridization step provides a higher specificity to the reaction. The supernatants were aspirated using sterile, nuclease-free pipette tips. A gel tip is used to remove all of the supernatant before the proper reverse transcription or PCR reaction mixes are added.

C. Step-by-Step Procedure

1. Follow steps 1–4 for FISHES procedure. (Go to step 4 of this procedure if amplification of only genomic DNA is desired.)

2. To the cell pellet, add 4.0 μl 4× SSC, 1.0 μl 5 mM phosphate buffer at pH 7.0, 1.0 μl 50 (units) placental RNase inhbitor, 2.0 μl (10 pmol) downstream primer, and 12.0 μl dH_2O. Incubate at 20–25°C below the melting temperature T_m for 1 hr.

3. Wash cells in 500 μl 1× RT buffer (Gibco BRL) and centrifuge 1500g. Resuspend cells in 50 μl of the same buffer and centrifuge in the microcentrifuge at the same speed. Aspirate the supernatant with the gel tip and add the reverse transcription mix. Incubate at 37°C for 2–3 hr.

4. For DNA amplifications only: Wash cells and resuspend in 100 μl 1× PCR reaction buffer containing ribonucleases (RNase A or a combination of RNase A/T1) after being reconstituted so that any mRNA sequences that could possibly be transcribed by the Taq polymerase are digested by the ribonucleases. Incubate at 37°C for 30 min and wash two times with the PCR buffer containing the nfBSA.

5. Wash cells 1× in 500 μl 1× PCR reaction buffer with 500 μg/ml nfBSA and resuspend in 100 μl of the same. Transfer cells to a 0.65-ml siliconized microcentrifuge tube. Centrifuge in microcentrifuge at 1500g for 3 min, then aspirate supernatant with a gel tip. Add 20 μl of the PCR reaction mix, including 40 pmol each of the upstream and downstream primers, and overlay with mineral oil. Run on the thermocycler for the desired number of cycles. Typical temperatures are 94°C—1 min melting, 55°C—2 min annealing, 72°C—2 min extension. Depending upon the nature of the sequence, alternative temperatures can be used.

6. After the last step cycle, hold for 5 min at 72°C and then reduce to 4°C until cells are recovered from under mineral oil and transfer the suspension to a clean 1.7-ml microcentrifuge tube.

7. Centrifuge and remove supernatant (may be saved for further analysis) and wash in 0.5 ml 1× SSC containing 500 μg/ml of BSA two times. At this point, the DNA can be extracted from the cells for analysis or the cells can be used for the FISHES procedure to detect the amplified product.

8. Follow FISHES procedure steps 4–8 to complete. To denature the PCR product within the cells, incubate the cells at 72°C for 10 min after adding the probe mix, and then transfer the cell suspension to 37°C for the hybridization. Do all stringency washes at the same temperature.

D. Critical Aspects

As for the FISHES protocol, the use of the proper controls is important. The following are some that should be included:

1. Cells that have been cycled only and then run through the FISHES protocol without a probe (there is also another increase in cellular autofluorescence after thermal cycling).
2. Cells that have been cycled, hybridized without the probe, and then followed with a secondary reagent if a hapten-tagged probe is used for the protocol.
3. The use of nonsense primers or no primers during the PCR.
4. Cells treated with RNases before an RT-PCR.

The reverse transcription protocol is essentially accomplished with the same reactants used for an *in vitro* reverse transcription. The length of time needed for the reverse transcription needs to be tested for each target to determine the optimal conditions for production of the cDNA *in situ*. Remember, the mRNA is trapped within the matrix of the fixed cell and the reverse transcriptase enzyme will need time to reach the target mRNA.

The amount of time required for the reaction to proceed can go from 30 min to 3 hr at 37°C. For sequences that have a high amount of secondary structure, the rTth polymerase (Perkin–Elmer Cetus, Norwalk, CT) can be used following the instructions from the company. This enzyme provides for reverse transcription activity in the range of 55 to 75°C. When using the rTth polymerase, follow the instructions provided for the reverse transcription as the reaction volume is already 20 μl.

E. Expected Results

Figure 12, distribution 4 shows cells from the same line in which the cytoskeletal actin mRNA was used as a template for the *in situ* RT-PCR. The reverse transcription was done using untagged nucleotides and the *in situ* PCR was performed using untagged nucleotides. The resulting product was detected within the cells using the same 195-base digoxigenin-tagged probe that was used for the mRNA FISHES protocol. When comparing the mean fluorescence intensities for distributions 3 and 4, it is evident there is much more signal after the *in situ* amplification.

Acknowledgments

This work is supported by USPHS Grant CA-05981 and CA-60201 awarded by the NCI, Departments of HHS. We thank Rose M. Budnick and David L. Sheedy for their assistance and contributions in the technical aspects of flow cytometry.

References

Bakkus, M. H. C., Brakel-Van Peer, K. M., Karin, M. J., Adriaansen, H. J., Wierenga-Wolf, A. F., van den Akker, T. W., Dicke-Evinger, M.-J., and Benner, R. (1989). *Oncogene* **4,** 1255–1262.

Bauman, J. G. J., and Bentvelzen, P. (1988). *Cytometry* **9,** 517–520.

Bauman, J. G., J., and van Dekken, H. (1989). *Acta Histochem., Suppl.* **37,** 65–69.

Chomczynski, P., and Sacchi, N. (1987). *Analy. Bioch.* **162,** 156–159.

Clementi, M., Menzo, S., Bagnarelli, P., Manzin, A., Valenza, A., and Varaldo, P. E. (1993). *PCR Methods Appl.* **2**(3), 191–196.

Devereaux, J., Haeberli, P., and Smithes, O. (1984). *Nucleic Acids Res.* **12**(1), 387–395.

Guatelli, J. C., Whitfield, K. M., Kwoh, D. Y., Barringer, K. J., Richman, D. D., and Gingeras, T. R. (1990). *Proc. Natl. Acad. Sci. U.S.A.* **87,** 1874–1878.

Haase, A. T., Retzel, E. F., and Staskus, K. A. (1990). *Proc. Natl. Acad. Sci. U.S.A.* **87,** 4971–4975.

Harlow, S. P., Timm, E. A., Jr., Lewis, D. E., and Stewart, C. C. (1993). *In* "Clinical Flow Cytometry: Principles and Applications" (K. D. Bauer, R. E. Duque, and T. V. Shankey, eds.), pp. 557–572. Williams & Wilkins, Baltimore, MD.

Higushi, R. (1989a). *In* "PCR Technology: Principles and Applications for DNA Amplifications" (H. A. Erlich, ed.), pp. 31–38. Stockton Press, New York.

Higuchi, R. (1989b). *In* "PCR Technology: Principles and Applications for DNA Amplifications" (H. A. Erlich, ed.), pp. 61–70. Stockton Press, New York.

Innis, M. A., Geland, D. A., Sninsky, J. J., and White, T. J. (1990). *In* "PCR Protocols: A Guide to Methods and Applications" (M. A. Innis, D. H. Gelfand, J. J. Sninsky, and T. J. White, eds.). Academic Press, San Diego.

Kallioniemi, A., Kallioniemi, O. P., Waldeman, F. M., Chen, L.-C., Yu, L.-C., Fung, Y. K. T., Smith, H. S., Pinkel, D., and Gray, J. W. (1992). *Cytogenet. Cell Genet.* **60,** 190–193.

Kawasaki, E. S., Clark, S. S., Coyne, M. Y., Smith, S. D., Champlin, R., Witte, O. N., and McCormick, F. P. (1988). *Proc. Natl. Acad. Sci. U.S.A.* **85,** 5698–5702.

Kuo-Ping, C., Cohen, S. H., Morris, D. W., and Jordan, G. W. (1992). *J. Histochem. Cytochem.* **40**(3), 333–341.

Li, B. D. L., Bauer, K. D., Carney, W. P., and Duda, R. B. (1993a). *J. Surg. Res.* **54,** 179–188.

Li, B. D. L., Harlow, S. P., Budnick, R. M., Sheedy, D. L., and Stewart, C. C. (1993b). *Breast Cancer Res. Treat.* **27**(1/2), 144.

Long, A. A., Komminoth, P., Lee, E., and Wolfe, H. J. (1993). *Histochemistry* **99,** 151–162.

Lupu, R., and Lippman, M. E. (1993). *Breast Cancer Res. Treat.* **27**(1/2), 83–93.

Maguire, H. C., Jr., and Greene, M. I. (1989). *Semin. Oncol.* **16,** 148–155.

Nederlof, P. M., van der Flier, S., Raap, A. K., Tanke, H. J., Ploem, J. S., and van der Ploeg, M. (1990). *Cytometry* **11,** 126–131.

Nuovo, G. J., MacConnell, P., Forde, A., and Delvenne, P. (1991). *Am. J. Pathol.* **139**(4), 847–854.

the plasma membrane to lysosomes (endocytosis). Our understanding of traffic along the secretory and endocytic pathways has benefited dramatically from the isolation and study of mutants defective in various steps of these pathways. The genetic disruption of specific membrane traffic processes has allowed researchers to define steps in these processes and to provide tools to biochemically isolate the factors involved. The isolation of membrane traffic mutants has enhanced and often surpassed the study of these processes with inhibitors. It is possible to design genetic selections and screens to identify mutants defective in poorly understood processes while the use of inhibitors or antibodies requires prior knowledge of the process. Inhibitors often cause unrelated deleterious effects and it is difficult to conduct cell-free replacement experiments containing drug-inhibited elements. Thus, isolation and characterization of mutants has become an important part of modern cell biology.

Yeast has been a particularly valuable system for studies of the secretory pathway. In *Saccharomyces cerevisiae,* at least 23 complementation groups have been isolated that affect specific stages in the secretory pathway (for review, see Whitters *et al.,* 1993). Yeast, however, are not ideal for the study of endocytosis or lysosomal hydrolase delivery. They do not use the mannose 6-phosphate delivery mechanism that is used by mammalian cells, and endocytosis, while present in *S. cerevisiae,* is rather limited, making it more difficult to study and less well characterized than in mammalian cells. In addition, the process of receptor recycling does not appear to occur in this organism.

Another genetically tractable organism that has been used to study the endocytic pathway is *Dictyostelium discoideum.* Several mutants in the endocytic and secretory pathways have been isolated, and several genes encoding *rab* proteins (small GTP binding proteins required for membrane traffic) have been cloned (for review, see Cardelli, 1993). However, like *S. cerevisiae,* the delivery of hydrolases does not rely on mannose 6-phosphate; it is thought to occur via some as yet unknown receptor-mediated system. Non-mannose phosphate receptor-dependent delivery of lysosomal hydrolases also occurs in mammalian cells and may be similar to the mechanisms used by *Saccharomyces and Dictyostelium.*

Since unicellular eukaryotes can serve as model systems for the study of many processes that occur in mammalian cells, and due to the difficulties inherent in carrying out genetic analyses on somatic cells, mutants of mammalian cells have not been used as extensively as mutants from more genetically tractable organisms. However, many processes are fundamentally different between mammalian cells and unicellular eukaryotes, and these processes are not suitable for dissection by classical genetics. A number of aspects of endocytic traffic fall into this category, and isolation and characterization of somatic cell mutants provide an alternative. The recent availability of a wide range of mammalian molecular biological methods makes this approach even more attractive.

Mammalian cell mutants with recessive defects have in fact been used to dissect steps in the secretory pathway and to characterize steps in the endocytic pathway. For example, a Chinese hamster ovary (CHO)[1] cell mutant (clone 15B), isolated by the Kornfeld laboratory (Gottlieb *et al.*, 1975; Tabas and Kornfeld, 1978) played a crucial role in the development of cell-free assays of Golgi transport in the Rothman laboratory (Fries and Rothman, 1980; Balch *et al.*, 1984). This assay has led to the identification of a number of proteins required for Golgi traffic, including the *N*-ethylmaleimide sensitive-fusion protein NSF (Block *et al.*, 1988). In addition, mutants with endocytically deficient phenotypes have been isolated by several laboratories (see Colbaugh and Draper, 1993, for review). These mutants have been divided into six complementation groups (End1 through End6) and continue to be useful in delineating steps along the endocytic pathway.

An underutilized, yet powerful, technique for isolating mutants of mammalian cells is fluorescence-activated cell sorting. This technique has the advantage of being able to examine large numbers of cells quickly and does not require that cells of undesirable phenotype be killed. It only requires a suitable fluorescence assay to mark cells of the desired or undesired phenotype. Examples of the use of cell sorting to isolate somatic cell mutants and variants are presented in Table I (this table does not include uses of cell sorting for cDNA expression cloning or isolation of normal cell types such as T lymphocyte subsets). The purpose of this chapter is to describe cell sorting methods which have been used in our laboratory to isolate somatic cell mutants with defects in specific aspects of the endocytic apparatus.

II. General Considerations

A. Mutagenesis

An important consideration in a mutant isolation scheme is the expected frequency of the desired phenotype. Since spontaneous mutation rates are low for cultured cells, the mutation rate is routinely increased by treating cells with a mutagen such as ethylmethane sulfonate (EMS). EMS is an alkylating agent that adds ethyl groups to all four bases. The alkylation of position 6 of guanine leads to pairing with thymine, resulting in GT to AT transitions (see Drake and Baltz, 1976, for review).

A guiding principle in the use of mutagens is that a balance should be struck between low extents of mutagenesis which lead to low isolation frequencies and high extents of mutagenesis which result in most or all cells having unconditional lethal mutations. A mutagen concentration that results in 10–50% survival is thought to provide the appropriate balance (Thompson and Baker, 1973). A dose–response curve should be generated for the particular cell line being used.

[1] Abbreviations used: bRPMI, RPMI containing 1 mg/ml bovine serum albumin; CBZ, carbobenzyloxy; CHO, Chinese hamster ovary; Cy5, cyanine 5.18-OSu; EMS, ethylmethane sulfonate; FITC, fluorescein isthiocyanate; MNA, methoxy-β-naphthylamine.

Table I
Somatic Cell Mutants/Variants Isolated by Cell Sorting

Parental cell type	Mutant/variant phenotype	Spontaneous/ mutagenized/ exposed	Number of rounds of sorting	Estimated frequency	Reference
I. Changes in surface macromolecules					
A. Surface immunoglobulin					
Myeloma X63 (mouse)	IgG class switch	Spontaneous	3–5	1.6×10^{-7} to 5.6×10^{-6}	Radbruch *et al.* (1980)
Hybridoma B1-8 (mouse)	Altered IgD	Spontaneous	6	10^{-8} to 10^{-6}	Brüggemann *et al.* (1982)
Lymphoblastoid T5-1 (human)	altered λ Ig	Spontaneous and mutagenized[a]	5	na	McFarland *et al.* (1992)
Bone marrow (chick)	Inc./dec. Ig expression	Spontaneous	1	0.03 to 0.5	Benatar *et al.* (1991)
B. Major histocompatibility antigens					
T lymphoma LDHB (mouse)	Various	Spontaneous	9–13	$<10^{-6}$ to 10^{-3}	Holtkamp *et al.* (1983)
T lymphoma HK13 (mouse)	Altered H-2K	Spontaneous	5	10^{-7} to 10^{-5}	Weichel *et al.* (1985)
C. Natural antibody binding site(s)					
Lymphoma (mouse)	Inc./dec. Ab binding	Mutagenized[b]	3	na	Tough and Chow (1988)
D. Mannose receptor					
Macrophage J774 (mouse)	Increased expression	Exposed[c]	1	na	Diment *et al.* (1987)

E. Fibronectin receptor					
CHO	Altered receptor	Spontaneous	3–5	3×10^{-7}	Schreiner *et al.* (1989)
II. Changes in metabolism					
A. Benzo(*a*)pyrene					
Hepatoma (mouse)	Inc./dec. catabolism	Spontaneous	3–4	na	Miller and Whitlock (1981)
B. Glutathione					
ovarian tumor (human)	Inc./dec. glutathione	Spontaneous	4	na	Lee and Siemann (1989)
III. Changes in Membrane Traffic					
A. Delivery of newly synthesized proteins to the plasma membrane					
CHO	Dec. cell surface expression of influenza HA	Mutagenized[d]	1–6	na	Hearing *et al.* (1989)
B. Receptor recycling					
CHO	Dec. Recept. recycling	Mutagenized[d]	3	10^{-7}	Cain *et al.* (1991)

Note. na, not available
[a] Treated with *N*-methyl-*N'*-nitro-*N*-nitrosoguanidine.
[b] Treated with 12-*O*-tetradecanoyl-phorbol-13-acetate.
[c] Methylation inhibited with 5-azacytidine.
[d] Treated with ethylmethane sulfonate.

In order to "fix" the mutations in the genome and to allow the cells to recover from the mutagen, the cells should be grown in the absence of selection for at least one doubling time after removal of the mutagen. The optimum length of time for recovery will vary depending on the locus and must be empirically determined (Thompson and Baker, 1973). Using the protocols described here, the minimum recovery time is dictated by the ratio between the number of cells mutagenized and the number desired for sorting, on the order of 10^7 cells. For example, starting with 10^7 cells before mutagenesis and allowing a 1- to 2-week recovery period provides approximately 10^7 cells for sorting (most animal cell lines have a doubling time of near 1 day). In our experience with the mouse B lymphoma line A20, 4 days are required after removal of EMS before the cells begin doubling again.

B. Feasibility of Isolating Recessive Mutations in Diploid Cell Lines

Many examples of recessive mutations isolated in diploid cell lines have been reported. The high frequency of isolation (often between 10^{-4} and 10^{-3}) and the fact that recessive mutations have been isolated suggest that the genes in question are present in one copy. Two hypotheses for the presumed presence of only one functional copy of many genes in diploid cell lines have been put forth. The first hypothesis is that many genes are disrupted by rearrangement. Even though CHO cells have near diploid DNA content (Deaven and Petersen, 1973), their genome has nonetheless undergone a significant amount of rearrangement. Siminovitch (1976) hypothesized that during rearrangement, one copy of a gene may have become disrupted, leaving the cell "functionally hemizygous" for that locus. A second hypothesis is that many genes are silenced by methylation. There are now several documented examples of reversion of a mutant simply by treating the cells with 5-azacytidine (Holliday, 1987,1990), a compound that inhibits the maintainance methylase and leads to hypomethylation of the genome (Jones and Taylor, 1981; Taylor and Jones, 1982). As hypomethylation is thought to allow transcription; Holliday has proposed that the mutant cells that reverted upon 5-azacytidine treatment were in fact heterozygous at the relevant locus, but that one copy of the gene had been rendered silent by *de novo* methylation (Holliday, 1987,1990).

C. Isolation of Temperature-Conditional Mutants

When designing a mutant selection scheme, the advantages of isolating mutants with temperature-conditional defects should be considered. Mutations in essential genes can be isolated by this approach, and the mutant cell line at the permissive temperature can serve as an important control during characterization. Use of a protocol that allows for the expression of temperature-conditional defects does not exclude the detection and isolation of mutants with constitutively expressed defects (Thompson and Baker, 1973).

Since mammalian cell growth is typically restricted to between 32 and 40°C, the temperature range available for selecting permissive and nonpermissive temperatures is limited (Thompson and Baker, 1973). A permissive temperature different from that supporting optimal growth temperature is often required. Permissive and nonpermissive temperatures of 33–34°C and 39.5–41°C, respectively, are most commonly used. A subline of the parental cells conditioned to the permissive temperature and viable at the nonpermissive temperature should be used for mutant isolations. This ensures that any lethality at the nonpermissive temperature observed in mutants after selection is not due to inability of the parental line to grow at this temperature. Stocks of the parental subline should be frozen along with any mutants, and the parental subline should be maintained at the permissive temperature to allow proper comparison with mutants. Cells should be maintained at the permissive temperature after selection.

We generally use an expression scheme designed to isolate temperature-conditional mutants with acute defects in the process under study. To that end, we use a short expression time (1–4 hr) at a moderate nonpermissive temperature (39.5°C). It is always possible to use a higher temperature (e.g., 41°C) than was used to isolate the mutant in order to accelerate expression (although not all cells may grow above 39.5°C). Longer expression times may also be used, but this may lead to the isolation of defects indirectly involved with the process of interest.

D. Signal to Noise Requirements

The observed frequency of somatic cell mutants typically ranges from 10^{-6} to 10^{-3}, depending on the extent of mutagenesis, the number of mutations which can give rise to the desired phenotype, and the sensitivity of those loci to mutation. In order to allow isolation of low-frequency mutants without many rounds of sorting, a high degree of enrichment for mutants is desired in each round. Our laboratory typically uses sort gates which select the 0.1% of mutagenized cells closest to the desired phenotype (i.e., those cells with the highest fluorescence when sorting for positives). In order for this approach to be successful, it must be possible to set a sort gate that includes most of the desired cells while excluding all but the tail of the parental distribution. Methods for estimating whether this is the case are discussed below.

As is the case for many flow cytometric applications, we recommend the use of logarithmic amplifiers when acquiring the fluorescence parameters. The histograms of log fluorescence typically observed are close in shape to Gaussian distributions, suggesting that a log-normal distribution can be used to model their behavior. For illustration, we assume that the desired population has a lower fluorescence than the normal population and that the standard deviation of both populations is the same. The number of cells falling inside a sort gate extending from zero fluorescence to a given position, **u,** can be estimated using the error function, erf (the integral of the Gaussian distribution).

$$\text{Fraction in gate} = 0.5 + \text{erf}\,[\mathbf{u} - \mu)/\sigma],$$

where μ and σ represent the mean and standard deviation of the log-normal distribution of desired cells (in channels from the logarithmic amplifier). For simplicity, we can normalize **u** and μ by division by σ and define **x** as the difference between **u** and μ in units of σ. We define Δ as the displacement between the means of the mutant and normal populations, again in units of σ. Given values for the minimum acceptable percentage retention of desired cells, **r,** and the minimum acceptable contamination by normal cells, **c,** the minimum difference between the distributions can be determined by solving the following two equations:

$$\mathbf{r} = 0.5 + \text{erf}\,(\mathbf{x})$$
$$\mathbf{c} = 0.5 + \text{erf}\,(\mathbf{x} - \Delta).$$

For **r** = 0.95 and **c** = 0.001, Δ is 3.35. For a four-decade log amplifier and analog–digital conversion at 10-bit (1024 channel) resolution, a typical value of σ for cultured cell populations is 64 channels (a log standard deviation of 0.25). This gives a minimum displacement between the means of the desired and normal populations of 214 channels or a linear signal to noise ratio of 6.9. Large coefficients of variation are often observed for cell populations labeled with endocytic tracers. For σ of 128 channels (a log standard deviation of 0.5), a linear signal to noise ratio of 47 is required for the same **r** and **c** values. Attaining a signal to noise ratio as high as this may require the use of recently developed probes with high extinction coefficients, large quantum yields, and excitation wavelengths in regions of low cellular autofluorescence (Southwick *et al.,* 1990; Mujumdar *et al.,* 1993).

It is recommended that a feasibility study be done to estimate the signal to noise ratio of a protocol before sorting is attempted. Fluorescence histograms should be acquired for a sample expected to mimic the desired mutants, as well as parental cells under the projected labeling conditions. For example, the maximal signal expected for mutants which fail to recycle transferrin (see Section III) would be the amount of labeled transferrin present before the chase period. This sample should be compared to parental cells after pulse and chase to determine the maximum signal to noise ratio expected and the fraction of parental cells anticipated to fall in the sort gate.

E. The Importance of Gating Using Light Scatter during Sorting

One of the truisms in genetics is that "you get what you select for." Our experience indicates that this is no less true when using cell sorting. In the example discussed in Section III, sorting for receptor recycling mutants, the desired phenotype was retention of more than the normal amount of labeled transferrin after a pulse and chase. Our initial sorting efforts using only a fluorescence gate resulted in the isolation of cells with dramatically increased size. These cells had increased numbers of receptors and took up more labeled transferrin during a pulse. They therefore met the sort criterion of elevated fluorescence after chase while still retaining the same *fraction* of internalized transferrin as parental cells (and therefore not being receptor recycling mutants at all). This problem was overcome by adding narrow gates on forward and side scatter to select cells of normal size, a practice which is recommended for all mutant selections using cell sorting.

III. Isolation of CHO Cells Defective in Receptor Recycling

A. Rationale

A combination of genetic, biochemical, and cell biological analyses during the past 15 years has provided significant insight into the processes by which ligands are internalized by receptors, by which receptor–ligand complexes are delivered to endosomes, and by which dissociated ligands are degraded in lysosomes (for reviews, see Kornfeld and Mellman, 1989; Hubbard, 1989; Rodman *et al.*, 1990; Murphy, 1993). Much less information is available regarding the process by which endocytosed receptors return to the plasma membrane, a process known as receptor recycling. To provide new tools for the analysis of this process, we developed a protocol for the isolation of mutants specifically defective in receptor recycling and successfully isolated such mutants (Cain *et al.*, 1991). The protocol is illustrative of one of the strengths of cell sorting, selection for a phenotype for which it is difficult to design a selective growth advantage. Since the iron carrying protein transferrin is normally recycled along with its receptor, it provides a convenient marker for the recycling pathway (Dautry-Varsat *et al.*, 1983; Klausner *et al.*, 1983). When normal cells (e.g., the CHO WTB line) are incubated with a fluorescent transferrin conjugate, they become fluorescent due to binding and internalization. During a chase in the presence of unlabeled transferrin, the fluorescent transferrin conjugate recycles to the plasma membrane and dissociates (making the cells nonfluorescent). We reasoned that mutants that were able to internalize transferrin but not recycle it would remain fluorescent after this chase, allowing them to be sorted from normal cells. For the reasons discussed above, we chose a temperature-conditional protocol. In order to avoid isolating mutants with gross metabolic defects (or other mutations which would affect both recycling and internalization) we added a second fluorescent criterion, ability to internalize FITC dextran. This probe is nonspecifically internalized in the fluid phase.

B. Protocol

1. CHO WTB cells (obtained from Dr. Brian Storrie, Virginia Polytechnic Institute and State University) are maintained as subconfluent monolayers in α-MEM supplemented with 10% calf serum, 0.29 g/liter L-glutamine, 4 g/liter proline, and, optionally, 100 units/ml penicillin and 100 μg/ml streptomycin. Cells should be conditioned to the permissive temperature (typically 33°C) for several generations.

2. Subconfluent cells (10^7) are exposed to 250 μg/ml EMS for 24 hr and washed free of EMS with α-MEM. The mutagenized cells are allowed to recover for 2 weeks with frequent media changes to remove dead cells.

3. Cells are preincubated for 2 hr at the nonpermissive temperature (typically 39.5–41°C) to allow partial expression of temperature-conditional defects. Since an expected phenotype of receptor recycling mutants is loss of surface receptors, full expression would result in inability to internalize labeled transferrin (and therefore little transferrin retention after the chase period). During the original isolation of recycling mutants (Cain *et al.*, 1991), various preincubation times from 0.5 to 2 hr were used, with only the 2-hr preincubation yielding any mutants.

4. Cells are washed twice with α-MEM salts and incubated for 30 min at the nonpermissive temperature with between 2 and 8 μg/ml Cy5 transferrin in α-MEM. This incubation is done in medium without serum to avoid competition by unlabeled transferrin present in the serum. The relatively short labeling period is chosen to permit saturation of intracellular compartments involved in recycling without allowing significant accumulation of labeled transferrin in lysosomes (a fraction of recycled ligands and receptors are missorted to lysosomes during each round of internalization and recycling). Cells are extensively labeled after this incubation (Fig. 1A).

5. Cells are washed twice with α-MEM salts and incubated for 1 hr at the nonpermissive temperature with 1 mg/ml unlabeled transferrin and 2 mg/ml FITC dextran in normal growth medium. Cells are washed six times with ice-cold α-MEM salts; careful washing is required to effectively remove the highly concentrated FITC dextran. After washing, cells are harvested by scraping into α-MEM and kept on ice until and during sorting. The vast majority of the cells should be well labeled with FITC dextran and have lost the bulk of Cy5 transferrin (Fig. 1C).

6. Steps should be taken to ensure sterility of the sheath and sample tubing. A 0.45-μm sterile filter is installed in line after the PBS reservoir with a three-way valve on the outlet side (toward the sorter). If sheath fluid containing NaN_3 is normally used, any residual NaN_3 should be removed by flushing with PBS without azide for 15 min. Then 5 ml of 70% ethanol is gently pushed through the sheath tubing using a syringe at the three-way valve. PBS is run for 15 min to remove residual ethanol from the sheath tubing. These measures should

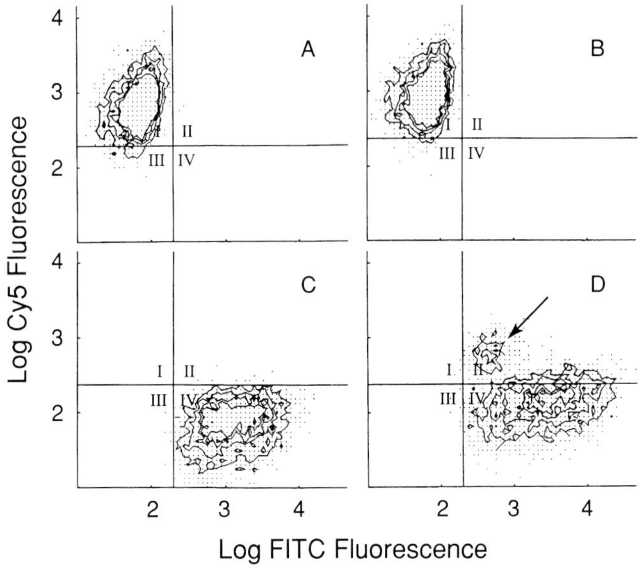

Fig. 1 Isolation of CHO cells defective in receptor recycling. Mutagenized CHO WTB cells were shifted from 33 to 41°C for 2 hr to allow expression of possible temperature-conditional defects and labeled for 30 min with 2.5 μg/ml Cy5 transferrin (A). The labeled transferrin was then chased by incubation for 1 hr with 1 mg/ml unlabeled transferrin and 2 mg/ml FITC dextran (C). Under these conditions, the vast majority of cells move from quadrant I (Cy5 transferrin positive, FITC dextran negative) to quadrant IV (Cy5 transferrin negative, FITC dextran positive) due to recycling of Cy5 transferrin and endocytosis of FITC dextran. The small number of double-positive cells (quadrant II) were sorted as described in the text. The population of cells obtained after two rounds of sorting was labeled (B) and chased (D) as for WTB cells. A distinct double-positive population, comprising approximately 9% of the total number of cells, was observed (arrow). Contours are drawn at 5, 10, 15, and 20 events/bin (5000 events total). Reprinted with permission from Cain *et al.* (1991).

sterilize (and keep sterile) the sheath line. To sterilize the sample tubing, 2 ml of 70% ethanol is placed in a sample tube and run through the sorter. PBS should be allowed to backflush through the sample tubing for 15 min to ensure that all ethanol is removed.

7. A first sort gate is set to include only cells with normal forward and side scatter values. A second sort gate is used to collect the 0.1% of mutagenized cells with the highest Cy5 transferrin fluorescence and a normal amount of FITC dextran uptake (quadrant II in Fig. 1C). Only cells passing through both gates are sorted.

8. Sorted cells are collected into tubes filled within a few millimeters of the top with ice-cold growth medium containing 100 units/ml penicillin, 100 μg/ml streptomycin, 0.1% gentamicin, 1% fungizone, and 0.1% nystatin. Between 10^6

and 10^7 total cells are analyzed and a minimum of 1000 sorted cells collected. After sorting, cells are plated at the permissive temperature.

9. After they have expanded sufficiently (typically 3 weeks), the sorted cells are resorted using the same procedure. No visible enrichment for desired cells is expected after the first round of sorting, but a detectable mutant population should be visible after the second or third sort (Fig. 1D). This population should be sorted and cloned by limiting dilution.

10. To clone by limiting dilution, cells are suspended at a concentration of 3 cells/ml and 100-μl aliquots are plated in 96-well trays (a maximum of one-third of the wells should develop colonies). Unused cells should be frozen to permit repetition of the cloning, if necessary. After 2–3 weeks, individual wells can be expanded into 12-well trays and then into 10-cm dishes as warranted by cell growth.

11. Individual clones should be frozen as soon as possible and tested at both the permissive and nonpermissive temperatures for the transferrin trapping phenotype.

The protocol presented above was used to isolate CHO cell lines with reduced ability to recycle endocytosed transferrin (Cain *et al.,* 1991). These lines are given the prefix TfT to denote their isolation by virtue of *T*rans*f*errin *T*rapping). One of these lines, TfT1.11, has been extensively characterized and shown to have a temperature-conditional defect in receptor recycling which affects the transferrin, α_2-macroglobulin, and low-density lipoprotein receptors. This line also has a pleiotropic defect in lysosome biogenesis, suggesting a link between recycling of receptors and transfer of ligands to lysosomes (Wilson *et al.,* 1993). Analysis of fusions with representatives of previously identified End complementation groups reveals that TfT1.11 defines a new complementation group, End6 (Bucci *et al.,* 1993).

IV. Isolation of B Cells with Reduced Cysteine Protease Activity

A specialized function of the endocytic pathway in certain cells of the immune system is antigen presentation via class II proteins of the major histocompatibility complex. Within the endocytic pathway, it is thought that cysteine proteases play an important role in processing both the antigen to be presented and the class II molecule itself (Streicher *et al.,* 1984; Blum and Cresswell, 1988; Diment, 1990). In order to establish a new system for analysis of the steps involved in antigen presentation, we have used cell sorting to isolate mutants with reduced cysteine protease activity from an antigen presenting cell line (the murine B cell lymphoma, A20). In addition to enabling study of the role of specific proteases in antigen presentation (without the use of inhibitors), these lines should be useful for the study of hydrolase delivery and lysosome biogenesis in B cells.

A. Methoxy-β-naphthylamine-Based Substrates

We initially explored the feasibility of using the fluorogenic substrate for cathepsin B (and cathepsin L) CBZ-ala-arg-arg-4-methoxy-β-naphthylamine (CBZ-ala-arg-arg-MNA) following the methods of Dolbeare and Smith (1977) and Dolbeare and Vanderlaan (1979). We have previously used this substrate (with either fluorometry or flow cytometry) to determine the kinetics with which endocytosed material encounters active cathepsin B activity (Roederer *et al.,* 1987; Bowser and Murphy, 1990). The substrate must be endocytosed in order to permit exposure to proteases in endocytic compartments, as it is minimally membrane permeant. Unfortunately, the product of its cleavage, MNA, is membrane permeant and does not accumulate in cells. To enable use in conjunction with flow cytometry, the substrate is coincubated with 5-nitrosalicylaldehyde. The latter is coendocytosed with the substrate and reacts with MNA to yield an insoluble product which can be detected with fluorescein optics (Dolbeare and Smith, 1977; Dolbeare and Vanderlaan, 1979). Two problems made this approach undesirable. First, the combination of substrate and 5-nitrosalicylaldehyde significantly alters the forward and side scatter properties of A20 cells (making setting of appropriate scatter gates problematic). Second, and more importantly, the signal resulting from labeling of A20 cells with 5 m*M* CBZ-ala-arg-arg-MNA and 2 m*M* 5-nitrosalicylaldehyde for 2 hr was only 5.7-fold higher than the autofluorescence from unlabeled cells (S. A. Brockman and R. F. Murphy, unpublished observations). Since the standard deviations of the distributions were also large (log standard deviations of approximately 0.5), the signal to noise ratio was too low to allow sufficient separation between labeled and unlabeled cells (see Section II,D).

B. Rhodamine 110-Based Substrates

To overcome these difficulties, we explored using the rhodamine 110-based substrates described by Leytus *et al.* (1983). $(CBZ\text{-phe-arg})_2$-rhodamine 110 is specific for cathepsin L (Rothe *et al.,* 1990,1992) and its product, rhodamine 110, has excitation and emission properties similar to those of fluorescein. A major advantage for use with cell sorting is that the product is well retained by cells. The substrate is not toxic to cells and does not appear to affect forward and side scatter properties (unpublished observations). $(CBZ\text{-phe-arg})_2$-rhodamine 110 was synthesized as described previously (Leytus *et al.,* 1983) and used at a concentration of 50 μg/ml to label cells for 30 min. Higher concentrations are insoluble (Assfalg-Machleidt *et al.,* 1992) and the signal to noise ratio reaches a maximum of 12.7 at 30 min (unpublished observations).

As was the case for isolating receptor recycling mutants, we wanted to avoid isolation of mutants defective in fluid-phase endocytosis. Since endocytosis of the substrate appears to be required for cleavage, such mutants would exhibit reduced product fluorescence (due to reduced endocytosis of the substrate) and be sorted along with the desired cells having reduced protease activity. A second fluorescence criterion was therefore incorporated in the selection protocol. A marker of fluid-phase endocytosis, Cy5 dextran, was included along with the substrate during labeling. This permitted sorting of cells with normal levels of Cy5 fluorescence (indicating that they were competent for endocytosis) and low levels of rhodamine 110 fluorescence (indicating inability to cleave the substrate) (Fig. 2A).

The following scheme, designed to allow selection of temperature-conditional mutants, was used to isolate A20 mutants with reduced cathepsin L activity (Brockman and Murphy, 1994).

C. Protocol

1. A20 cells (purchased from American Type Culture Collection, Rockville, MD) are maintained in RPMI containing 10% heat-inactivated fetal calf serum and 0.05 mM β-mercaptoethanol at 33°C in a humidified, 5% CO_2 environment for several generations prior to mutagenesis.

2. A total of 10^7 exponentially growing cells are exposed to 500 μg/ml EMS for 18–24 hr. The cells are washed free of EMS and replated in fresh medium at a density of 1×10^6 cells/ml. This routinely results in 50% killing. After 1–2 weeks of recovery, the cells are prepared for sorting.

3. On the day of the sort, the cells are washed twice in PBS (140 mM NaCl, 2.6 mM KCl, 8.1 mM Na_2PO_4, 1.5 mM KH_2PO_4, pH 7.4), resuspended at a concentration of 2×10^6 in RPMI containing 1 mg/ml bovine serum albumin (bRPMI), and incubated for 1 hr at 39.5°C. A parallel sample of parental cells is prepared for use in setting sort gates (see below). The washes and incubation in serum-free medium are included to avoid hydrolysis of the substrate by hydrolases in serum. This allows addition of a labeling cocktail directly to cells at the nonpermissive temperature.

4. To label the cells, bRPMI containing Cy5 dextran and $(CBZ\text{-phe-arg})_2$-rhodamine 110 is added to cells in suspension to give a final concentration of 0.5 mg/ml and 50 μg/ml, respectively. A sample of both mutagenized cells and parental cells are left unlabeled in order to assess the level of autofluorescence (see below). A stock solution of substrate is made in DMF at a concentration of 80 mg/ml. This stock can be kept at −20°C for several months. A working stock is made immediately before use by slowly diluting the stock to 4 mg/ml with distilled water at room temperature. If the diluent is added too quickly, the substrate will not stay in solution. Cells are incubated in the labeling medium for 30 min at 39.5°C, washed twice with PBS, and then resuspended at 3–5 $\times 10^6$ cells/ml in bRPMI. Cells are kept on ice before and during sorting.

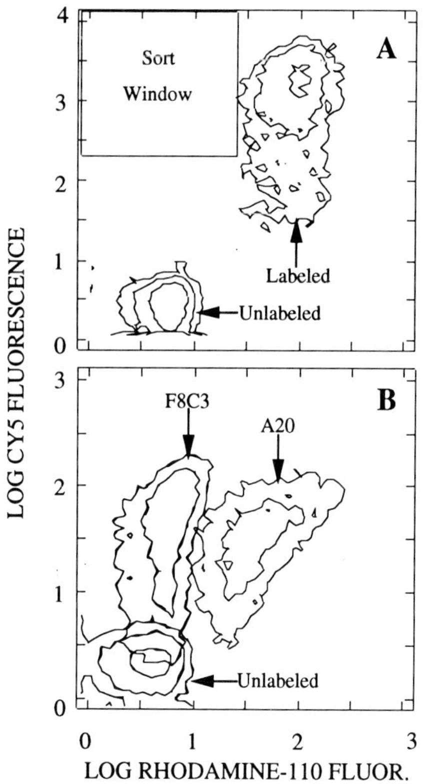

Fig. 2 Isolation of A20 cells with reduced cathepsin L activity. Cells were labeled with Cy5 dextran and $(CBZ\text{-phe-arg})_2$-rhodamine 110 as described in the text. (A) The fluorescence criteria used for sorting are shown superimposed on contour maps for labeled and unlabeled parental cells (unlabeled mutagenized cells were indistinguishable from their parental counterparts). This sort gate contains the lowest 0.1% of the parental cell population. (B) Contour maps from a subsequent experiment for mutant clone F8C3 are superimposed on those for labeled and unlabeled parental cells (unlabeled F8C3 was indistinguishable from unlabeled A20). F8C3 shows less than 10% of the parental level of rhodamine 110 fluorescence. Note that F8C3 exhibits normal levels of Cy5 dextran uptake, indicating that its reduced rhodamine 110 fluorescence is not due to failure to endocytose the substrate. Data from Brockman and Murphy (1994).

5. The sheath and sample tubing are sterilized as described in Section III,B.

6. To isolate cells that are the same size as parental cells, a strict gate is set in the forward and side scatter parameters around the labeled parental cell sample (unmutagenized). Therefore, only those cells that fall within that gate are considered for sorting. Once the scatter gate is set, a second gate is set in the rhodamine 110 and Cy5 parameters to collect the 0.1% of mutagenized cells that fall in the wild-type range of Cy5 fluorescence yet appear essentially unlabeled with rhodamine 110 (Fig. 2A).

7. Sorted cells are collected into tubes filled within a few millimeters of the top with growth medium containing 1% penicillin/streptomycin solution (Gibco, Grand Island, NY). At least 500 cells are collected per tube, plated in a 6-cm tissue culture dish, and allowed to expand at 33°C. When the cells have grown sufficiently to replate, medium without antibiotics can be used.

8. In roughly 3 weeks the sorted cell cultures should expand sufficiently to conduct another round of sorting. Typically no enrichment is seen in the sort window after one round of sorting; therefore, sorting is repeated using the same criteria. A noticeable population of cells meeting the sort criteria (perhaps as a tail extending from the normal population) should be visible after the second (or possibly third) round of sorting (assuming 0.1% sort gates). Clones may then be isolated by limiting dilution (see Section III) either before or after an additional round of sorting.

9. Once expanded sufficiently, individual clones should be frozen and tested for their ability to cleave the rhodamine 110 substrate. It is good practice to reclone any lines with the desired phenotype and retest their phenotype to ensure that they are truly clonal lines.

The protocol described above was used to isolate A20 lines with reduced ability to cleave the rhodamine 110 substrate (Fig. 2B). A significant enrichment was observed in the sort region after two rounds of sorting, and clonal lines were isolated by limiting dilution after a third sort. As expected, these lines were observed to be defective in presentation of antigen to T helper cells (Brockman and Murphy, 1994).

Ideally, this protocol should be followed for a number of separate mutageneses and one clone selected from each. This procedure ensures that separate clones are not derived from a single original mutant. If more than one clone is kept from a single mutagenesis/sort protocol, complementation analysis may be used to determine whether mutants in more than one complementation group have been obtained.

V. Conclusions

The protocols presented in this chapter are illustrative of many of the critical concepts of somatic cell mutant isolation by cell sorting and have been used to isolate mutants with defects in two aspects of endocytosis. Characterization of these mutants has already provided new information about the affected process (Wilson *et al.*, 1993). In addition to permitting the isolation of additional mutants of these types, it is hoped that these protocols will be useful as a starting point for isolation of other classes of mutants by cell sorting.

Acknowledgments

We thank Clark Brown, Sheree Rybak, Bruce Taillon, and Victor Yankiwski for helpful suggestions. The original research described in this article was supported by an institutional predoctoral training grant to S.A.B. from NIH (GM08067) and by research grants to R.F.M. from NIH (GM32508), NSF (DCB-8903657), and the Western Pennsylvania Chapter of the Arthritis Foundation.

References

Assfalg-Machleidt, I., Rothe, G., Klingel, S., Banati, R., Mangel, W. F., Valet, G., and Machleidt, W. (1992). *Biol. Chem. Hoppe-Seyler* **373,** 433–440.

Balch, W. E., Dunphy, W. G., Braell, W. A., and Rothman, J. E. (1984). *Cell (Cambridge, Mass.)* **39,** 405–416.

Benatar, T., Iacampo, S., Tkalec, L., and Ratcliffe, M. J. H. (1991). *Eur. J. Immunol.* **21,** 2529–2536.

Block, M. R., Glick, B. S., Wilcox, F. T., Wieland, F. T., and Rothman, J. E. (1988). *Proc. Natl. Acad. Sci. U.S.A.* **85,** 7852–7856.

Blum, J. S., and Cresswell, P. (1988). *Proc. Natl. Acad. Sci. U.S.A.* **85,** 3975–3979.

Bowser, R., and Murphy, R. F. (1990). *J. Cell. Physiol.* **143,** 110–117.

Brockman, S. A., and Murphy, R. F. (1994). Submitted for publication.

Brüggemann, M., Radbruch, A., and Rajewsky, K. (1982). *EMBO J.* **1,** 629–634.

Bucci, M., Moyer, T. W., Brown, C. M., Wilson, R. B., and Murphy, R. F. (1993). *Somatic Cell Mol. Genet.* **20,** 47–54.

Cain, C. C., Wilson, R. B., and Murphy, R. F. (1991). *J. Biol. Chem.* **266,** 11746–11752.

Cardelli, J. A. (1993). *Adv. Cell Mol. Biol. Membr.* **1,** 341–390.

Colbaugh, P. A., and Draper, R. K. (1993). *Adv. Cell Mol. Biol. Membr.* **1,** 169–198.

Dautry-Varsat, A., Ciechanover, A., and Lodish, H. F. (1983). *Proc. Natl. Acad. Sci. U.S.A.* **80,** 2258–2262.

Deaven, L. L., and Petersen, D. F. (1973). *Chromosoma* **41,** 129–144.

Diment, S. (1990). *J. Immunol.* **145,** 417–422.

Diment, S., Leech, M. S., and Stahl, P. D. (1987). *J. Leukocyte Biol.* **42,** 485–490.

Dolbeare, F. A., and Smith, R. E. (1977). *Clin Chem. (Winston-Salem, N.C.)* **23,** 1485–1491.

Dolbeare, F., and Vanderlaan, M. (1979). *J. Histochem. Cytochem.* **27,** 1493–1495.

Drake, J. W., and Baltz, R. H. (1976). *Annu. Rev. Biochem.* **45,** 11–37.

Fries, E., and Rothman, J. E. (1980). *Proc. Natl. Acad. Sci. U.S.A.* **77,** 3870–3874.

Gottlieb, C., Baenziger, J., and Kornfeld, S. (1975). *J. Biol. Chem.* **250,** 3303–3309.

Hearing, J., Hunter, E., Rodgers, L., Gething, M., and Sambrook, J. (1989). *J. Cell Biol.* **108,** 339–353.

Holliday, R. (1987). *Science* **238,** 163–170.

Holliday, R. (1990). *Development Suppl., (Cambridge, UK)*, 125–129.

Holtkamp, B., Cramer, M., and Rajewsky, K. (1983). *EMBO J.* **2,** 1943–1951.

Hubbard, A. L. (1989). *Curr. Opin. Cell Biol.* **1,** 675–683.

Jones, P. A., and Taylor, S. M. (1981). *Nucleic Acids Res.* **9,** 2933–2947.

Klausner, R. D., Ashwell, G., van Renswoude, J., Harford, J. B., and Bridges, K. R. (1983). *Proc. Natl. Acad. Sci. U.S.A.* **80,** 2263–2266.

Kornfeld, S., and Mellman, I. (1989). *Annu. Rev. Cell Biol.* **5,** 483–525.

Lee, F. Y. F., and Siemann, D. W. (1989). *Int. J. Radiat. Oncol. Biol. Phys.* **16,** 1315–1319.

Leytus, S. P., Patterson, W. L., and Mangle, W. F. (1983). *Biochem. J.* **215,** 253–260.

McFarland, R. D., Vincent, J. L., and Smith, G. J. (1992). *Environ. Mol. Mutagen.* **19,** 297–303.

Miller, A. G., and Whitlock, J. P., Jr. (1981). *J. Biol. Chem.* **256,** 2433–2437.

Mujumdar, R. B., Ernst, L. A., Mujumdar, S. R., Lewis, C. J., and Waggoner, A. S. (1983). *Bioconjugate Chem.* **4,** 105–111.

Murphy, R. F. (1993). *Adv. Cell Mol. Biol. Membr.* **1,** 1–17.

Radbruch, A., Liesegang, B., and Rajewsky, K. (1980). *Proc. Natl. Acad. Sci. U.S.A.* **77,** 2909–2913.

Rodman, J. S., Mercer, R. W., and Stahl, P. D. (1990). *Curr. Opin. Cell Biol.* **2,** 664–672.

Roederer, M., Bowser, R., and Murphy, R. F. (1987). *J. Cell Physiol.* **131,** 200–209.

Rothe, G., Oser, A., Assfalg-Machleidt, I., Machleidt, W., Mangel, W. F., and Valet, G. (1990). *Cytometry Suppl.* **4,** 77.

Rothe, G., Klingel, S., Assfalg-Machleidt, I., Machleidt, W., Zirkelbach, C., Banati, R., Mangel, W. F., and Valet, G. (1992). *Biol. Chem. Hoppe-Seyler* **373,** 547–554.

Schreiner, C. L., Bauer, J. S., Danilov, Y. N., Hussein, S., Sczekan, M. M., and Juliano, R. L. (1989). *J. Cell Biol.* **109,** 3157–3167.

Siminovitch, L. (1976). *Cell (Cambridge, Mass.)* **7,** 1.

Southwick, P. L., Ernst, L. A., Tauriello, E. W., Parker, S. R., Mujumdar, R. B., Mujumdar, S. R., Clever, H. A., and Waggoner, A. S. (1990). *Cytometry* **11,** 418–430.

Streicher, H. Z., Berkower, I. J., Busch, M., Gurd, F. R. N., and Berzofsky, J. A. (1984). *Proc. Natl. Acad. Sci. U.S.A.* **81,** 6831–6835.

Tabas, I., and Kornfeld, S. (1978). *J. Biol. Chem.* **253,** 7779–7786.

Taylor, S. M., and Jones, P. A. (1982). *J. Mol. Biol.* **162,** 679–692.

Thompson, L. H., and Baker, R. M. (1973). *In* "Methods in Cell Biology" (D. M. Prescott, ed.), Vol. 6, pp. 209–281. Academic Press, New York.

Tough, D. F., and Chow, D. A. (1988). *Cancer Res.* **48,** 270–275.

Weichel, W., Liesegang, B., Gehrke, K., Goettlinger, C., Holtkamp, B., Radbruch, A., Stackhouse, T. K., and Rajewsky, K. (1985). *Cytometry* **6,** 116–123.

Whitters, E. A., Skinner, H. B., and Bankaitis, V. A. (1993). *Adv. Cell Mol. Biol. Membr.* **1,** 307–339.

Wilson, R. B., Mastick, C. C., and Murphy, R. F. (1993). *J. Biol. Chem.* **268,** 25357–25363.

CHAPTER 9

Measurement of Micronuclei by Flow Cytometry

Michael Nüsse, Wolfgang Beisker, Johannes Kramer, Beate M. Miller, Georg A. Schreiber, Silvia Viaggi, Eva Maria Weller, and Jurina M. Wessels

GSF-Forschungszentrum für Umwelt und Gesundheit
Institut für Biophysikalische Strahlenforschung
D-85758 Oberschleissheim, Germany

I. Introduction

The induction of micronuclei in cell cultures or human lymphocytes exposed to ionizing radiation or chemicals can be used as a measure of both structural and numerical chromosome aberrations. Micronuclei represent genetic material that is lost from the genome of the cell during mitosis. They may contain one or several acentric chromosome fragments (clastogenic action), one or several whole chromosomes (by interference of the inducing agent with the mitotic spindle apparatus), or even combinations of both. Usually, the frequency of

METHODS IN CELL BIOLOGY, VOL. 42

micronuclei in cells is measured by microscopic scoring of micronuclei in several hundred or thousand cells. Some attempts have therefore been made to automate micronucleus scoring by image analysis (Pincu *et al.*, 1985; Fenech *et al.*, 1988; Romagna and Staniforth, 1989; Tates *et al.*, 1990) or by flow cytometry (for micronucleus scoring in mouse erythrocytes: Hutter and Stöhr, 1982; Hayashi *et al.*, 1992; Grawe *et al.*, 1992; for micronucleus scoring in cell cultures or human lymphocytes: Nüsse and Kramer, 1984; Nüsse *et al.*, 1992a,b; Schreiber *et al.*, 1992a,b; Miller and Nüsse, 1993).

We describe here a technique for obtaining a suspension of micronuclei and nuclei for flow cytometric measurement of the frequency of micronuclei per main nuclei in cell cultures or human lymphocytes exposed to ionizing radiation or chemicals. With this technique it is additionally possible to measure the DNA content of micronuclei and main nuclei for a simultaneous analysis of the cell-cycle distribution (main nuclei) and the DNA distribution of micronuclei. Analysis of the DNA distribution of micronuclei can give additional information on the mechanisms of action of the inducing agent (Nüsse *et al.*, 1992a; Miller and Nüsse, 1993). The two-step method includes a treatment of the cells with salt solution I, which contains a detergent, followed by an additional treatment with solution II containing citric acid and sucrose. Both solutions contain ethidium bromide (EB) as a DNA-specific fluorescent dye; in some cases Hoechst 33258 (HO) is used additionally. By this two-step treatment the cellular membrane and the cytoplasm are destroyed and nuclei and micronuclei are released in suspension. No mechanical treatment or centrifugation steps are necessary to obtain a solution of main nuclei and micronuclei, because mechanical treatment would induce cellular or nuclear debris that could overlap the small micronuclei during flow cytometric measurement. The nuclear membrane is maintained and mitotic cells that could release chromosomes in the suspension are usually not destroyed (Nüsse *et al.*, 1990).

Flow cytometric measurements of DNA content of micronuclei and nuclei are performed simultaneously using log mode, because the differences in DNA content between micronuclei and nuclei can be a factor of 100. Light scatter signals (forward light scatter and side scatter) are measured simultaneously for a discrimination of unspecific debris from micronuclei.

II. Application

The primary applications of the technique are measurement of the frequency of micronuclei in cell cultures exposed to ionizing radiation or chemicals (Nüsse and Kramer, 1984; Nüsse *et al.*, 1992a,b; Miller and Nüsse, 1993), measurement of the DNA distribution of micronuclei (Nüsse *et al.*, 1992a; Miller and Nüsse, 1993; Weller *et al.*, 1993), and also sorting of micronuclei with the same DNA content for further analysis of the chromosomal content of micronuclei with fluorescence *in situ* hybridization using for example centromeric DNA probes (Miller and Nüsse, 1993). The technique was also used to analyze micronuclei in primary cultures of rat hepatocytes exposed to chemicals *in vitro* (Cao *et al.*, 1993). With a modification of the technique we have studied micronucleus induction in human lymphocytes exposed to ionizing radiation (Schreiber *et al.*, 1992b). We are currently studying the applicability of the technique for a biological dosimetry in humans exposed to ionizing radiation. The method can probably not be used with *in vivo* material as for example tumor cells (with the exception of rat hepatocytes and human lymphocytes), because *in vivo* cells usually show a high amount of debris that will overlap the micronuclei during flow cytometric measurement and will therefore not allow an unambiguous identification of micronuclei.

III. Materials

A. Solutions

Two solutions are used for the preparation of a suspension of nuclei and micronuclei.

Solution I (prepare 500 ml stock):

584 mg/liter NaCl
1000 mg/liter Na-citrate
10 mg/liter RNase A from bovine pancreas (Serva, Heidelberg, Germany)
0.3 ml/liter Nonidet P-40 (NP-40)

Solution II (prepare 500 ml stock):

15 mg/liter citric acid
0.25 *M* sucrose

These solutions are made up using distilled water and should be filter sterilized. They can be stored at 4°C for at least several months. Ethidium bromide is added before use from a stock solution of 1 mg/ml in distilled water. Both solutions are used at room temperature.

B. Cells

Micronucleus induction has been studied with this technique in several cell lines including Chinese hamster embryo cells, various strains of V79 cells, mouse Ehrlich ascites tumor cells growing in suspension, mouse NIH-3T3 cells, human lymphoma K37 cells, rat hepatocytes, and human lymphocytes. The cells are usually treated and analyzed during exponential growth.

IV. Cell Preparation and Staining

For flow cytometric analysis of micronuclei, an exponentially growing cell culture should be used, because the cells have to progress through the cell cycle for the induction of micronuclei after (ionizing radiation, chemicals) or during (chemicals) treatment. About 1×10^6 cells are centrifuged at about 100 *g* (around 800–1000 rpm) for 5 min. The supernatant medium should be removed completely by aspirating or carefully pouring off. The small cell pellet should be shaken slightly (no vortexer) to resuspend the cells in the remaining medium (less than 50 μl).

Solution I (1 ml) containing 10 μg/ml EB (prepared before use and filtered again) is added to the cell pellet, and the cell suspension can be vortexed for a short time interval (2 sec).

After about 1 hr at room temperature, 1 ml of solution II containing also 10 μg/ml EB (prepared before use and filtered again) is added to the cell suspension, and the suspension is vortexed again (2 sec) and stored at 4°C before flow cytometric measurement. The suspension of nuclei and micronuclei can be measured directly after preparation or can be stored at 4°C for at least 1–2 weeks before measurement.

If the DNA distribution of nuclei has to be measured additionally it is recommended to use 25 μg/ml EB in solution I and 40 μg/ml EB in solution II. A lower coefficient of variation (CV) of the G_1 peak is obtained with these concentrations. The CV of the G_1 peak in several different cell cultures was usually found to be between 2 and 3%.

For a better discrimination of debris from micronuclei, especially in human lymphocytes, 1.5 μg/ml Hoechst 33258 can be added to solution II. With this double-staining technique two lasers are necessary to excite EB and HO simultaneously (see below).

V. Critical Aspects of the Procedure

Several aspects have been found to influence the preparation of a suspension of nuclei and micronuclei containing a low frequency of unspecific debris:

1. If more than 1.5×10^6 cells are treated with solution I and II (1 ml each), nuclear isolation is incomplete and a fraction of the cells could still have cytoplasm and cellular membranes. These cells can contain micronuclei that were not released in suspension thus influencing the results of micronucleus frequency. In this case a large amount of debris was usually observed. It is therefore recommended not to use too many cells.

2. Before measurement, the suspension of nuclei and micronuclei should be

checked in a fluorescence microscope using additionally phase-contrast optics. All nuclei should be free of cytoplasm and membranes, and micronuclei should be separated from nuclei. If this is not the case:

a. The cell number was too high (see 1).

b. Solution I should be prepared again. Because of the high viscosity of Nonidet P-40 it could be that solution I did not contain the proper concentration of the detergent.

c. The cell pellet contained too much medium. In this case try to remove the medium completely or use 2 or 3 ml of solution I and II instead of 1 ml each.

3. A large amount of unspecific debris always hampers the simultaneous measurement of micronuclei and nuclei. Usually, most of the debris can be gated out using forward or side scatter measurements if the debris is produced by fragments of cytoplasm or membranes (see below). However, dying cells in the cell culture, as for example necrotic or apoptotic cells, could lead to DNA fragments that show fluorescence intensities and light scatter signals similar to those of micronuclei but are clearly not micronuclei as verified by sorting of the respective particles. If plateau-phase cultures are studied or if chemicals are used for the induction of micronuclei, this effect has to be taken into consideration. It is therefore recommended to sort the particles considered to be micronuclei for visual examination before using the technique extensively.

4. Some chemicals (i.e., colcemid or 2-chlorobenzylidene malonitrile, a component of the tear gas CS) induce polyploid cells containing several nuclear fragments. Because the cellular membrane is destroyed, these nuclear fragments are released in suspension and will overlap the micronuclei during flow cytometric measurement. It cannot always be expected therefore that the frequency of micronuclei measured by flow cytometry agrees with results obtained by microscopic scoring. The criteria for the definition of micronuclei are not the same for microscopic scoring and for flow cytometry (see Nüsse *et al.,* 1992, and below).

VI. Standards

It is recommended that a good preparation of a suspension of micronuclei and nuclei is used as a standard before starting the flow cytometric measurements of freshly prepared samples. Because the suspensions of nuclei and micronuclei can usually be stored for 1–2 weeks, such a sample could be helpful for adjusting the flow cytometer.

VII. Instruments

A. Flow Cytometry

EB fluorescence (pulse height and pulse area) and side scatter as well as forward scatter of micronuclei and nuclei were measured simultaneously in list mode using a FACSstar$^+$ cell sorter (Becton–Dickinson). Excitation of EB was provided by the 488-nm line (500 mW) of an argon laser; EB fluorescence was detected with a long-pass filter (combination of KV 550 and OG 590). All parameters were registered in log mode (four decades). This is usually necessary, because micronuclei can have relative DNA contents between 0.5 and 10% of the DNA content of G_1-phase nuclei. The trigger was set between 0.5 and 1% of the mean EB fluorescence of the G_1 peak.

It is also possible to measure both types of particles in linear mode (as done earlier with a Cytofluorograf, Ortho Instruments). In this case usually not all micronuclei are registered together with the nuclei, because, due to the electronic threshold, particles with DNA content lower than about 2% of G_1 nuclei will be cut off.

If nuclei and micronuclei were double stained with EB and HO, dual-laser flow cytometry had to be performed (FACStar$^+$). In this case the first laser was adjusted to 488 nm (500 mW) and the second laser to the UV multilines (351.1–363.8 nm, 200–500 mW). Pulse height of forward light scatter and side scatter was recorded from the first laser as well as direct EB fluorescence (long-pass filter as above). EB fluorescence was used as a main trigger parameter. The second laser was used for excitation of HO. Direct fluorescence emission of HO was registered at wavelengths around 424 nm (band pass filter 424DF20, Becton–Dickinson). Fluorescence emission of EB excited by energy transfer from HO-fluorescence was measured at wavelengths around 630 nm (band pass filter 630DF20, Becton–Dickinson). Both fluorescences were split by a dichroic mirror with transmission between 340 and 490 nm and reflection between 590 and 640 nm.

B. Data Analysis

For analysis of the frequency and DNA distribution of micronuclei stained with EB alone (single-laser flow cytometry), a first gate has to be set to exclude unspecific debris according to side scatter and forward scatter (see Fig. 1). The frequency of micronuclei (number of micronuclei, N_{mn}) per nuclei (number of nuclei, N_n), N_{nm}/N_n, is calculated using additional gates for micronuclei (usually between 1 and 10% of the mean DNA content of G_1-phase nuclei) and nuclei (all nuclei are included).

For analysis of micronuclei that were double-stained with EB and HO, quotients of two of the three fluorescences measured had to be calculated for all particles included in the first gate. Only particles with the same quotients as the main nuclei are considered to be DNA containing particles (micronuclei).

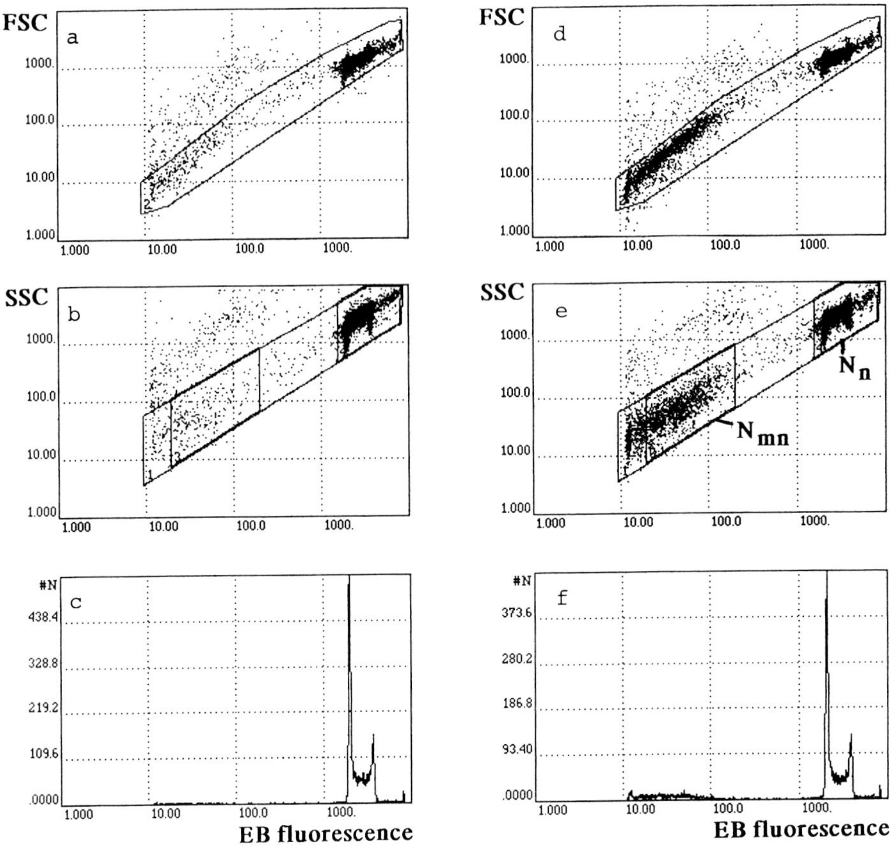

Fig. 1 Flow cytometric measurement of a suspension of micronuclei and nuclei. Human lymphoma cells K37 were irradiated with gamma radiation (D = 1 Gy). (a–c) control, (d–f) irradiated sample. The dot plots (10,000 events measured) show: (a and d) Forward scatter (FSC) against EB fluorescence (relative DNA content). The window shows the discrimination between micronuclei and debris according to forward scatter. (b and e) Side scatter (SSC) against EB fluorescence (relative DNA content). The window shows the discrimination between micronuclei and debris according to side scatter. The windows for micronuclei (N_{mn}) and nuclei (N_n) are indicated additionally. (c and f) DNA distribution of the suspension of nuclei and micronuclei. All particles found in both large windows are displayed.

Details of this more complicated technique are published elsewhere (Schreiber *et al.*, 1992a,b).

VIII. Results and Discussion

Figures 1a–1f show typical measurements of a suspension of nuclei and micronuclei prepared as described above and stained with EB. In this example, human lymphoma cells (K37 cells growing *in vitro*) were irradiated with ionizing radiation and analyzed 36 hr after exposure, when all cells have divided. Figures 1a–1c show the unirradiated control; Figs. 1d–1f, the irradiated sample (1 Gy). In the first two dot plots (Figs 1a and 1d) forward scatter (FSC) is plotted against EB fluorescence; in the second two dot plots (Figs. 1b and 1e) side scatter (SSC) is plotted against EB fluorescence. All parameters were measured in log mode. The large windows include nuclei and micronuclei and exclude unspecific debris; the two smaller windows (in Figs. 1b and 1e) are used for the calculation of N_{mn}/N_n. Micronuclei were counted in the region between 1 and 10% of the mean DNA content of G_1-phase nuclei. Figures 1c and 1f show the DNA distributions of micronuclei and nuclei.

Figure 1 shows that with both light scatter signals micronuclei can be separated from unspecific debris as verified by sorting of the respective particles. We usually measure both light scatter signals simultaneously with EB fluorescence and use the two gates (Figs. 1a and 1b or 1d and 1e) to exclude debris.

After cell cultures were treated with ionizing radiation, the results (frequency of micronuclei: N_{mn}/N_n) obtained with this flow cytometric technique agreed well with results from microscopic scoring. This is caused by the fact that radiation-induced micronuclei are mainly formed by acentric chromosome fragments that have a well-defined size distribution (for details, see Nüsse *et al.*, 1992a). This is not always the case, especially when certain chemicals are used to induce micronuclei. In this case micronuclei can sometimes be formed by whole single chromosomes or even by several chromosomes due to spindle damage (for details, see Nüsse *et al.*, 1992b; Miller *et al.*, 1992; Miller and Nüsse, 1993). In this case micronuclei can be found that have DNA contents between 1 and more than 50% of the mean DNA content of G_1-phase nuclei. Therefore, depending on the size of the window used to count the number of micronuclei N_{mn}, different values of N_{mn}/N_n can be obtained that not necessarily agree with results obtained by microscopic scoring.

Besides measurements of the frequency of micronuclei, the DNA distribution of micronuclei can be measured and analyzed. We have shown recently that several factors can influence the DNA content of micronuclei (Nüsse *et al.*, 1992a): The size distribution of the chromosomes, DNA synthesis in micronuclei (also see Kramer *et al.*, 1990), and the presence of chromosome fragments or whole chromosomes in micronuclei. Flow cytometric measurement of the DNA distribution of micronuclei alone cannot be used to completely understand the mechanisms that lead to the formation of micronuclei. Other techniques, as for example the use of DNA probes (chromosome-specific probes or a combination of telomeric and centromeric DNA probes) and fluorescence *in situ* hybridization are necessary as shown recently by us (Miller *et al.*, 1992; Nüsse *et al.*, 1992a; Miller and Nüsse, 1993).

Figure 2 shows two examples to illustrate this point. The left DNA distribution of micronuclei and nuclei (Fig. 2b) was obtained after Chinese hamster embryo cells (CCHE) were treated with the tear gas CS. The peaks in the micronucleus distribution show that the majority of micronuclei induced by CS was formed

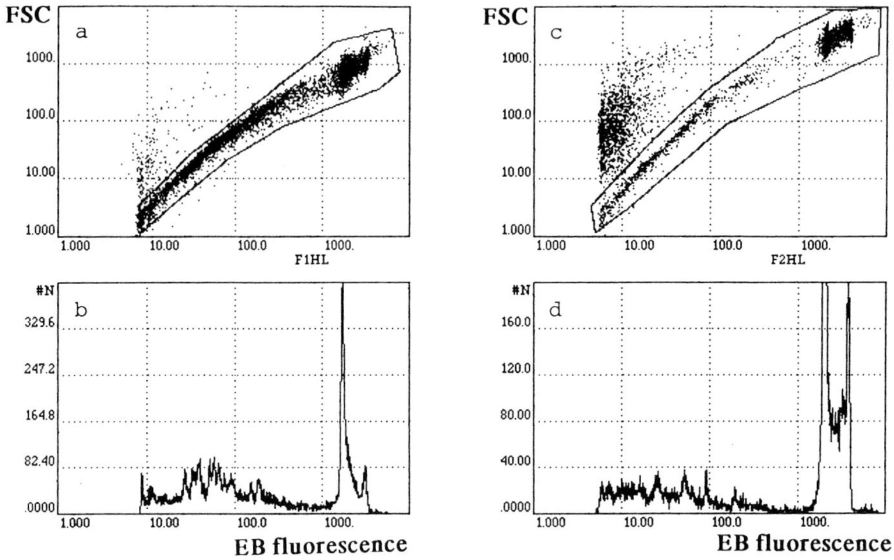

Fig. 2 Flow cytometric measurement of a suspension of micronuclei and nuclei. (a,b) Micronuclei induced by the tear gas CS (16 hr, 60 μM). (c,d) Micronuclei induced by bromodeoxyuridine (16 hr, 100 μM). (a,c) Forward scatter against EB fluorescence. (b,d) DNA distribution of micronuclei and nuclei from the particles defined by the gate in a and c.

by whole single chromosomes as verified by anti-kinetochore staining of micronuclei with the CREST serum (for details, see Nüsse *et al.*, 1992b). Similar results were obtained in mouse NIH-3T3 cells. Here the presence of whole single chromosomes in micronuclei was independently verified using fluorescence *in situ* hybridization with centromeric and telomeric DNA probes (Miller and Nüsse, 1993). Figure 2d shows in contrast to these results the DNA distribution of micronuclei induced in CCHE cells by bromodeoxyuridine (BrdUrd). In this case most BrdUrd-induced micronuclei (more than 95%) were formed by acentric chromosome fragments as verified by anti-kinetochore staining of micronuclei (Weller *et al.*, 1993). The peaks in this micronucleus distribution were caused by a nonrandom breakage of specific chromosomes (for details, see Weller *et al.*, 1993).

In conclusion, flow cytometry can be used to analyze frequency and DNA distribution of micronuclei induced in cell cultures by ionizing radiation or chemicals. For this measurement, a suspension of micronuclei and nuclei has to be prepared. Light scatter signals in combination with fluorescence intensity measurements are used to discriminate micronuclei from unspecific debris. With the same preparation the DNA distribution of nuclei can additionally be measured showing the effects of the treatments on the cell-cycle progression.

References

Cao, J., Leibold, E., Beisker, W., Schranner, T., Nüsse, M., and Schwarz, L. R. (1993). *Toxicol. In Vitro* **7,** 447–451.

Fenech, M., Jarvis, L. R., and Morley, A. A. (1988). *Mutat. Res.* **203,** 33–38.

Grawe, J., Zetterburg, G., and Anneus, H. (1992). *Cytometry* **13,** 750–758.

Hayashi, M., Norppa, H., Sofuni, T., and Ishidate, M. (1992). *Mutagenesis* **7,** 251–256.

Hutter, K.-H., and Stöhr, M. (1982). *Histochemistry* **75,** 353–362.

Kramer, J., Schaich-Walch, G., and Nüsse, M. (1990). *Mutagenesis* **5,** 491–495.

Miller, B. M., and Nüsse, M. (1993). *Mutagenesis* **8,** 35–41.

Miller, B. M., Werner, T., Weier, H.-U., and Nüsse, M. (1992). *Radiat. Res.* **130,** 177–185.

Nüsse, M., and Kramer, J. (1984). *Cytometry* **5,** 20–25.

Nüsse, M., Beisker, W., Hoffmann, C., and Tarnok, A. (1990). *Cytometry* **11,** 813–821.

Nüsse, M., Kramer, J., and Miller, B. M. (1992a). *Int. J. Radiat. Biol.* **62,** 587–602.

Nüsse, M., Recknagel, S., and Beisker, W. (1992b). *Mutagenesis* **7,** 57–67.

Pincu, M., Callisen, H., and Norman, A. (1985). *Int. J. Radiat. Biol.* **47,** 423–432.

Romagna, F., and Staniforth, C. D. (1989). *Mutat. Res.* **213,** 91–104.

Schreiber, G. A., Beisker, W., Bauchinger, M., and Nüsse, M. (1992a). *Cytometry* **13,** 90–102.

Schreiber, G. A., Beisker, W., Braselmann, H., Bauchinger, M., Bögl, K. W., and Nüsse, M. (1992b). *Int. J. Radiat. Biol.* **62,** 695–709.

Tates, A. D., van Welie, M. T., and Ploem, J. S. (1990). *Int. J. Radiat. Biol.* **58,** 813–825.

Weller, E. M., Dietrich, I., Viaggi, S., Beisker, W., and Nüsse, M. (1993). *Mutagenesis* **8,** 437–444.

CHAPTER 10

Sperm Chromatin Structure Assay: DNA Denaturability

Donald Evenson and Lorna Jost

Olson Biochemistry Laboratories
Department of Chemistry
South Dakota State University
Brookings, South Dakota 57007

Studies have shown that exposure to chemicals that alkylate free -SH groups on protamine molecules (e.g., methylmethane sulfonate, MMS) in late testicular or early epididymal sperm caused nearly 100% of sperm to have altered chromatin structure by 3 days postexposure (Evenson *et al.*, 1993b). This was 8 days prior to altered sperm head morphology. Of greater interest, the maximum chromatin alterations, measured by the SCSA, corresponded to the temporal pattern of sperm produced that resulted in maximal dominant lethal mutations. Since dominant lethal mutations are caused by chromosomes breaks, the SCSA was likely detecting damage to chromosomes about 8 days prior to maximum dominant lethal mutations. In another experiment (Estop *et al.*, 1993), mouse sperm aged *in vitro* showed alterations of chromatin structure by the SCSA at only 2 hr incubation; however, when these aged sperm fertilized mouse oocytes *in vitro*, pronuclear chromosomes from *in vitro* aged sperm did not show chromosome breaks until sperm had been aged for 6 hr. The SCSA appears to detect early stages of chromatin alterations that likely lead to whole chromosome breaks. Thus, the SCSA is viewed as a potentially important method to assay for early events of toxicant-induced chromosome damage.

In addition to screening for toxicant-induced damage to sperm, an equally important impact of the SCSA technique may be for assessment of animal and human subfertility (Evenson, 1986; Ballachey *et al.*, 1987,1988). In both types of studies it is important to know what the normal variation of chromatin structure is over time and how it relates to other typically measured semen parameters.

A longitudinal study of human sperm chromatin structure was made on monthly semen samples from 45 men over 8 consecutive months (Evenson *et al.*, 1991). The study showed that although the SCSA data often differed between donors, there was a remarkable homogeneity within a donor from month to month. In fact, the repeatability of the positions of just a few dots to major clusters of dots on the FCM cytograms from month to month were suggestive that particular stem cells had a consistent abnormality leading to very particular levels of abnormality in their progeny sperm. The SCSA data had higher repeatabilities and lower CVs than the classical measures of semen volume, sperm count, motility, morphology, and viability. Since sperm chromatin structure is a more repeatable feature of sperm and yet is responsive to environmental toxicants, it is looked upon favorably as a valuable biomarker for human toxicology studies.

Data for the relationship between human sperm chromatin structure and fertility are limited. Two major studies are in progress. Both confirm that semen samples from human infertility clinics demonstrate a high degree of chromatin structural heterogeneity within the sperm population (Evenson *et al.*, 1993a). Unresolved yet is the degree of heterogeneity that is compatible with fertility and normal embryo development.

III. Materials

A. Acridine Orange Staining Solutions

1. *AO stock solution:* Chromatographically purified AO (Polysciences, Warrington, PA) is dissolved in double-distilled water to a final concentration of 1 mg/ml. Nonpurified AO is not acceptable for this technique. AO is a toxic chemical and considerable care should be used when weighing out the powder. Typically a 15-ml glass scintillation vial is tared on a balance pan and about 5 to 10 mg of AO powder carefully transferred with a cupped spatula into the vial. An equivalent number of milliliters of water are then added to the vial which is then capped, covered with aluminum foil to minimize light exposure, and placed in refrigerator. This solution can be kept at 4°C for several months.

2. *Acid/detergent treatment solution for step 1 of AO staining procedure:* 0.15 *M* NaCl, 0.1% Triton X-100 (Sigma Chemical Co., St. Louis, MO), and 0.08 *N* HCl in double-distilled water. For 500 ml, admix 20 ml 2.0 *N* HCl, 4.39 g NaCl, 0.5 ml Triton X-100, and 480 ml double-distilled water. This solution will keep up to several months at 4°C. The working solution is kept in a 16-oz glass amber bottle containing an Oxford adjustable, 0.20- to 0.80-ml automatic dispenser (Lancer Division of Sherwood Medical, St. Louis, MO).

3. *Stock 0.1 M citric acid buffer:* To 21.01 g citric acid monohydrate (FW = 210.14) add double-distilled water to 1 liter. Store at 4°C.

4. *Stock 0.2 M Na_2PO_4 buffer:* To 28.4 g sodium phosphate dibasic (FW = 141.96) add double-distilled water to 1 liter. Store at 4°C.

5. *Stock AO staining solution:* Mix 370 ml 0.1 *M* citric acid buffer, 630 ml 0.2 *M* Na_2PO_4 buffer (due to crystallization at 4°C, this stock must be warmed prior to use), 372 mg EDTA, and 8.77 g NaCl. Mix well. Adjust to pH 6.0.

6. *AO staining solution:* 0.60 ml AO stock solution is added to each 100 ml of stock AO staining solution. This AO staining solution, kept in a 16 oz. glass amber bottle containing an Oxford adjustable, 0.80- to 3.0-ml automatic dispenser, is made fresh biweekly.

B. Buffers and Other Materials

1. *TNE buffer:* 0.01 *M* Tris, 0.15 *M* NaCl, and 1 m*M* EDTA, pH 7.4. Be cautious that this buffer remains free of bacterial contamination as this may cause problems with sample interpretation; it is preferable that this buffer is filtered through a 0.22-μm filter (Corning sterile filter unit No. 25932, Cat. No. 210-963, Curtin Matheson Scientific, Inc., Houston, TX) to remove any debris and bacteria and then stored in a sterile tissue culture flask at 4°C.

2. *Hanks' Balanced Salt Solution (HBSS):* (Gibco Laboratories, Grand Island, NY).

3. Polypropylene microtubes, 0.5 or 1.0 ml (Sarstedt, Inc., Princeton, NJ).

4. Polystyrene 12 × 75 mm, 4.5 ml, conical tubes, (Cat. No. 57.477, Sarstedt, Inc.).

5. Falcon No. 3033, 16 × 125 mm, tissue culture tubes (Becton Dickinson Labware, Lincoln Park, NY).

6. Nylon filters, 153-μm mesh, 1 in. diameter (Tetko, Elmsford, NY).

7. Dental plugger (Henry Schein, Inc., Port Washington, NY).

8. Corning No. 25702 cryogenic vials, with internally threaded caps, 2 ml (Cat. No. 237-347, Curtin Matheson Scientific, Inc.).

9. Tuberculin syringes, 1 cc, Becton Dickinson No. 9602 (Cat. No. 262-247, Curtin Matheson Scientific, Inc.).

IV. Cell Preparation

A. Fresh, Frozen, and Fixed Sperm Samples

Human semen samples are obtained by masturbation into plastic clinical specimen jars preferably after 2–3 days abstinence. For safety against potential infectious agents, e.g., hepatitis and HIV, samples are handled with disposable gloves in a biological safety cabinet. After 30 min has been allowed for semen liquefaction, aliquots of semen can be frozen directly in an ultracold freezer (−70 to −110°C) or placed into LN_2. A non selfdefrosting refrigerator freezer can be used for short-term freezing and storage and then the sample can be transferred preferably that same day to an ultracold freezer. In field situations, an ice chest containing dry ice may be used. Care should be taken to freeze the samples in an upright position using a test tube rack in the freezer. This is especially important if snap cap tubes are used because if the tube is inverted when frozen, the freezing pressure may partially open the snap cap. Furthermore, samples frozen at the bottom of a tube are thawed in the water bath with greater ease and safety. The easiest sample to work with is one that has been diluted into TNE buffer prior to freezing which reduces the viscosity for handling at the time of FCM measurements. Tall, narrow plastic snap cap tubes are preferred to minimize the surface to volume ratio of the sample in order to reduce cell "freezer burn" that may occur during long-term storage (months to years); freezing in an upright position helps to minimize this potential problem. A 1-ml microtube is favorable to store a mixture of 100 μl semen + 400 μl TNE buffer. Since 200 μl of semen/TNE mixture is used for preparation of the sample for FCM measurement, this sample can be used directly if the sperm count is low or diluted to about 1–2 × 10^6 if the sperm count is too high. If the first sample is too concentrated, the remaining 300 μl can be used for dilution or a repeat measurement. Since the AO/DNA phosphate ratio is very important for appropriate staining (Darzynkiewicz, 1979; practically applied as the number of cells/0.2 ml aliquot sample used for AO staining), it is very useful to know the sperm count in advance so that an aliquot can be diluted very closely to the desired range without a trial and error process. For severe oligospermic samples, undiluted semen can be used directly; the acid/detergent solution used in the fist step dramatically reduces any semen viscosity (Evenson and Melamed, 1983).

Commercial animal semen is often diluted with a variety of extenders which serve as a cryoprotectant and as a diluent for increasing the number of samples from a single ejaculate. These extenders generally do not interfere with the SCSA measurements of sperm chromatin. However, bull semen extended in *nonclarified* egg yolk citrate extender often causes some background noise which may or may not show up as debris noise extending into the sample signal region.

Frozen samples are preferred and the easiest to work with. However, in some cases it may be desirable to fix samples. Data from fixed samples are essentially similar to that obtained on fresh material (Evenson *et al.,* 1986a). In this case, sperm are centrifuged out of semen and resuspended in HBSS at a concentration of about 10^7/ml, and 1 ml of this suspension is forcefully pipetted into 10 ml cold (−20°C) 80% ethanol. Prior to analysis, the sperm are again pelleted, washed once in TNE buffer, and then processed for SCSA, always keeping the sample at 4°C.

B. Epididymal or Vas Deferens Sperm

For animal studies, a specific segment of the epididymis can be surgically removed from a killed animal and minced in TNE buffer with a curved pair of scissors in a petri dish set on crushed ice (Evenson *et al.,* 1986a). The mixture is transferred to a 12 × 75-mm tube and the larger fragments are allowed to settle. "Home-made" filtering systems are made by mounting a 1-in. 153-μm nylon mesh between the end of a 1-cc tuberculin syringe and its cap with its end cut off. The supernatant suspension is then passed through this filter. Do not apply pressure to the plunger. The vas deferens may also be excised and placed in a 60-mm petri dish containing TNE buffer and the sperm removed by pressing a blunt probe, a dental plugger works very well, along the length of the organ. In order to easily visualize the white "cord" of sperm being expressed, the petri dish is placed on a black Teflon-coated plate of steel set on the surface of crushed ice.

C. Sonication of Sperm Cells

Our laboratory has measured thousands of animal and human sperm samples by the SCSA. In previous years, there has been concern about whether any residual cytoplasmic droplets potentially containing RNA would add an artifact to the measurement of single-stranded DNA. Thus, in earlier studies, all samples were sonicated or some unsonicated samples were compared with their sonicated counterpart. The results have been so nearly identical (Evenson *et al.,*

1991) that sonication is no longer a routine procedure and this saves a great deal of time and effort. Investigators must remain aware of this potential problem and, if there is a reason to be concerned, then some or all samples should be sonicated. Sonication is preferred over RNase incubation which has the potential of causing incubation-related changes in chromatin structure possibly due to protease digestion of chromatin proteins. As an exception, rat sperm samples often used in toxicology experiments, *must* be sonicated because the long, fibrous tails tend to clog the flow cell. The broken tails in the sonicate can cause a problem in SCSA analysis. This can be corrected by electronic gating (see Section VII, Results and Discussion) or purification through a sucrose gradient (Evenson *et al.*, 1985). Mouse sperm, with much shorter tails, do not require sonication relative to the problem with rat sperm.

Sperm suspended in TNE buffer in a Falcon 3033 test tube immersed in an ice water slurry are sonicated for 30 sec at a setting of 50 on low power (Bronwill Biosonik IV Sonicator, VWR Scientific, Inc., Minneapolis, MN), cooled for 30 sec, and sonicated again for 30 sec. The half-inch probe is placed just above the bottom of the tube. Optimal time and power required for sperm head–tail/cytoplasm separation varies between species and needs to be tested for each sonicator to achieve an approximate ≥95% head/tail separation. Allow the sonicate to set 2 min on ice before preparing for the SCSA.

Human semen samples, potentially containing infectious agents such as hepatitis or HIV, must be sonicated only in a closed tube. Our laboratory utilizes a Branson Sonifier II, Model 450, coupled to a Branson Cup Horn (VWR Scientific, San Francisco, CA). Sample temperature is kept cold by 4°C water flowing through the cup horn. This is derived by using a Masterflex peristaltic pump (Cole Parmer Instrument Co., Chicago, IL) that drives water (21 ml/min) through approximately 3 ft of copper tube coil ($\frac{1}{4}$ in. id) set in a 4-liter flask containing an ice water slurry. Place 0.5 ml of TNE buffer containing $\leq 1 \times 10^6$ sperm cells into a 2-ml Corning cryogenic screw cap vial. The top end of this capped vial is inserted into the bottom side of a No. 11 rubber stopper which has a 12-mm hole drilled through it that will hold the vial securely. The rubber stopper, holding the vial, is then placed on top of the cup horn with the vial protruding down into the cup horn so that the bottom of the vial is just off the bottom of the cup. Samples are sonicated for 30 (rats) to 40 (humans) sec using 70% of 1-sec cycles at a setting of 3.0 output power. The cup horn sonicator is preferred over the probe method for case of use, uniformity between samples, and safety precautions.

V. Cell Staining and Measurement

In contrast to many procedures where a large batch of samples can be prepared at a lab distant from the flow cytometer, the SCSA procedure requires that samples are thawed and prepared in the immediate vicinity of the flow cytometer. Elapsed times for various components of the procedure are very important. A frozen sample is held by the top of the test tube which is mostly immersed in a 37°C water bath, just until the last remnant of ice disappears. After thawing, the sample is either diluted (all buffers and staining solutions are kept on crushed liquid ice) and prepared further or sonicated first.

Prior to measuring experimental samples, the instrument must obviously be checked for alignment. Very importantly, especially if the sample tubing has been bleached clean, an AO buffer mixture (0.4-ml acid detergent solution and 1.2-ml AO staining solution) needs to be passed through the instrument sample lines for at least 30 to 45 min prior to setting the photomultiplier tubes (PMTs) with the reference sample and measuring samples. This ensures that AO is equilibrated with the sample tubing. In contrast to the rumor spread by representatives of some commercial companies and uninformed flow operators, using AO in a flow cytometer does not ruin it for other purposes! The sample lines do not need to be replaced after measuring AO-stained samples! While the system does need to be equilibrated with AO (i.e., AO adheres to the sample tubing), this is FULLY rectified by rinsing the system for about 10 min with a 50% filtered bleach solution after finishing the AO measurements. Our laboratory has measured many other dyes and sample types after measuring AO-stained sperm without any associated problem.

A 0.20-ml aliquot of sample is placed into a 12 × 75-mm conical tube. A 0.40-ml aliquot of the first step low pH buffer is added with an automatic dispenser. This dispenser needs to be accurate and to have a maximum capacity near the amount being dispensed. A stopwatch is started immediately after the first buffer is dispensed. Exactly 30 sec later the AO staining solution is added. The sample tube is then placed into the flow cytometer sample chamber in a 30-ml beaker containing an ice water slurry. Although it is preferable to have the sample setting in an ice bath, the configuration of some FCM sample chambers may not permit this; since the sample is measured shortly after being removed from ice, this should not cause a significant difference in the data. The sample flow is started immediately after it is placed in the sample holder. Using the same stopwatch that was started with the addition of the first step buffer, the acquisition of the data is started at 3.0 min from the time of step one buffer addition. This allows for equilibration and stabilization of the sample, as well as uniformity between samples, all important points for AO staining. Also, the flow rate is checked during this time and if it is too fast, i.e., >300 cells/sec, a new sample is made at an appropriate dilution. The original sample cannot be diluted with AO buffer. It is implied that the sample and sheath flow valve settings of the instrument are never changed during these measurements and that the liquid flow rate is constant. Thus, a change in count rate is a function of sperm cell concentration only. A critical part of SCSA measurements is the use of a reference sample to monitor instrument stability throughout any experiment. (See Section VII, Results and Discussion, for more detail.)

VI. Instruments

Blue laser light (488 nm) excitation of AO-stained cells at a power of >35 mW is optimal. Fluorescence of individual cells is measured at wavelengths of red (630- to 650-nm long-pass filter) and green (515- to 530-nm band pass filter). Both green and red fluorescences are processed in *peak* mode of signal rather than area mode. Since mature, AO-stained mammalian sperm have very little red fluorescence, due to lack of RNA and single-stranded DNA, the red PMT gain may need to be set high enough that electronic noise may result with some instruments. Ortho Diagnostics engineers made a slight modification of the red fluorescence preamplifier circuit board to reduce background noise on the Cytofluorograf II.

Aliquots of the same semen samples from humans, stallions, and mice were measured on Cytofluorograf, Becton Dickinson FACScan, and Coulter Elite flow cytometers. The scattergram patterns for all instruments were similar indicating that any of these instruments can be used with the SCSA protocol. However, neither of the latter two instruments is capable of generating the very important ratio of red/red + green) fluorescence (α_t). In this case, list-mode files were transferred to an IBM compatible computer and processed using LISTVIEW software (Phoenix Flow, Inc., San Diego, CA) which gave

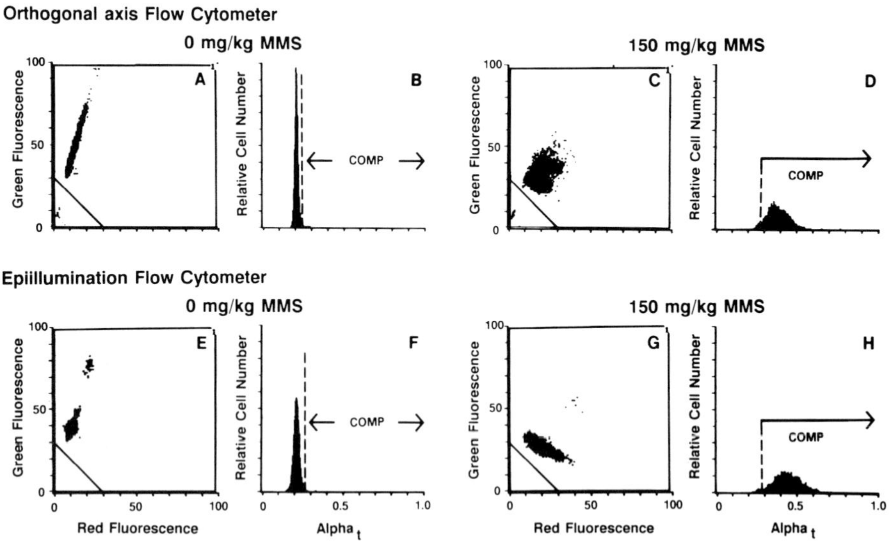

Fig. 1 Green vs red fluorescence cytograms of SCSA data obtained by measuring caudal mouse sperm on a Cytofluorograf (A,C) or an ICP22A (E,G) flow cytometer. Sperm were obtained from mice treated with 0 or 150 mg/kg MMS and killed at 13 days. (B,D,F, H) α_t frequency histograms; cells to the right of the vertical lines are cells with denatured DNA (COMPα_t; Evenson *et al.*, 1993b. Reprinted with permission of Wiley–Liss).

desired SCSA data. It is not appropriate to use a software or hardware configuration that defines α_t as red (>630 nm)/total (515 long-pass) fluorescence; this adds the unknown component of the 530- to 630-nm wavelengths into the denominator.

As an excellent alternative to the more expensive laser-based instruments, the Hg-arc lamp-driven Ortho ICP22A flow cytometer interfaced to the 2150 Data Handling system has been successfully used (Jost and Evenson, 1993; Evenson *et al.*, 1993b). Alternatively, the ICP22A can be interfaced to an IBM compatible 386 or 486 personal computer with the ACQCYTE system (Phoenix Flow, Inc.) installed. Figure 1 compares AO-stained mouse sperm measured by an orthogonal and an epiillumination flow cytometer. The epiillumination ICP22A does not produce an optical artifact (see Evenson *et al.*, 1993b, for discussion) as seen with orthogonal instruments when measuring sperm. Although the cytograms do not have the same cell fluorescence distribution, the pertinent data are the same and equally useful. Unfortunately, the ICP22A is no longer commercially available but other Hg-arc systems should work. The Argus flow-cytometer (Skatron, Bio-Rad, Italy) is known to work satisfactorily (P. De Angelis, O. P. F. Clausen, and D. P., Evenson, unpublished); however, it is hoped that a small, relatively inexpensive Hg-arc system will be produced in the near future that would accommodate the SCSA technique.

VII. Results and Discussion

A. SCSA Parameters

Two parameters are of particular importance in the evaluation of the SCSA-derived data. The first is the mean green fluorescence (X Green) which is related to the condensation of the sperm chromatin and extent of restricted access of DNA dyes. The sperm nuclear condensation process normally produces a five-fold reduction of DNA stainability relative to round spermatids (Evenson *et al.*, 1986b). Lack of appropriate sperm maturation results in an increased DNA stainability. Studies have shown that patients attending an infertility clinic often have an increased DNA stainability (Engh *et al.*, 1992; Evenson and Melamed, 1983). This can be visualized by univariate analysis (Engh *et al.*, 1992) as well as by the SCSA bivariate analysis. The SCSA has an advantage however in distinguishing debris from sperm signals.

The second parameters of particular importance are those of α_t. Interestingly, AO-stained samples that show high X Green usually do not exhibit extensive DNA denaturation and thus both abnormalities can be studied via the SCSA. SCSA analysis also includes mean total fluorescence (X Total = X Green + X Red), and the α_t parameters of mean α_t (Xα_t), standard deviation α_t (SDα_t), and cells *o*utside the *m*ain *p*opulation α_t (COMPα_t). Practically, COMPα_t indicates the percentage of abnormal cells and SDα_t describes the extent of the abnormalities. Note that the defined value range of α_t is from 0.0 to 1.0 (i.e., all green and no red fluorescence to all red and no green fluorescence; Darzynkiewicz *et al.*, 1975), but for practical considerations it is expressed in 1000 channels of fluorescence.

Harsher physical conditions have been applied to sperm with the SCSA (i.e., increasing AO concentration, more acidic conditions, longer incubation times) resulting in a higher percentage of COMPα_t and higher SDα_t; however, parallel dose–response curves are observed with the standardized procedure when compared with increased physical conditions for DNA denaturation. Thus, new information is not gained with increased physical conditions for denaturation. The current view is that the abnormalities observed by the COMPα_t and SDα_t are a "tip of the iceberg" effect and perhaps reflect abnormalities that may be present at lesser levels in the main population. Each component has valuable information which sometimes needs to be interpreted with special regards to the results; likewise, a specific component may be of more value when the response differs. For example, if the total sperm population shifts from green to red fluorescence, Xα_t and COMPα_t may be the most valuable descriptors. However, in many toxicology experiments, only a small to moderate percentage of cells have shifted to various degrees; in this case, SDα_t has had the highest correlation with dosage of toxicant or some other parameters.

B. Debris Exclusion

Figure 2 compares SCSA-derived data from a fertile male with those from a patient attending an infertility clinic. The distribution for DNA stainability (green fluorescence) of the normal population is broad due to a known optical artifact (see discussion in Evenson *et al.*, 1993b) but which has no effect on the α_t distribution of interest in this technique. Note that the distribution is much more homogeneous for the fertile individual than for the clinic patient. A very important, but sometimes difficult point, is deciding where to make the computer gates to cut out cellular debris and to distinguish between a normal population of sperm and cells with denatured DNA. The problem is accentuated with low sperm count samples and especially those samples derived from patients on chemotherapy that may result in debris from killed cells. For most animal and relatively normal human samples, a box is drawn first very near to the perimeters of the cytogram boundaries to exclude those events that are beyond the full channel limits, e.g., channels 2 to 254 inside a box of 0 to 256. Next, debris that falls to the lower left hand corner is dealt with in one of two ways. With samples having very little debris a near 45° angle line is drawn just below the bottom of the sperm signal as seen in Fig. 2. The 45° line is based on the premise that cells gain red fluorescence at the expense of green fluorescence at a near 45° angle. Human SCSA data are often more complicated and in some cases an elliptical circle has been used to exclude what was considered to be debris from the data (Evenson *et al.*, 1991). There is no perfect answer

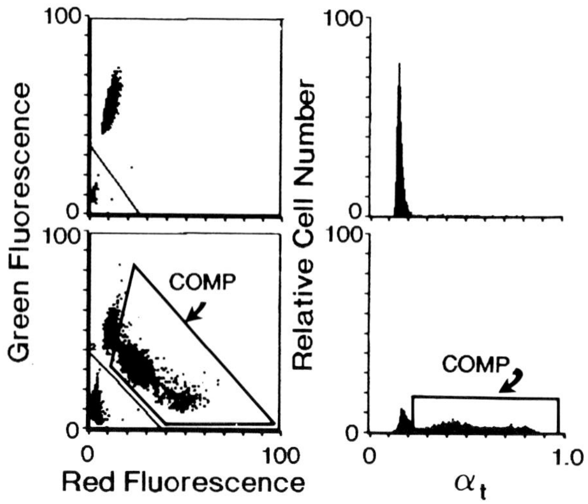

Fig. 2 FCM cytogram of two-parameter green (double-stranded DNA) vs red (single-stranded DNA) fluorescence distribution of sperm from a fertile human and a patient from an infertility clinic. The box marked COMP shows the cells outside the main population with abnormal chromatin structure. The α_t distribution shows the extent of the abnormality. (Evenson, 1990. Reprinted permission of Academic Press).

on this matter. After inspection of the data set a decision is made on what best fits the experiment. Whatever method is chosen, the important point is to be consistent throughout the entire data set and preferably between current and future experiments.

Because of their long tails and the need for sonication, rat sperm pose a particularly difficult problem when excluding debris. The rat tails break up into many pieces of debris that are seen as fluorescent signals by the flow cytometer. This problem can be overcome by using the flow cytometer's electronic signal processing capabilities. In Fig. 3, cytograms A and D show the regular peak mode, green and red fluorescence signals before debris is gated out. The sperm signal is not resolved from the debris. By gating out the debris in the green area vs peak cytogram (B), the resulting cytograms (C and E) are relatively "debris free" and analyzable. This processing technique may be useful in other species and situations where a large debris to sperm ratio exists.

C. Placement of the COMPα_t Line

The computer gate defining COMPα_t is easy to set if a significant percentage of the cells have fluorescence values equal to a normal population. This was true for all early experiments. However, when the entire population shifts (Evenson *et al.*, 1993b) then COMPα_t cannot be defined as the cells to the right of the main population in that cytogram. In that case, the computer gate must

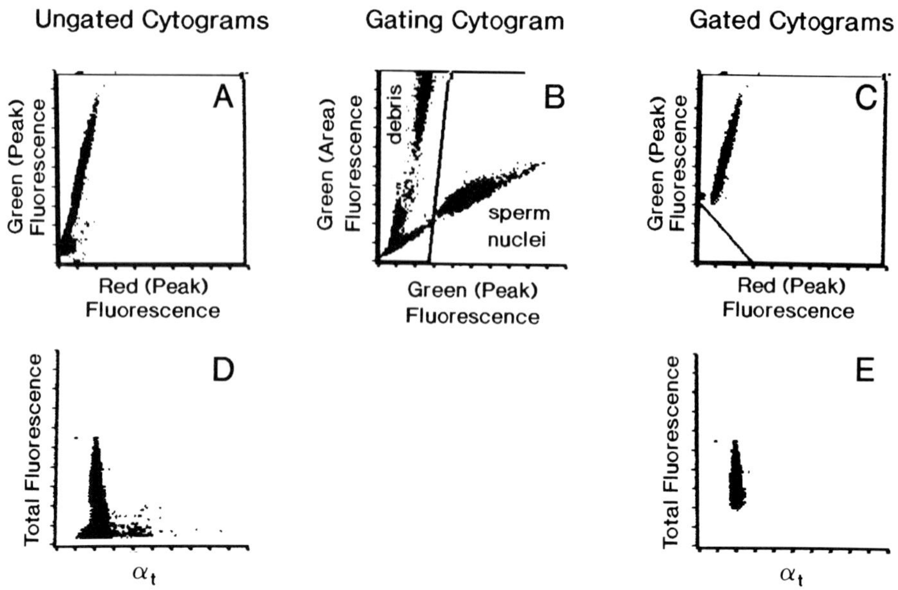

Fig. 3 FCM cytograms of ungated vs gated AO-stained rat sperm nuclei. Cytograms A and D (debris and sperm signals are not resolved) are prior to gating out the debris using the green fluorescence area vs peak mode cytogram (B), while cytograms C and E have the debris gated out and the sperm signal is resolved from the debris.

be set to the right side of the *control sample* population and all cells to the right of that line are defined as COMPα_t cells which in some cases may equal 100%. This procedure points out very strongly again the absolute need to have a reference sample to precisely set the instrument variables.

D. Reference Samples

SDα_t has been the most useful and most closely correlated parameter related to known fertility ratings (Ballachey *et al.*, 1987,1988). Because α_t measures, particularly SDα_t, are very sensitive to small changes in chromatin structure, studies using this parameter require very precise, repeat instrument settings for all comparative measurements whether done on the same or different days. These settings are obtained by using aliquots of a single semen sample that demonstrates heterogeneity of α_t. A semen sample is identified as a reference sample and then diluted with TNE buffer to a working concentration of 2×10^6 cells/ml. Several hundred 250-μl aliquots of this dilution are placed into small snap cap vials and frozen (-70 to -100°C) immediately. These reference samples are used to set the red and green PMTs to the same X Red, X Green, and α_t values (Xα_t and SDα_t) from day to day. The PMTs are set so that X Green is at about 50/100 channels and X Red at about 13/100 channels. Whatever channel numbers are established by a laboratory should be used consistently thereafter. All reference sample X Red and X Green values should fall within ±5 channels of a value that the flow operator decides upon. Strict adherence to keeping the reference values in this range should be maintained throughout the experiment. One or two freshly thawed reference samples are typically run after every 5 or 10 regular samples to ensure that there is no instrument drift. If the reference sample values are out of the range, first run a second reference sample, then check focus, and if that does not correct the problem, finally adjust PMT settings. If the first samples run during a day are seen to drift, it usually implies that AO was not equilibrated with the sample tubing.

E. Sampling Order

If the intent of an experiment is to determine the smallest amount of change measurable over time or increasing toxicant dosages, the most ideal situation is to measure all experimental samples of a particular set at one time period. Statisticians often do not like this approach and prefer placing all samples randomly coded in a box to be measured blindly. However, if the best "truth" of the experiment is desired, and recognizing that the flow cytometer randomly and objectively measures the samples, it is preferable to measure sets of samples in a single time frame. Each set is randomized within itself. However, if totally random measurements are preferred, it has been shown that very careful repeat settings of the red and green PMTs allow measurements to be made of compared samples over an extended period of time with nearly identical results. This includes samples that were measured fresh, then frozen, and the frozen/thawed samples measured up to 3 years later.

F. SCSA and Fertility

At this time, it is difficult to define what values are incompatible with normal fertility since the interpretation of α_t parameters is still being explored. However, from measurements of thousands of sperm samples derived from a variety of mammals, some of which had known fertility potential, the evidence strongly suggests that a broadly heterogeneous pattern is indicative of sub- or infertility. It is important to note however that fertility may occur with sperm with high α_t values and this may lead to early embryo death (Evenson *et al.*, 1993b).

VIII. Critical Points

A. Freezing and Thawing

Repeated evidence from various species shows that freezing and thawing a sperm or semen sample *once* does not cause significantly altered SCSA data relative to fresh samples. However, it is strongly emphasized that the samples must stay frozen in an ultracold freezer or in dry ice (−70°C or colder). Refrigerator freezers with automatic defrosters should be avoided due to the rise and fall of temperatures on a daily basis. Likewise, investigators must be aware that removal of samples from an ultracold freezer into ambient air shifts the sample to an approximate 100°C temperature change that may not immediately turn the frozen sample into a liquid sample but the strong physical forces of the temperature increase may damage the chromatin structure and cause an artifact. Thus, when handling any sample, e.g., moving from box to box in the freezer, DO NOT pick up the small test tube by the body of the tube; a warm human hand will produce microthawing of the sample very quickly. Use gloves or forceps or at least grasp the tube by the very top lip of the tube. If samples boxes must be manipulated, place them into a deep ice chest containing dry ice that will keep the box and the samples at least dry ice cold. Since ultracold freezers are often shared by a number of personnel, place the sample box in a rack that others are instructed not to remove. It is safest to place the sample box near the bottom of the freezer so that it is not subjected to ambient air when others may keep the door open for extended periods of time. When a group of samples are to be analyzed by flow cytometry, place the samples in an approximate 18-in. deep styrofoam box containing dry ice with a good cover and place it near the flow cytometer where the samples are prepared. Individual samples are removed, placed into a 37°C water bath to thaw, and then processed immediately.

B. Shipping of Samples

Typically semen samples are shipped by Federal Express (or similar overnight carrier) in $10\frac{3}{4} \times 7\frac{1}{2} \times 10\frac{1}{2}$ (L × W × H in., inside dimensions) commercial insulated shipping containers (i.e., FreezSafe Insulated Containers, Polyfoam Packers Corp., Wheeling, IL; Cat. No. 272-524, Curtain Matheson Scientific, Inc.). Small chunks of dry ice are first placed on the bottom of the shipping container, then the sample box is placed near the center of the shipping box, and then dry ice placed over and around the box. Ten pounds of dry ice added to the shipping container is satisfactory for Priority One overnight shipments from any point in the United States during any season. This amount will keep for at least 2 days. Shipments are made only Monday through Wednesday due to the rare case where the shipment box is "miss-shipped" somewhere else or it is held up for a day in a storm.

C. Reference Samples

Very few FCM protocols are as particular as the SCSA for using a reference sample. Fresh aliquots of this sample are run after approximately every 10 experimental samples to ensure that the instrument has not drifted or lost focus. Strict adherence to keeping the reference values in the range set by the flow operator should be maintained throughout the experiment.

D. Sample Flow Rate

The rate of sperm cell measurement is important and should be about 150 cells/sec. Samples that run over 300/sec are discarded and a new aliquot is diluted and stained to produce that approximate rate.

Acknowledgments

This work was supported in part by NSF Grant EHR-9108773 and the South Dakota Futures Fund. It is Publication No. 2749 from South Dakota State University Experiment Station. We gratefully acknowledge the skilled collaborative efforts and technical assistance of Donna Gandor and colleagues (Becton Dickinson, Inc. San Jose, CA); Ole Petter Clausen, Kenneth Purvis and Paula De Angelis (The National Hospital, Oslo, Norway); Richard Coico and Andrew Daley (CUNY, NY); and Barbara Stanton (Coulter Corporation, Hialeah, FL).

References

Ballachey, B. E., Hohenboken, W. D., and Evenson, D. P. (1987). *Biol. Reprod.* **36,** 915–925.

Ballachey, B. E., Saacke, R. G., and Evenson, D. P. (1988). *J. Androl.* **9,** 109–115.

Clausen, O. P. F., and Abyholm, T. (1980). *Fertil. Steril.* **34,** 369–373.

Darzynkiewicz, Z. (1979). *In* "Flow Cytometry and Sorting" (M. R. Melamed, P. F. Mullaney, and M. L. Mendelsohn, eds.), pp. 285–316. Wiley, New York.

Darzynkiewicz, Z., Traganos, F., Sharpless, T., and Melamed, M. R. (1975). *Exp. Cell Res.* **90,** 411–428.

Darzynkiewicz, Z., Traganos, F., Sharpless, T., and Melamed, M. R. (1976). *Proc. Natl. Acad. Sci. U.S.A.* **73,** 2881–2884.

Engh, E., Clausen, O. P. F., Scholberg, A., Tollefsrud, A., and Purvis, K. (1992). *Intl. J. Androl.* **15,** 407–415.

Estop, A. M., Munne, S., Jost, L. K., and Evenson, D. P. (1993). *J. Androl.* **14,** 282–288.

Evenson, D. P. (1986). *In* "Clinical Cytometry" (M. Andreeff, ed.), pp. 350–367. New York Academy of Sciences, New York.

Evenson, D. P. (1989). *In* "Flow Cytometry: Advanced Research and Clinical Applications" (A. Yen, ed.), Vol. 1, pp. 217–246. CRC Press, Boca Raton, FL.

Evenson, D. P. (1990). *In* "Methods in Cell Biology" (Z. Darzynkiewicz and H. A. Crissman, eds.), Vol. 33, pp. 401–410. Academic Press, San Diego.

Evenson, D. P., and Jost, L. K. (1993). *Cell Proliferation* **26,** 147–159.

Evenson, D. P., and Melamed, M. R. (1983). *J. Histochem. Cytochem.* **31,** 248–253.

Evenson, D. P., Darzynkiewicz, Z., and Melamed, M. R. (1980). *Science* **240,** 1131–1133.

Evenson, D. P., Darzynkiewicz, Z., and Melamed, M. R. (1982). *J. Histochem. Cytochem.* **30,** 279–280.

Evenson, D. P., Higgins, P. H., Grueneberg, D., and Ballachey, B. (1985). *Cytometry* **6,** 238–253.

Evenson, D. P., Baer, R. K., Jost, L. K., and Gesch, R. W. (1986a). *Toxicol. Appl. Pharmacol.* **82,** 151–163.

Evenson, D. P., Darzynkiewicz, Z., Jost, L., Janca, F., and Ballachey, B. (1986b). *Cytometry* **7,** 45–53.

Evenson, D. P., Baer, R. K., and Jost, L. K. (1989a). *J. Environ. Mol. Mutagen.* **14,** 79–89.

Evenson, D. P., Janca, F. C., Jost, L. K., Baer, R. K., and Karabinus, D. S. (1989b). *J. Toxicol. Environ. Health* **28,** 67–80.

Evenson, D. P., Jost, L., Baer, R., Turner, T., and Schrader, S. (1991). *Reprod. Toxicol.* **5,** 115–125.

Evenson, D. P., De Angelis, P., Jost, L. K., Purvis, K., and Clausen, O. P. F. (1993a). *Congr. Int. Soc. Anal. Cytol., 16th,* Colorado Springs, CO. *Cytometry,* Suppl. 6, March 21–26, p. 73.
Evenson, D. P., Jost, L. K., and Baer, R. K. (1993b). *J. Environ. Mol. Mutagen.* **21,** 144–153.
Evenson, D. P., Jost, L. K., and Gandy, J. G. (1993c). *Reprod. Toxicol.* **7,** 297–304.
Garner, D. L., Gledhill, B. L., Pinkel, D., Lake, S., Stephenson, D., Van Dilla, M. A., and Johnson, L. A. (1983). *Biol. Reprod.* **28,** 312–321.
Johnson, L. A., Flook, J. P., and Hawk, H. W. (1989). *Biol. Reprod.* **41,** 199–203.
Jost, L. K., and Evenson, D. P. (1993). *Congr. Int. Soc. Anal. Cytology, 16th,* Colorado Springs, CO. *Cytometry,* Suppl. 6, March 21–26, p. 17.
Otto, F. J., and Hettwer, H. (1990). *Cell. Mol. Biol.* **36,** 225–232.
Spano, M., and Evenson, D. P. (1993). *Biol. Cell.* **78,** 53–62.
Ward, W. S., and Coffey, D. S. (1991). *Biol. Reprod.* **44,** 569–574.

CHAPTER 11

Fine-Needle Cytopuncture and Flow Cytometry in the Characterization of Human Tumors

Frédérique Spyratos* and Marianne Briffod†

* Department of Biology
† Department of Pathology
Centre René Huguenin
92211 St. Cloud
France

METHODS IN CELL BIOLOGY, VOL. 42

I. Introduction

Cytologic diagnosis by means of fine-needle cytopunctures (with or without aspiration) is increasingly used in most body sites to obtain evidence of malignancy and to facilitate therapeutic decision making, whether the lesion is superficial or deep, and palpable or not. When cytologic samples are obtained from various parts of a lesion, the cellular material withdrawn is generally highly representative of the tumor and provides prognostic information such as the grade of malignancy. Furthermore, fine-needle cytopuncture is an easy sampling method to obtain tumor cells for the study of cell behavior during treatment.

The cytologic samples are particularly well suited for FCM-DNA analyses in that isolated cells are easily obtained without time consuming tissue disruption (Johnson *et al.*, 1988; Fuhr *et al.*, 1992).

II. Materials and Methods

A. Fine-Needle Cytopuncture Technique

Even though cells can be obtained by fine-needle cytopuncture from nonpalpable lesions throughout the body under special guidance (ultrasound, stereotaxic mammography, computed tomography) and submitted to FCM-DNA analysis with good results, the technique is usually applied to palpable lesions.

Samples can be obtained from palpable lesions by fine-needle aspirations (FNA), but we prefer the fine-needle cytopuncture technique without aspiration, first described in 1982 (Briffod *et al.*, 1982; Spyratos *et al.*, 1987), that is simpler and provides samples of better quality than FNA.

1. Fine-Needle Cytopuncture without Aspiration for FCM Analysis

The sampling method without aspiration is relatively simple and requires only a plastic syringe and a 23-gauge needle. Without local anesthesia, after careful palpation and immobilization of the tumor, the needle is inserted into the lesion, rotated gently, and moved to-and-fro, in several directions according to the consistency of the lesion (Fig. 1). As soon as a droplet is seen in the hub, the needle is withdrawn. Part of the extracted material is carefully ejected onto slides using a syringe filled with air, smeared, and stained for routine cytologic examination. The remaining material in the needle is ejected into buffer for FCM analysis. Three cytopunctures are carried out in different areas of the tumor and the slides are numbered accordingly.

A rapid visual assessment of the sample quality is made at the time of smearing. Quick staining and microscopic examination during the procedure are performed in the case of nonpalpable lesions.

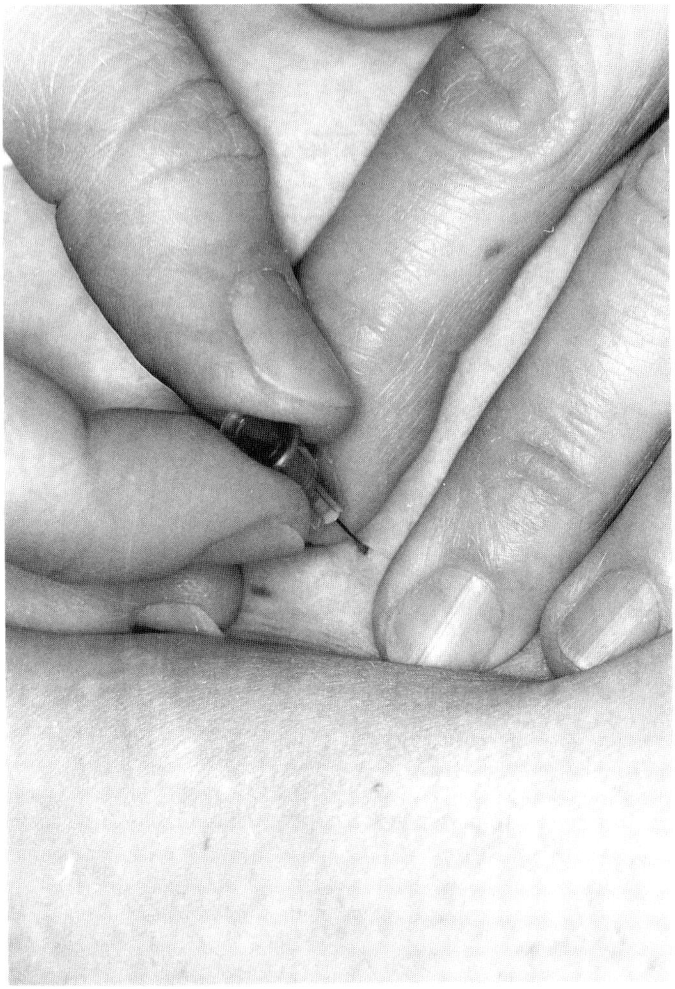

Fig. 1 Technique of cytopuncture with aspiration.

2. Advantages of the Procedure without Aspiration

This technique is not only simpler for the operator but also gives a better appreciation of the consistency of the mass being sampled. This reduces the number of insufficient samples. The needle, held directly between two fingers, provides tactile information unavailable when syringes or syringe holders are employed.

This procedure provides sufficient representative material for thorough characterization of the tumor. The cellular material obtained without aspiration is more concentrated, less traumatized, and less likely to be contaminated by blood. The presence of blood dilutes the tumor cells and may artifactually reduce the proliferation rate of the tumor. Excellence of cellular material is a prerequisite for diagnosis and adequate FCM information.

B. Conservation of Samples

For FCM study, the remaining material in the needle is ejected into a polypropylene tube containing citrate buffer with dimethyl sulfoxide (DMSO), according to Vindeløv *et al.* (1982a), and stored at −80°C until analysis. The FCM tube contains the samples from the different areas.

C. Preparation of Samples for FCM and FCM-DNA Analysis

The samples are thawed rapidly in a water bath at 37°C and rinsed before preparation for FCM analysis. Suspensions can then be stained according to classical protocols applied to nuclei or intact cells, after counting the cells. Samples obtained by our method produce high-quality FCM-DNA histograms with less blood and debris.

An additional measure of the quality of the histograms generated is the percentage of cells gated in for analysis. Because of the consistently high level of gated events (85% on average), a large amount of debris is rather a comment than a reason for rejection. The samples with the lowest gated percentage are associated with necrosis.

D. Standards

Several external and internal standards have been described in the literature. Ideally, with respect to the use of internal standards (added to the sample tube prior to fluorochrome staining), a sample should be run in duplicate, where one sample contains the internal control and the other does not. However, running duplicate samples is not always practical or feasible in the case of cytopunctures, because a large quantity of cells is necessary.

We suggest the use of hypodiploid references (chicken red blood cells, CRBC) for instrument alignment, setting the fluorescence threshold and verifying linearity. Unstimulated human PBL are used as an external DNA reference standard to adjust the diploid peak position. The peak position of CRBC can be somewhat variable in the low-channel regions, and CRBC are generally not recommended to identify the DNA diploid peak.

E. Comparison with Surgical Samples Obtained in the Same Patients

The reliability of FCM-DNA measurements on samples obtained by cytopuncture can be assessed by comparison with results obtained on tissues obtained at surgery from the same patients. In our experience with breast cancer

(Spyratos *et al.*, 1987), the aspirates contain a higher number of nondiploid nuclei than those obtained at surgery. Similar results have been reported by Greenebaum *et al.* for lung tumors (1984), Sneige *et al.* for lymph nodes (1991), and Russo *et al.* for liver hepatic tumors (1992). This may be due to mild cohesion of the tumor cells, which can therefore be more easily collected by cytopuncture. Moreover, the multiple samples and sampling in different areas of the tumor performed with the cytopuncture technique indicate that FCM samples from fine needle can be more representative of the overall tissue than excisional biopsies.

III. Critical Aspects

A. Cellularity

A good sampling technique is a prerequisite for FCM-DNA analysis and this takes experience. The best results are obtained when the cytopathologist routinely performs the sampling procedure. This should not be considered as a simple test able to be done by people with little experience.

When samplings are correctly performed, cytopuncture samples are adequate for FCM-DNA analysis in nearly 90% of cases (Spyratos *et al.*, 1987; Remvikos *et al.*, 1988). In such cases, the number of cells varies from thousands to several million. It is difficult to state a minimum number of cells needed: more is always better. Analyzable DNA histograms have been obtained on samples containing as few as 50,000 total cells (consisting of almost all tumor cells), but this can be influenced by cell type and integrity of the nuclear membrane.

B. Cytologic Quality Control

Quality control of samples by cytologic evaluation of cellular components on routine cytologic staining (May-Grünwald-Giemsa or Papanicolaou) is a necessity for the correct interpretation of DNA histograms. Confrontation of FCM data and cytologic examination gives a good estimation of the cellular components submitted to FCM analysis and permits the ratio between benign and malignant cells in the FCM samples to be evaluated. Indeed, single-parameter FCM cannot distinguish between benign and malignant cells in the case of diploid tumors. Furthermore, this control is very useful for assessing the malignancy of diploid populations, especially in the case of a small tetraploid peak which is difficult to differentiate from a diploid G_2M (after minimization of doublets by using "bit map" gating on integrated versus peak red fluorescence) or when benign inflammatory cells are numerous. It is also of particular interest to evaluate FCM-DNA changes more precisely during treatment.

The DNA histogram is considered uninterpretable when the cytological examination does not show tumor cells. The presence of necrosis should be recorded.

standard (lymphocytes) are used to establish diploid peak position. The Feulgen reaction applied to unstained slides permits nonmalignant cells to be eliminated visually. Indeed, although the Feulgen reaction can be done on archival stained slides after discoloration, in our experience better histogram quality is obtained by using unstained slides: previous staining induces a high level of background and broader peaks.

1. Advantages and Disadvantages

FCM-DNA analysis provides histograms of better resolution and a more reliable peak position and is better for detecting near-diploid tumors. Since FCM analyzes much larger numbers of cells, the results have more statistical significance. Furthermore, it is a rapid method (a few minutes to analyze 10,000 cells by FCM versus 30 to 60 min per slide for image analysis). Image analysis, by morphologic correlation and selection of malignant cells, permits associated tumor diploid peaks to be detected in multiploid tumors (Fig. 2). In these cases, analysis of slides from different areas of a tumor is frequently helpful to identify different populations. Moreover, image analysis enables the detection of small aneuploid populations diluted by a much larger diploid population, which are frequently missed by FCM (Fig. 3). Image analysis is more effective in detecting highly aneuploid populations to the right tail of the histogram (including endoreduplicated cells), but the CVs are approximately twice FCM values, making it difficult in some cases to estimate the S phase; it is also highly time consuming. In fact, these two methods are complementary in evaluating DNA content and proliferation.

B. Dual-Parameter FCM

The addition of antibodies recognizing cytokeratins may be used in an attempt to discriminate tumor from host cells to better characterize diploid and tetraploid epithelial tumors. Double labeling (keratin + DNA) may improve the sensitivity of DNA histogram classification and the evaluation of the S phase fraction in various cases: in particular, it permits the identification of the epithelial character of the diploid population. For reasons of cost effectiveness, it may be best to perform double-labeling after initial single-parameter analysis.

Histograms with broad CVs and/or high G_2M are good candidates for the double-labeling technique (Fig. 4). The overall gain in assessing DNA aneuploidy has been estimated by Van der Linden (1992) as 14% in a series of 311 primary epithelial tumors. The choice of the keratin is also important. The only limitation is the number of cells required for the controls.

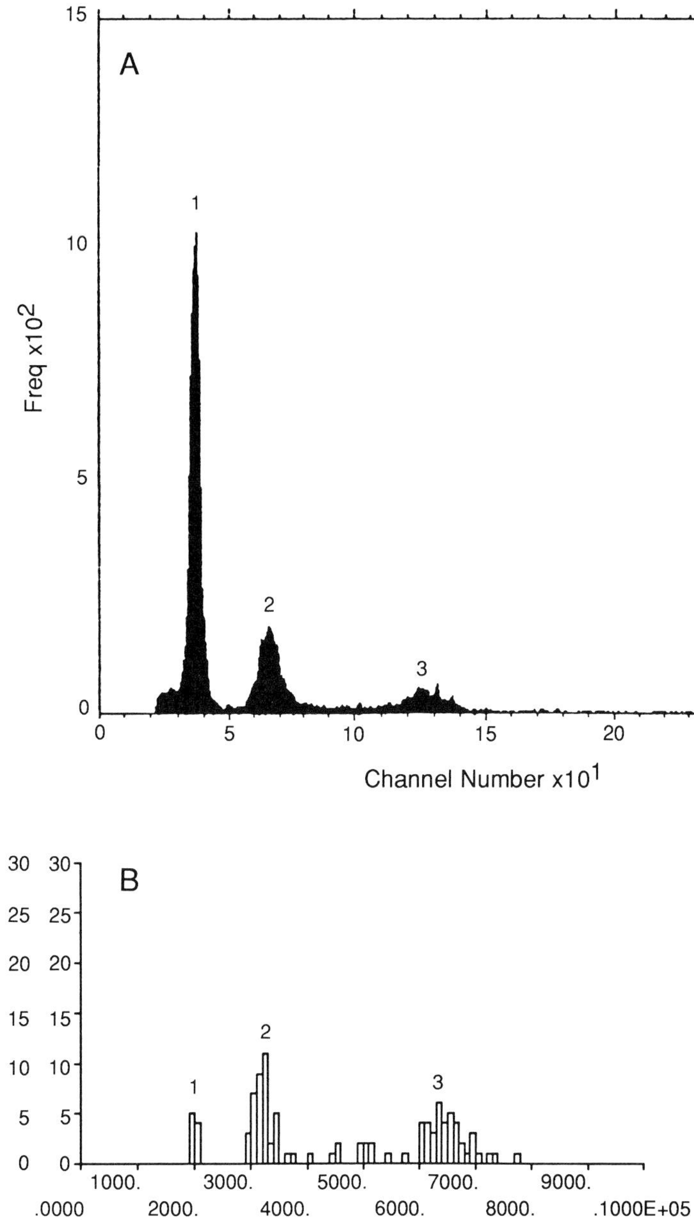

Fig. 3 Comparison of DNA histograms obtained by flow cytometry (A) and image analysis (B) from a metastatic axillary lymph node. The diploid peak consisted of lymphocytes (peak 1). Image analysis: lymphocytes used as internal standards; peak 1, FCM, G_0/G_1 diploid population; peak 2, G_0/G_1 aneuploid population; peak 3, G_2M aneuploid population.

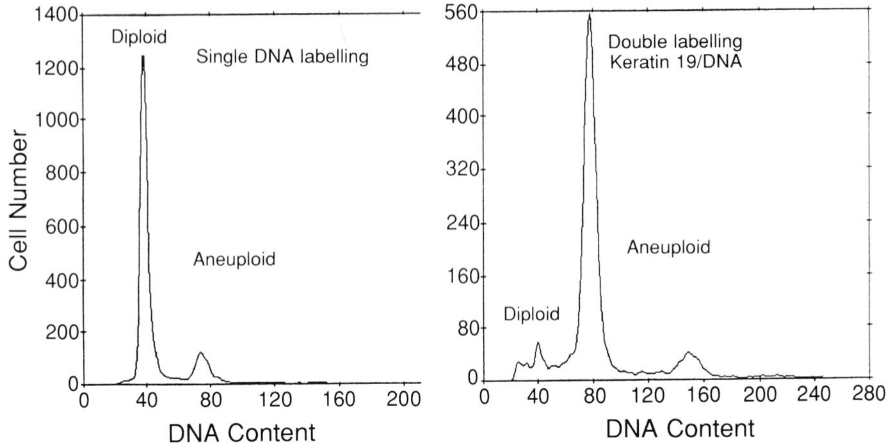

Fig. 4 FCM-DNA histogram of a metastatic axillary lymph node from a breast cancer patient with "questionable" ploidy status in FCM (small population of cells in the tetraploid region). The keratin 19-positive cells clearly revealed the tetraploidy of the tumor population.

C. Others Approaches to Evaluate Proliferation

The different methods used to quantify cell proliferation do not necessarily measure the same thing. These methods still are not standardized, although this will be required to minimize intra- and interlaboratory variations. Comparisons of these different approaches is still incomplete and they cannot be considered as interchangeable.

A number of monoclonal antibodies are now available that recognize cellular proteins associated with proliferation (Ki-67, P105, P120, PCNA, etc.) and which can be measured by FCM or immunocytochemical or immunohistochemical techniques (Linden *et al.*, 1992). This may overcome several problems associated with the interpretation of FCM-DNA histograms. The monoclonal antibody Ki-67 identifies an antigen present in the late G_1, S, and G_2M phases of the cell cycle that resting cells do not express. This pattern of cell staining identifies the growth fraction of the tumor and this contrasts with other techniques. Ki-67 staining is heterogeneous in most tumors. The correlation between SPF and Ki-67 staining appears to be better in tumors with a high rather than a low percentage of S phase cells.

Bromodeoxyuridine labeling indices directly reflect the proportion of cells in S phase only. Since bromodeoxyuridine (BrdUrd), like thymidine, is incorporated into nuclear DNA during DNA synthesis, the labeling indices obtained by incorporation of BrdUrd *in vivo* or *in vitro* provide a rough estimate of proliferative activity.

Biparametric BrdUrd/DNA FCM study after *in vitro* BrdUrd incorporation by breast tumor cells obtained by cytopuncture has also been proposed

(Remvikos *et al.*, 1991) to increase the proportion of patients for whom proliferative status could be determined.

Information on a single compartment of the cell cycle may be considered insufficient. The estimation of potential doubling time after injection of BrdUrd *in vivo* (Wilson, 1991) may prove to be fundamental to treatment schedule, particularly radiation therapy. This method is generally applied to tumor biopsies, but we have recently used cytopunctures of breast carcinomas, with better results than those obtained using biopsies (Fig. 5).

V. Clinical Applications

Some authors have used *ex vivo* cytopuncture of the excised tumor as an alternative to mince the tissue. The interest of fine-needle cytopunctures performed *in vivo* is to obtain information at the time of diagnosis or during treatment (Spyratos *et al.*, 1987; Remvikos *et al.*, 1988).

A. Different Locations

Although most applications concern breast cancer (Fuhr *et al.*, 1992), which is frequent and more accessible to cytopuncture, some studies have been undertaken on primary lesions of thyroid (Greenebaum *et al.*, 1985), prostate (Tribukait, 1991), lung (Vindeløv *et al.*, 1982b), pancreas, or liver (Russo *et al.*, 1992) and on metastatic lesions (Spyratos *et al.*, 1992).

B. Value for the Diagnosis of Malignancy

In malignant solid tumors, FCM-DNA analysis of cytopunctures is of little interest for diagnosis of malignancy. In general, it does not improve the cytologic diagnosis and, in our experience with breast and thyroid cancers, we have never found aneuploid populations when cytologic results were false negative or even suspicious. Nevertheless, in thyroid gland, in which the cytologic and even histologic differential diagnosis between follicular adenoma and follicular carcinoma is frequently difficult, the presence of an aneuploid population, reported in some adenomas (Greenebaum *et al.*, 1985), may lead to tumor excision when an aneuploid population is found in FCM cytopuncture samples.

C. Value for Prognosis

Published data suggest that both ploidy and S phase fraction can provide information that can be used in the clinical management of various tumors. Diploid tumors have a slightly lower risk of relapse than aneuploid tumors, and low S phase tumors have a more favorable prognosis regardless of ploidy status. Most results concern assays performed on frozen or formalin-fixed paraffin-

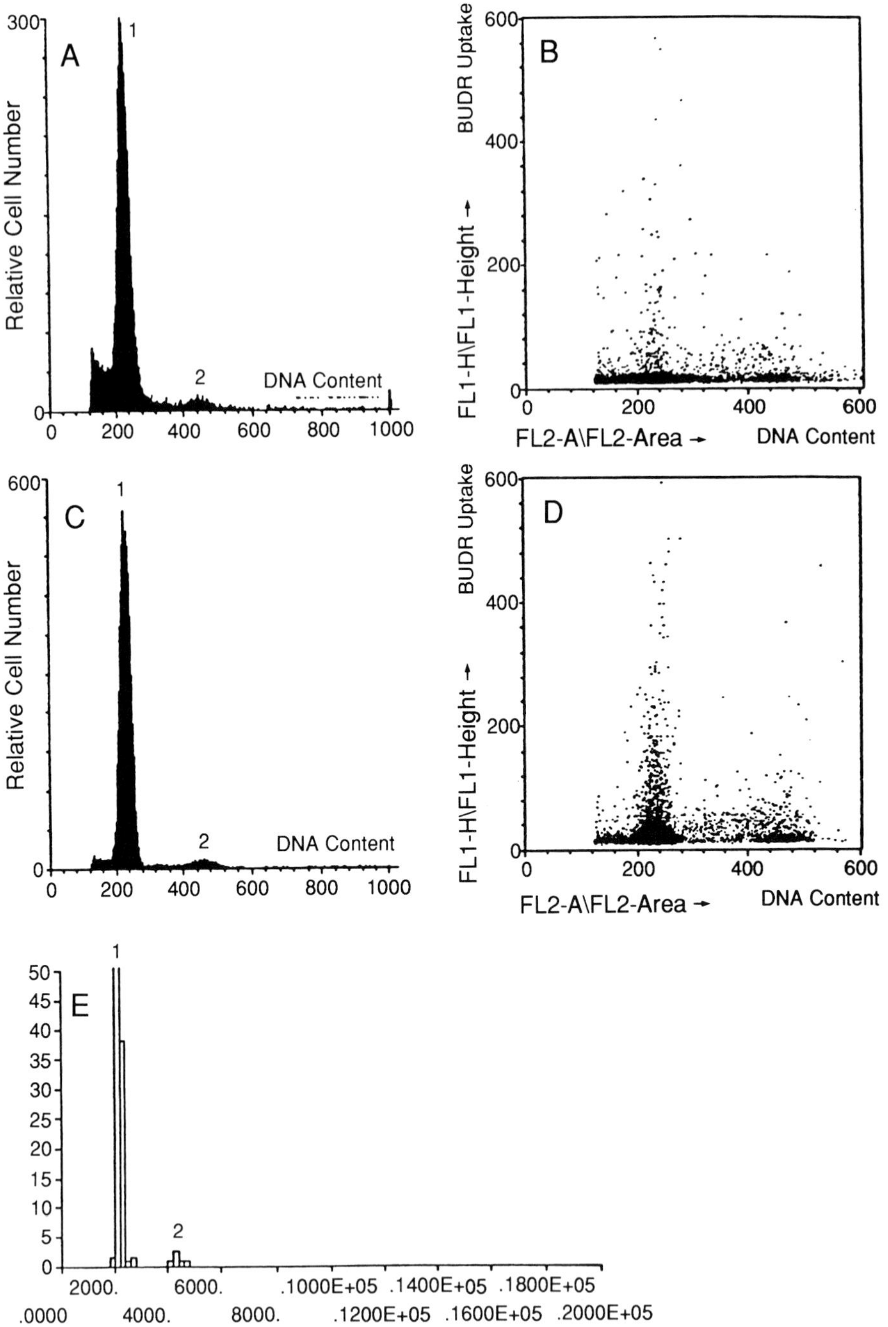

Fig. 5 FCM-DNA profiles and BrdUrd vs DNA cytograms of a diploid breast tumor (1, G_0/G_1; 2, G_2M) obtained 5 hr after injection of BrdUrd showing cell-cycle redistribution of BrdUrd. (A,B) Surgical sample difficult to interpret due to excessive background. (C,D) *In vivo* cytopuncture of the same lesion. (E) Image analysis after Feulgen reaction confirming the malignancy of the diploid population.

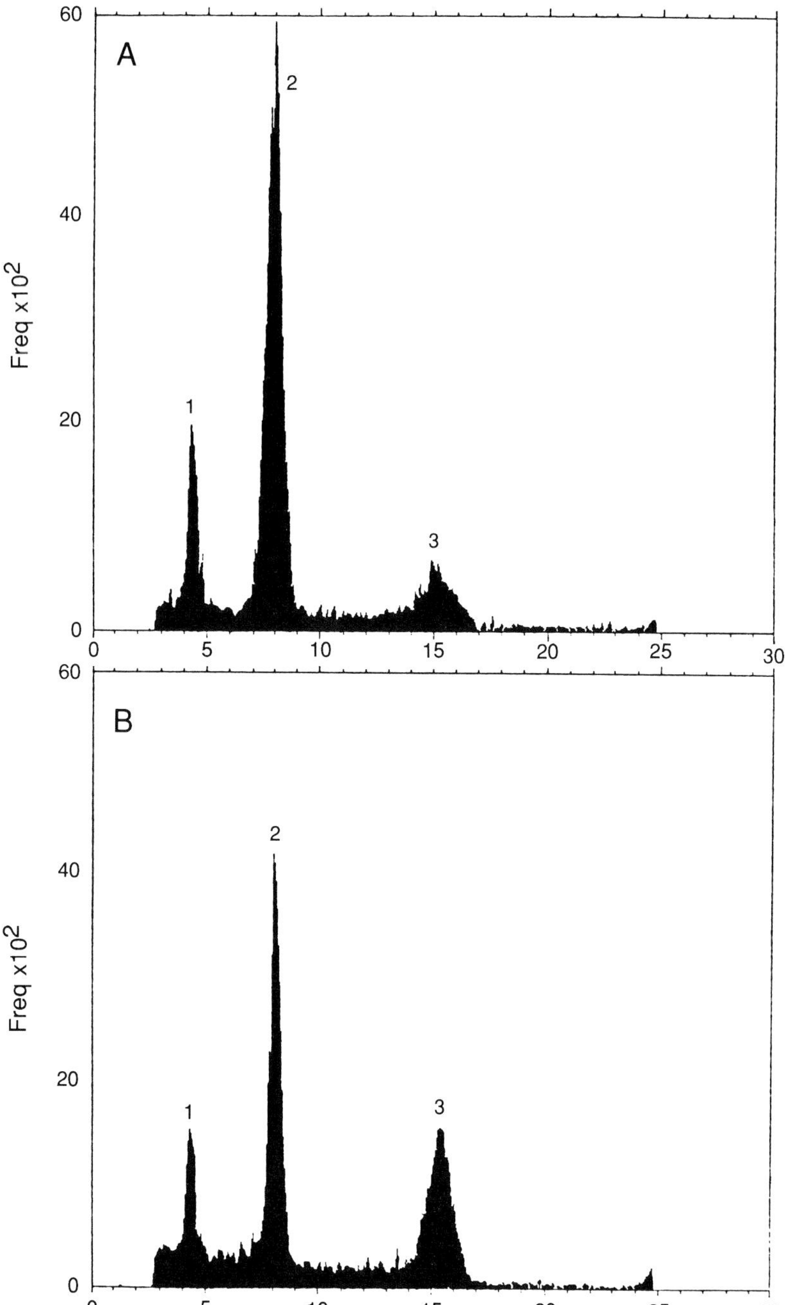

Fig. 6 FCM-DNA histogram of an aneuploid breast carcinoma before treatment (A) and after one cycle of chemotherapy (B). In (B), there is an increase in the aneuploid G_2M peak. Peak 1, G_0/G_1 diploid population; peak 2, G_0/G_1 aneuploid population; peak 3, G_2M aneuploid population.

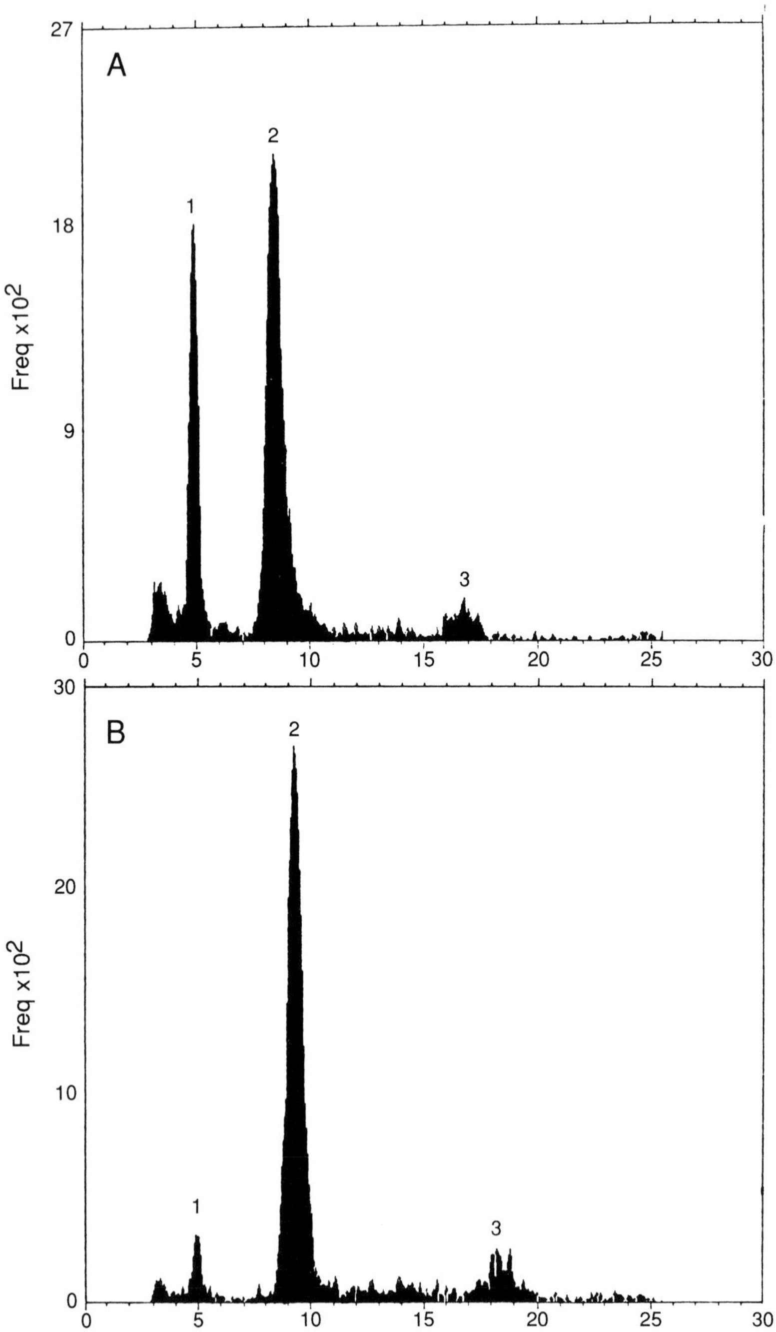

Fig. 7 FCM-DNA histogram of an aneuploid breast carcinoma with no change in FCM-DNA parameters before treatment (A) and after one cycle of chemotherapy (B). Peak 1: G_0/G_1 diploid population; peak 2, G_0/G_1 aneuploid population; peak 3, G_2M aneuploid population.

embedded tissues. However, improvements in methodology, data analysis, and quality controls are required before FCM analysis can be applied to routine determination of the prognosis.

Results of prospective studies on the prognostic value of FCM-DNA content and proliferation performed on cytopunctures are not yet available. The better quality of DNA histograms obtained by cytopuncture, leading to more reliable results, could better define the exact role of these parameters in the prognosis.

D. Evaluation of Tumor Response to Treatment

Early indicators of tumor response to treatment (chemotherapy, radiotherapy, or hormonotherapy) are needed by clinicians. By analogy to *in vitro* experiments, modifications in the cell-cycle distribution are to be expected (for instance accumulation of cells in S or G_2M). This has been shown in the pioneering work of Vindeløv *et al.* (1982b) and confirmed in breast cancer. Some studies have demonstrated, particularly for primary breast carcinomas (Briffod *et al.*, 1989; Remvikos *et al.*, 1989; O'Reilly *et al.*, 1992; Spyratos *et al.*, 1992), that FCM-DNA analyses of cytopuncture material can provide useful information regarding the likely tumor response to primary chemotherapy. Tumors with nondiploid DNA content and those with high S phase show a significantly better response to chemotherapy than diploid and low S phase tumors. Furthermore, since cytopuncture is an easy sampling method (permitting repeated sampling), it is particularly suited to studying tumor cell behavior during treatment.

Sequential flow DNA analyses have shown that tumors presenting early changes in DNA histograms (Fig. 6) (after one cycle of chemotherapy) have a significantly better response than other tumors (Fig. 7) (Briffod *et al.*, 1992; Spyratos *et al.*, 1992).

Similar findings have been reported using sequential cytopuncture and FCM-DNA analyses during radiation therapy (Zbieranowski *et al.*, 1992).

Sequential samplings can be very useful even if difficulties have been encountered with some multiploid tumors. Indeed, heterogeneity may still cause problems in assessing cellular DNA characteristics in a minority of cancers. If cytopuncture samples are not necessarily representative of the whole tumor mass, samples from different areas of the same tumor enhance the accuracy of DNA assessment. When Mullen and Miller (1989) reported changes in DNA profiles at 1- or 3-week intervals in untreated tumors, some of these discrepancies may have been linked to a subpopulation missed during sampling.

VI. Conclusion

The measurement of proliferation on cytological samples presents a number of advantages. Fine-needle cytopunctures can be obtained from patients at any stage of the disease, including inoperable tumors. Information obtained at the time of diagnosis can be useful for therapeutic decision making. The procedure is easily performed, allows rapid interpretation, and is highly accurate in the hands of a skilled and proficient cytopathologist.

References

Auer, G. U., Askenstein, U., Erhardt, K., Fallenius, A., and Zetterberg, A. (1987). *Anal. Quant. Cytol. Histol.* **9,** 138–146.
Briffod, M., Gentile, A., and Hebert, H. (1982). *Acta Cytol.* **26,** 195–200.
Briffod, M., Spyratos, F., Tubiana-Hulin, M., Pallud, C., Mayras, C., Filleul, A., and Rouëssé, J. (1989). *Cancer (Philadelphia)* **63,** 631–637.
Briffod, M., Spyratos, F., Hacène, K., Tubiana-Hulin, M., Pallud, C., Gilles, F., and Rouëssé, J. (1992). *Cytometry* **13,** 250–258.
Fuhr, J. E., Kattine, A. A., and Nelson, H. S. (1992). *J. Natl. Cancer Inst.* **84,** 1272–1276.
Greenebaum, E., Koss, L. G., Sherman, A. B., and Elequin, F. (1984). *Am. J. Clin. Pathol.* **82,** 559–564.
Greenebaum, E., Koss, L. G., Sherman, A. B., Elequin, F., and Silver, C. E. (1985). *Cancer (Philadelphia)* **56,** 2011–2018.
Johnson, T. S., Katz, R. L., and Pershouse, M. (1988). *Anal. Quant. Cytol. Histol.* **10,** 423–458.
Lee, A. K. C., Willey, B., Dugan, J. M., Hamilton, W. H., Loda, M., Heatley, G. J., Cook, L., and Silverman, M. L. (1992). *Pathol. Res. Pract.* **188,** 428–432.
Linden, M. D., Torres, F. X., Kubus, J., and Zarbo, R. J. (1992). *Am. J. Clin. Pathol.* **97,** S4–S13.
Mullen, P., and Miller, W. R. (1989). *Br. J. Cancer* **59,** 688–691.
O'Reilly, S. M., Camplejohn, R. S., Rubens, R. D., and Richards, M. A. (1992). *Eur. J. Cancer* **28,** 681–683.
Remvikos, R., Magdelénat, H., and Zajdela, A. (1988). *Cancer (Philadelphia)* **61,** 1629–1634.
Remvikos, Y., Beuzeboc, P., Zajdela, A., Voillemot, N., Magdelénat, H., and Pouillart, P. (1989). *J. Natl. Cancer Inst.* **81,** 1383–1387.
Remvikos, Y., Vielh, P., Padoy, E., Benyahia, B., Voillemot, N., and Magdelénat, H. (1991). *Br. J. Cancer* **64,** 501–507.
Russo, A., Bazan, V., Plaja, S., Cajozzo, M., and Bazan, P. (1992). *J. Surg. Oncol.* **51,** 26–32.
Sneige, N., Dekmezian, R., El-Naggar, A., and Manning, J. (1991). *Cancer (Philadelphia)* **67,** 1003–1007.
Spyratos, F., Briffod, M., Gentile, A., Brunet, M., Brault, C., and Desplaces, A. (1987). *Anal. Quant. Cytol. Histol.* **9**(6), 485–494.
Spyratos, F., Briffod, M., Tubiana-Hulin, M., Andrieu, C., Mayras, C., Pallud, C., Lasry, S., and Rouëssé, J. (1992). *Cancer (Philadelphia)* **69,** 470–475.
Tribukait, B. (1991). *Acta Oncol.* **30**(2), 187–192.
Van der Linden, J. C., Herman, C. J., Boenders, J. G. C., van de Sandt, and Lindeman, J. (1992). *Cytometry* **13,** 163–168.
Vindeløv, L. L., Christensen, I. J., Keiding, N., Spang-Thomsen, M., and Nissen, N. I. (1982a). *Cytometry* **3,** 317–322.
Vindeløv, L. L., Hansen, H. H., Gersel, A., Hirsch, F. R., and Nissen, N. I. (1982b). *Cancer Res.* **42,** 2499–2505.
Wilson, G. D. (1991). *Acta Oncol.* **30**(8), 903–910.
Zbieranowski, I., Le Riche, J. C., Jackson, S. M., and Olivotto, I. (1992). *Anal. Cell. Pathol.* **4,** 13–24.

Note Added in Proof

FCM-DNA histograms were obtained on an EPICS C (Coultronics, Hialeah, FL) and on a FACSTAR-Plus (Becton Dickinson, Mountain View, CA). Image analysis histograms were obtained on a SAMBA 2005, TITN, France).

CHAPTER 12

Functional NK Assays Using Flow Cytometry

Stefano Papa★ and Massimo Valentini†

★ Istituto di Scienze Morfologiche
Universitá di Urbino
61029 Urbino, Italy

† Laboratorio Analisi
San Salvatore Hospital
61100 Pesaro, Italy

METHODS IN CELL BIOLOGY, VOL. 42

I. Introduction

The immune system *in vivo* displays basically two types of cellular cytotoxicity that are, to a certain extent, mediated by different effectors that act in different ways. These two types of cells are: (i) cytotoxic T lymphocytes (CTL), which mediate an MHC-restricted lytic activity that needs to be preceded by sensitization toward the target; and (ii) natural killer (NK) cells, which display an MHC-unrestricted, nonadaptive cytotoxicity and are therefore termed "natural." Unlike CTL, NK cells are not clonally distributed, do not show restriction for MHC products at the target cell surface, and do not display any immunological memory. They represent a distinct lineage from T and B lymphocytes and originate directly from bone marrow. They are defined as $CD3^-$ large granular lymphocytes expressing CD16 and CD56 surface molecules. Note, however, that this apparently rigid classical distinction between CTL and NK cells is becoming less definite with the evidence of natural cytotoxic activities displayed *in vitro* by $CD16^-$ effectors. However, the NK function is defined as the ability of freshly isolated PBL to lyse specific targets (K562 cells for human NK, YAC cells for mouse NK) *in vitro*, without prior sensitization, in short-term assays. Since NK cells bear the FcR, they are also able *in vitro* to perform an ADCC (antibody-dependent cellular cytotoxicity) against antibody-coated target cells (P815). The cytotoxic activity of both NK cells and CTL is preceded by an effector–target cell contact, a molecular recognition, and binding. It has been also demonstrated that both NK and CTL are able to kill targets "in the distance" by secreting toxic factors, but this is not their main function. On the contrary, both NK cells and CTL have a cytoplasmic apparatus of granuli containing cytotoxic factors that are synthesized to be secreted in the intercellular spaces after binding to the target.

During this past decade a number of methodologies (based on optical microscopy or flow cytometry) have been developed to quantitatively determine the lytic activity of lymphocytes against specific targets (Papa *et al.*, 1988; Kimberley *et al.*, 1986) and/or the recognition and qualitative evaluation of cells forming conjugates with tumor targets (binding) (Segal and Stephany, 1984; Storkus *et al.*, 1986; Vitale *et al.*, 1992).

These techniques use single-cell assays (SCA), and thus, are more reliable and controlled tests compared to total cell assays (TCA) (e.g., ^{51}Cr release assay). Unlike TCAs, SCAs distinguish among effector cells, living and dead (killed) targets, and conjugates. This distinction can be achieved either by optical discrimination of the cell types (light microscopy) or by evaluation of their scattering and fluorescent properties. In particular, light scattering properties (or propidium iodide incorporation) of K562 can be used to detect cell death whereas specific autofluorescence can be used to detect the conjugates. Flow cytometry is therefore able to measure both the killing activity and the binding capacity of NK cells to K562. Not all the published methodologies use these parameters to identify dead target or conjugates (Table I).

Table I
Application of Cytometry to the Analysis of Cell–Cell Adhesion and Cytotoxicity

Reference	Instrument	Parameters employed to detect killed targets (T,E)	Parameters employed to detect conjugates (T,E)	Effector cells
Cavarec *et al.* (1990)	Flow cytometry		Hydroethidine (red)-stained T CFDAME (green)-stained E	LAK
Garcia-Pennarubia *et al.* (1989)	Light microscopy		Viewed by phase contrast	NK
Grimm and Bonavida (1979)	Light microscopy	Trypan blue staining		CTL
Kimberley *et al.* (1986)	Flow cytometry	Release of FDA from T		NK
Levow *et al.* (1986)	Flow cytometry		Hoechst 33342-stained E (blue) FITC-stained T (green)	CTL
Papa *et al.* (1988)	Flow cytometry	PI (red) staining		NK
Rubin *et al.* (1982)	Light microscopy	Trypan blue staining		NK
Segal and Stephany *et al.* (1984)	Flow cytometry		FDA (green)-stained T Rodamine ICT (red)-stained E	Spleen cells
Storkus *et al.* (1986)	Flow cytometry		FDS (green)-stained E Hoechst (blue)-stained T	NK
Vargas-Cortes *et al.* (1983)	Light microscopy	Trypan blue staining		NK
Vitale *et al.* (1989a)	Flow cytometry	SSC plus PI (red) staining		NK
Vitale *et al.* (1992)	Flow cytometry		Staining of conjugates (E) with fluorescent mAbs (G,O,R)	NK
Zarcone *et al.* (1986)	Flow cytometry	PI (red) staining		NK

Light scattering has certain advantages in avoiding any possible interference with lytic activity due to previous sample preparations, while the use of K562 autofluorescence allows the conjugates to be detected and the lymphocyte subset to which the effector belongs to be identified concurrently.

These techniques clearly discriminate effectors from targets and conjugates. While this is obtained on the scattering dot plot for the evaluation of cytotoxicity, in the case of conjugate evaluation this discrimination is made on the fluorescence dot plot, making this methodology more flexible for use with different targets. The cytotoxic assay (Papa *et al.*, 1988; Vitale *et al.*, 1989a) is performed after side scatter has been gated on the targets, and the function is evaluated in the fluorescence cytogram by propidium iodide (PI) staining of dead targets.

Also, the "binding assay" permits the different effectors attached to the targets (the conjugates) to be identified through the labeling of effectors with different monoclonal antibodies. Therefore this technique resembles a typical dual- or triple-fluorescence analysis as in a normal flow phenotype.

II. Application

These methodologies have been set up, primarily, to investigate non-MHC-restricted cytotoxicity (Herberman *et al.*, 1986). Their applicability is not restricted to this area and can be extended to ADCC or other adhesion mechanisms. Our experience demonstrates its usefulness (data not shown) with different targets, provided that the effector has clearly distinguishable scatter signals (cytotoxic assay) or autofluorescence (binding assay). This condition does not represent a real limitation, since most of the targets used *in vitro* are tumor cells that usually have different scattering and autofluorescence characteristic from lymphocytes.

A review of the literature on this matter is given in Table I.

III. Materials

A. Cytotoxicity Assay (SCCA)

A stock solution is prepared with 10 μg/ml of propidium iodide (Sigma Chemical, Co) in RPMI 1640 (Gibco, BRL) supplemented with 10% FCS (Flow Laboratories) and stored in aliquots at 4°C.

The propidium iodide is used, in this case, for the detection of dead cells, so that its dim degradation will not affect the test. Note that this product is known to be a tumorigenic agent.

The working solution is 0.5 μg/ml PI in RPMI 1640 supplemented with 10% FCS.

B. Binding Assay

This technique does not involve the use of any chemical; only freshly isolated (or cultured) lymphocytes and target cells cultured under standard conditions are needed. The conjugates will be stained by fluorescent monoclonal antibodies (mAbs). The choice of mabs depends only on the experimental protocol.

IV. Cell Preparation and Staining

A. Preparation of Cells for Binding and Killing Assays

1. Dilute 1 : 1 human blood with PBS.

2. Collect mononuclear cells using a Ficoll-Hypaque or Percoll density gradient (S = 1.077).

3. Wash twice with RPMI 1640 supplemented with 10% FCS and deplete residual monocytes by adhesion onto plastic flask walls for 30 min at 37°C and/or carbonyl iron phagocytosis.

4. Target cell lines are usually maintained in continuous culture in RPMI 1640 + 10% FCS, penicillin 100 U/ml, and streptomycin 0.1 mg/ml.

5. Harvest target cells (K562 or other) during the log phase of growth in order to have the minimum number of dead targets (if target cells grow adherent, use them before confluency).

B. Cytotoxic Procedure

6. Wash target and effector cells in PBS and mix them in an effector : target (E : T) ratio ranging from 6.25 : 1 to 25 : 1; in order to obtain 10,000 events during flow analysis 50,000 targets must be mixed for every *E*/*T* ratio.

7. Centrifuge the *E* : *T* mixed samples for 7 min at 250*g* at room temperature and gently place samples in a water bath at 37°C for 10 min to promote cell–cell adhesion.

8. Gently resuspend conjugates, to avoid recycling of effector cells, and bring them to a concentration of 10^5 cells/ml with stock solution reaching the final PI concentration of 0.5 μg/ml (working solution: dilute propidium iodide from stock solution with RPMI 1640 + 10% FCS).

9. Incubate samples for 1.5 hr at 37°C in a 5% CO_2 atmosphere, and then cool samples at 4°C. At the end of incubation cells must be brought back to a concentration of 10^6/ml by centrifuging for 7 min at 250*g* at 4°C and analyzed in flow cytometry.

10. Control samples, targets alone, must be incubated for 1.5 hr in working solution at a concentration of 10^5/ml in order to monitor spontaneous target death.

C. Binding Procedure

6. Mix target and effector cells in an $E:T$ ratio of 1 : 1. Seed a total of 2×10^5 cells in each tube, in RPMI 1640.

7. Centrifuge for 7 min at 250*g* at room temperature and incubate 10 min in water bath at 37°C to promote conjugate formation. Then leave 30 min at 4°C to prevent lytic activation.

8. Gently resuspend samples and stain with fluorochrome-labeled monoclonal antibodies (FITC, PE, APC, and/or others), which are known not to be present on the target cell surface for 20 min at 4°C. Be particularly gentle in this phase of the procedure to avoid disruption of the conjugates. It is preferable to use directly stained mAbs to reduce the number of sample washes. Indirect immunostaining yields good results as well.

9. Wash once in PBS and perform flow analysis.

V. Critical Aspects of the Procedures

A. Depletion of Monocytes

Both methodologies require an accurate depletion of monocytes. Monocytes will affect the cytotoxicity assay, even if the green autofluorescence displayed by targets (when present), which is the case of K562, could prevent their contamination in counting, because the scattering pattern of monocytes is relatively similar to that of little targets (Figs. 1A and 1B). This also depends on

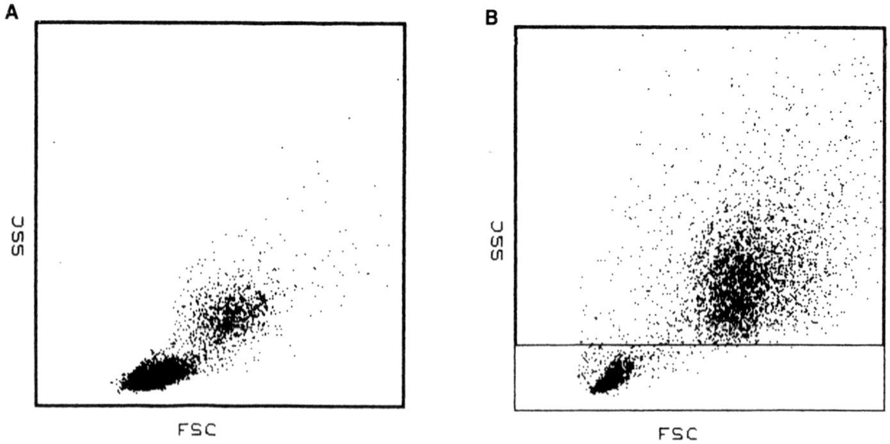

Fig. 1 (A) Scattering distribution of lymphocytes and monocytes in a mononuclear cell suspension not cleared by monocytes. (B) Scattering distribution of lymphocytes and K562 in a 1 : 1 ratio. The line corresponds to the SSC threshold level. Note the overlap between the edges of target and the monocyte areas. (FSC, forward scatter; SSC, side scatter).

the $E:T$ ratio employed in the two techniques. Those used in the SCA produce a relatively high contamination of monocytes if they are not cleared from the mononuclear cell preparation, while the 1 : 1 ratio of the binding assay will not interfere with counting if the mAbs or other fluorescent markers used to identify effectors are not coexpressed on the monocyte membrane (Fig. 2B).

B. Spontaneous Death of Target Cells

Regarding targets, when cell lines are employed it is necessary to work with cells in log phase of growth thus minimizing the amount of spontaneous death present in the samples. Dead cells will affect both assays: the SCCA, due to the high level of dead cells present in controls, and the binding assay, due to the increased autofluorescence of dead cells. This last problem can be avoided by gating on living cells on the scattering cytogram (Fig. 3A).

C. Instrument Quality Control

As with other flow techniques, these two need an accurate setting of the cytometer, since all the parameters are equally essential, for the exploitation of the assays. Moreover when these samples are run on analyzers (such as FACScan) remember to correct the amplification of scatter signals.

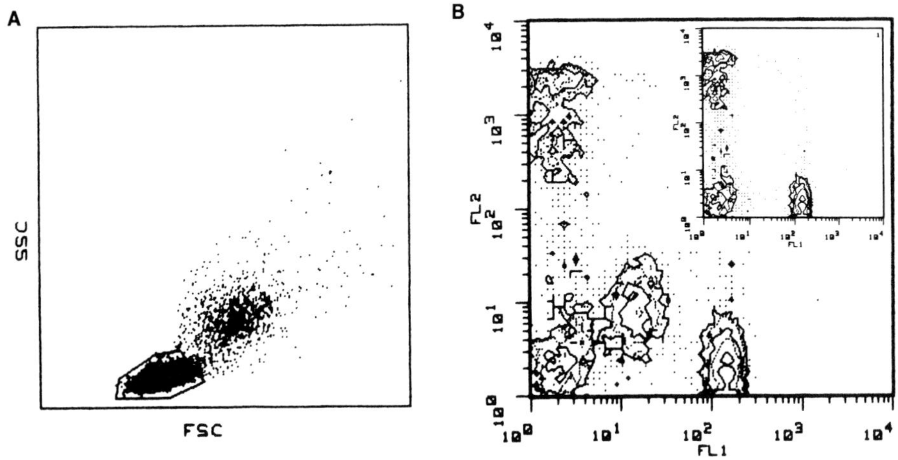

Fig. 2 Autofluorescence displayed by monocytes. (A) Scatter profile with gate drawn on the lymphoid area (Vitale, 1987). (B) Fluorescence dot plot of the whole sample previously incubated with anti-CD8 PE and anti-CD4 FITC. (Inset) Cytogram obtained after activation of the lymphoid scatter gate (A). (FL1, green fluorescence; FL2, orange fluorescence).

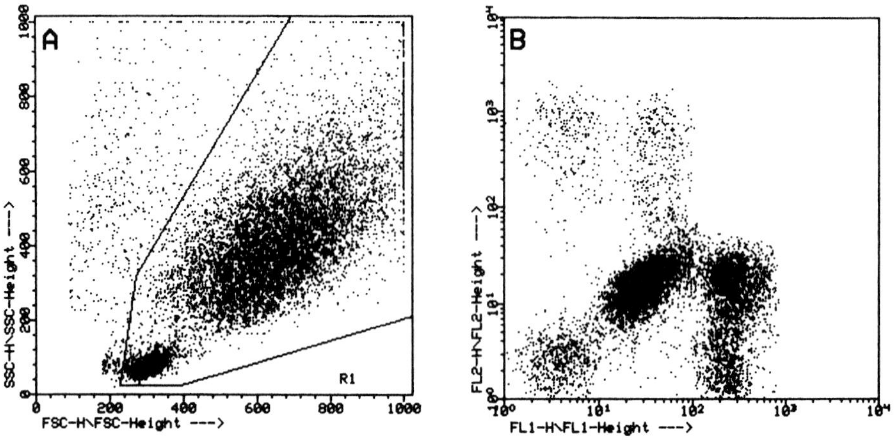

Fig. 3 (A) Scatter gate drawn in the binding assay to avoid contamination by conjugates made of dead targets. (B) Fluorescence cytogram of a binding assay after scatter gating as in A. (FL1, green-labeled lymphocytes with anti-CD3 FITC; FL2, orange-labeled mononuclear cells with anti-CD16 PE).

VI. Standardization

A. Scatter Optimization

For setting the proper signal amplification a 1 : 1 ($E:T$) mixed sample must be run in order to optimize the cytogram position of effectors and targets as shown in Fig. 3A. This is of the utmost importance in the case of the cytotoxic assay in which the position of the side scatter (SSC) threshold is crucial to avoid contamination with effectors in the fluorescence matrix.

B. Fluorescence Compensation

It is preferable to proceed with a dual-parameter compensation setting made by means of directly labeled cells (for example, lymphocytes labeled with anti-CD8 PE and anti-CD4 (FITC) (Fig. 2B) obtained by the mononuclear samples under investigation. In this way, the sample can be used for setting both the proper signal amplification for the scatter and then for the amplification and compensation of the fluorescence channels (PE for red and FITC for green).

VII. Instruments

For both assays, the set of filters employed is composed of a 488-nm BP on the orthogonal scatter, a 530/15-nm BP filter on the green channel, and a 585/30-nm BP on the red channel.

The optics and hydrodynamics of analyzers such as FACScan, ORTHO Absolute, and Coulter Profile are perfectly optimized to perform these assays. Moreover, as mentioned in the standardization section, compensation should be employed in order to perform accurate calculations.

A. Cytotoxicity Assay

In order to avoid clodging of the nozzle during analysis, sorters must be equipped with nozzles having a diameter of not less than 90 μm due to the high diameter of target cells.

The laser should be tuned at 200 mW in the 488-nm line which is the most typical and which produces in cells maximum autofluorescence spread in all the visible spectrum.

While both scatters are usually collected with a linear amplification, both fluorescences are amplified in log scale in order to enhance distinction between effectors and targets on the green channel (Papa *et al.*, 1988). Calculations are produced on the fluorescence cytogram with a previous control from the scatter cytogram, by gating on the area in which dead targets are localized (Fig. 4C and D).

B. Binding Assay

In order to perform this assay sorters must be equipped with a nozzle tip of diameter not less than 100 μm to avoid damage to the conjugates when they pass through the orifice (Vitale *et al.*, 1989b). Sample analysis must be performed at a slow flow rate with a sheath pressure around 10 psi in order to obtain a good orientation of conjugates within the jet.

In the case of a five-parameter instrument the setting of the third fluorescence channel should be obtained depending on instrumentation. While analyzers are fixed machines, sorters could be adjusted with proper filters for different fluorescent probes; for example, a dual-laser instrument will need for APC and Tricolor a 660-nm BP, while for Red 613, a 630-nm BP is needed.

VIII. Results and Discussion

As described in other chapters of this book the behavior of effector and target cells can be detected, in flow cytometry, by scatter signals (Mullaney and Dean, 1970). Due to their homogeneous morphology lymphocyte scatter signals are restricted to a very narrow area in both forward scatter (FSC) and SSC.

On the contrary, virtually all of the cells which derive from tumors and cell lines are much larger with a big cytoplasmic compartment rich in granules which enhance their SSC and autofluorescence signals. As shown in Figs. 1B and 3 the typical scatter and fluorescence dot plots encountered while doing these assays clearly reflect this situation. Consequently, in the cytotoxic assay, the cluster of targets, for ratios greater then 12.5 : 1, will be resolved (visualized) after the gating out of the lymphocytes (Figs. 4a and 4b). Moreover, this first step will show immediately whether a killing process has happened during incubation, by the appearance of a second cluster with lower FSC and slightly increased SSC values, compared to control, as shown in Figs. 4B and D.

This gate, used as a "live gate" in list-mode acquisition, will permit fluorescence analysis by which the NK lytic activity will be detected and estimated through the fluorescence produced by dead cells after PI incorporation. In order

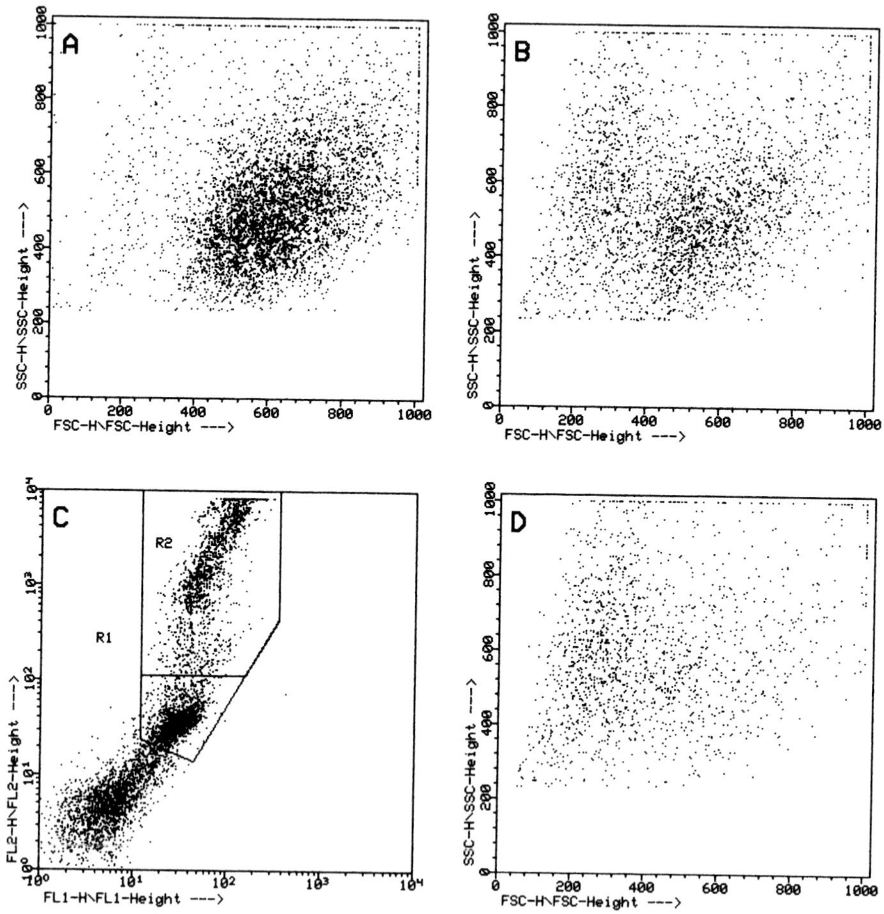

Fig. 4 Cytotoxic assay. (A) Scatter cytogram of control samples after SSC threshold. (B) Scatter cytotgram of a 25 : 1 *E*/*T* sample after incubation, previously gated with R1 (C). (C) Fluorescence cytogram showing the gate drawn to avoid contamination of lymphocytes and monocytes that can escape the SSC threshold. R1, gate which includes all K562, R2, subgate detecting dead targets. (D) Scatter dot plot of cells comprised within R2 gate. (FL1, green autofluorescence; FL2, red autofluorescence and PI fluorescence).

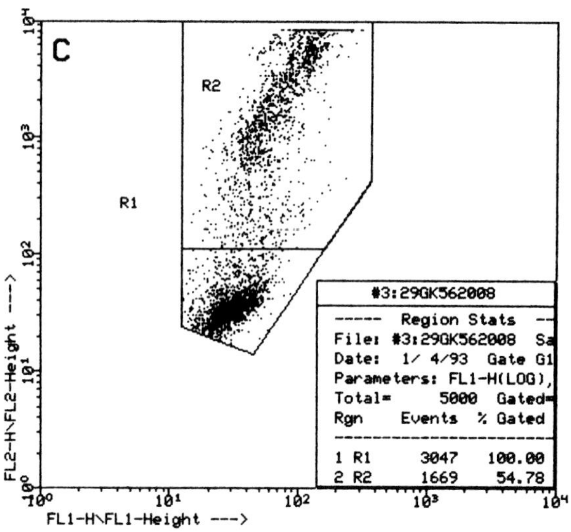

Fig. 5 Calculation of cytotoxicity as in Fig. 4C. R1 is the analysis gate corresponding to almost all K562 present in the sample. R2 is the window set to establish the rate of cytotoxicity as displayed in the insert.

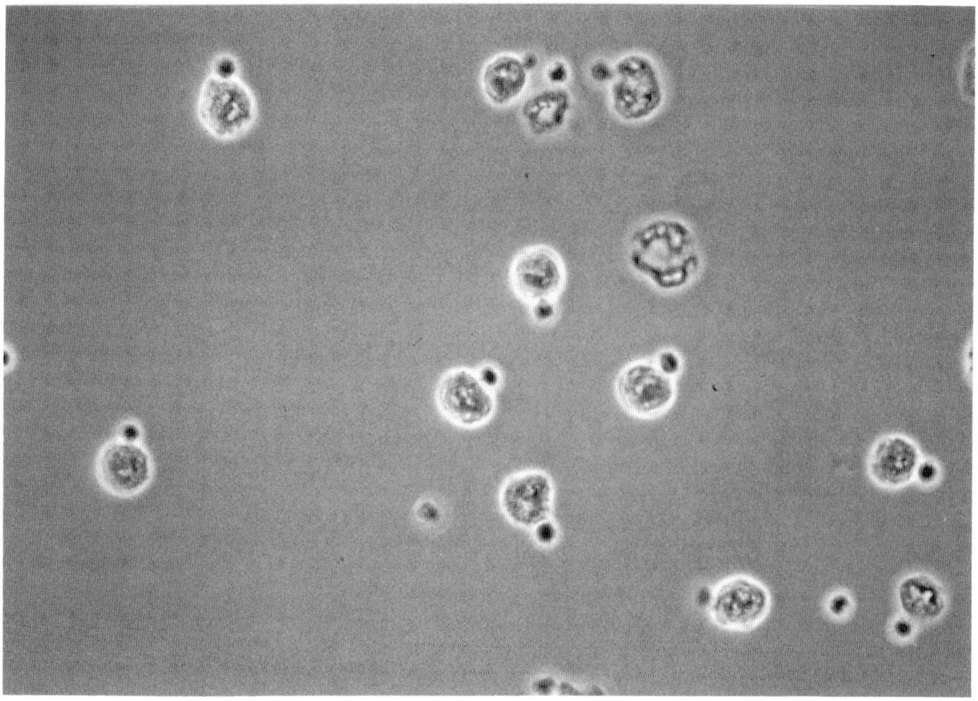

Fig. 6 Sorted CD16–K562 conjugates were placed onto a microscope slide and observed by phase-contrast microscopy. More than 90% are monoconjugates as expected.

to perform this second step a first control sample (a 1 : 1 *E* : *T* mixed sample) should be used both to set the SSC threshold and then to draw the fluorescence gate, which will be set on the basis of the autofluorescence and PI fluorescence displayed by control targets (compensation) (Fig. 4C). The morphological alterations induced in the target by NK cytotoxic activity clearly correlate with the changes in scatter signals and consequently with the progressive permeability to PI as previously reported (Vitale *et al.*, 1989a) (Fig. 4D). The evaluation of cytotoxicity requires the collection of 10,000 gated events in list mode (Fig. 5), with log green and red and linear scatters. Cytotoxicity is then expressed as the percentage of red positive targets in the whole target cluster.

Regarding the binding assay, autofluorescence of targets is the most important function of the assay and at the same time causes its limitation: this property of the target is necessary to distinguish bound from unbound lymphocytes and, therefore, to detect the conjugates and to assess the type of effector involved. In the presence of nonautofluorescent target cells, their specific mAb staining can be used theoretically to distinguish them from effectors. This implementation of the method is now under study in our laboratories.

This assay was designed to reveal the lymphoid subpopulation able to bind to targets in a multiparametric analysis, thus focusing also on small subsets (Fig. 6), which can be distinguished only by double-staining (with directly labeled mAb).

The discovery of new fluorochromes excitable in the 488-nm laser line, which emit in the far red region of the spectra, such as Red 613, has given more boost to the application of this technique.

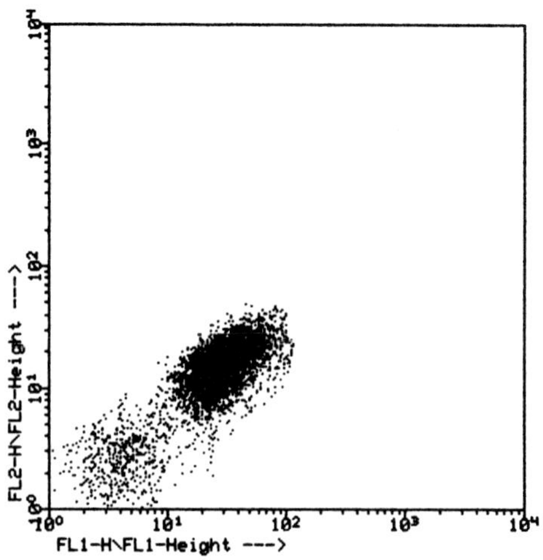

Fig. 7 Binding assay. Fluorescence cytogram of unlabeled lymphocyted and targets to show the autofluorescence displayed by K562.

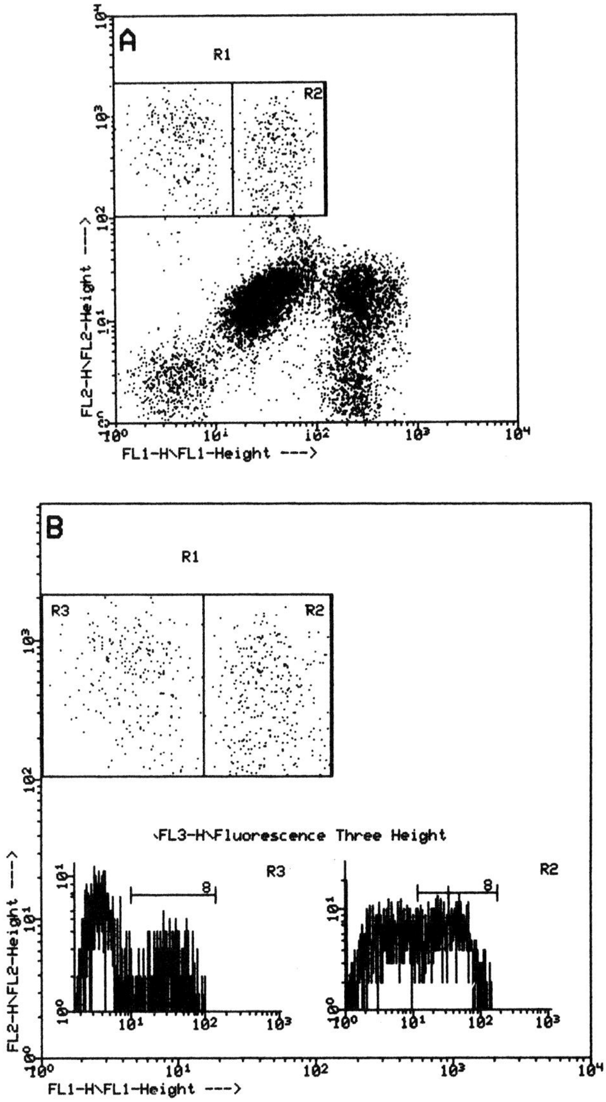

Fig. 8 Binding assay. (A) Flow cytometric analysis of PBL-K562 conjugates. Determination of the rate of binding of CD16 cells (R2/R1). (B) Gated CD16 cells (bound and unbound) were assayed for the expression of the third antigen, CD8, as shown in the two inserted histograms. Calculations were performed as reported in the text (color compensation was set on the phenotype sample as in Fig. 2B).

As the targets are clearly distinguished on the fluorescence cytogram in Fig. 7, in order to work with a two- or three-color analysis, the best condition is represented by two antigens not coexpressed, a shown in Figs. 2B and 3B, while the third, excited by a spatially separated second laser beam, can be coexpressed by one or both the previously defined major conjugate subsets (Vitale *et al.*, 1989b). The third could also display a dim fluorescence because it will be analyzed in a spectral region where the target autofluorescence does not exist (Fig. 8B).

The evaluation of the binding percentages is directly obtainable in the fluorescence cytogram for the major subsets (Vitale *et al.*, 1989b), by gating on the whole subset population (bound and unbound lymphocytes) and then calculating the value of the bound one (Fig. 8A).

The second step is a bit more complex for the calculation procedure: to determine the percentage of a double-stained subset which binds to the target two gates should be drawn: the first will determine the amount of double-stained cell in the sample (as bound and unbound cells collected together) (Fig. 8B); the second will determine how many of these double-stained cells are bound to the target. Both these numbers must be extracted as absolute in order to obtain a percentage. Consequently, dividing the second by the first value, we are now able to find the percentage of binding of a double-stained subset (Vitale *et al.*, 1992).

Acknowledgments

This work was supported by MPI 60% and National Research Council (CNR) grants 92.02735.04.

References

Cavarec, L., Quillet-Mary, A., Fradelizi, D., and Conjeaud, H. (1990). *J. Immunol. Methods* **130,** 251–261.

Garcia-Pennarubia, P., Koster, F. T., and Bankhurst, A. D. (1989). *Natl. Immunol. Cell Growth Regul.* **8,** 57–65.

Grimm, E., and Bonavida, B. (1979). *J. Immunol.* **123,** 2861–2867.

Herberman, R. B., Reinolds, C. W., and Ortaldo, J. R. (1986). *Annu. Rev. Immunol.* **4,** 651–664.

Kimberley, M. G., Chapman, G., Marks, R., and Penny, R. (1986). *J. Immunol. Methods* **86,** 7–13.

Lebow, L. T., Stewart, C. C., Perelson, A. S., and Bonavida, B. (1986). *Natl. Immunol. Cell Growth Regul.* **5,** 221–237.

Mullaney, P. F., and Dean, P. N. (1970). *Biophys. J.* **10,** 764–772.

Papa, S., Vitale, M., Mariani, A. R., Roda, P., Facchini, A., and Manzoli, F. A. (1988). *J. Immunol. Methods* **107,** 73–78.

Rubin, P., Pross, H. F., and Roder, J. C. (1982). *J. Immunol.* **128,** 2553–2558.

Segal, D. M., and Stephany, D. A. (1984). *Cytometry* **5,** 169–174.

Storkus, W. S., Balber, A. E., and Dawson, J. R. (1986). *Cytometry* **7,** 163–169.

Vargas-Cortes, M., Hellström, U., and Perlmann, P. (1983). *J. Immunol.* **62,** 87–99.

Vitale, M., Papa, S., Mariani, A. R., Facchini, A., Rizzoli, R., and Manzoli, F. A. (1987). *J. Immunol. Methods* **96,** 63–68.

Vitale, M., Neri, L. M., Comani, S., Falcieri, E., Rizzoli, R., Rana, R., and Papa, S. (1989a). *J. Immunol. Methods* **121,** 115–120.

Vitale, M., Rizzoli, R., Mariani, A. R., Neri, L. M., Facchini, A., and Papa, S. (1989b). *Cytotechnology* **2,** 59–62.

Vitale, M., Zamai, L., Neri, L. M., Manzoli, L., Facchini, A., and Papa, S. (1991). *Cytometry* **12,** 717–722.

Vitale, M., Zamai, L., Papa, S., Mazzotti, G., Facchini, A., Monti, G., and Manzoli, F. A. (1992). *J. Immunol. Methods* **149,** 189–196.

Zarcone, D., Tilden, A. B., Cloud, G., Friedman, H. M., Landay, A., and Grossi, C. E. (1986). *J. Immunol. Methods* **96,** 247–252.

CHAPTER 13

Antibodies to Intermediate Filament Proteins as Probes for Multiparameter Flow Cytometry of Human Cancers

T. Vincent Shankey

Departments of Urology and Pathology
Loyola University Medical Center
Maywood, Illinois 60153

I. Introduction

Flow cytometry has found increasing applications in clinical laboratories over the past 10 years. At present, the majority of clinical applications involve measurements of cell-surface antigen expression in the analysis of hematopoietic cells or DNA content analysis of human malignancies. Clinical applications involving the use of antibody staining technologies to measure intracellular antigens have not been extensively employed to this point. In part, this is due to the physical barriers presented by the cytoplasmic (and nuclear) membranes which restrict the passage of large molecules such as antibodies and some dye conjugates into the cell interior. An additional limitation is presented by the difficulties in isolating intact cells from many solid human tumors in a form suitable for multiparameter flow cytometric analysis.

While DNA content analysis (DNA ploidy and tumor S phase) of a variety of tumors has been reported over the past 10 years (reviewed in Barlogie *et al.*, 1983; Bauer *et al.*, 1993; Duque *et al.*, 1993; Hedley *et al.*, 1993; Shankey *et al.*, 1993; Wheeless *et al.*, 1993), such single-parameter measurements have several limitations. These include: difficulty in detecting near-diploid DNA aneuploid (DNA Index or DI ~ 1.1–1.2) and sometimes DNA tetraploid (DI 1.8–2.1) tumors; inability to detect rare tumor cells; (particularly in the presence of significant amounts of debris, aggregates, or reactive cells), and the inability to calculate tumor-specific S phase for DNA diploid tumors. While the presence or absence of DNA aneuploid tumor cell populations has been of value in predicting disease course for large groups of patients, or for individuals with specific stage or grade tumors (i.e., grade 2, superficial transitional cell bladder cancers), some studies have suggested that tumor proliferation (or S phase) may provide a more powerful predictor of disease course or response to therapy (see Tribukait *et al.*, 1982; Visakorpi *et al.*, 1991; Clark *et al.*, 1992; Bauer *et al.*, 1993; Hedley *et al.*, 1993; Shankey *et al.*, 1993). As discussed below, cytoplasmic markers, such as intermediate filament proteins, allow the identification of specific cell populations including tumor cells and may increase the clinical utility of tumor S phase estimates for specific types of tumors.

Cytokeratins represent the major intermediate filament protein found in all cells of epidermal origin, which include adenocarcinomas (roughly 70% of all solid tumors). Vimentin, the intermediate filament expressed by nonmuscle cells of mesenchymal origin, is found in many malignant hematopoietic cells (lymphomas), many sarcomas, and other tumors as detailed below. Antibodies to cytokeratins have found increasing use in flow cytometric analysis of breast, bladder, endocervical, and squamous cell tumors. Here, intermediate filament proteins are discussed as markers for multiparameter flow cytometric analysis of human tumors.

II. Intermediate Filament Proteins

Eukaryotic cells contain distinct classes of filamentous proteins which are characterized by their size. The intermediate filaments (10 nm) are named for their intermediate size between small actin filaments (6 nm) and large filaments of myosin (15 nm) and microtubules (25 nm). The intermediate filaments form the cytoskeleton of eukaryotic cells, extending as a network between the cell membrane and the nuclear membrane. They are believed to function as a cellular support network and as a mechanism to anchor the nucleus and cell organelles (Lazarides, 1980). Unlike other types of filamentous proteins which are similar from one cell type to another (actin, nonmuscle myosin, mictotubules), the intermediate filaments are different for cells of different embryologic origin. As indicated in Table I, there are five distinct types of intermediate filament proteins. All show a common organization scheme (see Fig. 1), with a central α helix (roughly 310 amino acids), flanked by amino-terminal and carboxyl terminal domains. The central rod-like domains show specific patterns of helical and nonhelical regions and are highly conserved. The N-terminal "head" and C-terminal "tail" nonhelical domains are heterogeneous in amino acid sequence (in number and type of amino acids) and are believed to define structurally unique regions for different intermediate filaments (Steinert *et al.,* 1985). All intermediate filaments are composed of two polypeptides which form a helical protofilament, which in turn aggregate into a six- or eight-stranded protofibril, which further aggregates into 10 nm filaments with either four or three protofibrils, respectively.

As summarized in Table I, neurons in the central or peripheral neural system (including neuroendocrine cells) express neurofilaments, while astrocytes and glial cells express glial fibrillary acidic protein (GFAP). All muscle cells express desmin, while nonmuscle cells of mesenchymal origin (including most hematopoietic cells) express vimentin. Cytokeratins (expressed by cells of epithelial origin) are unique in that there are 20 different types of cytokeratin molecules identified to date (Moll *et al.,* 1982,1990).

The majority of intermediate filament proteins are formed from two identical polypeptides forming a homodimer which assembles into specifically organized higher-order filamentous structures. In contrast, molecules of cytokeratins are

Table I
Subclasses of Intermediate Filament Proteins

Subclass	Molecular mass (kDa)	Normal tissue	Representative tumors
Cytokeratins	40–70	Skin, glandular epithelium	Adenocarcinomas, adenomas, transitional cell cancers, mesotheliomas, epithelioid, and synovial sarcomas
Desmin	53	Skeletal, smooth, and cardiac muscle cells	Rhabdomyosarcomas, spindle cell tumors (some)
GFAP[a]	51	Glial cells, astrocytes	Medulloblastoma, gliomas, and astrocytomas
Neurofilaments	68, 150, 200	Neurons	Neuroblastomas and pulmonary small cell Ca
Vimentin	58	Mesenchymal cells (Fibroblasts, lymphocytes, and endothelial cells)	Nonmuscle sarcomas, fibrosarcomas, liposarcomas, histiocytomas, lymphomas

[a] Glial fibrillary acidic protein.

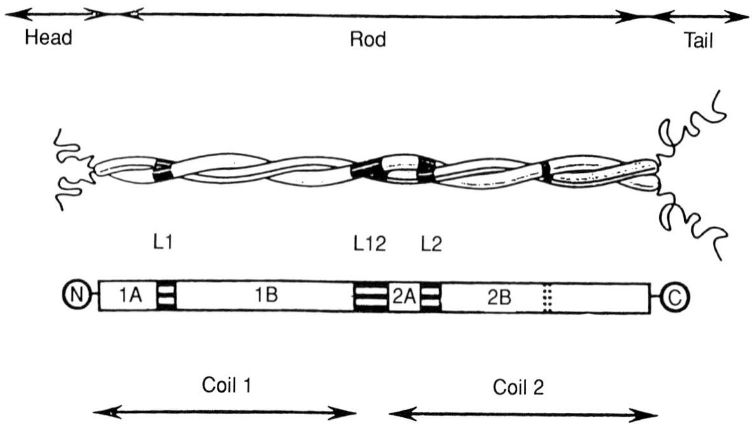

Fig. 1 Schematic representation of the structure of intermediate filaments. A central domain, composed of roughly 310 amino acids, forms a rod-like structure, flanked by an N-terminal "head" and C-terminal "tail." The central domain contains long sequences of α helix (regions 1A, 1B, 2A, and 2B) interrupted by nonhelical regions (L1, L12, L2). The N- and C-terminal regions show considerable variability in amino acid composition and size, and it is believed that these regions provide the majority of epitopes as well as the functional specificity found in different intermediate filament proteins. Two intermediate filament proteins form a coiled-coil dimer. For cytokeratins, two different cytokeratin proteins (one type I, one type II) form a heterodimer. For other intermediate filaments, two identical proteins form a homodimer. (Figure reproduced with permission from Nagle, 1988.

formed from two different polypeptide chains: one with an acidic (type I subfamily) and one with a neutral to basic (type II subfamily) isoelectric point. Ten different type I polypeptides (from 40 to 56.5 kDa) and 10 different type II polypeptides (from 52 to 67 kDa) have been identified. As shown in Table II, cytokeratin polypeptides pair in a specific pattern (i.e., cytokeratin 10 with 1 or 2, cytokeratin 18 with 8, etc.). Specific cytokeratin pairs are identified in different tissues. In addition, as cells within a tissue undergo normal differentiation, the pattern of cytokeratin pair expression changes. As a general rule, as cells progress from undifferentiated normal states (stem cells) to more differentiated cells, the cytokeratin polypeptides shift from higher- to lower-molecular-weight forms. In addition, the cytokeratins seen in complex epithelium (i.e., skin) are generally higher molecular weight than the peptides found in simple epithelia (Steinert *et al.*, 1985). Table II summarizes the patterns of expression of some cytokeratin pairs and the normal expression for these pairs in representative tissues.

As a model for the changes in cytokeratin pair expression in a normal tissue, the pattern of expression at different layers of the normal bladder urothelium is summarized in Table III (summarized from Cooper *et al.*, 1985; Schaafsma *et al.*, 1989; Moll *et al.*, 1990). The normal bladder contains cells expressing cytokeratins 4, 8, and 7 (type II peptides) and 13, 17, 18, 19, and 20 (type I

Table II
Examples of the Expression of Cytokeratin Pairs

Type I (acidic) subfamily			Type II (basic) subfamily			
Moll[a] number	Molecular mass[b]	pI[c]	Moll[a] number	Molecular mass	pI[c]	Normal expression
10	55.5	5.3	1/2	65/67	6.8	Keratinized epithelium
12	55	4.9	3	64	7.5	Corneal epithelium
13	51	5.1	4	59	7.3	Nonkeratinized stratified epithelium (esophagus, exocervix, tongue)
14/15	50	5.3/4.9	5	58	7.4	Stratified epithelium (keratinocytes)
16	48	5.1	6	56	7.8	"Hyperproliferative" keratinocytes
17	46	5.1	7	54	6.0	Simple, some stratified epithelium
18	45	5.7	8	52	6.1	Simple, some glandular epithelium
19	40	5.2	—	—	—	Simple epithelium

[a] Cytokeratin number, according to Moll classification (Moll *et al.*, 1982); Adopted from Nagle (1988).
[b] In kiloDaltons.
[c] Isoelectric point.

peptides). The pattern of cytokeratin pair expression at different cell layers of the bladder urothelium is controversial and may be complicated by cross-reactions of some monoclonal antibodies with more than one cytokeratin. As shown in Table III, the basal cells predominantly express the cytokeratin 4, 13 pair, while intermediate cell layers express the cytokeratin 7, 17 pair (Cooper *et al.*, 1985). All cell layers of the normal urothelium express cytokeratins 8, 18, and 19, with the most superficial cells containing only the 8, 18 pair (Schaafsma *et al.*, 1989), and cytokeratin 20 (Moll *et al.*, 1990,1992).

Table III
Cytokeratin Pairs Found in Normal Bladder Urothelium

Acidic peptides			Basic peptides			
Moll number	Molecular mass (kDa)	pI	Moll number	Molecular mass (kDa)	pI	Cells expressing cytokeratin pairs
13	51	5.1	4	59	7.3	Basal cell layer
17	46	5.1	7	54	6.0	Predominantly basal and intermediate
Cell layers						
18	45	5.7	8	52	6.1	Umbrella and intermediate cells
19	50	5.2	—	—	—	All cell layers
20	46					Umbrella cells

Note. Pattern of reactivity dependent on monoclonal antibody used. Adapted from Cooper *et al.* (1985), Schaafsma *et al.* (1989), and Moll *et al.* (1992).

III. Cell Isolation and Fixation

The simultaneous analysis of intracellular antigen expression and DNA content in hematopoietic cells is greatly simplified by the ease in obtaining single-cell suspensions from these tissues. Solid tumors present a sizable problem in isolating intact cells representative of the tumor cell population. A detailed review of the different techniques used to isolate cells from different human tumors is beyond the scope of this presentation (see Hitchcock and Ensley, 1993; Pallavachini *et al.*, 1990; Cerra *et al.*, 1990). These have included enucleation techniques which digest or dissolve the cytoplasmic membrane and leave nuclear membranes intact, or techniques which remove intact cells from tissues by mechanical means (i.e., scrapping), by enzymatic (trypsin, collagenase, etc.) digestion of intercellular matrices, or by a combination of techniques.

For the preparation of fresh solid tumor samples, the advantages of the enucleation techniques include speed, simplicity, very low coefficient of variation (CV) for DNA content measurements, and, for *some* types of tissues, recovery of nuclei from representative types of cells found in the original tissue. The disadvantages include loss of some or most of the cytoplasm (and loss of potentially important cell membrane and cytoplasmic antigens) and difficulty in recovering representative tumor nuclei from some tissues, including fibrosarcomas and prostate cancers. In contrast, techniques which yield intact cells can require prolonged incubation steps with enzymes, can provide a biased sampling of the cells present in the original tissue (i.e., preferential recovery of DNA diploid stromal cells), and may produce significant numbers of cells which have lost some or all of the cytoplasm. However, when sufficient numbers of representative, intact cells are isolated, multiparameter flow cytometric measurements are possible using antibodies to intracellular antigens in conjunction with DNA content measurements.

Although clinical applications of flow cytometry for the analysis of human tumors increasingly rely on the use of nuclei isolated from paraffin-embedded tumor samples (technique of Hedley *et al.*, 1983), this technique (0.5% pepsin digestion for 30–90 min) generally yields nuclei with little or no cytoplasm. This does not preclude the use of nuclear protein measurements, although possible digestion of nuclear proteins during the isolation procedure must be considered (Lincoln and Bauer, 1989). Although we have developed a techniquc to retain epitopes of cytokeratin proteins on some nuclei recovered from archival bladder tumors (Shankey *et al.*, 1990), only a minority of nuclei (generally <30%) retain immunochemically detectable cytokeratin following isolation by this technique.

While techniques have been reported which allow passage of antibodies into intact, living cells (Schroff *et al.*, 1984), these techniques can differentially lyse some cell populations and could allow some antigens to escape from the cell. The alternative approach requires fixation of the cell and permeabilization of the cell membrane to generate pores of large enough size to provide antibody molecules access to the cell interior.

Both fixation and permeabilization are critical steps for antibody-based measurements of intracellular antigens. It must be stressed that different techniques for fixation and/or permeabilization have different effects on different antigens and may also have different effects on the expression of the same antigen in different types of cells.

It is critical that the techniques used to fix and permeabilize cells be carefully studied and optimized for each cell type and for each antigen to be studied.

Although a number of different agents have been used to preserve cells or tissues for histologic study, most fall into two categories of chemical agents–dehydrating agents and crosslinking fixatives. Alcohols (usually methanol or ethanol) are generally used as aqueous solutions with a final alcohol concentration of 50–90%. All alcohols fix cells by dehydration and partial (or complete) denaturation of proteins. In general, alcohols act on proteins by changing their secondary and tertiary structure, stabilizing them in a nonaqueous or partially hydrated environment. Proteins have different susceptibility to irreversible denaturation, and a single concentration of alcohol will completely denature some proteins, partially denature others, and have no effect on others, all within the same cell. For antibody reactivity (retention of antigenicity) the ability of different antibody molecules to bind to different epitopes on alcohol-fixed proteins depends on the type of epitope (linear peptide versus conformational determinant) and the degree to which the epitope is altered. In addition, alcohols can "unmask" some epitopes by removing sterically inhibiting polypeptides, lipids, etc. from the vicinity of the epitope and may increase the fluorescence signal compared to other fixation techniques (Schimenti and Jacobberger, 1992). Alcohols extract some lipid from cell membranes and generate pores for antibodies to enter (and leave) cells. The multiple effects on intracellular antigens are perhaps better controlled by fixation of cells equilibrated to 4°C (below the membrane transition temperature) with cold alcohol. Carrying out alcohol fixation at −70°C may minimize many of the changes in cell morphology seen after alcohol fixation at higher temperatures (K. D. Bauer, unpublished).

Aldehyde fixatives act by crosslinking a variety of different cell constituents. One of the most commonly used fixatives, formaldehyde, reacts with functional groups on peptides, carbohydrates, and some lipids, to introduce methylene bridges between reactive groups. These reactions can link proteins to other proteins and to carbohydrates, or to nucleic acids. Unlike alcohol fixatives, the degree of aldehyde fixation is time and concentration dependent. Thus, fixation in 0.5, 1, or 4% paraformaldehyde can reach similar end points, at different rates.

Formaldehyde fixation of cells takes place either in an aqueous solution of formaldehyde gas (formalin) or using polymerized formaldehyde (paraformaldehyde) dissolved in water. Formalin is a 37–40% solution of formaldehyde, although the available fixative in solution is significantly lower in concentration.

All aqueous solutions of formaldehyde contain degradation or reaction products, including methanol and formic acid. The rate of formation of these compounds is affected by temperature, light, and pH. Significant levels of formic acid will irreversibly denature many proteins. Although buffered solutions (i.e., neutral-buffered formalin) have lower formic acid levels, the most advisable technique is to make up fresh fixative using paraformaldehyde dissolved in a neutral buffer, such as phosphate-buffered saline (PBS: pH 7.2–7.4). The solution should be prepared by carefully heating paraformaldehyde powder (use EM grade, available from several commercial supply houses) in buffer in a hot water bath, or by using a microwave oven (not the same one used to heat coffee or lunches), to depolymerize the powder and generate aqueous formaldehyde.

Crosslinking fixatives make cell membranes (for cells viable prior to fixation) relatively impermeable to large molecules such a antibodies. For immunochemical analysis, formaldehyde-fixed cells can be rendered permeable using lipophilic agents or detergents, such as Triton X-100, NP-40, Tween, or saponin, or with alcohols. Brief exposure (5–20 min) at low concentrations of these agents (0.1–0.5%) at 4°C is sufficient to extract lipid from the fixed membranes to generate pores of large enough size to allow antibody molecules (including 19S IgM) to pass into cells. It is important to avoid prolonged exposure to fixed cells to these agents, as this increases the likelihood that cellular proteins will diffuse out of the cell or can result in disintegration of lightly fixed cells.

IV. Multiparameter Flow Cytometric Analysis Using Antibodies to Cytokeratins

A. No-Wash Anti-cytokeratin Staining Technique

Our laboratory (Shankey, 1993; Clevenger and Shankey, 1993; Shankey *et al.*, 1994) and others (R. Duque, personal communication; R. Zarboe *et al.*, submitted for publication) have previously described techniques to label intermediate filaments, using directly conjugated monoclonal antibodies and a "no-wash" technique that does not involve any cell washing steps after the removal of cell fixative. These techniques are similar to that described by Larsen *et al.* (1991) for the staining of proliferating cells with anti-Ki-67 antibody. Important advantages of the no-wash technique include: (1) significant reduction in cell loss during multiple centrifugation and washing steps; (2) significant reduction in sample preparation times; (3) a quantitative determination of the intermediate filament content (provided directly labeled antibodies are used and antigen saturation is achieved); and, (4) reduced reagent costs.

The no-wash technique, as described below, can be used with either alcohol- or aldehyde-fixed cells. In either case, the cells are washed (once) following fixation, using cold (PBS) without Ca^{2+} or Mg^{2+} (pH 7.2) containing 5% bovine serum albumin (PBS/BSA).

No-wash technique:

1. Place 0.5–1.5 × 10^6 fixed and washed cells into two different test tubes (one for immunoglobulin isotype control and one for anti-cytokeratin antibody-stained sample).
2. Pellet cells by centrifugation at 400*g* for 10 min. Remove supernatant fluid. Add 25 μl normal goat sera (undiluted) to each tube, and resuspend pellet by flicking tube.
3. Add 250 μl RNase. A (chromatographically purified, DNase free, 200 Kunitz units/ml) in PBS/BSA containing 10 m*M* $MgCl_2$. Incubate for 30 min at 37°C.
4. Add 250 μl fluorescein isothiocyanate (FITC)-labeled anti-cytokeratin antibody (direct conjugate) to one tube; add concentration and isotope-matched FITC-labeled Ig (control) to other tube, both diluted in PBS/BSA. Mix, and incubate in the dark, on ice for 30 min.
5. Add 10 μl propidium iodide (PI) (stock, 2.5 mg/ml in PBS). Incubate in the dark, on ice, for a minimum of 1 hr before flow cytometric analysis.

This technique includes the addition of normal goat sera to minimize nonspecific Ig binding to the surface or cell interior (normal horse sera can also be used). The concentration of anticytokeratin antibody used should be determined to optimize staining for each type of cell and for different anti-cytokeratin antibodies (see Clevenger and Shankey, 1993). $MgCl_2$ is used, since previously published studies have indicated that RNase activity increases in low concentrations of Mg^{2+}.

The use of cytokeratin markers to distinguish different cell populations in heterogeneous cell mixtures (here, using the "no-wash" technique) is demonstrated in Fig. 2. Here, a mixture of diploid human peripheral blood lymphocytes and T-24 cells (bladder tumor cell line, ATCC, Camden, NJ) was analyzed by flow cytometry after the cells were stained with a monoclonal antibody which reacts with cytokeratins 8, 18, and 19 (Cam 5.2, Becton–Dickinson, San Jose, CA), in conjunction with DNA content analysis (after RNase treatment and propidium iodide staining). The top panel shows the single-parameter DNA content analysis of all cells in the mixture. The two-parameter measurement (DNA content plus cytokeratin) is shown in the middle panel, with the cytokeratin-positive, T-24 cells all contained in the region defined by the box. By gating the DNA content analysis on only the cells inside the box, it is possible to restrict the analysis to only the cytokeratin-positive T-24 cells (bottom panel). It is also possible to gate only the cytokeratin-negative peripheral blood lymphocytes in order to perform DNA content analysis on only the diploid lymphocyte (cytokeratin-negative) control population.

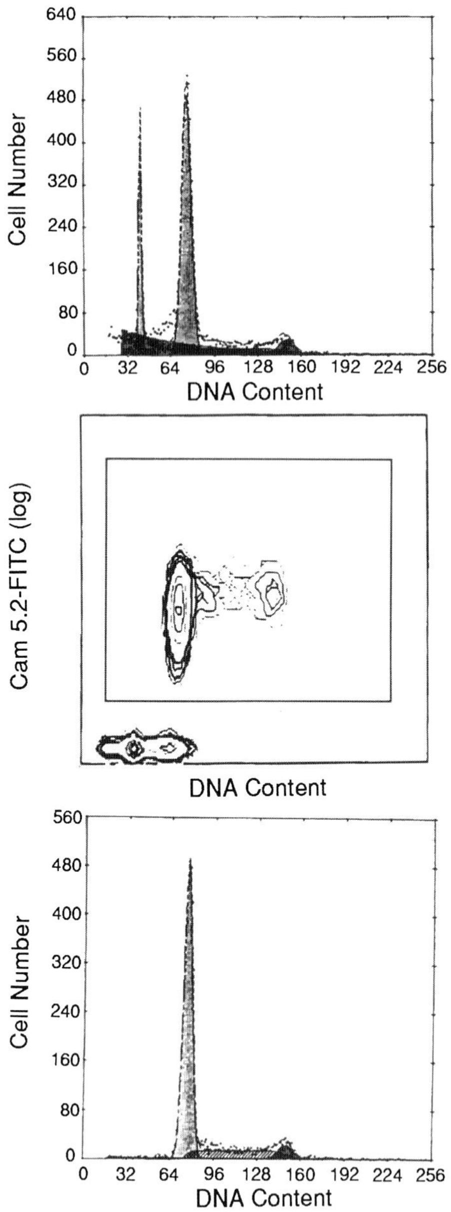

Fig. 2 Use of antibodies to intermediate filaments for simultaneous analysis of DNA content and cytokeratin expression in a mixture of human peripheral blood lymphocytes (cytokeratin negative) and human bladder tumor cell line, T-24. Cell mixture was fixed with 50% cold ethanol, washed, and processed using the "no-wash" technique. (Top) single-parameter DNA content analysis of all cells in mixture. (Middle) two-parameter histogram of DNA content versus cytokeratin expression, demonstrating clear separation of cytokeratin-positive cells (population in gated rectangular region) from diploid lymphocytes. (Bottom) single-parameter DNA content analysis of cytokeratin-positive (gated population) T-24 cells.

B. Technical Issues

The use of two (or more) fluorochromes to analyze complex cell populations found in clinical material by flow cytometry has the potential to generate artifacts caused by cell fixation, staining, or flow cytometry instrumentation. In a clinical setting, such artifacts could potentially impact on the determination of DNA ploidy and on the calculation of tumor S phase fraction.

Analysis of cytokeratin antibody-stained cells is generally performed using direct or indirectly labeled antibodies with FITC, and DNA staining performed with propidium iodide. Since cytokeratins represent a major intracellular protein (up to 20% of the total cellular proteins), the use of saturating concentrations of anti-cytokeratin antibodies can result in a large green (FITC) fluorescence signal. An additional problem is that populations of cells isolated from clinical specimens frequently contain cells with large variations in cytokeratin expression, necessitating the use of logarithmic amplification of the anti-cytokeratin (green) signal to include all cell populations. Inappropriate use of fluorescence signal compensation can result in artifactual signals, and the use of linear

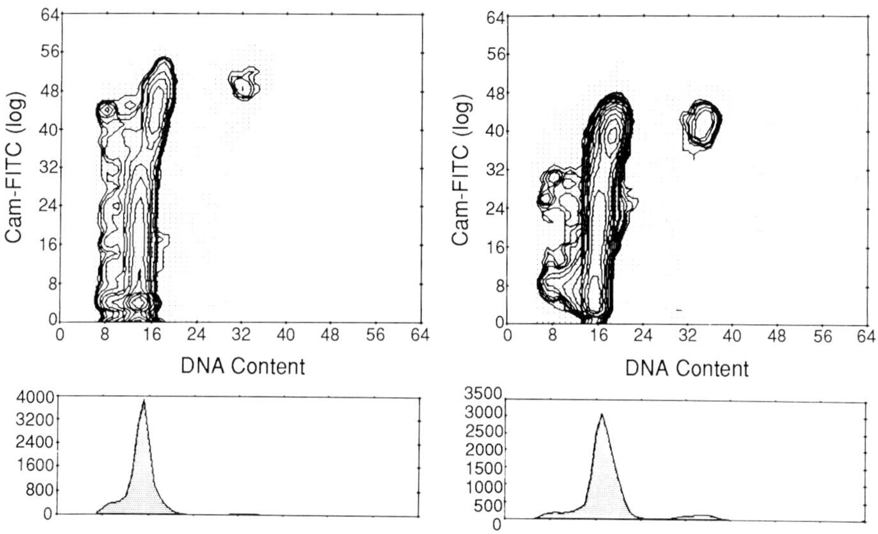

Fig. 3 Dual-parameter DNA content histograms of bladder washings stained with anti-cytokeratin antibodies and propidium iodide using the "no-wash" technique. Samples shown here contained only DNA diploid cells. Incomplete color compensation can result in cells brightly stained with FITC (anti-cytokeratin) showing increased red (DNA) fluorescence due to green fluorescence spill over into red photomultiplier. Single-parameter DNA content analysis could demonstrate a broad CV or a shoulder on the DNA diploid population, or if the DNA content of cytokeratin-positive and -negative cells were plotted separately, the results could indicate the presence of a near diploid DNA aneuploid tumor. These results indicate the importance of carefully analyzing the dual-parameter histograms before reaching a potentially incorrect conclusion based on an instrument-induced artifact.

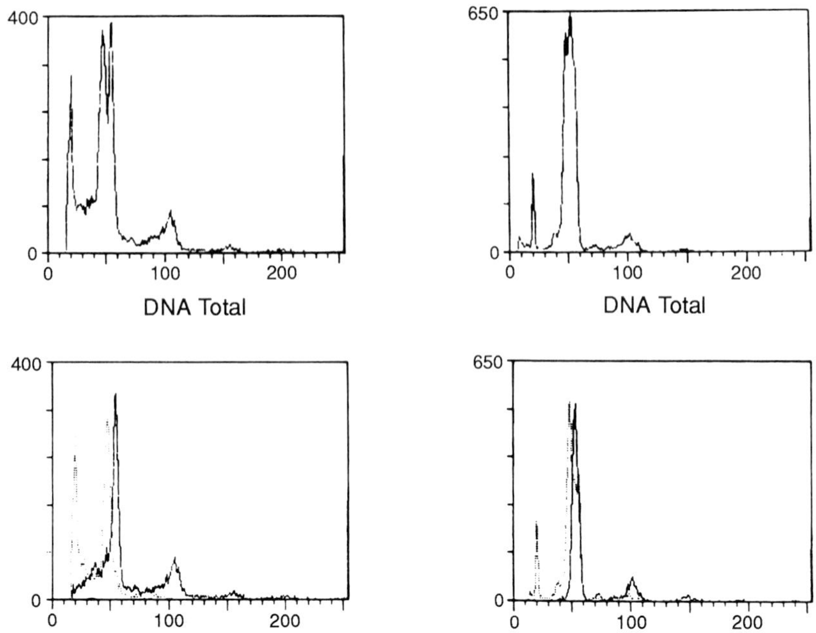

Fig. 5 DNA content histograms demonstrating the impact of cytokeratin gating. (Top) Ungated DNA content analysis showing a single DNA diploid peak with broad CV (right) or a peak–valley–peak pattern, suggestive of the presence of a near diploid DNA aneuploid population. (Bottom) DNA content of cytokeratin-positive (solid line) and cytokeratin-negative (dotted line) populations, demonstrating the higher DNA content of the cytokeratin-positive cells and suggesting the presence of near-diploid DNA aneuploid populations. (Reproduced with permission from van der Linden *et al.*, 1992.)

V. Clinical Application Using Multiparameter DNA Content Flow Cytometry

Multiparameter flow cytometric analysis of solid tumors using antibodies to intermediate filament proteins has been reported in a number of studies. As summarized in Table IV, studies published to date have used antibodies to cytokeratins, to vimentin, and to GFAP. The most frequently reported human cancer studied with simultaneous DNA content and anti-cytokeratin antibodies has been transitional cell cancer of the bladder. Over half of the studies listed in Table IV are authored by Drs. Frans Ramaekers, Peter Vooijs, and Frans Debruyne and reflect the sizeable impact that this group has had on the application of intermediate filament antibodies to clinical flow cytometry.

As discussed below, the use of anti-cytokeratin antibodies increases the sensitivity of flow cytometric detection of (DNA aneuploid) exfoliated bladder cancer cells and provides a very useful marker for the identification of DNA

Table IV
Flow Cytometric Applications of Immunofluorescence Measurements Using Antibodies to Intermediate Filaments

Antibody specificity	Application	Reference
Intermediate filaments	Lymphocytes	(Schroff *et al.*, 1984)
Cytokeratins (pan CK and CK18[a])	Bladder cell line	(Ramaerkers *et al.*, 1984)
Cytokeratins (pan CK)	Bladder TCC[b]	(Huffman *et al.*, 1986)
Cytokeratins (pan CK and CK18)	Bladder TCC	(Feitz *et al.*, 1985)
Cytokeratins (pan CK)	Bladder TCC	(Smeets *et al.*, 1987)
Cytokeratin (CK18)	Bladder TCC	(Hijazi *et al.*, 1989)
Cytokeratin (CK18)	Bladder TCC	(Konchuba *et al.*, 1992)
Cytokeratins (CK 10 and CK18)	Endometrial Ca	(Oud *et al.*, 1985)
Cytokeratin (CK18)	Endometrial Ca	(Oud *et al.*, 1986)
Cytokeratins (pan CK)	Squamous cell Ca	(Bijman *et al.*, 1986)
Cytokeratins (CK 5, 10, 14, and 16)	Psoriatic skin	(van Erp *et al.*, 1989)
Cytokeratin (CK8, 18, and 19)	Breast Ca	(Visscher *et al.*, 1990)
Cytokeratin CK7, 8, 18, 19)	Breast Ca	(Ferrero *et al.*, 1990)
Cytokeratin (CK8, 18, and 19)	Epithelial cancers	(van der Linden *et al.*, 1992)
Cytokeratin (CK7)	Teratoma	(Looijenga *et al.*, 1991)
Vimentin or Cytokeratin (pan CK)	Renal Cell Cancer	(Feitz *et al.*, 1986)
GFAP	Glioma cell line	(Ito *et al.*, 1989)

[a] Antibody specificity to cytokeratin polypeptide according to Moll classification (Moll *et al.*, 1982).
[b] TCC, transitional cell cancer.

diploid transitional cells. In breast cancers, the use of anticytokeratin antibodies has provided an important means to address the problem of tumor-specific S phase determinations for DNA diploid cancers.

A. Bladder Cancer

Antibodies to cytokeratins are useful for multiparameter flow cytometric analysis of exfoliated bladder cells for the detection of transitional cell cancers. By gating on cytokeratin-positive cells, it is possible to detect urothelial cells when present (both normal diploid urothelial cells and cancer cells). This gating technique makes it possible to "gate out" potentially interfering cells (red blood cells, granulocyte, debris, etc) and to detect bladder cancer cells in the presence of large numbers of noncancer cells. An additional advantage of this gating technique is that normal urothelial cells can be used as a diploid DNA content control and are usually present in bladder washings or mechanically dissociated tumor samples. The use of this internal diploid DNA content control generally eliminates problems caused by artifactual changes in DNA content of external DNA standards which can be generated by fixation and cell staining.

We currently use multiparameter flow cytometry (DNA content plus anti-cytokeratin) for routine analysis of bladder wash specimens. For these assays, we routinely stain samples with either a directly FITC-labeled monoclonal antibody to cytokeratins 8, 18, and 19 (Cam 5.2; Becton–Dickinson, San Jose, CA) or with a concentration-matched, isotope-matched FITC-labeled Ig control (to determine nonspecific Ig binding), using the "no-wash" technique described above. This "no-wash" technique greatly cuts down the amount of time necessary to stain samples and also reduces the cell loss associated with multiple washing and centrifugation steps. This is particularly useful for bladder washings, as they frequently contain few cells initially. Cells are fixed in 2% paraformaldehyde (containing lysolecithin), and are subsequently permeabilized using Triton X-100. Unlike many other cellular proteins, cytokeratin molecules are relatively resistant to denaturation by either alcohols or aldehyde-based fixatives, and our laboratory has extensive experience fixing bladder wash specimens with both types of fixatives.

A sample histogram for a bladder wash sample is shown in Fig. 6. The sample shown here came from an individual with biopsy-proven (grade 2, stage T2)

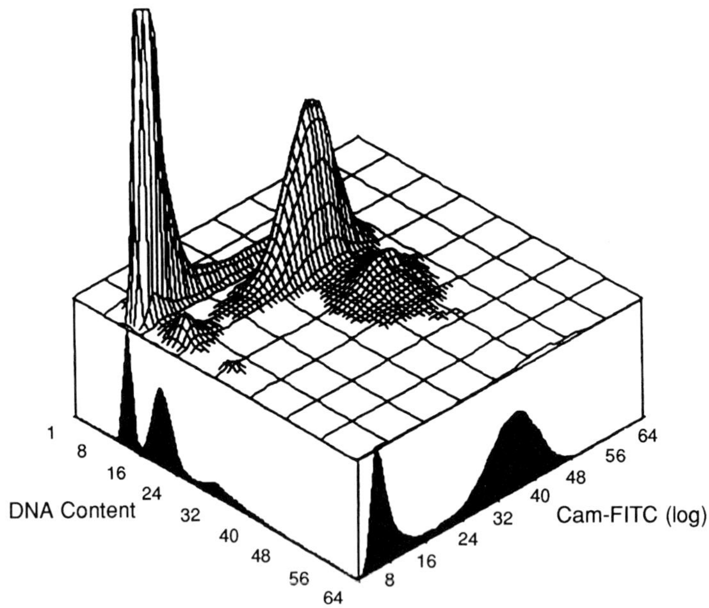

Fig. 6 Results of multiparameter flow cytometric analysis of bladder wash sample stained with anti-cytokeratin antibody (Cam 5.2-FITC, Becton–Dickinson, San Jose, CA) and propidium iodide. The sample contained approximately 50% DNA diploid, cytokeratin-negative cells and a large percentage of DNA aneuploid, cytokeratin-positive tumor cells. Analysis of dual-parameter histograms allows visual determination to be made of the percentage of DNA aneuploid cells that are cytokeratin negative (an indication of the presence of bare nuclei and the potential for poor cell fixation). As shown here, the small population of DNA diploid, cytokeratin-positive cells (presumably normal transitional cells) provides an internal control for diploid DNA content.

transitional cell cancer of the bladder. The bladder wash was stained with anti-cytokeratin (Cam 5.2, Becton–Dickinson, San Jose, CA) directly conjugated with FITC in conjunction with PI, using the "no wash" technique described above. As shown in this figure, the majority of DNA diploid cells are cytokeratin negative. It should be noted, however, that the DNA diploid, cytokeratin-positive cells provide an internal control for diploid DNA content. The figure also shows a predominant cytokeratin-positive, DNA aneuploid transitional cell tumor population (here DI 1.8). The 4C cytokeratin-negative population (most likely aggregates of diploid G_0/G_1 cytokeratin-negative cells) can be gated out of the analysis of the tumor population, making tumor-specific S phase estimates less likely to be affected by nontumor 4C cells.

B. Breast Cancer

Studies of DNA ploidy in breast cancer have established that although individuals with DNA diploid tumors have a favorable prognosis overall compared with DNA aneuploid tumors, DNA ploidy is a weak predictor of disease course (Hedley *et al.,* 1993). Several studies of breast cancers have indicated that tumor S phase does have predictive significance (Clark *et al.,* 1992; Hedley *et al.,* 1993). These studies have all been performed using single-parameter DNA content analysis, generally using archival paraffin-embedded tumor material and cell-cycle modeling programs which lack sophisticated debris or aggregate correction algorithms. While the association between prognosis and S phase will likely be increased with the increasing use of more sophisticated modeling programs, important limitations to all single-parameter analysis techniques remain.

In an important study using multiparameter flow cytometric analysis of breast cancer samples, Visscher and co-workers (1990) have demonstrated the impact of gating on cytokeratin-positive cell tumor S phase determinations. This study demonstrated that S phase measurements for DNA diploid tumors can be significantly increased using cytokeratin-positive cells, compared to S phase measurements on ungated populations. Using single-parameter DNA content measurement techniques, it is impossible to restrict the cell-cycle analysis to tumor cells only. The presence of large numbers of (cytokeratin-negative) noncycling diploid stromal or reactive cells artificially increases the G_0/G_1 population in single-parameter analysis of DNA diploid tumors and reduces the tumor S phase as a total of all diploid events. Such cytokeratin gating techniques could have a major impact on the importance of tumor S phase measurements in this and other types of tumors.

The potential impact of cytokeratin gating on S phase calculations is demonstrated in Fig. 7, taken from data provided by Dr. R. E. Duque (Lakeland Pathologists, Lakeland, FL). Shown here are two different tumor samples, stained with anti-cytokeratin and PI using a "no wash" technique. While the cytokeratin staining technique is useful in identifying samples with low percent-

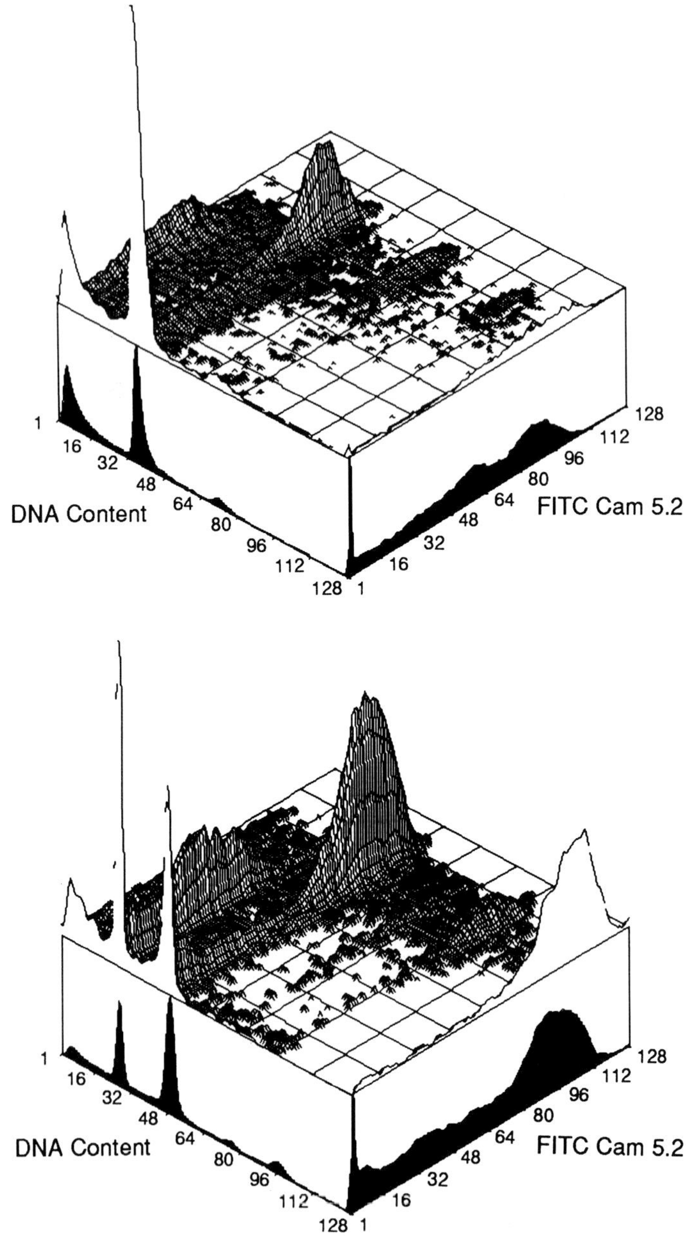

Fig. 7 Results of multiparameter flow cytometric analysis of fresh, fixed human breast tumors, showing DNA diploid (top) and DNA aneuploid (bottom) cytokeratin (Cam 5.2 anti-cytokeratin)-positive tumors. Gating the cell-cycle analysis had a significant impact on the S phase calculation of the DNA diploid tumor (top; here, 4.9% S phase ungated versus 15% S phase for the cytokeratin-gated cells). The S phase calculation of the DNA aneuploid tumor (bottom) was not significantly different for the cytokeratin-gated compared to the ungated analysis (data provided by Dr. R. E. Duque, Lakeland Pathologists, Lakeland, FL).

ages of DNA aneuploid cells, an important use is illustrated here. Comparing the S phase fraction of the ungated versus cytokeratin-gated population has little impact on DNA aneuploid tumors (Fig. 7, lower panel—here 14% using cytokeratin-gated events, 12.4% without). For the DNA diploid tumor (top panel), the calculated S phase was 4.9% for the ungated population and 15% for cytokeratin positive (gated) events—a threefold difference. As indicated previously by the study of Visscher *et al.* (1990), the removal of nontumor G_0 events can have a significant impact on the calculation of tumor-"specific" S phase for DNA diploid samples. Clearly, the impact these types of tumor-"specific" S phase calculations will have on predicting disease course on survival for DNA diploid tumors needs to be established by careful study.

VI. Conclusions

Single-parameter DNA content measurements using flow cytometry have been reported during the past 20 years for a large number of human tumors. While these studies suggest a useful role for DNA ploidy and S phase measurements for some types of tumors, the use of multiparameter flow cytometric analysis offers several important advantages. Tumor-related antigens such as intermediate filaments provide a means to identify relatively minor populations of DNA aneuploid cells and as indicated above could have a significant impact on tumor-specific S phase estimates for DNA diploid tumors. In addition, these markers stain both the tumor and its normal cellular counterpart, providing an internal diploid DNA content control for that sample. With the use of the "no wash" staining technique indicated here, samples can be processed in significantly less time, with lower cell loss and lower reagent costs (compared with techniques using multiple centrifugation steps).

The use of tumor-"specific" markers could provide a more sensitive means to detect DNA aneuploid tumors in multiparameter analyses. For DNA diploid tumors the use of intermediate filament markers such as cytokeratins still provides a potential problem. Here, the cytokeratin-positive population contains an undetermined percentage of normal (or reactive) cells which can potentially "dilute" the tumor G_0/G_1 population. If this cytokeratin-positive nontumor population represents a sizeable percentage of the total cells, the calculated S phase could still be significantly higher than that measured for the total cytokeratin-positive population. However, there are few tumor-specific antigens reported to date for human solid tumors. Lacking such markers, one potential approach that could prove useful is the use of additional tumor-"related" markers such as carcinoembryonal antigen (CEA), CA-125, etc. Ideally, the use of multiple markers, such as cytokeratin, plus CEA along with DNA content analysis could potentially improve detection and tumor-specific S-phase estimates and provide an internal diploid DNA content control. Such multiparameter flow cytometric analysis, using DNA content plus cytokeratins, plus a third parameter, such as drug resistance markers (p-glycoproteins), suppressor gene products (p53, MDM2, retinoblastoma protein), or escape from programmed cell death (bcl-2), could greatly expand the practical clinical application of flow cytometry.

Acknowledgments

The author gratefully acknowledges the contributions of Dr. Jia-Kuan Jin, Dr. Joseph Pyle, Ms. Sharon Graham, and Ms. Tina Lubrano. Work in the author's laboratory is supported by the National Cancer Institute (R55CA57910) and by grants from the American Cancer Society, Illinois Div. (No. 92-55), and from the Retirement Research Foundation (No. 93-49).

References

Barlogie, B., Raber, M., Schulmann, J., Johnson, T. S., Derwinko, B., Swartzendruber, D. E., Gohde, W., Andreff, M., and Freireich, E. J. (1983). *Cancer Res.* **43,** 3982–3997.

Bauer, K. D., Bagwell, C. B., Giaretti, W., Melamed, M., Zarbo, R. D., Witzig, T. E., and Rabinovitch, P. (1993). *Cytometry* **14,** 486–491.

Bijman, J. Th., Wagener, D. J. Th., Wessels, J. M. C., van den Broek, P., and Ramaekers, F. C. S. (1986). *Cytometry* **7,** 76–81.

Cerra, R., Zarbo, R. J., and Crissman, J. D. (1990). *Methods Cell Biol.* **33,** 1–11.

Clark, G. M., Mathieu, M. C., Owens, M. A., Dressler, L. G., Eudey, L. G., Tormey, D. C., Osborne, C. K., Gilchrist, K. W., Mansour, E. G., and Abeloff, M. D. (1992). *J. Clin. Oncol.* **10,** 428–432.

Clevenger, C. V., and Shankey, T. V.'(1993). *In* "Clinical Flow Cytometry: Principles and Application" (K. D. Bauer, R. E. Duque, and T. V. Shankey, eds), pp. 157–175. Baltimore: Williams & Wilkins.

Cooper, D., Schermer, A., and Sun, T-T. (1985). *Lab. Invest.* **52,** 243–256.

Duque, R. E., Braylan, R., Andreeff, M., Diamond, L., and Peiper, S. (1993). *Cytometry* **14,** 492–496.

Feitz, W. F. J., Beck, H. L. M., Smeets, A. W. G. B., Debruyne, F. M. J., Vooijs, G. P., Herman, C. J., and Ramaekers, F. C. S. (1985). *Int. J. Cancer* **36,** 349–456.

Feitz, W. F. J., Karthaus, H. F. M., Beck, H. L. M., Romijn C., van der Meyden A. P. M., Debruyne F. M. J., Vooijs, G. P., and Ramaekers, F. C. S. (1986). *Int. J. Cancer* **37,** 201–207.

Ferrero, M., Spyratos, F., Le Doussal, V., Desplaces, A., and Rouesse, J. (1990). *Cytometry* **11,** 716–724.

Hedley, D. W., Friedlander, M. L., Taylor, I. W., Rugg, C. A., and Musgrove, E. A. (1983). *J. Histochem. Cytochem.* **31,** 1333–1335.

Hedley, D. W., Clark, G. M., Cornelisse, C. J., Killander, D., Kute, T., and Merkel, D. (1993). *Cytometry* **14,** 482–485.

Hijazi, A., Devonec M., Bouvier, R., and Revillard, J-P. (1989). *J. Urol.* **141,** 522–526.

Hitchcock, C. L., and Ensley, J. F. (1993). *In* "Clinical Flow Cytometry: Principles and Application" (K. D. Bauer, R. E. Duque, and T. V. Shankey, eds), pp. 93–109. Baltimore: Williams & Wilkins.

Huffman, J. L., Garin-Chesa, P., Gay, H., Whitmore, W. F., and Melamed, M. R. (1986). *Ann. N.Y. Acad. Sci.* **468,** 302–315.

Ito, M., Nagashima, T., and Hoshino, T. (1989). *J. Neuropathol. Exp. Neurol.* **48,** 560–567.

Konchuba, A. M., Clemens, M. C., Schellhammer, P. F., Schlossberg, S. M., and Wright, G. L. (1992). *Cancer* **70,** 2879–2884.

Larsen, J. K., Christensen, I. J., Christiansen, J., and Mortensen, B. T. (1991). *Cytometry* **12,** 429–437.

Lazarides, E. (1980). *Nature (London)* **282,** 249–256.
Lincoln, S. T., and Bauer, K. D. (1989). *Cytometry* **10,** 456–462.
Looijenga, L. H. J., Oosterhuis, J. W., Ramaekers, F. C. S., de Jong, B., Dam, A., Beck, J. L. M., Sleijfer, D. Th., and Koops, H. S. (1991). *Lab. Invest.* **64,** 113–117.
Moll, R., Franke, W. W., Schiller, D. L., Geiger, B., and Krepler, R. (1982). *Cell (Cambridge, Mass.)* **31,** 11–24.
Moll, R., Schiller, D. L., and Franke, W. W. (1990). *J. Cell Biol.* **111,** 567–580.
Moll, R., Lowe, A., Laufer, J., and Franke, W. W. (1992). *Am. J. Pathol.* **140,** 427–447.
Nagle, R. B. (1988). *Am. J. Surg. Pathol.* **12** (Suppl. 1), 4–16.
Oud, P. S., Henderik, J. B. J., Beck, H. L. M., Veldhuizen, J. A. M., Vooijs, G. P., Herman, C. J., and Ramaekers, F. C. S. (1985). *Cytometry* **6,** 159–164.
Oud, P. S., Reubsaet-Veldhuizen, J. A. M., Beck, H. L. M., Pahlplatz, M. M. M., Hesselmans, G. H. F. M., Hermkens, H. G., Tas, J., James, J., and Vooijs, G. P. (1986). *Cytometry* **7,** 325–330.
Pallavachini, M. G., Taylor, I. W., and Vindelov, L. L. (1990). *In* "Flow Cytometry and Sorting" (M. R. Melamed, T. Lindmo, and M. L. Mendelsohn, eds.), 2nd Ed., pp. 187–194. New York: Wiley-Liss.
Ramaekers, F. C. S., Beck, H., Vooijs, G. P., and Herman, C. J. (1984). *Exp. Cell Res.* **153,** 249–253.
Schaafsma, H. E., Ramaekers, F. C. S., van Muijen, G. N. P., Ooms, E. C. M., and Ruiter, D. J. (1989). *Histochemistry* **91,** 151–159.
Schimenti, K. J., and Jacobberger, J. W. (1992). *Cytometry* **13,** 48–59.
Schroff, R. W., Bucana, C. D., Klein, R. A., Farrell, M. M., and Morgan, A. C. (1984). *J. Immunol. Methods* **70,** 167–177.
Shankey, T. V. (1993). *In* "Handbook of Flow Cytometyric Methods" (J. P. Robinson, ed), p. 87. New York: Wiley-Liss.
Shankey, T. V., R. C. Flanigan, and C. J. Herman. (1990). *J. Urol.* **143,** 272A.
Shankey, T. V., Kallioniemi, O-P., Koslowski, J. M., Lieber, M. L., Mayall, B. H., Miller, G., and Smith, G. J. (1993). *Cytometry* **14,** 497–500.
Shankey, T. V., Jin, J-K., Dougherty, S., Gandhi, K., and Pyle, J. (1994). *Cancer Mol. Biol.* **1,** 19–25.
Smeets, A. W. G. B., Pauwels, R. P. E., Beck, J. L. M., Geraedts, J. P. M., Debruyne, F. M. J., Laarakkers, L., Feitz, W. F. J., Vooijs, G. P., and Ramaekers, F. C. S. (1987). *Int. J. Cancer* **39,** 304–310.
Steinert, P. M., Steven, A. C., and Roop, D. R. (1985). *Cell (Cambridge, Mass.)* **42,** 411–419.
Tribukait, B., Gustafson, H., and Esposti, P. L. (1982). *Br. J. Urol.* **54,** 130–135.
van der Linden, J. C., Herman, C. J., Boenders, J. G. C., van de Sandt, M. M., and Lindeman, J. (1992). *Cytometry* **13,** 163–168.
van Erp, P. E. J., Rijzewijk, J. J., Boezman, J. B. M., Leenders, J., de Mare, S., Schalkwijk, J., van de Kerkhof, P. C. M., Ramaekers, F. C. S., and Bauer, F. W. (1989). *Am. J. Pathol.* **135,** 865–870.
Visakorpi, T., Kallioniemi, O-P., Paronen, I. Y., Isola, J. J., Heikkinen, A. I., and Koivula, T. A. (1991). *Br. J. Cancer* **64,** 578–582.
Visscher, D. W., Zarbo, R. J., Jacobsen, G., Kambouris, A., Talpos, G., Sakr, W., and Crissman, J. D. (1990). *Lab. Invest.* **62,** 370–378.
Wheeless, L. L., Badalament, R. A., deVere White, R. W., Fradet, Y., and Tribukait, B. (1993). *Cytometry* **14,** 478–481.

comprise most of the non-Hodgkin's lymphomas and may be either indolent or aggressive. This chapter addresses issues related to data analysis and display. Acute leukemias are considered very briefly. The emphasis is on B-cell lymphoproliferative disorders.

B. Pitfalls of Conventional Data Analysis

1. Determination of Positivity

In most laboratories that provide flow cytometric analysis of neoplastic hematopathology samples a typical procedure is as follows: a fraction of the mononuclear cells is gated based on forward scatter (FSC) and side scatter (SSC). The limits of this gate are placed in an arbitrary fashion since the fraction of larger normal lymphocytes frequently overlaps with the fraction of smaller blasts or neoplastic lymphocytes. The fluorescence of the gated population is then compared to a control. The fraction of cells that have a greater fluorescence intensity than an arbitrarily placed cursor is then considered "positive." This represents a methodological "carryover" from the more widespread application of CD4 cell counting in acquired immunodeficiency (AIDS) patients. Monitoring absolute CD4 levels in patients infected with HIV requires accurate technique in order to arrive at precise absolute CD4 cell levels. In neoplastic hematopathology, however, the detection of disease is more dependent on qualitative assessments that do not require precise cell counting techniques. Frequently, arbitrary and ambiguous (pseudorigorous) limits are set in order to separate normal from abnormal. The result of using this approach while attempting to be rigorous is that we are left with more questions than answers: in an acute leukemia, does 35% positive signify that there are 35% blasts in a cell suspension and they are 100% positive for a given marker? Does it mean that of all the blasts only 35% are positive? In the latter case, how would one describe the 65% of blasts that are negative? What is an acceptable false-positive percentage? Where does one place the cursor?

The acute leukemia literature is replete with reports that purport to distinguish "positive" from "negative" based on imaginary percentages above or below which events become (positive), or cease to be. The flow cytometric community has accepted this approach despite the glaring fact that no study has ever established what these limits are or that they even exist. The limits vary from one laboratory to another. This variance (20 vs 25 vs 30%, etc.) is a further indication of their imaginary nature. In the event that an arbitrary limit of 30% is established to distinguish positive events from negative, 31% would be "positive" whereas 29% would be "negative." The underlying concept is one of "all or nothing," implying that antigens may only be either fully expressed or not at all and that this is determined by an arbitrarily defined percentage limit. This approach ignores two fundamental issues: variations in fluorescence intensity of unimodal histograms (dim vs bright) and multimodal fluorescence (Duque and Braylan, 1991).

A more sophisticated and realistic method for separating dimly positive *unimodal* histograms from negative control histograms is the use of the statistical method of Kolmogorov-Smirnov (Young, 1977). This analysis determines the statistical probability of the existence of differences between two populations. Usually, however, the differences are obvious.

When the purported positive histogram is *bimodal* (or multimodal), the relative positivity is easily defined since there will be a negative subpopulation. In order to determine whether a histogram is unimodal, bimodal, kurtotic (wide), skewed (asymmetrical), etc., one must inspect it visually. Consequently, the visual analysis of histograms (single parameter or multiparameter) should be the first step in fluorescent data analysis.

2. Determination of Positivity in Acute Leukemia

In acute leukemias, the cell of interest (blasts) usually has combined forward scatter and immunofluorescence characteristics that allow them to be identified easily when they are simultaneously stained with antibody conjugates that are directed against lineage-associated cell-surface antigens. If one has an admixture of precursor B lymphoblasts and lymphocytes (mostly T cells), and the cell suspension is stained with anti-CD19, only the lymphoblasts and the few normal B cells will react. The presence of lymphoblasts is established both by their morphologic appearance and by their forward light scatter characteristics. Figure 1 shows the larger CD19-positive lymphoblasts and the smaller CD19-positive normal B cells. The number that should be used to describe this situation is not the "percent positive" but the percentage of cells that react with anti-CD19 that *also* have the forward scatter characteristics of blasts (i.e., larger than normal lymphocytes). In this case, 100% is the appropriate number since the fluorescence distribution of CD19 in the larger cells is unimodal. The more biologically meaningful numerical descriptor, however, is the number of blasts expressed as a percentage of total cells. The French–American–British (FAB) classification of acute leukemia requires the presence of at least 30% blasts in order to distinguish acute leukemias from myelodysplastic processes (Bennett *et al.*, 1976). The blastic population can be appropriately described in terms of its composite immunophenotype. An acute precursor B cell leukemia can be described as (composite immunophenotype) HLA-DR^+ $CD19^+$, $CD10^+$. Similarly, an acute myeloid leukemia may be described: HLA-DR^+, $CD13^+$, $CD33^+$, with the understanding that the disease has been defined by conventional (clinical and morphologic) criteria. Low-intensity expression of an antigen should be described as such. Again, this requires visual inspection of the histograms.

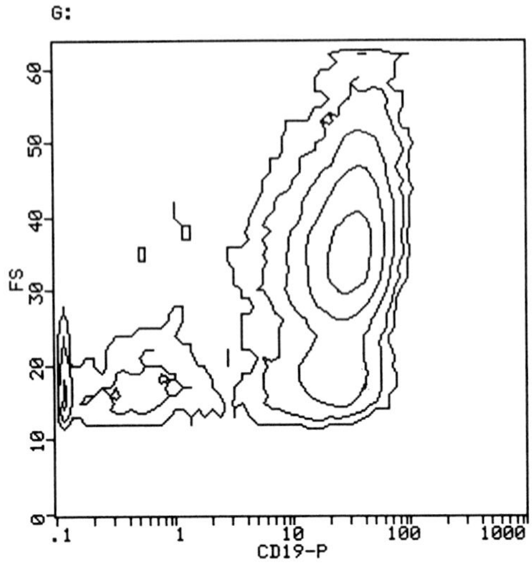

Fig. 1 Correlated display of forward scatter (FSC) vs fluorescence intensity (FL2). The histogram represents the mononuclear component of a suspension of bone marrow that contains 85% blasts. 100% of the blasts (larger cells) are positive for CD19. Additionally, there are a few small B cells that also react positively.

3. The $\kappa:\lambda$ Ratio

Other major pitfalls occur when one attempts to use "ratios" to distinguish normal from abnormal. If one has a specimen that is involved by a non-Hodgkin's B-cell lymphoma, the disease exists whether the entire specimen (lymph node, bone marrow, effusion, etc.) is replaced by a B-cell lymphoproliferative process (markedly altered $\kappa:\lambda$ ratio) or whether it is only partially replaced such as in a T cell-rich B-cell lymphoma (probably normal $\kappa:\lambda$ ratio) (Ramsay *et al.*, 1988). Using appropriate and interactive data display and analysis techniques, one can detect a contamination of 1% monoclonal B cells in a cell suspension composed of heterogeneous T and B cells (Braylan, 1993). In such a case, if one accepts a $\kappa:\lambda$ ratio of 1.5 ± 0.25 as normal, a 1% contamination of polyclonal B cells by monoclonal (κ) B cells would only alter the ratio by 2% ($\kappa:\lambda$ ratio of 1.53). In this example, the $\kappa:\lambda$ ratio would be insensitive to the presence of small (but pathologically significant) numbers of neoplastic B cells.

It is important, then, to reassess some of these issues in an effort to produce results that are biologically and clinically meaningful. In this chapter, I review more dynamic and intuitive approaches that require (a) a fundamental knowledge of neoplastic hematopathology, monoclonal antibodies, and flow cytometry and (b) a willingness to interact with graphic data rather than with numbers, which demands a familiarity with multiparametric data analysis and display techniques.

II. Correlated Multiparametric Data Analysis of B-Cell Neoplasia

A. Gating

When analyzed by modern flow cytometers, cells in suspension exhibit two *intrinsic* parameters (Shapiro, 1983): forward light scatter (FALS, FSC), an acceptable measure of size, and side scatter or 90° light scatter (SSC), a signal that correlates well with internal complexity. These features are routinely used to separate lymphocytes, monocytes, and polymorphonuclear leukocytes in peripheral blood without the need for staining procedures (Salzmann *et al.*, 1975). Depending upon whether one wishes to stain DNA, and/or surface and/or cytoplasmic antigens, there may be several *extrinsic* parameters. If the data are stored in "list mode" (Wood, 1993), one can then recreate the analysis and display scenario and correlate all appropriate parameters in any combination, as many times as desired. If there are four parameters (FSC, SSC, FL1, and FL2), there are 16 potential combinations of correlated display and analysis; they are useful in different ways. For example, a correlated display of FSC and SSC will allow us to separate debris, aggregates, neutrophils, etc., from the mononuclear fraction. This (mononuclear) fraction is usually the object of further analysis (Fig. 2). This same combination of parameters, however, is relatively inefficient in separating subpopulations of mononuclear cells that may exhibit overlapping forward scatter and/or side scatter characteristics (i.e.,

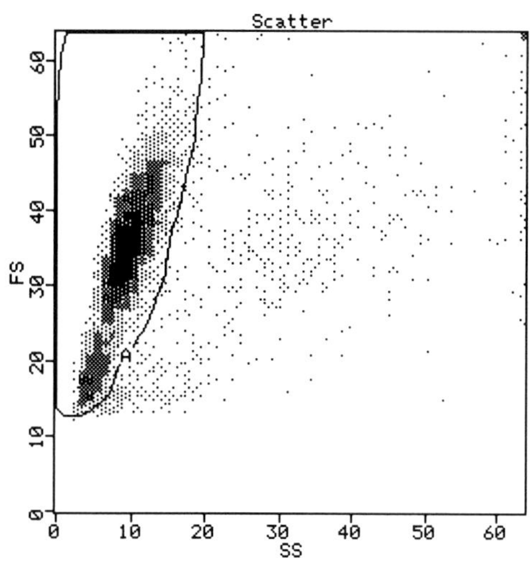

Fig. 2 Correlated display of forward scatter (FSC) vs side scatter (SSC) of a bone marrow cell suspension. There is mild peripheral blood contamination. The mononuclear component is contained within the gate. The majority of the cells are small and there is a subpopulation of larger cells. This should be the initial gate for further analysis.

"normal" cells may overlap with "abnormal" cells). Thus, attempting to "gate" a complex population of mononuclear cells solely on the basis of FSC and SSC in order to separate a subpopulation may prove to be frustrating since one can only guess where to place the gate.

Gating of flow cytometrically defined subcomponents of the mononuclear fraction of a cell suspension, then, should be performed after the cells have been allowed to visually separate themselves based on their intrinsic and/or extrinsic parameters and not *before*, as is usually done. Additional information can be obtained by staining the cell suspension with one or more antibodies conjugated to different fluorochromes (multicolor analysis).

B. Correlated Display of B-Cell-Associated Intrinsic and Extrinsic Parameters

It is more efficient to add an antibody directed against one of the presumed subpopulations (i.e., B cells) and allow the cells to visually separate themselves when appropriately displayed. This is shown in Fig. 1. In every cell suspension that is heterogeneous (and most cell suspensions from human sources are heterogeneous), the addition of an antibody directed against one of the subpopulations will invariably result in a bimodal distribution: positive and negative. In some cases, there may more than one positive population. For example, in a lymph node with partial involvement by a non-Hodgkin's B-cell lymphoma, there may be more than one CD20-positive population (bimodal or multimodal fluorescence) and there may be several populations based on size (forward scatter). This is depicted in Fig. 3.

The results can be described quite well in a binary fashion (i.e., small vs large; positive vs negative; dim vs bright, etc.). Once separated, these cellular subgroups can be probed for additional extrinsic parameters such as light-chain distribution on B cells, second surface or cytoplasmic antigens, and DNA content.

Ideally, multicolor analysis should be performed in response to specific questions. As an example, if the question is whether a patient with lymphocytosis has chronic lymphocytic leukemia (CLL), a combination of CD20 with anti-κ and a second tube with CD20 and anti-λ would determine the presence of light-chain restriction in the B-cell population (monoclonality). Since CLL usually coexpresses CD5, this may constitute a potential third color. The same information, however, can be obtained by inference by showing that a subcomponent of $CD5^+$ cells are $CD3^-$. The choice of two-color or three-color analysis is, obviously, optional. There is nothing magical about three- or four-color analysis except for the fact that these approaches decrease the required number of cells. It is unusual to have clinical situations in which there is an absolute need for three- or four-color analysis. There is, however, a need for multiparameter analysis.

With interactive analysis of graphic data, complex situations can be clearly defined without the need for pseudorigorous (but highly inexact) numerical

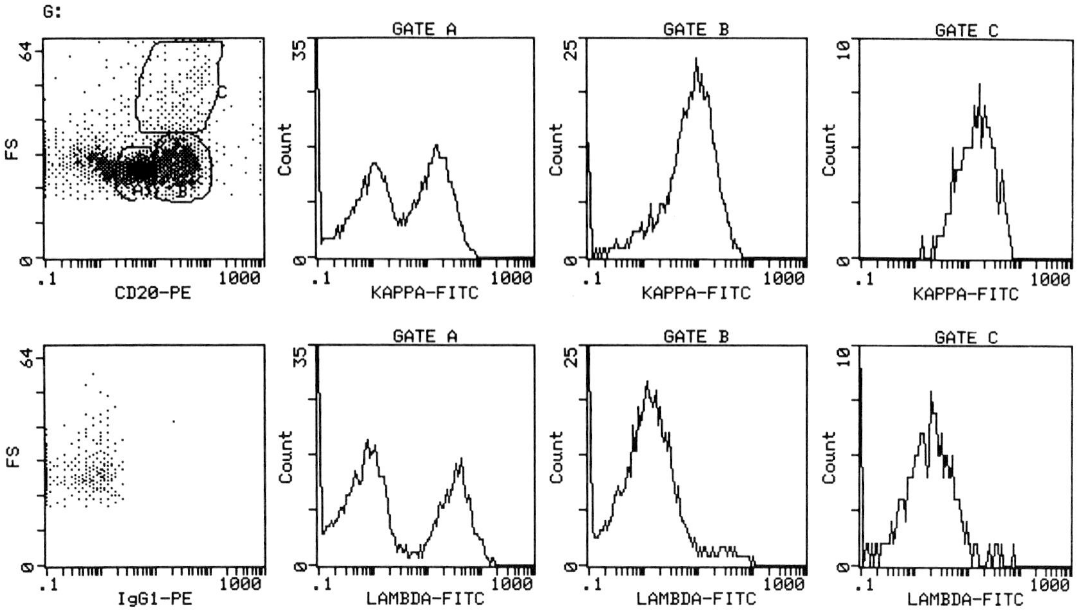

Fig. 3 Lymph node with partial involvement by a non-Hodgkin's B-cell lymphoma. See text for more details.

approaches. Figure 3 illustrates a complex situation wherein a lymph node is only partially involved by a B-cell lymphoma composed of small and large cells. Having excluded contaminating neutrophils, debris, etc., the mononuclear fraction is analyzed as a correlated display of size (FSC) vs fluorescence intensity of CD20. The first step is to determine how many (dual-parameter-defined) clusters are present. In this case there are four clusters as defined by the correlated display of FSC and CD20 reactivity. All can be defined descriptively in a binary fashion:

1. small, CD20-neative cells (presumably T cells);
2. small, CD20-positive cells (intermediate fluorescence);
3. intermediate (size), CD20-positive cells (bright fluorescence); and
4. large (size), CD20 positive cells (bright fluorescence).

The fluorochrome that anti-CD20 is conjugated to is phycoerythrin. Therefore, additional analysis can be performed with other compatible fluorochromes such as fluorescein isothiocyanate (FITC). The choice of antibodies for the additional analysis in this case is polyclonal anti-κ and anti-λ antibodies conjugated to FITC. Ideally, they should be $F(ab')_2$ to avoid non-specific binding to Fc receptors. Addition of blocking serum will further minimize nonspecific binding. If one gates the three positive (CD20 vs FSC-defined) subpopulations and examines the histograms of FLI (FITC-conjugated anti-κ and anti-λ), the results will show:

1. large, bright B cells that are monoclonal (kappa) (Gate C);
2. small and intermediate size, bright B cells that are monoclonal (kappa) (Gate B); and,
3. small, intermediate (fluorescence) B cells that are bimodal (and therefore polyclonal) (Gate A).

In this case we are taking advantage of the differences in fluorescence intensity (CD20) and size (FSC) that are associated with B-cell subpopulations. Fluorescence intensity is not usually addressed by most laboratories since numerical designation of positivity does not take fluorescence intensity into account.

Obviously, one does not always find these differences. For example, CD19 is less likely to exhibit B-cell subpopulation associated differences in intensity than CD20 (Almasri *et al.*, 1992). It remains to be seen whether other B cell-associated reagents exhibit these differences.

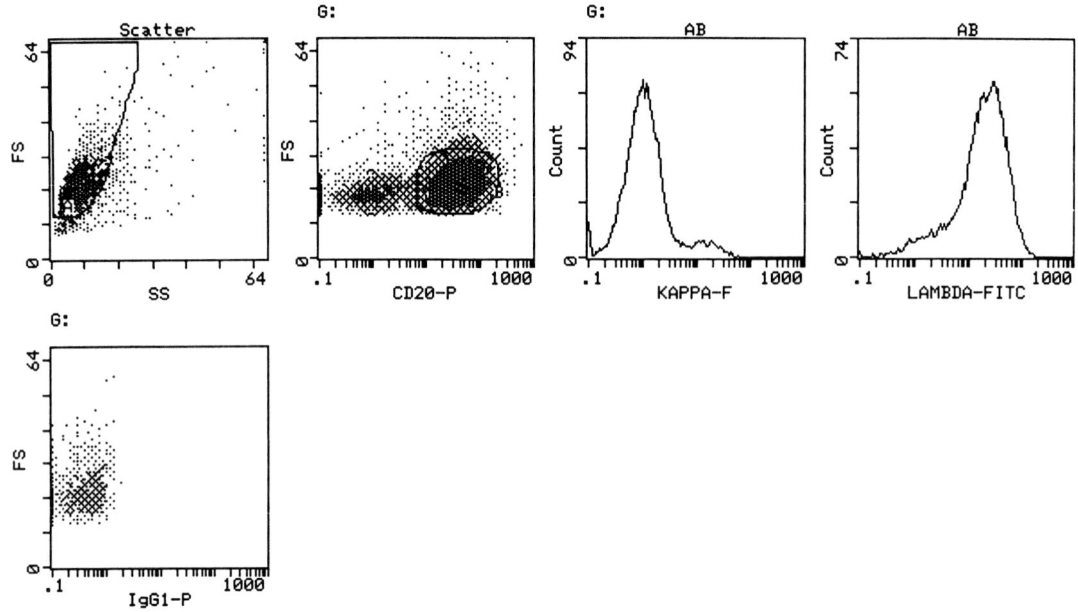

Fig. 4 The upper left frame shows the wide mononuclear gate. The lower left frame shows the appropriate isotype and fluorochrome control. The second frame from the left shows the CD20-positive cluster which is then gated (R2). The third and fourth frames show κ-FITC and λ-FITC fluorescence histograms of CD20-gated cells. There is a clear excess of κ-negative (lower left) and λ-positive (lower right) B cells. These monoclonal (λ) cells are admixed with a smaller proportion of polyclonal cells. For this reason, the histograms are bimodal.

In the event that the neoplastic and nonneoplastic B cells exhibit overlap in terms of size and/or intensity of B-cell antigens, large distortions in the proportion of $\kappa:\lambda$ cells ($\kappa:\lambda$ ratio) may be helpful but must be approached with caution. In normal circumstances, there are usually more κ-positive B cells than λ-positive B cells. Therefore, if one finds a clear excess of λ-positive cells, this may be taken as evidence that there is an admixture of polyclonal and monoclonal (λ) B cells (Fig. 4). If, however, the admixture is of polyclonal B cells with monoclonal (κ) B cells, it may be more difficult to obtain a high degree of certainty of the existence of a monoclonal *sub*population. Sometimes it may be useful to gate the brighter fraction of B cells and compare the shape of the $\kappa:\lambda$ histograms looking for obvious differences between the dimer and the brighter cells. If there is a difference in size (broad FLS distribution), but separate clusters cannot be discerned, one can gate the smaller B cells and compare the $\kappa:\lambda$ histograms with those of the larger B cells. In some cases, one must resort to molecular techniques to establish the presence of monoclonality.

Since monocytes have forward and side scatter characteristics that may overlap with those of large, neoplastic B or T cells or leukemic blasts, care must be taken to account for their presence by morphologic/immunologic correlation. Due to the presence of Fc receptors, monocytes may bind antibodies nonspe-

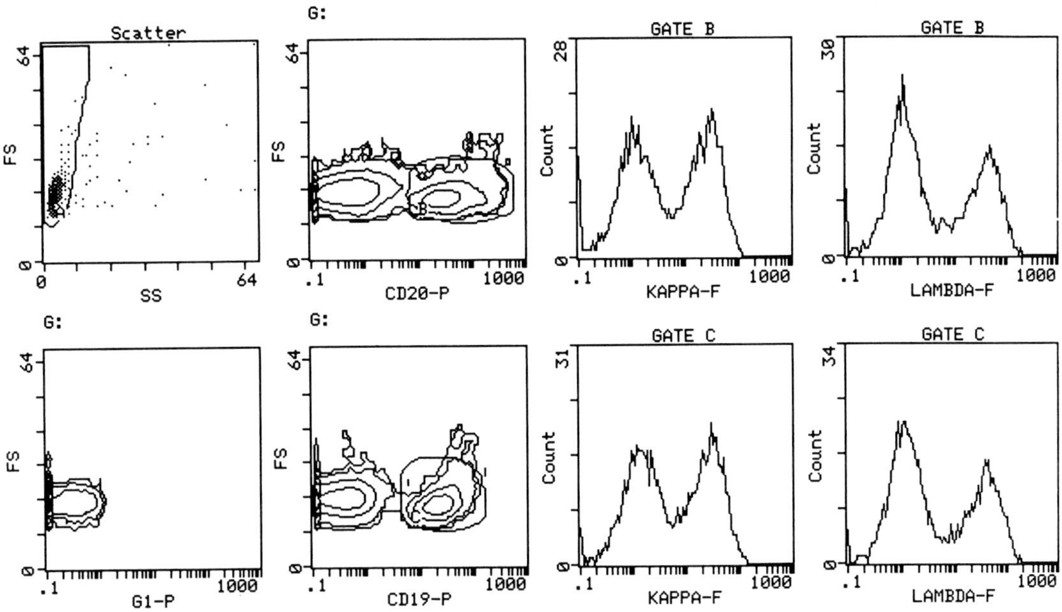

Fig. 5 The upper panel represents CD20-gated B cells; the lower panel represents CD19-gated B cells. The lower left frame represents fluorochrome and isotype controls. The light-chain histograms in the frames to the right are bimodal with a slight excess of κ, a normal finding.

cifically. To circumvent potential pitfalls, we routinely use horse serum in order to saturate Fc receptors and block nonspecific binding. In addition, it is also advisable to include CD14 to ascertain the presence of and relative numbers of these cells.

C. Analysis of B-Cell-Restricted κ/λ Histograms

Immunocompetent B cells express either κ light chains or λ light chains on their surface but not both. They are usually a component of an intact immunoglobulin molecule composed of two heavy chains linked by disulfide bonds to two light chains (Korsmeyer *et al.*, 1981). In normal circumstances, the proportion of B cells that express κ light chains exceeds that of those that express λ light chains by 1.5 to 2.0. Any cell suspension that contains normal B cells will show an admixture of κ-positive and λ-positive cells. If one restricts light-chain analysis to just B cells, a histogram display, with antibodies directed against either light chain, will be bimodal (Fig. 5). The negative control for each case is built in: the λ B cells will be the κ-negative control and vice versa. As illustrated in Fig. 5, the κ histogram should be approximately a mirror image of the λ histogram (60 : 40 and 40 : 60). As mentioned previously, one must take advantage of differences in fluorescence intensity and/or size to select subpopulations of B cells that only express one light chain but that would not necessarily disrupt a $\kappa : \lambda$ ratio if they are at a low frequency. Pure monoclonal populations of B cells exhibit *unimodal* κ or λ histograms. Several situations can explain a unimodal light-chain histogram gated on B cell reactivity:

1. If the κ histogram is brigher than the λ histogram, a κ-positive B-cell population is present.

2. If the λ histogram is brighter than the κ histogram, a λ-positive B-cell population is present.

3. If both κ and λ histograms are unimodal and overlapping (neither is brighter), there are two possibilities:

   a. The population is composed of precursor (surface immunoglobulin-negative) B cells, i.e., *blasts*.

   b. The population is neither κ nor λ positive but is surface membrane immunoglobulin (heavy-chain) restricted.

In lymph nodes are obtained from patients with florid follicular hyperplasia, there is usually an admixture of small and large B cells. The small B cells are usually clearly bimodal. However, the large B cells may exhibit κ and λ histograms that are not clearly bimodal and have a quasi-unimodal distribution that is kurtotic (wide histogram) and skewed to the left (equally for κ and for λ). Care must be taken to *not* identify these as monoclonal. The ambiguous $\kappa : \lambda$ distributions are a result of diminished expression of surface immunoglobulins by large, reactive germinal center cells (Picker *et al.*, 1987) (Fig. 6).

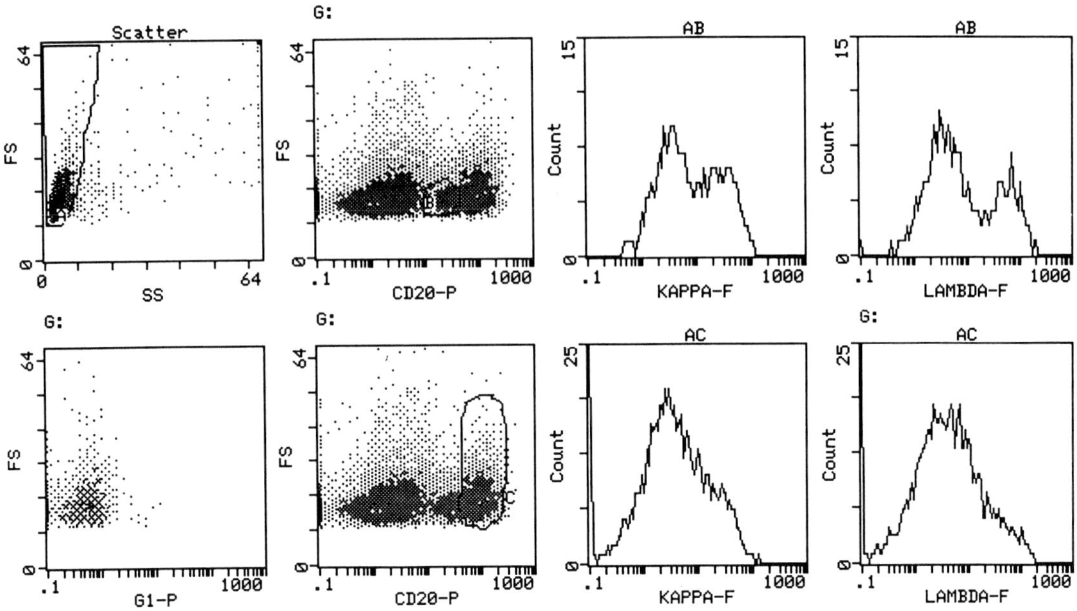

Fig. 6 Florid follicular hyperplasia. The upper panel illustrates CD20 gating of the smaller and dimmer B cells. The light-chain distribution is polyclonal (bimodal). The lower panel illustrates the kappa and lambda histograms following the gating of the CD20-positive, brighter and larger B cells. The kappa and lambda histograms in the lower right hand panels show an ambiguous fluorescence distribution. There is a large negative peak for both kappa and lambda. A smaller proportion of cells is represented by a smaller peak. This is produced by an overlap of polyclonal B cells.

In summary, there is a need for more interactive approaches to data analysis and display techniques to adequately detect and characterize B-cell neoplasia. This should result in an increase in detectability of diseases that may be otherwise masked by numerical designations that are inadequate and insufficient to portray phenomena that are more easily illustrated by simple graphics and described in plain language.

References

Almasri, N. M., Duque, R. E., Iturraspe, J. A., Everett, E. T., and Braylan, R. C. (1992). *Am. J. Hematol.* **40,** 259–263.

Bennett, J. M., Catovsky, D., Daniel, M. T., Flandrin, G., Galton, D. A., Gralnick, H. R., and Sultan, C. (1976). *Br. J. Haematol.* **33,** 451–458.

Braylan, R. C. (1993). *In* "Clinical Flow Cytometry: Principles and Application" (K. D. Bauer, R. E. Duque, and T. V. Shankey, eds.), pp. 203–234. Williams & Wilkins, Baltimore, MD.

Duque, R. E., and Braylan, R. C. (1991). *In* "Techniques in Diagnostic Pathology: Diagnostic Flow Cytometry" (J. S. Coon and R. S. Weinstein, eds.), pp. 89–102. Williams & Wilkins, Baltimore, MD.

Korsmeyer, S. J., Hieter, P. A., Ravetch, J. V., Poplack, D. G., Waldmann, T. A., and Leder, P. (1981). *Proc. Natl. Acad. Sci. U.S.A.* **78,** 7096–7100.
Picker, L. J., Weiss, L. M., Medeiros, L. J., Wood, G. S., and Warnke, R. A. (1987). *Am. J. Pathol.* **128,** 181–201.
Ramsay, A. D., Smith, W. J., and Isaacson, P. G. (1988). *Am. J. Surg. Pathol.* **12**(6), 433–443.
Salzmann, G. C., Crowell, J. M., Martin, J. C., *et al.* (1975). *Acta Cytol.* **19,** 374–377.
Shapiro, H. M. (1983). *Cytometry* **3,** 227–243.
Wood, J. C. S. (1993). *In* "Clinical Flow Cytometry: Principles and Application" (K. D. Bauer, R. E. Duque, and T. V. Shankey, eds.), pp. 71–92. Williams & Wilkins, Baltimore, MD.
Young, I. T. (1977). *J. Histochem. Cytochem.* **25,** 935–939.

CHAPTER 15

Analysis and Sorting of Hematopoietic Stem Cells from Mouse Bone Marrow

Jan W. M. Visser* and Peter de Vries†

* Laboratory of Stem Cells
New York Blood Center
New York, New York 10021
† Immunex Corporation
Seattle, Washington 98101

I. Introduction

All the different blood cell types are derived from a common ancestor, the pluripotent hematopoietic stem cell (PHSC). This cell not only has the capability to differentiate into committed progenitor cells of all the various blood cell lineages, but can also renew itself, thus maintaining the hematopoietic organ at a steady-state level throughout life. The production of mature blood cells out of stem cells and committed progenitors requires >10 cell divisions. As a consequence, the stem cells are rare cells, even in hematopoietic organs. Their incidence in adult mouse bone marrow is estimated to be between 5 per 1,000 and 2 per 100,000 (for reviews, see, e.g., Watt *et al.,* 1987; Visser and Van Bekkum, 1990; Spangrude *et al.,* 1991). No stem cell-specific cytochemical staining has been described yet. Stem cells are detected by examination of their offspring after *in vitro* culture or after transplantation into often lethally irradiated syngeneic or congenic recipients. There exists no consensus about the different culture and *in vivo* transplantation methods with regard to their specificity for pluripotent stem cells. Consequently, the quantification of stem cells by those methods may lead to widely different results. Also consequently, identification of PHSC by flow cytometry (FCM) should be combined with cell culture and transplantation assays for verification. Therefore, FCM methods should employ supravital staining and labeling, which can be combined with sorting, cell culture, and transplantation procedures.

Flow cytometry and cell sorting have contributed significantly to the purification of PHSC and, concomitantly, this has introduced uncertainty about the validity of existing methods for stem cell enumeration. For many years the spleen colony assay (CFU-S: colony forming unit spleen) has been regarded as the clonogenic test for stem cells. Colonies of hematopoietic cells, which appear on the surface of spleens of mice 8–14 days after lethal irradiation and bone marrow transplantation, were thought to be each derived from a pluripotent stem cell in the graft that based itself in the spleen. Magli *et al.* (1982) demonstrated that the spleen colony assay was more complicated than thought, because Day 8 and Day 12 spleen colonies were not necessarily related. Sorting experiments with Hoechst 33342 (Baines and Visser, 1983) and with antibodies against class I MHC antigens (Harris *et al.,* 1984; Visser *et al.,* 1984) have then for the first time indicated that early appearing spleen colonies (Day 8 CFU-S) are derived from other cells than late (Day 12-14) appearing ones. In addition, it could be shown that "true" stem cells with extensive repopulating ability in all blood cell lineages belong to the sorted population which also gives Day 12 CFU-S. Subsequently, Bertoncello *et al.* (1985) demonstrated that these Day 12 CFU-S can be sorted into two groups using rhodamine 123 (Rh123), a dye which stains the mitochondria. They observed that marrow repopulating ability (MRA), a quality of true stem cells, was only present among the Rh123-dull cells, which contain about one-third of the total amount of Day 12 CFU-S. Suspensions of Rh123-bright cells contained the other Day 12 CFU-S, all Day 8 CFU-S (Mulder and Visser, 1987), but no MRA. In addition, it was found that the stem cell type responsible for thymus repopulation was found in the Rh123-dull fraction (Mulder and Visser, 1987). Recently, Chaudhary and Roninson (1991) demonstrated that the low stainability of early human hematopoietic cells by Rh123 is significantly dependent on the presence of a membrane efflux pump, P-gp, that is encoded for by the MDR-1 gene. This indicates that the Rh123-dull cells and perhaps also the murine stem cells have a high expression of this efflux pump. On the other hand, we have demonstrated that the difference between the Rh123-dull and -bright cells is at least partly due to a difference in the sizes of their mitochondria (Visser *et al.*, 1991). The expression of an efflux pump in stem cells may explain another finding, viz., that stem cells do stain less readily with the dye Hoechst 33342 than other bone marrow cells (Baines and Visser, 1983; Pallavicini *et al.*, 1985; Neben *et al.*, 1991; Wolf *et al.*, 1993).

Recently, Jones *et al.* (1990) showed that fractions of cells can be obtained that contain PHSC without CFU-S. This suggests that even cells that give rise to late appearing spleen colonies (Day 12 CFU-S) are different from stem cells. Because of these observations the spleen colony assay has become a measure of stem cells which should be interpreted with much caution. However, no other clonal assays are available. The long-term repopulation assay as performed either with sex-mismatched transplantation and Y chromosome detection (Jones *et al.*, 1990; Visser *et al.*, 1991) or with antigenic markers (Spangrude *et al.*, 1988) is the only accepted assay system for PHSC at present. There is an urgent need for a more rapid stem cell detection and quantification assay. The flow cytometric detection of candidate stem cells employing a combination of markers is one of the most powerful alternatives to the current *in vitro* and *in vivo* assays.

Various protocols for identifying and sorting spleen colony forming cells have been described in recent years. Most of these employed FCM in combination with other physical and immunological cell separation techniques, which served to remove the bulk of mature cells. Since the successful stem cell separations by Van Bekkum *et al.* (1971), equilibrium density centrifugation is the method of choice for preenrichment of the PHSC from murine bone marrow. In the flow cytometer forward and perpendicular light scatter measurements have proved to be useful to further eliminate mature cells from analysis and sorting procedures. In addition, cells can be labeled in a viable state with fluorescent lectins and monoclonal antibodies (mAb) directed against cell-surface molecules, some of which are differentiation antigens. The fluorescence of individual cells due to the binding of lectins and antibodies can be determined quantitatively by FCM. We and others successfully employed the lectin wheat germ agglutinin (WGA) to sort the stem cells (Visser and Bol, 1981; Visser *et al.*, 1984; Lord and Spooncer, 1986; Ploemacher *et al.*, 1992). The use of WGA for this purpose was indicated by results obtained with free-flow electrophoresis and neuraminidase treatment to study membrane constituents of the stem cells and by subsequent analysis of WGA-labeled bone marrow cells using flow cytometry (Visser *et al.*, 1981). A number of antibodies have been successfully employed for stem cell separation. Two strategies could be followed here, either it was attempted to find a specific stem cell label (positive selection) or all committed and fully differentiated cells were labeled (negative selection). Useful antibodies for labeling stem cells were found to be anti-H-$2K^k$ (Visser *et al.*, 1984), and anti-Qa-m7 (Bertoncello *et al.*, 1986), both directed against class I MHC antigens. Also anti-Thy-1 (Muller-Sieburg *et al.*, 1986) and an antibody against a putative stem cell antigen (Sca-1; Spangrude *et al.*, 1988) have been reported to be of use for stem cell sorting. A diversity of cocktails of mAb against mature cells has been described and used to negatively select stem cells (Hoang *et al.*, 1983; Muller-Sieburg *et al.*, 1986; Spangrude *et al.*, 1988; Watt *et al.*, 1987; de Vries *et al.*, 1991; Migliaccio *et al.*, 1991). In addition, as mentioned above, supravital staining using Hoechst 33342 or rhodamine 123 was also found to be useful for stem cell identification and sorting. Recently, the expression of the protooncogene *c-kit* that encodes a transmembrane tyrosine kinase receptor for hematopoietic growth factor encoded by the steel locus and variably known as mast cell growth factor (MGF), stem cell factor, kit ligand, or steel factor, has been shown to be a useful marker to subdivide stem cell populations (de Vries 1991; Okada *et al.*, 1992; Ikuta and Weissman, 1992).

C-kit is expressed not only on stem cells with radioprotective and reconstitution capacity, but also on the large majority of Day 14 CFU-S and most of the *in vitro* colony forming cells and cells that can reconstitute the thymus (P. de Vries, unpublished data; de Vries *et al.*, 1992; Okada *et al.*, 1992; Ikuta and Weissman, 1992). Although the precise function of *c-kit*/MGF interactions in hematopoiesis is not yet clear it is obvious that these interactions must play an important role in the survival, proliferation, and differentiation of PHSC.

It has become clear that a combination of a number of labels and dyes is needed to unambiguously distinguish the PHSC from the other cells in the heterogeneous bone marrow. In this chapter we describe combinations that are of use to demonstrate heterogeneity of Day 12 CFU-S and that, therefore, may be of help to unravel further the true identity of the pluripotent stem cell.

II. Application

At present the analysis and sorting of PHSC is primarily employed in experimental bone marrow transplantation and gene therapy and to some extent in studies concerning the regulation of differentiation and proliferation.

The application for bone marrow transplantation arises from the concept that no malignant cells should be returned to the patient in the case of autologous transplantations and that preferably no lymphoid cells causing graft-versus-host reaction should be given in allogeneic grafts. As the malignant cells are different in most patients whereas the stem cells are probably similar, it is strategically of advantage to aim at developing a sorting procedure for the latter. Such a procedure would then also be of use for the separation of the lymphoid cells from the PHSC for allogeneic grafts. Transplantation of purified pluripotent stem cells which have self-renewal capacity should completely reconstitute the immunohematological system in patients after radiotherapy and chemotherapy, as well as in victims or radiation accidents. Application of flow cytometry for the engineering of human bone marrow grafts may have to wait until high-speed sorting techniques are implemented. It is envisaged, however, that a combination of cell separation techniques, including flow cytometry, could be successfully employed for that purpose. The present methodology for preparing those grafts already profits considerably from the analytical studies and preliminary flow cytometric sorting of murine and monkey cells. In addition, the analysis of the bone marrow by using specific stem cell staining and labeling and flow cytometry will be of use for monitoring the extent of effects of therapy on normal stem cells, which belong to the most radiosensitive cells in the body and which are essential for recovery after therapy.

A new application of purified stem cells concerns experimental gene therapy. Introduction or modification of genes in stem cells is expected to be of use for treating enzyme deficiencies and other malignancies by bone marrow transplantation. In order to evaluate the efficiency and extent of the DNA modifications in the stem cells, it is necessary to isolate them in sufficient quantities from the extremely heterogeneous suspension of bone marrow cells. Furthermore, to insert genes for therapy, it is also necessary that the stem cells remain pluripotential. Therefore, it is of importance to remove growth factor producing cells, which naturally occur in the bone marrow, before the stem cells are cultured, or, otherwise, the PHSC will differentiate and mature. In addition, enrichment of stem cells prior to transfection immediately improves the multiplicity of infection, which is important if low virus titers have to be employed, e.g., because of lengthy inserted genes. Growth factor producing cells should also be removed before stem cells are cultured if the effects of newly purified and recombinant growth factors on the regulation of PHSC are to be studied.

The hematopoietic organ is an attractive general model system for studying the regulation of differentiation and proliferation. Because of their pluripotency, the most interesting cells for this purpose are the PHSC, which are rare, and, which, therefore, are preferably analyzed by sorting them. The application of purified PHSC resulted in several findings. For instance, it was found that PHSC do not home in the thymus after transplantation, whereas some of the committed daughter cells of the PHSC, the prothymocytes, do so (Mulder and Visser, 1987 and 1988); that osteoclasts arise from PHSC; and that stem cells are sensitive to erythropoietin (Migliaccio *et al.,* 1988).

A new and fascinating application is the production of cDNA libraries from sorted stem cells. Several approaches are followed in different laboratories. RT-PCR techniques are applied to single-sorted candidate stem cells or to suspensions with low numbers (100–1000) of such cells. The amplified sequences are cloned and the resulting libraries are subtracted to select for stem cell-specific clones. Partial sequencing then reveals the new genes and hybridization with cDNA from different sources helps to screen for stem cell specificity of these genes. Alternatively degenerate primers for gene families can be used to screen the banks. Ershler *et al.* (1993) such discovered six new cdc-2-related genes in a cDNA bank obtained from a fraction of highly enriched stem cells from mouse bone marrow. Since the identity of the pluripotent stem cell is not established, it is important for this application that the flow cytometric properties of the sorted fractions are well documented. Overlapping properties are leading to contamination which is also amplified by the PCR and, therefore, clean separations based on nonoverlapping populations are to be recommended for this purpose.

It can be expected that the use of the analytical capabilities of flow cytometry for stem cell studies will become most important if methods for the specific labeling of those rare cells are improved. Concomitantly, the definition of pluripotent stem cells should be sharpened. The labeling method described below may be considered as one of the steps in this iterative process.

III. Materials

A. Buffers and Media

Bone marrow cell suspensions are prepared in Hanks' balanced salt solution (HBSS) (Laboratories Eurobio, Paris, France) buffered at pH 6.9 with 10 m*M* Hepes buffer (Merck & Co., Rahway, NJ). For mouse bone marrow cells the osmolarity has to be 300–305 mOsm/kg. In general, HBSS without phenol red is preferred in order to avoid optical side effects of the medium in the flow cytometer. This relates in particular to the sheath fluid in the sorter. Cell suspensions are washed before and between labeling with mAbs and conjugates with HBSS + Hepes, which we will further abbreviate as HH, containing 5% (v/v) of either fetal calf serum (FCS, Sera-lab, Ltd., Sussex, England) or newborn calf serum (NCS, Seralab) and 0.02% (v/v) sodium azide (Merck), which together will be further abbreviated as HSA (HH + serum + azide). After the final labeling, cells are resuspended in HH. Cells and media are kept at 0–4°C unless indicated otherwise.

B. Metrizamide

As the first step in stem cell purification procedures, bone marrow cells are separated using a discontinuous metrizamide (Nyegaard, Oslo, Norway) density gradient as described earlier (Visser *et al.*, 1984). The metrizamide (MA) is dissolved in HH + 1% (v/v) bovine serum albumine (BSA, fraction V, Sigma), adjusted first at pH 6.7 and, subsequently, at 300 mOsm/kg by appropriate dilution with water. Two solutions are prepared: a dense one of about 1.10 g/cm^3 and a light one of about 1.05 g/cm^3. MA solutions of wanted intermediate densities (1.078 g/cm^3) are prepared by mixing these two solutions.

C. Wheat Germ Agglutinin

Fluoresceinated wheat germ agglutinin (WGA/FITC; Polysciences, Inc., Warrington, PA) and Texas red-labeled wheat germ agglutinin (WGA/TxR: Molecular Probes, Eugene OR) were stored as a stock solution of 1 mg/ml at −20°C. Thawed aliquots can be stored at 0–4°C for up to several months. During prolonged storage (longer than 12 months) FITC and WGA may become released from each other. The solution is then processed on a Sephadex G-25 column to remove free FITC (Bauman *et al.*, 1985).

D. *N*-Acetyl-D-Glucosamine

WGA can be removed from cells by incubation with the competing sugar *N*-acetyl-D-glucosamine (*N*-ac-D-gluc). For this purpose, *N*-ac-D-gluc (Polysciences, Inc.) is dissolved at a final concentration of 0.2 *M* in HH which is diluted with water ($\frac{2}{3}$ HH + $\frac{1}{3}$ water) to compensate for the contribution of the sugar to the osmolarity. The solution is stored frozen; it is set at 37°C shortly before use.

E. Monoclonal Antibodies and Biotin and Avidin Conjugates

Monoclonal antibody 15-1.1 and its identical subclones 15-1.1.4 and 15-1.4.1 were raised by intrasplenic immunization of a Brown Norway rat with suspensions which were highly enriched for CFU-S. However, the mAbs do not bind to CFU-S but react strongly with bone marrow monocytes, granulocytes, and their immediate precursors. The antibodies are available upon request.

Aliquots of fluorescein isothiocyanate-labeled avidin (av/FITC, Sigma, Deisenhofen, FRG) and avidin conjugated to phycoerythrin (av/PE, Becton Dickinson) are stored frozen and used at a 100- and a 50-fold dilution, respectively. The dilutions are stored at 0–4°C for several weeks without noticeable negative effects on the labelings.

Murine recombinant mast cell growth factor (MGF: Immunex Corporation, Seattle, WA) was biotinylated (MGF/biotin) as described by Armitage *et al.* (1990). Briefly, biotin X NHS (Calbiochem–Novabiochem Corp., La Jolla, CA) and purified MGF were mixed at a molar ratio of 20 : 1 in 0.1 *M* NaHCO3, pH 8.4, for 30 min at room temperature, with inverting of the reaction tube every 5 min. At the end of the incubation period the mixture was centrifuged through a 1-ml Sephadex G-25 (Pharmacia) desalting column and the eluate adjusted to 100 mg/ml in PBS + 0.02% NaN3, assuming 100% recovery of MGF, and stored at 4°C.

F. Rhodamine 123

The supravital dye rhodamine 123 (Rh123; Eastman Kodak, Rochester NY) was dissolved in distilled water at a stock concentration of 1 mg/ml and stored at 0–4°C in the dark. Shortly before use 1 μl of this solution was diluted in 10 ml HH + 5% FCS. This dilution was prewarmed to 37°C. Alternatively, 1 μl Rh123 stock solution was added to 10 ml of the 0.2 *M* *N*-ac-D-gluc solution described above. FCS (5% v/v) was then also added to the solution.

IV. Instruments

Flow cytometry for the analysis and sorting of PHSC can be performed efficiently using an instrument with only one laser. Since multiple labelings of the cells are generally necessary, some time can be saved if a two- or three-laser instrument is employed. Good sorts have been obtained using standard B&D (FACS II, FACStar) and Coulter (Elite) instuments as well as using experimental flow cytometers in our own lab. It is of use to have a single-cell cloning device for the sorting and culturing of single stem cells in the wells of Terasaki trays.

V. Staining and Sorting Procedure

A. Density Gradient Centrifugation

Discontinuous gradients are prepared by first pipetting 1 ml of a high-density MA solution (1.100 g/cm^3) containing 5–6 $\times$ 10^7 bone marrow cells in a round-bottomed tube (Falcon, type 2057). On top of this solution, 3 ml of a MA solution with intermediate density (1.078 g/cm^3) is layered. To label the cells during the density separation, and to save time, this MA solution contains

WGA/FITC and WGA/TxR at a final concentration of 0.1 μg/ml. Finally, 1 ml of a low-density MA solution (1.055 g/cm^3) is put on top of the intermediate solution. The tube is centrifuged for 10 min, at 4°C, at 1000 *g*. The cells in the low-density fraction and from the interface between the top and intermediate layer (low-density cells) are collected, washed, centrifuged, and counted. This fraction typically contains 7–13% of the cells loaded onto the gradient.

B. WGA Sort

The low-density cells are subsequently analyzed and sorted using forward light scatter (FLS), perpendicular light scatter (PLS), and WGA/FITC fluorescence intensities as parameters. FLS and PLS are measured using linear amplification, and FITC fluorescence using logarithmic amplification. The laser is set at 488 nm, 0.5 W. The laser light blocking bar for FLS is set as narrow as possible (e.g., 1 mm in the FACS II), the one for PLS is taken broader than usual, viz., 2 mm in the FACS II, in order to block the optical disturbances generated by the droplet formation. The PLS is measured (preferably through a 488-nm laser line narrow band pass filter; e.g., Melles Griot, Irvine, CA) by an S11- or S20-type photomultiplier. The FITC fluorescence is measured by an S20-type photomultiplier which is blocked by a filter combination consisting of a 520-nm long wave pass (Ditric Research Optics, Hudson, MA) and a 520- to 550-nm band pass filter (Pomfret, Stamford, CT).

The FLS signal is used to trigger the flow cytometer. The threshold is set such that the erythrocytes are not detected, whereas the lymphocytes are (Fig. 1). The sort windows are set as follows. First, the fluorescence window is set

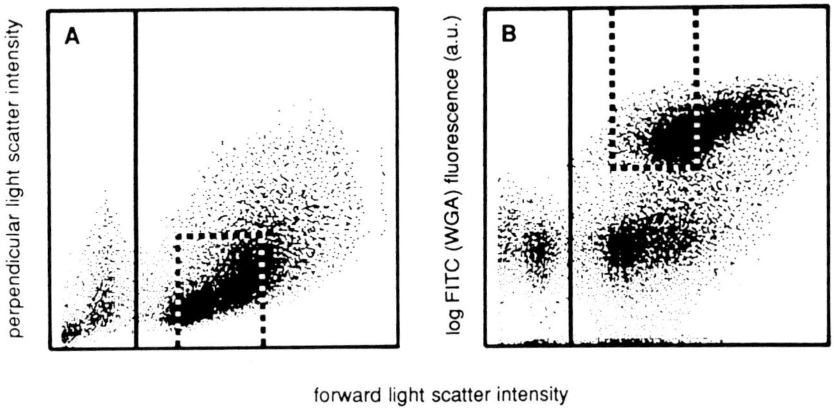

Fig. 1 Bivariate distribution of (a) the forward and perpendicular light scatter intensities and (b) the forward light scatter and WGA/FITC fluorescence intensities of mouse bone marrow cells. The straight vertical line indicates the FLS threshold for the triggering of the flow cytometer. The square window in (a) encloses the "blast" cells.

to include only positive cells. Those cells generally have a homogenous FLS distribution. The FLS window is then set to exclude the larger cells and aggregates (Fig. 1). The PLS distribution of the cells within the FLS/fluorescence windows is then examined; it is generally broad with a level top. The PLS window is set to exclude the half of the cells with the brightest PLS using that distribution (Fig. 1). Between 5 and 10% of the low-density cells normally are selected in this combination of windows. We call the FLS/PLS window the "blast window."

The cells are sorted with an analysis rate of 2500 to 3000 per second using a 50-μm nozzle (36,000 drops per second; 3 drops deflected per cell).

Deflected droplets are collected onto the wall of a 15-ml glass tube, which is coated with HH containing 5% BSA prior to the sorting. After about 2 hr 2×10^7 cells are analyzed and about 10^6 WGA-positive blast cells are collected. The sample and collection tubes are cooled to 2–4°C during the sorting.

C. Removal of WGA/FITC and Labeling with 15-1.1

After the WGA sort 10 ml of a 0.2 *M* solution of *N*-ac-D-gluc is added to the sorted cells (2–3 ml), and, subsequently, they are incubated for 15 min at 37°C in order to remove the WGA/FITC from the cells. Then they are centrifuged (10 min, 400*g*), and the pellet, resuspended in 1–2 ml of HSA, is transferred to a sample tube (Falcon 2058). The cells are centrifuged again and the pellet is now resuspended in 50–100 μl of the appropriate dilution of 15-1.1/FITC. This suspension is incubated for 30–45 min at 0–4°C and again centrifuged.

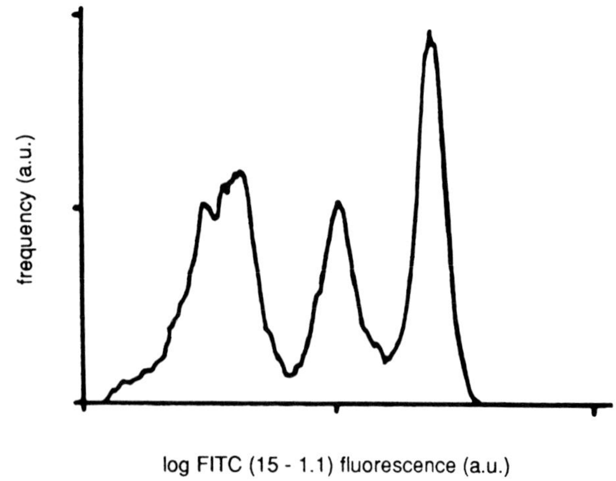

Fig. 2 Frequency distribution of the fluorescence intensities of sorted mouse bone marrow cells labeled with 15-1.1/FITC. "Blast" cell population after a density cut and selection of the brightly WGA fluorescent cells.

The pellet is resuspended in HH (1–2 ml) and again analyzed by the flow cytometer.

D. 15-1.1 Sort

Bound antibodies are detected using 0.5-W 488-nm argon laser light for FITC conjugates and 0.3-W 590-nm R6G dye laser light for Texas red and 600-nm light for allophycocyanin (APC). The photomultiplier which measured these latter conjugates is provided with 2-mm 630-nm long-pass filters (Schott, Duryea, PA) for Texas red and with a 1-mm 660 nm one (Schott) if APC was used. The signals are logarithmically amplified. Cells are now analyzed with the FLS trigger threshold just above the noise and debris level, permitting detection of the erythrocytes which passed the WGA sort as passengers in deflected droplets containing wanted cells. The FLS/PLS blast cell window is maintained for the sorting. The fluorescence window is set to sort the 15-1.1-negative cells (Fig. 2). At this stage the window can be set using the log fluorescence histogram which shows the wanted cells as a separate subpopulation.

E. Rhodamine 123 Sort

After the WGA sort, subsequent incubation in *N*-ac-D-gluc, and the 15-1.1 labeling and sorting, the cells (generally between 10^5 and 3×10^5) are incubated in 2 to 3 ml of a solution of 0.1 μg/ml Rh123 in HH plus 5% FCS for 15 min at 37°C. The cells are then washed once and resuspended in HH (1 ml) for a final sort run. The flow cytometer is equipped with the same filters as for the FITC measurements, only the photomultiplier power supply is reduced somewhat (800 V for Rh123 instead of 900 V). The FLS/PLS blast window now contains more than 50% of the cells. Two populations of cells can at this stage be recognized: a dull fluorescent one and a bright fluorescent one, which are clearly separated in the fluorescence histogram (Fig. 3). Both fractions are sorted and further studied. If stem cells are sorted for cDNA, the sheath fluid during this last sorting step consists of diethyl pyrocarbonate (DEPC)-treated HH and the sorted cells are collected in tubes that had been rinsed with a DEPC-treated BSA solution to prevent loss of RNA. For the same reason, sample and sorted cells are kept at 2–4°C as much as possible during the procedure.

F. Double and Triple Labeling

If a two- or three-laser flow cytometer is used the lectin and antibody labelings can be combined. WGA/1-pyrene-butyril (WGA/pyrene; Molecular Probes, Junction City, OR U.S.A.) instead of WGA/FITC is then used if UV lines (351–363 nm argon laser) are additionally available. And if a dye laser is added, WGA/TxR (Molecular Probes, Eugene, OR) can be used instead.

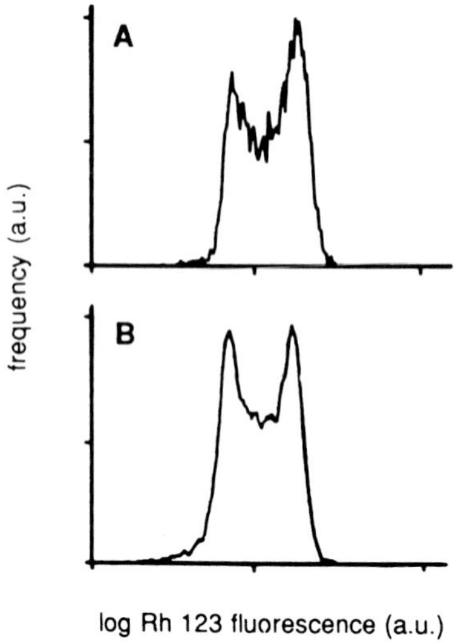

Fig. 3 Frequency distributions of the fluorescence intensities of sorted mouse bone marrow cells labeled with rhodamine 123. "Blast" cell population after a density cut and (a) double WGA/FITC sort (Lord and Spooncer, 1986), and (b) WGA/FITC sort and subsequent 15-1.1/FITC (Visser and de Vries, 1988; de Vries *et al.*, 1991).

G. MGF Sort

Instead of using two or three subsequent sorts, stem cells can also be purified in a single sort run as described earlier (de Vries *et al.*, 1992). After a density gradient separation and simultaneous labeling with WGA/TxR instead of WGA/FITC, the low-density cells are stained with 15-1.4.1/FITC and anti-B220/FITC (CD45R, PharMingen, San Diego, CA), and MGF/biotin. After 1 hr incubation

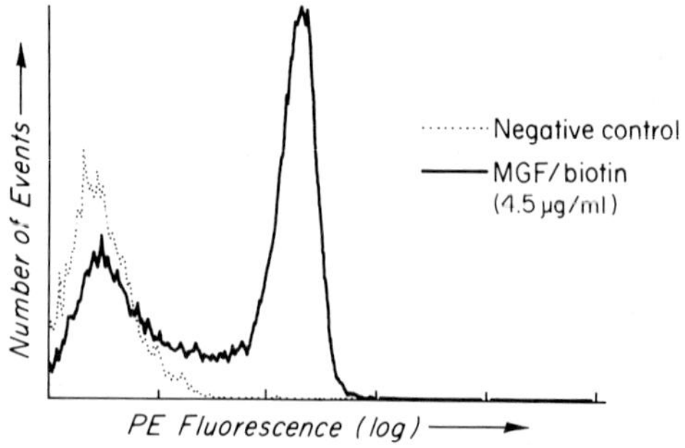

Fig. 4 Frequency distributions of sorted WGA$^+$/15-1.4.1$^-$/B220$^-$ low-density cells labeled with biotinylated MGF (solid line) and control cells that were incubated with excess unmodified MGF (dotted line), both after further incubation with avidin/PE.

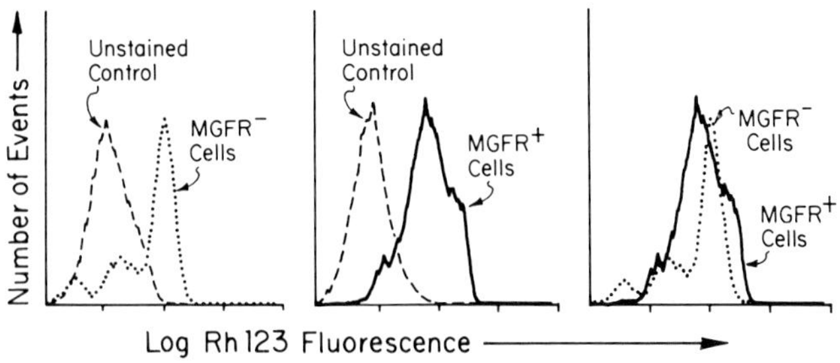

Fig. 5 Rh123 fluorescence intensity distributions of *c-kit*$^+$ (MGFR$^+$) and *c-kit*$^-$ (MGFR$^-$) sorted fractions of WGA$^+$/15-1.4.1$^-$/B220$^-$ low-density cells.

on ice the cells are washed once in HSA and are then incubated with UltraAvidin/PE (UA/PE: Leinco Technologies, Inc., St. Louis, MO) for 30 min on ice. The cells are washed once and are resuspended in HH plus 5% FCS at 2–3 $\times$ 10^6 cells/ml.

Subsequently, the *c-kit* distribution of the WGA$^+$/15-1.4.1$^-$ + B220$^-$/LD cells is analyzed using the LYSYSII software package (Becton–Dickinson) and the WGA$^+$/15-1.4.1$^-$ + B220$^-$/LD/*c-kit*$^-$ and WGA$^+$/15-1.4.1$^-$ + B220$^-$/LD/*c-kit*$^+$ cells are sorted (Fig. 4) using the FACStar Plus (normal R mode).

VI. Critical Aspects of the Procedure

A. Osmolarities, pH Values, and Densities

Since it is necessary to analyze and sort living cells, the osmolarity of the media and reagents is of importance. It is therefore useful to have easy access to an osmometer, so that the reagents can be checked regularly. This remark concerns all reagents. The right osmolarity is of utmost importance for the metrizamide of the density gradient, since the cells may swell or shrink due to osmolarity changes and, thereby, their density may change yielding a different result after centrifugation. Metrizamide, however, tends to dimerize at higher concentrations, so that its contribution to the osmolarity is not linearly related to its concentration. Therefore, we prepare two densities of isotonic metrizamide solutions and mix those to obtain intermediate densities. The osmolarity gradient is then negligible.

If the number of bone marrow cells recovered in the low-density fraction is less than 7% or more than 13% of the loaded cells, the metrizamide solution of the intermediate density is adjusted by adding aliquots of the high-density or the low-density solution, respectively, such that the recovery will be between 7 and 13% again, ideally 10%, in later experiments with the same reagents.

Other materials may be used for the density gradient, such as BSA, Nycodenz

(Nyegaard, Oslo, Norway), or Percoll (Pharmacia Fine Chemicals, Uppsala, Sweden), as long as they yield about 10% of the loaded cells in the low-density fraction. With BSA care should be taken that the osmolarity is correct (use dialysed BSA). Furthermore, with BSA the pH of the gradient will be low (near values of 5.2–5.5), and this affects the density of hematopoietic cells, such that lower densities of BSA solutions should be used to obtain 10% of the cells in the top fraction. With Percoll the pH is increased and in our hands not stable during storage, it tends to approach a value of 7.8–8.0 after storage for one night, whereas we prefer a pH of about 7.0. A pH of more than 7.5 is toxic for the *in vitro* clonogenic hematopoietic progenitor cells. Percoll solutions should, therefore, be freshly prepared or its pH adjusted shortly before use.

The amount of cells loaded on the density gradient may also affect its physical properties. If we load the bone marrow cells in the bottom fraction (with density 1.10 g/cm^3), only 10% of the cells will move upward, and up to 5×10^8 cells can be loaded without causing streaming. If the cells are loaded in the top fraction (of 1.05 g/cm^3), 90% of the cells will move downward, so that only 5×10^7 cells can be loaded per gradient tube. It is advantageous to divide the bone marrow cells at the start of the separation procedure over two or more density gradients, so that occasional small accidents with one of the tubes do not necessarily result in a complete failure.

Erroneous osmolarities which have caused us trouble in the past, apart from those in density materials, were found in the sheath fluid (commercially available salt solutions, whether or not supplemented with azide, and adjusted for human cells), in antibody solutions both from ascites and from hybridoma cell supernatant, and in *N*-ac-D-gluc solutions.

B. Sterility of Materials and Instruments

Since, most of the time, the cells have to be cultured after the sorting, contamination of the suspension should be avoided. Normal precautions are taken during the preparation and handling of the cell suspensions. In addition, the sorter is sterilized: the sheath fluid is first replaced by water, which is run through the tubing to the nozzle and backward to the sample tube position, in order to remove salt or other crystals. Next, 70% ethanol is run through the tubing and, preferably, is kept in the tubing overnight. Subsequently, the ethanol is replaced by sterile water, and, finally, this in turn is replaced by the sheath fluid of choice (sterile, colorless HH).

C. Avidin and the Opsonization of Transplanted Stem Cells

If the stem cells are studied *in vivo* after the sorting, they should not contain labels which trigger the immune system to remove them. A finding by Bauman *et al.* (1985) is of much help for the study of sorted stem cells: *in vivo* opsonization of biotinylated antibody-labeled stem cells can be prevented by additional labeling with avidin prior to transplantation. This is particularly useful if, e.g., anti-H-$2K^k$ biotin is employed (Visser *et al.,* 1984). For some mouse strains the appropriate antibody against the H-$2K^k$ haplotype is hard to obtain, or sometimes the biotinylation is suboptimal, causing loss of stem cells upon transplantation. One should avoid these conditions and, if necessary, compensate them by additional incubation with avidin (Bauman *et al.,* 1985). Negative selection of the stem cells using the mAb 15-1.1 avoids these problems.

D. WGA/FITC, WGA, and FITC

After prolonged storage FITC may be released from WGA/FITC. If bone marrow cells are incubated with free FITC, they will all become fluorescent. Therefore free FITC has to be removed from old solutions of WGA/FITC. A Sephadex G-25 column is used for that purpose (Bauman *et al.,* 1985). This does not, however, separate WGA/FITC from WGA, so that after several runs through the column, a higher concentration of the solution has to be added to the cells in order to obtain the same fluorescence intensity. This, however, may lead to concentrations of WGA which are sufficient for agglutination. Such high concentrations should be avoided for flow cytometry.

E. Coated and Cooled Collection Tubes

The sorting of sufficient numbers of stem cells takes several hours. In addition, the frequency of stem cells is low. Therefore, it is useful to coat the wall of the collection tube with serum or BSA, so that the deflected droplets efficiently slide to the bottom of the tube and the sort recovery is close to 100%. The added serum or BSA will also help to maintain the viability of the sorted cells. On the other hand, it prevents the adherence of monocytes and other adherent cells to the wall of the tube, and, therefore, these have to be removed by other methods, e.g., using the mAb 15-1.1. Furthermore, although the stem cells are physically relatively stable cells which are not killed by keeping them for up to 8 hr at room temperature, it is useful to cool the sample and collection tubes, in order to preserve cellular molecules with a short half-life, if those are to be studied after the sorting, or to shorten the lag phase at the culturing of the sorted cells.

VII. Results and Discussion

A. Sorting of Stem Cells Using WGA, 15-1.1, and Rh123

Table I gives the numbers of bone marrow cells and of CFU-S at the consecutive separation steps. The CFU-S have long served as a measure for pluripotent stem cells. It was assumed, on fair grounds, that each of the spleen colonies

Table I
Enrichment for CFU-S

Suspension sorted	Number of cells (% of unfractionated bone marrow cells)	Number of spleen colonies per 10^5 transplanted cells	
		Day 8 CFU-S	Day 12 CFU-S
Unfractionated bone marrow	100	28.3	30.6
LD cells	10	121	128
WGA^+LD cells	1	1830	2040
WGA^+/WGA^+/LD cells	0.5	2340	3870
$WGA^+/15 - 1.1^{-/+}$/LD cells	0.1	3900	5300
$WGA^+/WGA^+/Rh123^+$/LD cells	0.1	240	3910
$WGA^+/WGA^+/Rh123^{2+}$/LD cells	0.4	4810	3790
$WGA^+/15 - 1.1^{-/+}/Rh123^+$/LD cells	0.05	1060	8650
$WGA^+/15 - 1.1^{-/+}/Rh123^{2+}$/LD cells	0.05	6330	7840

in irradiated and transplanted mice represented between 10 and 20 transplanted pluripotent stem cells. It was assumed that only 5 to 10% of the transplanted stem cells homed in the spleen to form a colony (Visser *et al.*, 1984). This seeding efficiency factor (the so-called f factor) was deduced from serial transplantation experiments. According to these values some of the sorted fractions summarized in Table I should be more than 100% pure stem cells. However, the f factor was not determined for these fractions. Besides, the fractions differ with respect to the time of appearance of the spleen colonies. The Rh123-dull cells give virtually no colonies at 8 days after irradiation and transplantation, but high numbers at 14–16 days. In addition, the Rh123-dull fraction contains the cells with MRA, whereas the Rh123-bright fraction does not. In these two aspects the Rh123-dull fraction resembles pluripotent stem cells in mice 1–2 days after 5-fluorouracil treatment. These have been shown to initially home to the bone marrow after transplantation and not to the spleen. Part of their offspring may then migrate to the spleen to form late appearing colonies. The spleen colony assay and the seeding efficiency factor, therefore, do not seem to be the appropriate tools to quantitate these stem cells. Since no other colony assay for PHSC is established, the absolute purity of the Rh123-dull fraction cannot be determined at present.

The Rh123-bright fraction contains cells which form many Day 8 and many Day 12 spleen colonies. Enrichment factors for those types of CFU-S of 200 to 250 are regularly obtained in this fraction. Although this seems to be inferior to those reported by Spangrude and co-workers (1988), the final result is similar: the mice used by Spangrude have a very low content of CFU-S, about 3 times less than the mice in most other laboratories, including ours. They report an enrichment factor of 750, three times more than we do; consequently, the Rh123-bright fraction contains ultimately a similar frequency of Day 12 CFU-S as the cell suspensions sorted by Spangrude *et al.* (1988). Unfortunately, the MRA is completely lost in this fraction, whereas also the 30-day radioprotective activity of pluripotent stem cells is not found in the Rh123-bright fraction. The Day 12 CFU-S in this Rh123-bright fraction, therefore, do not represent pluripotent stem cells. However, the MRA and the radioprotective activity are found in the Rh123-dull fraction, which, therefore, should be further studied to elucidate the nature of the hematopoietic stem cell.

B. Sorting with the Biotinylated Growth Factor MGF

We previously examined the effect of MGF on the WGA^+/15-1.1$^-$/$Rh123^+$ and WGA^+/15-1.1$^-$/$Rh123^{2+}$ cells in liquid culture and found that these cells do moderately respond to MGF alone and strongly to MGF in combination with IL-3 or IL-1 (de Vries *et al.*, 1991). These results indicated that at least some of the cells must express *c-kit*. We decided to look at the expression of *c-kit* on stem cells by using biotinylated MGF. Figure 4 clearly shows the bimodal distribution of *c-kit* on the WGA^+/15-1.4.1$^-$/LD cells. Based on these results, the sort procedure was changed into a three-color single-sort procedure. Stem cells are being selected using WGA/TxR instead of WGA/FITC and 15-1.4.1/FITC + B220/FITC and MGF/biotin + UA/PE in a single sort after a density cut. B220/FITC was included in the purification procedure because it was shown that some of the previously selected cells expressed the B220 marker. The present procedure therefore is very similar to the one in which the whole cocktail of antibodies against lineage markers was used (Spangrude *et al.*, 1988) with the exception that it is not mouse strain dependent.

The Day 14 CFU-S content of the *c-kit*$^-$ and *c-kit*$^+$ subpopulations is shown in Table II. There is an almost absolute separation of Day 14 CFU-S between the *c-kit*$^+$ and *c-kit*$^-$ cell fractions. The *c-kit*$^+$ cells are 260-fold enriched for Day 14 CFU-S to unit purity, if a spleen seeding efficiency of 5 to 10% is assumed. In contrast, the *c-kit*$^-$ fraction contained only two times more Day 14 CFU-S than unfractionated bone marrow. Most of the *in vitro* colony forming

Table II
Enrichment for Day 14 CFU-S in MGF Sorts

Suspension sorted	Number of cells (% of unfractionated bone marrow cells)	Number of Day 14 CFU-S per 10^5 transplanted cells
Unfractionated bone marrow	100	41.3 ± 5.1
WGA^+/15 − 1.4.1$^-$/B220$^-$/*c-kit*$^-$ LD cells	0.01 ± 0.01	96 ± 22
WGA^+/15 − 1.4.1$^-$/B220$^-$/*c-kit*$^+$ LD cells	0.04 ± 0.04	10,747 ± 1,029

Note. Data obtained from de Vries *et al.* (1992).

precursor cells were also found in the *c-kit*$^+$ fraction, which is in agreement with data reported by others.

Labeling the *c-kit*$^-$ and *c-kit*$^+$ cells with Rh123 showed different staining intensities in both populations (Fig. 5), indicating there is still heterogeneity in these very small subpopulations of cells. Sorting the *c-kit*$^+$ cells on the basis of Rh123 retention into Rh123$^+$ and Rh123^{2+} subpopulations did not result in a further separation of Day 14 CFU-S.

We have shown earlier that the *c-kit*$^+$ cells upon intrathymic injection could reconstitute the thymus of sublethally irradiated mice with high efficiency (de Vries *et al.*, 1992). Intravenous injection of approximately 9000 *c-kit*$^-$ or *c-kit*$^+$ cells resulted in reconstitution of both the B and T lymphoid compartments of irradiated SCID mice. These data, together with the Day 14 CFU-S data, indicate that at least a subpopulation of the *c-kit*$^+$ cells must be pluripotent cells because they give rise to reconstitution of both the myeloid (CFU-S, CFU-C) and lymphoid lineages. This was also indicated by the results of MRA assays, where we looked at secondary CFU-M-derived colonies and secondary Day 14 CFU-S and results from long term reconstitution assays. It was found by using Ly 5 congenic mice that 6–8 months after injection of 100 *c-kit*$^+$ cells, donor-derived cells could be found in the bone marrow, thymus, spleen, and blood of lethally irradiated recipient mice. The results with the *c-kit*$^-$ cells are less clear. Occasionally, when relatively large numbers of *c-kit*$^-$ cells were used we found *in vivo* reconstitution in the different assays (Day 14 MRA, SCIDS, thymic reconstitution), but never in the long-term reconstitution assay. We therefore believe that the occasional reconstitution was due to the presence of low numbers of contaminating *c-kit*$^+$ cells in the *c-kit*$^-$ fraction.

Although the exact function of *c-kit*/MGF interactions in the earliest events in hematopoiesis is not yet clear, the above results indicate that they must play an important role and might contribute to a better understanding of these events.

References

Armitage, R. J., Beckman, M. P., Idzerda, R. L. Alpert, A., and Fanslow, W. C. (1990). *Int. Immunol.* **2,** 1039–1045.

Baines, P., and Visser, J. W. M. (1983). *Exp. Hematol.* **11,** 701–708.

Bauman, J. G. J., Mulder, A. H., and Van den Engh, G. J. (1985). *Exp. Hematol.* **13,** 760–767.

Bertoncello, I., Hodgson, G. S., Bradley, T. R., Hunter, S. D., and Barber, L. (1985). *Exp. Hematol.* **13,** 999–1006.

Bertoncello, I., Bartelmez, S. H., Bradley, T. R., Stanley, E. R., Harris, R. A., Sandrin, M. S., Kriegler, A. B., McNiece, I. K., Hunter, S. D., and Hodgson, G. S. (1986). *J. Immunol.* **136,** 3219–3224.

Chaudhary, P. M., and Roninson, I. B. (1991). *Cell (Cambridge, Mass.)* **66,** 85–94.

de Vries, P., Brasel, K. A., Eisenman, J. R., Alpert, A. R., and Williams, D. E. (1991). *J. Exp. Med.* **173,** 1205–1211.

de Vries, P., Brasel, K. A., McKenna, H. J., Williams, D. E., and Watson, J. D. (1992). *J. Exp. Med.* **176,** 1503–1509.

Ershler, M. A., Nagorskaya, T. V., Visser, J. W. M., and Belyavsky, A. V. (1993). *Gene* **124,** 305–306.
Harris, R. A., Hogarth, P. M., Wadeson, R. J., Collins, P., McKenzie, I. F. C., and Pennington, D. G. (1984). *Nature (London)* **307,** 638–641.
Hoang, T., Gilmore, D., Metcalf, D., Cobbold, S., Watt, S., Clark, M., Furth, M., and Waldmann, H. (1983). *Blood* **61,** 580–588.
Ikuta, K., and Weissman, I. L. (1992). *Proc. Natl. Acad. Sci. U.S.A.* **89,** 1502–1506.
Jones, R. J., Wagner, J. E., Celano, P., Zicha, M. S., and Sharkis, S. J. (1990). *Nature (London)* **347,** 188–189.
Lord, B. I., and Spooncer, E. (1986). *Lymphokine Res.* **5,** 59–72.
Magli, M. C., Iscove, N. N., and Odartchenko, N. (1982). *Nature (London)* **295,** 527–529.
Migliaccio, G., Migliaccio, A. R., and Visser, J. W. M. (1988). *Blood* **72,** 944–951.
Migliaccio, G., Migliaccio, A. R., Valinsky, J., Langley, K., Zsebo, K., Visser, J. W. M., and Adamson, J. W. (1991). *Proc. Natl. Acad. Sci. U.S.A.* **88,** 7420–7424.
Mulder, A. H., and Visser, J. W. M. (1987). *Exp. Hematol.* **15,** 99–104.
Mulder, A. H., and Visser, J. W. M. (1988). *Thymus* **11,** 15–27.
Muller-Sieburg, C. E., Whitlock, C. A., and Weissman, I. L. (1986). *Cell (Cambridge, Mass.)* **44,** 653–662.
Neben, S., Redfearn, W. J., Parra, M., Brecher, G., and Pallavicini, M. G. (1991). *Exp. Hematol.* **19,** 958–967.
Okada, S., Nakauchi, H., Nagayoshi, K., Nishikawa, S., Miura, Y., and Suda, T. (1992). *Blood* **80,** 3044–3050.
Pallavicini, M. G., Summers, L. J., Dean, P. N., and Gray, J. W. (1985). *Exp. Hematol.* **13,** 1173–1181.
Ploemacher, R. E., van der Loo, J. C. M., van Beurden, C. A. J., and Baert, M. R. M. (1992). *Leukemia* **7,** 120–130.
Spangrude, G. J., Heimfeld, S., and Weissman, I. L. (1988). *Science* **241,** 58–62.
Spangrude, G. J., Smith, L., Uchida, N., Ikuta, K., Heimfeld, S., Friedman, J., and Weissman, I. L. (1991). *Blood* **78,** 1395–1402.
Van Bekkum, D. W., Van Noord, M. J., Maat, B., and Dicke, K. A. (1971). *Blood* **38,** 547–558.
Visser, J. W. M., and Bol, S. J. L. (1981). *Stem Cells* **1,** 240–249.
Visser, J. W. M., and de Vries, P. (1988). *Blood Cells* **14,** 369–384.
Visser, J. W. M., and Van Bekkum, D. W. (1990). *Exp. Hematol.* **18,** 248–256.
Visser, J. W. M., Bol, S. J. L., and Van den Engh, G. (1981). *Exp. Hematol.* **9,** 644–655.
Visser, J. W. M., Bauman, J. G. J., Mulder, A. H., Eliason, J. F., and De Leeuw, A. M. (1984). *J. Exp. Med.* **59,** 1576–1590.
Visser, J. W. M., de Vries, P., Hogeweg-Platenburg, M. G. C., Bayer, J., Schoeters, G., Van den Heuvel, R., and Mulder, A. H. (1991). *Semin. Hematol.* **28,** 117–125.
Watt, S., Gilmore, D., Davis, J. M., Clarke, M. R., and Waldmann, H. (1987). *Mol. Cell. Probes* **1,** 297–326.
Wolf, N. S., Koné, A., Priestley, G. V., and Bartelmez, S. H. (1993). *Exp. Hematol.* **21,** 614–622.

ripheral blood circulation. Reticulocytes continue a maturational process over a 1- to 3-day period in the peripheral blood compartment, which is marked by continued hemoglobin synthesis and loss in cellular RNA levels, volume, and surface expression of transferrin receptor (CD 71), resulting in the creation of a mature erythrocyte or red blood cell (Houwen, 1992). Thus, it is through peripheral blood reticulocyte analysis, both enumeration and maturational assessment, that erythropoietic activity can be gauged, allowing for the clinical evaluation and classification of anemias. The degree of peripheral blood reticulocytosis can be used to segregate anemic patients into categories of aplastic, hypoproliferative, and compensated types (Table I), which then allows for more focused testing for specific etiologies.

The traditional laboratory methodology of reticulocyte analysis for the past century has been the manual microscopic counting of erythrocytes or red cells on supravitally stained blood samples. Although slide-based methods, such as new methylene blue staining, do provide clinically useful information, the lack of precision in these subjective assays is well documented (Savage *et al.*, 1985). In particular, the poor precision of manual counting methods does not allow for clinically useful sequential monitoring of patients with low reticulocyte levels or for therapeutic monitoring of drug therapies for either erythropoietic inhibition or enhancement. Furthermore, assessment of different maturational stages, such as the four stages proposed over 50 years ago by Heilmeyer using grading of RNA content (Heilmeyer and Westhaeuser, 1932), is difficult to reproduce and is impractical for the clinical laboratory (Crouch and Kaplow, 1985). Flow cytometric technology and nucleic acid binding fluorescent dyes are ideally suited to overcome most of the limitations of manual reticulocyte analysis. The rapid flow cytometric counting of 20,000–50,000 cells in seconds,

Table I
Classification of Anemias by Reticulocyte Analysis

Type of anemia	Reticulocyte count	Reticulocyte maturity index
Aplastic anemia	Low	Low
Toxic drug reaction	Low	Low/normal
Bone marrow regeneration	Low	Normal/high
Chronic disease	Low/normal	Normal
Iron deficiency	Low/normal	High
Thalassemia	Normal/high	Normal/high
Myelodysplastic syndrome	Low/normal	Normal/high
Pernicious anemia	Low/normal	High
Hemoglobinopathy	Normal/high	High
Compensated blood loss	Normal/high	High
Hemolytic anemia	High	High

rather than the usual 10 min to manually count 1000 cells with microscopic methods, improves both analysis time and statistical precision, since the relevant cell population is normally in low frequency (0.5–2%). Additionally, the reticulocyte fluorescence distribution can be exploited to derive a reticulocyte maturation index (RMI). Reticulocyte fluorescence is proportional to the amount of cellular RNA, hence the newly produced or more immature reticulocytes are present in the highly fluorescence fraction of the reticulocyte distribution. The fluorescence distribution of the reticulocyte population can be measured in various ways to derive a quantitative reticulocyte maturity index (Davis *et al.,* 1993a; Davis and Bigelow, 1989,1990; Tanke *et al.,* 1983).

Methods of flow cytometric reticulocyte analysis for both clinical and experimental applications have been reported over the past 15 years utilizing a variety of fluorescent dyes. The relative advantages and disadvantages of the various dyes, including thioflavin T, acridine orange, cyanine dye Di03C6, pyronin Y, propidium iodide, ethidium bromide, and thiazole orange, are now of historic interest and have been reviewed elsewhere (Davis and Bigelow, 1993; Davis, 1993; Tanke, 1992). Thiazole orange (TO) has distinct advantages over other nucleic acid binding dyes for reticulocyte analysis using commercial multipurpose FCM instruments and is currently utilized by the vast majority of clinical laboratories. TO reticulocyte analysis offers the ability to use easily reticulocyte enumeration and a stable, reproducible fluorescence intensity measurement over a 60-min period (Davis and Bigelow, 1989; Lee *et al.,* 1986; Chin-Yee *et al.,* 1991; Ferguson *et al.,* 1990; Schimenti *et al.,* 1992). Furthermore, the fluorescence signal provides a high quantum yield or good separation of the reticulocyte (positive) from the mature red blood cell (negative) populations. TO is commercially available only in a preformulated reagent solution from a single source (Becton–Dickinson Immunocytometry Systems, San Jose, CA) at the time of this writing. Dedicated and more automated reticulocyte analysis on instruments using flow cytometric principles is also currently available. The Sysmex R-1000 and R-3000 instruments (TOA Medical, Inc., Kobe, Japan) use the fluorescent dye auramine O, which by employing automated sample staining and operator-independent data analysis provides good counting precision and the ability to derive an RMI (Tichelli *et al.,* 1990). Most recently hematology analyzers have been adapted to employ flow cytometry technology using absorbance of laser light measurements following reticulocyte staining of blood samples with dyes. These methods include the RNA binding dyes of oxycine (Technicon H3, Miles, Inc., Tarrytown, NY) and new methylene blue (Coulter MAXM, Coulter Corp, Hialeah, FL). Thus, in a very short time the clinical reticulocyte assay has come full circle in leaving the clinical hematology laboratory to the flow cytometry laboratory, only to return to the hematology laboratory with the advent of integration of semiautomated reticulocyte counting on automated blood cell counters.

II. Reticulocyte Maturity Index

The importance of a stable and reproducible fluorescence (or absorbance) signal in reticulocyte analysis relates to the ability to derive a RMI. The improved precision of reticulocyte counting afforded by FCM analysis over manual microscopic methods can be viewed as a significant, but small advancement for laboratory hematology. The ability to provide additional measurement parameters, beyond simple quantitation, promises to be a greater dividend of applying flow cytometric technology to clinical reticulocyte analysis. The RMI is one such parameter. The potential for clinical utility of an RMI derived from FCM reticulocyte analysis was first suggested by Tanke *et al.*, 1983,1986) in their studies using Pyronin Y reticulocyte analysis and the demonstration of bone marrow suppression in patients undergoing cancer chemotherapy. Subsequent work using thiazole orange methods on commercial flow cytometers and the Sysmex reticulocyte analyzer has indicated clinical utility of the RMI in the monitoring of bone marrow engraftment following bone marrow transplantation (Ball *et al.*, 1990; Davies *et al.*, 1992; Davis *et al.*, 1989; Lazarus *et al.*, 1992), and distinguishing different types of anemia (Tsuda and Tatsumi, 1989; Wells *et al.*, 1992), and serves as a therapeutic monitor for patients receiving erythropoietin treatment (MacDougall *et al.*, 1992). Additionally, the combination of absolute reticulocyte counts and the RMI offers the potential for better classification and diagnosis of anemias (Table I).

Despite these reports of clinical utility of using fluorescence intensity information to derive a RMI, widespread acceptance and implementation into clinical practice have been hampered by the inability to standardize the measurement between laboratories. Initial studies with the RMI using the thiazole orange method of reticulocyte analysis utilized mean channel fluorescence measurements to derive the measurements (Davis and Bigelow, 1989; Davis *et al.*, 1989; Wells *et al.*, 1992). This becomes problematic given the variety of commercial models of flow cytometers employing various means of quantifying the fluorescence intensity (3 vs 4 logs, 256 vs 1024 channels, etc.) Furthermore, there is currently no method, such as involving the use of calibrated beads, that allows for interinstrument or interlaboratory standardization of the RMI. We have proposed a method of RMI measurements expressed as a fraction of highly fluorescent thiazole orange-stained reticulocytes, which uses normal healthy individuals to define the cursor settings (Davis *et al.*, 1993a). This method of RMI calculation has the advantage over mean channel fluorescence measurements of allowing for a common quantitative range (0.00–1.00), independent of instrument model, data analysis software versions, or staining conditions between laboratories. Initial studies with the highly fluorescent fraction approach to RMI calculation with interlaboratory correlations were encouraging with correlation coefficients as high as $R^2 = 0.68$ achieved (Davis *et al.*, 1994). Additional validation is required, but the method has the additional merit of not only being referenced to a normal population, but also having conceptual similarities with the approach used by the Sysmex, Technicon, and Coulter instruments of dividing the reticulocyte populations into distinct regions of staining intensity or RNA content.

III. Data Analysis

Data analysis in FCM reticulocyte analysis can be problematic and arbitrary with many instruments. In particular, the RMI measurements require repeated cursor movements with most software programs available on the common multipurpose FCM instruments. Second, the definition of cursor placement to distinguish the mature red cell population from the more fluorescent reticulocyte population lacks consensus. Some authors advocate using unstained controls (autologous blood in PBS) and applying the channel position on this control sample where only 0.1–0.2% of the gated red cells are "positive" to the TO-stained sample (Davis and Bigelow, 1989; Lee *et al.,* 1986). Others favor a fixed channel position applied to all TO-stained samples, previously determined by analysis of random unstained blood samples (Chin-Yee *et al.*, 1991; Schimenti *et al.,* 1992). In clinical practice the results of the reticulocyte percentage using either approach are not clinically significantly different in the majority of samples. However, we favor the use of autologous autofluorescent controls for cursor settings, despite the additional time required for sample analysis, for several reasons. Most importantly, the rule-based approach of cursor placement from an autologous sample improves the reproducibility or precision of the RMI measurements and compensates for the increase in autofluorescence seen with storage of blood samples. Additionally, analysis of autologous unstained controls allows for the ready identification of unusual clinical states where red cells might exhibit increased autofluorescence, such as patients receiving drugs with fluorescent properties or the disease porphyria where normal heme synthesis is altered, resulting in the accumulation of fluorescent molecules in the red cells (Brun *et al.,* 1988).

Some of the variability in data analysis can be minimized through the use of software programs specifically designed for reticulocyte analysis. These programs remove the operator subjectivity involved in the cursor placement defining the positive reticulocyte population. Laboratories equipped with the FACScan instruments can purchase the ReticCount software (Becton–Dickinson Immunocytomentry Systems, San Jose, CA). Off-line data analysis in laboratories equipped with IBM clone personal computers running Windows (Microsoft Corp., Redmond, WA) can utilize the Reticfit software (Verity Software House, Topsham, ME). Reticfit, unlike ReticCount, can analyze FCM data from any instrument using FCS format and provides parameters of reticulocyte enumeration and RMI (Fig. 1).

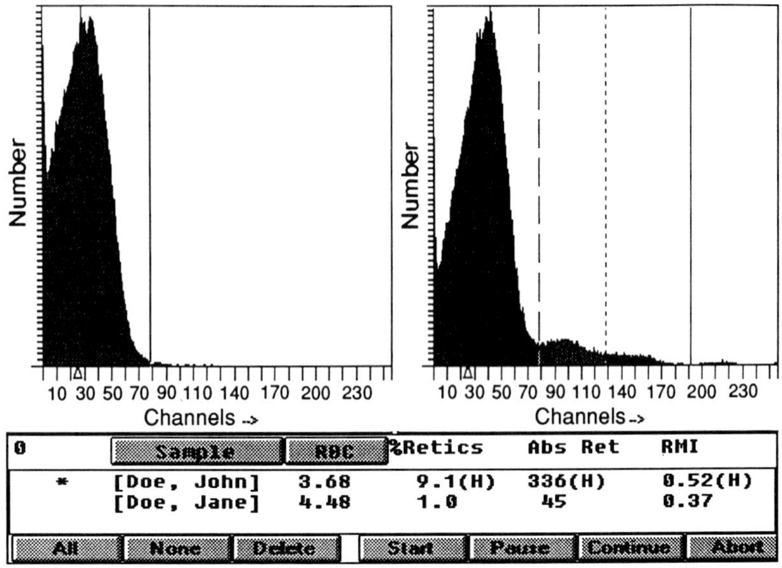

Fig. 1 Thiazole orange-stained whole blood (right) and unstained autofluorescence histogram (left) as analyzed by ReticFit software (Verity Software House, Topsham, ME). The cursor position for defining reticulocytes is set automatically from the unstained sample at the channel position where 0.2% of the events is to the left of the cursor and applied to the same channel position on the thiazole orange-stained histogram (long dash line cursor). Leukocytes and nucleated red cells (small peak left of solid line cursor) are excluded from calculations. The reticulocyte maturity index is calculated by the software and derived from the ratio of the number of highly fluorescent reticulocytes (between short dash line cursor and solid line cursor) to the total number of reticulocytes. Results are displayed with abnormal values flagged, requiring the operator to only manually enter the red cell count (RBC) to enable the software to calculate the absolute reticulocyte count (Abs Ret).

Flow cytometric reticulocyte analysis does have some potential pitfalls or sources of error that should be familiar to laboratories performing this testing (Table II). A common theme to most of these problematic clinical situations is the fact that reticulocyte dyes, such as TO, will stain both RNA and DNA. Hence, if red cells included in the gated population of analysis are nucleated or contain residual fragments of nuclear material (Howell Jolly bodies), a falsely elevated reticulocyte count and RMI value can result. Leukocytes may also be included in the gated population, also representing a potential source of falsely elevated results, unless data analysis is performed to exclude such cells from reticulocyte quantitation. Elimination of nucleated cells from contamination of reticulocyte analysis can readily be handled with an upper fluorescent cursor (Fig. 1). The fluorescence intensity of TO-stained leukocytes and nucleated red cells can be determined by using simple light scatter gating around leukocytes. Alternatively, two-color analysis with TO and either negative gating

Table II
Clinical Conditions with Potential Interference with Flow Cytometric Reticulocyte Analysis

Leukocytosis	Parasitic infections
Leukemia	Malaria
Leukemoid reaction	Babesiosis
Thrombocytosis	Cold agglutinin syndromes
Large platelet syndromes	Drugs (autofluorescence)
Bernard-soulier syndrome	
May-Hegglin anomaly	Porphyria (autofluorescence)
Alport's syndrome	
Gray platelet syndrome	Red cell inclusions
"Mediterranean" macro thrombocytopenia	Nucleated RBCs
	Howell-Jolly Bodies

of CD 45-positive leukocytes or positive gating of glycophorin A-positive red cells can be performed to further evaluate leukocyte contamination within the red cell gates (Davis and Bigelow, 1993; Serke and Huhn, 1993). However, the use of an upper fluorescence gate is equally effective in excluding leukocytes from reticulocyte analysis, avoiding the additional methodologic complexity and cost of adding monoclonal antibodies, and simultaneously ensures the exclusion of nucleated red cells from analysis (Fig. 2). Exclusion of platelets as a source of error can usually be readily performed by careful light scatter gating, as the majority of platelets are significantly smaller than red cells.

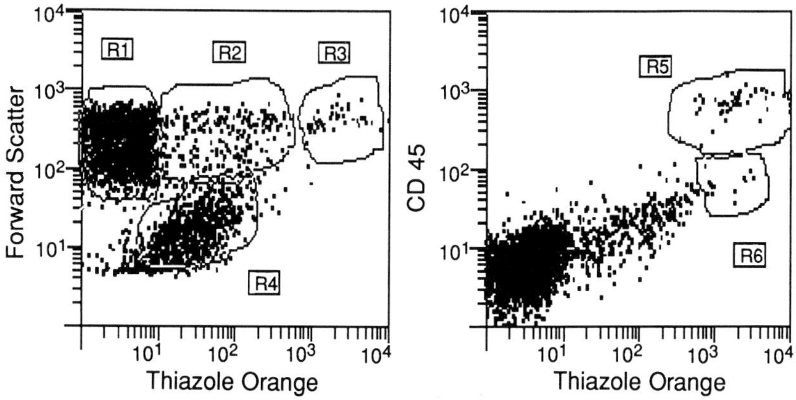

Fig. 2 Thiazole orange staining of whole blood. Bivariate display of forward scatter versus thiazole orange (left) without gating on the red cell population clearly demonstrates subpopulations of mature red cells (R1), reticulocytes (R2), leukocytes and nucleated red cells (R3), and platelets (R4). Dual staining of whole blood with CD 45 percp and thiazole orange light scatter gating on the red cell cluster (right) demonstrates the highly fluorescent thiazole orange CD 45-positive leukocyte (R5) and CD 45-negative nucleated red cell (R6) subpopulations.

Visualization of the distinct platelet population on TO-stained whole blood can be achieved in a two-parameter display of forward light scatter and fluorescence intensity (Fig. 2). Recognition of intracellular parasites, such as malaria, and Howell-Jolly body containing red cells, as commonly seen in patients following splenectomy, remains problematic. Recognition of these conditions requires correlation with routine morphologic studies of peripheral blood smears by the clinical hematology laboratory. Cold agglutinin disease, although a rare condition, can also present a problem for flow cytometric reticulocyte analysis secondary to the temperature-dependent antibody-mediated red cell clumping phenomenon. These conditions that potentially interfere with flow cytometric reticulocyte analysis are infrequent, representing less than 5% of any clinical laboratory practice, but require the laboratory to default to manual microscopic reticulocyte counting methods.

Another source of error with flow cytometric reticulocyte analysis is the phenomenon of falsely elevated reticulocyte percentage that can result from data acquisition at high flow rates. The sample flow rates should be adjusted to ensure that sample analysis does not exceed 4000 cells per second. Otherwise, the potential of coincidence (two cells traversing the laser beam path simultaneously) increases, which favors undercounting of the predominant mature red cell population and results in overcounting the proportion of reticulocytes.

IV. Quality Control

Quality control (QC) methods should be adopted by each laboratory to ensure continuous monitoring of results for accuracy and precision. Suggested guidelines for flow cytometric reticulocyte counting are forthcoming in the National Committee for Clinical Laboratory Standards (NCCLS) document H44-P. Stabilized reticulocyte control material in the normal and abnormal ranges can be obtained from several commercial sources (Becton–Dickinson Immunocytometry Systems, San Jose, CA, and Strack Laboratories, Omaha, NE). These materials function effectively for long-term monitoring of instrument performance drift in reticulocyte enumeration, but may not effectively serve as QC for RMI measurements. Rabbit blood (Corash *et al.*, 1988) and citrated red cell suspensions (Tsuda and Tatsumi, 1990) stored under refrigerated conditions have also been proposed as suitable reagents for QC material. Correlation with manual microscopic counts can serve as a QC method, albeit time consuming, subject to bias due to lower sensitivity compared to the TO method (Schimenti *et al.*, 1992; Pappas *et al.*, 1992) but cannot function to address QC of the RMI measurements. We advocate the use of retained patient specimens as an effective, inexpensive short-term QC material for both reticulocyte counting and RMI measurements (Davis and Bigelow, 1989,1990; Davis, 1993). Since patient samples can be stored refrigerated for up to 72–96 hr without significant change in the reticulocyte parameters (Davis and Bigelow, 1989; Chin-Yee *et al.*, 1991), selected samples of both normal and abnormal values can be stored and reanalyzed in successive analytical batches to detect any significant shift in assay performance.

V. Method for Thiazole Orange Reticulocyte Analysis

A. Reagents/Equipment Required:

1. Phosphage-buffered saline with 0.1% sodium azide (PBS), pH 7.4.
2. Retic-COUNT (thiazole orange, Becton–Dickinson Immunocytometry Systems, San Jose, CA. Cat. No. 92-0004). Store refrigerated and in the dark. Warning: The reagent solution contains a nucleotide-reactive material and may be carcinogenic. Handle with care. Avoid contact with skin and mucous membranes.
3. 12 × 75-mm polystyrene test tubes.
4. Vortex or similar mixer.
5. Pipettors accurate at 5 μl and 1.0 ml.
6. Flow cytometer with an argon ion laser tuned to 488-nm excitation equipped with fluorescence detector with filter combination to collect green light (e.g., 525 band pass) and light scatter (forward and wide angle) detectors. Light detectors should be connected to log amplifiers.

B. Specimen Requirements: EDTA-Anticoagulated Whole Blood

1. Specimens held at room temperature must be analyzed within 8 hr postcollection. Preferably specimens are refrigerated within 4 hr of collection and stored at 2–8°C until analysis.
2. Specimens held at refrigerator temperature (2–8°C) must be analyzed within 72 hr postcollection. Thoroughly mix and warm specimens to room temperature immediately prior to analysis.

C. Staining Procedure

1. For each specimen to be analyzed, set up two 12 × 75 tubes. Label one tube "control" with patient identification. Place 1.0 ml of PBS in this tube. Label the second tube "test" with patient identification. Place 1.0 ml of ReticCount solution in this tube.
2. To each of the two tubes add 5 μl of well-mixed blood specimen and mix gently.
3. Incubate at room temperature in the dark for 30 min.

4. Analyze on the flow cytometer within 60 min postincubation if measuring RMI along with reticulocyte counts. Continue to protect the tubes from bright light. Mix tubes gently prior to flow cytometric analysis.

D. Flow Cytometric Analysis:

1. Adjust log amplifier gains to allow light scatter gating around red cell population and preferably to allow identification of the platelet population. Adjust green fluorescent log amplifier gains to visualize the control autofluorescence samples within the first decade or log of fluorescence scale.

2. Acquire up to 50,000 events from within the red cell-gated region at a flow rate of approximately 4,000 cells per second.

3. Data results can be stored as either list-mode files or single-parameter green fluorescence histograms of the gated red cell population.

E. Data Analysis

1. Use the control sample for each specimen to define the lower threshold channel position of the reticulocyte population. Determine the cursor position at which less than 0.2% of the unstained cells are to the right of the unstained autofluorescence sample histogram.

2. Place the lower threshold cursor determined above on test or thiazole orange-stained sample histogram. The upper threshold cursor position is placed on histogram to exclude leukocytes and nucleated red cells (10× fluorescence intensity of reticulocytes) from analysis (this cursor position can be determined by previously examining the fluorescence intensity of gated leukocyte populations). These two cursor positions define the "positive interval" or reticulocyte population.

3. The reticulocyte percentage result is calculated as follows: "Test" % in positive interval − "control" % in positive interval = % reticulocytes. The anticipated normal range is 0.5–2.5%.

4. The absolute reticulocyte count is calculated using the red blood cell (RBC) count obtained from the automated blood cell counter as follows: Number of RBC $\times$ 10^{12}/liter $\times$ % reticulocytes $\times$ 10 = reticulocytes $\times$ 10^{9}/liter. The anticipated normal range is 20–135 10^{9}/liter.

5. The RMI calculation requires initial analysis of a normal or reference population to determine the central 95% interval of the mean fluorescence intensity (Davis *et al.*, 1993a). A highly fluorescent reticulocyte region is established between the upper limit channel number of the 95% interval of the normal range and the upper cursor position for the reticulocyte region. The RMI is then calculated as follows: % highly fluorescent reticulocytes ÷ % reticulocytes. The anticipated normal range is 0.20–0.50.

F. Procedural Quality Control:

1. Select two specimens (<24 hr old) from the current day's run and label as QC1 and QC2. Specimens are selected to include both normal and abnormal values. Record the results for reticulocyte %, reticulocyte absolute count, and RMI on the QC worksheet.

2. Refrigerate the QC specimens at 2–8°C.

3. On the next working day, warm the QC specimens to room temperature and perform reticulocyte analysis by flow cytometry after primary instrument QC has been successfully completed and prior to running patient specimens.

4. Record the results on the QC worksheet and calculate the percentage difference between the 2 days on the three reticulocyte parameters. All values must be within the laboratory's established accepted range (≤10% for reticulocyte percentage and count; ≤20% for RMI).

5. If any value falls outside the established range, repeat the QC run and evaluate. If the second run is outside the established range, notify the laboratory supervisor before proceeding with patient specimens.

6. Alternatively, commercial reticulocyte control material, formulated at multiple levels of reticulocyte percentage, can be purchased from either Bectin–Dickinson Immunocytometry Systems (San Jose, CA) or Streck Laboratories (Omaha, NE).

References

Ball, E. D., Mills, L. E., Cornwell, G. G., Davis, B. H., Coughlin, C. T., Howell, A. L., Stukel, T. A., Dain, B. J., McMillan, R., Spruce, W., Miller, W. E., and Thompson, L. (1990). *Blood* **75,** 1199–1206.

Brun, A., Steen, H. B., and Sandberg, S. (1988). *Scand. J. Clin. Lab. Invest.* **48,** 261–267.

Chin-Yee, I., Kenney, M., and Lohmann, R. C. (1991). *Clin. Lab. Haematol.* **13**(2), 177–188.

Corash, L., Rheinschmidt, M., Lieu, S., Meers, P., and Brew, E. (1988). *Pathol. Immunopathol. Res.* **7,** 381–394.

Crouch, J., and Kaplow, L. (1985). *Arch. Pathol. Lab. Med.* **109,** 325–329.

Davies, S., Cavill, I., Bentley, N., Fegan, C., Poynton, C., and Whittaker, J. (1992). *Br. J. Haematol.* **81,** 12–17.

Davis, B. H. (1993). *In* "Clinical Flow Cytometry: Principles and Applications" (K. B. Bauer, R. E. Duque, and T. V. Shankey, eds.), pp. 373–386. William & Wilkins, Baltimore, MD.

Davis, B. H., and Bigelow, N. C. (1989). *Arch. Pathol. Lab. Med.* **113**(6), 684–689.

Davis, B. H., and Bigelow, N. C. (1990). *Pathobiology* **58**(2), 99–106.

Davis, B. H., and Bigelow, N. (1993). *Ann. N.Y. Acad. Sci.* **677,** 281–292.

Davis, B. H., Bigelow, N., Ball, E. D., Mills, L., and Cornwell, G. (1989). *Am. J. Hematol.* **32**(2), 81–87.

Davis, B. H., DiCorato, M., Bigelow, N., and Langweiler, M. (1993a). *Cytometry* **14**(3), 318–326.

Davis, B. H., Bigelow, N., Koepke, J., Borowitz, M., Houwen, B., Jaccobberger, J., Pierre, R., Corash, L., Ault, K., and Batjer, J. (1994). *Am. J. Clin. Path.* (in press).

Ferguson, D. J., Lee, S. F., and Gordon, P. A. (1990). *Am. J. Hematol.* **33**(1), 13–17.

Heilmeyer, L., and Westhaeuser, R. (1932). *Z. Klin. Med.* **121,** 361–365.

Houwen, B. (1992). *Blood Cells* **18,** 167–186.

Lazarus, H. M., Chahine, A., Lacerna, K., Wamble, A., Iaffaldano, C., Straight, M., Rabinovitch, A., Schimenti, K. J., and Jacobberger, J. (1992). *Am. J. Clin. Pathol.* **97**(4), 574–583.

Lee, L. G., Chen, C. H., and Chiu, L. A. (1986). *Cytometry* **7**(6), 508–517.

MacDougall, I. C., Cavill, I., Hulme, B., Bain, B., McGregor, E., McKay, E., Sanders, P., Coles, E., and Williams, J. D. (1992). *Br. Med. J.* **304,** 225–226.

Pappas, A. A., Owens, R. B., and Flick, J. T. (1992). *Ann. Clin. Lab. Sci.* **22**(2), 125–132.

Savage, R., Skoog, D., and Rabinovitch, A. (1985). *Blood Cells* **11,** 97–112.

Schimenti, K., Lacerna, K., Wamble, A., Maston, L., Iaffaldano, C., Straight, M., Rabinovitch, A., Lazarus, H., and Jacobberger, J. (1992). *Cytometry* **13,** 853–862.

Serke, S., and Huhn, D. (1993). *Clin. Lab. Haematol.* **15,** 33–44.

Tanke, H. J. (1992). *In* "Flow Cytometry in Hematology" (O. D. Laerum and R. Bjerknes, eds.), pp. 75–93. Academic Press, San Diego.

Tanke, H. J., Rothbarth, P., Vossen, J., Koper, G., and Ploem, J. (1983). *Blood* **61,** 1091–1097.

Tanke, H. J., van Vianen, P., Emiliani, F. M., Neuteboom, I., de Vogel, N., Tates, A. D., de Bruijn, E., and van Oosterom, A. (1986). *Histochemistry* **84**(4–6), 544–548.

Tichelli, A., Gratwohl, A., Driessen, A., Mathys, S., Pfefferkorn, E., Regenass, A., Schumacher, P., Stebler, C., Wernli, M., Nissen, C. *et al.* (1990). *Am. J. Clin. Pathol.* **93**(1), 70–78.

Tsuda, I., and Tatsumi, N. (1989). *Eur. J. Haematol.* **43**(3), 252–254.

Tsuda, I., and Tatsumi, N. (1990). *Am. J. Clin. Pathol.* **93**(1), 109–110.

Wells, D. A., Daigneault-Creech, D. A., and Simrell, C. R. (1992). *Am. J. Clin. Pathol.* **97,** 130–134.

CHAPTER 17

Analysis of Platelets by Flow Cytometry

Kenneth A. Ault and Jane Mitchell
Maine Medical Center Research Institute
South Portland, Maine 04106

I. Introduction

Platelets are anucleate cellular fragments which circulate in large numbers in blood (2 to 4 × $10^5/\mu l$). They are primarily responsible for maintaining the integrity of the vasculature by plugging leaks as they occur. In order to do this they have the capability to respond to changes in blood flow (shear stress) and to the subendothelial matrix of vessels. When they detect an abnormality they are able to respond in basically three ways. They can undergo adhesion to the vessel wall, they can aggregate with other platelets to form a "platelet plug," and they can release the contents of their cytoplasmic granules which results in the recruitment of more platelets, the initiation of an inflammatory response, and the initiation of the coagulation cascade. Interestingly this entire functional repertory of platelets can be evaluated by flow cytometry. Platelet dysfunction can result in an increased tendency to bleeding and may very well play an important role in a variety of cardiovascular diseases such as myocardial infarction or stroke.

Only in the past 5 years has flow cytometry assumed any importance in the evaluation of platelets. However, it has now become clear that flow cytometric techniques are extremely useful in the study of platelets. The flow cytometer can overcome some very difficult problems that had slowed progress in understanding platelets. For this reason flow cytometric techniques for the study of platelets have achieved considerable and still rapidly growing importance.

This chapter attempts to outline the methods used for analysis of platelets by flow cytometry. Because there are a number of different flow cytometric applications associated with platelets, we do not attempt to detail each of the applications but rather cover the general principles that are important whenever platelets are being studied by flow cytometry. The biology and physical properties of platelets are sufficiently different from those of other cell types in that they frequently require different methods of preparation, labeling, and analysis than are used for flow cytometric evaluation of other cell types. These special considerations are dealt with in detail. The specific applications will be mentioned and whatever unique aspects they entail will be outlined, but the reader is referred to more detailed accounts of specific applications in the literature for discussion of the rationale, alternative methods, and interpretation of individual tests. There have been several recent reviews which cover the biological and medical issues rather than strictly technical issues (Ault *et al.*, 1991; Ault, 1992; Ault and Mitchell, 1993).

The flow cytometric evaluation of platelets has a historical component which is both of some interest and relevant to some of the discussion to follow. Much of the early work with flow cytometry was directed toward the study of nucleated cells of the blood, particularly lymphocytes. Until the past 5 years no one had used flow cytometers for the study of platelets, and in fact most workers referred to the small particles found in blood as "debris" with little thought that the debris consisted almost entirely of platelets. Thus one can find extensive discussions of methods to eliminate debris and of the confusion created when debris contaminated the lymphocyte gates.

It has now become clear that the study of platelets by flow cytometry is a fruitful area of research and clinical applications. They have been elevated above the category of debris. However, as we discuss below, there is now a new category of debris which those of us interested in platelets take care to eliminate from our platelet gates. It is likely that there is much useful information in this new debris which includes fragments of red cells, white cells, and platelets as well as large immune complexes and perhaps other interesting objects. Thus this chapter might be subtitled The Flow Cytometric Analysis of Debris.

II. Applications

There are several distinct applications of flow cytometry to the study of platelets which differ to some extent in the methods of platelet preparation, labeling, and analysis, and of course in the specific markers used. These are summarized in Table I and described briefly below.

Phenotyping: Platelets can be characterized according to their expression of a considerable variety of surface glycoproteins that are identified by monoclonal antibodies. The surface of the platelet is at least as well understood as that of the lymphocyte. In the case of most of the major platelet surface glycoproteins their function is known and monoclonal antibodies are readily available. Unlike lymphocytes, however, there are very few subsets of platelets that have been identified. Thus the major utility of platelet phenotyping is to determine unambiguously that one is studying platelets. Markers such as CD41 and CD42 can be used to reliably distinguish platelets from "debris." In addition, there are a few, relatively rare diseases in which there are abnormalities of platelet surface structures. For example, in Glanzman's thrombasthenia the GPIIb/IIIa molecule is absent or markedly reduced and this can be easily detected using a CD41 monoclonal.

Platelet-associated immunoglobulin (PAIg): There are abundant data suggesting that the ability to measure the amount of antibody on the surface of platelets provides useful information in diagnosing the relatively common disorders in which antibodies directed against platelets result in thrombocytopenia. This has been a controversial area due in large part to technical problems associated with the measurement of PaIg. Recently flow cytometry has been used to an increasing extent to reliably measure the amount of Ig (and sometimes C3) on the surface of platelets. The flow cytometric method has several advantages, such as the ability to perform the measurement in patients who are severely thrombocytopenic with relatively small volumes of blood; the ability to be absolutely certain that one is measuring platelet-associated Ig rather than Ig on other types of "debris" such as red cell fragments and immune complexes; and the ability to restrict the measurement to the platelet surface thus avoiding the problem of a large amount of Ig inside the platelet which may not be relevant to the process of immune platelet destruction.

Platelet activation and aggregation: Numerous investigators have recently demonstrated that it is possible to elegantly study various aspects of platelet function through the use of activation-specific monoclonal antibodies and the use of light scatter to measure platelet aggregation. These methods of studying

standard reagent, its use in the fixation of platelets requires more care than is usually necessary. Paraformaldehyde solutions should be checked for pH (should be near 7.0) and careful attention has to be paid to the concentration. Overfixation is a common cause of poor labeling.

Monoclonal antibodies directed against platelet-specific markers are becoming more widely available. The most complete catalog of these reagents (at the time of this writing) is that of AMAC, Inc. (160B Larrabee Road, Westbrook, ME 04092; 1-800-458-5060).

There are a number of reagents used to either activate platelets (agonists) or to prevent their activation. Some of these are listed in Table II with their usual concentrations and stability. Sources for these reagents include standard chemical supply companies and the manufacturers of hematology instruments used to evaluate platelet function.

IV. Cell Preparation

There are three basic sources for preparation of platelets for flow cytometry. These are whole blood, platelet-rich plasma, and washed platelets. The use of whole blood is recommended because it greatly reduces the artifactual platelet activation that is inevitable in any process which manipulates the platelets. Platelets can be studied in whole-blood samples which have been fixed using paraformaldehyde. The protocol used in our laboratory is to place 50 μl of fresh blood directly into a tube containing 1 ml of 2% (w/v) paraformaldehyde in phosphate-buffered saline. This results in rapid fixation of the platelets and preserves them in a state of activation and aggregation that accurately reflects their status in the blood (Ault *et al.*, 1989). The great advantage to fixation is that the effects of subsequent artifactual changes in the platelets are removed. This procedure lends itself particularly well to clinical studies in which the samples cannot be analyzed immediately. The major disadvantage of fixation is that it may destroy some immunological determinants that are of interest. For example, the very interesting activation marker on GPIIb/IIIa which is identified by the monoclonal antibody PAC-1 is altered by fixation. Alternatively, whole blood may be labeled without fixation by placing a small amount of blood into tubes containing premeasured amounts of the antibodies (Shattil *et al.*, 1987). This results in rapidly labeling the platelets, which are then diluted to a final volume of 0.5 to 1 ml. The result is that the effective concentration of the antibodies is reduced so that the pace of subsequent labeling is slowed. When such samples can be analyzed within a short period of time, perhaps an hour or less, the results are excellent. However, aggregation is not observed in such protocols due to the dilution of the platelets and deaggregation which occurs.

Table II
Reagents Used to Activate Platelets and to Prevent Activation

	Agent	Concentration	Stability	Comments
Agonists	Adenosine diphospate	5 to 20 μM	Stable	Works poorly on washed platelets
	Epinephrine	100 μM	Stable	A very weak agonist, frequently combined with ADP
	Thrombin	0.5–1 unit/ml	Unstable, freeze aliquots	A strong agonist, difficult to use in whole blood due to clotting
	Arachidonic acid	500 μg/ml	Unstable, freeze aliquots	
	Collagen	0.2 mg/ml	Stable	
Antagonists	Prostoglandin E1	1–10 μg/ml	Unstable, freeze aliquots	The best single antagonist
	Theophyllin	1 mM	Stable	
	Heparin	10–100 u/ml	Stable	Blocks thrombin-induced activation
	Hirudin	0.5–25 u/ml	Stable	Blocks thrombin-induced activation

Platelet-rich plasma (PRP) is prepared from blood by a slow centrifugation which removes most of the erythrocytes and results in plasma containing large numbers of platelets. Because the platelets have not been pelleted there is not a great deal of artifactual activation. However, there is a time penalty in that there is progressive activation during the time required for the centrifugation. Nevertheless the use of PRP is the customary method for platelet function studies and flow cytometric techniques based on PRP are still widely used. In studies which do not depend upon platelet activation, such as PAIg or reticulated platelets, PRP may be an excellent preparative step. It has the advantage that the platelets are still in their normal physiologic plasma.

Finally, washed platelets may be prepared in a variety of ways, most of which require centrifugation to pellet the platelets. Unless inhibitors of activation are used this will always result in considerable platelet activation. In our hands it is not possible to obtain washed platelets with less than 20–40% expression of CD62 unless one uses inhibitors such as PGE1. Even in the presence of such inhibitors the levels of activation obtained are highly variable. Recently, it has been suggested that it is possible to obtain washed platelets without pelleting them using a discontinuous density gradient. This approach has not been explored extensively to our knowledge. Washed platelets are best used in protocols which are not affected by platelet activation. For example in measuring PAIg it is necessary to wash the platelets out of the plasma in order to label with anti-immunoglobulin reagents (Ault, 1988).

A. Sample Protocol for Preparation of Fixed Whole-Blood Samples for Measuring Platelet Activation and Aggregation

Draw blood using minimal application of a tourniquet and a clean venipuncture.

Use a heparinized syringe or a green top (heparinized) Vacutainer tube.

Transfer 50 μl of blood immediately into a tube containing 1 ml of 2% paraformaldehyde in PBS.

Transfer 0.45 ml of blood into a tube containing 50 μl of 200 μM ADP.

Mix the tube and incubate 5 min at room temperature.

Remove 50 μl of blood from the tube and transfer to a tube containing 1 ml of 2% paraformaldehyde in PBS.

Allow both fixed samples to fix for 1 hr at room temperature.

Wash the cells three times with 1 ml of sterile filtered Tyrode's buffer. Centrifuge at 1800g for 5 min. During washing there may be considerable lysis of erythrocytes.

Resuspend the cells in 1 ml of Tyrode's buffer and store at 4°C until labeled.

B. Sample Protocol for Preparation of Platelet-Rich Plasma Containing Resting and Activated Platelets

Draw blood in purple top Vacutainer tubes (EDTA anticoagulant).

Centrifuge tubes 200*g* for 10 min.

Remove the supernatant platelet-rich plasma from each tube and pool (the volume of PRP will be about 25% of the blood volume).

Transfer 4.5 ml of PRP to a tube containing 0.5 ml of 100 μM PGE 1 (to keep platelets resting).

Transfer 4.5 ml of blood to another tube containing 0.5 ml of 200 μM ADP (to induce activation). Incubate 5 min at room temperature.

The PRP can be used immediately for labeling or can be fixed in an equal volume of 2% paraformaldehyde in PBS.

Wash fixed platelets and store as above.

C. Sample Protocol for Preparation of Washed Platelets for Measurement of PAIg

Collect blood in purple top Vacutainer tubes (EDTA anticoagulant).

Centrifuge blood at 200*g* for 10 min.

Remove supernatant platelet-rich plasma.

Centrifuge PRP at 1800*g* for 10 min.

Resuspend platelet pellet in 1 ml Tyrode's buffer containing 0.3% EDTA and 0.5% bovine serum albumin.

Wash platelets again with Tyrode's EDTA/BSA.

Add washed platelets to an equal volume of 2% paraformaldehyde in PBS.

Wash fixed platelets and store as above.

V. Staining

Labeling fresh or fixed platelets with fluorescent monoclonal antibodies follows exactly the same protocol used for labeling other cells with only a few additional considerations. If fixed platelets are being labeled it is critically important in preliminary experiments to confirm that the epitope detected by the antibody is still detectable after fixation. Thus any new labeling procedure should be first tested on both fresh and fixed platelets. In our experience, the commercially available antibodies for CD41, CD42, and CD62 all work well on fixed platelets. One should obviously perform a titration of a new antibody on a fixed number of platelets in order to determine the level of antibody which is saturating. It is theoretically possible for an antibody to cause agglutination of the platelets which will remove them from the normal platelet region of the flow cytometer. We have encountered this difficulty only when using antiplatelet antiserum, not with monoclonal antibodies.

Platelets have an Fc receptor (FcRII, CD32) and can thus bind antibodies nonspecifically. However, this is not, in our experience, a sufficient problem to require the use of blocking for the Fc receptors. In general the nonspecific labeling of activated platelets will be somewhat higher than that of resting platelets, perhaps due to increased expression of the Fc receptor.

Labeling platelets with thiazole orange to detect RNA can be done with either fresh or fixed platelets. In our experience, the dye penetrates fixed platelets more rapidly and reaches equilibrium within 15 min at room temperature (Ault *et al.*, 1992). Fresh platelets may require longer incubation to achieve full labeling; we generally allow 1 hr at room temperature. The labeling of fixed platelets is stable for about 1 hr at room temperature but will then gradually decrease. The labeling of fresh platelets is more stable. The dye should not be washed out as it will leak back out of the platelets.

A. Sample Protocol for Labeling of Fixed Whole-Blood Samples for Platelet Activation and Aggregation

If the samples have not been washed out of fixative before, the fixative must be removed before labeling.

If the platelet count is normal it should be possible to divide one sample containing 50 μl of fixed whole blood into at least five tubes for labeling and still have enough platelets in each tube to analyze, (50 μl of normal blood contains 1.2×10^7 platelets).

Pellet the platelets by centrifugation at 1800*g* for 5 min.

Add 10 μl of FITC-CD41 at a dilution determined in preliminary experiments.

Add 10 μl of biotinylated-CD62 at a dilution determined in preliminary experiments.

Mix and incubate 15 min at room temperature.

Wash once with Tyrode's buffer.

To the pelleted platelets add 10 μl of phycoerythrin–streptavidin at the manufacturer's recommended dilution.

Incubate 15 min at room temperature.

Wash once and resuspend in 0.5 ml Tyrode's buffer for analysis.

B. Sample Protocol for Labeling of Fixed Washed Platelets for PAIg

Transfer approximately 1×10^6 fixed platelets to each of 5 tubes.

Add the following reagents (using the antibodies at predetermined dilutions):

Tube	Purpose	Add	Add
1	Control	10 μl Biotin CD41	Nothing
2	Two-color anti-Ig	10 μl Biotin CD41	FITC–anti-human Ig
3	IgG	10 μl Biotin CD41 (optional)	FITC–anti-human IgG
4	IgM	10 μl Biotin CD41 (optional)	FITC–anti-human IgM
5	C3	10 μl Biotin CD41 (optional)	FITC–anti-human C3

We have not found it cost effective to add CD41 to every tube. The anti-immunoglobulin reagents are all $F(ab')_2$ rabbit anti-human antibodies. The use of IgM, C3, and IgA are optional, see Ault (1988).

After 15 min at room temperature the tubes are washed once with Tyrode's buffer and phycoerythrin–streptavidin is added to each tube containing biotin CD41.

After an additional 15 min the tubes are washed again and the pellet is resuspended in 0.5 ml Tyrode's buffer for analysis.

C. Sample Protocol for Labeling of Fixed Platelets for Reticulated Platelets

Place 1×10^6 fixed platelets in a tube in a volume of 50 μl.

Resuspend the platelets in 0.5 ml of thiazole orange solution (50 μg/ml in saline).

Incubate 15 min to 1 hr at room temperature.

Analyze.

VI. Critical Aspects

One of the most critical aspects for successful flow cytometric analysis of platelets is the resolving power of the flow cytometer. Even when platelets are properly prepared and labeled one will not obtain good data unless the platelets can be clearly resolved. The size distribution of normal platelets is from 1–5 μm. Thus a good test of the flow cytometer is the ability to clearly resolve 1-μm beads using forward light scatter. Platelet microparticles are smaller, in the range of 0.1 to 1 μm, thus the instrument must be capable of resolving smaller particles if they are of interest. It is particularly important to realize that there are a large number of particles in this size range other than platelets. Cellular debris, as discussed more fully below, can be easily confused with platelets. Even more problematic is the fact that bacteria which may contaminate reagents or buffers can fall exactly in the forward scatter region of platelets. If there are even small numbers of bacteria in a buffer used for washing they will be concentrated with every wash step and may contaminate the sample in very large numbers making resolution of platelets impossible.

THUS IT IS CRITICAL THAT ALL REAGENTS AND BUFFERS USED IN THE PREPARATION AND LABELING OF PLATELETS BE FILTERED THROUGH AN 0.2μm FILTER AND THAT THEY BE STORED AND USED UNDER STERILE CONDITIONS.

As an additional guarantee that the particles in the size range of platelets really are platelets it is advisable to include in every procedure a sample labeled with at least one platelet-specific marker such as CD41 or CD42. In some cases this may be used simply to qualify the gate used for analysis of platelets, in other cases it may be desirable to trigger the instrument on the fluorescent signal for the platelet-specific marker thus excluding nonplatelet particles.

Another critical aspect of successful analysis of platelets has to do with the wide size distribution of normal platelets. They vary in size under normal conditions from 1 to 5 μ. In some disease states they can be considerably larger, and of course they have a marked propensity to form aggregates which may range in size up to 20–30 μm. This is much greater size heterogeneity than is seen in other cell types such as lymphocytes. For this reason, it is customary for platelets to span about one decade on a log light scatter scale and more than a decade on log fluorescent scales. The result is a unique signature for platelets on the flow cytometer which allows the experienced operator to be fairly certain one is dealing with platelets. However, regardless of what label is used with platelets there will always be strong correlation between fluorescence and size which must be taken into account. Thus it is usually desirable when studying platelets to use log amplification of all parameters, including light scatter, and to present results in a two-parameter fashion with light scatter versus fluorescence (see examples below). The definition of a platelet as "positive" for a particular marker must frequently take into account the size of the platelet as well as its fluorescence intensity. Analysis of platelets using single-parameter histograms will result in very wide distributions with poorly resolved "positive" populations.

VII. Standards

There are no standards that are specific to the study of platelets. It is advisable to check the ability of the flow cytometer to resolve small particles with the routine use of 1-μm beads. Since fluorescent signals from platelets are generally of considerably lower magnitude than those from larger cells, good fluorescence sensitivity should also be assured through the use of calibrated fluorescent beads.

In studies of platelet activation it is necessary to have negative (resting) controls and positive (activated) controls. Fresh blood platelets fixed immediately should have a level of CD62 labeling that is less than 5% of the population.

The most common problem is the inability to obtain good resting samples. The use of PGE1 to help ensure quiescence was described above. Platelets activated with thrombin or high doses of ADP should be greater than 90% positive for CD62. Samples of fixed resting and activated platelets can be kept in the laboratory for many weeks and can even be kept frozen with little change in their light scatter or labeling properties.

VIII. Instrument

The setup of a flow cytometer for the analysis of platelets is similar to that for other types of immunofluorescence measurements. All parameters, including light scatter, should be log amplified. The forward scatter gain should be set high enough to clearly resolve 1-μm particles, and the alignment and "tuning" of the instrument should be such that instrument noise falls well below the platelet cluster. Gains for fluorescence channels should be set higher than is normal for analysis of lymphocytes. This can only be determined by running positive and negative controls.

The setting of compensation for FITC and PE is identical to that for any immunofluorescence protocol and can be adequately done with beads designed for the purpose.

If the platelets have been enriched in the process of preparation (PRP or washed platelets) the instrument will probably be triggered on light scatter with the threshold set just below the level of the platelet cluster. If the platelets are in whole blood, or if platelet microparticles are of interest, the samples will have been labeled with a platelet-specific antibody and the instrument must be triggered on the fluorochrome corresponding to that antibody, with the threshold set below that of the platelet cluster. This is best done by looking at a dot display of forward scatter versus fluorescence while adjusting the threshold. It is usually not possible to exclude all of the leukocytes with a single threshold. Variable numbers of leukocytes will fall above the threshold due to either increased autofluorescence or binding of platelets to the leukocytes. Platelet microparticles are seen as particles that are positive for the platelet-specific marker but distintly smaller than normal platelets. They are best resolved on a display of forward or side scatter versus fluorescence.

IX. Results and Discussion

Figure 1 shows the typical platelet cluster with a strong positive correlation between the two light scatter measurements. This cluster is usually easily distinguished from the leukocyte and erythrocyte clusters which do not show a strong size correlation. Figure 1 also illustrates that it may be difficult to accurately separate platelets from leukocytes using light scatter gating alone.

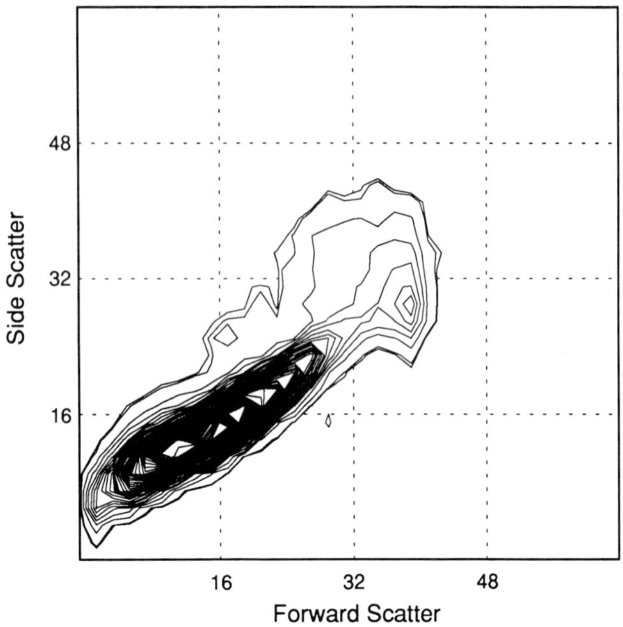

Fig. 1 The typical platelet cluster in whole blood. A fixed whole-blood sample labeled with FITC-CD41 analyzed by triggering on FL1 so as to include only those particles that express CD41. The result is the typical platelet cluster showing a more than one decade variation in light scatter. A small cluster of leukocytes can be seen to the right of the platelet cluster. Both parameters are four-decade log scales.

This is especially true if the platelets have become aggregated in the preparation process. Thus, in Figure 2, a more useful display of forward scatter versus fluorescence is shown. Here the platelet cluster is much easier to distinguish from that of leukocytes, and it is on this display that platelet gating is best performed.

Figure 3 shows an analysis of resting platelets gated as shown in Fig. 2. There is minimal expression of CD62. The control is not shown here; however in our laboratory control labeling is always within the first decade (below channel 16 in the figure). Figure 4 shows an analysis of platelets which were intentionally activated and allowed to aggregate. There are several features of interest. First is the strong expression of CD62 both on normal-sized (single) platelets and on the aggregates which extend up to sizes that somewhat exceed that of leukocytes. Of interest is the fact that in this whole-blood system, the streak of aggregates does not extend to infinity. Thus we do not observe the appearance of macroscopic aggregates such as would be formed in PRP. It is likely that the size of the aggregates is self-limited by the marked fall in the effective concentration of platelets and the presence of large numbers of erythro-

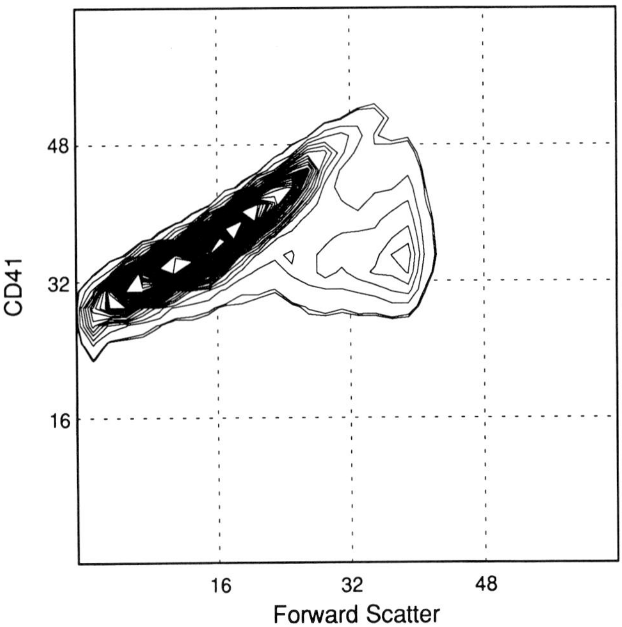

Fig. 2 Use of CD41 to identify platelets in whole blood. The same data as shown in Fig. 1 illustrating the CD41 labeling of the platelets. The trigger threshold is located at the boundary of the first and second decade. The labeling of the leukocytes is to a large extent due to the binding of platelets to monocytes and granulocytes. These parameters are the best ones for placing a gate around the platelet cluster for subsequent analysis of platelet activation.

cytes which inhibit further aggregation. Also of note is the presence of a small number of platelets that do not express CD62 even with maximal stimulation. These apparently defective platelets have not been well studied.

A difficulty arises when one attempts to quantitate platelet aggregation in this system. We have used the percentage of platelet particles that are larger than resting platelets as judged by forward light scatter (Ault *et al.*, 1989). This is a useful parameter which correlates fairly well with the degree of platelet aggregation estimated in a traditional platelet aggregometer, which also uses light scatter to detect aggregation (Carmody *et al.*, 1990). However, it grossly underestimates the proportion of platelets that are actually participating in the aggregation. The maximal value for the percentage of platelet particles larger than resting platelets is usually about 50–70%. At that level of aggregation we estimate that in excess of 99% of the platelets are in aggregates. Each leukocyte-sized aggregate contains at least several hundred platelets as judged by their staining characteristics. Most platelet aggregation in whole blood is reversible, thus unfixed samples which are not analyzed for more than 30 min after preparation will not show much if any platelet aggregation.

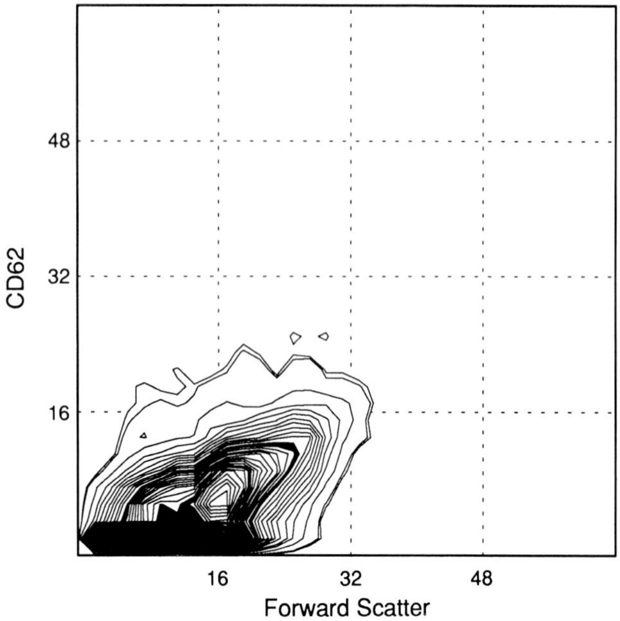

Fig. 3 Analysis of platelet activation and aggregation in resting platelets. Shown here are the same data shown in Figs. 1 and 2, now illustrating the labeling of the gated platelets with CD62. In resting blood the proportion of platelets labeled above control levels (approximately the boundary of the first decade) is less than 5%.

Typical results for the measurement of PAIg are shown Figs. 5 and 6. Figure 5 shows a sample which has normal levels of PAIg. In this sample nearly all of the particles gated as platelets (using light scatter gating in this case) were indeed CD41 positive. It can be seen that there is some heterogeneity in the level of anti-Ig labeling of these normal platelets. Sometimes in patients and occasionally in normals it is possible to resolve two distinct populations of platelets having considerably different levels of Ig labeling. The explanation for this phenomenon has not been clearly defined. It most likely reflects differences in permeability of the platelets to the anti-Ig antibody, perhaps due to differences in activation status. It does not appear to correlate with platelet size nor with the presence of reticulated platelets. Figure 6 shows data from a patient with immune thrombocytopenia. The platelet gate now contains several distinct populations of particles. Those that are CD41 positive have a clearly increased level of anti-Ig labeling. The mean fluorescence intensity of this population can be used as measure of PAIg. In this particular example there are three additional populations of particles that do not express CD41. It is important to realize that all of these populations copurify with platelets by centrifugation, and all fall within the same "platelet" light scatter gate on the

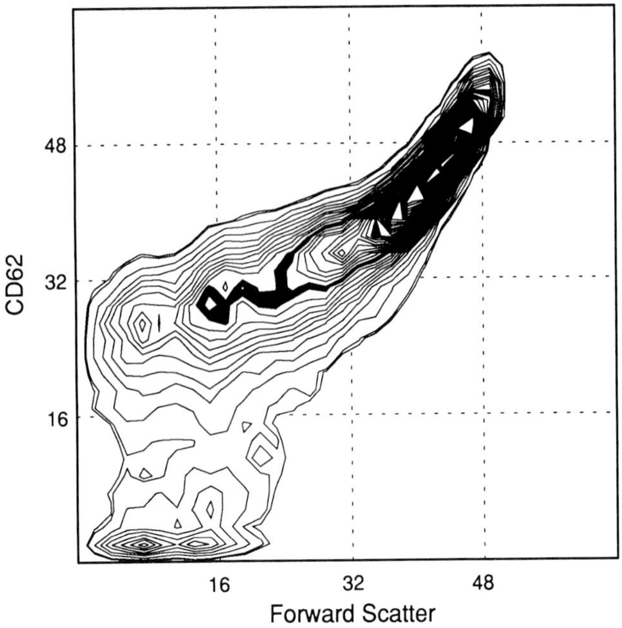

Fig. 4 Measurement of platelet activation and aggregation in stimulated whole blood. This sample was prepared from blood which had been stimulated with 20 μM ADP prior to fixation. There is now extensive positive labeling with CD62 as well as dramatic change in the forward scatter distribution. The peak of large CD62-positive particles are platelet aggregates approximately the size of leukocytes.

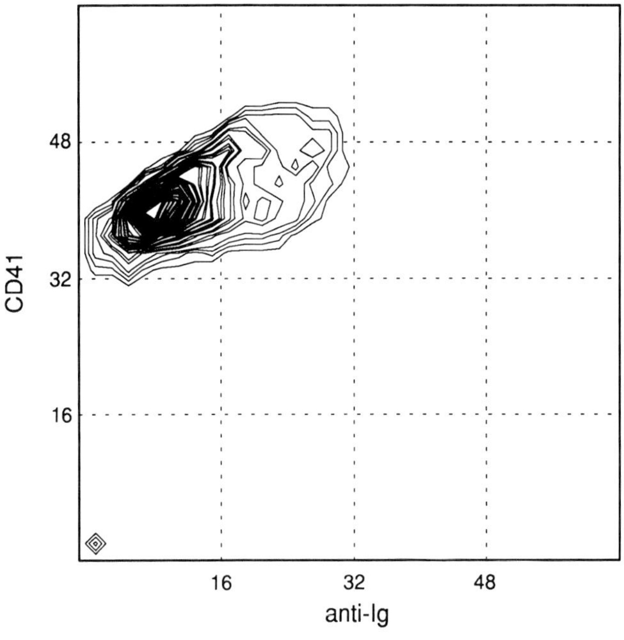

Fig. 5 Labeling of platelets for immunoglobin. Illustrated here is the result of an assay for PAIg which is within the normal range. The platelets are identified by the presence of CD41 and the mean level of PAIg is determined from the anti-Ig staining.

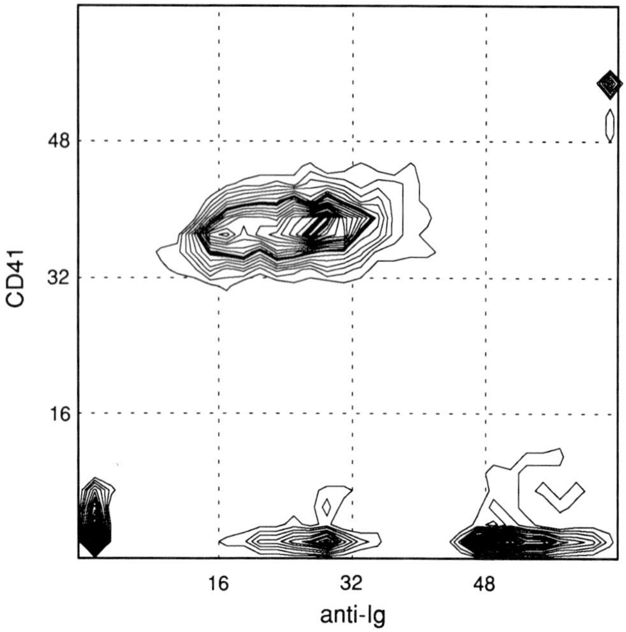

Fig. 6 A sample with elevated PAIg and nonplatelet particles. A sample from a patient with elevated PAIg was processed the same as that in Fig. 5. Here the platelets which are CD41 positive have elevated levels of PAIg. In addition to the platelets the light scatter gate contains three additional populations of CD41-negative particles which have no immunoglobulin, low levels, and high levels, respectively. Although not illustrated here, further workup usually suggests that these are red cell fragments and immune complexes.

flow cytometer. In our experience, the particles which are negative for Ig frequently label with anti-glycophorin and probably represent red cell fragments or "microspherocytes." Non-platelet particles which label brightly for Ig can usually be shown to also label very brightly for IgG, IgM, and C3 and, we feel, represent large immune complexes. The intermediate population in this particular sample is unexplained.

The frequent presence of these nonplatelet particles in patient samples raises several important issues. First is the one discussed above, that many particles fall in the platelet size range. Thus light scatter gating alone cannot be relied upon to identify platelets, especially in abnormal samples. Second, the presence of significant numbers of nonplatelet particles in preparations of "washed platelets" from patients could have a major impact on the measurement of PAIg. We feel that this is one of the strongest arguments in favor of the flow cytometric approach to PAIg measurement. As shown in Fig. 6 the flow cytometer permits one to reliably restrict the measurement to platelets, ignoring particles which are not platelets and which may have 10 to 100 times more or less immunoglobulin per particle than the platelets. It is likely that this is one of the difficulties which has plagued previous attempts to measure PAIg by other techniques.

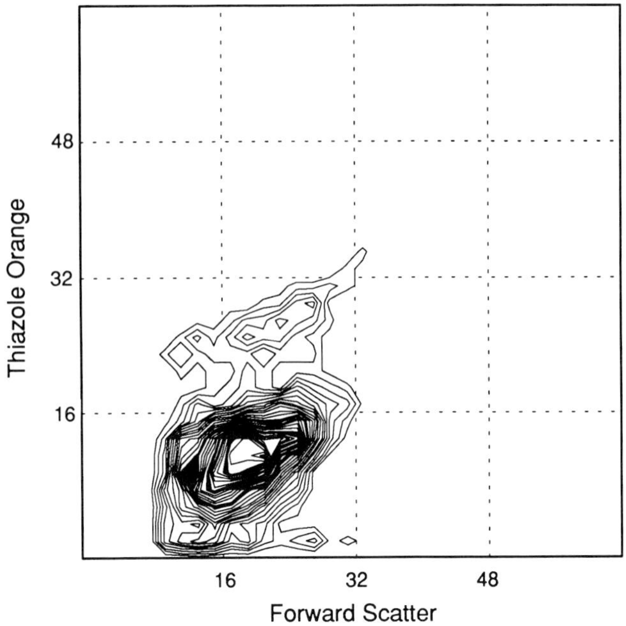

Fig. 7 Reticulated platelets. This shows results from a whole-blood sample labeled with Thiazole orange and gated on platelets. A subpopulation of platelets, about 11% in this case, have increased thiazole orange labeling and have been referred to as reticulated platelets. They have the properties of platelets newly released into the circulation.

Figure 7 shows a typical analysis of reticulated platelets. This sample was obtained by Thiazole orange labeling of whole blood followed by light scatter gating on platelets. Similar results can be obtained by Thiazole orange labeling of fixed washed platelets. All of the platelets label to some extent with Thiazole orange; however, a subset can usually be resolved which has increased labeling and which, on average, is slightly larger than the average normal platelet. This is a good example of a situation in which a population can be fairly clearly identified in two-parameter, light scatter versus fluorescence display but is very poorly resolved in a one-parameter fluorescence histogram. The normal size heterogeneity of platelets can easily mask a small population with increased fluorescence for their size. In order to enumerate these reticulated platelets we have used a line which is parallel to the platelet cluster and defines the upper bound of the cluster (Ault *et al.,* 1992). Platelets falling above this line are considered "positive." Thus the definition of "positive" includes consideration of the size of the platelet.

References

Abrams, C. S., and Shattil, S. J. (1991). *Thromb. Haemostasis* **65,** 467–473.
Abrams, C. S., Ellison, N., Brudzynski, A. Z., and Shattil, S. J. (1990). *Blood* **75,** 128–138.

Adelman, B., Michelson, A. D., Handin, R. I., and Ault, K. A. (1985). *Blood* **66,** 423.

Ault, K. A. (1988). *Pathol. Immunol. Res.* **7,** 395–408.

Ault, K. A. (1992). *In* "Clinical Flow Cytometry" (A. L. Landay, K. A. Ault, K. D. Bauer, and P. S. Rabinovitch, eds.), pp. 387–403. New York Academy of Sciences, New York.

Ault, K. A., and Mitchell, J. (1993). *In* "Flow Cytometry of the Platelet-Megakaryocyte System" (R. E. Scharf and K. J. Clemeson, eds.). Elsevier, Amsterdam (in press).

Ault, K. A., Rinder, H. M., Mitchell, J. G., Rinder, C. S., Lambrew, C. T., and Hillman, R. S. (1989). *Cytometry* **10,** 448–455.

Ault, K. A., Mitchell, J. G., Rinder, H. M., Rinder, C., and Hillman, R. S. (1991). *In* "Flow Cytometry in Hematology" (O. D. Laerum and R. Bjerknes, eds.), pp. 153–163. Academic Press, London.

Ault, K. A., Rinder, H. M., Mitchell, J., Carmody, M. B., Vary, C. P. H., and Hillman, R. S. (1992). *Am. J. Clin. Pathol.* **93,** 637–646.

Berman, C. L., Yeo, E. L., Wencel-Drake, J. D., Furie, B. C., Ginsberg, M. H., and Furie, B. (1986). *J. Clin. Invest.* **78,** 130–137.

Carmody, M., Ault, K. A., Mitchell, J. G., Rote, N. S., and Ng, A. (1990). *Hybridoma* **9,** 631–641.

Coller, B. S. (1985). *J. Clin. Invest.* **76,** 101.

Corash, L. (1990). *Blood Cells* **160,** 97–108.

Corash, L., and Rheinschmidt, M. (1986). *In* "Manual of Clinical Laboratory Immunology" (N. R. Rose, H. Friedman, and J. L. Fahey, eds.). Am. Soc. Microbiol., Washington, DC. pp. 254–257.

Corash, L., Mok, Y., and Rheinshmidt, M. (1986). *In* "Applications of Fluorescence in the Biomedical Sciences," pp. 567–584. Liss, New York.

Davies, T. A., Drotts, D., Weil, G. J., and Simons, E. R. (1988). *Cytometry* **9,** 138–142.

de Bruijne-Admiraal, L. G., Modderman, P. W., von dem Borne, A. E. G. K., and Sonnenberg, A. (1992). *Blood* **80,** 134–142.

Ginsberg, M. H., Frelinger, A. L., Lam, S. C., Forsyth, J., McMillan, R., Plow, E. F., and Shattil, S. J. (1990). *Blood* **76,** 2017–2023.

Ingram, M., and Coopersmith, A. (1969). *Br. J. Haematol.* **17,** 225–228.

Johnson, P. C., Ware, J. A., Cliveden, P. B., Smith, M., Dvorak, A. M., and Salzman, E. W. (1985). *J. Biol. Chem.* **260,** 2069–2076.

Johnston, G. I., Pickett, E. B., McEver, R. P., and George J. N. (1987). *Blood* **69,** 1401–1403.

Kienast, J., and Schmitz, G. (1990). *Blood* **75,** 116–121.

Lazarchick, J., and Hall, S. A. (1986). *J. Immunol. Methods* **87,** 257–265.

Marti, G. E., Magruder, L., Schuette, W. E., and Gralnick, H. R. (1988). *Cytometry* **9,** 448–455.

Nieuwenhuis, H. K., van Oosterhout, J. J. G., Rozemuller, E., van Iwaarden, F., and Sixma, J. J. (1987). *Blood* **70,** 838–845.

Rinder, C. S., Bohnert, J., Rinder, H. M., Mitchell, J., Ault, K. A., and Hillman, R. S. (1991a). *Anesthesiology* **75,** 388–393.

Rinder, C. S., Mathew, J. P., Rinder, H. M., Bonan, J., Ault, K. A., and Smith, B. R. (1991b). *Anesthesiology* **75,** 563–570.

Rinder, C. S., Bonan, J. L., Rinder, H. M., Mathew, J., Hines, R., and Smith, B. R. (1992). *Blood* **79,** 1201–1205.

Rinder, H. M., Murphy, J., Mitchell, J. G., Stocks, J., Hillman, R. S., and Ault, K. A. (1991a). *Transfusion (Philadelphia)* **31,** 409–414.

Rinder, H. M., Bonan, J., Rinder, C. S., Ault, K. A., and Smith, B. R. (1991b). *Blood* **78,** 1730–1737.

Rinder, H. M., Bonan, J., Rinder, C. S., Ault, K. A., and Smith, B. R. (1991c). *Blood* **78,** 1760–1769.

Rinder, H. M., Munz, U., Smith, B. R., Bonan, J. L., and Ault, K. A. (1993). *Arch. Pathol. Lab. Med.* **117,** 606–610.

Rosenfield, C. S., Nichols, G., and Bodensteiner, D. C. (1987). *Am. J. Clin. Pathol.* **87,** 518–522.

Shattil, S. J., Cunningham, M., and Hoxie, J. A. (1987). *Blood* **70,** 307–315.

Sims, P. J., Faioni, E. M., Wiedmer, T., and Shattil, S. J. (1988). *J. Biol. Chem.* **263,** 18205–18212.

CHAPTER 18

Flow Cytometry in Malaria Detection

Chris J. Janse and Philip H. Van Vianen

Laboratory of Parasitology
University of Leiden
2300 RC Leiden
The Netherlands

METHODS IN CELL BIOLOGY, VOL. 42

I. Introduction

Malaria is a parasitic disease found in tropical and subtropical regions which is caused by unicellular organisms of the genus *Plasmodium*. In man, 4 different species are responsible for the disease, *Plasmodium falciparum, P. vivax, P. ovale,* and *P. malariae*.

A large part of the life cycle of these parasites takes place in the blood circulation, where these organisms invade red blood cells in which they grow and multiply. Failure of existing methods to control malaria, the lack of an effective vaccine, and increasing drug resistance of the parasites are factors which play a role in the increase of malaria cases. Both for laboratory research aimed at the development of new control strategies and for monitoring the effects of existing control projects in the field, the availability of rapid, sensitive, and reproducible techniques for the detection and analysis of blood infection are required. The blood stage infection is the most relevant part of the life cycle; the blood stages cause the clinical symptoms and are targets for a number of drugs. Moreover, the demonstration of the presence of parasites in the blood is used for diagnosis and treatment of the disease.

Flow cytometry has proven to be a useful tool for the analysis of blood infection by malaria parasites. Analysis of blood-stage development (Janse *et al.*, 1987; Mons and Janse, 1992) and determination of susceptibility to drugs by flow cytometry (van Vianen *et al.*, 1990a,b) are reproducible and rapid and detection of blood-stage parasites appears to be sensitive and reproducible (van Vianen *et al.*, 1993; P. H. van Vianen, unpublished results).

The analysis and detection of malaria infection by flow cytometry makes almost exclusive use of fluorescent dyes which are specific for nucleic acids (for review, see Mons and Janse, 1992). DNA-specific dyes are especially useful since the parasites multiply inside the red blood cell (RBC) population of the blood cells. Since RBC do not contain DNA, DNA-specific fluorescence from infected RBC can only be due to fluorescence of dyes bound to parasite DNA. Consequently, infected cells can be discriminated from noninfected cells based on their fluorescence intensity. In addition, since parasites multiply within the RBC by several mitotic divisions, the fluorescence intensity of stained parasites increases during development of the parasites. This can be analyzed by flow cytometry and used to determine the developmental stage of the parasite. The total DNA content of a parasite is 100–200 times less than that of nucleated blood cells. Therefore, the nucleated blood cells can easily be distinguished from parasites on the basis of the difference in fluorescence intensity.

The vast majority of the reported flow cytometric studies on malaria parasites have been carried out with the A/T-specific DNA dyes, Hoechst 33258 and Hoechst 33342. These dyes give a strong fluorescence with parasite DNA after fixation of infected blood cells (Myler *et al.*, 1982; Bianco *et al.*, 1986) or free parasites and after vital staining of parasites (Franklin *et al.*, 1986). Moreover, the relative fluorescence intensity of different blood stages after Hoechst staining corresponds closely to the relative DNA content of these stages (Janse *et al.*, 1987). A/T-specific dyes are particularly suited since the A/T content of DNA of malaria parasites is extremely high, ranging from 70 to 82% in different species. These dyes are now routinely used in studies to determine parasite development and DNA synthesis in parasites and for determination of the level of drug resistance in parasites obtained from patients. Moreover, it has been shown that these dyes can be used for sensitive detection of parasites in blood samples obtained from patients in clinical and in epidemiological studies (van Vianen *et al.*, 1993; P. H. van Vianen, unpublished results).

Other dyes have been used for parasite detection (for review, see Mons and Janse, 1992), such as acridine orange (Hare, 1986; Whaun *et al.*, 1983), propidium iodide (Saul *et al.*, 1982; Pattanapanyasat *et al.*, 1992), and thiazole orange (Makler *et al.*, 1987). These dyes have the disadvantage that they bind both to DNA and to RNA. The reticulocyte population of blood cells contains RNA and the fluorescence of parasite-infected cells can be in the same range as that of noninfected reticulocytes which hampers the discrimination between infected cells and noninfected cells.

Below we describe flow cytometric methods for (1) measurement of parasite development and DNA synthesis by parasites, (2) determination of susceptibility of parasites to drugs, and (3) detection of low numbers of parasite-infected cells in blood samples from patients.

II. Applications

A. Flow Cytometry and the Developmental Cycle of the Parasite

Malaria parasites have a complex life cycle which alternates between two different hosts: mosquitoes and vertebrates such as reptiles, birds, rodents, nonhuman primates, and humans. An infection starts with a bite of an infected mosquito, which injects parasites into the blood of the vertebrate host. These parasites penetrate liver cells and, after one developmental cycle, the parasites are released in the blood and invade RBC.

For flow cytometry the blood stages (see Fig. 1) are the most important (see below). After entering the RBC the small haploid parasites (called ring forms) develop and grow until they nearly fill the RBC (these growing stages are the so-called trophozoites). After this growth phase most of the parasites synthesize DNA and enter mitosis. In each parasite, now called schizont, three to five rapid mitotic divisions follow each other, resulting in the production of 8–32 haploid merozoites. After rupture of the RBC, these merozoites can penetrate new RBC and the multiplication cycle starts again. One multiplication cycle takes 24–72 hr, depending on the species. This mitotic multiplication results in a rapid increase in the number of infected red blood cells.

A small percentage of the merozoites do not continue the asexual multiplication but differentiate within the RBC into precursor cells of the gametes, the so-called gametocytes. These stages develop into gametes when they are ingested by mosquitoes, in which further development takes place. No flow cytometric studies on the mosquito stages of the parasite have been reported.

B. Flow Cytometry and Determination of Parasite Development, DNA Synthesis, and Drug Susceptibility

The blood stages of several species of malaria parasites can grow and multiply under culture conditions. These *in vitro* cultures allow the effect of new drugs to be studied on parasite development and DNA synthesis under standardized

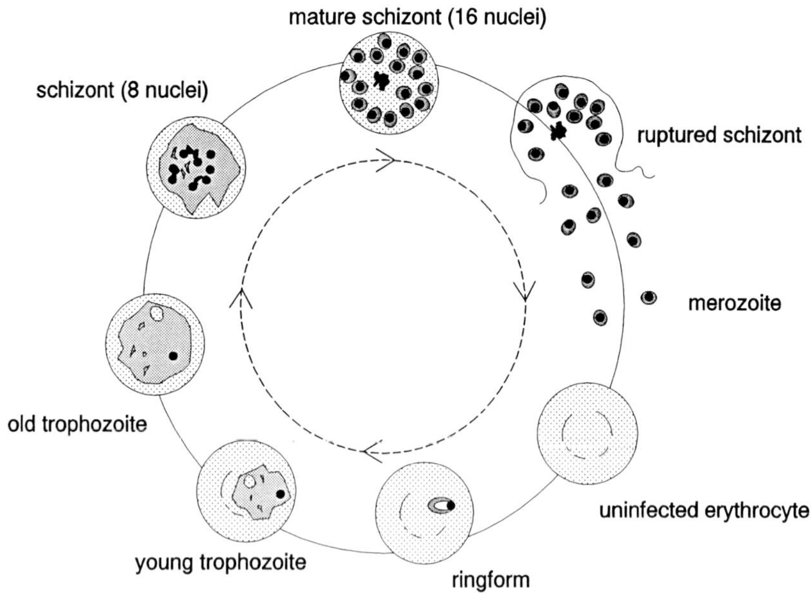

Fig. 1 Schematic representation of blood-stage development of malaria parasites. Only part of the life cycle, the asexual erythrocytic development, is shown.

conditions. The effect of new drugs on parasite development in culture is routinely monitored by microscopic counting of the number of infected cells or by measuring the incorporation of radioactive precursors into the nucleic acids of the parasites. Flow cytometry is a very good alternative for the determination of the effect of new drugs on parasite development in culture. Using DNA-specific dyes the increase/decrease of the number of infected cells can be measured rapidly and reproducibly. In addition, the amount of DNA synthesis can be determined precisely which is a reliable characteristic of the development of the parasites (Janse *et al.*, 1987,1989; van Vianen *et al.*, 1990a). The advantages of flow cytometry are the speed of measurement, the accuracy, reproducibility, and the large number of parasites analyzed.

For the determination of drug susceptibility of parasites from field isolates the World Health Organization (WHO) (1982) has developed a standard *in vitro* microtest. In these tests haploid small blood stages (ring forms, trophozoites) are cultured for short periods in the presence of different concentrations of drugs. Giemsa-stained slides are made from these cultures to monitor development of the parasites from ring forms into the DNA synthesizing stages (schizonts) by light microscopy. Determination of parasite development by light microscopy is time consuming and the results can easily be influenced by human errors. Since development and DNA synthesis of parasites can accurately be assessed by flow cytometry using DNA-specific dyes, flow cytometry is therefore very useful for determination of drug susceptibility. Recently, a method has been developed for the fully automated reading of the WHO microtests by flow cytometry (van Vianen *et al.*, 1990b).

C. Flow Cytometry and Detection of Blood Stages

The demonstration of the presence of blood stages is used for diagnosis and treatment of malaria. The "gold standard" is the microscopic detection of blood stages in thin or thick blood smears which are stained with Giemsa. However, sensitive detection in large numbers of blood samples collected from patients under primitive field conditions using this method poses problems. The sensitivity is highly influenced by local circumstances and working conditions and depends on the availability of experienced people for recognition of parasites. The quality of microscopic slides is not always optimal, due to improper preparation of the smears, applying contaminated staining solution, or suboptimal use of the staining procedure. The sensitivity is very much dependent on the experience of the microscopist and the time spent reading the slides. Because microscopy is labor intensive, human factors such as loss of concentration, especially when large numbers of samples need to be screened with a low percentage of positives, can account for misreading of samples (Payne, 1988).

In the past few years, alternatives to microscopic detection of malaria parasites have been investigated, such as immunological methods to demonstrate antibodies or antigens (Tharavanij, 1990), detection of parasites by fluoresence microscopy (Rickman *et al.*, 1989; Kawamoto and Billingsley, 1992), and the use of malaria-specific radioactively labeled DNA and RNA probes (Tharavanij, 1990). At present none of these techniques appears to be superior to microscopic examination of Giemsa-stained blood smears. Flow cytometry has been shown to be potentially suited to overcome most of the above-mentioned problems with parasite detection (van Vianen *et al.*, 1993). Although several fluorescent dyes have been reported to be useful for detection of parasite-infected cells most studies have been performed with the DNA-specific Hoechst dyes (Mons and Janse, 1992). Detection and counting of the number of infected cells can be performed using Hoechst 33258-stained RBC which are fixed by glutaraldehyde. However, this method is not sensitive enough to detect very low numbers of infected RBC (<0.1%). It appears to be necessary to first free the parasites from the RBC by lysing these blood cells (van Vianen *et al.*, 1993). This reduces the sample volume and the number of cells to be analyzed.

III. Materials

Materials for the *in vitro* culture of *P. berghei* have been described by Janse *et al.* (1989). For materials for *in vitro* cultures of *P. falciparum* see references in Trigg (1985).

Phosphate-buffered saline (PBS) tablets (Flow Laboratories) are dissolved in double-distilled deionized (demi) water and HCl is used to adjust the pH at 7.2. PBS for fixation solution or staining solution is filtered through a 0.22-μm filter to remove small particles. PBS is stored at 4°C.

Hoechst 33258 (Janssen Chimica) is dissolved in demi-water at a stock concentration of 500 μM. The stock is stored at −20°C. Final concentration for cell staining is 2 μM in PBS.

Propidium iodide (Sigma) is dissolved in demi-water at a stock concentration of 1 mg/ml, which is stored at 4°C. Final concentration for cell staining is 1 μg/ml in PBS.

Glutaraldehyde (Zeiss, high grade, 70%) is diluted with PBS at a stock concentration of 25% and stored at −20°C. Final solution is made by diluting the stock $\frac{1}{100}$ with filtered PBS (to 0.25%) and stored at 4°C.

Lysis solution (Becton–Dickinson Immunocytometry Systems, 10× stock) is stored at room temperature. Final solution is made by diluting the stock $\frac{1}{10}$ with demi-water and filtering through a 0.22-μm filter. Storage is at 4°C.

IV. Cell Preparation and Staining

A. Collection of Blood Samples from Infected Humans and Laboratory Animals

For flow cytometric analysis only small blood samples are required. Samples from human patients can be taken from blood collection tubes treated with heparin or EDTA as anticoagulants, which are routinely used for collection of blood. Alternatively small samples can be drawn using heparinized capillaries from the finger after a finger prick.

Malaria parasites which infect nonhuman primates and rodents are regularly used as models for the study of malaria. Small blood samples (20–200 μl) from infected rodents, such as mice and rats, are usually collected from the veins at the end of the tail using heparinized capillaries and resuspended in PBS or culture medium. Cells are collected by centrifugation for 5 sec (15,000*g*) in an Eppendorf centrifuge or at 450*g* for 10 min. When larger amounts of blood are needed (for example for cultures of the blood stages; see below), a cardiac puncture under etheranesthesia is performed and blood is collected either in PBS containing heparin (20 IU/ml) or in culture medium RPMI 1640 (see below) containing heparin (20 IU/ml). Heparin is added as an anticoagulant. Cells are collected by centrifugation at 450*g* for 10 min.

B. Collection of Samples from *in Vitro* Cultures of the Blood Stages of Malaria Parasites

In vitro cultures of the blood stages of two species are regularly used for the study of parasite development and drug susceptibility. These are the human parasite *P. falciparum* and a parasite which infects rodents, *P. berghei*. Culture methods for both species have been described extensively (Trigg, 1985; Janse *et al.*, 1989). Here the methods are described very briefly.

Infected RBC are obtained either directly from humans and rodents or from liquid nitrogen storage. Cultures are normally started with young stages of the parasite, the ring forms. Infected RBC are incubated in culture medium RPMI 1640 containing Hepes buffer (5.94 g/liter) $NaHCO_4$, serum (10–20%), and antibiotics. This cell suspension at an RBC concentration ranging from 0.5 to 10% is incubated at 37°C in culture plates, flasks, petri dishes, or Erlenmeyers and gassed with a mixture of 10% CO_2, 5% O_2, and 85% N_2. In these cultures parasites develop from ring forms into schizonts, after which invasion of new RBC takes place. For both parasite species methods have been described to synchronize the asexual development of the blood stages, so that all parasites are at the same stage of development during the complete cycle. Samples from the culture are centrifuged for 5 sec (15,000*g*) in Eppendorf centrifuges or at 450*g* for 10 min to remove culture medium.

C. Collection of Samples from Standard Drug Susceptibility Tests

For determination of drug susceptibility of *P. falciparum* parasites obtained from patients, the WHO (1982) has developed a standardized microtest. Infected blood, obtained from patients, is incubated in complete culture medium (RPMI 1640 + 10% human serum) in standard 96-well microtiter plates for 26–30 hr, according to the WHO procedure, at an RBC concentration of 10%. The plates are predosed with different concentrations of drugs. Samples from the microtest can be prepared as described for the *in vitro* cultures of the blood stages of the parasites. However, for flow cytometric analysis the samples can remain in the culture wells at the end of the culture period and be fixed and stained in the wells after removal of the culture medium (see below).

D. Fixation of Infected Erythrocytes

Infected RBC can be fixed with glutaraldehyde, paraformaldehyde, or a combination of these two. Both glutaraldehyde and (para)formaldehyde have the disadvantage that they crosslink cell membrane components, which hampers the penetration of high-molecular-weight molecules such as monoclonal antibodies, DNA probes, or large fluorochromes. In addition these fixatives can have a significant quenching effect on the emission of fluorescence from certain DNA-bound fluorochromes (Crissmann *et al.*, 1979). Despite these disadvantages aldehyde-type fixatives appear to be very useful for fixation and staining of infected RBC with Hoechst dyes. They induce no significant cell aggregation or lysis, which are frequently observed when for example ethanol and methanol are used as fixative.

The usual procedure for fixation is as follows: Samples of infected blood cells obtained from humans, laboratory animals, or cultures are washed once in PBS before fixation. Typically these samples consist of 1 ml of blood cell suspension of 0.5–10% (10^7–10^9 cells), collected in eppendorf tubes. The blood cells are centrifuged for 5 sec at 15,000*g* and the supernatant is removed. Subsequently 1 ml of 0.25% glutaraldehyde in PBS is added and the sample is mixed vigorously. Fixation is done at 4°C for 15 min. The cell suspension in the glutaraldehyde solution ranges between 0.5 and 10%. After fixation, cells are washed twice with PBS. Fixed cells are stored in PBS at 4°C until being stained for flow cytometry. Cells can be kept for more than a year at 4°C without deteriorating. We have found that the washing steps with PBS both before and after fixation are not necessary for accurate flow cytometric readings. Blood cells can be added directly to the fixative and can be stored without removal of the fixation solution.

E. Fixation of Free Parasites

To detect very low numbers of parasites in blood samples from patients it is beneficial to lyse the RBC before fixation in order to reduce the sample volume and the number of cells to be analyzed (van Vianen *et al.*, 1993). For this method samples of 50 μl of blood (1–5 $\times$ 10^8 cells) are collected in eppendorf tubes with 1 ml of FACS lysing solution (Becton–Dickinson Immunocytometry Systems, San Jose, CA) containing 1.5% formaldehyde as a fixative. For proper lysis of the RBC, the blood cells are added directly to the lysis solution and the sample is mixed well in the solution. This treatment ruptures red blood cells, releasing the malaria parasites which are subsequently fixed by the formaldehyde. The white blood cells (WBC) remain intact and are fixed as well. The samples are lysed and fixed for 30 min at room temperature and are stored in the lysing solution at 4°C. We found that samples can be stored up to a year in this way.

F. Fixation of Infected Blood Cells from Drug Susceptibility Microtests

Microtests are performed in 96-well microtiter plates. To fix the cells in the wells, culture medium is carefully removed from the cells using a micropipette leaving the blood cells at the bottom of the wells. The cells are fixed immediately by adding 200 μl 0.25% glutaraldehyde in PBS. The plates containing the fixed samples can be sealed and stored at 4°C until analysis. In this way material can be kept in fixing solution or in PBS for over 6 months at 4°C without significant deterioration.

G. Staining of Infected Blood Cells and Free Parasites with Hoechst 33258 after Fixation

Samples of fixed infected blood cells or free parasites are stained for 1 hr at 37°C in the dark in 1–2 μM Hoechst 33258 in PBS. In case of the samples containing fixed RBC, part of the sample is diluted with PBS to a volume of 1 ml (approximately 10^6–10^8 RBC/ml). To this suspension 2–4 μl of a 500 μM stock solution of Hoechst 33258 is added. In case of samples containing free parasites, 200–500 μl of the sample is centrifuged for 1 min in an eppendorf centrifuge (15,000g) and the supernatant is carefully removed to prevent loss of cells. To the pellet 0.2 to 0.5 ml of a staining solution containing 1 μM Hoechst 33258 in PBS is added, and the suspension is mixed. In the final cell suspension at least 10^5 WBCs/ml should be present. The cells remain in staining solution until analysis, which is usually performed within 0–3 hr after staining.

The same procedure can also be applied to samples from drug susceptibility tests. Fixation solution in the wells is carefully removed using a micropipette and the cells are resuspended in 200 μl staining solution which contains 2 μM Hoechst 33258 in PBS. Cells are stained in the plates at 37°C for 1 hr in the dark.

H. Staining of Free Parasites with Hoechst 33258 in Combination with Propidium Iodide

When low numbers of infected cells are present in blood samples (<0.1%), red blood cells are lysed before staining and analysis by flow cytometry. We have found that staining of the parasites with Hoechst 33258 in combination with propidium iodide improves the capability to distinguish parasites from background fluorescence (see Section III, Results and Discussion) (P. H. van Vianen, unpublished results).

haploid parasites, as well as schizonts containing more than 30 nuclei, need to be presented in the same histogram, the fluorescence gain setting is set in a logarithmic scale. The lower threshold in the FSC is set so that free merozoites, which are much smaller than the RBC, are still included.

The fluorescence intensity and FSC of 10,000–50,000 cells per sample are measured, collected in list mode, and stored using the standard BD Consort 30 software. Data analysis can be performed using the Consort 30 software, but for analysis of malaria parasite development in culture, specialized software is developed (see also Section VIII, Results and Discussion).

B. Analysis of Samples Containing Free Parasites Stained with Hoechst 33258 in Combination with Propidium Iodide

To detect and count free parasites, samples are analyzed using a FACStar equipped with a Coherent Innova 90 laser tuned to UV excitation (351 nm, 50 mW). A dichroic mirror (DM560), a BP 485/22 for the blue Hoechst fluorescence, and a LP 620 for the red propidium iodide are selected as emission filters. Calibration of the FACStar is done using calibration beads containing defined amounts of fluorescent dye (Hoechst). A lysed blood sample containing sufficient numbers of parasites is used for the initial settings of the machine. Because both parasites and WBC have to be included in the same fluorescence histograms, the fluorescence gain settings are set in the logarithmic scale. To eliminate small weakly fluorescing particles which are not of interest, an electronic threshold is set in the red fluorescence signal, just below the fluorescence signal of parasites from the control sample. All samples are analyzed using these fixed settings. Of each sample, 5000 events are analyzed and the data are stored using the standard Becton–Dickinson Consort 30 software. For data analysis, parasite and WBC populations are identified and selected by setting a gate in a two-dimensional dot plot of FSC and blue fluorescence. This gate is set using the data from the control sample.

C. Analysis of Samples from Microtests Containing Hoechst-Stained Parasites

The analysis of samples from microtests is as described for the analysis of Hoechst-stained infected RBC by the FACStar as described above. However, the samples are not fed into the flow cytometer by hand, but sampling from the wells is performed automatically using an AutoMATE (Becton–Dickinson). This enables the fully automatic analysis and storage of samples from a complete 96-well microtiter plate. The acquisition is performed using the AutoMATE Control Program (ACP). With the ACP, samples of interest can be selected. Furthermore information for each sample can be added and stored together with the flow cytometric data. This can be used for identification purposes and for data processing by the program described below (Reinders *et al.*, 1994).

The program selects and identifies the events of interest, such as uninfected

RBC, infected RBC, and free parasites by their FSC and fluorescence characteristics. This is used to determine the percentage of infected RBC whereas parasite development is calculated from the fluorescence distribution of the infected RBC which can be expressed as the increase in the mean number of nuclei per parasite (growth) and as the total number of nuclei synthesized, respectively. Results from several samples from the same culture series are combined in graphs and tables (see Fig. 5). Alternatively, results from single samples can also be displayed.

VIII. Results and Discussion

A. Blood Stages: Development and DNA Synthesis

The nuclei of parasites stained with Hoechst dyes show a strong specific fluorescence. Therefore, infected RBC are clearly separated from uninfected cells on the basis of Hoechst-DNA fluorescence intensity (Fig. 2). The assess-

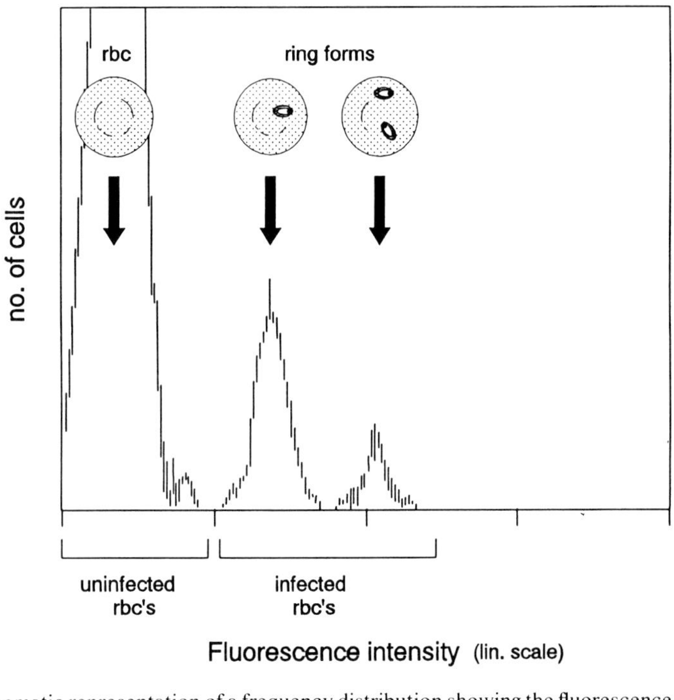

Fig. 2 Schematic representation of a frequency distribution showing the fluorescence distribution of a blood sample containing uninfected RBC and malaria-infected RBC containing haploid non-DNA synthesizing parasites, the ring forms. The RBC were fixed with glutaraldehyde before being stained with Hoechst 33258. Two peaks represent the infected cells: the first peak consists of RBC with one haploid ring form and the second peak consists of RBC infected with two ring forms.

ment of the percentage of infected cells by flow cytometry based on this difference in fluorescence intensity corresponds closely to the assessments by microscopic examination of Giemsa-stained slides (Janse *et al.*, 1987).

Frequency distributions of the fluorescence values of young ring forms and merozoites show narrow, symmetrical Gaussian distributions (see Fig. 2). These stages of the parasite are haploid and non-DNA synthesizing. In the fluorescence histograms of these stages often a small second peak is observed of cells with a double-fluorescence intensity. This peak represents infected cells containing two ring forms (double-infected RBC) or is caused by the simultaneous measurement of two infected cells or two free parasites.

During development of the ring forms into the old trophozoites, parasites increase in size but do not synthesize DNA. Old trophozoites show about 10% higher fluorescence intensity than ring forms, which is due to an increase of a non-specific background fluorescence of the cytoplasm of the parasite (Janse *et al.*, 1987). In the schizont stage of development a rapid increase in DNA content and number of nuclei occurs as the result of three to five mitotic divisions, resulting in the production of 8–32 merozoites per parasite. The increase in the number of nuclei is proportional to the increase of the fluorescence intensity of Hoechst-stained schizonts (Fig. 3). Therefore, the frequency distributions of the fluorescence intensities of a population of (dividing) stages at different time points are representative of the development and degree of DNA synthesis of the parasites (Janse *et al.*, 1987). These frequency distributions can be used to determine the inhibition of parasite development and DNA synthesis by antimalarial drugs (see below).

B. Determination of Drug Susceptibility of Parasites

Determination of antimalarial activity of drugs *in vitro* which inhibit development of the blood stages can be performed routinely using flow cytometry. Here we describe results from experiments using two different species, *P. falciparum* and *P. berghei*.

1. *P. berghei*

Young ring forms are cultured for 24 hr under standard culture conditions in RPMI 1640 medium to which different concentrations of the drugs are added. Blood-stage development of *P. berghei* from haploid ring forms to the mature schizonts containing 8–24 nuclei takes 22–24 hr. RBC containing the mature schizonts do not burst spontaneously in culture, but remain intact and viable for several hours. Samples for flow cytometry are taken from these cultures, before the culture is started and after 24 hr. Samples at the start of the cultures contain ring forms which show a narrow symmetrical frequency distribution of their fluorescence intensities (Fig. 2). The mean fluorescence intensity of ring forms/young trophozoites represents the haploid DNA content and can be used

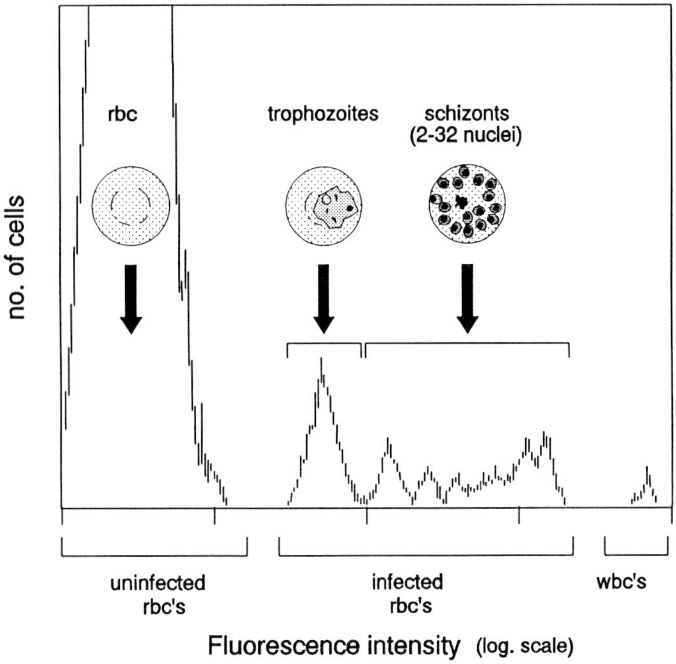

Fig. 3 Schematic representation of a frequency distribution showing the fluorescence intensity of a sample containing uninfected RBC and infected RBC. Parasites range from haploid non-DNA synthesizing trophozoites to mature schizonts containing 16–32 merozoites with immature schizonts in between. The RBC were fixed with glutaraldehyde before being stained with Hoechst 33258. The fluorescence gain setting is set in a logarithmic scale. The small peak with the highest fluorescence intensity represents the white blood cells (WBC). From van Vianen *et al.* (1990b), with permission.

as an internal standard to calculate the number of nuclei in the schizonts. In the 24-hr sample from cultures without antimalarial drugs parasites show fluorescence values between 1 and 24 times the haploid amount. This comprises mature schizonts with 8–24 nuclei, immature schizonts in the process of DNA synthesis, and some degenerated parasites and free merozoites which are liberated from the RBC during handling of the samples. Figure 3 shows a schematic representation of a histogram of the fluorescence distribution of a sample containing trophozoites and schizonts. Figure 4 shows a two-parameter dot plot representation of flow cytometric data showing fluorescence intensity and FSC of a sample containing trophozoites, schizonts, and free merozoites.

Based on the mean fluorescence intensity of the cells, the software developed for this purpose calculates the percentage of infected cells and the total number of nuclei and the average number of nuclei per parasite present. Additionally the parasite growth and DNA synthesis of the whole series are calculated and

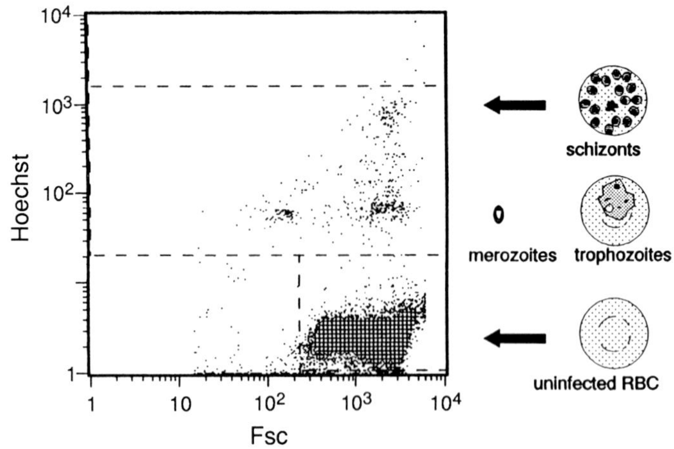

Fig. 4 Two-parameter dot plot representation of fluorescence intensity and FSC from a sample containing uninfected RBC, free merozoites and infected RBC containing trophozoites, and immature and mature schizonts. The cells were fixed with glutaraldehyde before being stained with Hoechst 33258.

presented in a graph and table. Parasite growth is defined as the average number of nuclei/parasite in a sample divided by the maximum average number of nuclei/parasite in the cultures. DNA synthesis is defined as the total number of nuclei in a sample divided by the maximum total number of nuclei in the cultures. Figure 5 shows an example of the calculation of parasite growth and DNA synthesis in samples obtained from cultures containing different concentration of an antimalarial drug. Figure 6 gives an example of the flow cytometric comparison of inhibition of parasite growth/DNA synthesis by three related antimalarial drugs. The results obtained by flow cytometry are comparable with results obtained by microscopic examination of parasites in Giemsa-stained slides (Janse *et al.*, 1987; van Vianen *et al.*, 1990a) or by measurement of the incorporation of radioactive precursors into RNA/DNA (C. J. Janse, unpublished results).

2. *P. falciparum*

Different *in vitro* tests have been described for determination of drug susceptibility of *P. falciparum* parasites. Parasites, often ring forms or young trophozo-

---

Fig. 5 Results from the computer program which calculates parasite inhibition by an antimalarial drug in a series of cultures. Each histogram represents the fluorescence distribution of parasites cultured without drug (control) or in the presence of different concentrations of the drug ranging from 0.3–100 ng/ml. In this experiment *P. berghei* ring forms were cultured for 22 hr in the presence of sodium artesunate after which samples of the cultures were fixed with glutaraldehyde and stained with Hoechst 33258. From the histograms it is clear that ring forms still develop into (mature)

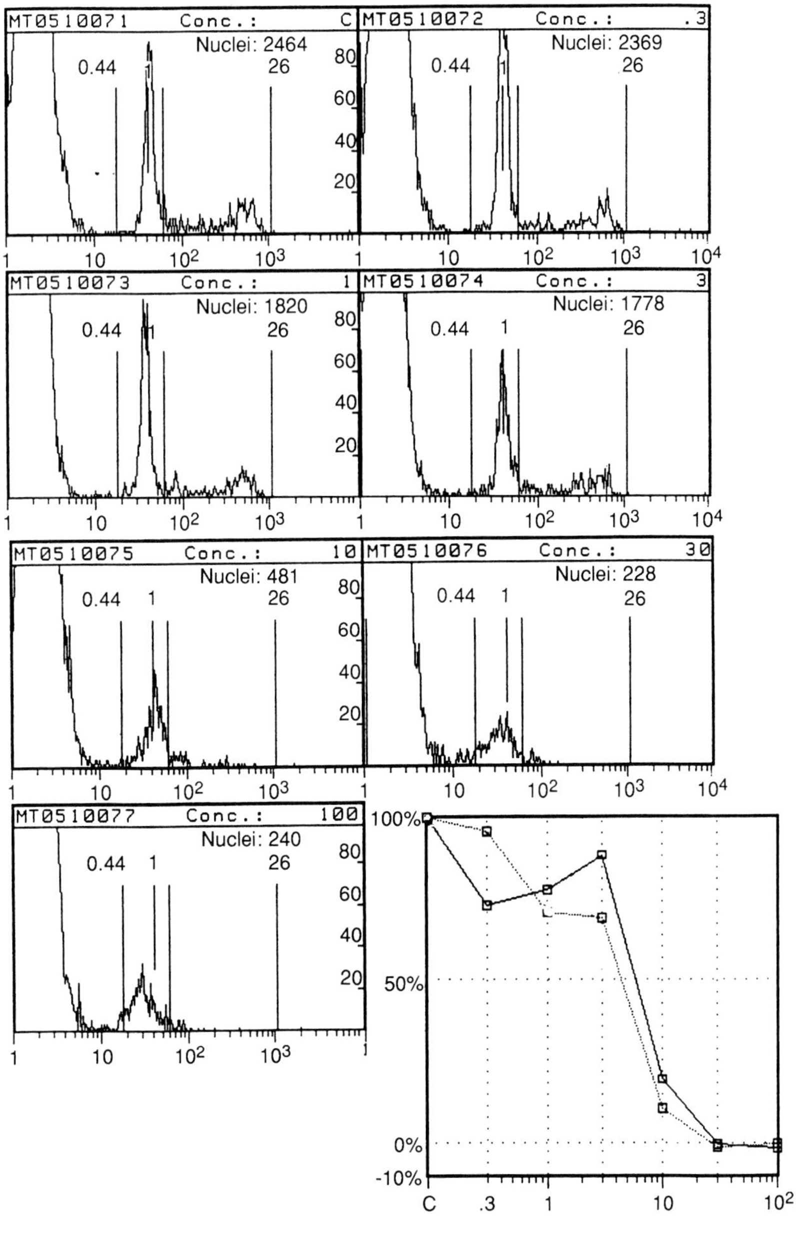

schizonts at low concentrations up to 3 ng/ml. No schizonts can be detected at 30–100 ng/ml. The first peak in the histograms contains ring forms/free merozoites and trophozoites and the mean fluorescence intensity of these cells represents the haploid amount of DNA (one nucleus). This value is used to calculate the total number of nuclei in the different samples which is shown in the upper right corner of the histograms. The graph shows the inhibition of growth (. . .) and DNA synthesis (———). (See also Section VII.B.)

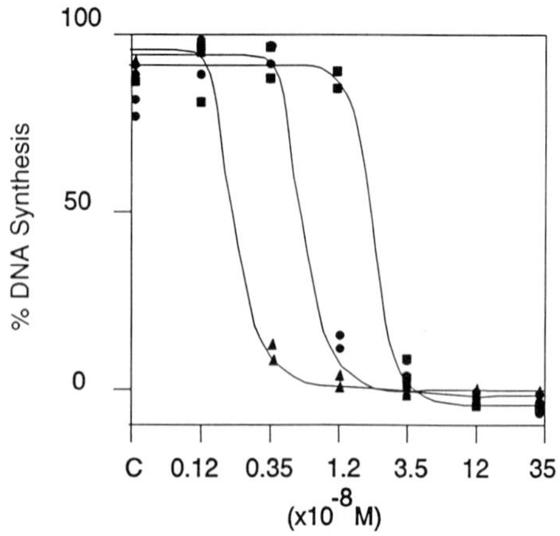

Fig. 6 An example of comparing the antimalarial activity of three related drugs, artemisinin (■), dihydroartemisinin (▲), and sodium artesunate (●) as measured using flow cytometry. In this experiment ring forms of *P. berghei* were cultured for 22 hr in the presence of different concentrations of the three drugs. At the start of the cultures and after 22 hr, samples were taken from the cultures, fixed with glutaraldehyde, and stained with Hoechst 33258. The total number of parasite nuclei in the samples is calculated as described in the text of Section VII,B. DNA synthesis is defined as the increase in number of parasite nuclei during the culture period.

ites, are cultured for prolonged periods (48–96 hr) in the presence of different concentrations of drugs. The length of the culture periods depends on the process studied (e.g., reinvasion or schizont development) and on the method used to determine parasite development. In Giemsa-stained slides differences in the number of infected cells compared to the control culture are counted or the number of DNA synthesizing parasites (schizonts) is determined as parameters for parasite development. Alternatively the incorporation of radioactive precursors into the DNA of parasites can be measured. The development of ring forms into mature schizonts takes 48 hr, after which schizonts burst spontaneously and the free merozoites enter new RBC. In cultures without drugs this will result in the increase of the number of infected cells. Figure 7 gives an example of determination of the chloroquine susceptibility of *P. falciparum* in the "extended 72-hr test" by flow cytometry.

A standardized test to determine drug susceptibility is the WHO microtest. Here parasites are cultured for 26–30 hr in 96-well microtiter plates, which are predosed with different concentrations of antimalarial drugs. During this culture period ring forms or young trophozoites develop into (immature) schizonts. The percentage of schizonts after culture is used for the assessment of susceptibility. Since flow cytometry can rapidly and reproducibly determine the increase of the number of nuclei a method has been developed for automated flow cytometric analysis of drug susceptibility of *P. falciparum* in microtests. For this purpose the AutoMATE is used for automatic sampling from the plates for flow cyto-

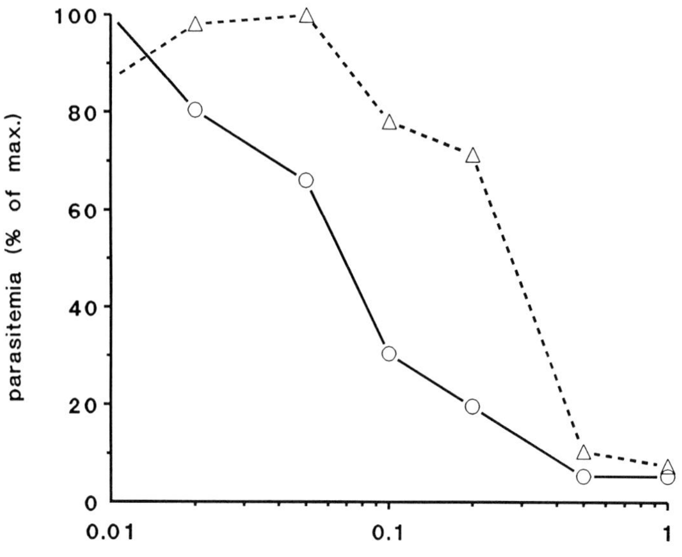

Fig. 7 Comparison of the chloroquine susceptibility of *P. falciparum* isolates (○ = clone T9/94; △ = isolate TM 152) in the "extended 72-hr test" using flow cytometry. In this test parasites are cultured for 72 hr in the presence of different concentrations of chloroquine and inhibition of development is determined by measurement of the increase/decrease of the number of infected cells during the culture period. Here the number of infected cells is determined by flow cytometry after fixed (infected) cells are stained with Hoechst 33258. In 63 tests it was found that the results obtained by flow cytometry closely corresponded with results obtained by microscopic examination of Giemsa-stained slides (van Vianen *et al.*, 1990b). From van Vianen *et al.* (1990b), with permission.

metric measurements. Automatic sampling and flow cytometric analysis of a complete 96-well plate takes 2 hr. Data analysis is performed with the described software which calculates the percentage of infected cells, the total number of nuclei per sample, and the average number of nuclei per parasite in each sample. Parasite growth and DNA synthesis for a complete culture series is calculated as described above. Figure 8 shows an example of the determination of chloroquine susceptibility of *P. falciparum* in microtests by flow cytometry compared to results by microscopic examination. We have shown in several experiments that automatic reading of a microtest by flow cytometry gives results comparable with those of microscopic examination of Giemsa-stained slides (van Vianen *et al.*, 1990b; P. H. van Vianen, unpublished results).

C. Detection and Counting of Low Numbers of Parasites

Since infected cells show a higher fluorescence intensity than noninfected cells, the percentage of infected cells can be established by flow cytometry.

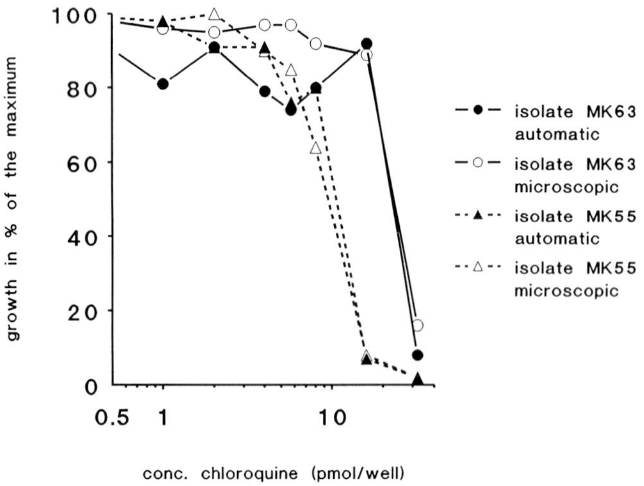

Fig. 8 Comparison of chloroquine susceptibility of two *P. falciparum* isolates in standard drug susceptibility microtests. Inhibition of development was determined either by flow cytometric analysis or by microscopic examination of Giemsa-stained thick smears. Parasites were incubated in 96-well microtiter plates under standard culture conditions for a period of 30 hr. These plates were predosed with different concentrations of chloroquine. After 30 hr samples were taken to prepare slides for microscopic examination. The rest of the culture material was fixed in the plates with glutaraldehyde and stained with Hoechst 33258. Flow cytometric reading of the plates was performed using the AUTOmate and parasite growth and DNA synthesis was calculated as described in Section VII,B. Calculation of parasite growth by microscopy is done by dividing the number of schizonts counted by the total number of parasites counted. From van Vianen *et al.* (1990b), with permission.

When this percentage is higher than 0.1%, reproducible counts are obtained in samples where the blood cells are fixed before staining and measurement. However, less reproducible results are obtained with fixed RBC when the percentage is lower than 0.1%, due to the presence of low numbers of RBC or reticulocytes in the blood samples, which show aspecific background fluorescence. Their fluorescence intensity is in the same range as that of infected cells and they may interfere with the measurement. Therefore, to detect low numbers of parasites in blood samples, the RBC are first lysed before fixation of the parasites. Figure 9 shows the fluorescence intensity and the FSC of a blood sample, which is lysed before fixation and staining with Hoechst 33258. Three populations can be distinguished: WBCs, platelets, and parasites. From dilution experiments, in which infected blood was diluted with uninfected blood, we have found that *P. falciparum* parasites were reproducibly detected at a percentage of about 0.005% (Fig. 10). Using this method, the detection of *P. berghei* is somewhat less sensitive due to interference by residual bodies of the nucleus in a low percentage of rodent RBC. These residual bodies do not lyse and can

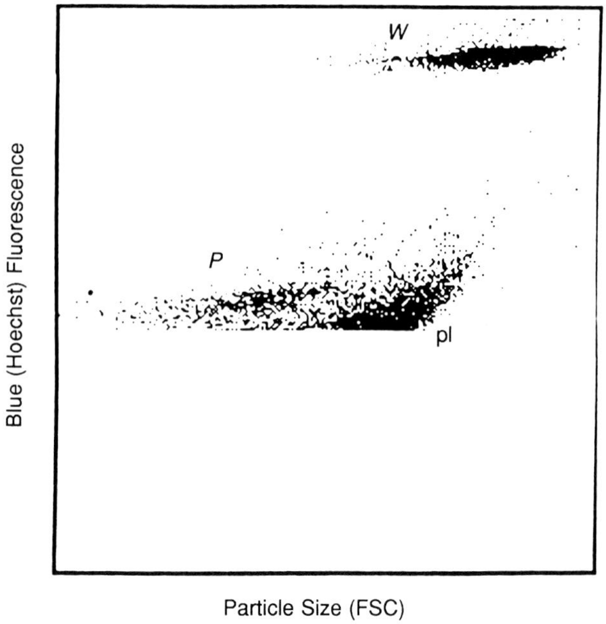

Fig. 9 Two-parameter dot plot representation of fluorescence intensity and FSC from a blood sample after lysis of the RBC in FACS lysing solution, fixation with formaldehyde, and staining with Hoechst 33258. The blood sample is from a patient infected with *P. falciparum* with a percentage of infected RBC of about 0.01%. Three populations of cells can be distinguished: white blood cells (W), platelets (pl), and parasites (P). From van Vianen *et al.* (1993), with permission.

show fluorescence intensities comparable to that of parasites (see the relative high number of "parasites" in noninfected blood samples in Fig. 10).

The reproducible detection of lower numbers of parasites is hampered by the fact that the fluorescence intensity and FSC of a low percentage of platelets fall in the same range as those of the parasites. We found that the combination of Hoechst 33258 and propidium iodide staining of the lysed blood samples allows a better separation of platelets and parasites. When the samples are stained with propidium iodide alone parasites cannot be separated from platelets on the basis of the red fluorescence. In combination with Hoechst 33258 staining, however, the red fluorescence of the parasites significantly increases while the fluorescence intensity of the platelets remains the same, allowing a better separation of those two populations. It is suggested that energy transfer from DNA-bound Hoechst 33258 to propidium iodide takes place, by which the red fluorescence of the parasites is more enhanced than that of platelets which do not contain DNA. This staining method has been used in clinical and epidemiological studies to detect low numbers of parasites in blood samples from patients. We

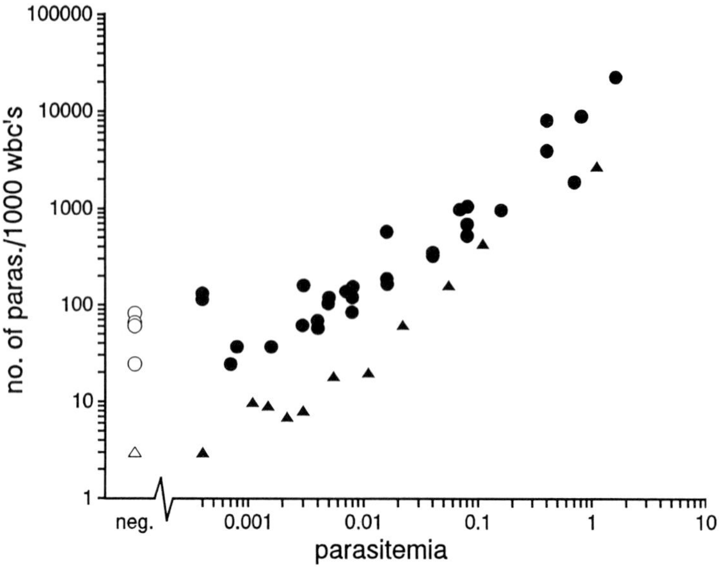

Fig. 10 Relationship between the number of parasites counted by flow cytometry and the expected percentage of infected cells (parasitemia) in blood samples. In these experiments infected blood was serially diluted with noninfected blood after which RBC were lysed and the parasites fixed and stained with Hoechst as described in Fig. 9. Parasite numbers counted by flow cytometry (in parasites per 1000 WBCs) are compared with the expected parasitemia calculated from the starting parasitemia and the dilution factor. (*P. berghei*, O; *P. falciparum*, △; open symbols are from blood samples containing no parasites). From van Vianen *et al.* (1993), with permission.

found that as few as 20 parasites per μl of blood (equivalent to 0.0004% infected RBC) could be detected (P. H. van Vianen, unpublished results).

IX. Comparison of Methods

A. Drug Susceptibility of Parasites

The use of an instrument such as a flow cytometer for the analysis of samples collected under primitive field conditions in developing countries seems contradictory to the necessity to keep field procedures uncomplicated. However, the collection of samples in the field is very simple. For drug susceptibility tests only the addition of fixative to the culture wells is essential. In fact the procedure is more easy than making smears of the culture material. Compared to other techniques, flow cytometric analysis of microtests provides more information. With microscopic examination of thick smears only the percentage of schizonts is counted and with the use of radioactive precursors only the total amount of precursor in the nucleic acids per sample is used to establish parasite development. Flow cytometry enables the combination of a number of different parameters. Information on the percentage of infected cells, the schizogonic development of individual parasites, extra- versus intraerythrocytic parasites, and invasion of new RBC can be generated in one analysis. An additional improvement is the automation of the analysis and processing of the data. All data are processed in a standardized way, maintaining objectivity, and the data remain available for reexamination. Data from large numbers of tests can be combined.

B. Detection and Counting of Parasites

Routine diagnosis of malaria is generally based on examination of Giemsa-stained blood smears for blood-stage parasites. Since this disease requires a rapid treatment, ideally within several hours after attending the hospital, a rapid sensitive and simple diagnostic method is required. This is provided by the Giemsa-stained smear examination. Since flow cytometric analysis as described in this chapter requires a FACStar flow cytometer, the method is less suitable for direct diagnosis of malaria, especially in hospitals in developing countries. However, flow cytometric analysis has proven to be a sensitive, rapid, and reproducible method when large numbers of samples have to be screened for malaria parasites in epidemiological studies or clinical studies. It allows for the reproducible detection of less than 20 parasites per microliter of human blood (0,0004% infected RBC). This is comparable to the sensitivity obtained under optimal laboratory conditions by the use of radioactively labeled DNA probes specific for malaria parasites. Nonradioactive methods are less sensitive. The sensitivity of microscopic examination is reported to be better. In optimal circumstances, 1 parasite per μl blood (0.00002% parasitemia) can be detected, although 10–20 parasites per μl is probably more realistic. When both the probe and the microscopic method were tested under field conditions their sensitivities appeared to be much lower. In these circumstances, both radioactive DNA probes and microscopy were reported to approach their limits of reproducible detection at parasitemias of 0.016% in the field (Barker *et al.*, 1989). Flow cytometric analysis is not affected by the circumstances in which the samples are collected and handled, and the same sensitivities have been found in the laboratory and with field samples. Furthermore flow cytometry allows for quantitation of the number of parasites. This is not easily performed with other methods. In conclusion, flow cytometry seems to be the best technique available for the detection and quantitation of parasites in large numbers of blood samples. Especially in view of the recent interest in vaccine development and the use of newly developed drugs this method is suitable for the follow-up of patients after treatment or the follow-up of vaccine trials and for large-scale epidemiological research in malaria.

Acknowledgments

This research was financed in part by the Netherlands' Ministry for development Cooperation (Project ID. OSAM-08775). Flow cytometry was performed within the Department of Haematology, AZL, Leiden by M. van der Keur and P. P. Reinders. We thank Dr. H. J. Tanke (Department of Cytochemistry and Cytometry, Medical Faculty, University of Leiden) for critical reading of the manuscript and stimulating discussions and P. P. Reinders for developing the software for the calculation of parasite growth.

References

Barker, R. H., Jr., Suebsang, L., Rooney, W., and Wirth, D. F. (1989). *Am. J. Trop. Med. Hyg.* **41,** 266–272.
Bianco, A. E., Battye, F. L., and Brown, G. V. (1986). *Exp. Parasitol.* **62,** 275–282.
Crissmann, H. A., Stevenson, A. P., Kissane, R. J., and Tobey, R. A. (1979). *In* "Flow Cytometry and Sorting" (M. R. Melamed, P. F. Mullaney, and M. L. Mendelsohn, eds.), pp. 234–262. Wiley, New York.
Franklin, R. M., Brun, R., and Grieder, A. (1986). *Z. Parasitenkd.* **72,** 201–212.
Hare, J. D. (1986). *J. Histochem. Cytochem.* **34**(12), 1651–1658.
Janse, C. J., van Vianen, P. H., Tanke, H. J., Mons, B., Ponnudurai, T., and Overdulve, J. P. (1987). *Exp. Parasitol.* **64,** 88–94.
Janse, C. J., Boorsma, E. G., Ramesar, J., Grobbee, M. J., and Mons, B. (1989). *Int. J. Parasitol.* **19,** 509–514.
Kawamoto, F., and Billingsley, P. F. (1992). *Parasitol. Today* **8,** 81–83.
Makler, M. T., Lee, L. G., and Recktenwald, D. (1987). *Cytometry* **8,** 568–570.
Mons, B., and Janse, C. J. (1992). *In* "Flow Cytometry in Hematology" (O. D. Laerum and R. Bjerkness, eds.), pp. 197–211. Academic Press, London.
Myler, P., Saul, A., Mangan, T., and Kidson, C. (1982). *Aust. J. Exp. Biol. Med. Sci.* **60,** 83–89.
Pattanapanyasat, K., Webster, H. K., Udomasangpetch, R., Wanachiwanawin, W., and Yongvanitchit, K. (1992). *Cytometry* **13,** 182–187.
Payne, D. (1988). *Bull. W. H. O.* **66,** 621–626.
Reinders, P. P., van Vianen, P. H., van der Keur, M., van Engen, A., Mons, B., Tanke, H. J., and Janse, C. J. (1994). *Cytometry* (in press).
Rickman, L. S., Long, G. W., Oberst, R., Cabanban, A., Sangalang, R., Smith, J. I., Chulay, J. D., and Hoffman, S. L. (1989). *Lancet* **1,** 68–71.
Saul, A., Myler, P., Mangan, T., and Kidson, C. (1982). *Exp. Parasitol.* **54,** 64–71.
Tharavanij, S. (1990). *Southeast Asian J. Trop. Med. Public Health* **21,** 3–16.
Trigg, P. I. (1985). *Bull. W. H. O.* **63,** 397–398.
van Vianen, P. H., Klayman, D. L., Lin, A. J., Lugt, C. B., van Engen, A. L., van der Kaay, H. J., and Mons, B. (1990a). *Exp. Parasitol.* **70,** 115–123.
van Vianen, P. H., Thaithong, S., Reinders, P. P., van Engen, A. L., van der Keur, M., Tanke, H. J., van der Kaay, H. J., and Mons, B. (1990b). *Am. J. Trop. Med. Hyg.* **43,** 602–607.
van Vianen, P. H., van Engen, A., Thaithong, S., van der Keur, M., Tanke, H. J., van der Kaay, H. J., Mons, B., and Janse, C. J. (1993). *Cytometry* **14,** 276–280.
Whaun, J. M., Rittershaus, C., and Ip, S. H. C. (1983). *Cytometry* **4,** 117–122.
World Health Organization (WHO) (1982). "WHO/MAP/82.1." WHO, Geneva.

CHAPTER 19

Large-Scale Chromosome Sorting

John J. Fawcett, Jonathan L. Longmire, John C. Martin, Larry L. Deaven, and L. Scott Cram

Life Sciences Division and Center for Human Genome Studies
Los Alamos National Laboratory
University of California
Los Alamos, New Mexico 87545

I. Introduction

Several areas of research have benefited from the availability of sorted chromosomes including clinical cytogenetics, cell biology, and molecular genetics. Within these areas, some applications require only a single sorted chromosome (Rabinovitch, 1994), some typically require a few hundred (Telenius *et al.*, 1992, Carter *et al.*, 1992, and Voijs *et al.*, 1993), and others, such as the construction of genomic libraries, require several million sorted chromosomes (Longmire *et al.*, 1993). The focus of this chapter is on large-scale chromosome sorting for constructing large insert chromosome-specific libraries using cloning vectors such as Charon 40, cosmids, and yeast artificial chromosomes (YACs). Up to 2 μg of sorted chromosomal DNA is commonly used for constructing YAC libraries (McCormick *et al.*, 1993). This requires the accumulation of 4 to 20 million chromosomes (Mayall *et al.*, 1984). Requirements for production sorting differ from and are more demanding than the requirements of analytical sorting which typically requires only a few minutes of actual sorting time. For successful production sorting, additional precautions are needed to assure high purity, uninterrupted sorting, and recovery of high-molecular-weight DNA. This chapter assumes the reader is knowledgeable about general sorting techniques and chromosome isolation and staining. Several manuscripts and reviews are available which describe basic principles (Cram *et al.*, 1990a,b; Gray, 1989; Deaven *et al.*, 1986; Trask and Pinkel, 1990; and Van Dilla *et al.*, 1986). Additional prerequisite sorting information can be acquired from instrument manufacturer's documentation and customer support services. Many of the techniques described here are also useful for large-scale sorting of cells or other cellular constituents.

Chromosome sorting is typically accomplished using chromosomes suspended in a stabilizing buffer and stained with two DNA binding fluorochromes; Hoechst 33258 and chromomycin A3. The binding specificity of each fluorochrome allows one to measure the ratio of adenine–thymine base pairs (A : T) to guanine–cytosine base pairs (G : C) and thereby resolve many more chromosomes in complex karyotypes such as human or mouse than is possible with single fluorochrome staining (Gray, 1989; Breneman *et al.*, 1993; Weier *et al.*, 1994). Individual chromosomes flow through spatially separated argon ion laser beams tuned to the appropriate wavelengths; 351/364 nm to excite Hoechst 33258 and 457 nm to excite chromomycin A3. When the ratio of the two fluorescence intensities uniquely defines a chromosome, an electronic sorting window is set to sort that chromosome. As a fluid stream containing chromosomes jets from the analysis chamber, the stream is broken into droplets and a charge is applied to the droplet containing the chromosome to be sorted. The charged droplet passes through an electric field where the charged droplets are separated from the uncharged droplets.

II. Sample Preparation

A polyamine buffer is used for production chromosome sorting to stabilize chromosome structure and to maintain high-molecular-weight DNA (Sillar and Young, 1981; Cram *et al.*, 1990a). Alternative buffers for other applications have been published (Gray, 1989; Cram *et al.*, 1990b). An optimal chromosomal suspension contains a high proportion of individual chromosomes and a negligible number of contaminating particles such as nuclei, clumps of chromosomes, or single chromatids (Telenius *et al.*, 1993). A typical sample concentration is 3×10^7 chromosomes/ml.

A. Chromosomal Purity

To improve sort purity and reduce contamination from chromosomes having a similar A : T to G : C ratio, rodent–human hybrid cell lines containing only one, or a small number of, human chromosomes are often used as sources of isolated chromosomes. Contamination of sorted human chromosomes by a small number of rodent chromosomes can be enumerated using fluorescence *in situ* hybridization (FISH) techniques (Trask and Pinkel, 1990).

B. Interphase Nuclei and Clumps

Contamination from interphase nuclei is monitored by sorting 10,000 fluorescent events in a spot on a nitrocellulose filter. The resulting spot is then examined under a fluorescence microscope and the number of nuclei are counted. Nuclei present in a sample of flow-sorted chromosomes are regarded as a serious contaminant due to their genomic DNA content. The presence of large numbers of nuclei in a chromosomal suspension may also affect purity because nuclei

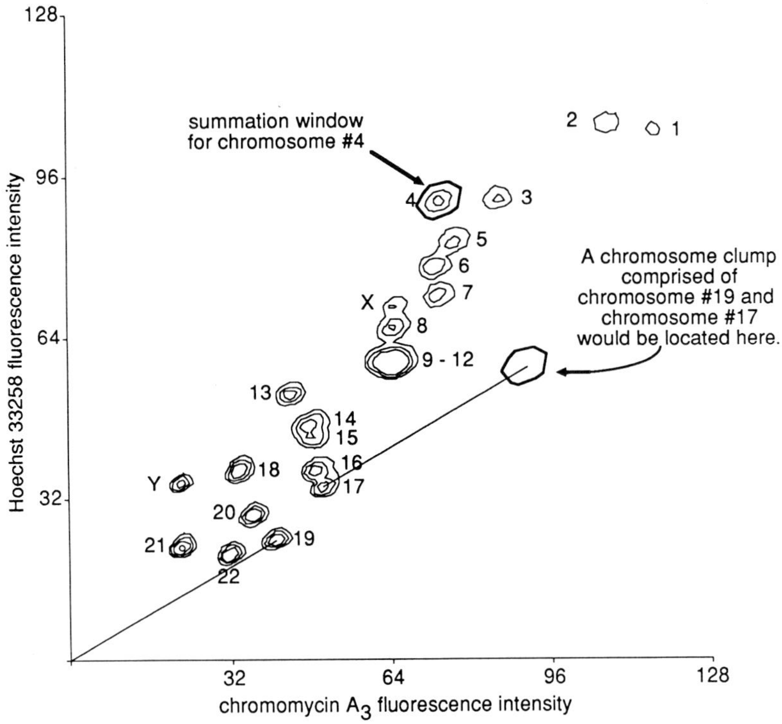

Fig. 1 Bivariate flow karyotype for GM 130 cells illustrating the expected location of a clump of two human chromosomes (numbers 17 and 19). For a high-quality sample such as illustrated here, only 0.15% of the total number of events were located in an area equal to a window which defines human chromosome 4.

perturb the flow stream and the formation of droplets. A preparation of chromosomes is regarded as unsatisfactory for production sorting if more than one nucleus is observed in every 10,000 sorted events.

Clumps of chromosomes or chromosome fragments can imitate the fluorescence properties of single chromosomes. The amount of clumping present in a sample is qualitatively monitored by observing pulse shapes on the instrument's oscilloscope and by quantitative analysis of the flow karyotype. Figure 1 illustrates the expected location of a clump containing one copy of chromosome 17 and one copy of chromosome 19. The vector distance from the origin to the 19 peak is added to the centroid of the 17 peak. The position of the resulting peak is well resolved from the peaks which contain only single chromosomes. For this cell line, the number of events in this area is an indicator of the number of clumps present in the sample. This type of analysis is used to compare multiple chromosome preparations.

C. Chromatids

Chromatids are a source of contamination when their fluorescence properties are similar to that of normal intact chromosomes. Because a chromatid contains half the amount of DNA found in a whole chromosome, its position on a bivariate flow karyotype can be estimated by drawing a straight line half the distance from a chromosomal peak to the origin. Figure 2 shows two views of a flow karyotype of cell line WAV-17. This mouse–human hybrid cell line

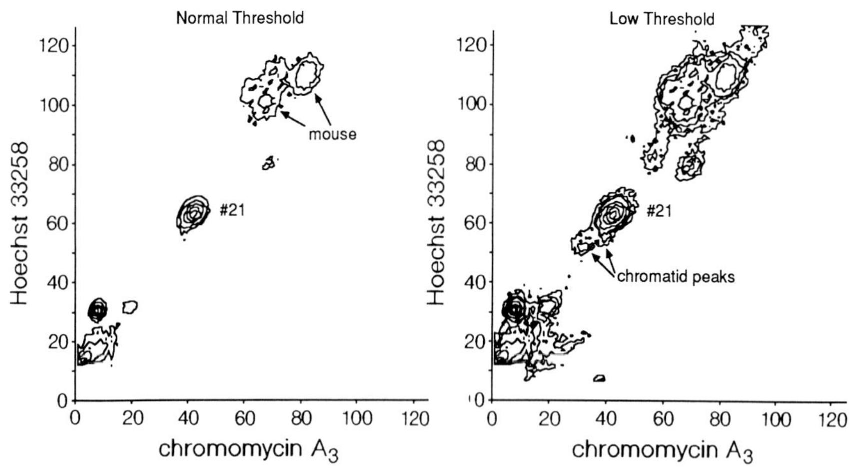

Fig. 2 Bivariate flow karyotype for a mouse–human somatic cell hybrid (WAV-17) containing three copies of human chromosome 21. The presence of events arising from mouse chromatids on the shoulder of the human 21 chromosome peak is revealed by contours displayed at a very low level of counts.

contains three copies of human chromosome 21. In the first view, the chromosome 21 peak appears to be well resolved from other chromosomes. The second view of the same flow karyotype was displayed using lower contour levels and reveals events attributable to murine chromatids on the shoulder of the human chromosome 21 peak. These chromatids overlap the low-fluorescence side of the peak of human chromosome 21. Unless the sort window is set up to be very restrictive, both the human chromosome 21 and the mouse chromatids would be sorted together. The frequency of chromatids in a chromosome preparation increases when a high colcemid concentration is used to synchronize cells in preparation for chromosome isolation (Telenius *et al.*, 1993).

D. Chromosome Staining and Debris Removal

The following information explains modifications which have been made to a previously published polyamine chromosome isolation procedure (Cram *et al.*, 1990a). A typical isolation of chromosomes for a production sort yields from 12 to 24 one-milliliter samples held in 15-ml tubes. These samples are combined and 5 ml of sample aliquoted into however many 15-ml tubes is required to accommodate the total volume. The pooled samples are centrifuged at 30*g* for 3 min to pellet cellular debris and nuclei. The clear supernatents are collected from each tube and set aside. The sample tubes are vortexed again (without additional buffer) and recentrifuged at 30*g*, and the clear supernatent is added to the previous samples. It has been found that fresh samples (less than 1 day old) require less centrifugation time (45 sec) than samples held at 4°C for several days. Throughout the procedure the samples are kept on ice to prevent degradation.

Chromosome samples are stained with chromomycin A3 (CA3) 16 hr prior to being sorted by the addition of 150 μl of stain (500 μg CA3/ml MacIlvaine's buffer) per ml of sample. After 3 hr, 4.5 μl of Hoechst 33258 (500 μg/ml of distilled water) per ml of sample is added, the suspension is vortexed, and 3- to 6-ml aliquots are distributed into 15-ml conical polypropylene centrifuge tubes. The sample tubes are left undisturbed for 16 hr (4°C) after which (without vortexing) they are centrifuged for 1 min at 280*g*. The resulting supernatents are carefully removed, pooled in a 50-ml tube, and gently vortexed.

Sample filtration is achieved by drawing the sample into a 1- or 3-cc sterile disposable syringe using a blunt 18G × 1-$\frac{1}{2}$″ disposable needle (Becton–Dickinson Co., PO Box 854, Rutherford, NJ, 07070; special order product 30260). The needle is discarded and a new needle is placed on the syringe with a 8 × 8-mm piece of 62-μm sterilized nylon mesh (Small Parts, Inc., PO Box 4650, Miami Lakes, FL 33014-0650; part CMN-62-D) wedged between the needle and the syringe. The sample is then dripped through the mesh into the final sample holder where it is stored at 4°C until used.

III. Preparation Evaluation

A. Molecular Weight Determination

Experience has shown that chromosomal preparations that have high-molecular-weight DNA prior to sorting retain their molecular weight through the sorting process. To assure that all samples have a high molecular weight prior to sorting, each sample is checked as a precaution prior to its use for a long-term sorting experiment. A simple and rapid molecular weight assay was developed which consists of observing DNA viscosity following treatment of the chromosomal suspension with a nuclear lysis buffer and sodium dodecyl sulfate (SDS) as outlined below.

Procedure: Molecular Weight Estimation of Chromosomal DNA

a. Nuclear Lysis Buffer

100 m*M* Tris-HCl, pH 8

100 m*M* EDTA

10 m*M* NaCl.

(Store at 4°C).

b. Procedure

1. Place 75 μl of the chromosomal preparation into a microcentrifuge tube (assuming a chromosome concentration of about 3–4 × 10^7 chromosomes/ml).
2. Add 25 μl of nuclear lysis buffer.
3. Add 2.5 μl of 20% SDS (sodium dodecyl sulfate).
4. Wait several seconds and check the viscosity of the sample by stirring the preparation, drawing some of it into a pipette tip, and withdrawing the tip from the preparation. If the sample contains high-molecular-weight DNA (greater than 50 Kb), there will be a "string" of DNA between the pipette tip and sample surface as the tip is withdrawn from the solution. If a string of DNA does not form, then the sample probably should not be used for chromosome sorting if the goal is to construct genomic libraries containing large DNA inserts.

The rate of DNA molecular weight degradation is dependent on the cell line from which the chromosomes are isolated. Chromosomes isolated from human or mouse cell lines generally retain their molecular weight longer than preparations from hamster cell lines. When sorting chromosomes from a human–hamster hybrid cell for a partial enzymatic digest, large insert library, the sorting should occur within a week of chromosome isolation. Chromosomes isolated from human cell lines or from mouse–human somatic cell hybrids can be used up to 3 weeks after isolation.

B. Microscopic Evaluation

Chromosome morphology, concentration, and debris level is evaluated by fluorescence microscopy prior to sorting. To avoid chromosome shearing, which can occur when a drop of sample is placed on a microscope slide and cover-slipped, draw a few microliters of stained sample into a square microcapillary pipette (Vitro Dynamics, Inc., 114 Beach St., Rockaway, NJ 07866; Cat. No. 5010, 0.1 mm path length and 50. mm long) and place it on a microscope slide. Microscopic evaluation of multiple samples can be made by placing several pipettes side by side on a microscope slide and viewing them in a single field of view. High-quality chromosomal preparations will appear very bright and will have a high contrast against a black background. Preparations in which the background appears milky or diffuse will frequently produce flow karyotypes with an unacceptably high level of debris. The relative number of nuclei in a sample can also be determined by this procedure.

IV. Production Chromosome Sorting

A. Chromosome Analysis Rate

The maximum flow analysis rate depends on several factors and varies with the application. The relationships between analysis rate, droplet frequency, fraction of a population to be sorted, coincidence rate, dead time, sort rate, sort purity, and other factors have been described (Watson, 1991; Albright *et al.,* 1991). Documentation from the instrument manufacturers also addresses some of these issues. The fastest sorting rates are possible in experiments where peak resolution is not critical (low-resolution requirements) and the fraction of the population to be sorted is high. Slow analysis rates are used when maximum resolution is required, when sample supply is limited, or when the concentration of chromosomes is low (less than 4×10^6 chromosomes/ml). Dilute samples run at high analysis rates result in sample stream widening and loss of resolution. The maximum analysis rate for high-quality chromosome samples is 1000–2000/sec. The theoretical sorting rate for sorting a human chromosome from a euploid cell line is 4.3% of the analysis rate or 43–86 chromosomes/sec. The actual sort rate is significantly less than the theoretical rate due to the presence of chromosome clumps, nuclei, and debris, all of which fluoresce and contribute to the total fluorescent event rate. When high-quality samples are analyzed using an electronic discrimination level set to exclude events with less than half the fluorescence of human chromosome 21, 70–85% of the remaining events will be chromosomes. Coincidence detection and two-droplet charging will typically lower the sort rate by an additional 10%.

Typical analysis criteria are

analysis rate	1500/sec
background count rate[1]	150 events/sec
sample consumption rate	150 μl/hr
sorting rate[2]	50 chromosomes/sec.

B. Specific Instrument Requirements

For large-scale/production chromosome sorting, a flow cytometer should have the following features.

• Intense and stable lasers with power output in the UV (351 and 364 nm) of greater than 100 mW and, at 457 nm, greater than 200 mW are recommended. Accurate centering of the laser aperture is important both for high-resolution measurements and for laser output stability. Laser "current regulation" (as opposed to "light regulation") has been found to have advantages for stable laser operation.

• A low coefficient of variation (CV) for both the UV and the visible integrated fluorescence signal is required. Alignment is checked using 2-μm calibration beads (Polysciences, Inc., Warrington, PA 18976-2590; Cat. No. 18604); a full peak CV of less than 1.65% with UV excitation and less than 0.9% with 457-nm excitation indicates acceptable instrument performance.

• The flow chamber or exit orifice, whichever is used, must be capable of efficient and stable droplet formation. A highly stable droplet breakoff position is more critical than normally expected. Flow tips capable of producing stable droplets at minimum drive amplitude produce the best resolution. Evaluation of several flow tips is recommended to identify the best one or two.

• A large-volume (5 liter), autoclavable, refrigerated, sheath tank provides long-term (33 hr) uninterrupted sheath flow (assuming a 76-μm tip, 12.75 psi). Heat sterilization is required to reduce the potential for bacterial contamination. Pseudomonad contamination of the tubing is of particular concern (Anderson *et al.*, 1990).

• Instrument delay timing is adjusted until greater than 97% of the sorted calibration beads are recovered. Recoveries greater than 99% are not uncommon.

• A circulating chiller is used to cool the sheath tank, the sample reservoir, and the sorted sample. Temperature drift in the sample or sheath can result in a shift in peak position on the bivariate flow karyotype during sorting.

[1] Includes interphase cells, small debris, and other triggering particles.

[2] Assumes sorting is from a normal human diploid cell where one chromosomal type is sorted from a background of 24 different chromosomal types.

Bubbles in the fluidic system are a major problem for long-term stable sorting experiments since they can cause turbulence in the flow chamber which in turn degrades resolution and affects sorting recovery. The presence of bubbles in the fluidics can be detected by pinching the sheath supply line several centimeters from the flow cell. If pinching the sheath supply line stops the flow suddenly, the system is probably free of bubbles. If the sample stream continues to flow, then gas in the line is expanding, pushing sheath from the nozzle until the bubble(s) reaches equilibrium with atmospheric pressure. If the system is making droplets more efficiently than usual (an unusually low droplet drive setting), then a partial plug could be causing a restriction in the exit stream resulting in an increase in the stream velocity. Both of these factors affect droplet breakoff stability.

For long-term production sorting, presort precautions include changing the in-line sheath filter (Millipore, Bedford, MA 01730; SterivaxGS 0.22-μm filter unit, Cat. No. SVGS010115), cleaning the deflection plates to remove any hidden salt bridges, and replacing the sample tubing to avoid carryover contamination of DNA. When sorting begins, the droplet stream is monitored closely for the first 30 min and then intermittently for any changes in appearance or position of the droplet breakoff point.

The opening of the tube into which the sorted sample is collected, is covered with Parafilm (American Can Company, Greenwich, CT) leaving a small hole in the center of the tube opening into which the deflected droplets enter. The Parafilm provides two features: it reduces evaporation and contamination by restricting the size of the tube opening and it allows one to quickly discover if there is a problem with the trajectory of the deflected stream by the appearance of liquid on the outside surface of the Parafilm.

A dedicated laboratory that is free from vibration and that has well-controlled lighting and temperature is critical for long-term sorting. Careful attention to details has allowed us to sort continuously, around-the-clock for 7 to 10 days, stopping only to change sample, add sheath solution [a modified polyamine buffer (see Appendix)], or make minor adjustments to the amplifier gains.

C. System Shutdown

In addition to the usual procedures recommended by the instrument manufacture for shutting down the sorter, we have found the following steps to be necessary. First, 70% ethanol is run through the sample line at a high flow rate to flush the line, and then a small rubber cup (eye dropper bulb, cut to fit) containing 70% ETOH is placed over the flow tip if the instrument is being shut down for 1 day or less. If the system is shut down for more than 1 day, the sheath line from the exit of the sheath line filter is removed and 70% ethanol is forced through the sheath tubing, flow cell, and sample tubing to remove all traces of the polyamine sheath buffer. Crystals of dried polyamine sheath buffer are difficult to clear and will cause plugs in the system unless the system is well rinsed.

Appendix: Polyamine Sheath Buffer

For all solutions, use deionized, triple-distilled water. Use sterile glassware and keep everything as clean as possible.

Stock Solutions

Solution A EGTA

Place approx. 1.8 liters of water in a 2-liter flask.

Add 3.8 g EGTA, anhydrous, MW 380.4; Sigma Cat. No. E-4378.

Place flask on a magnetic stirrer and begin dissolving the EGTA.

Add pellets of NaOH to dissolve the EGTA. Often four pellets can be added initially. If the EGTA is not in solution after 15 min, add one pellet of NaOH every 5–10 min until EGTA is dissolved.

Add additional water to make 2.0 liters and mix well.

Cover with Parafilm and store in the refrigerator.

Solution B EDTA

Place approx. 0.9 liters of water in a 1-liter flask.

Add
- 119.28 g KCl
- 47.28 g Tris–HCl, anhydrous, MW 157.6; Sigma Cat. No. T-3253
- 23.38 g NaCl
- 15.21 g EDTA, anhydrous, MW 380.2; Sigma Cat. No. ED4SS.

Place flask on a magnetic stirrer until dissolved.

Add additional water to bring up to 1.0 liter and mix well.

Cover with Parafilm and store in the refrigerator.

Spermine

Dissolve 13.93 g spermine tetrahydrochloride in 100 ml water (Calbiochem, Behring Diagnostics, La Jolla, CA 92037; Cat. No. 5677).

Spermidine

Dissolve 25.46 g spermidine trihydrochloride in 100 ml water (Calbiochem, Cat. No. 56766).

Spermine and spermidine are aliquoted into 2.0-ml portions and stored in the freezer. The required 2.0-ml aliquots of each are thawed just before use.

Sheath Solution

To make 4 liters of sheath solution:

Combine the following in a 4-liter beaker while mixing: 400 ml solution A (EGTA), 200 ml solution B (EDTA), and 3.4 liters of water.

Adjust the pH to 7.2 using concentrated solutions of HCl or NaOH.

Add 2 ml each of spermine and spermidine stock solutions.

Mix the sheath solution well before use.

Acknowledgments

This work was performed under the auspices of the Department of Energy and the National Center for Research Resources of NIH (Grant RR01315). We thank Mrs. Fawn Gore for her assistance in preparation of the manuscript.

References

Albright, K. L., Cram, L. S., and Martin, J. C. (1991). *In* "Cell Separation Science and Technology" (D. S. Kompala and P. Todd, eds.), Vol. 464, pp. 73–88. American Chemical Society Symposium Series.

Anderson, R. L., Holland, B. W., Carr, J. K., Bond, W. W., and Favero, M. S. (1990). *Am. J. Public Health* **80,** 17–21.

Breneman, J. W., Ramsey, M. J., Lee, D. A., Eveleth, G. G., Minkler, J. L., and Tucker, J. D. (1993). *Chromosome* **102,** 591–598.

Carter, N. P., Ferguson-Smith, M. A., Perryman, M. T., Telenius, H., Pelmear, A. H., Leversha, M. A., Glancy, M. T., Wood, S. L., Cook, K., Dyson, H. M., Ferguson-Smith, M. E., and Willatt, L. R. (1992). *J. Med. Genet.* **29,** 299–307.

Cram, L. S., Campbell, M. L., Fawcett, J. J., and Deaven, L. L. (1990a). *In* "Methods in Cell Biology" (Z. Darzynkiewicz and H. A. Crissman, eds.), Vol. 33, pp. 377–382. San Diego, CA: Academic Press.

Cram, L. S., Ray, F. A., and Bartholdi, M. F. (1990b). *In* "Methods in Cell Biology" (Z. Darzynkiewicz and H. A. Crissman, eds.), Vol. 33, pp. 369–376. San Diego, CA: Academic Press.

Deaven, L. L., Van Dilla, M. A., Bartholdi, M. F., Carrano, A. V., Cram, L. S., Fuscoe, J. C., Gray, J. W., Hildebrand, C. E., Moyzis, R. K., and Perlman, J. (1986). *Cold Spring Harbor Symp. Quant. Biol.* **51,** 159–167.

Gray, J. W., ed. (1989). "Flow Cytogenetics." San Diego, CA: Academic Press.

Longmire, J. L., Brown, N. C., Meincke, L. J., Campbell, M. L., Albright, K. L., Fawcett, J. J., Campbell, E. W., Moyzis, R. K., Hildebrand, C. E., Evans, G. A., and Deaven, L. L. (1993). *Genet. Anal. Tech. Appl.* **10,** 69–76.

Mayall, B. H., Carrano, A. V., Moore II, D. H., Ashworth, L. K., Bennett, D. E., and Mendelsohn, M. L. (1984). *Cytometry* **5,** 376–385.

McCormick, M. K., Campbell, E. W., Deaven, L. L., and Moyzis, R. K. (1993). *Proc. Natl. Acad. Sci. U.S.A.* **90,** 1063–1067.

Rabinovitch, P. S. (1994). Personal communications. *Proc. Natl. Acad. Sci. U.S.A.*, in press.

I. Introduction

A. Background

Analysis and sorting of rare cells has proceeded relatively slowly because its problems appear daunting to the average researcher. However, due to the invention of polymerase chain reaction (PCR) and its possibilities for working with as few as one sorted rare cell as described later in this chapter, the analysis and sorting of rare ($<0.1\%$) and even ultrarare ($<0.001\%$) cells will become increasingly important and of interest to more and more researchers. Some important applications include the detection and isolation of rare fetal cells from maternal blood for prenatal diagnosis (Cupp *et al.*, 1984), detection of rare metastatic breast (Leslie *et al.*, 1990) or neuroblastoma (Frantz *et al.*, 1988) cells in bone marrow, monitoring of minimal residual disease (Ryan *et al.*, 1984), detection of stem cells (Visser and de Vries, 1990), detection of rare HIV-infected cells in peripheral blood (Cory *et al.*, 1987), and mutation frequencies in genetic toxicology (Jensen and Leary, 1990). The importance of rare event analysis by flow cytometry and rare cell sorting, including clinical applications, can be expected to grow in the years ahead (Ashcroft, 1988).

Other "rare cell" applications may be less evident. There are, in fact, numerous "rare event" applications that are not recognized as such. For example, many people presume that it is necessary to have on the order of 10^6 cells to perform a flow cytometric analysis. On the contrary it is quite feasible to examine a few thousand or even a few hundred cells if these cells are surrounded by distinguishable (on the basis of at least one flow cytometric parameter) "carrier" cells. Thus a thousand total (now "rare") cells constitute an approximate 0.1% subpopulation when mixed in among 10^6 carrier cells and can be analyzed by conventional flow cytometry and even sorted down to the single-cell level. Rare cell analysis and sorting requirements push most conventional flow cytometry and cell sorting methodologies at every stage to their limits. It is important to understand both the problems encountered and their theoretical basis. Many people launch into rare event applications failing to recognize them as extremely difficult or impossible given the requirements of the application and the technology available. With a little foresight many of these problems can be either circumvented or diminished to manageable levels. Much can be accomplished with conventional technology if the problems are recognized and addressed in suitable ways. But methods quite acceptable for non-rare problems become unacceptable when applied to the detection or isolation of rare cell subpopulations. The important thing is to be neither naive nor intimidated but to solve the problem at hand in the best way you can with the resources you have available. This chapter primarily addresses in a fairly nontechnical, practical, and pragmatic way what you can do with conventional flow cytometry and cell sorting instrumentation to analyze and sort rare cells. It also attempts to point out some of the methodological pitfalls. Discussed briefly in this chapter are more advanced technologies that now exist in some research laboratories; perhaps they will someday be available commercially when the simpler methods on currently available instrumentation prove insufficient to enough researchers.

B. Theory and Practice of Rare Cell Detection

Obviously having a high-speed flow cytometer/cell sorter optimized for rare cell analyses would greatly enhance one's capabilities of performing detection and isolation of rare cell types. However, despite the unfortunate absence of commercially available high-speed, rare events instrumentation, much can be done with conventional flow cytometer/cell sorters provided that the researcher understands what the instrument is doing to his/her experiments.

Most flow cytometers use cell light scatter as a default system trigger for obvious reasons. All cells should have a light scatter signal, whereas only some cells will be fluorescent. Light scatter can also be used to eliminate debris at the analog level prior to digital signal processing which can greatly reduce the effective dead time of the instrument. Unfortunately, most flow cytometers have upper practical limits to analysis or sorting rates (typically 2,000–10,000 trigger events/sec which translates to 2,000–10,000 total cells/sec if light scatter is used as a trigger signal). One way people have tried to get around the signal processing rate limitation problem is to use the rare fluorescent signal rather than light scatter as the system trigger (McCoy *et al.*, 1991). While this is a good method to try, one should be aware of its limitations. First, original frequency information will be lost. However, by knowing the overall cell concentration, sample volume flow rate, and time for collection of a number of fluorescent-positive cells, one can at least estimate the original frequency of the rare cells. More importantly, since the system trigger is seeing only fluorescence (and not the light scatter) of the rare cells it is effectively blind to the presence and identity of all other cells and debris. Frequently debris and dead or damaged cells are the brightest objects in the sample. For this reason sorting blindly for all fluorescent-positive cells is very unwise. The real problem is with cell sorting anti-coincidence; since this method makes the instrument blind to nearby fluorescent-negative cells, anti-coincidence is effectively disabled. However, if accurate original frequency information is not important, or if an impure but enriched sort is adequate, this method can be very useful to researchers with conventional instrumentation.

Instrument dead times are important to understand since they will determine the rates at which rare cells can be processed. Conventional instruments have typical deadtimes of 10–25 μsec, meaning that when a cell is detected above trigger threshold, the instrument will spend the next 10–25 μsec processing that cell. If the instrument has multiple laser beams the dead time may be even greater if the instrument has a dead time for the entire period that the cells pass between the two (or more) laser beams. Some flow cytometers do not allow several cells to be processed through the first beam before being processed through the second or third beams. During this dead time single-queue signal processing systems in most commercial instrumentation will ignore any other cells coming through the flow cytometer. More modern instrumentation allows, to various degrees, cells processed by the initial beams to be stored in analog pipe delays, shift registers, or other devices such that the entire system is not locked up in dead time while those cells traverse the remaining beams, greatly improving the situation (Parson *et al.*, 1985; Peters *et al.*, 1985; van den Engh and Stokdijk, 1989; van Rotterdam *et al.*, 1992; Leary *et al.*, 1993). Just as supercomputing has rapidly headed in the direction of multiprocessor parallel systems, future flow cytometry signal processing will operate with either multiqueue or full parallel processing.

Classical definitions of instrument dead time provide an easily calculable estimate of the number of cells that will be missed,

$$N = \frac{n}{1\text{-}nd},$$

where N is the actual number of cells per second, n is the number of cells observed, and d is the instrument deadtime.

Although this is often a good approximation it is actually incorrect because it assumes equal spacing of signals rather than the Poisson arrival statistics of cells through laser beams or distribution in sort droplets. You can easily determine the dead time of your instrument under the same conditions that you would be acquiring data for your application (depending on how the instrumentation is designed, dead times may vary considerably depending on how the equipment is run!). Do the dead time experiment described below, customized to be as close as possible to your application in terms of number of parameters, signal levels, signal processing, gating, and number of cells (input to the instrument) per second. A particular important variable in many systems is the time that it takes for a given signal to return to a baseline of zero volts. Nonintegrated ("pulse height") signals, either log peak or linear peak, will return to baseline more quickly than either linear-amplified integrated or log-amplified integrated signals. While integrated signals sometimes appear to give greater sensitivity they are costly in terms of instrument dead time and they are very susceptible to integration of noise, particularly if those integrated signals are log amplified.

These dead time experiments are fairly simple to perform. First one needs a high throughput rate of particles or cells. For example, one can use red blood cells and calculate (by any of a variety of measurements) the number of cells per unit volume. Then determine under these conditions the volume of sample passing through your flow cytometer by weighing a volume of fluid in your sample vial before and after (several minutes depending on the accuracy of the result desired) on an analytical balance of sufficient accuracy. You will typically need a series of cell samples of concentrations from about 5×10^6 to about 1×10^8 cells/ml. Most flow cytometers use about 30–60 μl/min of sample, but this may vary widely with the instrument model. Then simply record the total number (n) of cells per second counted by your instrument over that number of minutes and compare with the number (N) of cells per second that should have passed through the instrument based on the product of the number of cells per milliliter times the sample volume used. Then use the above equation to calculate the dead time. Rearranging the equation yields

$$d = \frac{N-n}{Nn}.$$

If N and n are cells per second, the dead time d is in seconds. For example, if you expected on the basis of cell concentration and volume a throughput of 20,000 cells/sec with your system and you observed only 15,000 cells/sec, your instrument dead time (under those conditions!) would be

$$d = \frac{20{,}000-15{,}000}{20{,}000 \times 15{,}000} = 16.7 \times 10^{-6} \quad \text{or} \quad 16.7\ \mu\text{sec}$$

Sample graphs (generated using the software package Mathematica) which show the effects of dead time d on the cell counting loss are shown in Fig. 1A. Although this approximation is good enough for most uses, you should know that a more exact formulation depends on Poisson statistics as derived from queuing theory. Queuing theory describes the basis of arrival statistics of cells to the excitation beam and to the sort droplets (Gross and Harris, 1985). This chapter gives you an elementary introduction to queuing theory to the point where you can calculate with a simple calculator the percentage counting losses and sort purity of your experiments. Again using Mathematica the more exact losses due to instrument dead time can be calculated as shown in Fig. 1B. For most people the analysis dead times are not an issue if one operates at less than about 10,000 cells/sec. However, as can be seen from Fig. 1A, dead time becomes an increasingly important issue at rates beyond 10,000 cells/sec. True high-speed systems presently in research laboratories have through a variety of methods reduced the dead time for analysis of rare cells to less than 2 μsec (Leary *et al.*, 1993), roughly the time it takes for cells to traverse a typical laser excitation beam. On the other hand, the effect of cell coincidence in sorting is readily apparent to most researchers because the queue length (i.e., the sorting unit of 1, 2, or 3 droplets typically) is much longer than that in the case of cell coincidence in the excitation source.

C. Theory and Practice of Rare Cell Sorting

In cell sorting most people are pushing the limits of instrument throughput. Sorting dead time is a factor which affects everyone trying to sort rare cells. One can quickly estimate the amount of time it will take to isolate a number of cells of known frequency. For example, to sort 10^4 (ten thousand) rare cells of frequency 10^{-4} (1 rare cell per 10,000 total cells) at a rate of 2000 total cells/

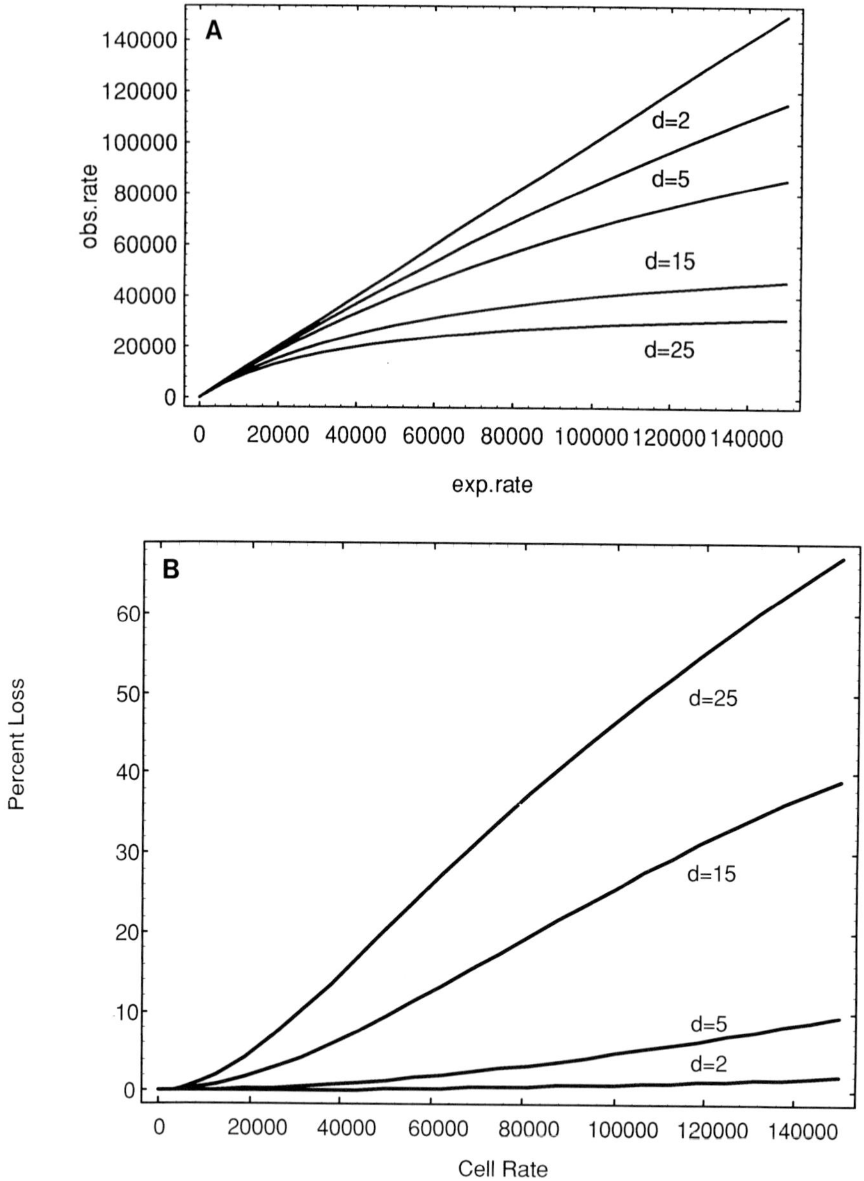

Fig. 1 (A) A Mathematica-generated graph of observed (obs) versus expected (exp) number of cells per second as a function of instrument dead time (d) based on the conventional formula $N = n/(1/nd)$ versus cell throughput rate. (B) A similarly generated graph of the percent loss in the number of observed cells per second versus expected number of cells per second as a function of instrument dead time (d) based on queueing theory dead time calculations.

sec will require the experimenter to spend almost 14 hr processing 10,000 times 10,000 cells or 10^8 (100 million) total cells! Even at 5000 cells/sec (the practical sorting rate limit for most cell sorters, especially if one requires anticoincidence), the process would take over 5 hr! If the cells must also be maintained in a viable state this may represent a significant problem. This assumes, of course, that both the sorter and the researcher can also remain in a "viable" state for this period of time! The researcher must also be able to obtain this many total cells and have the reagents sufficient to process them. This example shows why rare event researchers to date have not often attempted to sort this many rare cells. If only small numbers, potentially only 1 cell which can be cloned and expanded in tissue culture or whose DNA or RNA can be amplified by PCR, are needed the problem becomes practical. The one remaining problem is to be able to visualize enough rare cells to be able to isolate them from undesired cells and debris. Unfortunately, as the reader will see later in the chapter, it is difficult to label rare cells specifically without labeling many other cells and dead/damaged cells nonspecifically. Thus to really do the job right requires that the experimenter process many controls and obtain large enough distributions on each control to determine the optimal region to sort. If multiple parameters have been used to try to separate the positive rare cells of interest from background there are a small number of cells spread out over many dimensions. There are ways to deal with some of these problems as discussed later in the chapter.

Sorting of enriched but impure rare cell subpopulations is an interesting case to think about. While at first counterintuitive, it is easy to show that cell sorters are actually far more efficient (in terms of enrichment factor) for isolating rare cell subpopulations at very high speeds rather than at low speeds as shown in Fig. 2. The more rare the cell of interest, the greater the potential enrichment factor of a high-speed enrichment sort!

The mathematics necessary to calculate the degree of purity of your sort are not difficult. A sample calculation of the Poisson arrival statistics of cells to sorting units as predicted by queuing theory, including use of a simple recursion relationship, greatly simplifies the calculations making them easy to perform with a hand-held scientific calculator.

$$P(x) = \frac{\lambda^x}{x!} e^{j\lambda},$$

where $x!$ is "x factorial,"

$$\text{for example, } 0! = 1$$
$$1! = 1$$
$$2! = 2 \times 1 = 2$$
$$3! = 3 \times 2 \times 1 = 6, \text{ etc.},$$

and where λ is the length of the queue.

$$P(0) = e^{-\lambda}.$$

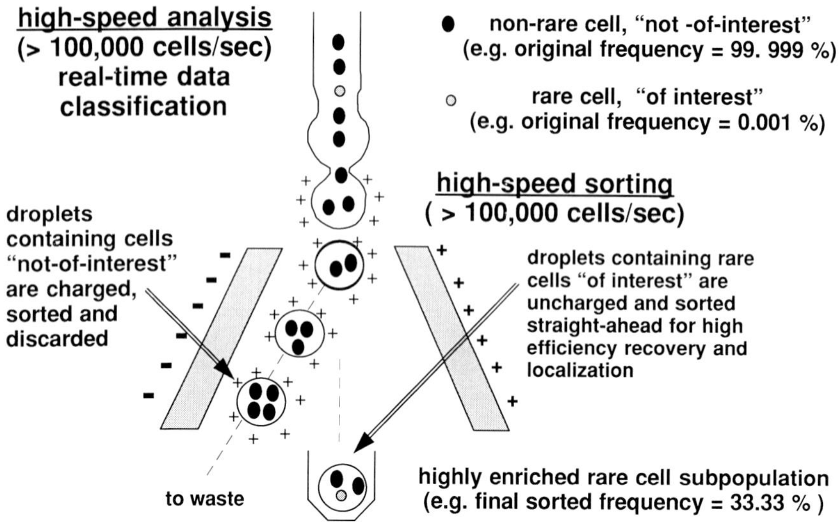

Fig. 2 High-speed sorting for enrichment of rare cells is actually more efficient than lower speed sorting of non-rare cells. In this example, a high-speed sort of a 0.001% subpopulation of rare cells at 100,000 cells/sec yields an average of 3 cells/droplet. However, one of these cells is of interest leading to a 30,000-fold enrichment which may be based on multiple parameters.

Using a recursion relationship to simplify calculations

$$P(x) = \frac{\lambda}{x} P(x\text{-}1),$$

so the probability of one cell in the interval λ is

$$P(1) = \frac{\lambda}{1} P(0),$$

the probability of two cells in the interval λ is

$$P(2) = \frac{\lambda}{2} P(1),$$

etc.

Now we compute an actual example. Assume you are trying to sort 5000 cells/sec with two droplets per sorting unit at a droplet frequency of 32,000 droplets/sec. The calculations for determining the probability of finding a cell within a (2 droplets/sorting unit) queue length λ are as follows:

$$\lambda = f\,t,$$

where $f = 5000/\text{sec}$ and where

$$t = \frac{2 \text{ droplets}}{\text{sorting unit}} \times \frac{1}{32{,}000 \text{ droplets/sec}} = \frac{0.0000625 \text{ sec}}{\text{sorting unit}}$$

so that $\lambda = 5000 \times 0.0000626 = 0.313$ and the probability of finding no cells within two sorted droplets is $P(0) = e^{-\lambda} = e^{-0.313} = 0.732$. Meaning that 73.2% of these two-droplet sorting units are empty and contain no cells! Using the recursion relationship, the probability of finding one cell within these two droplets, $P(1)$, is

$$P(1) = \frac{\lambda}{1} P(0) = \frac{0.313}{1} \times 0.731 = 0.229$$ (i.e., 22.9% of the sorting units contain 1 cell)

$$P(2) = \frac{\lambda}{2} P(1) = \frac{0.313}{2} \times 0.229 = 0.036$$ (i.e., 3.6% of the sorting units contain 2 cells),

etc., for cases of 3 or more cells/sort interval.

To calculate the purity of the total sort one must calculate the purity of each sort unit (1, 2, or 3 droplets/sorting unit depending on choice of user and capability of the instrument). It should be obvious that the longer the queue length the greater the probability that two or more cells will be found in the queue length defined by the sorting unit. While three-droplet sorting is the easiest to set up and maintain, it leads to the greatest amount of contamination. In contrast, two-droplet sorting is still fairly easy to set up and maintain yet leads to a considerable drop in contamination due to cell coincidence in the sorting unit. For very high-speed sorting, one-droplet sorting can and has been performed in our laboratory. However, it is not only difficult to set up but inherently unstable in terms of sort fluidic stability and must be performed very carefully!

In the above example, the sort purity of rare cells is

$$= 100 - 50 \times P(2) - 67 \times P(3) - 75 \times P(4) - \ldots$$ (continue to desired accuracy),

since a sort unit containing two cells [one of which is the desired cell while the other cell is undesired, $P(2)$] has a sort purity of only 50%; a sort unit containing three cells (one desired, two undesired) has a sort purity of only 33% (hence a 67% contamination level); a sort unit containing four cells (one desired, three undesired) has a sort purity of only 25% (hence a 75% contamination level); etc. Obviously, this very simple calculation considers the probability of a sorting unit containing two or more positive cells of interest to be statistically insignificant, which is true for rare cell sorting but not for the general case! For the non-rare case coincidences of positives occur frequently and can decrease sort yield if a conventional full-anticoincidence sort strategy is used.

Most conventional sorters have a feature known as "anticoincidence" to eliminate all sort units containing two or more cells. Depending on your sorting needs it may or may not be a good idea to use this feature. If the goal is to enrich a rare cell subpopulation and obtain as many rare cells as possible (and possibly further process by a subsequent re-sorting or by another method of micromanipulation of sorted cells, as discussed later in this chapter), it may be better to not use anticoincidence since the yield of rare cells drastically decreases as a function of sort rate when anticoincidence is active. The problem with conventional anticoincidence strategies is that if anticoincidence is engaged, nearby cells are rejected regardless of whether they are "friend" or "foe," (meaning "desired" or "undesired," respectively). What this means in practical terms is that nearby desirable cells are thrown away, vastly lowering sort yield. Also, beyond a certain point (typically beyond a few thousand cells/sec on many instruments), anticoincidence totally shuts down cell sorting, as many experimenters will attest. Again these problems can be addressed by employing more sophisticated "friend/foe" "flexible" sorting strategies (Corio and Leary, 1993).

A graph of sorting purity versus rate as a function of the number of droplets/sorting unit (with no anti-coincidence) is shown in Fig. 3. The effect of an all-or-nothing "full" anti-coincidence (which rejects all sorting units containing greater than one cell, regardless of identity) is to sharply reduce the yield of sorted cells at higher sorting rates. If purity (and not yield) is the only concern, full anti-coincidence should be used with a total cell parameter such as forward angle light scatter chosen as the trigger input to the anti-coincidence circuitry.

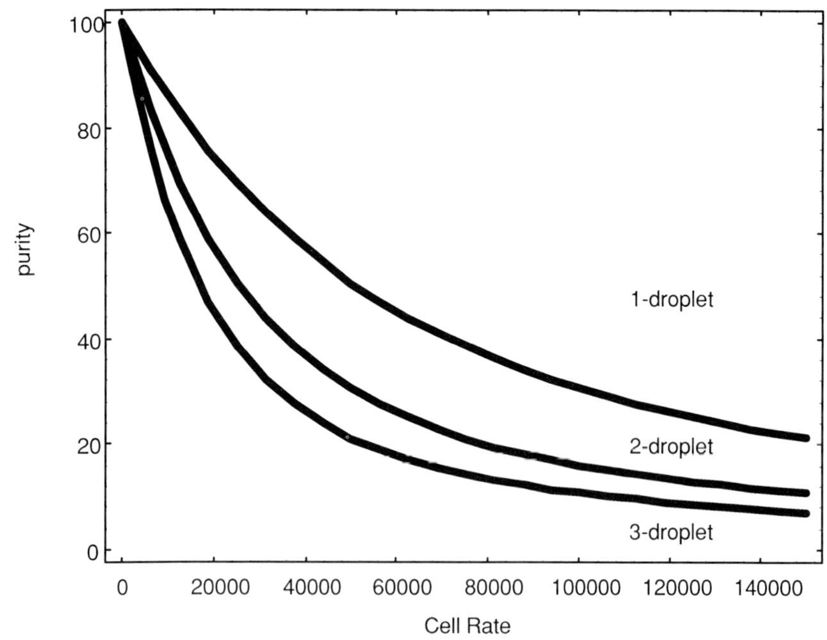

Fig. 3 A Mathematica-generated graph of sorting purity versus cell throughput rate as a function of the number of droplets per sorting unit assuming no anti-coincidence.

For this to work, the cell rate must be reduced to have an average time between cells greater than the dead time of the anti-coincidence circuitry of the experimenter's cell sorter. While the optimal cell rate can be calculated from dead time rate and queuing theory models, it is far easier for most experimenters to determine this empirically. High-speed impure sorting enrichment can be accomplished by more sophisticated "flexible sorting" algorithms and hardware that allow the experimenter to optimize the yield/purity ratio based on probable identity and desirability of neighboring cells (Corio and Leary, 1993).

The Mathematica calculations used to generate Figs. 1 and 3 are essentially the same but include the graphics capability of viewing the results as a function of cell analysis or sort rate and allowing us to parameterize the results in terms of particular variables such as the number of droplets/sorting unit. While not discussed in this chapter, other non-droplet sorting methods such as "zapper" elimination of unwanted cells (Herweijer *et al.*, 1988; van Rotterdam *et al.*, 1992) can also be described in the same manner with queuing theory although in those cases the queue length becomes the "zap" queue length rather than the sort droplet interval.

II. Methods

A. General Strategies for Rare Cell Detection and Isolation

To develop intelligent strategies for detection and isolation of the rare cells of interest to you, first analyze your requirements. Does your application require accurate original frequency information? If so, flow cytometry is probably the best technique to employ. Do you require a small or substantial number of isolated rare cells? If you need as many rare cells as possible then consider the use of one or more preenrichment procedures, e.g., selective lysis of unwanted red blood cells, gradients (e.g., percoll or ficoll-hypaque), magnetic bead sorting (using one or more antibodies), affinity columns (with attached ligands or antibodies), sedimentation (e.g., unit gravity sedimentation, velocity sedimentation), centrifugal elutriation, aqueous polymer phase separation, or cell electrophoresis. For a concise review of many of these cell separation methods refer to Kompala and Todd (1991).

If you require accurate original frequency information, and/or need only a few isolated cells, then consider using flow cytometry and cell sorting with various helpful hints provided in this chapter. This includes high-speed impure sorting followed by resorting by flow cytometry (FCM) or image analysis with micromanipulation re-sorting of sorted cells by "cookie cutter" sorting (Wade, 1987) or optical trapping (Buican, 1991) particularly if imaging or another non-FCM method is needed to separate false positives from true positives. For example, this can be particularly useful when using fluorescence *in situ* hybridization to determine correct cell identity in cases of aberrant number of chromosome copy number as in prenatal diagnosis or detection of tumor cells. Or, alternatively, high-speed analysis can be followed by careful slower speed sorting of only a few cells if these few cells can be expanded in tissue culture or analyzed by PCR.

B. Setup Procedures for Rare Cell Detection and Isolation

One problem facing rare event researchers is that of instrument setup and calibration. Usually it is not possible to have a positive control available for instrument setup. Likewise it is frequently difficult to have a perfect negative control since rare cell analysis frequently works at levels close to background noise. Indeed, it is the negative control which determines the levels of detection of rare cells possible. A functionally useful negative control is usually, but not always, available and is critical to the success of the experiment.

In the case of rare cell subpopulations it is impossible to have enough cells to see and to use for electronic compensation of rare cells labeled with multicolored probes as discussed in the following sections. One way to deal with this problem is to use standardized beads (available from several vendors) with the same fluorescent probes present in the same quantities. Correct color compensation is essential; improper compensation can lead to false-positive or false-negative subpopulations which can make rare cell detection difficult or impossible.

As with other features of rare cell analysis, methodological errors which can be easily tolerated for analysis of non-rare cells quickly become fatal to the outcome of rare cell experiments. For example, compensation should be done with more than one probe of each color so that you can be confident that the brightest probes are compensated and the dimmer ones are not overcompensated. Depending on the instrument used, overcompensated signals can become negative (inverted pulses which can cause dim positives to be clipped to zero). One way to check for the existence of these miscorrelations is to dope into a negative sample beads which are fluorescent on all fluorescent parameters to be used in the experiments. If subsequent fluorescence results reveal any signals which are fluorescent for some, but not all, fluorescent parameters then miscorrelations are occurring for those signal processing conditions. For rare event analysis even a small number of miscorrelated signals can lead to false positives or false negatives. For weak rare fluorescent signals it is important to not use log integrated signal processing which can lead to integration of noise with subsequent production of false-positive signals.

It is important to include a negative control probe for detection of rare cells. Sorting double-positives without a negative control is dangerous! Uniformly positive samples are frequently false-positive labeling dead or damaged cells. Use of propidium iodide to exclude dead or damaged cells does not, in general, work well for the detection of rare cell subpopulations because many dead or damaged (or other nonspecific antibody binding cells) not recognized by propidium iodide will bind antibodies nonspecifically. We do not really care whether cells are live or dead or have leaky membranes. We care whether cells arer binding antibody nonspecifically. The logical solution is to attempt to label the cells with one or more antibodies (''negative selectors'') which should not label cells of interest. Use of negative selectors to eliminate false-positive labeling cells, including dead or damaged cells, is an important function of the negative selection probes described in the next section. Other parameters such as forward and side light scatter can also be used as additional negative selectors to eliminate unwanted cells which may, in fact, be staining specifically for the positive selectors and negative for the fluorescent negative selector markers.

C. Rare Cell Detection by Different Flow Cytometry Strategies

1. Use of Multiple Positive and Negative Selection Parameters

In order to accurately and efficiently identify and isolate rare cells it is essential to develop an appropriate system of positive and negative selection markers. Multiple positive selection markers, each inadequate by itself, can in combination become much more accurate. Sometimes these positive or negative selection markers can make use of physical properties of the cells as measured by forward angle light scatter or side (90°) light scatter. For fluorescent markers, if each positive selection marker has its own color of fluorescence the number of colors can quickly grow beyond the maximum number on typical flow cytometers. However, if cocktails of monoclonal antibodies are used, a large number of positive selection markers can be used with the same color reporter molecule. However, using multiple markers of the same color will only work for the case of a logical OR (whereby a cell may label with one or the other marker and you cannot tell, and do not really care, which one) condition between the markers. In most cases we wish to use logical AND (whereby a cell is known to be simultaneously positive for two different markers of different colors) conditions to make use of the full power of multiparameter flow cytometry. It is equally important to use negative selection markers to eliminate other cell types which may cross-react with some of the positive selection markers. A number of negative selection markers can be labeled with the same color reporter molecule (but of a different color than that of the positive selection markers!). Thus powerful combinations of positive and negative selection markers can be used with a total of only two or three colors of fluorescence making these experiments possible on most commercial instrumentation. If one of the positive selection marker colors is used as a trigger, conventional flow cytometers can be used at high speeds to analyze rare cells for at least some applications (McCoy *et al.*, 1991). Positive and negative selection marker strategies can be used either with conventional low-speed flow cytometry/cell sorting or with nonconventional high-speed, multiparameter flow cytometry methods (Leary *et al.*, 1993) which perform real-time data classification into cells ''of interest,'' ''not of interest,'' and ''not sures'' at rates in excess of 100,000 cells/sec (Fig.

4). It should be emphasized that while the rate in cells/sec is "high-speed," the actual velocity of the cells through the system is the same as in conventional instruments. Hence the cells are more closely spaced as they travel through the system. The important consequence of this similar velocity is that viable cells can be obtained without the increased damage caused by systems which do increase the cell velocity. Unlike the method of McCoy *et al.*, all cells are counted so original frequency information is preserved. Since all cells are detected (and hence potentially contaminating nearby cells are detected and identified) this information can be used by sophisticated "flexible sorting" (Corio and Leary, 1993) strategies which can provide much higher numbers of sorted cells of higher purity in a given time. For example, if small numbers of contaminating red blood cells are also sorted they may not affect a subsequent PCR amplification of DNA sequences in rare nucleated cells of interest because the red blood cells are not nucleated and the presence of small amounts of hemoglobin will not interfere with the PCR reaction. Results dramatically illustrating the importance of positive and negative selection marker strategies are shown in Fig. 5. The labeling strategy is shown in Fig. 5A. The negative selection gate is shown in Fig. 5B. As shown by comparing Fig. 5C with its isotype

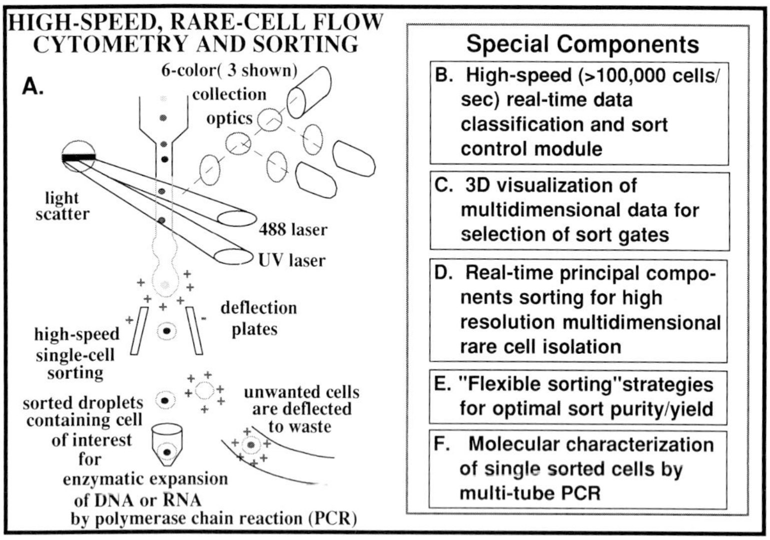

Fig. 4 Example of a high-speed rare cell analysis and sorting system in the author's laboratory which incorporates many of the features described in this chapter. The system has an acquisition system dead time of less than 2 μsec and is capable of performing multiparameter high-speed analysis of ultrarare cell subpopulations at rates in excess of 100,000 cells/sec. Real-time principal component sorting in hardware and sophisticated "flexible" sorting anti-coincidence strategies permit isolation of much greater numbers of rare cells at higher purity than can be obtained on conventional cell sorters.

control in Fig. 5E it would be difficult or impossible to sort out rare cells of interest due to the high nonspecific background. However, when only one negative selection marker, in this case CD45, coupled to a phycoerythrin–cyanine-5 (PE-Cy5) tandem conjugate energy transfer probe which labels mature maternal leukocytes as well as dead/damaged and nonspecifically binding cells, is used the situation changes dramatically. Comparison of Figs. 5D and 5F reveals an easily identifiable subpopulation of "putative" fetal cells that can be sorted. Rare cell subpopulations should always be thought of as "putative" until the rare cells are sorted and characterized, preferably by unequivocal molecular means such as PCR-amplified cell-specific DNA sequences. While tandem conjugate energy transfer probes are extremely useful for multicolor fluorescence flow cytometric experiments, they can sometimes lead to problems when cells are dimly positive (in this case for PE) due to energy transfer "leakage." An energy transfer probe that "leaks" fails to achieve energy transfer and gives off donor rather than acceptor fluorescence (in this case it would falsely give off PE rather than Cy5 fluorescence). In that case it is difficult to distinguish between cells with leaky energy transfer probes and true fluorescence-positive cells (in this case PE-positive cells). However, even somewhat leaky energy transfer probes can still be used successfully as negative selector probes since any cells which are positive for the donor fluorescence will also be positive for acceptor fluorescence and all acceptor fluorescence-positive cells are eliminated. This is analogous to the situation of not worrying about propidium iodide (PI) fluorescence in dead/damaged cells spectrally overlapping with fluorescein fluorescence since all PI-positive cells will be gated out of the distribution. This is why we have used the tandem conjugate probe on the negative selector in the above example.

2. Data Analysis Methods

Rare event data analysis presents some unique problems. To have enough rare cells to be statistically significant over background one must sample a large number of total cells. If all of these data are stored as correlated list-mode data the data quickly become unmanageably large (in the tens of megabytes or more per data file). While modern electronics can digitize at very high speeds and computer storage has become less of a problem, one should question the wisdom of approaching the situation in a brute force fashion. If 99.999% of the cells are clearly not of interest, does it make sense to store all of this data as a correlated list-mode file? A more sensible policy is to digitize and store as correlated list-mode data only cells "of interest" and "not sures" performing a "real-time data classification" at speeds in excess of 100,000 total cells/sec prior to digitization. All other cells "not of interest" can be quickly counted but not stored as list-mode data. This allows very large total sample sizes to be quickly and easily counted to preserve original frequency information and all data of interest and not sures to be stored as list-mode data where more

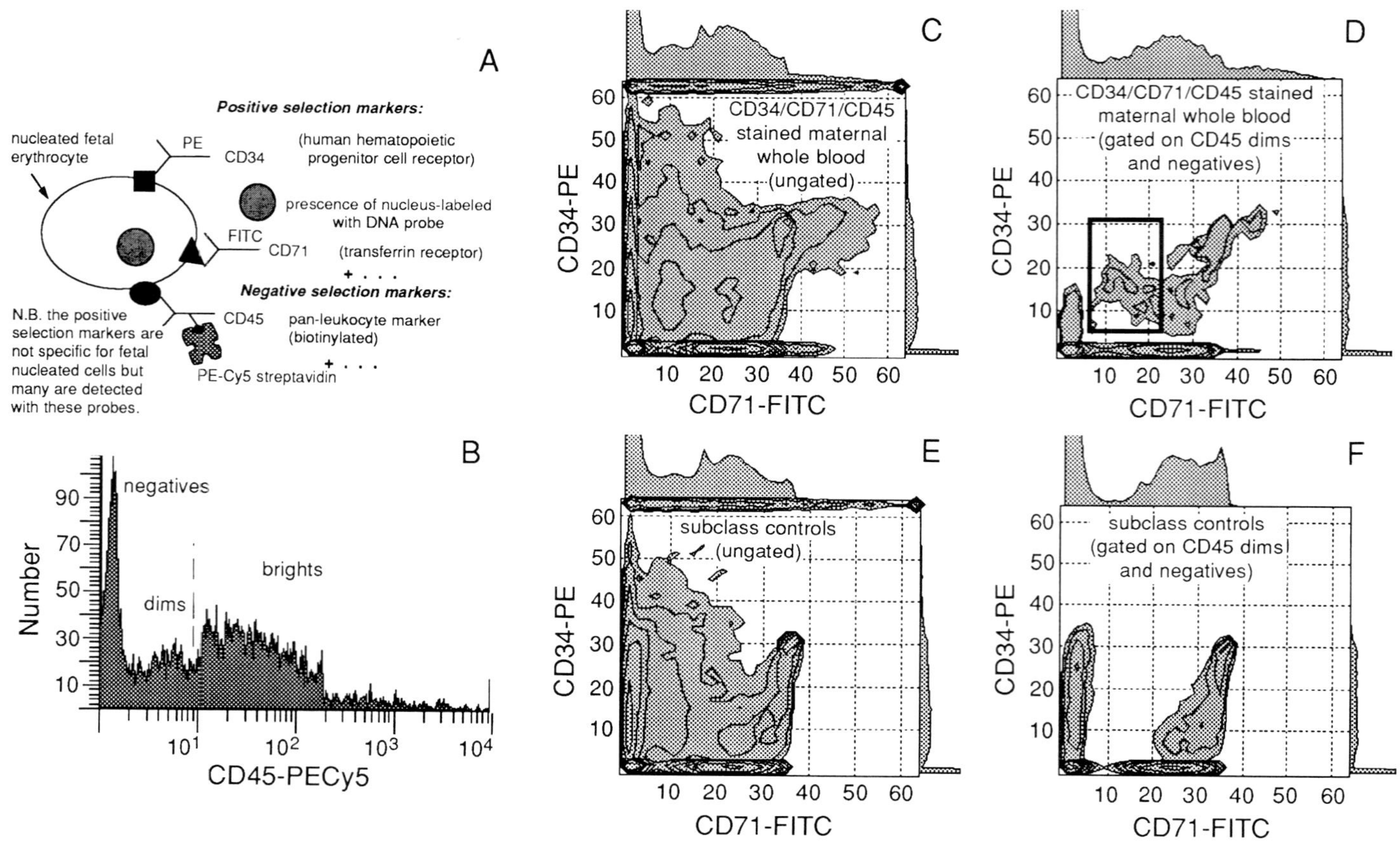
A
Positive selection markers:
nucleated fetal erythrocyte
PE
CD34
(human hematopoietic progenitor cell receptor)
prescence of nucleus-labeled with DNA probe
FITC
CD71
(transferrin receptor)
+ . . .
Negative selection markers:
CD45
pan-leukocyte marker (biotinylated)
PE-Cy5 streptavidin
+ . . .
N.B. the positive selection markers are not specific for fetal nucleated cells but many are detected with these probes.
B
Number
negatives
dims
brights
CD45-PECy5
C
CD34/CD71/CD45 stained maternal whole blood (ungated)
CD34-PE
CD71-FITC
D
CD34/CD71/CD45 stained maternal whole blood (gated on CD45 dims and negatives)
CD34-PE
CD71-FITC
E
subclass controls (ungated)
CD34-PE
CD71-FITC
F
subclass controls (gated on CD45 dims and negatives)
CD34-PE
CD71-FITC

powerful secondary stage data processing can be used to help further examine the not sures. These methods have been used in our laboratory for a number of years with considerable success (Leary *et al.*, 1991,1993). If wide enough gates are chosen on positives to include the not sures, the method of McCoy *et al.* (1991) can be used. An indirect calculation of original frequency can be performed by comparing the fluorescence distributions of the positive triggered samples with other distributions triggered by total signals on aliquots of the same samples, but this calculation is difficult to perform particularly on very rare cells.

The problem of sufficient sample size is addressed in Section III,C. Sampling statistics can become critical to the correct estimation of rare cell frequencies. They can also be critical in determining the amount of sample that must be processed to obtain the needed number of rare cells to sort.

3. Rare Cell Isolation by Development of Different Sorting Strategies

When sorting rare cells, one must ask a series of questions to develop a sensible sorting strategy which will be designed to meet the experimenter's goals. First, do you need as many cells as possible? If so, you should probably perform a high-speed preenriching sort with full anti-coincidence disabled. On the other hand, is sort purity more important than cell number? If so, you should consider doing one or more of the following: (a) perform extensive analysis of multiparameter data before sorting to get the best possible gates, (b) do high-precision cell sorting with full anti-coincidence, (c) do single-cell (per sort unit with full anti-coincidence) sorting for multitube PCR if you can get the information you want from specific cellular DNA or RNA, (d) do sorting followed by subsequent image analysis and micromanipulation re-sorting in order to examine the cells of interest by methods not possible or efficient by flow cytometry/cell sorting. The proper strategy must be developed according to the needs of the researcher. How to develop such a strategy is the central subject of this chapter.

Fig. 5 An example of a high-speed (85,000 cells/sec) analysis of ultrarare human fetal cells in whole (unlysed) maternal blood based on three-color fluorescence analysis of $CD34^+$ + $CD71^+$ + CD45 processed through the instrument described in Fig. 4. Cells were labeled as shown in A and gated on CD45-negative and dim cells as part of a negative selection parameter as shown in B. The positively labeled sample yielded results before CD45 gating as shown in C and after CD45 gating as shown in D. These gates can be applied in hardware at rates in excess of 100,000 cells/sec permitting the sorting hardware to see the distribution shown in D. The importance of this can be seen by comparing the isotype-matched negative control sample before and after gating on CD45. By high-speed hardware gating prior to sorting one is then comparing D to F. A rare cell (1.8×10^{-7} frequency, approximately 2 rare cells per 10 million total cells) subpopulation invisible in C and E, is now clearly visible in D. Figure reprinted with permission of *Annals of the New York Academy of Sciences* (Leary, 1994a).

III. Critical Aspects of the Procedure

A. Cell Preparation

Several aspects of rare cell analysis make cell preparation critically important. Analysis of non-rare cell subpopulations is forgiving in that dead or damaged cells which frequently bind cellular probes nonspecifically can be tolerated, and partial loss of cell subpopulations of interest can be tolerated since there are many more cells available than are needed for flow cytometric analysis or sorting. But in the event of rare cell subpopulations, the presence of even a 1% subpopulation of dead or damaged cells can be hundreds or thousands of times greater in number than the rare cell subpopulation of interest! The greater the degree of cell death or damage the greater the detection problem and flow cytometric methods adequate for non-rare cells are totally inadequate. An example is the use of viability dyes such as PI commonly excluded from dead cells. However, PI only labels a fraction of the dead or damaged cells (those with the most leaky cell membranes) and by no means labels all dead or damaged cells which nonspecifically bind cellular probes.

Many cell preparation techniques, particularly gradients or other preenrichment techniques, may lose more than 90% of the rare cells of interest. Indeed, one problem of cell preparation is the washing centrifugation steps whereby low cell concentrations (<100,000 cells/ml) can typically lead to high rates of cell loss during these steps prior to flow cytometric analysis. Losses of rare cells can lead to problems of finite sampling statistics not normally encountered in flow cytometry, which will be discussed briefly later in this chapter. In many cases the best solution is to keep cell manipulation to a minimum and to keep rare cells rare, i.e., surrounded by large numbers of "carrier" cells, provided that these cells can be readily distinguished from the rare cells during FCM analysis and that they do not interfere with subsequent analyses to be performed on sorted rare cells.

B. Marker Specificity

At the level of rare or ultrarare events, no markers are ever completely specific! Coupled with the problems of nonspecific labeling of dead or damaged cells which are impossible to completely remove, it is essential to think in terms of multiple-positive selection markers and multiple-negative selection markers. It is somewhat paradoxical that the advent of monoclonal antibodies has led to many modern researchers being almost completely untrained in good immunocytochemical techniques. It is mistakenly thought that the marvelous specificity of monoclonal antibodies will solve the problem. Unfortunately, monoclonal antibodies can identify epitopes as small as seven amino acids. Similar epitopes on other cell types are frequently undetected during routine, non-rare screenings of these antibodies against other cell types. Add to this the problem of binding of antibodies to cells via Fc receptors, phagocytosis of antibodies (largely, but not completely stopped at 4°C or in the presence of phagocytosis inhibitors), and even the best monoclonal antibodies may have high enough nonspecific background to preclude the detection of rare cells. One important negative selection marker is an antibody which should not bind to any rare cells. Many rare cell studies attempt to select rare cells on the basis of two positive selection markers. For rare cells the largest contributor is frequently dead/damaged or otherwise false-positive cells which also tend to label as positive with each antibody. Depending on the application, a cocktail of negative selection markers can be used to exclude most dead and damaged cells which nonspecifically bind antibodies as well as specific cell types not of interest. To minimize the number of fluorescent colors that need to be detected, all negative selectors can be labeled with the same fluorescent molecule. A good paradigm without excessive complexity is for detection of rare cells with two positive selection markers, each with its own fluorescent color, and a third fluorescent color encompassing all negative selection markers.

While titration of antibodies and cytochemical stains should always be performed, it is critically important in rare event analysis. Do careful titrations of reagents to select optimal reagent concentration for the optimal signal to noise ratio (*S*/*N*), which can be easily and quantitatively determined by flow cytometry. A somewhat facetious but frequent truism is that if it looks "good" to you by microscopy it is probably overstained to the point where its *S*/*N* is poor. Our eyes are good peak detectors and we are excellent at pattern recognition which can eliminate many artifacts. Unfortunately, flow cytometry is a "zero-resolution" technology which performs integrated fluorescence detection very well but is not sensitive to the spatial distribution of that fluorescence. Provided that there is adequate signal to measure, *S*/*N* is far more important than signal. This should be remembered particularly when immunocytochemical amplification methods are used. While the signal may be considerably amplified making it more pleasing to the human eye, the amplification process may actually make the *S*/*N* worse! Some sample *S*/*N* determinations for a particular application using a variety of immunofluorescent labeling methods were previously reported by our laboratory (Cupp *et al.*, 1984). For this particular application the use of fluorescent immunobeads proved to be by far the best method both in terms of signal *S* and *S*/*N*. But this is not always the case. For example, many cells phagocytose immunobeads and it is difficult to prevent completely even at 4°C and with inhibitors. Steric hindrance of immunobeads in the surface of cells leads to nonstoichiometric measurements of antigens/cell. Unfortunately, except for delayed fluorescence probes, fluorescent dyes tend to self-quench at high concentrations. Thus a losing game is frequently played because the steric hindrance problem varies as the second power of the bead diameter (cross-sectional area) whereas the amount of dye per bead, and hence signal level, varies with the volume (third power of the diameter) of the bead.

C. Sorting of Rare Cells

Two-step sorting (a high-speed enrichment sort followed by a slower sort for increased purity) can reduce the amount of time it takes to sort a given number of cells at a given purity by more than 90%. This shortening of sort time may be critical for the sorting of live cells where they may not be kept viable for long sorting operations. On the other hand the viability of sorted cells can vary considerably. It is not the journey through the sorter that leads to lower viability in some cell types; it is the impact at 10 m/sec into the sort vessel that causes most of the damage. Never sort cells onto hard surfaces such as microscope slides without providing some king of cushioning fluid to allow the sorted cells to decelerate before impact. And remember that all sorted cells will be diluted with sheath fluid by a factor equal to the relative volumes of the sample and sheath streams of your instrument (usually about a 60-fold dilution depending on flow cell orifice size and sample stream diameter and flow rate). Since the sorted sample will be highly diluted with sheath fluid, make certain that the cells can tolerate the sheath fluid. While phosphate-buffered saline (PBS) (and many cells dislike PBS!) is commonly used for sheath fluid in cell sorters other fluids can be used provided that they provide proper ion charges for cell sorting and do not disturb the viscosity for fluidic requirements or refractive index for optical requirements. For example, it is common to suspend the sample in a suspension containing 1% bovine serum albumin (BSA). However, if this is done with the sheath fluid it can cause problems with the viscosity which can disturb the droplet breakoff position. A sheath solution of 0.25% BSA will work nicely in most systems. Cell culture medium (minus serum) can also be used but phenol red in many media can lead to increased levels of autofluorescence. Some cell sorters have 0.2-μm in-line filters on the sheath lines that can clog on particulate matter and cause flow variations. These in-line filters must be kept clean to minimize fluidic variations. Interestingly, a high-speed sort at 100,000 cells/sec will yield a sorted sample at a dilution appropriate for a regular-speed second step sorting at a few thousand cells/second.

For high-speed enrichment sorting, the cells are sufficiently close together that they themselves represent a variation in local viscosity and cells occurring in the necks of the droplet breakoff points can cause a variable charge to be placed on the droplet to be sorted. This will result in "fanning" of the sorted cell stream, which can lead to reduced recovery and increased contamination if considerable care is not used. It is always best to do "straight-ahead" sorting when at high sort rates. In straight-ahead sorting the sorted droplets have no charge placed on them. Droplets containing undesired cells are charged and electrostatically diverted. However, in this case it is not important to collect these cells, only to separate them from the desired cells which are sorted straight ahead. Straight-ahead sorting can yield close to 100% sort recovery and should always be used for sorting rare cells.

To obtain pure cells, cell sorting must be done at lower rates whereby a significant fraction of sorting units contains only one cell and whereby sorting units containing more than one cell are completely rejected by the anti-anticoincidence system. Since there are no perfect markers, or in many cases combinations of markers, it is frequently not possible to sort rare cells to 100% purity by conventional methods. A way around this limitation is to sort to single-cell level and deal with this uncertainty factor by performing a number of single-cell sorts sufficient to capture at least one true rare cell, despite the presence of many false-positive cells with that combination of markers. This has been done for many years in the case of sorting live, single-cell clones and then expanding these isolated clones in tissue culture, but it is now possible through PCR to do single-cell molecular characterizations in the same manner. High-speed analysis can be followed by slower speed, full anti-coincidence "high-resolution sorting" to sort rare single cells. This "single-cell, multitube PCR" sort (Fig. 6), unlike enrichment sorting described in Fig. 2, allows us to examine pure rare single cells at the molecular level. In this method we ensure that the sort is pure by sorting one, and only one, cell per PCR tube and perform PCR analysis of DNA or RNA only on a single-cell level. It may be necessary in some circumstances to use "nested primers" to obtain the necessary amount of specific amplification especially if the cells have been treated with fixatives.

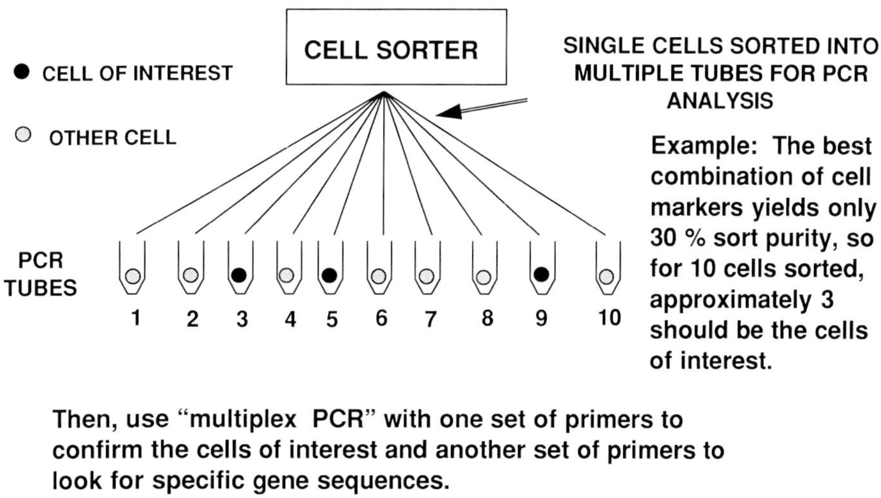

Fig. 6 Concept of single-cell, multitube PCR. In this example, the best combination of cell markers yields only 30% sort purity; so for 10 cells sorted, approximately 3 should be the cells of interest. Then use "multiplex PCR" with one set of primers to confirm the cells of interest and another set of primers to look for specific gene sequences. While the method will in theory always work, if the sort purity is poor the necessary number of tubes quickly becomes impractical. Molecular characterizations of sorted single cells can now be performed on specific DNA or RNA sequences by PCR amplification of these sequences.

Such a strategy will, in theory, always be successful but if the sort purity is poor the number of tubes necessary to isolate and screen pure rare cells can become prohibitive or impractical.

Recovery of small numbers of pure sorted rare cells requires development of strategies appropriate to each application but there are a number of general principles for rare cell sorting in addition to the usual ones you should always perform in any sorting experiment. Always, if possible, perform straight-ahead sorting of desired rare cells to obtain maximum recovery. Never sort small numbers of cells directly onto uncleaned and uncoated slides. Cells can be sorted into high-viscosity droplets or gel material coatings preexisting on the slide surface to keep the sorted cells from air-drying. Due to the extreme sensitivity of the method sorting cells for PCR requires ultraclean fluidics free from the contamination of even a single cell from a previous sample. It is best to sort directly into PCR tubes so that no further direct manipulation of the sorted cells is required. Cells from previous samples frequently appear hours later as they are dislodged from instrument tubing. If it is practical, the best thing to do is to replace as much of the sample tubing as possible between samples. However, it should be kept in mind that to achieve very stable sorting, the new tubing must be thoroughly wetted for approximately half an hour before sorting is attempted. In cases where it is not practical to replace the tubing we typically flush the entire sample delivery system with freshly diluted Clorox (bleach) for a minimum of 20–30 min. Other groups have used different cleansing agents including a 0.07 *M* NaOCl/0.05 *M* NaOH solution for a 20-min interval (Gross *et al.*, 1993).

D. Strategies for Data Acquisition, Data Storage, and Data Analysis for Rare Cells

The user should think very carefully about the problems of data acquisition and storage when deciding to perform flow cytometric analysis of rare cell subpopulations. A list-mode data file of 10 million cells [a modest sample size for a rare cell subpopulation of, say, 10^{-4} frequency (i.e., 0.01%)] with six parameters measured per cell and with each parameter stored as two 8-bit bytes takes 120 megabytes of disk space for one file and contains only 1000 rare cells of interest! A more practical way for most users, who lack the real-time high-speed front-end data classification described later in this chapter, is to first collect enough total cells in a parameter-uncorrelated histogram to allow at least an approximate determination of the relative frequency of the rare cells of interest. Then use "live" acquisition gates in which a rare parameter is used as a system trigger to the flow cytometer to collect rare cells of interest as list-mode data with all parameters stored in correlated fashion (McCoy *et al.*, 1991). The 1000 cells of interest then would require only 12 kilobytes (plus file header space) of storage space (a relatively small data file!).

A problem not encountered in most flow cytometric analyses but one of potentially critical importance in the analysis of rare cell subpopulations is that of finite sampling statistics. One normally assumes that any flow cytometric sample is a statistically representative sample of the total cell population (or in the case of a gated acquisition of rare cells, a representative sampling of the total distribution of rare cells). This may not be the case when very rare cells are analyzed. In some "ultrarare" (e.g., 10^{-6} frequency or, equivalently, 0.0001%) cell applications, one cannot be certain either that the frequency of rare cells measured is accurate or that the distribution of properties is representative of the entire rare cell subpopulation. Finite sampling theory involving binomial statistics can be applied to allow confidence limits on measured rare cell frequencies. We have been using a software program SSIZE ("sample size") to perform these calculations (Blumenson, 1992). The problem of finite sampling theory is not an area of major concern to most flow cytometry researchers, but it is of critical importance if one is to obtain accurate assessments of the number of rare cells in a total cell population where typically only a very small number of rare cells are detected or recovered. In most experiments only a small aliquot of the total cell sample is processed. The frequency of the cell subpopulation of interest is large enough that one can effectively consider that (1) the sample size is infinite, (2) removing some cells during a sampling does not alter the probability of finding the cells of interest in subsequent samplings, and (3) the frequency obtained in an aliquot is representative of what would be obtained if the entire sample were processed.

The reverse process can be performed with actual flow cytometric data and/or sort results to obtain an estimate with confidence limits for the actual frequency of the number of fetal cells of that type in maternal blood. Two examples of input and output from SSIZE are given as follows. If one wanted to obtain confidence bounds on measurements of fetal cell subtype frequencies, and one estimated that in 10^9 total maternal and fetal cells a particular fetal cell subtype was present at a relative frequency of 10^{-6}, and if one wanted to have a 95% probability of being able to isolate 100 of these fetal cells for further growth or characterization, one would need to process at least 1.16×10^8 cells. If you performed the experiment and subsequently found 150 cells of the desired subtype within that sample size of 1.16×10^8, the most likely fetal cell frequency is 1.29×10^{-6} with a 95% confidence that the correct frequency of this fetal cell type is between 1.10×10^{-6} and 1.50×10^{-6}. For a second case consider an estimated fetal cell subtype frequency of 10^{-6} where you wish to have a 95% probability of sorting a fetal cell within a total available sample size of 10^7 total cells. From SSIZE calculations you would need to plan to process as many as 2.59×10^6 total cells during a sorting experiment. If in doing this actual experiment you found out that you actually were able to obtain 3 fetal cells, then the most likely estimate of the frequency of fetal cells of that type in maternal blood becomes 1.16×10^{-6} with 95% confidence that the frequency of fetal cells is between 0.3×10^{-6} and $3.10 \times {}^{-6}$.

Another problem frequently encountered with rare cell analysis is the problem that multiparameter measurements on rare cells while important and useful result in a spreading of a small number of data points over a large multiparameter data space, the so-called "tyranny of dimensionality." This makes viewing of the rare cell data difficult. One way to look at the data is to reduce the dimensionality of the rare cell data using principal components. Instead of looking through multiple bivariate displays to view the data, one can look at one display which tries to provide a good (in this case based on choosing a projection plane that maximizes the statistical variance of the data rather than choosing a standard bivariate view which must be orthogonal to some original list-mode parameters) view of all of the multiparameter data within a single bivariate principal component display. An even better way to view the data is to look at a trivariate display of original data or, better yet, of the first three principal components, as we do using home-built software in our laboratory which projects 3D views of the data either as stereo pairs on a conventional IBM compatible personal computer or as hologram-like images on an autostereoscopic display (Dimension Technologies, Inc., Rochester, NY) which also allows a 3D mouse to interact with the data. Commercially available software packages such as DataDesk available on Macintosh computers, or S-plus on IBM-compatible personal computers, can take ASCII format list-mode data from other sources and display it dynamically in rotating 3D displays of the first three principal components. One sees "clouds" of data points representing each cell subpopulation. Again a detailed description of this is beyond the scope of this chapter but may be of interest to some readers. Principal component hardware sorting has been implemented in the author's laboratory which permits sorting of cell subpopulations not visible in any of the possible bivariate displays of the data. A brief account of some of these methods including use of biplots (Gabriel, 1971) is described in Leary *et al.* (1991).

New software developed in our laboratory permit listmode data mixing of positive and negative cells in any proportion to test (1) limits of detectability, (2) ability to successfully locate rare positive cells during analysis, and (3) new sort strategies (Leary *et al.*, 1994b). Such software allows us to more rapidly develop rare cell analysis and sorting strategies for new applications.

IV. Instrumentation

A. Conventional Instrumentation

Conventional instrumentation will not be described in detail except to point out important features that should be used where available for rare cell analysis. The most important point to consider is the choice of a trigger signal. All subsequent signal processing will depend on this choice as will all of your results! If one chooses a trigger for all cells (e.g., forward light scatter since all cells, and unfortunately all debris, will have a light scatter signal), then the subsequent signal processing of your instrument will attempt to process all cells subject to the dead time of your instrument. This trigger signal will also be used to determine cell coincidence within sorting units should you decide to employ the anti-coincidence feature, if present, on your cell sorter. Should you decide to use a property only of rare cells (e.g., a relative amount of one or more fluorescent signals corresponding to markers used to identify rare cells), the instrument will be blind to all other cells both in terms of measuring the relative frequency of the rare cell subpopulation and in terms of sort anti-coincidence detection of neighboring contaminating cells (which will not be seen). If a total cell trigger (e.g., forward light scatter) is used, logical gates around other parameters can be used during data acquisition to reduce the number of cells stored as list-mode data and to employ anticoincidence, but the instrument will still be blind with respect to any cells below this trigger level and to the extent of its dead time at the rate cells are being processed. Depending on how your instrument is configured, this dead time may or may not be reduced by using nonintegrated "peak" signals which are available on most commercial instruments.

B. High-Speed Flow Cytometry

A multiparameter hardware/software system was developed in our laboratory which allows multiparameter analysis of cells at rates in excess of 100,000 cells/sec (Fig. 4). This high-speed system, an outboard module attachable to either commercial or home-built flow cytometers, allows processing of 100,000,000 cells in less than 17 min as opposed to approximately 6 hr at 5000 cells/sec on the same instrument without this module. The high-speed system performs high-speed counting, logic gating, and count rate error checking. The count rate error checking requires that the cells flow at an even rate according to Poisson arrival statistics. On a millisecond basis the sample rate is checked to prevent "bursts" of cells coming through the system which can cause false-positive rare cells to be detected as shown by another group (Gross *et al.*, 1993). Indirectly, by acting as a high-speed front-end filter of signals, the system can be used to control cell sorting.

Actual instrument dead time depends on the pulse widths of the signals as well as delay lines, if used. The actual throughput rate is limited not by the signal and software processing times, but rather by the in-excitation-beam cell coincidence caused by asynchronous cell arrival times in the cell sorter or similar type of device. Use of thresholds and logical gating from total and rare cell signals with other nonrare signals allows multiparameter rare event list-mode data to be acquired reasonably both in terms of signal processing speeds and total amount of data to be stored by a conventional second-stage data acquisition system. Analysis of these multiparameter rare event data also permits, with further data processing techniques, reduction or elimination of many "false positives," an important problem in the analysis of rare cell subpopulations.

V. Results: Some Rare Cell Applications and Strategies

Several examples of rare cell analyses as performed in this author's laboratory illustrate different strategies for flow cytometric analysis and/or sorting of rare cell subpopulations. These strategies were formulated for specific situations and subject to subsequent characterizations of sorted rare cells by different techniques.

As an example of results which can be obtained with a high-speed cell sorter, cells were processed in our laboratory at 85,000 cells/sec for three-color immunofluorescence using two positive (but not fetal cell unique) selection markers (CD34 and CD71) and one negative selection marker (CD45). This application used for selection of human fetal cells from the blood of a pregnant woman involves flow cytometric analysis of early erythroid cells (highly enriched but not specific for rare fetal cells) in unlysed maternal whole blood (processed at a concentration of approximately 5×10^8 cell/ml). Data are shown in Fig. 5. The region of interest contains a calculated frequency of 1.8×10^{-7} based on a total sample size of 10^8 cells. Finite sampling theory gives the 95% confidence range of the rare cell frequency to be between 1.8×10^{-7} and 2.0×10^{-7}. To have 95% probability of being able to sample (and sort in the absence of other possible losses) 10 cells from a sample aliquot containing 10^8 cells would require us to process at least 71% of the sample. Such calculations can obviously be used to determine how much sample can be first analyzed to set sorting gates while still allowing for enough rare cells to remain for subsequent sorting. This application also shows why it is difficult to use depletion cell separation technologies (e.g., affinity columns and magnetic bead sorting). While it is theoretically possible to allow sorting on the basis of levels of antigen expression with these technologies, these sorting levels are much more difficult to control. In the sample shown in Fig. 5, approximately half of the fetal cells are CD45 dim rather than CD45 negative. It is also very difficult to recover cells from affinity columns or through magnetic sorting devices when there may be only a few rare cells, although such methods are useful as cell isolation or preenrichment procedures to obtain larger numbers of not very rare cells (e.g., 1% subpopulations) when the total number of rare cells available is large (greater than a few thousand cells). We are now attempting to perform single-cell multitube PCR (cf. Fig. 6) to isolate and characterize at the molecular level pure single human fetal cells from maternal blood. Multiple sets of PCR primers can be used on single cells to simultaneously confirm the identity of the cell and to check for specific gene sequence abnormalities.

Sometimes flow cytometry by itself is not able to completely distinguish between true- and false-positive rare cells. Other complementary technologies such as image analysis and confocal microscopy must be used in a second-stage processing of sample. An example of an application in progress in our laboratory is the sorting of rare fetal cells on the basis of the number of copies of a chromosome labeled by fluorescence *in situ* hybridization (FISH) (e.g., using a Y chromosome-specific probe or a chromosome 18- or 21-specific probe) to detect either a male fetal cell or a trisomic 18 or 21 fetal cell in maternal blood for noninvasive prenatal diagnosis (Leary *et al.,* 1991). FISH is only semiquantitative in that staining intensity can vary enough from cell to cell to prevent the complete separation of normal disomic cells from aberrant trisomic cells. While methods such as ratiometric FISH (Nederlof *et al.,* 1992) are helpful, it is still necessary to follow a first-step sort by a second-stage examination of sorted cells by fluorescence microscopy or fluorescence image analysis/confocal microscopy to properly distinguish between true positives with an aberrant number of true chromosome spots (as opposed to fluorescent debris within or outside the nucleus). Subsequent micromanipulation [e.g., by cookie cutter sorting (Wade, 1987) of cells on a Meridian ACAS 570 image analysis/confocal microscope at our institution] of true-positive rare cells for subsequent molecular characterizations can further confirm their identity and identify other aberrant properties.

A third application involves the search for rare metastatic breast cancer cells in patients for minimal residual disease detection. High-speed analysis and sorting techniques are being developed to detect and to isolate rare metastatic cells for subsequent analysis of specific mRNA levels to see if putative metastatic genes are "turned on" in these cells. In addition high-speed flexible sorting strategies are being developed to purge rare metastatic breast cancer cells from bone marrow prior to autologous bone marrow transplants in patients undergoing chemotherapy.

VI. Discussion

This chapter demonstrates that with sensitivity to the problems of rare cell analysis and sorting, it is possible to analyze rare and, in some cases (with suitable diligence!), ultrarare cells. Impure sorts may suffice for many situations. The theoretical discussions in this chapter are designed to enable experimenters to determine whether their experiments are at all feasible. Each biological applications should be approached pragmatically; the analysis or sort may not need to be perfect, only good enough to answer the question at hand.

Acknowledgments

This work was performed under the auspices of NIH Grants GM38645, HD20601, and CA61531. I thank Mr. Scott McLaughlin, Ms. Janet Gram, and Mr. Stefan Burde for their critical reading of the manuscript and Ms. Christine Taillie for her help in preparing the manuscript and figures for publication. Flow cytometric sample preparation was performed by Janet Gram. Rare cell sorting was performed by Scott McLaughlin. Peripheral blood from pregnant women was provided by collaborator, Dr. Donald Schmidt, at Children's Hospital in Buffalo, NY.

References

Ashcroft, R. G. (1988). *Cytometry* **3,** 85–88.

Blumenson, L. (1992). SSIZE Computer Software (finite sampling theory).

Buican, T. N. (1991). *ACS Symp. Ser.* **464,** 59–72.

Corio, M. A., and Leary, J. F. (1993). U.S. Pat. 5,199,576.

Cory, J. M., Ohlsson-Wilhelm, B. M., Brock, E. J., Sheaffer, N. A., Steck, M. E., Eyster, M. E., and Rapp, F. (1987). *J. Immunol. Methods* **105,** 71–78.

Cupp, J. E., Leary, J. F., Cernichiari, E., Wood, J. C., and Doherty, R. A. (1984). *Cytometry* **5,** 138–144.

Frantz, C. N., Ryan, D. H., Cheung, N. V., Duerst, R. E., and Wilbur, D. C. (1988). *Prog. Clin. Biol. Res.* **271,** 249–262.

Gabriel, K. R. (1971). *Biometrika* **58,** 453–467.

Gross, D., and Harris, C. M. (1985). "Fundamentals of Queuing Theory," 2nd ed. Wiley, New York.

Gross, H.-J., Verwer, B., Houck, D., and Recktenwald, D. (1993). *Cytometry* **14,** 519–526.

Herweijer, H., Stokdijk, W., and Visser, J. W. (1988). *Cytometry* **9,** 143–149.

Jensen, R. H., and Leary, J. F. (1990). *In* "Flow Cytometry and Sorting" (M. R. Melamed, T. Lindmo, and M. L. Mendesohn, eds.), 2nd ed., pp. 553–562. Wiley-Liss, New York.

Kompala, D. S., and Todd, P., eds. (1991). "Cell Separation Science and Technology," ACS Symp. Ser. No. 464. Am. Chem. Soc., Washington, DC.

Leary, J. F., Ellis, S. P., McLaughlin, S. R., Corio, M. A., Hespelt, S., Gram, J. G., and Burde, S. (1991). *ACS Symp. Ser.* **464,** 26–40.

Leary, J. F., Schmidt, D. F., Gram, J. G., McLaughlin, S. R., Dalla Torre, C., Burde, S. (1994a). "High-Speed Flow Cytometric Analysis and Sorting of Human Fetal Cells from Maternal Blood for Molecular Characterization." *In* Annals of the New York Academy of Sciences, Vol. 731. In press.

Leary, J. F., Schmidt, D., Gram, J., McLaughlin, S., Della Torre, C., Burde, S., Ellis, S. (1994b). "Isolation of Rare Cells by High-Resolution Cell Sorting for Subsequent Molecular Characterization—Applications in Prenatal Diagnosis, Breast Cancer and Autologous Bone Marrow Transplantation." 24th Annual Cancer Symposium: Cytometry 2000. Detroit, Michigan. Norwell, Massachusetts: Kluwer Academic Publishers. In press.

Leary, J. F., Corio, M. A., and McLaughlin, S. R. (1993). U.S. Pat. 5,204,884.

Leslie, D. S., Johnston, W. W., Daly, L., Ring, D. B., Shpall, E. J., Peters, W. P., and Bast, R. C. (1990). *Am. J. Clin. Pathol.* **94**(1), 8–13.

McCoy, J. P., Jr., Chambers, W. H., Lakomy, R., Campbell, J. A., and Stewart, C. C. (1991). *Cytometry* **12,** 268–274.

Nederlof, P., van der Flier, S., Raap, A. K., and Tanke, H. J. (1992). *Cytometry* **13**(8), 831–838.

Parson, J. D., Hiebert, R. D., and Martin, J. C. (1985). *Cytometry* **6,** 388–391.

Peters, D., Branscomb, E., Dean, P., Merrill, T., Pinkel, D., Van, D., and Gray, J. (1985). *Cytometry* **6,** 290–301.

Ryan, D. H., Mitchell, S. J., Hennessy, L. A., Bauer, K. D., Horan, P. K., and Cohen, H. J. (1984). *J. Immunol. Methods* **74,** 115–128.

van den Engh, G., and Stokdijk, W. (1989). *Cytometry* **10,** 282–293.

van Rotterdam, A., Keij, J., and Visser, J. W. (1992). *Cytometry* **13,** 149–154.

Visser, J. W., and de Vries, P. (1990). *In* "Methods in Cell Biology" (Z. Darzynkiewicz and H. Crissman, eds.), Vol 33, pp. 451–468.

Wade, P. (1987). "'COOKIE CUTTER' Method of Cell Selection," Appl. Note C-1. Meridian Instruments.

CHAPTER 21

Cell Sorting of Biohazardous Specimens for Assay of Immune Function

Janis V. Giorgi
Laboratory of Cellular Immunology and Cytometry
Department of Medicine
Division of Clinical Immunology and Allergy
University of California, Los Angeles School of Medicine
Los Angeles, California 90024

I. Introduction

This chapter describes general precautions that must be taken when sorting potentially biohazardous specimens. A specific method, modified from a previously published technique, is detailed to verify the efficacy of aerosol containment by sorters (Merrill, 1981). The precautions presented in this chapter are directed toward sorting cells in specimens potentially harboring pathogens such as human immunodeficiency virus (HIV) that are transferred primarily parenterally in a laboratory situation. Those pathogens transferred primarily through aerosols, e.g., *Mycobacterium tuberculosis*, require additional precautions that are not addressed here. The chapter also outlines techniques found useful in maintaining the viability and function of activated lymphocytes during sorting and provides one example of how sorting has been used to study the immunology of the acquired immunodeficiency syndrome (AIDS).

II. Applications

Flow cytometry sorting can be used in studies of human infectious diseases to separate leukocyte and lymphocyte subsets in the peripheral blood or lymphoid tissue on the basis of cell-surface phenotype. The sorted cells can be used to examine immune responses to the pathogen under investigation, characterize the cellular basis of immunopathogenesis caused by the organism, and characterize the cells that are infected with the pathogen. Methods for sorting leukocytes for studies of immune function have been described and are beyond the scope of this chapter (Parks *et al.*, 1986). Survival of some human and animal pathogens have been reviewed (Sattar and Springthorpe, 1991; Schoenbaum *et al.*, 1990). Extensive literature is available on aerobiology (Dimmick, 1969) and on methods of providing protection against biohazards in a laboratory settings (Hambleton and Dedonato, 1992; Centers for Disease Control, 1992).

Specific guidelines to prevent transmission of HIV and other blood-borne pathogens in a laboratory setting have been set forth (Centers for Disease Control, 1987,1988a,b,c). These later references are of great importance to the flow cytometry laboratory that undertakes sorting of samples potentially infected with HIV or hepatitis B virus. In all cases, extraordinary care needs to be taken in handling specimens from humans regardless of whether they are known to harbor a particular infectious organism.

III. Critical Aspects of the Procedure

Critical issues related to the instrument and to verification of aerosol containment by the sorter are covered in Sections IV and V below. Other critical issues of maintaining the laboratory and initiating the procedures are outlined in this section. Sorting potentially biohazardous specimens is appropriate only in laboratories (i) with prior experience sorting healthy control material of the type that contains the pathogen to be sorted and (ii) with biosafety precautions already in place for performing analytic flow cytometry on the pathogen under investigation. Most institutions have a biosafety advisor or committee who can help establish a safe environment and from whom approval will most likely be needed to carry out potentially biohazardous sorting. The critical aspects of safe sorting of potentially biohazardous specimens are summarized in Table I.

Cell sorting of biohazardous specimens must consider (i) the room, (ii) the

Table I
Assurance of Safety during Biohazard Sorting

Place physical barriers between hazard and operator
- Disposable lab coats, gloves
- Surgical masks, glasses, face shields

Control and test for aerosol containment
- Precautions for containment
  - Standard biosafety features of the sorter
  - Custom-designed modifications are required
- Test for complete aerosol containment
  - Test every 1–3 months

Establish proper air flow in room
- Negative pressure
- 20 changes per hr

Limit access to room during the sort
- Allows operator to concentrate on safe techniques
- Maintains negative pressure in room

Provide adequate training to sorter operators
- Train in the biology of infectious organism
- Train in flow cytometry and sorting (e.g., 2–4 years of flow cytometry experience with 1–2 years sorting experience)

instrument (see Section IV, below), and (iii) the operator. With regard to the room, P3 containment is desirable, but not essential. Most biosafety professionals require that the room have negative air pressure and a minimum of 20 changes of air per hour (hr). Furthermore, access to the room must be limited during the sorting of HIV-positive samples in order not to interrupt the negative pressure environment of the room and to allow the operator to concentrate on safe techniques throughout the sort.

As a general precaution during sorting, an apparatus that consists of a stainless steel box with an ultraviolet (UV) light inside and an intake exhaust fan (American Ultraviolet Company, Murray Hill, NJ) can be used during the sort to sterilize room air. Some experts advise this will provide some protection against exposure, but how much is controversial. Although the effectiveness of UV light in inactivating pathogens varies, it is traditional to use UV light to sterilize hoods and rooms, e.g., when one leaves at the end of the day. This can be done in the sorter room with a UV light in the ceiling.

Protection of the operator and others in the room is the ultimate concern. It is suggested that an operator have 2–4 years of flow cytometry experience, including 1–2 years of experience with sorting, before that technician undertakes sorting that is potentially biohazardous. The experience should include working with the pathogen under study. This experience provides both knowledge regarding safe operation of the flow cytometer and a general knowledge of the characteristics of the pathogen and its infectious properties. With adequate experience working with potentially biohazardous specimens and knowledge regarding the organism, an operator is more likely to be able to work safely.

Physical barriers, known as personal protection, should also be used to protect the operator. These barriers should focus on the hands and mucous membranes of the face. Double gloves should be used. The nose and mouth should be covered with a surgical mask and the eyes with safety glasses, then a full face shield should be placed over these. A disposable laboratory coat and a disposable head covering should also be used. Use of a sort-stream viewing camera is advisable as it allows the operator to keep his/her face away from the area of the instrument that holds the greatest potential biohazard.

IV. Instruments

A deflected droplet (jet-in-air) sorter is usually used to sort cells for immune function studies because, in general, greater numbers of cells can be obtained more quickly than is possible using a fluidic switching (closed flow cell) system. These sorters produce aerosols as part of their normal operation. Aerosol containment by the cell sorter should be complete both during routine operation and when a clog is simulated (see Section V below). Standard biosafety features on most sorters include an efficient vacuum containment system and an enclosed sample collection chamber. If aerosols escape from the sorter during the test sort, modifications must be made to the sorter containment system. No potentially biohazardous sort should be attempted until aerosol containment is complete. Check the seals around the sort chamber door, and modify them if necessary. Additional custom biosafety features can be envisioned but are probably not critical unless aerosol containment is not complete. The waste vacuum evacuation system can be improved on some systems by increasing the negative pressure on the waste evacuation system, for example, by attaching an auxiliary vacuum pump to the sorted-sample collection chamber. If this is done, a bleach trap must be added to the line connecting the collection chamber to the additional vacuum pump to filter the aerosol through the bleach. If negative pressure is increased too much, however, the sorting streams will be affected.

Note that the sorter vacuum system is designed to contain aerosols only when the sort chamber door remains closed. Air flow will be disrupted as soon as the door is opened. A 3-min delay after sorting has stopped before the door is opened will probably allow the chamber to clear on most sorters. The time can be checked with bottled smoke (Lab Safety Supply, Inc., PO Box 1368, Janesville, WI, Cat. No. WA-4811). An auxiliary pump, turned on for a few moments immediately prior to opening the sort chamber door, can be configured for use as an additional safety measure.

It is critical to meticulously maintain the sorter. A leak in a worn flow line might cause pathogen-contaminated liquid to leak into the room. This type of rare accident probably represents the greatest potential biohazard. The operator should check frequently during the sort and following the sort to assure that no leak has occurred and immediately clean up any material with bleach if it is discovered. Keep adsorbent pads and a container to dispose of the contaminated waste.

Care must be taken to help users prepare samples to avoid formation of cell aggregates. These enhance clog formation, aerosol formation, and other machine problems. Even if the instrument is spotless, a bad sample will clog the tip. Prefiltering the sample with nylon mesh to remove clumps should be done if the sample is suspect.

V. Verification of Aerosol Containment

Verification of aerosol containment using the method outlined below, or a similar approach, should be performed every 1–3 months. This method is adapted from a previously published technique (Merrill, 1981).

A. Materials

Materials used for phage sorting to verify aerosol containment are listed together with potential vendors. Methods for preparing, storing, and expanding these biologic materials are described in the ATCC Catalogue of Bacteria and Phages (Gherna *et al.*, 1989). A basic review of relevant methods and properties of bacteria and bacteriophage can be found in many introductory microbiology laboratory manuals (e.g., see Miller, 1992) or in older books on the subject (e.g., see Adams, 1950).

1. *Escherichia coli* (*E. coli*): The American Type Culture Collection (ATCC), 12301 Parklawn Dr., Rockville, MD, Cat. No. 11303.
2. Bacteriophage T4 (T4 phage): ATCC No. 11303-B4.
3. Prepared nutrient agar plates: Becton–Dickinson, Inc., Cockeyville, MD, Cat. No. 97801.
4. Nutrient agar: Difco, Detroit, MI, Cat. No. 0001-01-8.
5. Nutrient broth: Difco, Cat. No. 0003-01-6.
6. Sheath fluid (10× Dulbecco's phosphate-buffered salt solution): Mediatech, Washington DC, Cat. No. 20-030-LV.

B. Preparation for the Test Sort

1. Prepare Bacteria and Bacteriophage Stocks

a. *E. coli:* grow 10–20 ml of *E. coli* to turbidity while still maintaining log phase growth (Gherna *et al.*, 1989, p. 403; Miller, 1992, p. 21).

b. T4 phage: stock at 10^9 phage/ml or higher (Gherna *et al.*, 1989, p. 289).

2. Prepare *E. coli* Lawns

Twenty-six nutrient agar plates are needed. Set aside 1 as a control (step 7), and 4 each for steps 2 and 3. Use the other 17 plates to prepare *E. coli* lawns as follows: 2–4 hr prior to the T4 phage sort, overlay each with 50–100 μl of *E. coli* (prepared in B.1.a) in 5 ml of 0.7% soft agar; incubate the plates for 2 to 4 hr at 37°C. The *E. coli* on these plates must be in log-phase growth but not confluent when T4 phage is added (steps 4–5) so that plaques will form.

C. Method for Verification of Aerosol Containment

Step 1. Calculate the Sample Flow Rate

A sample is run for 10 min and the amount of volume consumed is determined. The sample flow rate should approximate the maximum rate of sample consumption during routine cell sorting, usually 20–30 μl/min. Record the value on a log sheet such as that shown in Fig. 1.

Step 2. Titrate the T4 Phage Stock

Plates: four agar plates.

The T4 phage stock is retitered during each test to verify a concentration of at least 10^9 phage/ml. Titrations are determined by adding 0.9 ml of phage dilutions in nutrient broth to 50–100 μl of *E. coli* in 5 ml soft agar and overlayering on culture plates using dilutions of 10^{-5}–10^{-8} (10,000–10 plaque forming units, PFU/plate). The initial concentration of the phage stock should be as high as possible to increase the sensitivity of the test.

Step 3. Determine the Concentration/Min of T4 Phage Emerging from the Flow Cytometer

Plates: four agar plates.

Run the T4 phage stock through the sorter and collect the sheath stream in nutrient broth for 1 min. Dilute 10^{-1}–10^{-4} in nutrient agar (dilute more if the stock is at a concentration higher than 10^9) and add 0.9 ml of the dilutions to 50–100 μl of *E. coli* in 5 ml soft agar and pour over the *E. coli* lawns. Experience suggests the T4 phage stock is diluted about 2 logs as a result of mixture with the sheath fluid when it passes through the flow cytometer and loses about 2 additional logs of activity as a result of the physical stress of sorting. The phage should emerge from the cytometer at 10^5/ml or greater, indicating it is sufficiently concentrated to be of value to test aerosol containment.

DATE: 11/30/93

T4 PHAGE SORT FOR TESTING OF AEROSOL CONTAINMENT

STEP 1. **CALCULATION OF SAMPLE RATE (μl/min)**

Volume fed to machine (μl): 344μl Feeding time (min): 10min Sample rate (μl/min): 34.4

Does flow meet established criteria (20-40μl/min)? YES [✓] NO [ ]

STEP 2. **RE-TITRATION OF T4 PHAGE STOCK (MINIMUM 10^9 PFU/ML)**

Plate ID	Dilution	# of plaques	Dilution factor	Vol on plate	PFU/ml
2.1	$1/10^5$	>1000			
2.2	$1/10^6$	302			
2.3	$1/10^7$	98	10^7	0.9	1.08×10^9
2.4	$1/10^8$	32			

Does the phage stock meet minimal concentration requirement (10^9 PFU/ml)? YES [✓] NO [ ]

STEP 3. **CONCENTRATION OF T4 FLOWING/MIN**

Collect sheath stream for 60sec

Volume collected (ml): 2.12ml Sheath rate (ml/min): 2.12ml/min

Plate ID	Dilution	# of plaques	Dilution factor	Vol on plate	Sheath rate	PFU/min
3.1	$1/10^1$	Confluent				
3.2	$1/10^2$	>1000				
3.3	$1/10^3$	508				
3.4	$1/10^4$	63	10^4	0.9	2.12	1.48×10^6

Does the phage stock meet optimal established criteria for flow rate through flow cytometer (>10^5 PFU's/min)? YES [✓] NO [ ]

STEP 4. **AEROSOL CONTAINMENT DURING ROUTINE OPERATION**

Time: 2 hr

Plate ID	Plate location	# of plaques
4.1	inside right	0
4.2	inside left	14
4.3	door left	0
4.4	door center	0
4.5	door right	0
4.6	Y adjust	0
4.7	at vacuum	0
4.8	sample area	0

Is aerosol contained under optimal sort conditions? YES [✓] NO [ ]

STEP 5. **AEROSOL CONTAINMENT DURING SUBOPTIMAL OPERATION**

Time: 15min

Plate ID	Plate location	# of plaques
5.1	same as above	16
5.2		110
5.3		0
5.4		0
5.5		0
5.6		0
5.7		0
5.8		0

Is aerosol contained under suboptimal sort conditions? YES [✓] NO [ ]

REMARKS: Successful test Sort.

SIGNED: [signature]

Fig. 1 Log sheet from a typical T4 phage sort illustrating the multistep process of verifying aerosol containment during routine and simulated-clog sorting conditions.

Step 4. Test Aerosol Containment during Routine Operation

Plates: eight *E. coli* lawns.

Place two plates inside the sort chamber, three just outside the sort chamber door, and three wherever other aerosols may escape. Choose these later places by examining the pathways of the vacuum and waste lines and selecting areas of potential aerosol escape.

Run the T4 phage stock through the sorter for 2 hr under ordinary sort conditions (1000 sort decisions/sec each left and right). The goal is to obtain at least a few plaques on the inside of the sorting chamber as a positive control. Maintain tight side streams and the door closed. Place 15-ml conical tubes into the sort block to catch the side streams. These tubes will have to be periodically changed if they become full. Do not allow them to overflow. It is important to ensure that the sheath fluid does not kill the phage. Certain sheath fluids, especially those that contain detergents, cause the phage titer to drop markedly. A recommended buffer is PBS with Ca^{2+} and Mg^{2+}; Mg^{2+} stabilizes the T4 phage. Note that antibiotics in the sheath may aerosolize into the *E. coli* lawn, especially on the plates inside the sorting chamber, and prevent growth of the *E. coli*.

Step 5. Test Aerosol Containment during Suboptimal Sorter Operation

Plates: eight *E. coli* lawns placed at the same positions as those in step 4.

The sorter is modified to produce as much aerosol as possible either by positioning the sheath stream to splash against the waste collection vessel (Merrill, 1981) or by turning off the droplet drive which creates fanning side streams (L. E. Hultin, personal observation). Run the T4 phage stock through the sorter for 15 min. The goal here is to obtain 20–200 plaques on plates placed inside the sort chamber.

Step 6. Prepare Control Plate

Plate: one *E. coli* lawn.

Pour 5 ml of soft agar over agar plate. This is used to verify that the bacterial lawn grows to confluence when no phage are added.

Step 7. PFU Calculation

Cover, invert, and incubate all 26 petri dishes at 37°C for 18 hr. Count the plaques on all plates and calculate the PFU/ml for step 2, the PFU/min for step 3, and total PFU/plate for steps 4–6. The agar plate from B.2 should be blank and that from step 6 should have a confluent *E. coli* lawn.

D. Results from a Typical Test

Data from a typical T4 phage sort are shown in Fig. 1. We generally obtain anywhere from a few to 20–200 plaques on the inside of the sorting chamber when verifying that aerosols are indeed formed as a result of sorting. Picking up even a few phage may take 2 hr when the instrument is in routine operation. Picking up 20–200 plaques should take no more than 15 min when a clog is simulated. The plates placed outside the flow chamber at various places to detect aerosol if it were escaping from the instrument have been negative in all the phage sorts we have performed to date on our FACStar^Plus (Becton–Dickinson Immunocytometry System, Inc., San Jose, CA) for both routine sorting mode and simulated failure mode. This indicates that aerosols should be contained inside our flow cytometer under routine flow cytometer operation even if a clog were to occur.

VI. Staining and Cell Separation of Potentially Biohazardous Specimens

Cell separation and staining are carried out using the same methods for separating and staining cells from healthy control individuals except that appropriate biosafety precautions are taken. In cases where the cells being sorted are activated cells that have specific immune function, e.g., anti-HIV-directed cytotoxic T cells, the cells may survive poorly outside the body. Cytokines are

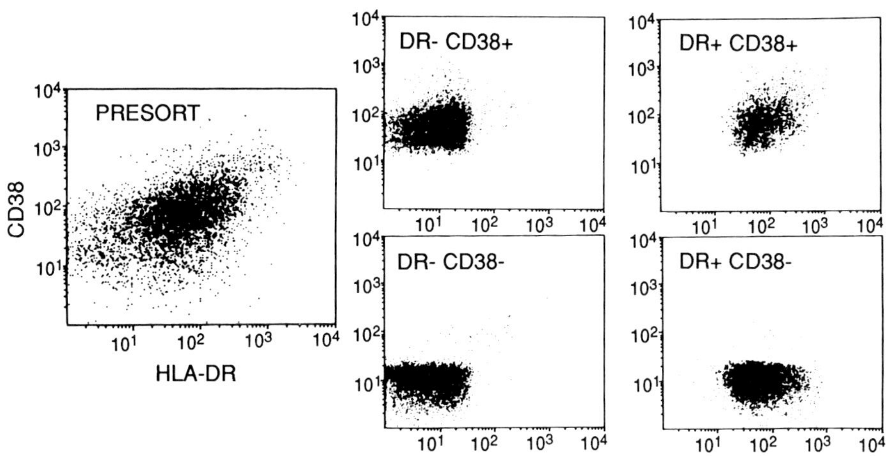

Fig. 2 Anti-HLA-DR FITC (*x*-axis) and CD38 PE (*y*-axis) staining of the presort sample (left) and postsort analysis of four populations sorted on the basis of expression of one, both, or neither activation Ag (right). This figure is reproduced from Ho *et al.* (1993).

often needed to maintain the viability of such cells and these are seldom present in the medium used for sorting the cells. Although it may be possible to add cytokines or other nutrients to improve survival and permit maintenance of function, we found that addition of IL-2 to maintain the viability of cytotoxic T cells simply caused clumping of the cells. It is essential to work as quickly as possible with the cells to prepare them for the test of cell function. Maintain the cells suspended at 15°C using a refrigerated circulator (Brinkmann Instruments, Inc. Westbury, NY). Use tissue culture medium (TCM) such as Hanks' balanced salt solution (HBSS), 2 μM L-glutamine, and 25 mM Hepes supplemented with 10% human serum (HS). Use TCM without serum as the sheath fluid. Collecting the sorted cells into TCM plus 20% HS and 50 μg/ml gentamycin, as well as completing the processing within 5-hr, permitted the anti-HIV cytotoxic activity of MHC-restricted T cells to be maintained, as described in Ho *et al.* (1993). Maintenance of the functional activity of healthy control individuals' lymphocytes does not generally require such stringent handling.

VII. Results and Discussion

An example of a three-color sort of $CD8^+$ T cells separated from peripheral blood on Ficoll-Hypaque then stained with allophycocyanin (APC)-labeled CD8 (Leu 2) mAb, and mAb against two activation antigens (Ag) are shown in Fig. 2. The fluorescence profiles of anti-HLA-DR FITC and CD38 PE of the four $CD8^+$ populations obtained are illustrated. Between 1 and 2×10^6 cells of each of the four $CD8^+$ populations were generally obtained within the 5-hr window that was allowed for the sorts. These fractions were tested for their ability to lyse autologous EBV-transformed lymphoblasts infected with HIV protein expressing vaccinia viruses in a chromium-51 release assay. $HLA\text{-}DR^+CD38^+$ lymphocytes were found to have the highest CTL activity (Ho *et al.*, 1993).

VIII. Alternate Technologies for Cell Separation

From a biohazard standpoint, it would be advantageous to use alternate methods for separating viable cells and to sort fixed cells whenever possible. For example, polymerase chain reaction analysis to determine DNA and RNA content, as well as most other molecular biologic work, can be done on cells that are fixed first and then sorted. Non-flow cytometric methods of separation of major cell subsets including B cells, T cells, $CD4^+$ cells, $CD8^+$ cells, and NK cells are available. These methods include flasks coated with mAb (Applied Immune Sciences, Menlo Park, CA) and two mAb-based magnetic bead separation approaches, Dynabeads (Dynal, Oslo, Norway) and MACs separation,

(Miltenyi Biotech, GmbH, Bergisch Gladbach, Germany). Cells can generally be positively or negatively selected.

These other methods are recommended rather than flow sorting if the cells of interest can be selected with sufficient purity. More cells can be obtained in a shorter time and the potential hazard of sorting is avoided. As many as 10^9 cells can quickly be separated at a reasonable cost. Such large cell numbers would be impractical using flow current sorting technology. However, these alternate methods cannot generally select subpopulations defined on the basis of multiple Ag the way sorting can. In addition, the purity of sorted cells can reach 99.9% whereas the alternate methods seldom exceed 97%. An example of a situation in which three-color cell sorting was required because alternate methods did not work is described above (Ho *et al.*, 1993).

Acknowledgment

Supported by the UCLA AIDS Institute through funding from CFAR AI-28697 and the Jonsson Comprehensive Cancer Center through funding from CA-16042.

References

Adams M. H. (1950). *In* "Bacteriophages" (J. H. Comroe, ed.), pp. 443–522. Yearbook Publ., Chicago.

Centers for Disease Control (1987). *Morbid. Mortal. Wkly. Rep.* **36,** Suppl. 2S, 3S–18S.

Centers for Disease Control (1988a). *Morbid. Mortal. Wkly. Rep.* **37**(S-4), 1–17.

Centers for Disease Control (1988b). *Morbid. Mortal. Wkly. Rep.* **37**(S-4), 19–22.

Centers for Disease Control (1988c). *Morbid. Mortal. Wkly. Rep.* **37**(24), 378–388.

Centers for Disease Control (1992). "NIOSH Recommended Guidelines, September," pp. 1–55. CDC, Atlanta, GA.

Dimmick, R. L. (1969). *In* "An Introduction to Experimental Aerobiology" (R. L. Dimmick, A. B. Akers, R. J. Heckly, and H. Wolochow, eds.). Wiley (Interscience), New York.

Gherna, R., Pienta, P., and Cote, R. (1989). "American Type Culture Collection: Catalogue of Bacteria and Phages." American Type Culture Collection, Rockville, MD.

Hambleton, P., and Dedonato, G. (1992). *BioTechniques* **13,** 450–453.

Ho, H.-N., Hultin, L. E., Mitsuyasu, R. T., Matud, J. L., Hausner, M. A., Bockstoce, D., Chou, C.-C., O'Rourke, S., Taylor, J. M. G., and Giorgi, J. V. (1993). *J. Immunol.* **150,** 3070–3079.

Merrill, J. T. (1981). *Cytometry* **1,** 342–345.

Miller, J. H. (1992). "A Short Course in Bacterial Genetics: A Laboratory Manual and Handbook for *Escherichia coli* and Related Bacteria." Cold Spring Harbor Lab., Cold Spring Harbor, NY.

Parks, D. R., Lanier, L. L., and Herzenberg, L. A. (1986). *In* "Handbook of Experimental Immunology" (D. M. Weir, L. A. Herzenberg, C. Blackwell, and L. Herzenberg, eds.), pp. 29.1–29.21. Blackwell, Oxford.

Sattar, S. A., and Springthorpe, V. S. (1991). *Rev. Infect. Dis.* **13,** 430–447.

Schoenbaum, M. A., Zimmerman, J. J., Beran, G. W., and Murphy, D. P. (1990). *Am. J. Vet. Res.* **51,** 331–333.

CHAPTER 22

High-Speed Photodamage Cell Sorting: An Evaluation of the ZAPPER Prototype

Jan F. Keij,[*] Ad C. Groenewegen,[†] and Jan W. M. Visser[*]

[*] Laboratory of Stem Cells
New York Blood Center
New York, New York 10021

[†] Department of Molecular Pathology
TNO Medical Biological Laboratory
2280 HV Rijswijk
The Netherlands

I. Introduction

Photodamage cell sorters were first suggested as a high-speed alternative to droplet sorters more than a decade ago (Martin and Jett, 1981; Shapiro, 1983). After cells are detected in a first laser, undesired cells are "sorted out" by exposing them to a lethal pulse of laser light from a second laser. Unlike droplet sorters, photodamage sorters are not limited in sort rate by the droplet frequency. Photodamage sorters can accomplish high rates for the sorting mechanism, i.e., the modulation of the lethal laser beam can be achieved in 100 nsec. Furthermore, nozzle tips with large orifice diameters can be used to sort very large cells, such as filamentous algae.

Generally the photodamaging effect of focused laser light (350–633 nm) in flow cytometers is minimal at the normally applied power densities. However, cells can be made extremely sensitive to light after uptake of photosensitizers. Based on their mode of action photosensitizers are divided into two classes. Excited Type I photosensitizer molecules react with acceptor molecules in their environment directly or via a free radical mechanism. For example, excited 5-bromo-2′-deoxyuridine (BrdUrd) can liberate a highly reactive bromine radical. Type II photosensitizers, such as rhodamine 123, generate reactive oxygen species, such as singlet oxygen and the superoxide radical from the excited triplet state.

Photosensitized cell killing is used clinically in the treatment of skin diseases and certain cancers where exposure of the treated area to low-power laser light lasts for several minutes. For photodamage cell sorting, the choice of a photosensitizer is primarily determined by the excitation optimum, the ability to penetrate viable cells, and the ability to kill cells after only a brief exposure to a high-power laser flash. Herweijer *et al.* (1988) applied a photosensitizer combination of BrdUrd and Hoechst 33342 in a flow cytometer. The plating efficiency of photosensitized L1210 was reduced by four decades after exposure to a 5-μsec pulse of 300 mW of 351/364 nm (UVA) laser light. Although successful, this method has the drawback that it requires actively dividing cells and several days of culturing with BrdUrd to ensure that all cells are labeled.

Experience with type II photosensitizers has been less successful. Protoporphyrin-labeled L1210 cells (Herweijer, 1988) and 10-dodecylacridine orange (DAO)-labeled yeast cells were highly photosensitive to a 514-nm laser fluence of 10 kJ/m^2. In these experiments, this fluence was delivered over a period of several seconds. However, when exposed to an identical fluence in the flow cytometer both cell types survived completely. The mechanism behind this failure is currently being investigated with computer models of the photon-induced reactions in the irradiated cells. Preliminary data suggest that slowly decaying triplet state photosensitizer molecules accumulate during high fluence rate exposures, which leads to a reduced generation of the toxic singlet oxygen molecules during flash irradiations. Diffusion of oxygen into the cells was found to be the second parameter that limited photoinactivation in the flow cytometer.

This finding aborted our search for type II photosensitizers and led us to investigate the possibilities of exploiting the cell's intrinsic photosensitizer, DNA.

The literature revealed that most mammalian cells are effeciently killed after exposure to a 50 J/m^2 fluence of short wavelength ultraviolet light (UVC). Irradiation of DNA with UVC results in DNA damage. The prevalent DNA photoproducts are the pyrimidine dimers and the (6-4) pyrimidine-pyrimidone photoproducts, both of which are known to be the cause of cell death (Mitchell, 1988). Also, it could be assumed that UVC killing would work in a flow cytometer as UVC killing is fluence rate independent up to fluence rates of 10^{12} W/m^2 (Zavilgelsky *et al.,* 1984). Another advantage of UVC killing is that there are no restrictions on the probes used to identify (un)desired cells in the sort samples.

Most existing and commercially available dual-beam flow sorters can be converted to photodamage sorters with minor modifications. Recently, we have completed the design and construction of such an instrument (the ZAPPER), which was designed to operate at sort rates of 100,000 cells/sec. Some of the design-related studies are presented in this chapter.

II. Application

In our study of the hemopoietic system, flow cytometric purification of murine and human progenitor cells has played a pivotal role (Visser *et al.,* 1984). The desire for high-speed sorting of these and other rare cells has been the main drive behind the development of the ZAPPER. For example, the sorting of stem cells from a bone marrow graft containing 10^9 cells for transplantation purposes would take 111 hr at a droplet sort rate of 2500 cells/sec. The ZAPPER running at 50,000 cells/sec would process such a sample in 5.5 hr. Furthermore, bone marrow fractions preenriched for stem cells, using antibody (CD34)-coated beads or flasks, could be rapidly processed by the ZAPPER to a desired level of purity.

Other applications involving rare cells are the purging of leukemic cells from bone marrow grafts, the isolation of hybrid cells obtained through fusion procedures, and the isolation of hybridoma class switches and mutant cells. Photodamage sorting can also be employed when large numbers of sorted cells are required. Sorting of X or Y chromosome bearing sperm cells for insemination (Johnson *et al.,* 1989) is an interesting possibility.

III. Materials

A. Lasers and Modulators

For BrdUrd/Hoechst 3342 photodamage sorting, a 2025/5 argon ion laser [Spectra Physics (SP), Palo Alto, CA] tuned to 351/364 nm was used. The 300-mW laser beam was modulated with an acousto-optic modulator (AOM),

Model 70A (IntraAction Corp., Bellwood, IL). The AOM consists of a transparent fused quartz crystal connected to a piezoelectric transducer. A driver that can be externally triggered completes the system (Fig. 1). When triggered, the transducer produces sound waves centered around 70 MHz. The sound waves induce periodic changes of the crystal's refractive index. These changes lead to deflection of the incoming laser beam. The intensity ratio of the undeflected and deflected beams is referred to as the contrast ratio or extinction rate (ER). The first-order deflected beam was used to zap the undesired cells. By blocking all non-first-order beams from irradiating the cells, this system resulted in an ER larger than 1000 : 1.

The first-order beam is deflected at an angle of about 4.2 mrad. The zap beam could be completely deflected within 200 nsec, resulting in a zap pulse of 400 nsec. There was a 2-μsec propagation delay between the triggering of the AOM and the initiation of the deflection. The transmission of the AOM was 93% and the first-order beam contained 78% of the incoming power.

In an alternative arrangement, the same 2025/5 laser was used to pump a 395B frequency doubler unit (SP), which resembles a dye laser. However, the critical component is not a dye ribbon but a potassium dihydrogen phosphate (KDP) frequency doubling crystal. In this double-pass unit the crystal is cooled by a Peltier element. Excess heat is removed from the 395B by a circulating waterbath set at 15°C. An output–feedback system assures a stable 257-nm output over time. A maximum output of 100 mW 257-nm light could be achieved, but an output of 50 mW was typical.

An electro-optic modulator (EOM), Model LM13P (Gsänger, Planegg, FRG), fitted into an adjustable mirror mount, was used to modulate the 257-nm laser output power. The EOM consists of a cylinder-shaped potassium dideuterium phosphate (KD*P) birefringent crystal with electrodes fixed perpendicular to the optical propagation direction. Attached to the exit side of the EOM is a calcite polarizing prism, which splits the laser beam into two mutually orthogonal, linearly polarized beams. The extraordinary beam (e-ray) exits the prism at an angle, while the ordinary beam (o-ray) goes straight through. A driver that can be externally triggered is used to supply a voltage over the electrodes, thereby inducing a 90° rotation of the laser beam polarization inside the EOM (Fig. 1). As a result of the rotation, the intensities of the o-ray and e-ray are reversed. Using the o-ray, rise times of 50 nsec were observed, allowing zap pulses of 100 nsec. A propagation delay of 300 nsec was observed between triggering the EOM driver and the initiation of the deflection. The EOM transmission for 257-nm laser light was 50%, and a maximum ER of 150 : 1 was achieved.

B. Optics

In the experiments performed by Herweijer *et al.* (1988), the beams of the 351/364-nm zap laser and 488-nm detection laser were focused to round spots with a diameter of 54 μm using planoconvex glass lenses (f = 12.5 cm).

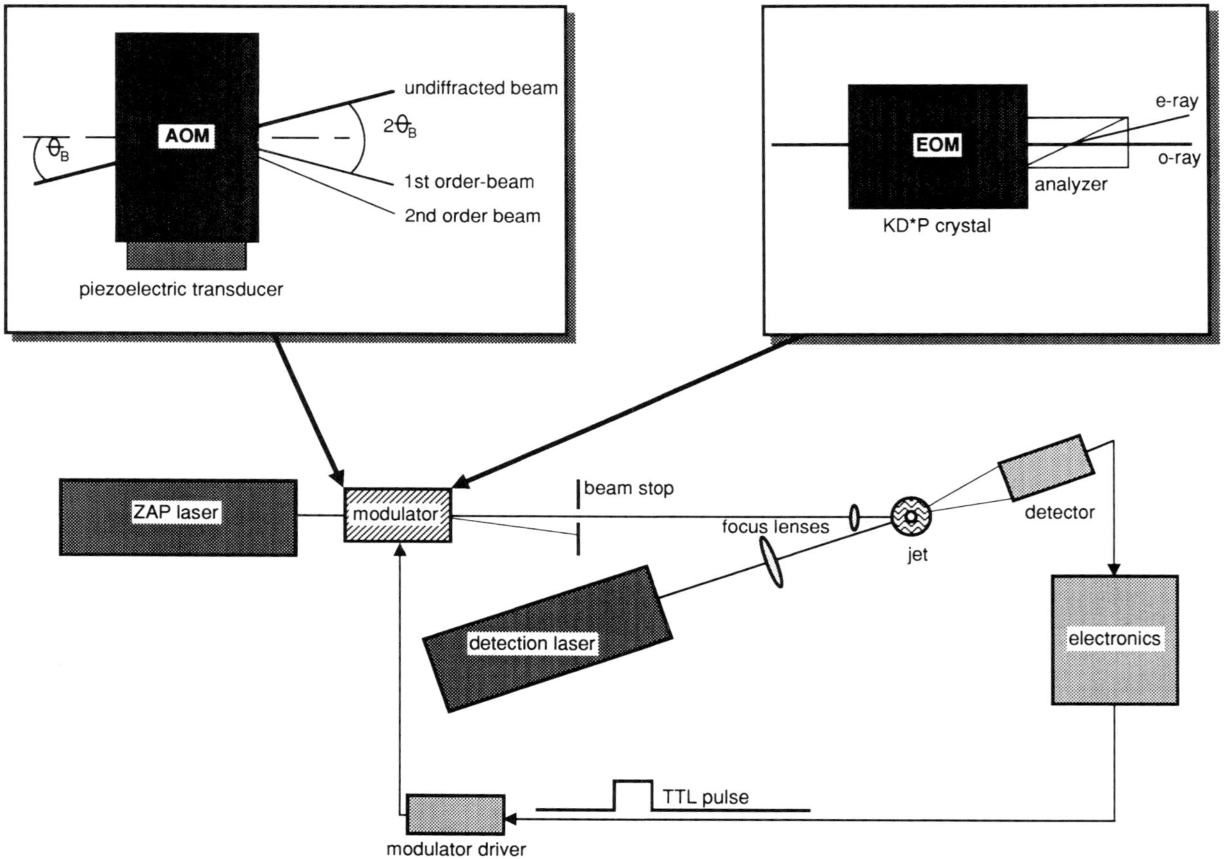

Fig. 1 A schematic (top view) representation of a photodamage cell sorter, highlighting the critical components. (Upper left) An acousto-optic modulator, illustrating propagation directions of the incoming beam and the diffracted beams, where θ_B represents the Bragg angle. (Upper right) An electro-optic modulator, showing the propagation directions of the unmodulated o(rdinary)-ray and the deflected e(xtraordinary)-ray. (Bottom panel) Overview of the ZAPPER design. After cells are detected in the detection laser beam, the electronics process the data and determine whether the cell should be killed. If so, a TTL pulse is sent to the modulator driver and the laser is deflected onto the jet as the cells pass.

In the ZAPPER the 488-nm detection laser beam was focused to a round spot with a diameter of 20 μm using a glass lens (f = 4 cm). The 257-nm beam was focused to a round spot using a synthetic fused silica lens (f = 2.5 cm), resulting in a spot diameter of 45 μm.

Side scatter and fluorescence were collected through a synthetic fused silica lens (f = 1.5 cm). All optics were from Melles Griot (Irvine, CA). Emissions were collected onto photomultipliers (Hamamatsu, Middlesex, NJ); Model R1477 was used for visible emissions and Model R166UH for 257-nm side scatter.

C. Reagents and Cells

Sheath fluid was infusion quality 0.9% NaCl (NPBI, Emmer-Compascuum, the Netherlands). Alignment of the 257-nm laser was done with 4.8-μm beads (Polysciences, Warrington, PA) which have excitation and emission optima at 273 and 340 nm, respectively.

Cells tested for their ability to survive passage through the nozzle tip at increased sheath fluid velocities were: Ng 227 (baker's yeast), lymphocytes (human), GM-CFU (murine granulocyte-macrophage colony forming units), 3T3 (murine fibroblasts), LT 12 (rat promyelocytic leukemia), 15.1.4 (murine hybridoma), HL-60 (human promyelocytic leukemia), Jurkat (human T cells), and WEHI (murine promyelocytic leukemia). Cells suspended in lymphocyte separation medium (LSM, Organon Teknika, Durham, NC) were delivered to the nozzle using a mechanically driven syringe after all air bubbles were removed from the sample inside the syringe. Cell survival was determined either by plating the cells in the appropriate media or by microscopic live/dead evaluation after the samples were resuspended in 0.1% eosin.

Cells tested for their ability to survive passage through the 257-nm laser beam were *Escherichia coli* SURE (Stratagene, La Jolla, CA), *E. coli* WT (isolated wild type from monkey intestines), Ng 227, AB 1380 (auxotrophic yeast), GM-CFU, LT 12, CTLL-2 (murine IL-2-dependent T-cells), and 15.1.4. Cell survival was determined by plating the cells in the appropriate media.

For the photodamage sorting, a sample of *E. coli* SURE was transformed with the pBluescriptII SK plasmid (Stratagene), providing α complementation for the blue/white color selection and ampicillin resistance. It was designated BS. Another batch of *E. coli* SURE cells was transformed with the pGEM 1 plasmid (Promega, Madison, MI), which encodes ampicillin resistance. It was designated pGEM 1.

Prior to photodamage cell sorting, both BS and pGEM 1 cells were separately grown overnight in 10 ml liquid medium. One strain was stained by supplementing the medium with 1.5 μg DAO. Harvested cells were washed three times and aliquots were mixed before sorting.

Sort windows were defined using DAO fluorescence and time of flight. Cells in the sort window were to be spared when zapping in the KILL mode; in the LIVE mode cells in the sort window were to be zapped. At a rate of 25,000 cells/sec, the percentage of detected doublets and clumps was below 3%. Equal numbers of cells of controls and sorted samples were plated on selection medium. After incubation at 37°C for 24 hr, blue colonies were scored as BS cells and white colonies as pGEM 1 cells. The number of surviving cells in each sample was calculated from the observed number of colonies. The plating efficiency of the control samples (passed through the ZAPPER but not UVC irradiated) was only 30–60%, due to the ampicillin in the selection medium.

One liter of *E. coli* SURE selection medium (Luria broth) contained 10 g Bacto tryptone (Difco, Detroit, MI), 5 g yeast extract (Difco), 5 g NaCl (Sigma), and 10 g Bacto agar (Difco). After autoclaving this medium, a sterile-filtered solution containing 80 mg 5-bromo-4-chloro-3-indolyl-β-D-galactopyranoside [X-gal, Molecular Probes (MP), Eugene, OR], 120 mg isopropyl-*b*-D-thiogalactopyranoside (IPTG, Sigma, St. Louis, MO), and 50 mg ampicillin (Sigma) was added. DAO was obtained from MP.

IV. Critical Aspects of the Procedure

A. High-Speed Processing of Viable Single Cells

Sort rates in a photodamage cell sorter are limited by coincident arrival of cells during the laser beam transit time, the electronics processing time, and the sort execution time. When comparing a droplet sorter to a photodamage sorter it is obvious that the execution of the sort decision is the rate limiting parameter. For example, it takes 25 μsec to execute the sort decision in a droplet sorter deflecting 1 droplet per sorted cell at the droplet frequency 40,000/sec. A photodamage sorter could execute a sort decision in 100 nsec. Taking a repeat delay of 100 nsec into account, a theoretical sort rate of 5,000,000/sec could be achieved.

However, this would require that the detection and the processing of the cells proceed at a comparable rate. At high rates coincident arrival of cells in the laser beam becomes a critical issue (Lindmo and Fundingsrud, 1981). Coincident arrival of cells in the laser beam was significantly reduced by reducing the laser beam transit time: the focused beam height was reduced from 50 to 20 μm and the velocity with which the cells travel through the beam was increased from 10 to 30 m/sec. A simple equation was found to describe the fraction of detected events which represented coincident arrivals for cell rates up to 50,000/sec,

$$F_{\text{coincidence}} = 0.5\ PR,$$

where P represents the measured average pulse width of a single cell passing through the laser beam and R represents the real cell rate. Employing the 20-μm laser spot, a sheath fluid velocity of 33 m/sec, and electronics with a dead time of 4 μsec, practical processing rates of 100,000 cells/sec were achieved (Keij *et al.*, 1991).

However, at a sheath velocity of 50 m/sec, it had previously been shown that cells did not survive passage through the orifice jewel (Peters *et al.*, 1985). As cell survival is of critical importance in sorting, several cell types were tested for their ability to survive increased sheath velocities (Table I). Sheath fluid was delivered to the nozzle tip by either gas pressure or mechanical pressure. Rapidly expanding gas in the sheath fluid had no effect on the cell viability. This could be concluded from experiments where the sheath fluid was delivered to the nozzle tip mechanically. From these data it could be concluded that cell death was caused by the shear forces endured during the acceleration in the nozzle tip.

B. Cell Preparation and Labeling Considerations

Sorting performance is greatly influenced by the quality of the sorted sample. Undetected doublets and clumps may lead to incorrect sort decisions and reduce the sort rate and sort purity. It is of critical importance to start with a suspension of single cells. For example, we noted that *E. coli* colonies taken from agar plates could not be completely resuspended in saline, while diluted liquid cultures resulted in almost complete single-cell suspensions.

The staining protocol should consist of as few steps as possible to retain good viability and to avoid massive clumping due to extracellular DNA. Thus,

Table I
Effect of Average Sheath Fluid Velocity on Cell Viability[a]

	Viability (%)						
Cell type $\bar{v}_s$ (m/sec)[b]	14	20	24	28	31	34	36[c]
Ng 227	100	—	—	—	—	97	
lymphocytes	100	100	100	100	100	100	
GM-CFU	109	97	85	92	96	86	
3T3	101	98	92	82	85	85	
LT12	106	98	92	90	87	60	
15.1.4	64	64	55	47	45	42	
HL60	99	86	69	55	39	43	50
Jurkat	81	75	72	67	32	17	15
WEHI	97	73	33	23	34	12	26

[a] Cells were processed through a 100-μm nozzle tip. Viabilities are expressed as percentages of the respective controls, which were samples of the same cell type kept on ice during the experiment.

[b] The average sheath fluid velocity ($\bar{v}_s$) was increased by increasing the gas pressure on sheath fluid.

[c] A pressure of 44 kg/5.7 cm^2 was applied to a 50-ml syringe plunger, resulting in a sheath fluid pressure of 7.7 atm. Friction of the plunger resulted in an average sheath fluid velocity of 36 m/sec instead of the expected 39 m/sec.

when antibodies are used to distinguish desired and undesired cells, directly tagged monoclonals are preferred. After staining, the sample quality can be retained by keeping it on ice, by adding DNase, and by gently stirring it during sorting. We routinely resuspend the samples in LSM to increase the settling time.

Another important issue is how cells should be stained. Depending on whether the yield or the purity of the desired cell is of paramount importance, the staining procedure should be adjusted to meet that requirement. If the sort yield is critical, emphasis should be put on a procedure that stains all desired cells and coincidence detection should be aborted. If purity is critical, the emphasis should be on excluding the possibility that desired and undesired cells appear in the sort window together. Ideally, several probes should be used simultaneously that stain desired and undesired cells in a mutually exclusive manner. Any double-stained doublets that are not detected by coincidence detectors will not meet the sort requirements and thus will not reduce the purity of the sorted cells.

For photodamage sorting of BrdUrd/Hoechst-labeled cells there are additional considerations. First, all cells should have incorporated BrdUrd. This may require several days of culturing the cells in the presence of BrdUrd at a concentration of 0.5 μg BrdUrd/ml. Second, after being stained with Hoechst 33342, the cells become extremely photosensitive. All sample handling should take place in a dimly lit room, for exposure of the cells to ambient light from fluorescent tubes will kill them.

C. Sterility

Great care must be taken to retain sterility in the sorting environment as all the sheath fluid passing through the cytometer during the sort will be collected. A typical 1-hr sort, using a 100-μm nozzle and a sheath velocity of 10 m/sec, results in a collected volume of 300 ml. A single included microbe could infect the sorted population. To reduce the probability of contamination, the ZAPPER optics, fluidics, and sheath fluid container are enclosed in a laminar flow cabinet. For critical applications, antibiotics could be added to the sheath fluid. Immediately after use, the fluidics are rinsed with diluted bleach, followed by extensive rinsing with 70% ethanol.

D. Alignment of the Zap Laser

Of the two types of modulators, the AOM is far easier to align. First, the AOM which is attached onto a simple rotating mount is adjusted so that the laser beam enters the AOM at the known Bragg angle. Then, the sound wavelength is adjusted to obtain an optimal first-order beam.

To align the EOM, it is important to first obtain a stable output from the frequency doubler, as tuning the 257-nm output influences the output angle significantly. The EOM, mounted in an adjustable mount is then positioned into the beam and transmission of the beam is optimized with the *X, Y,* and *Z* micrometers. Now, by rotating the EOM, the ER of the o-ray is optimized. The ER is further optimized by adjusting the tilt angle in the *X, Y,* and *Z* planes. Finally, the EOM is connected to the driver and the ER is fine tuned using the driver voltages LO and HI. After aligning the EOM, the driver is set to SIGNAL which allows external switching from HI to LO.

Before sorting, the zap laser needs to be carefully aligned onto the passing cells. This requires that scatter or fluorescence signals from exposed particles be monitored simultaneously with the delayed trigger pulse for the deflection of the zap laser. After aligning the zap laser onto the particles, the (delayed) zap trigger is adjusted in time to enfold the zap laser signal. In the ZAPPER we use a four-channel digital oscilloscope Model 54503A (Hewlett-Packard, Palo Alto, CA) to monitor the alignment of the zap laser.

To adjust the zap delay trigger pulse for the EOM or AOM propagation delay, a sort window is set around all passing events and the ZAPPER is set to zapping. The zap delay is moved forward in time until no signals (KILL mode) or maximum signals (LIVE mode) from the zap laser are visible on the oscilloscope.

E. Postsort Treatment

During and immediately after the sort, DNA repair in the undesired cells must be avoided or inhibited to obtain optimal purity of the desired cells. Exposure of the irradiated cells to bright light should be avoided, for this could trigger a repair mechanism known as photoreactivation. Photoreactivation has been demonstrated in Halobacterium (Iwasa *et al.,* 1988), yeast (Winckler *et al.,* 1988; Boyd and Harris, 1987), goldfish (Mano *et al.,* 1982), frog (Rosenstein and Setlow, 1980), and human leukocytes (Sutherland, 1974). The ability to photorepair DNA damage was shown to be enhanced in goldfish cells if they were preirradiated under bright white light (Yasuhira *et al.,* 1991). On the other hand, Tyrrell *et al.* (1984) found increased sensitivity of human lymphoblasts to 254-nm irradiation after preirradiation at 365 nm. To avoid the possibility of triggering the photoreactivation mechanism, all cell staining and processing is performed under dim yellow light.

Conditions relating to the growth of the irradiated cells can also influence recovery from UVC irradiation. Maintaining cells under nongrowth conditions (liquid holding) has been shown to enhance the recovery of irradiated yeast cells (Schall and Kiefer, 1988). Furthermore, growing *E. coli* cells on minimal media was also found to enhance recovery (Wang and Smith, 1984). Addition of DNA repair inhibitors, such as caffeine (Calkins *et al.,* 1988), could also be considered. However, these methods require careful evaluation for each individual cell type with respect to the plating efficiency of the desired cells and the increased killing of the undesired/zapped cells.

V. Instrumentation

A. Pulse Processing Electronics

The ZAPPER's parallel processing design has been described by Van den Engh and Stokdijk (1989). Their original design was modified for high-speed sample processing. All the analog electronic components were analyzed and adjusted if necessary. In many cases, active "pulse shaping" in the analog electronics was sacrificed for speed, resulting in a system that could process signals with pulse widths down to 1 μsec. A reduction in the digital electronics dead time was accomplished by replacing the 4.5-μsec analog-to-digital converters with 3-μsec versions. Combined, these modifications resulted in the maximum processing rate obtainable from the 16-parameter design.

B. Special ZAPPER Features

1. Digital Delay Line Timer

Photodamage sorting requires the setting of a delay time between detection and sorting. In the ZAPPER an end pulse is generated 24 μsec after the system is triggered. This end pulse triggers the bus controller to coordinate the transfer of the data onto the data bus. The sort decision is then formed, based on these data. This sort decision is stored into a first in–first out temporary memory (FIFO) while the cell travels to the zap laser position. An adjustable zap delay between 20 and 99 μsec can be set. This allows for ample alignment flexibility of the zap laser, in the ZAP mode.

Besides sorting, the ZAPPER is used for the analysis of rare events. In the ANALYSIS mode, the end pulse is generated 60 μsec after the system is triggered. This mode allows for high-speed analyses using two measuring lasers and for alignment of the zap laser optics.

2. Killer Beam Mode Selector

This module allows the operator to choose between several sort options. One switch controls the deflection mode of the zap laser. In the ALIGN mode, the position of the zap laser beam is fixed onto the jet to allow beam alignment with the beam steering optics. In the KILL mode, the zap laser beam is aimed at the jet and all passing cells are exposed to the photodamaging light. If a detected cell meets the sort criteria, a TTL pulse is sent to the modulator driver and the high-intensity beam is deflected from the jet allowing the desired cell to pass undamaged. This mode is intended for sorts where the killing of the undesired cells is of absolute importance, as is the case when purging leukemic cells from a bone marrow graft.

In the LIVE mode, the situation is reversed; the zap laser is only deflected onto the jet when an undesired cell passes. This mode is aimed at applications where the yield of the desired cells is critical, such as, the isolation of class switch variants.

Another switch controls the duration of the sort pulse. In the TOF mode, the zap pulse duration is equal to the time of flight of the sorted event. This mode allows for the exclusive sorting of the processed event, ignoring any additional cells in the electronics dead time. In the FIXED mode, the zap pulse duration can be set manually. This allows the operator to set a wider time window to assure complete sorting of the passing event(s).

VI. Results and Discussion

Previously, Herweijer *et al.* (1988) had demonstrated the feasibility of photodamage cell selection using UVA irradiation of BrdUrd/Hoechst-labeled cells. However, the disadvantages of this method led us to investigate the possibilities of UVC photodamage cell selection.

Prior to performing a variety of sort experiments with the ZAPPER, the feasibility of the new concept was verified. For this purpose several cell types were processed through the focused 257-nm beam at a velocity of 10 m/sec. The data, as illustrated in Fig. 2, confirmed that exposure to a 257-nm laser beam focused to a 45-μm round spot killed most cell types efficiently at relatively low laser power densities. For most of the cell types that were tested a 5 log cell kill was reached at 40 mW; the two yeast strains required 100 mW. From these data it could also be concluded that most cell types completely survived exposure to the low doses of 257-nm laser light which "leak" through the EOM in the LO setting.

The high-speed capabilities of the ZAPPER can be fully exploited for small and sturdy cells which survive passage through the nozzle at increased velocities (Table I). However, the received fluence of UVC is proportional to the time the cell spends in the laser. Thus, for optimal performance a compromise must be found between the processing speed and the killing efficiency for each application.

By the time the ZAPPER electronics were completed, the frequency doubling crystal output was reduced to 2 mW. In the presented preliminary tests of the ZAPPER performance, the extreme UV sensitivity of the DNA repair-deficient *E. coli* SURE strain was used. Samples containing mixtures of the BS and pGEM 1 transformants were sorted at a rate of 25,000 cells/sec. The results of a representative experiment are summarized in Table II. All cells survived passage through the sorter when the 257-nm laser beam was blocked or set to LO. However, only a small fraction of cells survived passage through the 257-nm laser beam set to HI. In the LIVE mode sorts, where the 257-nm laser beam was only deflected onto the jet when an undesired cell passes, the yields of the desired cells and the purity were good. In the KILL mode sorts, where the photodamage laser was only deflected from the jet to allow a desired

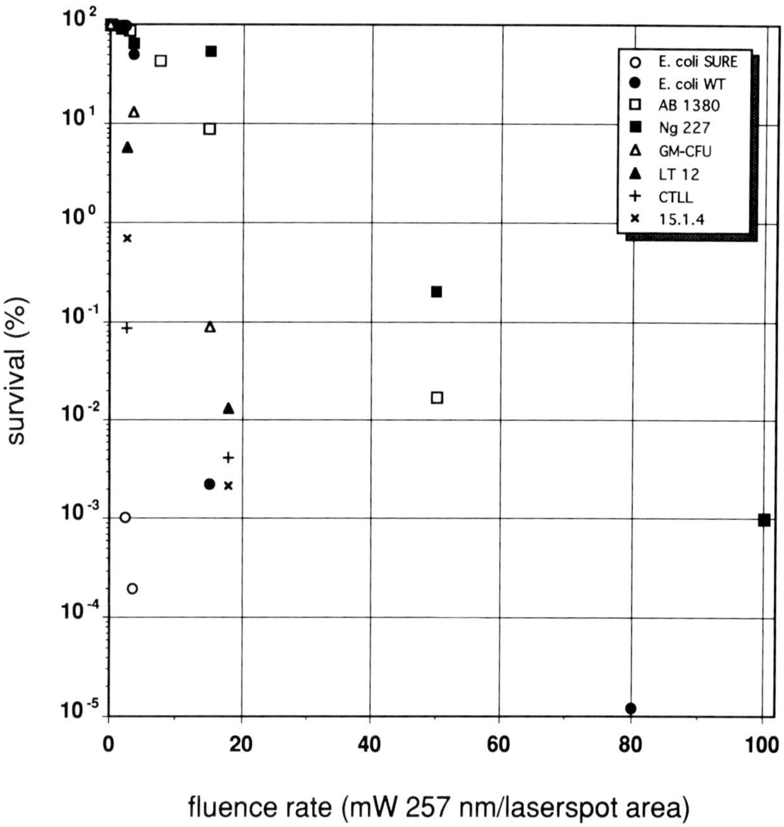

Fig. 2 Cell survival as a function of the zap laser power. Cells were processed through the zap laser that was focused to a round 45-μm spot, at a velocity of 10 m/sec. The DNA-repair deficient *E. coli* SURE strain was the most UV-sensitive, while the commercial baker's yeast strain Ng 227 was the least sensitive; the DNA-repair-competent mammalian cells revealed intermediate sensitivity.

cell to pass, the purity of the desired cells was good but the yield was only around 40%. This loss of desired cells was traced to the EOM.

It was found that during a sort in the KILL mode, the EOM did not completely return to the LO setting during the 6.5-μsec deflection. Increasing the deflection duration to 12 μsec almost completely restored the performance of the EOM. However, for high-speed cell sorting, this was not a satisfactory solution for it would greatly reduce the maximum sort rates.

In a KD*P-based EOM, the electro-optic effect and an elongation of the KD*P crystal occur when a voltage is applied. The combined effect of these phenomena determine the extinction ratio. At high switching frequencies, the

Table II
Effect of Sort Mode on Purity and Yield of Desired Cells[a]

	LO	HI	KILL_1	KILL_2	LIVE_1	LIVE_2
			Surviving cells[b]			
BS	22,997	4	129	*12,267	*22,613	2,816
pGEM 1	22,827	1	*7,339	233	1,565	*22,144
			Yield (%)[c]			
BS	100	0.017	0.6	*53.3	*94.0	12.2
pGEM 1	100	0.004	*32.2	1.0	6.8	*97.0
			Purity (%)[d]			
BS	50.18	—	1.7	*98.1	*93.5	11.3
pGEM 1	49.82	—	*98.3	1.9	6.5	*88.7

[a] Cells were processed through a 50-μm nozzle tip at 10 m/sec. The LO and HI samples were exposed to approximately 0.01 and 1.5 mW of 257-nm laser light, respectively. pGEM 1 cells were set in the sort window for the KILL_1 and LIVE_1 sorts, whereas for the KILL_2 and LIVE_2 sorts, the BS cells were set in the window. In the KILL and LIVE columns numbers preceded by an * indicate the desired cells.

[b] Cell survival for each sample was calculated for the observed number of colonies. Identical cell numbers were plated for all samples.

[c] Yields were based on the surviving cell counts in the LO samples. In the control samples, 24,192 (BS) and 18,688 (pGEM 1) cells survived passage through the blocked 257-nm laser beam, respectively.

[d] Purities were calculated from the combined numbers of surviving cells, obtained in the individual sorts.

piezoelectric effect cannot keep up with the electro-optic effect. The reduced performance of our EOM was probably caused by this piezoelectric effect.

It should be noted that in DNA repair competent cells, the reduced performance of the EOM is not of critical importance. The cellular DNA repair mechanisms offer sufficient protection against low-level UVC irradiation. However, errors made during the repair of DNA damage are considered the cause of mutations (Hutchinson, 1987). Thus, in applications such as stem cell purification for bone marrow transplantation purposes, optimal performance of the EOM is required. Replacing the EOM with an AOM and accepting the reduced transmission of the AOM for UVC may be preferable for such applications.

Of greater concern to the applicability of photodamage cell sorting were the frequency doubler characteristics. The doubling crystals rated at 100 mW, have a warranted lifetime of 3 months. When used daily at 100 mW, they did not meet this specification. Also, due to the beam characteristics, we were not able to focus the 257-nm beam to the desired small spot. Thus, in the current version of the ZAPPER the cell transit time through the 257-nm spot exceeds the electronics dead time, which, with high speed in mind, is an undesirable situation. Furthermore, the output stability over time at high power is also limited and, generally, it does not exceed an hour. In a future version of the ZAPPER, an alternative zap laser should be selected.

Recently an alternative was announced by Coherent: an intracavity frequency-doubled Innova 300 argon ion laser. The engineering model emitted 500 mW at 257 nm or 350 mW at 244 nm. Such a laser would solve all the problems mentioned above. Another option would be a large 20-W BeamLok 2085 argon ion laser (SP) which emits up to 350 mW at 275 nm. Emissions of 257 and 275 nm induce comparable levels of DNA damage (Matsunaga *et al.*, 1991). Although more expensive, application of such a laser would result in better transmission of the emitted beam through the modulator optics, and the EOM could be replaced with a more convenient AOM.

Despite the problems relating to the modulation of the zap laser, several advantages of the photodamage method should be mentioned. The day-to-day setup prior to sorting required less than 15 min. Large nozzle tip diameters can be used which enable the sorting of large cells, while also reducing the occurrence of plugs without reducing the sort rate; this in contrast to droplet sorters. Finally, the performance of the photodamage sorter was extremely stable. Factors that commonly reduce the long-term sort reliability of a droplet sorter, such as minor air bubbles in the nozzle which may influence the breakoff point and fanning of the deflected stream, have no effect on the performance of the ZAPPER.

In conclusion, we have shown that photodamage sorting in a flow cytometer is an alternative to droplet sorting and we have now constructed the first high-speed photodamage cell sorter. We have demonstrated that the concept works with low-power photodamage lasers. As indicated, further improvements are considered. A coincidence detector has recently been implemented to enhance sort purity. Currently, we are exploring the performance limits of the prototype using DNA repair competent cells. In the immediate future we intend to apply the instrument to the isolation of hybridoma class switches.

References

Boyd, J. B., and Harris, P. V. (1987). *Genetics* **116,** 233–239.

Calkins, J., Wheeler, J. S., Keller, C. I., Coley, E., and Hazle, J. D. (1988). *Radiat. Res.* **114,** 307–318.

Herweijer, H. (1988). Ph.D. Thesis, University of Leiden, Leiden, The Netherlands.

Herweijer, H., Stokdijk, W., and Visser, J. W. M. (1988). *Cytometry* **9,** 143–149.

Hutchinson, F. (1987). *Photochem. Photobiol.* **45,** 897–903.

Iwasa, T., Tokutomi, S., and Tokunaga, F. (1988). *Photochem. Photobiol.* **47,** 267–270.

Johnson, L. A., Flook, J. P., and Hawk, H. W. (1989). *Biol. Reprod.* **41,** 199–203.

Keij, J. F., Van Rotterdam, A., Groenewegan, A. C., Stokdijk, W., and Visser, J. W. M. (1991). *Cytometry* **12,** 398–404.

Lindmo, T., and Fundingsrud, K. (1981). *Cytometry* **2,** 151–154.

Mano, Y., Kator, K., and Egami, N. (1982). *Radiat. Res.* **90,** 501–508.

Martin, J. C., and Jett, J. H. (1981). *Cytometry* **2,** 114.

Matsunaga, T., Hicda, K., and Nikaido, O. (1991). *Photochem. Photobiol.* **54,** 403–410.

Mitchell, D. L. (1988). *Photochem. Photobiol.* **48,** 51–57.

Peters, D., Branscomb, E., Dean, P., Merill, T., Pinkel, D., Van Dilla, M., and Gray, J. W. (1985). *Cytometry* **6,** 290–301.

Rosenstein, B. S., and Setlow, R. B. (1980). *Photochem. Photobiol.* **32,** 361–366.
Schall, M., and Keifer, J. (1988). *Photochem. Photobiol.* **47,** 155–157.
Shapiro, H. M. (1983). U.S. Pat. 4,395,397.
Sutherland, B. M. (1974). *Nature* (*London*) **248,** 109–113.
Tyrrell, R. M., Werfelli, P., and Moreas, E. C. (1984). *Photochem. Photobiol.* **39,** 183–189.
Van den Engh, G., and Stokdijk, W. (1989). *Cytometry* **10,** 282–293.
Visser, J. W. M., Bauman, J. G. J., Mulder, A. H., Eliason, J. F., and De Leeuw, A. M. (1984). *J. Exp. Med.* **59,** 1576–1590.
Wang, T. V., and Smith, K. C. (1984). *Photochem. Photobiol.* **39,** 793–797.
Winckler, K., Golz, B., Laskowski, W., and Bende, T. (1988). *Photochem. Photobiol.* **47,** 225–230.
Yasuhira, S., Mitani, H., and Shima, A. (1991). *Photochem. Photobiol.* **53,** 211–215.
Zavilgelsky, G. B., Gurzadyan, G. G., and Nikogosyan, D. N. (1984). *Photochem. Photobiophys.* **8,** 175–187.

CHAPTER 23

High-Gradient Magnetic Cell Sorting

Andreas Radbruch,* Birgit Mechtold,* Andreas Thiel,*
Stefan Miltenyi,† and Eckhard Pflüger†

* Institut für Genetik
Universität zu Köln
50931 Köln
Germany

† Miltenyi Biotec GmbH
51429 Bergisch Gladbach
Germany

METHODS IN CELL BIOLOGY, VOL. 42

I. Introduction

Magnetic microparticles instead of fluorochromes can be coupled to specific ligands or antibodies against cell-surface structures. Such magnetic staining reagents can be used to isolate the labeled cells directly from complex cell mixtures. The absence of a natural magnetic background in most cell types and the ease by which extremely high magnetic forces can be produced using high-gradient magnetic technology permit the use of very small colloidal superparamagnetic particles (diameter < 100 nm) as magnetic labels. These tags do not interfere with optical analysis. Magnetically labeled cells can be stained simultaneously with fluorochromated antibodies to control and evaluate the quality of magnetic separation or to further analyze and sort the cells by flow cytometry or microscopy. Here, we describe magnetic cell separation based on staining with colloidal superparamagnetic particles and separation of stained cells in high-gradient magnetic fields (Molday and Molday, 1984; Miltenyi *et al.*, 1990; MACS; Fig. 1).

Although magnetism provides only one parameter for separation, magnetic cell sorting is superior to fluorescence-activated cell sorting (FACS) in several aspects and, in others, a valuable complementation of it. In FACS, the most powerful multiparameter sorting technology, cells are analyzed one by one and sorted one after the other. As a serial sorting device, its capacity is limited by the frequency of analysis and sorting, which is about 5000 cells per second, i.e., 10^8 cells in 6 hr. The capacity of MACS is only limited by the capacity of the magnetic matrix for labeled cells, which can be up to 10^9 cells. All cells are "analyzed" and sorted simultaneously allowing up to 10^{11} cells to be processed in about 30 min and giving it a leading edge in sorting of rare cells. Not only the fast processing but also the low physical forces of MACS sorting favors recovery of viable cells, compared to FACS, where the cells are submitted to considerable stress by acceleration in the nozzle. The sorting of rare cells, a combination of enrichment of the rare cells by gentle, fast, and parallel MACS followed by further enrichment by FACS, can yield enrichment rates of up to 10^8-fold in one experiment. Not many biological problems would require higher enrichment rates.

II. Application

Due to its origin in immunology, high-gradient magnetic cell sorting has been used primarily for the isolation of various cell types from human and animal hematopoietic systems (Hansel *et al.*, 1991; Schmitz and Radbruch, 1992; Kato and Radbruch, 1993; Vollenweider *et al.*, 1993; Irsch *et al.*, 1994); it has been used also for separation of plant cells and bacteria (Kronick and Gilpin, 1986). Basically, separation of organelles and chromosomes, as well as very large

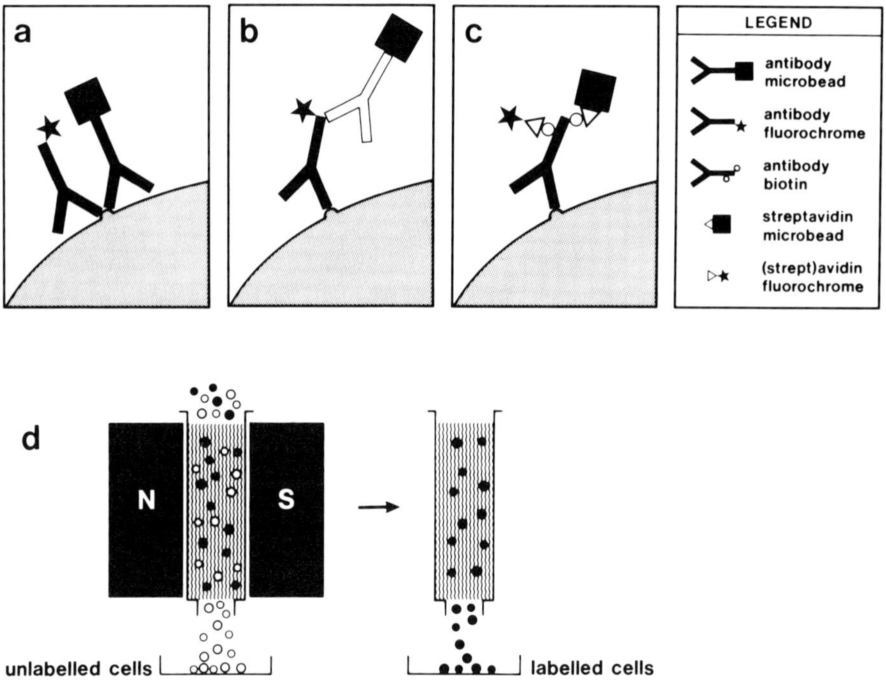

Fig. 1 Principle of high-gradient magnetic cell sorting. (a–c) Various principles of labeling cells with magnetic microbeads. (a) Competitive binding of directly microbead-conjugated and fluorochrome-conjugated antibodies; (b) piggyback of fluorochrome-conjugated first antibody and microbead-conjugated second anti-antibody; (c) piggyback of haptenized first antibody, anti-hapten-conjugated microbeads, and fluorochrome-conjugated anti-hapten. In (d) the differential behavior of magnetic and nonmagnetic cells in a separation column inside and outside of an external magnet is illustrated.

and fragile cells and cell aggregates, could be done by MACS, as long as immunomagnetic staining can be performed.

Designing the optimal strategy for any specific cell separation depends on

1. the availability of suitable cell-surface markers,
2. the composition of the original cell mixture,
3. frequency of wanted cells and the characteristics of the other cells,
4. specific requirements with regard to purity, recovery, and activation status of sorted cells.

Positive selection, i.e., *enrichment* of wanted cells, requires one or more cell-surface markers which are specific for the cells of interest. Since the separation process distinguishes only between nonmagnetic and magnetic cells, it is important that the negative cells really remain unlabeled, i.e., the staining reagent has to be titrated to eliminate background staining. Depending on the type of cell and on the surface marker used for staining, the sorted cells can be functionally influenced, e.g., by staining T lymphocytes with anti-CD3 (MacDonald and Erard, 1986). This is a problem inherent to staining with antibodies which recognize and crosslink physiological receptors on the surface of the target cell and thus may induce or suppress proliferation or differentiation of the cell. As far as is known, labeling with antibody-conjugated colloidal magnetic particles has no effect other than labeling with an unconjugated crosslinking antibody. The bound microbeads will be capped off or endocytosed and degraded, depending on cell type and antigen. Similar beads have been used to treat iron deficiency in humans and animals and have been shown to be nontoxic and biodegradable (Muir and Goldberg, 1961).

Negative cell sorting, i.e., labeling and *depletion* of unwanted cells from complex mixtures, will not yield homogenous, pure cell populations, unless antibodies against all unwanted cell types are available, which usually is not the case. However, the fraction of depleted, unstained cells is functionally "naive." A side effect of depletion is that dead cells, debris, and cell clumps are often removed as well from the cell suspension due to unspecific labeling.

The *isolation of rare cells* bears additional problems: First, discrimination of rare cells and cells of the major population is difficult because of the variation of immunofluorescence (%CV) which may lead to an overlap of the fluorescence distributions of frequent and rare cells, even in cases of otherwise acceptable staining. Often this may even make it difficult to identify the rare cells unambiguously for analysis and sorting. The overlap in magnetic staining is usually less significant, but is hard to access. Second, large cell numbers have to be processed in order to obtain rare cells in reasonable numbers. As a parallel enrichment method with high discriminatory power, MACS is predestined for sorting of rare cells, especially when combined with FACS (Weichel *et al.,* 1992; Kato and Radbruch, 1993; Irsch *et al.,* 1994).

III. Materials

A. Cells and Staining Reagents

1. Single-cell suspension of good viability (Ficoll-Hypaque centrifugation removes dead cells and debris, cotton wool column aggregates, and dead cells; Esser, 1992).

2. Phosphate-buffered saline (PBS) with 1% w/v bovine serum albumin (BSA) or 1% v/v fetal calf serum (FCS) (PBS/BSA or PBS/FCS).

3. Propidium iodide (PI), 1 mg/ml in water.

4. Immunofluorescent and magnetic staining reagents in PBS/BSA with 0.05% sodium azide. For sterile sorting, reagents can be sterilized by filtration or, in cases of fluorochrome conjugates, by centrifugation, which also removes aggregates (3 min, Eppendorf centrifuge). *Titration* of staining reagents is essen-

tial to prevent background staining, i.e., magnetism. Titration can be done by immunofluorescence and flow cytometric evaluation (Radbruch, 1992).

5. MACS CD14 microbeads (Miltenyi Biotec, Bergisch Gladbach, FRG; Fax +49-2204-85197).

6. CD14-FITC (Leu M3 fluorescein conjugate; Becton–Dickinson).

7. MACS ramG1 microbeads (MACS rat anti-mouse IgG1 microbeads; Miltenyi Biotec).

8. CD3-FITC (Leu 4 fluorescein conjugate, mouse IgG1; Becton–Dickinson).

9. CD34 isolation kit (A1, blocking reagent; A2, modified CD34 antibody, clone QBEND-10; B, MACS microbeads; Miltenyi Biotec).

10. CD34 staining antibody (HPCA-2 phycoerythrin conjugate; Becton–Dickinson).

B. Sorting Equipment

1. Standard cell culture plasticware, such as Eppendorf tubes, sample tubes for sorter, pipettes and tips, syringes, and needles.

2. Cell filters, i.e., nylon mesh with 30- to 50-μm pore size, either pieces of about 2 cm^2, glued to the bottom of a syringe (Esser, 1992), or commercially available (from Phoenix, San Diego, CA; Fax +1-619-259-5268), autoclaved for sterility.

3. Cell counting device: Coulter counter or Neubauer chamber.

4. Centrifuge for cells and Eppendorf centrifuge.

5. MACS equipment: MACS separator and separation columns (Miltenyi *et al.*, 1990; Miltenyi Biotec, Bergisch Gladbach).

6. Ethanol (70%) in water (to sterilize and fill up the MACS reusable columns air bubble free). Alternatively MACS ready-to-use columns may be used. They come sterile and can be filled directly with buffer.

7. Flow cytometer (optional) or fluorescence microscope for separation control.

IV. MACS: Staining and Sorting

A. Preparation and Staining of Cells

A single-cell suspension of good viability is a prerequisite for any good cell separation. Methods for obtaining such single-cell suspensions are described elsewhere (Esser, 1992). Since the magnetic separation process cannot distinguish between labeled single cells and aggregates containing labeled cells, it is important to avoid cell aggregates which might be composed of positive and negative cells. The cells should be resuspended carefully, e.g., by finger-flicking the sediment in the tip of the tube after centrifugation, before adding new buffer. Staining should be optimized by titration of the staining reagents (Radbruch, 1992; see also Section III,A) to maximize discrimination between positive and negative cells and, even more important, to avoid background staining, which could make magnetic separation impossible. Adapting the staining to variable cell numbers should be done by adjusting the volume of the staining reagent, leaving the concentration unchanged.

1. Indirect Staining

1. Spin down all cells (usually 10^5–10^{10}).

2. Remove supernatant carefully, e.g., by aspiration, and finger-flick the pellet.

3. Add a titrated amount of first staining reagent, e.g., unconjugated, fluorochromated, or biotinylated antibody, specific for the marker molecule in question, to a final volume corresponding to the number of positive cells, but exceeding the total cell volume at least twofold, e.g., 20 μl for 2×10^6 cells, containing 10^6 positive cells.

4. Stain for 10 min on ice.

5. Dilute the cells with PBS/BSA and spin them down (10 min, 300*g*).

6. Wash once: remove supernatant, resuspend the pellet by flicking, fill up with PBS/BSA, and spin down again.

7. For indirect staining, repeat steps 2 to 5 with second reagent, i.e., stain the cells marked with the biotinylated first antibody with MACS streptavidin microbeads (diluted 1 : 10 from stock) or stain the unconjugated or fluorochromated first antibody with MACS anti-isotype microbeads. Stain for 15 min at 6–12°C (in refrigerator).

8. Shake cells gently, then add a titrated amount of fluorochrome-conjugated streptavidin or, in the case of unconjugated first antibody, isotype-specific antibody or marker-specific antibody.

9. Dilute the cells with PBS/BSA and spin them down (10 min, 300*g*); resuspend them in PBS/BSA.

10. Check staining by microscopy or flow cytometry.

11. Store the cells on ice and in the dark, occasionally resuspending them, until they can be applied to the MACS column.

2. Direct Staining

1. Spin down all cells (usually 10^5–10^{10}).

2. Remove supernatant carefully, e.g., by aspiration, and finger-flick the pellet.

3. Add the specific MACS antibody microbeads as instructed by the manufacturer.

4. Stain for 15 min at 6–12°C (in refrigerator).

5. Add a titrated amount of the specific antibody conjugated to a fluorochrome to label the cells for sort control.

6. Stain for 10 min on ice.

7. Dilute the cells with PBS/BSA and spin them down (10 min, 300*g*); resuspend the cells in PBS/BSA.

8. Check staining by microscopy or flow cytometry.

9. Store the cells on ice and in the dark, occasionally resuspending them, until they can be applied to the MACS column.

B. Cell Sorting

1. Setting Up a MACS Separation Column

1. Select a column with a capacity of at least the number of positive cells you want to collect. Columns are available that can hold between 10^7 cells (type A) to 10^9 (type D) cells. Attach a three-way stopcock to the column of your choice.

2. For reusable columns: Attach a 10-ml syringe filled with 70% ethanol to the side plug of the stopcock and fill the column slowly from bottom to top with ethanol avoiding trapping air bubbles in the matrix by gently flicking the column. Ready-to-use columns are filled with sorting buffer from the bottom and then washed with buffer once.

3. Place the column in the MACS separator, close the stopcock, remove the ethanol syringe, and replace it with a syringe filled with PBS.

4. Wash the column with several volumes of PBS and then PBS/BSA from the top of the column, and clear the side plug of the stopcock of ethanol using PBS from the syringe.

5. Immediately before separation wash the column with ice-cold PBS/BSA to cool it down, then attach a needle for flow regulation at the bottom plug of the stopcock (small needle, low flow rate; large needle, high flow rate). The whole procedure can be performed in a sterile workbench.

2. Setting Up a MACS Rare Cell Column or a MiniMACS Separation Column

For isolation of up to 10^7 labeled cells and rare cells, a MACS rare cell column or a MiniMACS column will be advantageous because it is easier to handle. Separation columns are sterile. They just have to be unpacked and filled by adding 0.5 ml of PBS/BSA or PBS/FCS. A flow resistor (needle) can be attached to slow down the flow rate for separation of dimly labeled cells. The whole procedure can be performed in a sterile workbench.

3. Magnetic Sorting

1. Cells, stained as described above, are passed through the MACS column placed in the MACS separator at low flow rates (0.5–2 ml/min). The flow frequency of cells can be monitored, for example, by checking drops under the microscope from time to time. When the majority of cells has passed the column, the column is washed at higher flow rates depending on the intensity of staining and following the rule

$$\text{High relative fluorescence} = \text{high flow rate},$$

until only few more cells come with the effluent. The cells not retained on the column in the MACS magnet are the *negative fraction,* and, if washed off at higher flow rates, the *wash fraction.* Aliquots are analyzed for depletion by microscopy or flow cytometry.

2. The column is then closed at the bottom, carefully removed from the MACS separator, and eluted outside the MACS by flushing with PBS/BSA (3–5 volumes) from the top using a syringe plunger to increase the flow rate and thus wash off the cells from the ferromagnetic matrix. If the elution volume is not too high, the cells can be processed further directly; otherwise it is necessary to spin them down (10 min, 300*g*). This fraction is the *positive fraction.* An aliquot is used for analysis of enrichment, either by microscope or by flow cytometry. For live/dead cell discrimination, propidium iodide is added to a final concentration of 0.01 μg/ml.

3. The quality of cell sorting is defined by the enrichment and depletion rates for labeled cells, the recovery of the wanted cells, and their viability.

The *enrichment rate* can be calculated as

$$f_E = \frac{\%\ \text{negative in original sample}}{\%\ \text{positive in original sample}} \times \frac{\%\ \text{positive in positive fraction}}{\%\ \text{negative in positive fraction}}. \quad (1)$$

Accordingly, the *depletion rate* is

$$f_D = \frac{\%\ \text{positive in original sample}}{\%\ \text{negative in original sample}} \times \frac{\%\ \text{negative in positive fraction}}{\%\ \text{positive in positive fraction.}}. \quad (2)$$

The recovery is

$$\text{Recovery} = \frac{\text{absolute number of wanted cells in positive fraction}}{\text{absolute number of wanted cells in original sample}} \quad (3)$$

and the viability can be defined as

$$\text{Viability} = \frac{\%\ \text{live, wanted cells in positive fraction}}{\%\ \text{live, wanted cells in original sample}}, \quad (4)$$

although frequently it is given just as the frequency of live versus dead, wanted cells after the sort.

An essential figure in these calculations is the frequency of positive cells in the original sample. For rare cells it often simply cannot be determined with precision, making the calculation of enrichment rates difficult.

For calculation of the recovery, the cells have to be counted immediately before and after the sort. Otherwise, the sometimes considerable loss of cells during centrifugation steps will contribute to the calculated rate.

V. Critical Aspects of the Procedure

A. Staining

Like that in immunofluorescence, the critical aspects of immunomagnetic staining are specificity and handling of the labeling antibody by the labeled cells. In general, short staining times improve the specificity of staining and few washing steps improve the recovery of cells throughout the staining procedure. To avoid capping of the staining antibody from the cell surface before or during the sort, one should work fast, keep the cells cold, and, whenever possible, use the reversible inhibitor of capping sodium azide (0.05% in PBS/BSA). However, antibody-conjugated microbeads require prolonged incubation times for staining on ice, so that they are used at 6–12°C refrigerator temperature (15 min).

Problems with specificity and background usually can be improved by titration of the antibody, either in immunofluorescence and flow cytometry or in magnetic test sorts. Like any staining reagent, the magnetic microbeads can be titrated also. This can be done functionally using different concentrations of microbeads for one particular, standardized separation and determining the concentration with the best enrichment rate, i.e., low background and high specific labeling. Background staining is lethal for successful separation because even slightly labeled cells will be retained in the magnetic field. Commercial preparations of magnetic microbeads, however, usually come titrated.

Fluorescent labels should be introduced after magnetic staining to avoid competition. Since the magnetic labeling will usually be carried out at non-saturating conditions, nearly complete fluorescent labeling can be achieved.

B. Sorting

Ferromagnetic columns should be chosen according to their capacity for labeled cells. For optimal enrichment the capacity should correspond to the number of positive cells. For optimal depletion the capacity can be larger than that. Air bubbles trapped in separation columns can reduce the actual capacity of the column considerably and therefore have to be avoided. In reusable columns they are avoided by filling the column with 70% ethanol from the bottom (see Section IV,B). Ready-to-use columns can be filled directly with buffer.

Large cell aggregates and clumps of debris may clog the column, although this is less of a problem for MACS than for FACS. Nevertheless, they should be removed by filtration of the cell suspension through a nylon mesh of defined mesh width, usually 30 or 50 μm (see Section III).

The sensitivity of the separation can be influenced by varying the flow rate of cells through the column. For depletion a slow flow rate will be chosen, whereas for enrichment a faster flow rate is preferred. In addition, magnetic labeling and optimal flow velocities depend on each other, i.e., intensely labeled cells will also attach to the matrix at high flow rates, whereas dimly labeled cells will not. For individual antibodies it may be advisable to calibrate the speed of flow in test sorts for optimal sorting efficiency.

Dead cells and debris, some of which may be unspecifically labeled, in that case will attach to the MACS column and be recovered with the positive fraction. For depletion experiments, this may be a wanted effect. For enrichment of rare cells, however, one should keep in mind that unspecifically stained dead cells, even if they are rare, are also enriched and may comprise a major fraction of the positive fraction. Thus, for enrichment experiments, dead cells and debris have to be either accepted or removed before or after sorting, using Ficoll-Hypaque density gradients or cotton wool columns (Esser, 1992). For the sort control by flow cytometry, dead cells and debris can be gated out by scatter and PI fluorescence (FSC/SSC and F2/F3) (Weichel *et al.,* 1992; Kato and Radbruch, 1993).

In general, to improve purity-enriched cells, eluted from the first column, can be passed over another column, at either the same or slightly different flow rate in order to remove unspecifically bound cells and particles from the first enrichment. An example for a double-pass column is shown in Fig. 5.

C. Sterility

The most frequent sources of contamination are staining reagents, especially if they are used by several persons over some time and stored in a refrigerator. Yeast can be eliminated by fungicides such as nystatin, but it is advisable to separate yeast from cells by Ficoll-Hypaque gradient centrifugation beforehand. Contamination of cells with bacteria can be avoided by using an antibiotic which is not normally used, such as gentamycin.

With respect to the sorting procedure, ready-to-use columns are sterile upon delivery, while multiuse columns are sterilized upon filling them with 70% ethanol (see Section IV,B). The whole MACS separation can be performed in a sterile workbench.

VI. Controls

Monitoring the magnetic sort for recovery requires a cell counting device. This could be a simple Neubauer chamber or a more sophisticated Coulter counter.

The volume of the negative fraction and the volume required for washing, i.e., the time points for changing between the steps of the sorting procedure, can be either estimated, as indicated before (Section IV,B), or determined by collecting drops of the washing fluid from the column and judging the number of cells by phase-contrast microscopy. When no more cells leave the column, it is time to proceed to the next step. Usually, this will be after three void volumes of buffer.

With regard to specificity, the sort is monitored by immunofluorescence. In most systems tested so far, the magnetic label does not occupy all target molecules of the specific antibody on the cell surface. Sufficient target molecules are left over for labeling with the same antibody conjugated to a fluorochrome, directly or indirectly (Section III,B). Thus, the cells are labeled both magnetically and fluorescent, allowing the sorting process to be monitored by fluorescence microscopy or, preferably, by flow cytometry. Since the magnetic label does not interfere with the optical analysis, additional parameters can be included in this sort control, like propidium iodide for live/dead cell discrimination or fluorochromated antibodies specific for particular "unwanted" cells.

For cytometric analysis of rare cells, e.g., in biomedical research, aliquots of the positive fraction can be directly counterstained and analyzed for the parameters in question (Kato and Radbruch, 1993; Irsch *et al.*, 1994), a unique feature of high-gradient magnetic cell sorting.

In evaluating the frequencies of contaminating cells in positive and negative fractions, which, due to the lack of better evaluation software, is usually done by setting a statistical threshold between the fluorescence distributions of labeled and unlabeled cells (Figs. 3–5), one should keep in mind that fluorescence distributions in general have a broad variation (CV). Cells from the wanted population, crossing the threshold due to this variation will be falsely classified as contaminating cells. True contaminating cell populations are those which have the same mean fluorescence as the original unwanted population.

VII. Instruments

To magnetize the high-gradient magnetic separation columns, a strong magnet is required for generation of the magnetic fields. The design of the columns and the magnet has been described elsewhere (Miltenyi *et al.*, 1990). We have used the MACS, SuperMACS, and MiniMACS from Miltenyi Biotec (Fig. 2), together with columns from that manufacturer.

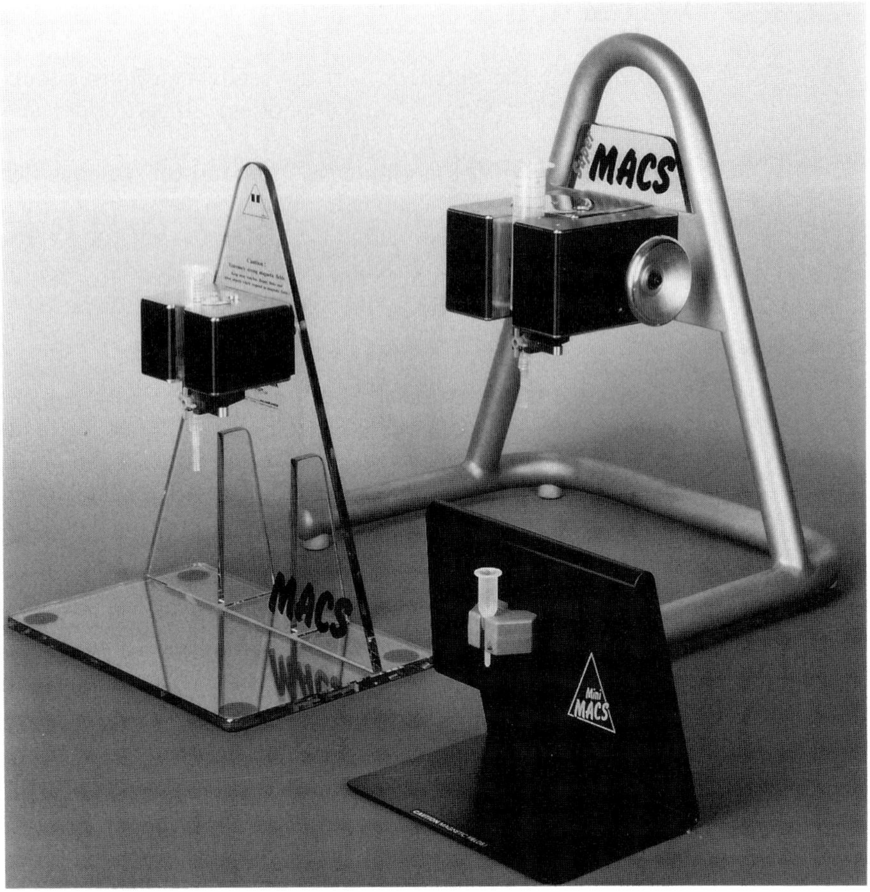

Fig. 2 Commercially available high-gradient magnetic cell sorters, the MiniMACS, the MACS, and the SuperMACS.

The SuperMACS can be used with columns with volumes of up to 50 ml and capacities of up to 10^9 labeled cells. It is optimal for depletion and for handling of large cell numbers. The MACS is a general purpose instrument which can be equipped with columns ranging in capacity between 10^7 and 2×10^8 cells. We have used this machine in most experiments (e.g., Section VIII,A). The MiniMACS is optimal for the efficient enrichment of up to 10^7 cells. We have used this device especially for the isolation of rare cells (Section VIII,B).

For sort control, we have used a FACScan (Becton–Dickinson) and for fluorescence-activated sorting of MACS preenriched cells a modified FACS IV (Weichel *et al.*, 1992).

VIII. Results

A. Isolation and Depletion of Monocytes

A typical MACS separation using direct magnetic labeling is shown in Fig. 3. Monocytes are separated from other peripheral blood mononuclear cells (PBMC). In this experiment, 2×10^8 PBMC had been stained with 2 ml of MACS CD14 microbeads, diluted 1 : 5 in PBS/BSA from the purchased stock. After 15 min in refrigeration, 200 μl of CD14-FITC (50 μg/ml) was added, and the mixture incubated for another 5 min in the cold. After one wash with 15 ml of cold PBS/BSA, the cells were resuspended in 2 ml PBS/BSA. An aliquot of about 10^5 cells was analyzed in a FACScan.

The remaining cells were applied to a B2 column with a 22-guage needle, as a flow resistor, inserted into a MACS magnet. The negative cells were collected in 20-ml. An aliquot of this *negative fraction* was analyzed in the FACScan. The column with the positive cells was removed from the MACS and the positive cells were pushed off the wires from the bottom to the top of the column using a syringe attached to the side of the three-way stopcock. The

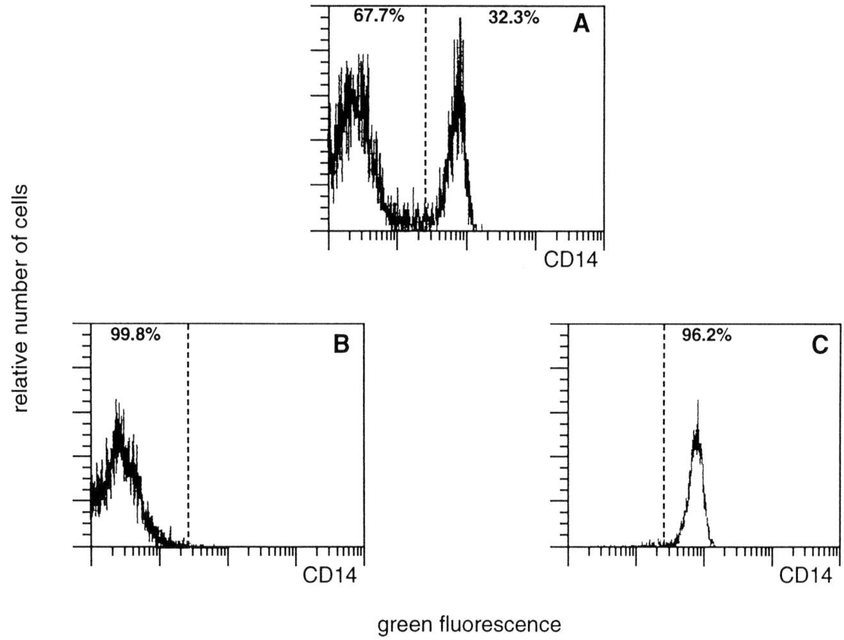

Fig. 3 Separation of CD14 monocytes from PBMC with directly conjugated MACS CD14 microbeads. For analysis the cells were stained with fluoresceine-conjugated CD14. Fluorescein-fluorescence histograms of live cells as gated by scatter and PI fluorescence. (A) Before sort: 67.7% negative cells, 32.3% positive cells; (B) negative fraction: 99.8% negative cells; (c) positive fraction after one washing cycle: 96.2%.

column was then reinserted into the MACS magnet, and cells which had bound unspecifically in the first round were washed off. The column was then washed with 20 ml cold PBS/BSA using a 20-gauge needle. With this little intermezzo the purity of the positive fraction can be considerably increased. After the column was removed from the magnet, the flow resistor was removed from the column, a 10-ml syringe with PBS/BSA was attached to the top of the column via the adapter, and the retained cells were flushed out. An aliquot of this *positive fraction* was analyzed in the FACScan as well. Absolute cell numbers were determined in a Neubauer chamber. In this experiment the recovery of CD14-depleted cells in the negative fraction was 89%, and the recovery of CD14-positive cells, 71%. The enrichment rate (f_E) was calculated to be 53, while the depletion rate (f_D) was 238. This MACS separation was optimized for efficient depletion. The depleted fraction could easily be used for further magnetic separations. The enrichment rate can be increased by repeating the washing cycle. We have used similar MACS separations leading to the discovery of the critical importance of murine monocytes/macrophages for the specific induction of interferon-γ expression in normal T helper lymphocytes (Schmitz and Radbruch, 1992).

B. Isolation and Depletion of T Lymphocytes

In an experiment similar to that described in Fig. 3, we separated T lymphocytes from PBMC (Fig. 4). The protocol is included to demonstrate the indirect magnetic labeling with MACS rat anti-mouse IgG1 microbeads. A total of 10^8 PBMC are suspended in 350 μl PBS/BSA with CD3-FITC (50 μg/ml) and incubated for 10 min at 6–12°C. Cells are washed with 15 ml PBS/BSA, the pellet is resuspended in 800 μl PBS/BSA, and 200 μl of MACS ramG1 microbeads are added. After 10 min at 6 to 12°C (or 20 min on ice), cells are washed and resuspended in 1 ml PBS/BSA. The cell suspension is applied to a B1 column with a 23-gauge needle. The passing cells are applied onto the column again. Then, the negative cells are washed out in 15 ml PBS/BSA. After the washing step with 15 ml PBS/BSA and a 21-gauge needle, the positive fraction is collected outside of the MACS magnet with 5 ml of PBS/BSA applied from the top. All fractions are evaluated for purity, recovery, and viability as described above.

C. Enrichment of Rare Cells

The isolation of rare cells is a challenge for any cell sorting technology. In principle, positive selection, i.e., isolation of the rare cells according to the marker molecules they express, is superior to negative selection, i.e., depletion of all other cells according to markers they express, since usually discriminative markers are not available for all other cells, especially in the case of malignant

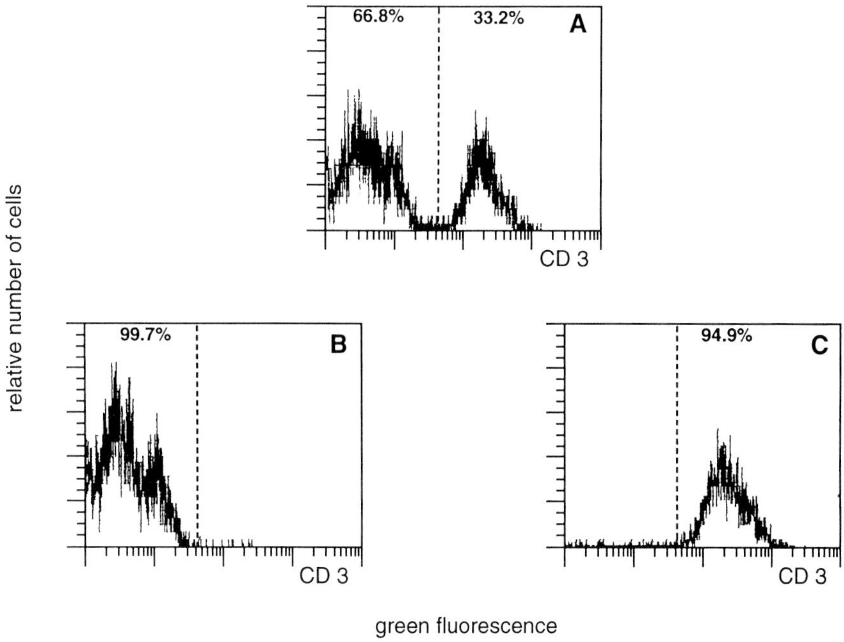

Fig. 4 Separation of CD3 T cells from PBMC with FITC-conjugated Leu 4 (IgG1) antibodies and rat anti-mouse IgG1 microbeads. Fluorescein-fluorescence histograms of gated live cells. (A) Before sort, (B) negative fraction f_D = 165, (C) positive fraction f_E = 37.

cells. Here, we describe the purification of rare $CD34^+$ hematopoietic stem cells from peripheral blood by MACS.

For the enrichment of CD34 cells we used an improved indirect magnetic labeling technique. PBMC (5×10^8) were incubated with blocking reagent (human Ig) and haptenized CD34 antibody (clone QBEND-10) in PBS/BSA/EDTA (5 m*M*). Cells were washed with PBS/BSA and labeled magnetically with hapten-specific MACS microbeads (10 min at 6–12°C) in PBS/BSA/EDTA (5 m*M*).

For flow cytometric analysis, phycoerythrine-conjugated HPCA-2 antibody was added. After 5–10 min incubation on ice, PBS/BSA/EDTA was added and cells were filtered through a 30-μm nylon mesh. The cells were washed, the pellet resuspended in 2 ml PBS/BSA/EDTA, and the cell suspension applied to a MiniMACS separation column which had been equilibrated with buffer before. The negative cells were washed off the column with PBS/BSA/EDTA. Retained cells were eluted from the column outside of the magnet by pipetting buffer onto the column and using the plunger supplied with the column.

An aliquot of the positive cells was used for flow cytometric analysis. The positive cells from the first column were now applied to a new separation

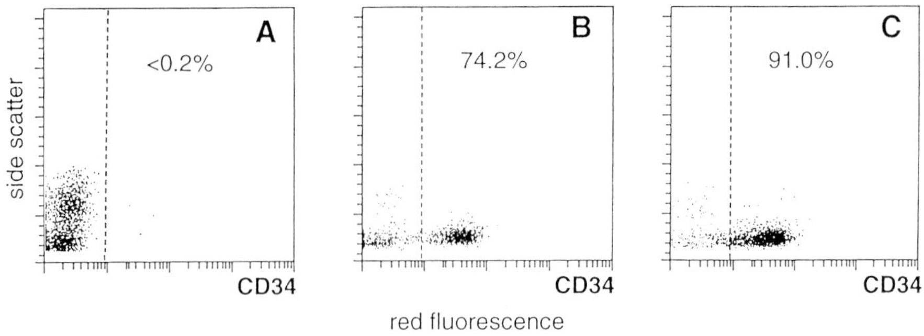

Fig. 5 Isolation of CD34 cells from PBMC with hapten-conjugated QBEND-10, hapten-specific microbeads, and physoerythrine-conjugated HPCA-2. Phycoerythrin-fluorescence versus side scatter dot plots of cells before (A) and after the first (B) and second (C) enrichment are shown. Live cells were gated according to forward scatter and PI fluorescence.

column and the magnetic separation was repeated. During the second enrichment, negative cells were again washed off the column. Positive cells were eluted in PBS/BSA. Figure 5 shows flow cytometric side scatter versus phycoerythrin-fluorescence dot plots of the original fraction before separation and of the cell fractions eluted from the first and second columns. PBMCs were taken from a fresh blood sample of a normal donor with a less than 0.2% population of CD34-positive cells. The calculated enrichment rate (f_E) in this experiment is more than 4000. The typical recovery of CD34 cells is between 65 and 80% of the calculated fraction in unseparated PBMC of normal donors. MACS-enriched CD34 cells show a normal colony formation potential in semisolid short-term colony assays. Immunophenotyping revealed that peripheral CD34 cells show a different pattern of surface markers compared to bone marrow-derived CD34 cells (Kato and Radbruch, 1993; Miltenyi *et al.*, 1994).

Acknowledgments

This work was supported by the German Ministry of Science through the Genzentrum Köln.

References

Esser, C. (1992). *In* "Flow Cytometry and Cell Sorting" (A. Radbruch, ed.), pp. 133–140. Springer, New York.

Hansel, T. T., DeVries, I. J., Iff, T., Rihs, S., Wandzilak, M., Betz, S., Blaser, K., and Walker, C. (1991). *J. Immunol. Methods* **145,** 2130–2136.

Irsch, J., Irlenbusch, S., Radl, J., Burrows, P., Cooper, M., and Radbruch, A. (1994). *Proc. Natl. Acad. Sci. U.S.A.* **91,** 1323–1327.

Kato, K., and Radbruch, A. (1993). *Cytometry* **14,** 384–392.

Kronick, P., and Gilpin, W. (1986). *J. Biochem. Biophys. Methods* **12,** 73–80.

MacDonald, H. R., and Erard, F. (1986). *Curr. Top. Microbiol. Immunol.* **126,** 187–194.

Miltenyi, S., Müller, W., Weichel, W., and Radbruch, A. (1990). *Cytometry* **11,** 231–238.

Miltenyi, S., Guth, S., Radbruch, A., Pflüger, E., and Thiel, A. (1994). *In* "The Mulhouse Manual" (E. Wunder, ed.), pp. 201–215. AlphaMed Press, Dayton, Ohio.

Molday, R., and Molday, L. (1984). *FEBS Lett.* **170**(2), 232–237.

Muir, A., and Goldberg, L. (1961). *Q. J. Exp. Physiol. Cogn. Med. Sci.* **46,** 290–298.

Radbruch, A. (1992). *In* "Flow Cytometry and Cell Sorting" (A. Radbruch, ed.), pp. 34–46.

Schmitz, J., and Radbruch, A. (1992). *Int. Immunol.* **4,** 43–52.

Vollenweider, I., Vrbka, E., Fierz, W., and Groscurth, P. (1993). *Cancer Immunol. Immunother.* **36,** 331–336.

Weichel, W., Irlenbusch, S., Kato, K., and Radbruch, A. (1992). *In* "Flow Cytometry and Cell Sorting" (A. Radbruch, ed.), pp. 159–167. Springer, New York.

CHAPTER 24

Contributions of Flow Cytometry to Studies with Multicell Spheroids

Ralph E. Durand
Medical Biophysics Department
British Columbia Cancer Research Centre
Vancouver, British Columbia V5Z 1L3

I. Introduction

Multicell spheroids, as introduced by Sutherland and his co-workers more than 20 years ago (Inch *et al.*, 1970; Sutherland and Durand, 1976), have now found widespread application. Much of the novelty and ultility of the system is, of course, a result of its multicellular composition: as the spheroids grow, peripheral cells remain in a rapidly proliferating state, whereas cells deeper into the spheroid become quiescent and eventually die to form a necrotic core. These characteristics have led to extensive studies of spheroids as *in vitro* models of nodular tumors, in addition to their use for examining more basic questions of cell interaction and growth control processes (Acker *et al.*, 1984; Mueller-Klieser, 1987; Sutherland, 1988; Carlsson and Nederman, 1989).

Spheroids, in essence, consist of an organized collection of heterogeneous cells. It is that heterogeneity which, at least in the opinion of this author, makes the system a natural and often necessary candidate for flow cytometry studies. This is illustrated in this chapter by studies from the author's laboratory that are representative of many of the general techniques and applications that have evolved.

II. Spheroid Growth and Analysis

A. "Gel Culture" Techniques

Several different methods have evolved for production of multicell spheroids. For any cell type that can grow under anchorage-independent conditions, a guaranteed technique for producing spheroids is to use "gel" culture (McAllister *et al.*, 1967; Dalen and Burki, 1971; Carlsson, 1978; Acker *et al.*, 1984). Soft agar or other materials such as methyl cellulose can be used as a supporting matrix, and cells introduced into the matrix before gelling will subsequently grow by cell division to form multicell nodules. While this technique has considerable flexibility in terms of the different types of cells which can be grown as spheroids, it is not widely used because of the technical difficulty of recovering the spheroids/cells from the supporting matrix for further study.

B. Liquid Overlay Cultures

The two predominant techniques now used for spheroid growth are liquid overlay cultures and stirred suspension cultures. For the former, cells are introduced into regular culture medium which overlays a surface to which cells will not adhere (Yuhas *et al.*, 1977; Acker *et al.*, 1984; Carlsson and Nederman, 1989). Typically, this consists of agar, agarose, or, in some cases, simply non-tissue culture plastic. Under these conditions, cells will generally aggregate and then grow by cell division to form multicell clusters. A variation of the technique is to grow individual spheroids in each well of a multiwell plate. The liquid overlay technique has the dual advantages of being applicable to a large variety of cell types and allowing repetitive observation of individual spheroids. Conversely, the static growth conditions may add additional complexity to analysis of results (diffusion of oxygen and nutrients may be "top-down" rather than "outside-in").

C. Suspension Culture Techniques

The stirred suspension culture method is applicable only to cell types which spontaneously aggregate and grow as spheroids; spheroids are produced simply by introducing monodispersed cells into the culture flask (Sutherland and Durand, 1976; Acker *et al.*, 1984). For such cells, this technique offers a number of advantages. Perhaps principal among these are the large number of cells/spheroids which can be cultured per flask, the relative uniformity of the spheroid population that can be produced, and the ease of manipulation, feeding, etc.

Combinations of the liquid overlay and suspension techniques are also often useful (Acker *et al.*, 1984). As an example, cells which will not spontaneously aggregate in stirred suspension cultures can be "initiated" as spheroids in liquid overlay culture, and when small aggregates have formed, those are transferred into stirred suspensions for continued growth. Conversely, spheroids can be harvested from stirred suspension cultures and placed in multiwell plates for repeated volume or size measurements.

D. Available Assays: An Overview

As was implicit in the previous paragraph, one of the principal advantages of the spheroid system is the multitude of end points which can be studied. As in a tumor growing in an animal, spheroid growth and regrowth or regression rates all provide useful information. Similarly, most spheroids can be nondestructively disaggregated to monodispersed cell suspensions using enzymatic treatment plus mechanical agitation. The individual cells recovered from spheroids in this manner can then be subjected to any of the traditional assays that are used for other cultured cells. Consequently, the system is unique in that the "net" response of the cell population as a whole can be assessed, as can the individual responses of the subpopulations of cells comprising the spheroid. Coupling these measurements has provided novel mechanistic information relevant to tumor growth and response to therapeutic interventions.

III. Flow Cytometry Techniques

A. Overview

Once spheroids are reduced to a single-cell suspension, the subsequent use of flow cytometry is limited only by the competence and imagination of the investigator. DNA, RNA, protein, thiol, and immunochemical analyses can all be performed as easily as analyses with cells from any other source.

B. Sorting Cell Subpopulations

As previously suggested, the cell sorting capabilities of flow cytometers have had a major impact on spheroid studies and, more recently, tumors grown in rodents (Chaplin *et al.*, 1985; Olive *et al.*, 1985). A cell sorter can be used to selectively recover cells from different regions of the spheroid. In theory, this requires only three factors: (1) a nontoxic fluorochrome which (2) penetrates slowly into spheroids to produce a gradient of staining intensity and which (3) remains localized within the cells even after spheroid disaggregation. Hoechst 33342 is one fluorochrome that comes close to meeting all of these criteria (Durand, 1982); other choices include carbocyanine derivatives and acridine orange analogues (Durand *et al.*, 1990). We have found the Hoechst compound particularly useful since a staining gradient of more than 1000-fold can be established between the brightest and dimmest cells of the spheroids and since this can be achieved at nontoxic concentrations of the dye. In our typical experiments, illustrated later in this chapter, we administer Hoechst 33342 at 2 μM for 20–30 min prior to spheroid disaggregation and then pass the cells through the cell sorter to recover 10 populations each containing 10% of the spheroid cells (Durand, 1986a). By sorting in this fashion, we essentially perform a "reconstruction" experiment in which subsequent analyses provide functional or structural information for cells which are located at *known* depths within the spheroid at the time of dye exposure. As shown in subsequent sections, this technique is particularly useful for experimental cancer therapy studies, where, for example, radiation or drug delivery and efficacy can be determined throughout spheroids of various types.

IV. Studies of Spheroid Oxygenation

A. Spheroid Radiosensitivity

The combination of flow cytometry/cell sorting techniques for spheroid reconstruction, plus clonogenicity assays of cell viability, has been used to advantage to study the response of multicell spheroids to radiation under various environmental conditions. Figure 1 shows the spectrum of responses that can be obtained in Chinese hamster V79 spheroids, as well as indicating the novel information available only with the use of flow cytometry and cell sorting. Cells that are sensitive to radiation when fully oxygenated are more resistant when anoxic; additionally, cycling cells near the spheroid periphery tend to be somewhat more radioresistant due to increased numbers of resistant S phase cells (Fig. 1). The sigmoid curves between the oxygenation extremes show the transition

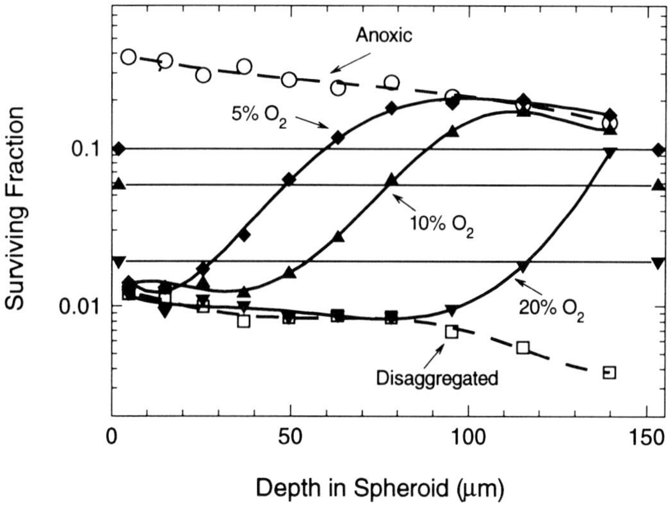

Fig. 1 Clonogenicity of cells from V79 Chinese hamster cell spheroids exposed to 12-Gy X-rays after equilibration with the indicated atmospheres. Spheroids were stained with 2 μM Hoechst 33342 for 20 min immediately postirradiation, then reduced to single cells by trypsinization and sorted into 10 consecutive fractions each containing 10% of the total cells. Horizontal lines indicate the net response of unsorted cells from each spheroid population; note the transition from oxygenated (radiosensitive) to anoxic (resistant) cells with increasing depth into the spheroids.

from aerobic to hypoxic conditions. Since 10 fractions were recovered from the spheroids, simply counting the number of aerobic versus hypoxic fractions gives an immediate indication of the percentage hypoxia present in the spheroids. To illustrate the extra information available from the cell sorting, Fig. 1 also shows the "net" response of the spheroids irradiated under the intermediate conditions of oxygenation: the horizontal lines. While there is a clear progression toward increased resistance as the level of oxygen was decreased, the "single-point" data provide no indication of the heterogeneity of response through the spheroid.

B. Role of Cellular Thiols

Our cell sorting techniques have contributed substantially to understanding the mechanism of action of thiol-dependent radiosensitizers and radioprotectors (Durand, 1984). An example is shown in Fig. 2, where modification of the radiation response of spheroids (under a 10% oxygen atmosphere) is examined after glutathione depletion with L-buthionine-S,R-sulfoximime (BSO) or after treatment with the aminothiol radioprotector, WR-2721. The similarity between Figs. 2 and 1 is striking; only a subset of the cells were markedly sensitized by thiol depletion or significantly protected by thiol addition. Since little effect

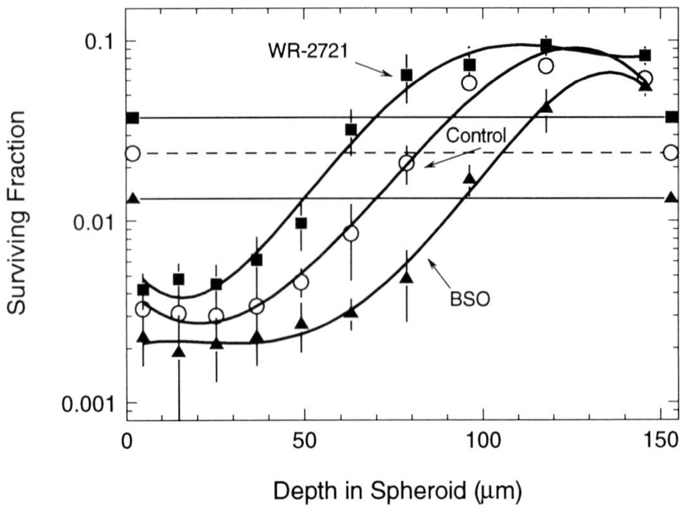

Fig. 2 Clonogenicity of cells from V79 spheroids exposed to 15 Gy X-rays in a 10% oxygen environment. Addition of 0.3 mg/ml WR-2721 30 min preexposure resulted in a net increase of resistance, whereas 24-hr incubation with 0.1 mM L-BSO produced a net sensitization. In both cases, however, the effect was largely due to changes in the size of the anoxic subpopulation of cells, in turn indicating that oxygen "consumption" rate was highly dependent upon the number of reduced (oxidizable) thiols present in the spheroids. (Data previously presented in a different format in Durand, 1984.)

was seen for either aerobic or hypoxic cells, we conclude that the primary effect of these agents is to modify the oxygenation status of the "borderline hypoxic" subpopulations of cells within the spheroids (in analogy to Fig. 1). Note that this conclusion directly resulted from the use of spheroids and cell sorting techniques; the horizontal lines included in Fig. 2 show the net survival of cells from the spheroids and, while providing a relative indication of modified sensitivity, certainly do not provide sufficient mechanistic information to lead to the insights just described.

V. Evaluation of Antineoplastic Drugs

A. Drug Activity in Spheroids

One of the major factors contributing to the utility of the spheroid system is, again, the three-dimensional (tissue-like) nature of the model. This has resulted in extensive studies of drug delivery and drug toxicity, since it is of obvious interest whether cancer chemotherapeutic agents are able to penetrate to, and be active in, clonogenic tumor cells not bordering the tumor vasculature (Durand, 1986b).

Experience in our own and other laboratories has led to the consensus that most cancer chemotherapy agents in common clinical use do not display significant "penetration" problems (Carlsson and Nederman, 1989; Durand, 1989a). Conversely, however, the activity of most drugs is markedly influenced by the microenvironment and cycle status of the target cells. Some illustrations are provided in Fig. 3, where the response of cells from various regions of V79 spheroids to representative, common cancer chemotherapeutic agents is shown. One of the unique features of the spheroid system, and of these studies, is that the exposure/sorting procedures can be conducted in either sequence; that is, cells can be exposed to the drug while "*in situ*" in the spheroid or, conversely, entirely equivalent cells can be sorted from the spheroids and then exposed to the anticancer agent in a Petri dish under known microenvironmental conditions. Figure 3 shows the two general results we have observed: (1) with the exception of the nitrosoureas, every drug that we have studied has shown marked differences in activity in the intact spheroid relative to cells exposed under monodis-

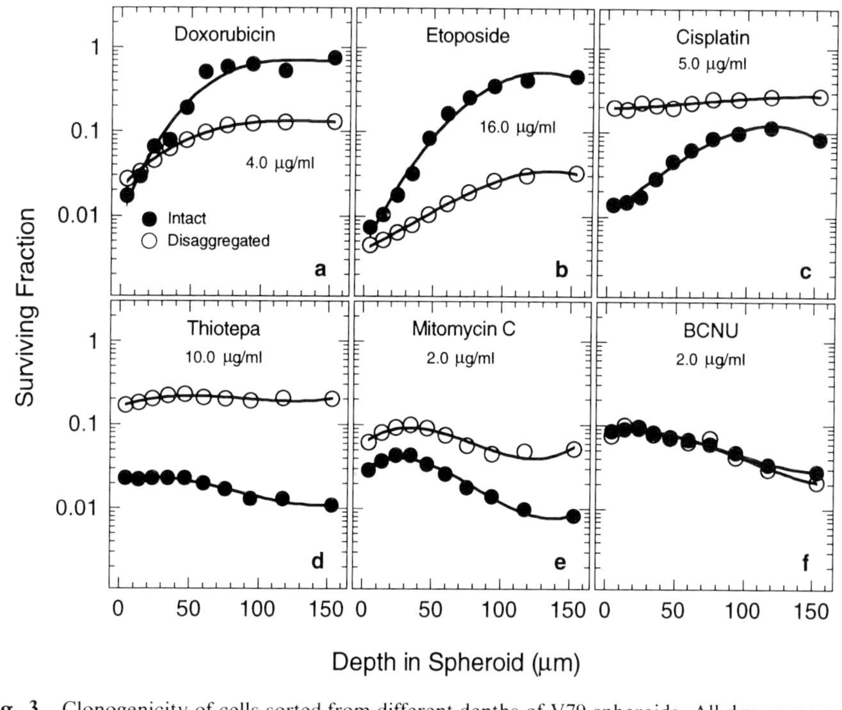

Fig. 3 Clonogenicity of cells sorted from different depths of V79 spheroids. All drug exposures were 2-hr duration; closed symbols represent cells exposed in intact spheroids, while open symbols show the response of cells sorted prior to drug exposure. All exposures produced about 99% toxicity to the most sensitive subpopulation of cells in intact spheroids; note, however, that different subpopulations were most sensitive to the different drugs. No consistent pattern for *in situ* versus *in vitro* exposure was observed. (Some of these data were previously presented in Durand, 1986b.)

persed conditions, and (2) every drug studied to date has shown considerable variation in activity throughout the spheroid.

One might expect that the external, rapidly cycling cells of spheroids would be more sensitive to chemotherapeutic agents than the internal quiescent cells, due to a combination of drug delivery and cycle-active processes. While this is certainly true for the antimetabolite/antibiotic compounds (for example, doxorubicin and etoposide in Fig. 3), not all agents behave this way. In fact, mitomycin C and perhaps somewhat surprisingly the nitrosoureas are preferentially toxic to the inner noncycling cells of spheroids. The practical significance of these observations is twofold: since clinical chemotherapy combinations are typically chosen on the basis of "non-cross-resistance," studies like these add yet another dimension to the standard clinical definition of cross-resistance. Additionally, data like these provide a more objective rationale for choosing and using chemotherapy combinations and for altering combinations as the status of the tumor/spheroid changes during (or in response to) therapy.

A final point must also be made: cisplatin, an agent with broad spectrum activity in the clinic, was much *more* effective against cells in intact spheroids than when identical cells were exposed after spheroid disaggregation. This demonstrably different activity "*in situ*" versus "*in vitro*" raises interesting questions regarding the adequacy and wisdom of *in vitro*-based drug development programs, as well as suggesting a need for considerable caution in predicating clinical therapy based on *in vitro* sensitivity assays of the tumor cells.

B. "Slow" Drug Penetration: A Therapeutic Advantage?

It is of considerable interest that a very large gradient of activity is seen for doxorubicin in Fig. 3. Since the compound is itself fluorescent, intracellular drug levels can be estimated using flow or static cytometry techniques (Krishan and Ganapathi, 1979; Durand and Olive, 1981). These have generally been consistent with the observed levels of cytotoxicity. However, such observations in spheroids lead to a major contradiction: only a small subset (the external cells) is sensitive to the drug. Consequently, the apparent penetration problem for doxorubicin suggests that it should have little if any clinical activity. Like others, we originally explained this inconsistency by the fact that clinical treatment is not delivered as a single administration of the drug. While that explanation is partially correct, detailed studies of doxorubicin uptake and retention in multifraction experiments provided further insight (Durand, 1990a).

Figure 4 shows data similar to those already presented for doxorubicin in Figure 3, but with a slight variation: Fig. 4a shows survival of cells sorted immediately after the 2-hr drug administration, using a protocol like that for the data shown in Fig. 3. However, in Fig. 4b, we show the survival observed if sorting and assay were delayed for 24 hr. In this case, the drug was administered for the same 2-hr period, but then removed from the growth medium and the spheroids were left intact rather than being immediately disaggregated for

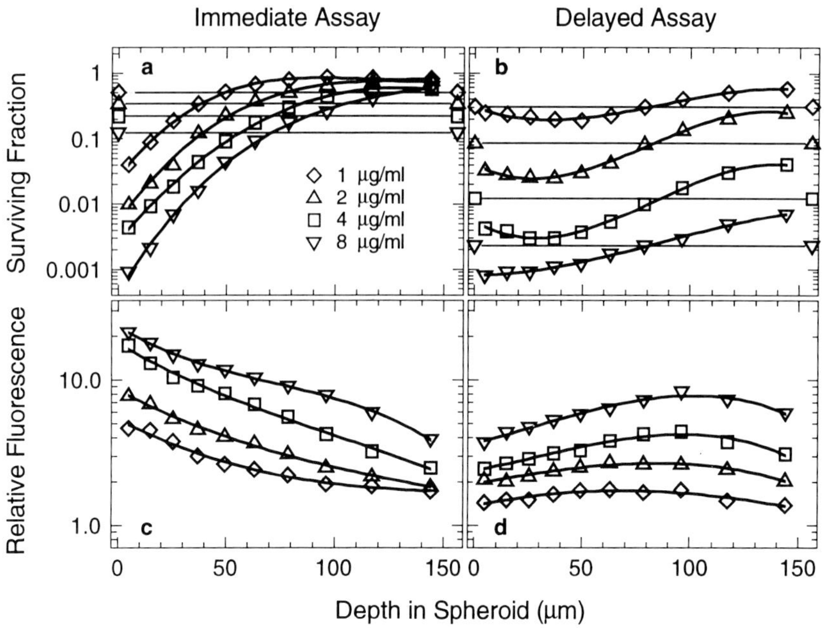

Fig. 4 Clonogenicity (top) or relative intracellular fluorescence (bottom) of cells recovered from V79 spheroids either immediately (left) or 24 hr (right) after a 2-hr exposure to the indicated concentrations of doxorubicin. Note that intracellular drug levels of the inner cells *increased* for the higher exposures during the 24-hr postexposure interval, indicating that the high levels of drug initially accumulated in the external cells diffused both outward *and inward* during the postexposure interval. (Data previously presented in a different format in Durand, 1990a.)

clonogenicity analysis. When interpreted in conjunction with the flow cytometry estimates of intracellular drug levels in the lower panels of Fig. 4, an interesting conclusion emerges. Doxorubicin effectively diffuses only slowly through the spheroid, but that slow transport also applies to the *outward* diffusion of accumulated drug. Clearly, it is a very different situation for a cell to remain *in situ* after drug exposure as opposed to being immersed in a sea of non-drug containing medium. In either case, the cell may efficiently "pump out the drug." However, in a tissue-like situation, this may produce a high local extracellular concentration favoring diffusion back into the cell. In the even more extreme scenario, it is perhaps possible that the drug is "pumped" directly into adjacent cells, resulting in a rather futile role for the P-glycoprotein molecule in cells of solid tumors.

Perhaps the most significant conclusion gained from Fig. 4 is the danger of extrapolating from the culture environment to that in a three-dimensional tissue. Quite different local microenvironments potentially surround subsets of cells in tissue, a situation totally unlike isolated cells in a Petri dish. "Excision" assays (whether from spheroids or animal tumors) should therefore be timed so that cell exposure to the drug is pharmacologically realistic; an obvious problem with this approach, however, is the cell turnover and loss which will necessarily occur in dynamic systems.

VI. Drug Interactions: Synergy in Cell Subpopulations

Drug interaction studies are a natural extension for the spheroid model. One such example is shown in Fig. 5, illustrating the interaction between cisplatin and etoposide. Figure 5a shows survival curves for each of the drugs when administered independently, as well as the predicted survival if the agents acted separately (the product of the two single-agent survival curves), and the observed surviving fraction when the drugs were administered together. Two features are evident: not only was more toxicity observed than expected (suggesting synergism), but, additionally, the degree of interaction was *not* uniform through the spheroids. Rather, the interaction was somewhat greater in the internal than in the external cells for the particular exposure conditions used (the relative difference between observed and expected was almost as great for the innermost cells as for those more peripheral, despite the fact that much less overall toxicity was observed).

Defining "synergism" presents continuing problems, since the numerous techniques which quantify the degree of drug interaction (Steel and Peckham, 1979; Chou and Talalay, 1984; Gessner, 1988; Berenbaum, 1989) often lead to somewhat different conclusions. We find the isobologram technique to be one of the more useful, since not only sub- or supraadditive interactions can be identified, but also an "envelope of additivity" is defined which in turn provides mechanistic information concerning the mode(s) of action of the drugs. The panels on the right in Fig. 5 show isobologram analyses (at the 95% level of cytotoxicity) for the 10% outermost and 10% innermost cells of spheroids. Several features are evident; one of the more important is that a supraadditive interaction is only seen at specific drug administration ratios (it is of considerable interest that these overlap the ratios and expected doses that are used clinically). It is also of note that increased interactions are seen when a higher proportion of toxicity is due to platinum than due to etoposide. To maximize the interaction in the internal (quiescent) cells of the spheroids, it is necessary to use an even higher platinum to etoposide dose ratio.

Figure 5 shows the interaction of two agents which have relatively similar patterns of toxicity through the spheroids; much of our other work has focused on combinations of drugs which have "complementary" rather than similar toxicity patterns. For example, based on Fig. 3, one might expect that cisplatin and mitomycin C would be logical candidates to use together, as would cisplatin and the nitrosoureas. We have in fact demonstrated synergistic interactions for

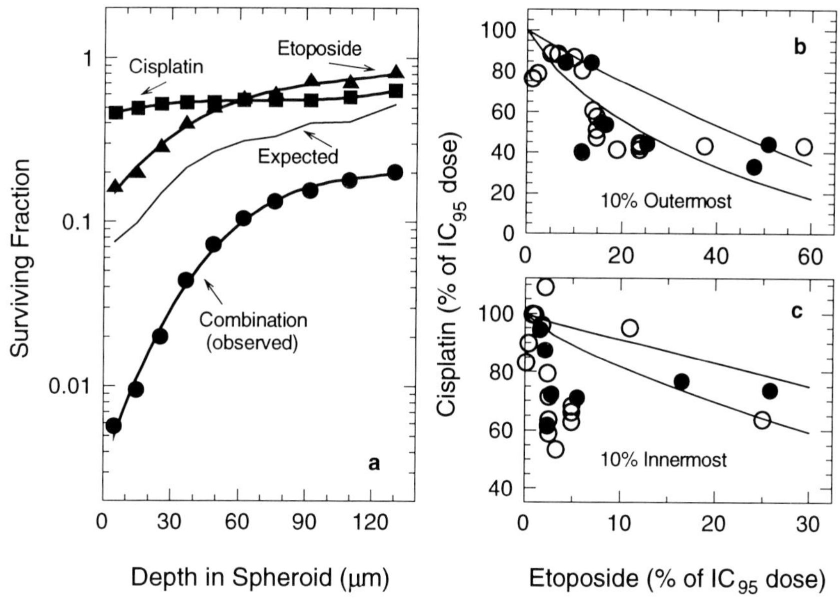

Fig. 5 (a) Clonogenicity of cells sorted from V79 spheroids exposed for 1 hr to 4.0 μg/ml cisplatin, 4.0 μg/ml etoposide, or the combination; the line without symbols shows the expected response for independent toxicity of the agents. From numerous experiments like those in (a) dose–response curves were produced for each subpopulation of cells, allowing isobologram analyses at various survival levels as in (b) and (c). Open symbols indicate 3- to 5-point dose–response curves, with closed symbols from curves defined by at least six values. Note that supraadditive responses were seen only for selected dose combinations; additionally, greater synergism (further displacement from the "envelope of additivity" shown by the bounding curves) was evident for the innermost cells.

both of these combinations (Durand, 1989b,1990b). While increased tumor cell kill is always desired, it must be remembered that increasing the "therapeutic ratio," that is, the ratio of damage to malignant versus normal tissue, is of paramount importance. While this cannot be directly addressed in spheroids, increased activity against the tumor-like internal cells is, we believe, an essential prerequisite for improved therapeutic efficacy in the clinic.

VII. Multifraction Treatments

A. Advantages of Spheroids

Spheroids growing in stirred suspensions can be easily (though not inexpensively) maintained and at larger sizes grow relatively slowly thus providing a quasi-steady-state system ideal for multifraction experiments with cytotoxic agents. It is rather surprising how little information is available for multifraction drug or radiation administrations; in most laboratory studies that have been performed with cell or tumor systems, observations have necessarily been made only for a few time points. Consequently, mechanistic information often can only be deduced from a judicious choice of agents and experimental protocols. This is rather unfortunate, in that clinical activity in therapeutic protocols is typically achieved only with multiple treatments.

B. Regrowth: A Dominant Variable

A major advantage of the spheroid system is that large numbers of identical spheroids can be propagated, thus making it feasible to repetitively determine the number of viable cells per spheroid *before and after every treatment* in a multifraction regimen. Typical experiments using treatments expected to be about equally toxic based on single-treatment experiments are shown for cisplatin, doxorubicin, and irradiation in Fig. 6; the drugs were administered daily and radiation twice daily. Interestingly, the spheroids responded to both drugs in a way reminiscent of many tumors in patients: an initial, rapid decrease of the number of viable cells per spheroid was seen, but eventually, regrowth began despite continued treatment. In the clinic, this is routinely described as "emergence of drug resistance." The data in Fig. 6, however, question the accuracy of that terminology. Since the number of viable cells was measured before and after every treatment, it can be seen from Figs. 6d–6f that the cell kill per treatment actually *increased* for each agent throughout the course of the regimen, yet was ultimately offset by sufficient regrowth to produce the apparent "resistance."

These observations raise a number of intriguing questions for cancer therapy. Clearly, they highlight the importance of regrowth or repopulation rate; the amount of regrowth is perhaps not surprising to investigators with a flow cytometry (or cell kinetics) background, since these spheroids, at the start of treatment, had growth fractions of only 30–40%. The depopulation induced by treatment led to an increase in the growth fraction to virtually 100%, with a concomitant ability of the system to more than compensate for the "per treatment" cell kill. If a similar phenomenon occurs in human tumors, one must question whether the "induced resistance" seen in the clinic has a genetic or, perhaps, a *kinetic* basis. If the former, we need better drugs and treatment strategies. If, however, any part of the resistance is due to changing cell kinetics, our current armamentarium of antiproliferative drugs is probably adequate, if used appropriately, to overcome that part of the problem. We find it intriguing that the argument just stated is entirely compatible with clinical experience which suggests that high-intensity treatments are often more successful, as are shorter treatment regimens.

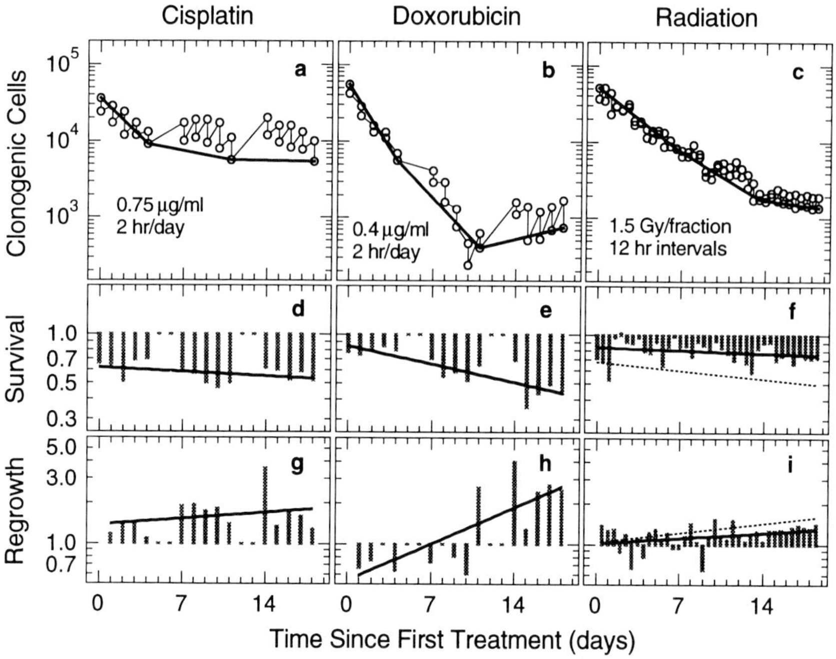

Fig. 6 Multifraction treatments of spheroids, where the number of viable cells per spheroid was determined immediately before and after each treatment (top). (Center) Net cell kill per treatment; (bottom) increase in viable cells per spheroid between exposures. The lines in the top panels indicate the (rather paltry) information that would have been forthcoming had only one assessment per week been performed; in the bottom six panels, regression lines showing the trends of the data are shown. Note that radiation was delivered twice as frequently as drugs; the dotted lines (f) and (i) show the net toxicity and regrowth that would have been measured for the comparable case of one assessment per day. In all cases, the cells became progressively more sensitive to treatment; regrowth rate was, however, always sufficient to confer a diminishing net response.

VIII. Cell Kinetics: Role of Pretreatment Kinetics

The data just described indicate the considerable role of cell kinetics in determining the response of the spheroid system and, by analogy, that of human tumors to multifraction therapy. Flow cytometry, with the new techniques that have evolved for measuring cell kinetics in human tumors (Begg *et al.,* 1985), has rekindled an interest in this important field. It is now possible and indeed commonplace, for the potential doubling time (T_{pot}) of human tumors to be determined on an individual basis with high precision and reliability (Begg *et al.,* 1985; Hall *et al.,* 1992; White and Terry, 1992). Even with the obvious question that must be raised about the adequacy and representativeness of any particular biopsy specimen, evidence is emerging that this type of kinetic data may identify patients more or less likely to respond to conventional therapy (Begg *et al.*, 1992). A question that remains, however, is whether *pretreatment* cytokinetic parameters are always relevant to ultimate outcome. Stated differently, is it the observed pretreatment growth kinetics which are of importance or, rather, tumor repopulation potential which will ultimately determine the response of the tumor and fate of the patient?

This question is addressed for spheroids in Fig. 7, where experiments like those shown in Fig. 6 have been reanalyzed to determine the actual doubling time of the clonogenic cells at various stages of spheroid treatment with cisplatin, radiation, or doxorubicin. Clearly, in all cases, the observed doubling time (T_d) of the clonogenic cell number within the spheroid decreased as the number of viable cells was decreased; the different patterns observed for the different agents reflect combinations of depopulation rate and growth inhibition induced by the treatment (cisplatin shows little if any cytostatic activity, whereas doxorubicin is markedly cytostatic as well as cytotoxic). Again, if the spheroid results are representative, one must question whether the pretreatment T_{pot} or T_d of tumors really will be predictive for the repopulation potential of the system.

Clearly, there are always problems in extrapolating from any laboratory system to the clinic. The spheroid system is only a model and therefore unlikely

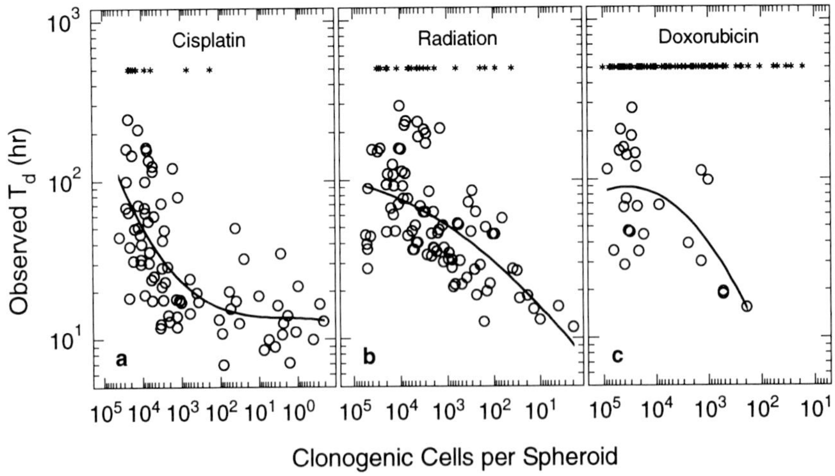

Fig. 7 Data from experiments like those in Fig. 6, where the doubling time of the clonogenic cells was calculated from the amount of regrowth observed in each time interval. In some cases, minimal regrowth (or, occasionally, cell loss) was recorded between treatments; those data were plotted as stars at an arbitrary doubling time of 500 hr and excluded from the regression analysis. In all cases, data are included for a variety of treatment intensities and interfraction intervals; the treatment scheme per se was clearly important only in terms of the rate or efficiency of spheroid cell depopulation.

to be quantitatively predictive for any type of human cancer. Nonetheless, the value of the spheroid system seems heightened by the qualitative similarities of spheroid and human tumor response to protocols like those just described. We also believe that it may be particularly noteworthy and informative that spheroid growth kinetics, at the time we begin multifraction studies, show T_{pot} values in the 3- to 7-day range, with T_d values of 14–40 days. Neither is grossly different than the averages emerging from clinical tumor cell kinetics studies (Rew *et al.*, 1991; Remvikos *et al.*, 1991; Begg *et al.*, 1992).

IX. Alternatives for Kinetic Measurements: Overcoming Heterogeneity

A common problem which arises when trying to deduce the cell kinetics of tumors is the degree of cellular heterogeneity observed. Like tumors, as spheroids enlarge, karyotypic heterogeneity occurs. In fact, the V79 Chinese hamster

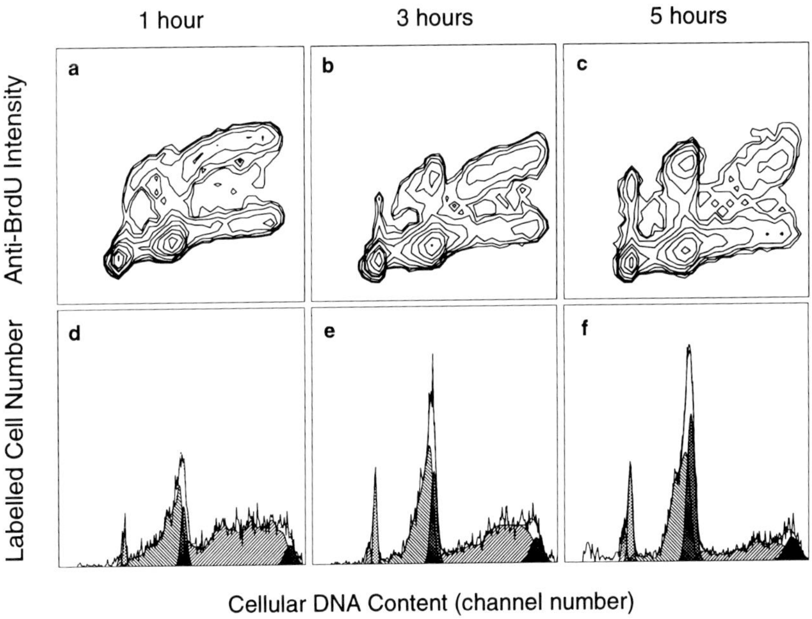

Fig. 8 An illustration of the difficulties in deducing cell kinetic information with a heterogeneous system. Large V79 spheroids were pulse labeled with BrdU for 10 min, then conventionally analyzed to determine T_s and T_{pot}. Note the difficulty (impossiblity!) of calculating the relative movement of the S phase cells, since the cells in the 4N (middle) peak cannot be classified as diploid versus tetraploid. Using objective analyses of the DNA distributions of the labeled cells (bottom panels, where the diploid G_1, diploid S, 4N, tetraploid S, and tetraploid G_2 cells were modeled by commercial software) this problem can be overcome (see text, and Durand, 1993).

cell spheroid system is perhaps the worst case example, since the initial, diploid starting cell population spontaneously and irreversibly becomes tetraploid as the spheroids enlarge (Olive *et al.*, 1982). This has some interesting consequences for flow cytometric estimations of T_{pot}, as shown in Fig. 8.

The bivariate plots of anti-BrdU intensity and cellular DNA content for BrdU pulse-labeled spheroids in the top panels of Fig. 8 show the difficulty in assigning BrdU labeled cells to the diploid or tetraploid component as a function of time after labeling. Both S phase populations are progressing, but it is not clear whether the 4N cells, for example, represent diploid G_2 cells or tetraploid G_1. To deal with that difficulty, we developed an alternate method for estimating T_{pot} based on objective estimates of the progression of labeled cells out of S phase (Durand, 1993). This is illustrated using a commercial software package for DNA histogram analysis (Modfit, from Verity Software House, Topsham, ME) in the lower panels of Fig. 8, leading to the new analysis which we have called the "%S" technique.

Unlike the conventional relative movement technique introduced by Begg *et al.* (1985), we do not measure the average DNA content of the proliferating, labeled cells. Rather, we use the number (or fraction) of initially labeled cells *remaining* in S phase at the time of measurement to indicate the rate of progression from S phase and, therefore, the duration of S phase T_s. Two advantages arise: (1) computerized DNA analysis gives objective estimates of cell distribution within the cycle, and (2) the fate of the labeled cells becomes irrelevant (one need not assume that they continue to progress).

X. General Conclusions

The topics that we have reviewed in this chapter have been chosen to highlight the unique types of data that can be obtained with the spheroid system and, additionally, to show the invaluable contribution of flow cytometry and cell sorting to those analytical techniques. Cellular heterogeneity is, arguably, the most unique and troublesome feature of spheroids. Understanding the responses of the heterogeneous cell subpopulations, which is a necessary prerequisite to understanding the response of the system as a whole, is only possible if one has the technology to separately recover and study those heterogeneous cells. Clearly, the "wedding" of multicell spheroids and flow cytometry techniques has been a productive and informative undertaking.

Often our studies have produced unexpected and controversial conclusions; equally often, our results have been categorized as merely "reinforcing the obvious." At the risk of continuing the latter, we briefly emphasize some of the more important conclusions illustrated by the data presented here. First, the action(s) of cytotoxic or cytostatic agents are unlikely to be identical in single-cell versus multicell systems. This raises obvious questions concerning the utility of any predictive assay based on the *in vitro* response of tumor cells and also questions the adequacy of conventional test systems for new drug development. Second, the presumption that each treatment in a multifraction tumor therapy regimen will be of equal efficacy is totally incompatible with prevailing concepts of tumor heterogeneity and tumor growth dynamics. Similarly, the presumption that the pretreatment T_{pot} will reflect the maximal tumor repopulation rate can only be valid for the rather unlikely situation where the tumor growth fraction is 100%! Finally, and still on the theme of growth kinetics, we believe it may be instructive that spheroids display "emergence of drug resistance" (according to standard clinical criteria) under conditions where only the repopulation potential has changed markedly. As previously argued, this has (very testable) implications for clinical radio- and chemotherapy protocols.

In conclusion, it is perhaps notable as well that many of the techniques just described for multicell spheroids can be applied, albeit with more difficulty, to the study of certain tissues and tumors in experimental animals (Chaplin *et al.*, 1985; Olive *et al.*, 1985; Yan and Durand, 1991). In particular, for human tumors xenografted into immunocompromised hosts, the ability to collect subsets of cells and to quantitatively assay their response to cytostatic and cytotoxic perturbations is, we feel, an important advantage. Without question, the further evolution of flow cytometry systems and techniques will have continued impact.

Acknowledgments

Supported by the Medical Research Council of Canada and the National Cancer Institute of Canada.

References

Acker, H., Carlsson, J., Durand, R. E., and Sutherland, R. M., eds. (1984). "Spheroids in Cancer Research." Springer-Verlag, Berlin.
Begg, A. C., McNally, N. J., Shrieve, D. C., and Kärcher, H. (1985). *Cytometry* **6**, 620–626.
Begg, A. C., Hofland, I., Van Glabekke, M., Bartelink, H., and Horiot, J. C. (1992). *Semin. Radiat. Oncol.* **2**, 22–25.
Berenbaum, M. C. (1989). *Pharmacol. Rev.* **41**, 93–141.
Carlsson, J. (1978). *In Vitro* **14**, 860–867.
Carlsson, J., and Nederman, T. (1989). *Eur. J. Cancer Clin. Oncol.* **25**, 1127–1133.
Chaplin, D. J., Durand, R. E., and Olive, P. L. (1985). *Br. J. Cancer* **51**, 569–572.
Chou, T. C., and Talalay, P. (1984). *Adv. Enzyme Regul.* **22**, 27–55.
Dalen, H., and Burki, H. J. (1971). *Exp. Cell Res.* **65**, 433–438.
Durand, R. E. (1982). *J. Histochem. Cytochem.* **30**, 117–122.
Durand, R. E. (1984). *Int. J. Radiat. Oncol. Biol. Phys.* **10**, 1235–1238.
Durand, R. E. (1986a). *Cancer Res.* **46**, 2775–2778.
Durand, R. E. (1986b). *J. Natl. Cancer Inst.*, **77**, 247–252.
Durand, R. E. (1989a). *J. Natl. Cancer Inst.* **81**, 146–152.
Durand, R. E. (1989b). *Int. J. Cancer* **44**, 911–917.
Durand, R. E. (1990a). *Cancer Chemother. Pharmacol.* **26**, 198–204.
Durand, R. E. (1990b). *Br. J. Cancer* **62**, 947–953.
Durand, R. E. (1993). *Cytometry* **14**, 527–534.

Durand, R. E., and Olive, P. L. (1981). *Cancer Res.* **41,** 3489–3494.
Durand, R. E., Chaplin, D. J., and Olive, P. L. (1990). *In* "Methods in Cell Biology" (Z. Darzynkiewicz and H. A. Crissman, eds.), Vol. 33, pp. 509–518. Academic Press, San Diego.
Gessner, P. K. (1988). *J. Am. Coll. Toxicol.* **7,** 987–1012.
Hall, P. A., Levison, D. A., and Nicholas, A. (1992). "Assessment of Cell Proliferation in Clinical Practice." Springer-Verlag, New York.
Inch, W. R., McCredie, J. A., and Sutherland, R. M. (1970). *Growth* **34,** 271–282.
Krishan, A., and Ganapathi, R. (1979). *J. Histochem. Cytochem.* **27,** 1655–1656.
MacAllister, R. M., Reed, G., and Huebner, R. J. (1967). *J. Natl. Cancer Inst.* **39,** 43–47.
Mueller-Klieser, W. (1987). *J. Cancer Res. Clin. Oncol.* **113,** 101–122.
Olive, P. L., Leonard, J. C., and Durand, R. E. (1982). *In Vitro* **18,** 708–714.
Olive, P. L., Chaplin, D. J., and Durand, R. E. (1985). *Br. J. Cancer* **52,** 739–746.
Remvikos, Y., Vielh, P., Padoy, E., Benyahia, B., Voillemot, N., and Magdelénat, H. (1991). *Br. J. Cancer* **64,** 501–507.
Rew, D. A., Thomas, D. J., Coptcoat, M., and Wilson, G. D. (1991). *Br. J. Urol.* **68,** 44–48.
Steel, G. G., and Peckham, M. J. (1979). *Int. J. Radiat. Oncol. Biol. Phys.* **5,** 85–91.
Sutherland, R. M. (1988). *Science* **240,** 177–184.
Sutherland, R. M., and Durand, R. E. (1976). *Curr. Top. Radiat. Res. Q.* **11,** 87–139.
White, R. A., and Terry, N. H. A. (1992). *Cytometry* **13,** 490–495.
Yan, R., and Durand, R. E. (1991). *Int. J. Radiat. Oncol. Biol. Phys.* **20,** 271–274.
Yuhas, J. M., Li, A. P., Martinez, A. O., and Ladman, A. J. (1977). *Cancer Res.* **37,** 3639–3643.

CHAPTER 25

Functional Measurements Using HL-60 Cells

J. Paul Robinson, Padma Kumar Narayanan, and Wayne O. Carter

Department of Physiology and Pharmacology
Purdue University
West Lafayette, Indiana 47907

I. Introduction

In 1977 Collins *et al.* described a cell line that was derived from the peripheral blood of a patient with acute myelogenous leukemia (Collins *et al.,* 1977). These peripheral blood leukocytes, designated HL-60, are derived from the progenitors of granulocytes and monocytes which have undergone neoplastic transformation. Hence HL-60 cells resemble the blast cells of their lineage phenotypically (myeloblastic and promyelocytic cells with azurophilic granules) (De La Maza *et al.,* 1985; Harris and Ralph, 1985). However, more mature myeloid cells (myelocytes, metamyelocytes, bands, and segmented neutrophils) may also be seen (Collins *et al.,* 1977). HL-60 cell lines are heterogeneous, proliferate continuously in suspension culture, and exist in an arrested, yet pliant state of maturation (Dufer *et al.,* 1989). Proliferation is exponential with a doubling time of 34 hr up to a density of 3×10^6 cells/ml (Collins *et al.,* 1977; Cowen *et al.,* 1991).

Exposure of HL-60 cells to a variety of agents induces terminal differentiation along either of two distinct pathways—one that bears many of the surface properties and functional characteristics of mature neutrophils and the other that resembles characteristics of monocytes/macrophages (Koeffler, 1983). This property of the cell line has been exploited by researchers in different laboratories to: (i) study differentiation of blast cells to granulocytes and macrophages, (ii) examine the genetic defects that may be responsible for the maturational arrest of leukemic cells *in vitro,* (iii) delineate the nature of their neoplastic transformation and define the conditions under which their malignancy can be reversed, and (iv) design therapy to induce maturation *in vivo* which might be expected to alter cell-cycle kinetics (Koeffler, 1983; Harris and Ralph, 1985; Ross, 1985; Thompson *et al.,* 1988; Dufer *et al.,* 1989).

Koeffler (1983) described many different types of maturation inducers. They can be classified as monocyte/macrophage-like differentiation inducers and granulocyte-like differentiation inducers. Agents that fall into the former category are phorbol diesters (Katagiri *et al.,* 1992), phospholipase C (PLC) (Cowen *et al.,* 1991; Madden *et al.,* 1992), teleocidins (Koeffler, 1983), cytokines, and vitamin D derivatives (Zhou *et al.,* 1991). Granulocyte-like differentiation inducers can be divided into physiologic and nonphysiologic inducers (Koeffler, 1983). The physiologic inducers are retinoids (Matzner *et al.,* 1987; Janick-Buckner *et al.,* 1991) and colony stimulating factors (CSFs) (Koeffler, 1983). The nonphysiologic inducers of differentiation are dimethyl sulfoxide (DMSO); purine and pyrimidine analogues; chemotherapeutic agents like actinomycin D, dibutyryl c-AMP bromodeoxyuridine, 5-azacytidine, 6-thioguanine, daunomycin, cytosine arabinoside, and vincristine; and P_2 purinergic receptor agonists (Koeffler, 1983; Matzner *et al.,* 1987; Thompson *et al.,* 1988; Cowen *et al.,* 1991). Calcium ionophores and phorbol diesters have been shown to act synergistically in inducing monocytic differentiation of HL-60 cells (Cowen *et al.,* 1991). Pilz *et al.* (1987) suggested that HL-60 differentiation can also be induced by starvation of a single essential amino acid. Granulocytic and monocytic differentiation inducers are known to increase the intracellular cAMP-dependent protein kinase C activity and decrease *c-myc* oncogene expression (Koeffler, 1983; Ross, 1985; Forsbeck *et al.,* 1985; Chaplinski and Niedel, 1986). Nitroprusside and $NaNO_2$, which activate cytosolic guanylate cyclase and increase the intracellular cGMP concentration, induce granulocytic differentiation of HL-60 cells (Boss, 1989).

II. Application

The properties described above can be used to establish a useful model for functional characterization of phagocytic cells. Parameters that are usually adopted as yardsticks of granulocyte differentiation are (i) increased oxidative burst response to soluble as well as particulate stimuli, (ii) increased intracellular calcium mobilization in response to formylated chemotactic peptides, (iii) a reduction in cell size (see Fig. 1) with a decreased nuclear/cytoplasmic ratio, (iv) a sharply reduced rate of proliferation, (v) a reduction in the number of cells expressing transferrin receptors, (vi) increases in the percentage of cells expressing both type 1 (CR1) and type 3 (CD11b) complement receptor, and (vii) decreased esterase content. Changes suggestive of the monocyte/macrophage pathway are increased adherence, growth inhibition, decreased levels of *c-myc* mRNA, and expression of monocyte cell-surface markers (CD14) (Cowen *et al.*, 1991). Induced and uninduced HL-60 cell populations are heterogenous in their stages of differentiation. As induction causes a shift to a much higher proportion of mature cell types, all stages from promyelocyte to PMN are present in the culture, at each step of differentiation. Virtually all of the above functions can be relatively easily evaluated using flow cytometry.

III. Materials and Methods

The HL-60 cell line (ATCC) is maintained in RPMI medium 1640 (Sigma Chemical Company, St. Louis, MO). To make 200 ml total volume, the medium is supplemented with 5% fetal calf serum (Harlan Bioproducts for Science, Indianapolis, IN); 5% newborn calf serum (Sigma); 20,000 U penicillin, 20 mg streptomycin, and 50 μg amphotericin B (Sigma); and 2 ml 200 mM L-glutamine (Sigma) at 37°C, 5% CO_2, and cultured in 25-cm^2 flasks. Cell viability is assessed by trypan blue dye exclusion or by propidium iodide (PI) dye uptake using flow cytometry.

A. Estimation of Cellular Esterases

Enzymatic activities of cells or tissues can be determined with flow cytochemistry which offers some special advantages as a method of measuring myeloid maturation (Ross, 1986). With absorptive histochemical stains one can distinguish the various stages of myeloid maturation and differentiate between the monocytic and granulocytic maturation (Malin-Berdel and Valet, 1980; Thompson *et al.*, 1988).

Principle: The presence and approximate quantity of cellular esterases can be estimated by adding a dye such as carboxyfluorescein diacetate, which is rapidly hydrolyzed by the cellular esterases, and measuring the fluorescence of cells. The quantity of esterases is directly proportional to the stage of maturation of the HL-60 cells, especially when this cell line is induced to differentiate along the monocytic pathway. Cells induced to monocytic maturation have increased quantities of esterase. On the other hand, cells induced to differentiate along the granulocytic pathway have a reduced esterase content.

1. Reagents

Phosphate-buffered saline (PBS): Use fluorescent antibody (FA) Bacto buffer (Difco Laboratories, Detroit, MI), 100 g (bottle) and make up to 10 liters, pH 7.20.

Carboxyfluorescein diacetate (CFDA-AM) (Molecular Probes) (Cat. No. C-1354): Prepare a 12 mM stock solution of pure AR grade CFDA-AM in 1.0 ml spectrograde acetone (Watson, 1993). Keep this in the dark at -10°C. Add 50 μl stock CFDA-AM to 250 μl PBS. This is a 2 mM solution. Use 1 μl per 1 ml of cells for a final 2 μM concentration.

Stock PBS gel:

EDTA (disodium salt) 0.2 M	7.604 g
Dextrose, 0.5 M	9.0 g
Gelatin, 10% (Difco)	10 g
Distilled water	100 ml.

To make PBS gel, heat water to 45–50°C and slowly add gelatin while mixing with a magnetic stirrer. Continue stirring and add EDTA and dextrose. Do not exceed 55°C because gelatin and glucose will "caramelize." Store in 1.2-ml aliquots at -20°C. Remove 1 ml to make 100 ml stock buffer.

Working solution PBS gel (make daily as needed): warm 1 ml gel to 45°C, add 95 ml warm PBS, and mix. Adjust pH to 7.4 and make up to 100 ml.

2. Instrumentation

Using appropriate beads calibrate linear green fluorescence (IGF) to the desired channel at a gain of 10 or 20. Collect forward light scatter (FALS) and IGF for each tube. Record settings each day. Select the population of interest using FALS-90° light scatter. Set the gates with unloaded cells (no CFDA-AM), then check the gates with CFDA-AM-loaded cells. Collect LOG green fluorescence if satisfactory log–lin conversions are available on the instrument. Excitation is 488 nm, and emission is collected with a 525-nm band pass filter.

3. Cell Preparation

Prepare a population of cells (1×10^6/ml) in PBS gel buffer. Warm the cells to the working temperature (37°C). Mix 1 ml of cells and 1 μl of dye and collect data for a specified time interval (e.g., 10 min).

4. Comments

When HL-60 cells are induced to granulocytic maturation, the number of cells that are positively stained for esterases will decrease daily. In contrast, their expression is upregulated during monocytic differentiation. Thus, esterases are inducible and their measurement offers a useful marker of cell maturation.

B. Expression of CD11b Receptors

Studies on the regulation of expression of CD11b (MAC-1) during differentiation of HL-60 cells represent a valuable approach for characterization of the expression of leukocyte adhesion molecules during human myeloid differentiation. Surface changes that occur during the differentiation of myelomonocytic precursor cells to granulocytes and monocytes/macrophages can be visualized through the adoption of suitable *in vitro* models like HL-60 which can be induced to differentiate along the myelomonocytic pathway. The use of monoclonal antibodies directed against the surface antigens that show marked changes during differentiation enables separation of mature from immature cells (Perussia *et al.*, 1981; Janick-Buckner *et al.*, 1991; Back *et al.*, 1991) and provides a method to track the differentiation.

1. Principle

Differentiation of HL-60 cells along the granulocytic pathway results in enhanced CD11b surface antigen expression, consistent with high levels of expression of CD11b on human granulocytes. This CD11b molecule is noncovalently associated with a common β or CD18 unit which increases with differentiation along both granulocytic and macrophage pathways. The CD11b molecule is a membrane glycoprotein present on mature granulocytes and monocytes and is expressed by less than 10% of undifferentiated cells. These glycoproteins are thought to mediate binding and migration of leukocytes through the vascular endothelium, and attachment of C3bi-coated particles as in serum complement opsonized phagocytosis.

2. Reagents and Methods

OKM1 (Ortho Diagnostics, Raritan, NJ)

MO-1/FITC (Coulter Immunology, Hialeah, FL)

GAM-FITC (Goat-Anti-Mouse antiserum)(Caltag, San Francisco, CA)

Phosphate-buffered saline: To a 100-g bottle of Bacto buffer add distilled water to 10 liters, and adjust pH to 7.40

2% paraformaldehyde (PF): In a 500-ml flask, combine 10 g PF, 5 g (FA) Bacto buffer, and 400 ml distilled H_2O. Stir until solids dissolve. This may take several hours; the materials can be left on the stirrer overnight. When the solution has cleared, adjust pH to 7.3–7.4 and make up to 500 ml with distilled water. Place in a 500-ml bottle, cover with foil to protect from light, and store at 4°C.

3. Comment

Normally use PF with equal volumes of reagent to be fixed (final concentration of PF is 1%); however, we have found better phenotyping with HL-60 cells if the 2% PF is added directly to the vortexed cell pellet. This reagent is good for 1 month.

4. Indirect Immunophenotyping

After 48 hr of culture, obtain a cell count. Centrifuge cultures at 200*g* for 5 min at room temperature. Resuspend in PBS at 3–5 × 10^6/ml. Aliquot 100 μl cells into 12 × 75-mm tubes. Add 10 μl OKM1 to the cells and incubate at 4°C for 30 min. Add 10 μl of 1 : 10 dilution of GAM-FITC after the incubation. Add 2 ml of PBS to each tube. Spin at 250*g* for 10 min at 4°C. Aspirate supernatant, cap tubes, and vortex. Add 250 μl 2% PF while vortexing. Cover and store in refrigerator until ready to run on cytometer. Collect ungated list-mode data (FALS, 90° scatter, log FITC). Control samples should be labeled with GAM-FITC only, plus pooled human serum.

5. Direct Immunophenotyping

An antibody which is directly conjugated to a fluorescent label like FITC or PE can be used for this method. Add 10 μl of Mo1-FITC (1 : 8) (or similar antibody) to 100 μl of cells. Incubate for 30 min at 4°C. Add 2 ml PBS to the tube and spin at 250*g* for 10 min at 4°C. Aspirate supernatant EdE and vortex. Add 250 μl of 2% PF while vortexing to fix the samples. Cover and store in refrigerator until ready to run on cytometer. Proper isotypic controls should be used.

6. Instrumentation

Set up a two-parameter histogram (FALS-90° LS) to gate differentiated and undifferentiated cells. Set bitmaps on the population of interest. For instance, it might be important for drug screening studies to discriminate between differentiated and undifferentiated cells based on scatter (see Fig. 1). Set histograms for LOG FITC gated on FALS-90° scatter plot and collect 5000 events.

C. Oxidative Burst Measurements

Differentiated HL-60 cells are capable of demonstrating a respiratory burst after stimulation with PMA, FMLP, or other activating agents. Either H_2O_2 or O_2^- can be measured using standard flow cytometric methods. The techniques are described in detail in Chapter 28 in volume 41.

1. Materials

a. Hanks' balanced salt solution (HBSS)

i. Stock HBSS, 10× concentrated: NaCl, 40 g; KCl, 2.0 g; Na_2HPO_4, 0.5 g; $NaHCO_3$ 0.5 g; add distilled water to 500 ml.

ii. Stock Tris, 1.0 *M:* Tris base, 8.0 g; Tris–HCl, 68.5 g; add distilled water to 500 ml; adjust pH to 7.3.

iii. Preparation of 100 ml HBSS:

Stock HBSS, 10×	10 ml
Distilled water	80 ml
Tris, 1.0 *M*	2.75 ml
$CaCl_2$, 1.1 *M*	170 μl
$MgSO_4$, 0.4 *M*	200 μl
Dextrose	220 mg.

Adjust pH to 7.4 and add distilled water to 100 ml.

b. PBS gel (see previous description).

c. 2′,7′-Dichlorofluorescin diacetate (DCFH-DA, MW 487.2) (Molecular Probes Inc., Eugene, OR), 20 m*M* solution:

i. Weigh 2–9 mg of DCFH-DA and place in a foil-covered 12 × 75-mm tube.

ii. Add absolute ethanol in a milliliter volume equivalent to the weight in milligrams of the DCFH-DA divided by 9.74. For example, if you add 4 mg DCFH-DA add 4/9.74 = 402 μl of ethanol.

iii. Cap the tube, mix, cover in foil, and store at 4°C until use.

d. Hydroethidine (HE, MW 315) (Molecular Probes Inc., Eugene, OR), 10 m*M* solution: Stock solution 10 m*M* in dimethylformamide (3.15 mg/ml).

e. PMA (phorbol 12-myristate 13-acetate) (Sigma Chemical Co., St. Louis, MO): PMA is toxic and carcinogenic; additionally DMSO is readily absorbed through the skin. Wear gloves while handling solutions, prepare solutions in a hood, and be extremely cautious!

i. Stock PMA (2 mg/ml in DMSO): Mix well and aliquot 15–20 μl of stock PMA in small capped polypropylene bullets. Store at −20°C.

ii. Working PMA solution (make daily as needed): 5 μl PMA stock in 10 ml PBS gel (1000 ng/ml PMA solution). A final PMA concentration of 100 ng/ml will predictably result in maximal cell stimulation (for example, use 900 μl of cells in solution and 100 μl of working PMA solution). It is actually better to add a very small volume of activation reagent so as not to dilute the dye concentration. Thus it is preferable to add 1 μl/ml if possible. However, be aware that addition of very small volumes of fluorochromes may result in greater error.

2. DCFH-DA Assay

i. Centrifuge cells from one large flask (25 cm^2) to remove medium. Resuspend in PBS gel at 2 × 10^6/ml.
ii. Wash (250*g*) for 10 min and resuspend in HBSS plus 1% BSA. Final cell concentration should be 2.0 × 10^6 cells/ml.
iii. Add 1 μl 20 m*M* DCFH-DA per ml of cell suspension to be loaded.
iv. Incubate loaded cells at 37°C for 15 min.
v. Stimulate cells with PMA: add 100 μl PMA (working solution) to 900 μl of cell suspension (final PMA concentration 100 ng/ml). Reserve some loaded, unstimulated cell suspension for a control.
vi. Maintain cell sample at 37°C and run stimulated and unstimulated samples every 15 min on the cytometer for a total of 60 min.

3. Hydroethidine Assay

Procedures i and ii are identical to those above.

iii. Add 1 μl HE per ml of cell suspension to be loaded.
iv. Incubate loaded cells at 37°C for 5 min.

Procedures v and vi are identical to those above.

4. Combined DCFH-DA and Hydroethidine Assay

Procedures i and ii are identical to those above.

iii. Add 1 μl 20 m*M* DCFH-DA per ml of cell suspension to be loaded.
iv. Incubate loaded cells at 37°C for 15 min.
v. Add 1 μl HE per ml of cell suspension to be loaded.
vi. Incubate loaded cells at 37°C for an additional 5 min.
vii. Stimulate cells with PMA: add 100 μl PMA (working solution) to 900 μl of cell suspension (final PMA concentration 100 ng/ml). Reserve some loaded, unstimulated cell suspension for a control.
viii. Maintain cell sample at 37°C and run stimulated and unstimulated samples every 15 min on the cytometer for a total of 60 min. It is important to establish a standard procedure for running oxidative burst assays. This can be achieved by finding fluorescent beads which fall generally within the range of fluorescence of activated cells. These beads are then used to set up the flow cytometer each time, setting the high voltage of the photomultipliers based upon the bead fluorescence. If a full calibration is performed, the mean channel fluorescence can then be equated with the quantity of H_2O_2 formed per cell.

5. Instrumentation

All studies are carried out using a 15-mW argon laser operating at a wavelength of 488 nm. Optical filters, 488-nm dichroic, 488-nm laser blocking, 550-nm dichroic, 525-nm band pass (for DCFH-DA), and 610-nm long pass (for HE), are placed in the fluorescence collection pathway. Gated list-mode data are collected for FALS, 90° scatter (90° LS), log 90° light scatter (log 90), FITC, log FITC, EB, log EB, and TIME.

6. Comment

When using esterase-dependent dyes, it is better to add as small a concentration of activating substance as possible so as not to dilute the dye concentration. Although there are many instances where this will not alter the results, especially if correct controls are used, try to maintain a constant dye concentration if at all possible. Additionally, since the DCF fluorescence emission extends into the red spectrum, it is better to use a 610-nm band pass and ensure adequate color compensation for the DCF in the EB channel when the two fluorochromes are used together.

7. Critical Aspects

Several aspects of HL-60 cell culturing must be kept in mind while interpreting experiments using repeated samples from a continously maintained culture of HL-60 cells. HL-60 cells which are heterogenous to start with become more homogenous when passaged for a very long period of time. Multiple impairments in stimulus–response coupling and cell regulation develop in association with increased rate of cell replication. Repeated passages lead to the predominance of a subset of cells that proliferate more rapidly and are more functionally impaired. Alterations in doubling time, impairment in calcium mobilization, decreased response to differentiation inducers, and decreased oxidative burst have been observed in cells beyond passage 60. Hence it is clearly important that changes in cell function be related to the age of the culture or the passage number.

IV. Results and Discussion

HL-60 cells show some unique light scatter changes with differentiation. Figure 1a shows this change as there is a dramatic reduction in cell size and increase in 90° light scatter. Since spontaneous differentiation is continually occurring, some differentiated cells are always present in a so-called "undifferentiated" populations. The reduction in size of differentiating HL-60 cells over a period of 5 days can be seen in Fig. 1b. A rapid method for checking the

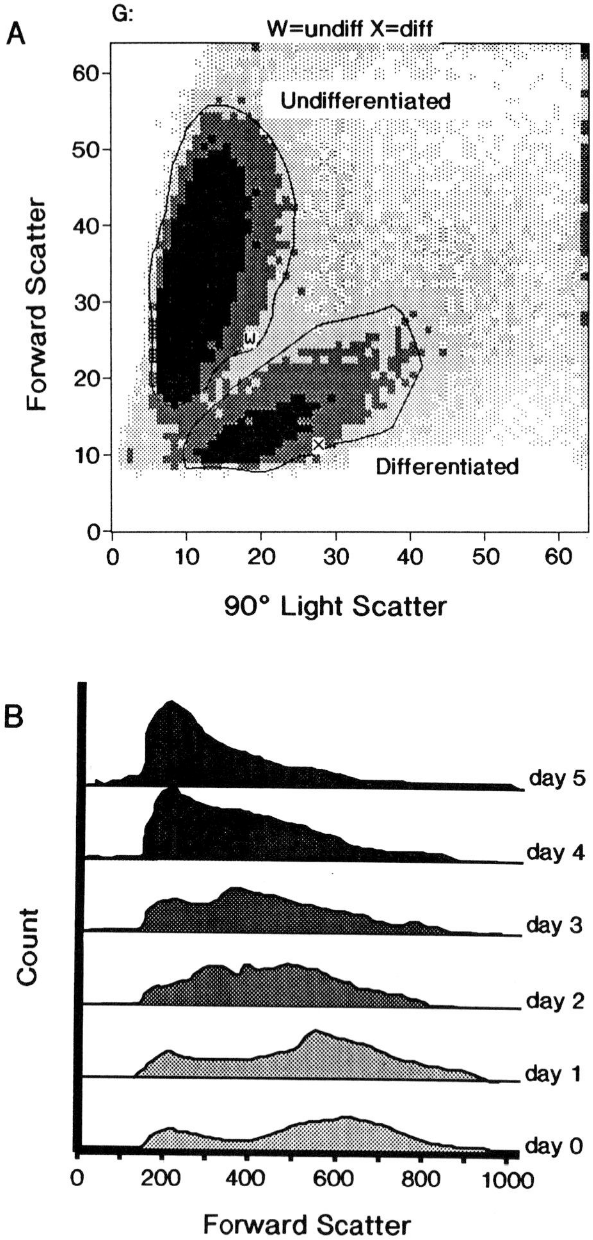

Fig. 1 A change in the scatter pattern of HL-60 cells when differentiated with DMSO. (a) Scatter plot of FALS versus 90° light scatter of both differentiated and undifferentiated HL-60 cells. Undifferentiated HL-60 cells (Day 0) have a significant cell population of large cells. This is mainly due to the large size of the immature promyelocytes and myeloblasts present in the culture. (b) On differentiation, cells become smaller and more granular (Day 5).

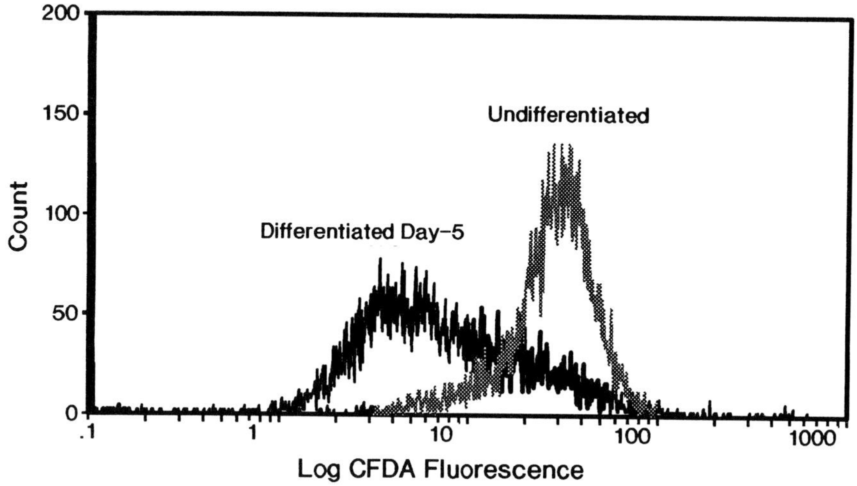

Fig. 2 Changes in esterase content of HL-60 cells when differentiated into granulocytes using DMSO. Undifferentiated HL-60 cells (Day 0) have high levels of esterase activity compared to differentiated cells. Fluorescence histograms (log CFDA fluorescence) were collected after the cells were incubated at 37°C for 30 min followed by flow cytometry.

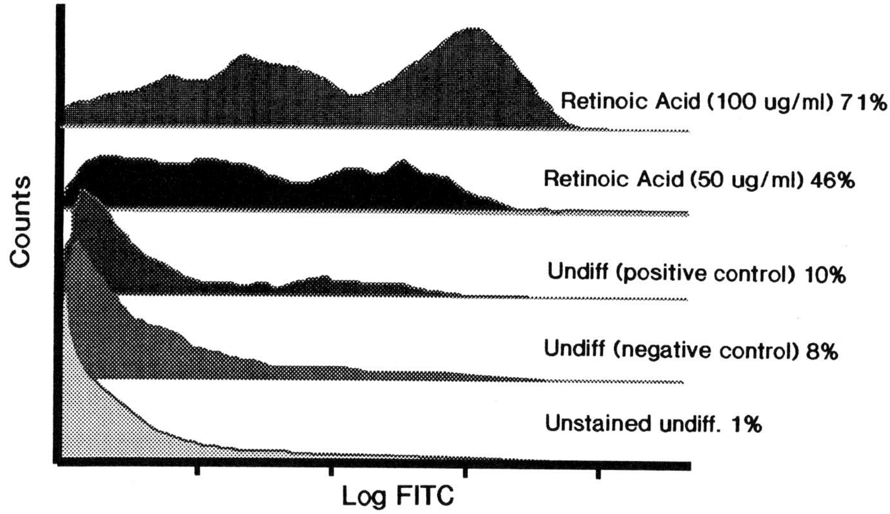

Fig. 3 A dose-dependent response of CD11b expression on HL-60 cells to retinoic acid (50–100 μg/ml). An increase in CD11b expression can be seen, with the highest dose of retinoic acid (100 μg/ml) eliciting maximum response. CD11b expression increases as HL-60 cells differentiate into mature granulocytes. Percentages of HL-60 cells expressing CD11b are shown beside each histogram. All measurements were made 48 hr after addition of differentiation agent.

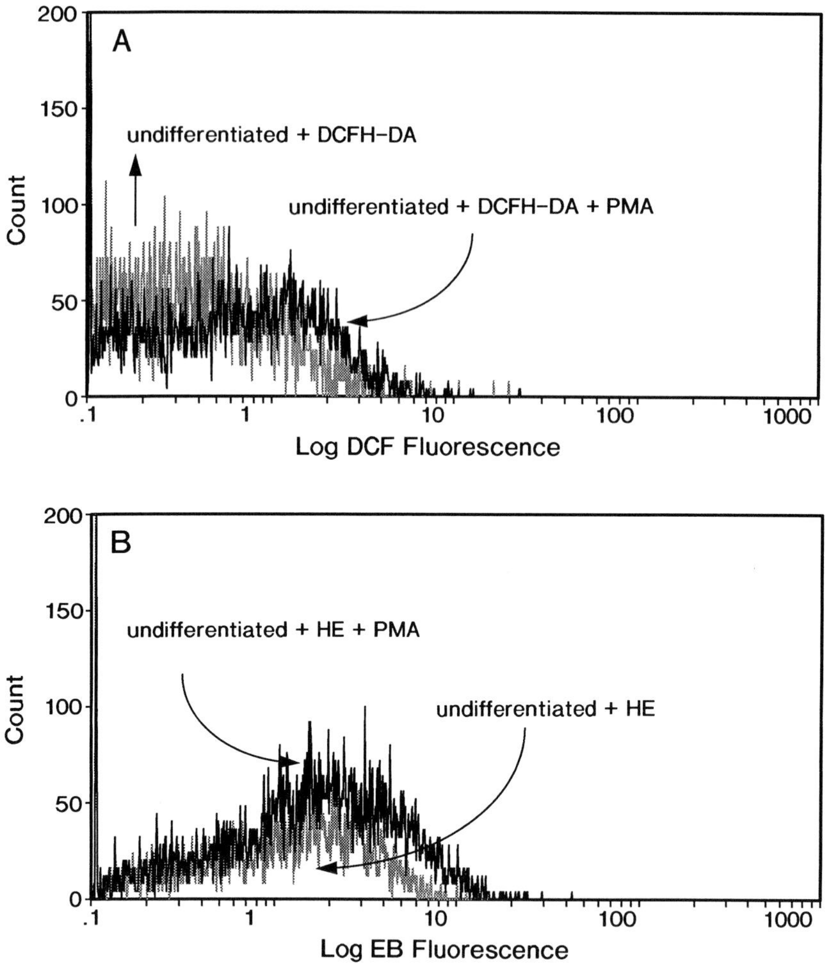

Fig. 4 Examples of PMA stimulation of HL-60 cells to produce H_2O_2 (DCF fluorescence) or O_2^- (EB fluorescence). Undifferentiated cells are shown for both the control (no stimulation) and the PMA-stimulated H_2O_2 and O_2^- responses. (A and B) H_2O_2 and O_2^- production, respectively, for undifferentiated HL-60 cells. (C and D) H_2O_2 and O_2^- production, respectively, for differentiated HL-60 cells (5 days). Responses were measured after 30 min stimulation with PMA (100 ng/ml).

differentiation status of HL-60 cells is to evaluate the esterase content as shown in Fig. 2. Undifferentiated cells have reasonably high esterase content, which decreases in granulocytic differentiation (Fig. 2) and increases in monocytic differentiation (not shown).

Differentiated HL-60 cells also express increased numbers of cell receptors such as CD11b. This is shown in Fig. 3, where the increased expression of CD11b is observed with maturation into granulocytic cells. This change in cell-

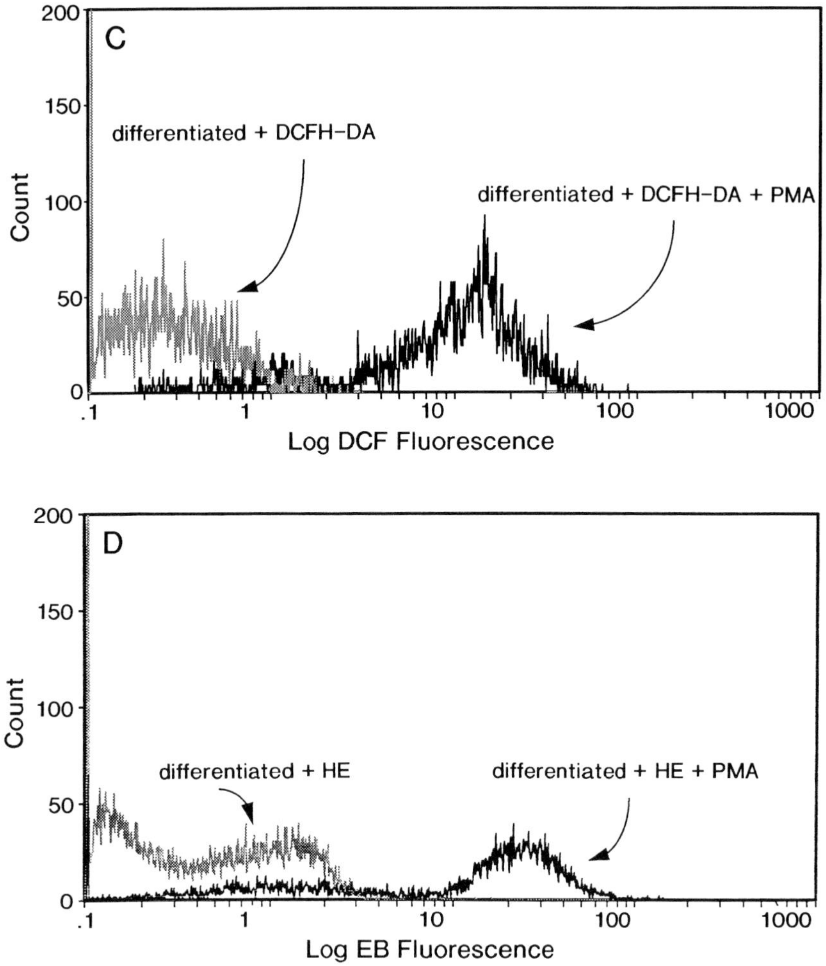

Fig. 4 Continued

surface expression is particularly useful for evaluation of chemicals thought to elicit an immune response, since screening for CD11b expression is a relatively simple assay. However, one must control for the high background fluorescence commonly observed in HL-60 cells.

An additional functional assay useful for evaluating HL-60 cells is the determination of their reactive oxygen species production. A substantial increase in both O_2^- and H_2O_2 production can be observed in differentiated HL-60 cells (Fig. 4) as they mature into granulocytic cells, indicative of the maturation of the respiratory burst components. Flow cytometry provides an excellent technique for monitoring this maturation. A significant advantage afforded by HL-60 cells is the ability to accurately reproduce differentiation conditions, thereby creating an efficient model for studying phagocytic cell function. Care in interpretation of studies must of course be made since the HL-60 cell line is leukemic in origin.

Acknowledgment

Funding for these studies was provided by Grants P42 ES04911 and GM38827 from the National Institutes for Health.

References

Back, A. L., Gollahon, K. A., and Hickstein, D. D. (1991). *J. Immunol.* **148,** 710–714.
Boss, G. R. (1989). *Proc. Natl. Acad. Sci. U.S.A.* **86,** 7174–7178.
Chaplinski, T. J., and Niedel, J. E. (1986). *J. Leuk. Biol.* **39,** 323–331.
Collins, S. J., Gallo, R. C., and Gallagher, R. E. (1977). *Nature (London)* **270,** 347–349.
Cowen, D. S., Berger, M., Nuttle, L., and Dubyak, G. R. (1991). *J. Leuk. Biol.* **50,** 109–122.
De La Maza, L. M., Peterson, E. M., Goebel, J. M., Fennie, C. W., and Czarniecki, C. W. (1985). *Infect. Immun.* **47,** 719–721.
Dufer, J., Biakou, D., Joly, P., Benoist, H., Carpentier, Y., and Desplaces, A. (1989). *Leuk. Res.* **13,** 621–627.
Forsbeck, K., Nilsson, K., Hansson, A., Skoglund, G., and Ingelman-Sundberg, M. (1985). *Cancer Res.* **45,** 6194–6199.
Harris, P., and Ralph, P. (1985). *J. Leuk. Biol.* **37,** 407–422.
Janick-Buckner, D., Barua, A. B., and Olson, J. A. (1991). *FASEB J.* **5,** 320–325.
Katagiri, K., Katagiri, T., Kajiyama, K., Uehara, Y., Yamamoto, T., and Yoshida, T. (1992). *Cell. Immunol.* **140,** 282–294.
Koeffler, H. P. (1983). *Blood* **62,** 709–720.
Madden, M. C., Becker, S., Koren, H. S., and Friedman, M . (1992). *J. Leuk. Biol.* **51,** 118–123.
Malin-Berdel, J., and Valet, G. (1980). *Cytometry* **1,** 222–228.
Matzner, Y., Gavison, R., Rachmilewitz, E. A., and Fibach, E. (1987). *Cell Differ.* **21,** 261–269.
Perussia, B., Lebman, D., Ip, S. H., Rovera, G., and Trinchieri, G. (1981). *Blood* **58,** 836–843.
Pilz, R. B., Van den Berghe, G., and Boss, G. R. (1987). *J. Clin. Invest.* **79,** 1006–1009.
Ross, D. W. (1985). *Cancer Res.* **45,** 1308–1313.
Ross, D. W. (1986). *Cytometry* **7,** 263–267.
Thompson, B. Y., Sivam, G., Britigan, B. E., Rosen, G. M., and Cohen, M. S. (1988). *J. Leuk. Biol.* **43,** 140–147.
Watson, J. V. (1993). Quantitation of esterase activity using FDA. *In* "Handbook of Flow Cytometry Methods." (J. P. Robinson, ed.), p. 194. New York: Wiley–Liss.
Zhou, J.-Y., Norman, A. W., Akashi, M., Chen, D.-L., Uskokovic, M. R., Aurrecoechea, J. M., Dauben, W. G., Okamura, W. H., and Koeffler, H. P. (1991). *Blood* **78,** 75–82.

CHAPTER 26

HIV Infection: Diagnosis and Disease Progression Evaluation

Janis V. Giorgi* and Alan Landay†

* Laboratory of Cellular Immunology and Cytometry
Department of Medicine
Division of Clinical Immunology and Allergy
University of California, Los Angeles School of Medicine
Los Angeles, California 90024

† Department of Immunology and Microbiology
Rush Presbyterian-St. Luke's Medical Center
Chicago, Illinois 60612

METHODS IN CELL BIOLOGY, VOL. 42

I. Introduction

The primary application of flow cytometry in diagnosis and disease progression evaluation in HIV disease is enumeration of circulating lymphocyte subset levels. Monoclonal antibodies (mAb) directed against cell-surface differentiation antigens (Ag) are used to distinguish lymphocyte subsets and subpopulations. The immune status of the patient is reflected in the circulating lymphocyte subset profile.

Routine immunophenotyping for HIV disease currently utilizes two-color immunofluorescence analysis. The state of the art is clearly moving to three-color analysis. One-color analysis is no longer acceptable because of increased awareness regarding Ag coexpression among distinct cell types. Monoclonal antibodies used for HIV disease immunophenotyping are almost always directly conjugated to a fluorochrome, generally to fluorescein isothiocyanate (FITC) or phycoerythrin (PE) or, for three-color immunofluorescence, to a dye that emits in the far red, such as peridinin chlorophyll protein or, alternatively, a PE–cyanine or PE–Texas red tandem conjugate. Generally flow cytometers with a single air-cooled argon ion laser, tuned to 488 nm, are used because they are versatile and easy to operate and large numbers of samples can be run relatively rapidly.

Most analysis routines set a gate around the lymphocytes, and values are usually provided by the flow cytometer as a percentage of the lymphocytes with the immunophenotype of interest. The percentage can be multiplied by the peripheral blood absolute lymphocyte count to determine the absolute number of each lymphocyte subset per volume of blood (mm^3 or liter). These values, especially the $CD4^+$ cell number (CD4 count), are used to stage disease, predict outcome, select subjects for trials of therapeutic vaccines, and decide on the timing of treatment with anti-retroviral drugs or drugs for prophylaxis against opportunistic infections. Flow cytometry immunophenotyping is also invaluable for investigations of the underlying basis of lymphocyte dysfunction and activation in HIV disease.

II. Application

The many uses of flow cytometry to enumerate lymphocyte subsets and study the cellular basis for immune deficiency as well as certain virologic aspects of HIV disease have been reviewed by us previously (Landay *et al.*, 1990). A recent review outlines the use of immunophenotyping to elucidate the cellular basis of immunologic abnormalities in adult and pediatric HIV disease and AIDS and to characterize the stages of $CD4^+$ cell loss during disease progression (Giorgi, 1993). Immunophenotyping has contributed to our understanding of the relationships between $CD8^+$ cell activation, soluble immune system activation markers including cytokines, and low $CD4^+$ cell numbers and function (Autran and Giorgi, 1992; Giorgi, 1992).

By far the greatest use of immunophenotyping in HIV disease is to measure $CD4^+$ cell numbers. $CD4^+$ cell numbers are used as a criterion for entry into clinical trials and initiation of treatment, to monitor the effects of therapy, and to predict prognosis. The Center for Disease Control (CDC) has revised their classification system for HIV infection to include as AIDS cases all HIV-infected persons with <200 $CD4^+$ T lymphocytes/mm^3 or a $CD4^+$ T lymphocyte percentage of <14% (Centers for Disease Control, 1993). Anti-retroviral drug therapy is currently initiated at 500 $CD4^+$ cell/mm^3 in adults and for prophylaxis for *Pneumocystis carinii* at 200 $CD4^+$ cells/mm^3. The median $CD4^+$ cell level in uninfected adults is around 850–1000/mm^3. Guidelines for prophylaxis in infants, who have about three times as many circulating lymphocytes as adults, have also been published (Centers for Disease Control, 1991). Increases in $CD4^+$ cell levels are used as an indicator of therapeutic efficacy of drugs used to treat HIV infection largely due to the strong relationship between low $CD4^+$ cell levels and progression of HIV disease to AIDS. With regard to vaccines, sequential $CD4^+$ cell measurements have been used to show that no decrease in $CD4^+$ cell levels results from HIV gp120 vaccination and to show that such vaccinations may actually lead to improvements in $CD4^+$ cell levels in HIV seropositive individuals. The relative merits of using $CD4^+$ cell absolute number, $CD4^+$ percentage, or CD4/CD8 ratio for prognosis in HIV-infected subjects have been discussed (Giorgi, 1993; Taylor *et al.*, 1989).

III. Materials

Some of the mAb, many available from multiple vendors, that are useful in studies of the immunopathogenesis of HIV disease are listed in Table I. The mAb panel known as the basic immunophenotyping panel (Schenker *et al.*, 1993; National Committee for Clinical Laboratory Standards, 1992), often used to enumerate the major circulating lymphocyte subsets, is provided in Table II. Most other essential materials including lysing reagents, some of which are

Table I
Monoclonal Antibodies Commonly Used for AIDS Immunophenotyping

Cluster	Cell distribution	Function/differentiation state marked
CD3	T	TCR complex
CD4	T subset; monocytes	Accessory molecule for interaction with MHC class II + Ag; HIV receptor
CD8	T and NK subsets	Accessory molecule for interaction with MHC class I + Ag
CD16	NK; PMN; T subset	IgG Fc receptor type III
CD19	B	B cell proliferation involvement
CD20	B; T subset	B cell activation and proliferation involvement
CD28	T subset	CD80, CD86 Ag receptor
CD38	NK; T subset; plasma cells	NAD glycohydrolase
CD45	Leukocytes	Common leukocyte Ag
CD45RA	Most T; B; NK	Marker of naive cells
CD45RO	T subset	Marker of activated/memory cells
CD56	NK; T subset	N-CAM
CD57	NK and T subsets	Carbohydrate Ag
CD62L	B; T subset; monocytes; NK	Lymph node homing receptor
anti-HLA-DR	B; T subset; monocytes	MHC class II

Note. Adapted from Barclay *et al.* (1993) and Lanier and Jackson (1992).

Table II
Basic Immunophenotyping Panel[a]

mAb Combination[b]	Cell type enumerated
1. CD45/CD14	Percentage lymphocytes in gating region[c]
2. IgG_1/IgG_2	Isotype control
3. CD3/CD4	CD4 T cells ($CD4^T$)[d]
4. CD3/CD8	CD8 T cells ($CD8^T$)[d]
5. CD3/CD19 (or CD19)	Total B cells[e]
6. CD3/CD16 + CD56	Total T cells; total NK cells[f]

[a] From Giorgi (1993), with permission.
[b] FITC/PE-labeled reagents.
[c] Lymphocytes will be $CD45^{bright}CD14^-$; this tube also evaluates monocyte contamination.
[d] Indicated cell type will be positive for both mAb.
[e] All cells that express CD19.
[f] Total T cells: all that express CD3, i.e., quadrant II + IV; total NK cells, $CD3^-$(CD16 and/or $CD56)^+$, i.e., quadrant I.

listed in Table III, are also available from the vendors who distribute mAb or from general laboratory supply houses.

Formulas for three common reagents used in immunophenotyping are as follows.

1. Staining buffer (PBS with 0.1% sodium azide and 2% serum): For 5 liters, to 4350-ml double-distilled H_2O, add 100 ml newborn calf serum (NCS), 500 ml 10× PBS w/o calcium and magnesium (Irvine Scientific, cat. no. 9242), and 50 ml of 100× NaAz (5g NaAz in 50 ml H_2O). Mix well and adjust to pH 7.2–7.4 using 0.1 *M* NaOH or 0.1 *M* HCl. Store covered at room temperature. Filter before use, e.g., in 500 ml aliquots, weekly.

2. Ammonium chloride (NH_4Cl) lysing solution: For 1 liter of 10×, to 850-ml double-distilled H_2O, add 82.9 g ammonium chloride, 10 g potassium bicarbonate, and 0.37 g ethylenediamine tetra acetic acid (EDTA) disodium salt. Bring to a final volume of 1.0 liter with double-distilled H_2O. Keep tightly closed and store at 4°C. Prepare 1× working solution daily. When NH_4Cl lysing solution is prepared in the laboratory from bulk dry chemical, keep in mind that especially ammonium chloride is extremely hydroscopic. Start with new bottles of the chemicals and keep the bottles tightly closed.

Table III
Lysing Reagents

Reagent name	Company[a]	Fixation included	Wash	Incubation time[b]	Sample hold time[c]	
					Without fixation	With fixation
Ammonium chloride solution[d]						
Standard recipe	n/a	No	Yes	5 min + 3 min	2 hr	1 day
Freeze-dried	Ortho	No	Yes	10 min	2 hr	1 day
FACS lysing solution[e]	BDIS	(Yes)	Yes	10 min	(2 hr)	1 day
Whole-blood lysing reagent kit[f]	Coulter	Yes	Yes	30 sec–2 min	n/a	1 day
Whole-blood lyse and fix reagent[f]	Gentrak	Yes	Yes	<4 min	n/a	1–2 day
OptiLyse	AMAC	Yes	No	10 min	n/a	2 day
ImmunoPrep reagent system[g]	Coulter	Yes	No	35 sec	n/a	1 day

[a] Ortho Diagnostics, Raritan, NJ; Becton–Dickinson Immunocytometry Systems, Inc., San Jose, CA; Coulter Corporation, Hialeah, FL; Gentrak, Plymouth Meeting, PA; AMAC, Westbrook, ME; n/a, not applicable.

[b] Time that reagent is incubated with the cells to either lyse or lyse and fix the sample.

[c] Maximum recommended time after the sample is fixed until it is analyzed.

[d] Samples lysed with ammonium chloride can be fixed in 0.5 or 1% paraformaldehyde solution.

[e] Product can be used without additional fixation in which case samples should be analyzed within 2 hr; manufacturer recommends samples be fixed in 0.5% paraformaldehyde.

[f] Products involve separate lysis and fixation.

[g] Used with the Coulter Q-Prep workstation.

3. Paraformaldehyde: For a 2% stock solution, add 2 g paraformaldehyde (Eastman–Kodak, Cat. No. 421-500) to 100 ml of PBS. Heat to 70°C (do not exceed this temperature) in a fume hood until the paraformaldehyde goes into solution (approximately 60 min). Allow the solution to cool to room temperature. Adjust to pH 7.4 using 0.1 *M* NaOH or 0.1 *M* HCl. Filter and store at 4°C, protected from light. Use at final concentration of 0.5 or 1%.

IV. Cell Preparation and Staining

A. Basic Immunophenotyping Panel

Whole blood is recommended for routine immunophenotyping. The basic panel of mAb provided in Table II, with results presented in Fig. 1, is recommended for immunophenotyping by the National Committee for Clinical Laboratory Standards (1992) and the CDC (Centers for Disease Control, 1992). NIAID has indicated that a shorter panel comprising the first four mAb combinations is appropriate for evaluating adult HIV infection, while the first five combinations are recommended for pediatric HIV disease (Calvelli *et al.*, 1993). The advantages of using this panel have been discussed elsewhere (Schenker *et al.*, 1993) and include biological accuracy of lymphocyte subset enumerations and several quality control checks among the values obtained from this panel that confirm the technical accuracy of the results.

B. Advantages of Whole-Blood Staining

The primary reasons for using whole blood rather than gradient-separated mononuclear cells (PBMC) for immunophenotyping are the simplicity and accuracy of the whole-blood method. Gradient purification of PBMC, e.g., using Ficoll-Hypaque, is more labor intensive and expensive and can alter lymphocyte subset analysis results. The major reason for this is that density gradient separation does not adequately separate the major hematological components in blood from HIV-infected subjects. In HIV disease, the interface band between the medium and the plasma or serum not only contains lymphocytes and monocytes, as it does in separations of blood from normal donors, but often also contains granulocytes and nucleated erythrocytes which have the same light scatter as lymphocytes and contaminate the flow cytometry analysis. Both the disease process and specific antiviral therapies including zidovudine probably contribute to these abnormalities. The result is that lower percentages of the stained lymphocyte populations are reported because the denominator of the equation, i.e., subset count ÷ lymphocyte count = subset percentage, is artificially inflated.

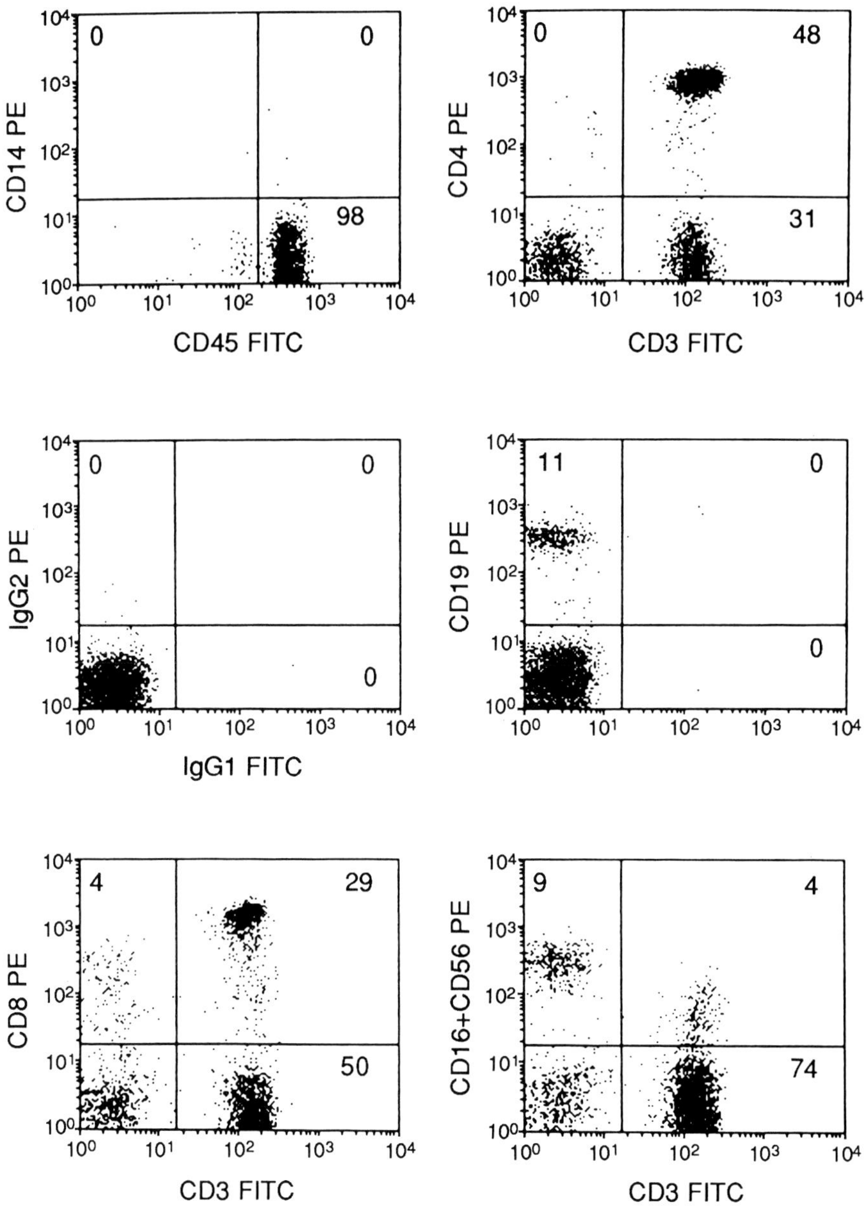

Fig. 1 Dual-color histograms from immunophenotyping whole blood using the basic immunophenotyping panel. Reprinted from Giorgi (1993), with permission.

C. Problem Antigens

Routine immunophenotypic analysis with the basic immunophenotyping panel should always be performed using a whole-blood method. Nevertheless, research studies on certain cell-surface Ag may benefit from Ficoll-Hypaque purification of the cells if care is taken to lyse residual lymphocytes prior to flow cytometric analysis. One of the Ag that may benefit from Ficoll-Hypaque purification is CD16, and anti-FcRIII. There are two reasons for this. First, soluble FcRIII in the serum, present at various concentrations in different people, can absorb CD16 mAb. In addition, CD16 is present at high levels on granulocytes. Density gradient separation of the lymphocytes will circumvent both problems. The latter problem can also be overcome by adding extra mAb or by using a CD16 mAb such as Leu11c that does not react as strongly with the form of CD16 expressed on granulocytes. Staining of a number of other cell-surface Ag, e.g., β_2-microglobulin, found in the serum in soluble form, also benefits from cell separation and consequent removal of the soluble Ag from the staining reaction.

Another problematic Ag is the CD11b molecule. This Ag is found at low levels on NK cells and putative T suppressor cells, but at such high levels on granulocytes that high concentrations of the mAb are needed if whole-blood samples, rather than density gradient separated lymphocytes, are stained to enumerate the percentage of lymphocytes that express this Ag.

One of the most problematic cell-surface Ag on lymphocytes is CD62L (Leu 8). This Ag is lost in the cold (4°C) after just a few hours and is lost quickly even at room temperature when reacted with CD62L mAb. Other Ag are also modulated upon reaction with mAb directed against them, only usually more slowly, e.g., in about 12 hr at 37°C for CD3, or even longer for CD8. It is well known that anti-Ig antibodies lead to capping of Ig within about an hour at temperatures above 4°C. Azide is added to staining buffers to retard capping, but has no effect on CD62L loss over time.

D. Basic Whole-Blood Staining Procedure

This procedure includes paraformaldehyde fixation and can be performed with materials, including NH_4Cl lysing solution, prepared from the recipes in Section III of this chapter.

1. Add 50 μl of staining buffer to 12 × 75 round-bottom tubes. Add appropriate mAb for staining 5×10^5 cells.
2. Add 50 μl of blood.
3. Vortex tubes individually in order to observe that all blood is mixed with mAb.
4. Incubate in the dark at 22°C for 15 min.
5. Vortex tubes in test tube rack.

(Commercial lysing and fixing reagents described in Table III can be used at this point)

6. Add 1 ml NH_4Cl lysing solution and vortex tubes in test tube rack again.
7. Place in the dark at 22°C for 5 min.
8. Spin at 4°C, 250*g* for 5 min.
9. Aspirate supernatant and vortex rack vigorously to break up pellet.
10. Repeat steps 6–9, but reduce second lysing to 3 min.
11. Wash in 2 ml staining buffer.
12. To fix in paraformaldehyde, resuspend while vortexing in 750 μl 1× PBS without protein, followed by 250 μl 2% paraformaldehyde.

Samples should be run as soon as possible if not fixed. They can be stored at room temperature in the dark. When samples cannot be run until 30 min or more after staining, they should be stored at 4°C and brought to room temperature again for 10 min before analysis. Each sample should be vortexed again individually before being analyzed. Samples fixed in paraformaldehyde can be held for 24 hr prior to analysis.

E. Three-Color Immunophenotypic Analysis

Three-color immunofluorescence analysis has been used for accurate $CD4^+$ cell enumeration in a single tube using CD45 and CD3 mAb to select the lymphocytes and T cells, respectively; CD4 mAb was used to select the $CD4^+$ cells (Nicholson *et al.*, 1993b). Other applications of three-color immunofluorescence extend the power of flow cytometry to enumerate distinct functional subpopulations of lymphocytes, e.g., by gating on $CD4^+$, $CD8^+$, or $CD19^+$ cells and then evaluating two additional markers on these cells. Many such applications can be found in the immunology literature from the past 5 years.

V. Critical Aspects of the Procedure

A. Biohazards

Immunophenotyping blood from humans requires precautions including use of gloves, no mouth pipetting, and no use of glass pipettes, needles, or other sharp laboratory supplies. Biosafety procedures must be rigorously followed. These have been described in detail elsewhere (Centers for Disease Control, 1988). The commercially available flow cytometers currently on the market have been manufactured to permit safe operation during analysis of human blood. The greatest risk to the technician is during sample preparation, and special care should be given to containing aerosols during centrifugation and vortexing of samples. Fixation with paraformaldehyde inactivates HIV but is generally performed after samples are stained. Obviously, this only provides protection once the samples are fixed (Lifson *et al.*, 1986; Nicholson *et al.*, 1993a).

B. Absolute Counts

Lymphocyte subset counts are generally determined using the estimates of the percent of positively stained cells from the flow cytometer and values from a complete blood count (CBC) performed by routine hematology methods according to the formula: Absolute lymphocyte subset number = percent of lymphocyte subset × white blood count (WBC) × percent of all lymphocytes. The standard for the maximum time a sample can be held prior to performing basic hematology is 6 hr and before staining for immunophenotype analysis is 24 hr. In practice, most investigators accept a hematology count performed within 24 hr on samples submitted for immunophenotype analysis.

Certain recently released flow cytometers such as the CytoronAbsolute (Ortho Diagnostic Systems, Inc., Raritan, NJ) determine absolute counts directly. Other devices such as the Cytek Absolute Count Module (Cytek Development, Fremont, CA) can be used with existing instruments like the FACScan (Becton–Dickinson Immunocytometry Systems, Inc., San Jose, CA; BDIS) or EPICS (Coulter Corporation, Hialeah, FL; Coulter) instruments. Direct determination of absolute counts is expected to reduce some of the measurement error in current absolute count methodologies.

C. Erythrocyte Lysis

Many products are available to lyse erythrocytes (Table III). These should be used as directed by the manufacturer. Even when following directions, attention must be paid to whether they are working properly. One of the most basic and essential aspects of lymphocyte immunophenotyping is that erythrocyte lysis must be complete. If it is not, the lymphocytes percentages will all be artificially reduced due to unlysed erythrocytes that fall in the lymphocyte gate and inflate the denominator of the percentage equation (counts of stained cells ÷ counts in the gating region). A simple way to judge whether this has occurred is to check that the lymphosum, i.e., the sum of the lymphocyte subset percentages of T + B + NK, is equal to 100 ± 5% (Schenker *et al.*, 1993). If a laboratory's values are generally <95% on normal subjects, or if the laboratory's median T cell value is <70%, the most likely reason is that their method for erythrocyte lysis is not optimal. The laboratory should work on their method of lysing erythrocytes to optimize the protocol.

D. Backgating

Backgating on the fluorescence of the CD45/CD14 mAb combination should be used to set the light scatter gate for each patient. This technique has been discussed in detail by Loken *et al.* (1990) who introduced it. The light scatter gate should be set liberally so as not to exclude lymphocytes. Backgating on the fluorescence of the $CD45^{bright}$ $CD14^{-}$ cells allows estimation of the percentage of lymphocytes included in the light scatter gate. Loken (Loken *et al.*, 1990) suggests that 98% of the lymphocytes be included in the analysis, but inclusion of 95% is more generally attainable in practice (Centers for Disease Control, 1992). The gate set on the CD45/CD14 mAb combination is then used for the remainder of the analyses on that patient.

Backgating on each mAb combination for each patient can also be performed to assess the percentage of cells stained with each mAb combination that has been included in the analysis. The percentage should average 95%. In certain cases, the percentage will be lower than 95%, sometimes significantly so, due to cell clumping. Large cell clumps that fall in the granulocyte gating region were first described by Jackson (1990) who referred to lymphocytes within these clumps as "escapees." These clumps represent a serious staining artifact that must be avoided. Figure 2 shows how to detect cell clumps that contain escapees during acquisition. A forward light scatter × fluorescence 1 dot plot

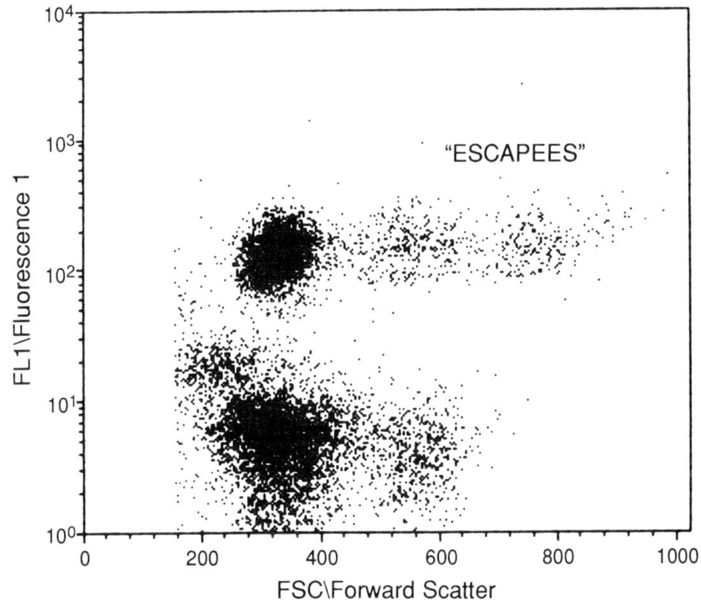

Fig. 2 Detection of "escapees" during acquisition. See text for details.

is created. Clumps of cells containing escapees can be observed as positively stained ($CD3^+$ in Fig. 2) with increased forward light scatter. These clumps will alter the lymphocyte subset measurements, leading to inaccurate results.

Escapees usually develop after staining with certain FITC (but not PE)-conjugated mAb, including CD8-FITC and CD3-FITC, but seldom with others, including CD45-FITC. Because they do not form equally among all the mAb combinations for a subject, only by backgating individually on each tube can escapees present in each sample be monitored. The percentage of the lymphocyte population included in the lymphocyte region may be 98% for the CD45/CD14 combination, but much lower, e.g., 85% for another combination such as CD3-FITC/CD8-PE. Backgating on every tube is practical if a computer program is created that automatically performs backgating as part of the routine analysis procedure.

Backgating can be performed, as shown in Fig. 3, to determine the percentage of the stained cells that were included in the analysis. An analysis screen can be set up with six dot plot (DP) windows. Window 1 is an ungated forward × 90° light scatter DP used to create a lymphocyte gate retion (R1). Window 2 is an FL1 × FL2 DP gated on R1 from which the T cell subset percentages are recorded. Window 3 is an ungated FL1 × FL2 DP; in this DP, three regions (R2, R3, and R4) are created to be used for the purpose of gating on the positively stained lymphocyte populations.

The example shown in Fig. 3 is useful for analyzing the NIAID mAb combination panel (Calvelli *et al.*, 1993), i.e., the first four mAb combinations in Table II. R2 is created to include $CD4^+$ $CD3^+$ and $CD8^+$ $CD3^+$ cells in their respective tubes (combinations 3 and 4 in Table II), and R3 for $CD4^-$ $CD3^+$ and $CD8^-$ $CD3^+$ cells in those tubes. R4 is used to gate on the CD45 brightly stained lymphocytes, when tube 1 (data not shown) is backgated. In an effort to make the analysis as automated as possible Regions 2, 3, and 4 are amply created so that it is usually not necessary to change them from tube to tube, although they should be changed if the lymphocyte clusters do not fall within them. Windows 4, 5 and 6, shown in the bottom 3 panels, are all forward × 90° light scatter plots. Window 4 displays the R1 lymphocyte gate for R2 ($CD8^+$ $CD3^+$ cells shown here, or $CD4^+$ $CD3^+$ cells on tube 3), while Window 5 displays R3 ($CD4^-$ $CD3^+$ or $CD8^-$ $CD3^+$ cells), and Window 6 shows R4 ($CD45^{bright}$ stained cells). The backgate percentage is obtained by requesting region statistics for DP 4, 5, and 6. Record the percentage of cells that fall within R1. It is also a good idea to check that the number of events in R2, R3, and R4 match the number of events in Windows 4, 5, and 6, respectively.

It is useful to create a report sheet that includes backgating percentages as well as T cell subset percentages as a way to monitor the occurrence of clumping. Backgating can be incorporated into the routine analysis of all lymphocyte staining to determine the frequency of clumping. It is especially useful to ensure accurate results by new or varying technical staff. Experienced laboratories should investigate the utility of this quality control check and incorporate it

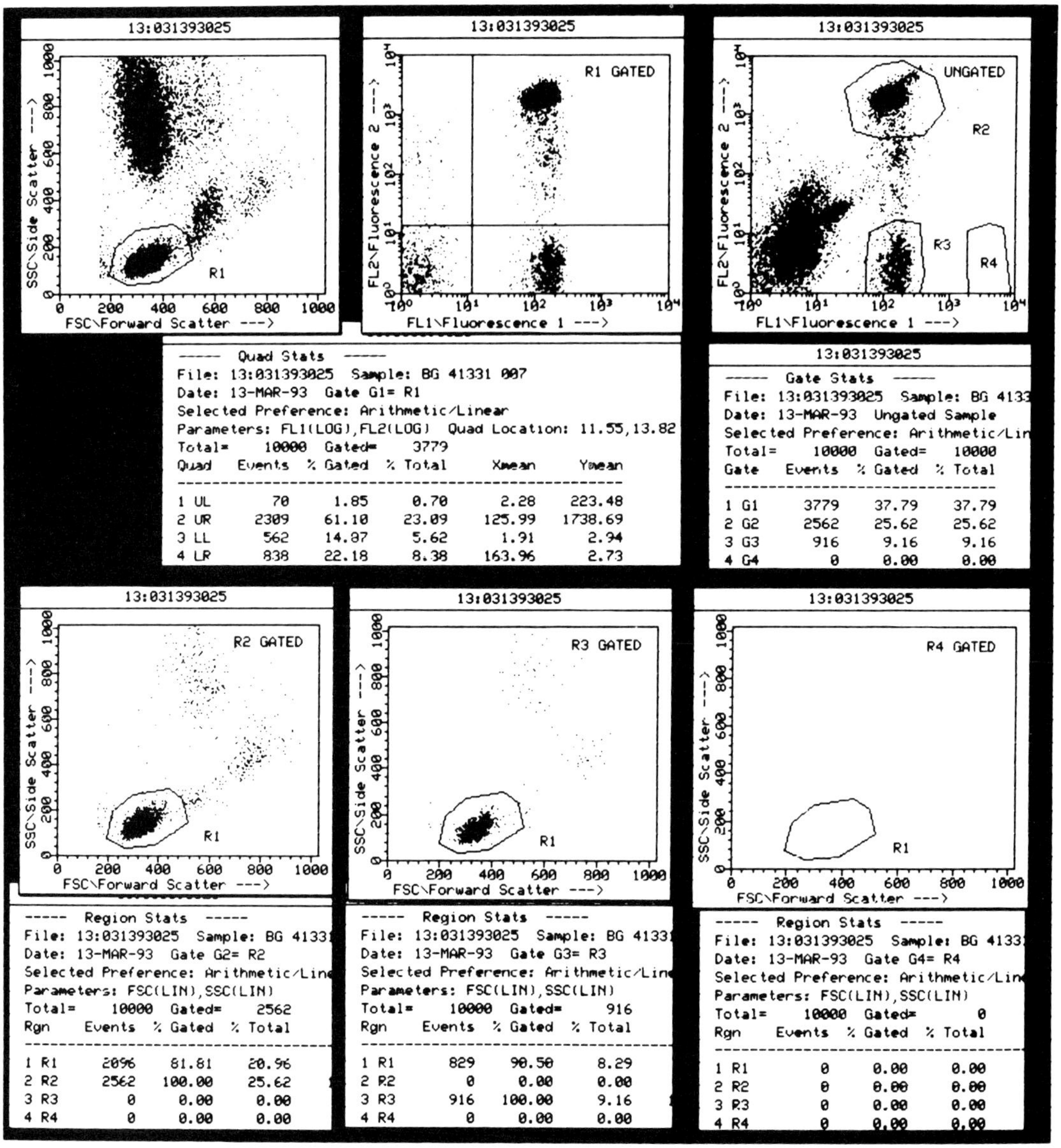

Fig. 3 Detection of "escapees" during analysis. See text for details.

into their routine if they find it useful. This technique should be continued until it is clear that "escapees" can be adequately recognized and steps taken to disperse clumps during acquisition (Fig. 2).

The exclusion of stained lymphocytes during analysis is an error that cannot be corrected mathematically. Instead, the sample should be revortexed and rerun. If this does not correct the problem, the sample should be restained and rerun. Although it would be desirable to have all backgate values be 100%, a practical lower cutoff for an acceptable backgate percentage is 90% on an individual sample, with an overrall average of 95%. Several steps can be taken during sample preparation to avoid formation of escapees: (i) perform incubations at room temperature, (ii) allow samples to warm to room temperature prior to flow cytometric analysis if they have been refrigerated, and (iii) vortex to break up cell clumps, e.g., by vortexing in the centrifuge racks thereby maximizing agitation by bumping the tube multiple times, as opposed to vortexing tubes individually, and centrifugation at no greater than 250*g*.

VI. Controls and Standards

A. Reference Range

Each laboratory should generate their own reference range for the lymphocyte subsets measured by that laboratory. Separate ranges for adults (Reichert *et al.*, 1991) and for children (Hannet *et al.*, 1992) must be generated. Only one specimen from each person should be used in the calculation. Specimens to generate the reference range should be processed in exactly the same way the test blood is processed. The number of specimens that should be tested to generate a statistically accurate reference range is large, around 70–200. In practice, about 40 individuals give a useful range, although outliers cannot be accurately identified. The laboratory director should consult a biostatistician for guidance on these issues.

B. Verification of the Linearity of the Instrument Log Amplifier

The laboratory should assess whether their flow cytometer's logarithmic amplifier is operating appropriately. Quantitative fluorescence microbeads (Flow Cytometry Standards Corporation, FCSC, Research Triangle Park, NC) and Immuno-Brite beads (Coulter) can be used. These products are mixtures of several fluorescence intensities of positive beads plus a blank bead. The channel number for each bead's intensity is plotted versus the fluorescence equivalent stated by the manufacturer. According to the manufacturers, a linear plot indicates linearity of the log amplifier. Alternatively, two beads of slightly different fluorescence intensities can be moved across the range of output voltages by altering the PMT settings, and the data analyzed as described previously (Schmid *et al.*, 1988).

C. Evaluation of Instrument Sensitivity

Quantitative fluorescence microbeads (FCSC), Immuno-Brite beads (Coulter), or Calibrite beads (BDIS) can be used. Analyze the mixed bead population at fixed instrument settings using PMT voltages similar to those used for stained lymphocytes. Measure the peak-to-peak separation of the blank and positive beads. The PMT voltage and peak-to-peak sensitivity should be charted and examined for changes and to document stability of the instrument sensitivity over time. Calibrite beads are used together with a software package that adjusts the fluorescence PMT voltages to place the blank bead in a similar channel from day to day. The software also measures peak-to-peak sensitivity of the FITC- and PE-labeled beads.

D. Standardization of Fluorescence Intensity Day to Day

Several products are available including fixed chicken erythrocytes (Riese Enterprises, Inc., Mountainview, CA), fluorospheres that are latex particles with fluorescent dye inside (available from many manufacturers including Polysciences, Warrington, PA), or microbeads that are surface labeled with FITC or PE (FCSC). These are used to adjust the PMT voltage to place the fluorescent particles in a specified target channel each day.

E. Titration of mAb Lots

Although most manufacturers titrate mAb lots and recommend an amount of mAb to use for each, good quality control requires that this be verified in the laboratory where the mAb will be used. Titrations of each lot should be done to verify reproducible reactivity (percent and median channel) against either a known lot of cryopreserved lymphocytes or one of the recently developed standardized preserved cell preparations.

Products available for this include CD-Chex (BDIS), a stabilized liquid leukocyte preparation, and CYTO-TROL (Coulter), a lyophilized lymphocyte preparation. Each new lot of mAb should be titrated and 2× the saturating amount or the manufacturer's recommended amount should be used.

F. Calculation of the Number of Molecules of Ag Expressed on the Cell Surface

This can be accomplished using either Quantum simply cellular microbeads (FCSC) or QIFIKIT microbeads (Biocytex, Marseille, France). Stain the beads with mAb. Plot the mean intensity of each stained bead versus the manufacturer's specified binding capacity. Use a calibration curve for each fluorochrome/mAb combination to determine the number of Ag per cell.

G. Proficiency Testing Programs

The laboratory may wish to enroll in a certification program such as that organized by the College of American Pathologists (CAP) or in a proficiency testing program such as that organized by FAST Systems (Gaithersburg, MD).

VII. Instruments

Modern analytic instruments with a single air-cooled laser are generally most convenient for AIDS immunophenotyping. They are generally easier to operate and samples can often be run more rapidly compared to research flow cytometers that incorporate both sorting and analytic applications. Standard filter packages for each of the several commercially available instruments are generally similar for FITC and PE. However, the standard filter pack on each instrument is optimized for certain specific third-color reagents, usually either for PerCP or for PE–cyanine and PE–Texas red. Certain instruments, most notably the FACScan (BDIS), are more sensitive than others for measurements of fluorescence intensity. Whereas this does not substantially affect most major lymphocyte subset percentage measurements, it does profoundly affect percentage measurements of Ag with low level or heterogeneous expression, e.g., CD38, CD25, HLA-DR, CD8, CD56, and most adhesion molecules.

VIII. Results and Discussion

A. Typical Results from the Basic Immunophenotyping Panel

A typical example of the flow cytometric results on the basic immunophenotyping panel are shown in Fig. 1. Note that the cursor settings from the isotype control tube are used to set the cursor on tubes 3–6. Less than 1% of the cells should be scored as positive with either IgG1 or IgG2. However, the cursors on the CD45/CD14 tube should be set to enumerate the $CD45^{bright}$ $CD14^{-}$ cells, because the purpose of this tube is to determine what percentage of cells in the gating region are lymphocytes. The cursor is set to exclude the granulocytes, which are $CD45^{dim}$, but must be set far enough to the left of the lymphocyte cluster that B cells, which can have slightly lower CD45 staining in AIDS, are not excluded. Although these mAb combinations do not need to be run routinely with the specimens, each laboratory should stain samples from a number of AIDS patients with CD45 FITC/CD19 PE and CD45 FITC/CD13 PE in order to discern how these populations appear in the analyses under conditions in use in each laboratory. This information can then be used to guide cursor placement on this tube so that most granulocytes are excluded and virtually all B cells are included in the analyses.

B. Conventions for Expressing Quantitative Immunophenotypic Data

Several different ways of expressing immunophenotyping data are available. These include:

1. Measurements from the flow cytometer

a. Percentage of lymphocytes (e.g., 45% of the lymphocytes are $CD4^{+}$): The denominator of the calculation is the number of events in the gating region.

b. Percentage of a subset of lymphocytes [e.g., 50% $CD5^+ \parallel CD19^+$, where " ‖ " is read "given" (Giorgi *et al.*, 1993); this means that 50% of the B cells are $CD5^+$]: The denominator of the calculation is the number of events with the marker indicated to right of the ‖ .

c. Absolute count of lymphocyte subsets (e.g., 200 $CD4^+$ cell/mm^3): Using either a CytoronAbsolute (Ortho) or using modified methods or attachments to other flow cytometers, absolute lymphocyte subset measurements per volume of blood can be determined.

d. Fluorescence intensity measurements (e.g., the median staining intensity of CD4 mAb on monocytes is 10× less than that on lymphocytes): The most common expressions for these measures are the arithmetic mean, which reflects the overall reactivity of a mAb with a cell type, or the median or the geometric mean, which both reflect how brightly the mAb would stain a typical cell. Fluorescence intensity measurements can be expressed as channel numbers or as relative linear fluorescence intensity measurements. As in the example, ratios of one intensity to another can be compared.

2. Flow cytometric measurements combined with hematology measurements (e.g., 200 $CD4^+$ cells/mm^3): The subset percentage measurement from the flow cytometer multiplied by the absolute numbers of lymphocytes/volume of blood determined by conventional hematology methods.

3. Flow cytometric measurements converted to numbers of molecules of Ag (e.g., approximately 50,000 molecules of CD3 Ag are expressed on a typical T cell): Estimates of numbers of molecules of Ag expressed on a cell can be made using quantum simply cellular microbeads (FCSC) or QIFIKIT beads (Biocytex).

IX. Alternative Technologies

The primary application of flow cytometry in HIV disease is to determine the $CD4^+$ T cell number (CD4 count). It is this measurement that has the greatest utility in clinical practice. Nevertheless, conventional flow cytometry technology is quite costly and usually also requires that separate hematology instrumentation be available to determine the absolute lymphocyte count so that the absolute CD4 count can be determined.

Several alternative technologies have been developed that allow absolute CD4 counts to be measured directly on whole blood at a reduced cost for instrumentation. All of these use simple to operate technology and do not require separate hematology instrumentation. These and other technologies are reviewed in detail elsewhere (World Health Organization, 1994). A few points are mentioned here about four of the systems because they may be of interest to a reader with a background in flow cytometry technology. The FACSCount (BDIS) instrument uses flow-based technology. It is equipped with a helium–neon laser and determines $CD4^+$ and $CD8^+$ T cell subsets by fluorescence and scatter measurements combined. The antibody reagent contains a microbead calibrator for obtaining the absolute counts. The Immuno VCS system (Coulter) is also a flow-based system although a manual nonflow counterpart is available for this system that uses the same reagents. The flow-based system is a modified hematology analyzer that determines the absolute CD4 count with the addition of antibody-coated latex beads. A third alternative technology, the TRAx CD4 assay (T Cell Diagnostics, Inc., Cambridge, MA), is an enzyme immunoassay that involves lysis of a whole-blood sample and subsequent analysis of total CD4 protein. A standard calibration curve is run in order to determine the absolute CD4 count. The Zymmune assay (Zynaxis Cell Science, Inc., Malvern, PA) uses a combination of fluorescent and magnetic bead technologies. A CD4 antibody coupled to the magnetic bead captures the $CD4^+$ cells. Subsequently, a fluorochrome-labeled CD4 antibody allows cell enumeration on a fluorometer. These various $CD4^+$ cell counting technologies are expected to be useful for the determination of routine CD4 counts in developing countries and in situations in which only the CD4 count is needed for clinical management.

Acknowledgment

Supported by the UCLA AIDS Institute through funding from CFAR AI-28697 and the Jonsson Comprehensive Cancer Center through funding from CA-16042.

References

Autran, B., and Giorgi, J. V. (1992). *In* "Immunodeficiency in HIV Infection and AIDS" (G. Janossy, B. Autran, and F. Miedema, eds.), pp. 171–184. Karger, Basel.

Barclay, A. N., Birkeland, M. L., Brown, M. H., Beyers, A. D., Davis, S. J., Somoza, C., and Williams A. F. (1993). "The Leukocyte Antigen Facts Book." Academic Press, London.

Cavelli, T., Denny, T. N., Paxton, H., Gelman, R., and Kagan, J. (1993). *Cytometry* **14,** 702–715.

Centers for Disease Control (1988). *Morbid. Mortal. Wkly. Rep.* **37**(24), 378–388.

Centers for Disease Control (1991). *Morbid. Mortal. Wkly. Rep.* **40**(RR-2), 1–13.

Centers for Disease Control (1992). *Morbid. Mortal. Wkly. Rep.* **41**(RR-8), 1–16.

Centers for Disease Control (1993). *Morbid. Mortal. Wkly. Rep.* **41**(RR-17), 1–19.

Giorgi, J. V. (1992). *In* "Immunodeficiency in HIV Infection and AIDS" (G. Janossy, B. Autran, and F. Miedema, eds.), pp. 1–17. Karger, Basel..

Giorgi, J. V. (1993). *Ann. N.Y. Acad. Sci.* **677,** 126–137.

Giorgi, J. V., Liu, Z., Hultin, L. E., Cumberland, W. G., Hennessey, K., and Detels, R. (1993). *J. Acquired Immune Defic. Syndr.* **6,** 904–912.

Hannet, I., Erkeller-Yuksel, F., Lydyard, P., Deneys, V., and DeBuyére, M. (1992). *Immunol. Today* **13,** 215–218.

Jackson, A. (1990). *Clin. Immunol. Newsl.* **10,** 43–55.

Landay, A., Ohlsson-Wilhelm, B., and Giorgi, J. V. (1990). *AIDS* **4,** 479–497.

Lanier, L. L., and Jackson, A. L. (1992). *In* "Manual of Clinical Laboratory Immunology" (N. R. Rose, E. C. deMacario, J. L. Fahey, H. Friedman, and G. M. Penn, eds.), pp. 157–163. Am. Soc. Microbiol., Washington, DC.

Lifson, J. D., Sasaki, D. T., and Engleman, E. G. (1986). *J. Immunol. Methods* **86,** 143–149.

Loken, M. R., Brosnan, J. M., Bach, B. A., and Ault, K. A. (1990). *Cytometry* **11,** 453–459.

National Committee for Clinical Laboratory Standards (1992). "Clinical Applications of Flow Cytometry: Quality Assurance and Immunophenotyping of Peripheral Blood Lymphocytes; Proposed Guideline," *NCCLS* Doc. H42-T. Villanova, PA.

Nicholson, J. K. A., Browning, S. W., Orloff, S. L., and McDougal, J. S. (1993a). *J. Immunol. Methods* **160,** 215–218.

Nickolson, J. K. A., Jones, B. M., and Hubbard, M. (1993b). *Cytometry* **14,** 685–689.

Reichert, T., DeBruyére, M., Deneys, V., Tötterman, T., Lydyard, P., Yuksel, F., Chapel, H., Jewell, D., Van Hove, L., Linden, J., and Buchner, L. (1991). *Clin. Immunol. Immunopathol.* **60,** 190–208.

Schenker, E. L., Hultin, L. E., Bauer, K. D., Ferbas, J., Margolick, J. B., and Giorgi, J. V. (1993). *Cytometry* **14,** 307–317.

Schmid, I., Schmid, P., and Giorgi, J. V. (1988). *Cytometry* **9,** 533–538.

Taylor, J. M. G., Fahey, J. L., Detels, R., and Giorrgi, J. V. (1989). *J. Acquired Immune Defic. Syndr.* **2,** 114–124.

World Health Organization Global Programme on AIDS (1994). *AIDS* **8**(1), WHO1–WHO4.

CHAPTER 27

Cell-Cycle Analysis of *Saccharomyces cerevisiae*

Bruce S. Dien, Marvin S. Peterson, and Friedrich Srienc

Department of Chemical Engineering and Materials Science
University of Minnesota
Minneapolis, Minnesota 55455
and Institute for Advanced Studies in Biological Process Technology
St. Paul, Minnesota 55108

METHODS IN CELL BIOLOGY, VOL. 42

I. Introduction

Flow cytometry has become an important tool for studying the cell-cycle kinetics of yeasts. It has been used to study the phenotypic effects of various mutations upon cell-cycle kinetics in the yeasts *Schizosaccharomyces pombe* and *Saccharomyces cerevisiae*, which are important models for studying the regulation of the eukaryotic cell cycle (Cross *et al.*, 1989; Ghiara *et al.*, 1991; Richardson *et al.*, 1989). Flow cytometry can be used to probe the physiological state of a culture or to measure the degree of synchrony (Van Doorn *et al.*, 1988). Flow cytometry is also a convenient tool for determining the total genome size of yeast strains such as *S. cerevisiae* and *Candida albicans*, which are found in nature at a variety of different ploidy levels (Dvorak *et al.*, 1987; Hutter and Eipel, 1979).

This chapter emphasizes flow cytometry measurement methods of the DNA content of *S. cerevisiae* populations, although the analysis methods might be applicable to other yeasts with minor modifications. Other papers detailing the DNA staining of this and other types of yeasts for flow cytometry are summarized in Table I. A recent review also discusses the application of flow cytometry to the analysis of the single-cell protein content of yeast (Alberghina and Porro, 1993).

Important differences exist between yeasts and mammalian cells which need to be considered when preparing, measuring, and analyzing the DNA distributions of yeasts. First, yeasts have approximately 300 times less chromosomal DNA than mammalian cells (Wehr and Parks, 1969). The fluorescence intensity of the stained nuclear yeast DNA is, thus, much lower than that for mammalian cells. The DNA histograms measured have coefficients of variations (CV) for the G_1 and the G_2 + M peaks which are appreciably higher than those typically obtained with mammalian cells.

Table I
References on DNA Staining of Yeasts

Yeast	DNA Stain	Reference
C. albicans	Mithramycin	Dvorak *et al.* (1987)
C. albicans	Propidium iodide	Olaiya and Sogin (1979)
S. cerevisiae	Acriflavine	Doran and Bailey (1986)
S. cerevisiae	DAPI	Munch *et al.* (1992)
S. cerevisiae	Mithramycin	Slater *et al.* (1977)
S. cerevisiae	Propidium iodide	Hutter and Eipel (1979)
S. cerevisiae	Olivomycin	Kuchenbecker and Braun (1985)
S. pombe	Propidium iodide	Beach *et al.* (1985)
S. pombe	Various dyes	Sazer and Sherwood (1990)

Second, yeasts have a much larger ratio of mitochondrial to nuclear DNA than mammalian cells (Wehr and Parks, 1969). Mitochondrial DNA usually comprises 5–20% of the total yeast DNA (Dujon, 1981). This interferes with accurate determination of the cell-cycle fractions because mitochondrial DNA is synthesized continuously within the cell cycle. Mitochondrial DNA consists of approximately 80% AT bases and, therefore, is expected to have a more pronounced effect with AT-specific stains, such as Hoechst and DAPI, than with other DNA-specific dyes such as mithramycin which is GC specific (Sazer and Sherwood, 1990). Furthermore, the large amount of nonnuclear material compared to the small genome size significantly contributes to nonspecific staining. Protocols for staining yeast DNA need to be carefully optimized to minimize these effects.

Third, yeasts have a cell wall. Its presence affects both the DNA staining procedure as well as the analysis of the cell-cycle kinetics. Many DNA staining procedures require the enzymatic digestion of RNA. The cell wall impedes the diffusion of this enzyme, ribonuclease, into the cell (Scherrer *et al.,* 1974) and makes the digestion step less efficient than that with mammalian cells. The lower efficiency of the RNA digestion step is usually compensated by increasing its length (Hutter and Eipel, 1979).

The presence of the cell wall also influences the interpretation of DNA distributions because newborn cells remain attached by their cell wall for a certain amount of time following cytokinesis (Section VII), which marks the end of mitosis. Though these newborn cells still attached to each other are G_1 phase cells, they are included in the DNA histogram together with the G_2 + M phase cells. The mitotic phase cells which have segregated their chromosomes and the newborn cells still attached to each other can be indentified by analyzing DNA signals in peak and area mode (Block *et al.,* 1990).

Fourth, for budding yeasts, the newborn daughter cells are normally much smaller than the newborn mother cells because they divide asymmetrically. Since the main cell-cycle size regulation point for budding yeast is at the end of the G_1 phase, the daughter cells spend (on average) more time in the G_1 phase than the mother cells. Therefore, the duration of the G_1 phase, which can be calculated from data obtained from a DNA histogram, is the average of two distinct populations, the mother and the daughter cells.

Section III, Methods, detailed protocols are presented for staining yeast DNA with mithramycin, propidium iodide, and acriflavin. The bromodeoxyuridine (BrdUrd) method (Dolbeare *et al.*, 1991), as adapted for yeast, is also presented in this section. A section has also been included which discusses the mathematical analysis of yeast DNA histograms. Representative data obtained with each of the protocols and the subsequent mathematical analyses are compared in Section VII, Results. In addition, an example is given in which the BrdUrd labeling technique is applied to measure the distributions of cell properties at specific points within the cell cycle.

II. Materials

All solutions should be filtered through a 0.2-μm filter before use.

A. Staining Solutions

1. *Acriflavin:* Dissolved 10 ml 0.5 *N* HCl, 500 mg $K_2S_2O_5$, and 20 mg acriflavin (Sigma, St. Louis, MO) in 100 ml dH_2O; store at 4°C in the dark.

2. *Mithramycin:* Mix 200 μl mithramycin stock solution, 200 μl $MgCl_2$ stock solution (250 m*M*, see Section II,C), and 9.5 ml phosphate-buffered saline (PBS, see Section II,C). Prepare mithramycin stock solution by dissolving 1 mg mithramycin A (Cat. No. M6891, Sigma) in 1 ml PBS. Stock solution can be stored at least 1 month in the dark at 4°C.

3. *Propidium Iodide:* Dissolve 5.0 mg propidium iodide (Sigma) in 100 ml PBS; store at 4°C in the dark.

B. Antibody Solutions

1. *Preabsorbed anti-BrdUrd IgG:* Dissolve 0.1 g acetone-dried *Escherichia coli* powder (Harlow and Lane, 1988) in 10 ml PBS. Incubate at 97°C for 15 min. Centrifuge at 4000*g* for 20 min at 4°C and discard supernatant. Resuspend pellet in 10 ml PBS/BSA. Add 20 μl of IU-4 ascites solution containing 1 mg/ml protein (lot number 400, Cat. No. MD5000, Caltag Laboratories, South San Francisco, CA). Incubate for 30 min at 37°C. Centrifuge cells at 4000*g* for 20 min at 4°C. Recover supernatant and discard pellet. Store antibody solution on ice.

2. *Secondary Ab solution:* 10 mg RNase A (Sigma), 20 μl affinity-purified FITC-conjugated goat anti-mouse $F(ab')_2$ fragment solution containing 150 μg/ml protein (Tago, Burlingame, CA) in 10 ml PBS/BSA.

C. Other

1. *Acid alcohol:* Mix 1 ml *concentrated* HCl (12 *N*) in 99 ml 70% EtOH.

2. *BrdUrd solution:* Dissolve 25 mg BrdUrd (Sigma) in 5 ml dH_2O; sterilize by filtration.

3. *Chromatin denaturation solution (CDS):* Dissolve 2.5 ml 4 *N* HCl, 0.5 g Triton X-100 (Sigma), 1.75 g NaCl in 97.5 ml dH_2O; heat to dissolve Triton X-100 and cool on ice before using to treat fixed cells.

4. *Formaldehyde solution (0.25% v/v):* Mix 0.156 ml formaldehyde solution (16% ultrapure formaldehyde, EM Grade, Polyscience, Warrington, PA) in 10 ml PBS.

5. *2-Mercaptoethanol solution (0.143 M):* Disssolve 100 μl 2-mercaptoethanol (Sigma) in 10 ml PBS.

6. *$MgCl_2$ stock solution (250 mM):* Dissolve 5.08 g $MgCl_2 \cdot 6H_2O$ in 100 ml dH_2O.

7. *Phosphate-buffered saline:* For a 0.05 *M* solution, dissolve 1.104 g $NaH_2PO_4 \cdot H_2O$, 5.965 g Na_2HPO_4, 8.75 g NaCl in 1.01 dH_2O; pH 7.5.

8. *PBS bovine serum albumin (BSA):* Dissolve 1.0 g BSA (Cat. No. A-7906, Sigma) and 0.1 g sodium azide in 100 ml PBS.

9. *PBS/sorbitol (2.6 M):* Dissolve 47.32 g sorbitol (Sigma) in 100 ml 0.1 *M* PBS.

10. *Zymolyase solution (40 U/ml):* Dissolve 20 mg 20T zymolyase (20 U/mg; ICN ImmunoBiologicals, Lislie, IL) in 10 ml PBS; prepare solution right before use and store on ice.

III. Methods

A. Staining DNA with Propidium Iodide

1. *Cell fixation:* From an exponentially growing culture of cells, remove a 1-ml sample, transfer to a 1.5-ml Eppendorf tube, and centrifuge for 5 sec at 16000*g* at room temperature. Discard the supernatant and resuspend the cells in 300 μl of ice-cold dH_2O. Cells are fixed by adding 700 μl of ice-cold 95% EtOH and vortexing. Cells can be stored at −20°C in the 70% EtOH fixative for at least 3 weeks without noticeably affecting the final results.

2. *Chromatin denaturation:* Centrifuge fixed cells for 5 min at 824*g* and 4°C. Wash cells twice with 200 μl cold PBS. Transfer 2×10^6 cells to a 1.5-ml nonstick Eppendorf tube (Rainin Instruments, Woburn, MA). Resuspend cells in 500 μl ice-cold CDS and incubate on ice for 10 min. Centrifuge cells for 5 min at 824*g* and 4°C and wash cells twice with 200 μl PBS.

3. *Formaldehyde incubation:* Incubate cells in 500 μl formaldehyde solution for 30 min at room temperature. Centrifuge cells for 5 min at 824*g* and 4°C and wash twice with 200 μl PBS.

4. *Cell wall permeation:* Incubate cells in 200 μl zymolyase solution at 30°C for 10 min. Centrifuge cells for 5 min at 824*g* and 4°C and wash twice with 200 μl PBS.

5. *RNA digestion:* Incubate in 1 ml RNase solution for 30 min at 37°C. Centrifuge cells for 10 min at 824*g* and 4°C and wash once with 200 μl PBS. For wash step, centrifuge cells for 5 min at 824*g* and 4°C. Cells at this point can be stored in 200 μl PBS for at least 4 days at 4°C.

6. *Propidium iodide staining:* Immediately before flow cytometric analysis, incubate cells in 200 μl of propidium iodide solution for 30 min at room temperature and centrifuge cells for 5 min at 824*g* and 4°C. Resuspend in 200 μl PBS and store on ice shielded from light. To analyze the sample, transfer 100 μl of

this cell solution to a new test tube and dilute with 1.9 ml of filtered distilled water.

B. Mithramycin Staining of DNA

1. *Cell fixation and chromatin denaturation:* Fix cells and denature chromatin as detailed in steps 1 and 2 of Section II,A.

2. *Mithramycin staining:* Incubate cells in 1 ml mithramycin staining solution at room temperature. Cells can be analyzed on the flow cytometry within $\frac{1}{2}$ to 4 hr. Cells need to be kept in the staining solution during analysis.

C. Acriflavine Staining (Block *et al.*, 1990)

1. *Cell fixation:* Fix cells as described in step 1 of Section II,A.

2. *RNA hydrolysis:* Transfer 5×10^6 cells to new 1.5-ml Eppendorf tube and centrifuge for 5 sec at 16000*g* at room temperature. Discard supernatant and resuspend the pellet in 0.7 ml at 4 *N* HCl and incubate for 20 min at room temperature.

3. *Acriflavine staining:* Centrifuge cells for 5 sec at 16000*g* at room temperature, resuspend in 0.7 ml of acriflavine staining solution, and incubate for 20 min at room temperature.

4. *Washing cells:* Centrifuge cells for 5 sec at 16000*g* at room temperature. Wash cells three times with 0.7 ml of acid alcohol solution. Disperse the pellet during these wash steps by repeated pipetting with a 1-ml pipetman. Resuspend the cells in distilled water and store at 4°C in the dark. Samples are stable for at least 1 week. For analysis samples can be diluted in distilled water.

D. BrdUrd Labeling and Staining of Cells

1. *Pulse labeling with BrdUrd:* After cells have grown to a density of 1×10^7 cells per ml of culture, add 1 ml of BrdUrd solution per 50 ml of medium (final BrdUrd concentration: 6.5 m*M*).

2. *Cell fixation:* After 10 min, remove 5 ml of cell solution and transfer to a glass test tube. Cool cells immediately for 10 sec in an ice water bath. Centrifuge cells at 4000*g* for 1 min at 4°C. Discard supernatant and resuspend cells in 1.5 ml of sorbitol solution. Add 3.5 ml of cold 95% EtOH while vortexing cells and then place on ice. Repeat sampling as desired during labeling period. Store cells at −20°C until ready for subsequent steps.

3. *Chromatin denaturation:* Wash fixed cells twice with 2 ml cold PBS. Centrifuge cells at 824*g* for 5 min at 4°C. Transfer 2×10^7 cells to a new 10×75-mm glass test tube (Fisher, Pittsburgh, PA). Incubate the cells in 2 ml of cold CDS, on ice, for 10 min. Wash cells twice with 2 ml of cold PBS.

4. *2-Mercaptoethanol incubation:* Incubate cells in 2 ml 2-mercaptoethanol solution for 30 min at 30°C. Centrifuge cells at 824*g* for 5 min at 4°C. Wash cells twice with 2 ml cold PBS.

5. *Formaldehyde incubation:* Incubate cells in 2 ml formaldehyde solution for 30 min at room temperature. Centrifuge cells at 824*g* for 5 min at 4°C. Wash cells once with 2 ml cold PBS and once with 2 ml cold dH_2O.

6. *Partial DNA denaturation:* Resuspend cells in 2 ml cold dH_2O. Incubate cells in a 98°C water bath for 10 min. Upon removal, cool samples immediately in an ice water bath.

7. *Cell wall permeation:* Centrifuge cells at 824*g* for 5 min at 4°C and resuspend cells in 2 ml of cold zymolyase solution for 12–16 hr on ice. Centrifuge cells at 824*g* for 5 min at 4°C and resuspend cells in 1 ml cold PBS. The cells are suitable for immunological staining for at least 2 days provided they are kept on ice.

8. *Primary antibody incubation:* Transfer 100 μl of cell solution to a cooled, 1.5-ml nonstick Eppendorf tube. Centrifuge cells for 5 min at 824*g* and 4°C, resuspend cells in 1.0 ml of primary antibody solution, and incubate for 1 hr at 37°C. After incubation, centrifuge cells for 10 min at 824*g* and 4°C. Wash cells twice with 200 μl PBS/BSA. For wash steps, centrifuge cells for 5 min at 824*g* and 4°C. Allow cells to incubate for 15 min at room temperature between last two wash steps.

9. *Secondary antibody incubation:* Centrifuge cells for 5 min at 824*g* and 4°C and resuspend cells in 1 ml of secondary antibody solution. Incubate cells at 37°C for 75 min. After incubation, centrifuge cells for 10 min at 824*g* and 4°C. Wash cells twice with 200 μl cold PBS. For wash steps, centrifuge cells for 5 min at 824*g* and 4°C. Cells can be stored on ice in the dark for at least 24 hr.

10. *Propidium iodide staining:* Immediately before flow cytometric analysis of cells, centrifuge cells for 5 min at 824*g* and 4°C, resuspend them in 200 μl of propidium iodide solution, and incubate them in the dark at room temperature for 30 min. Centrifuge cells and resuspend in 200 μl of cold PBS. To analyze the cells, transfer 100 μl of cell solution to a new glass test tube and add approximately 1.9 ml of distilled water.

IV. Critical Aspects

A. Growth of Cells

Meaningful comparison of cell-cycle parameters based on measured DNA distributions of two or more populations of cells requires that the cells in each population are measured in the same physiological state. Therefore, cells should be in balanced exponential growth before harvesting. Balanced growth is present when the DNA distribution of the population becomes time invariant. Exponential growth of the culture, as determined by monitoring cell number or optical density over time, does not guarantee that the culture is in balanced growth. The time a culture needs to grow before the DNA frequency distribution becomes stable is a complex function of the inoculum history, medium, and strain. Cultivating cells in a homogeneously mixed continuous culture with precisely controlled environmental conditions is ideal for obtaining balanced growth. Precise control of the culture conditions requires using a bioreactor equipped with controls for operating parameters such as aeration, pH, medium flow rates, and temperature.

When continuous cultivation is not possible, it is recommended that cultures be kept in exponential growth for at least 24 hr before the DNA distribution of the population is measured. Such extended exponential growth can be obtained in shaker flask cultures by diluting the culture with fresh medium. When grown in YPD medium (1% yeast extract, 2% peptone, and 2% dextrose), cells typically leave the balanced growth state two generations prior to the beginning of stationary phase. The DNA distribution should be measured before the culture has reached 25% of its maximum cell density.

B. Propidium Iodide Staining Protocol

During fixation, the cells need to be dispersed before ethanol is added or they will clump. If the water and EtOH used for cell fixation are not kept on ice and instead allowed to warm to room temperature, the autofluorescence of the cells will increase. Denaturation of the chromatin and permeabilization of the cell wall (steps 2–4 of Section II,A) significantly increase the quality of the measured DNA distributions compared to other protocols which do not include these steps (Darzynkiewicz *et al.*, 1984; Dien and Srienc, 1991). After the cells are treated with zymolyase, they need to be kept in either glass test tubes or nonstick Eppendorf tubes because they tend to stick to the surface of normal Eppendorf tubes. The formaldehyde incubation step is important for increasing the time cells can be stored on ice, presumably by diminishing the activity of native DNase and proteinase activity (Pearse, 1968). Finally, it is preferable not to measure the DNA of yeast cells directly in the propidium iodide solution because removal of the cells from the staining solution diminishes nonspecific staining.

C. BrdUrd Staining Technique

Yeasts are not able to take up and to incorporate extracellular thymidine or BrdUrd into the replicating DNA. The yeast strain used for the BrdUrd labeling has been genetically engineered to allow this (Sclafani and Fangman, 1986). The yeast strain is auxotrophic for thymidine and thymidine needs to be added to a final concentration of 100 μg/ml to all media. A staining protocol using these strains has been described in detail previously (Dien and Srienc, 1991). To avoid the appearance of petites, the strains were maintained on YPG plates [1% yeast extract, 2% peptone, 3% (v/v) glycerol, and 1.5% agar] with 100 μg/ml thymidine added. In steps 6–8 it is essential that the cells remain on ice. If the cells are allowed to warm up, presumably the rate of DNA renaturation will be enhanced. Since the BrdUrd antibody only recognizes BrdUrd on single-stranded DNA, the staining levels will be diminished (Beisker *et al.,* 1987). The specific composition of the *S. cerevisiae* cell wall is dependent on the strain, the growth conditions, and the growth rate. Under certain growth conditions a more efficient cell wall removal procedure might be required (Kuo and Yamamoto, 1975).

There are many different anti-BrdUrd antibodies available (Dobeare *et al.,* 1991). The type IU-4 was chosen because of its high sensitivity for incorporated BrdUrd (Vanderlaan *et al.,* 1986). The initial staining protocol was developed with an antibody preparation of high titer provided by F. Dolbeare, which gave excellent results. Unfortunately, the Caltag antibody ascites solution is synthesized in mice and appears to be contaminated with additional antibodies which bind to yeast. It has been observed that ascites solutions typically contain antibodies which cross-react with *S. cerevisiae* proteins (Pringle *et al.,* 1991). The level of background staining can be reduced by preincubating the primary antibody with DNase-free acetone-dried *E. coli* powder.

V. Flow Cytometry

Data were acquired using a Cytofluorograf IIs flow cytometer equipped with an Ortho 2151 data acquisition system (Ortho Diagnostic Systems). Data analysis was performed using the software packages Listview, Multi2d and Multicycle (Phoenix Flow Systems). The light source was an argon ion laser (Coherent Innova 90-5, Coherent, Palo Alto, CA) suitable for emission at wavelengths of 457 and 488 nm. The laser wavelength used to excite each dye and the filters used for collection of the emitted fluorescence are listed in Table II. In all cases, the laser power was set at 200 mW. For the BrdUrd method, the FITC fluorescence was measured in peak mode.

Table II
Flow Cytometric Setup for Various Dyes

Dye	Laser wavelength (nm)	Filter	Filter source
Acriflavine	488	530 nm long pass	Corning Glass
FITC	488	FITC band pass	Ortho Diagnostic
Mithramycin	458	530 nm long pass	Corning Glass
Propidium iodide	488	570 long pass	Corning Glass

VI. Analysis of DNA Histograms

Cell-cycle analysis of DNA histograms using parametric models allows for convenient and quantitative comparison of a series of DNA histograms. Two models commonly used to analyze the DNA distributions of mammalian cells are the Dean and Jett model (1974) and the Fried model (1976). We have found that the Dean and Jett model, with a few modifications, fits the yeast DNA distributions better than the Fried model (Peterson, 1991).

The Dean and Jett (1974) model fits the G_1 and G_2 + M DNA peaks with Gaussian distributions and the S phase with a broadened second-degree polynomial. The use of a broadened second-order or even first-order representation of the frequency distribution of the S phase subpopulation in yeast can lead to serious errors in estimating the cell-cycle fractions and may increase the computation time. It can frequently result in negative fractions of the subpopulations as the nonlinear optimization searches for the best fit. To prevent these problems, a constant or uniform distribution can be chosen for the S phase frequency distribution and placed into an exponential function (Peterson, 1991). To avoid negative fits of the G_1 or G_2+M phase distributions, the weight (w) of each Gaussian distribution should also be represented by an exponential form, i.e., exp (w). If the histogram is particularly noisy, the CVS (standard deviation/mean) of the G_1 and G_2+M phases can be set equal to one another and the mean of the G_2+M peak fixed at a value twice that of the G_1 peak.

The analysis of the data presented was performed with the Multicycle program (Phoenix), a commercial software package based on the Dean and Jett model. The S phase was fit with a constant frequency. The analysis gave in all cases similar CVs for the G_1 and G_2+M phases, ratios for the G_2+M to the G_1 mean DNA contents between 1.9–2.1, and only positive frequencies.

From the cell-cycle fractions, the average transit times in each cell-cycle phase have been determined by assuming an exponential age distribution as described by Slater *et al.* (1977).

VII. Results

A. Comparison of DNA Stains

Typical DNA histograms from a diploid strain of *S. cerevisiae* stained with mithramycin, acriflavine, and propidium iodide are displayed in Fig. 1. The cell-cycle fractions estimated from these histograms using Multicycle (Phoenix) are summarized in Table III. All data were obtained from the same sample from an exponentially growing yeast population. The percentages of mitotic cells and cell doublets listed in Table III were estimated directly from DNA area versus DNA peak cytograms such as the one displayed in Fig. 2.

The populations stained with propidium iodide seem to have a larger fraction

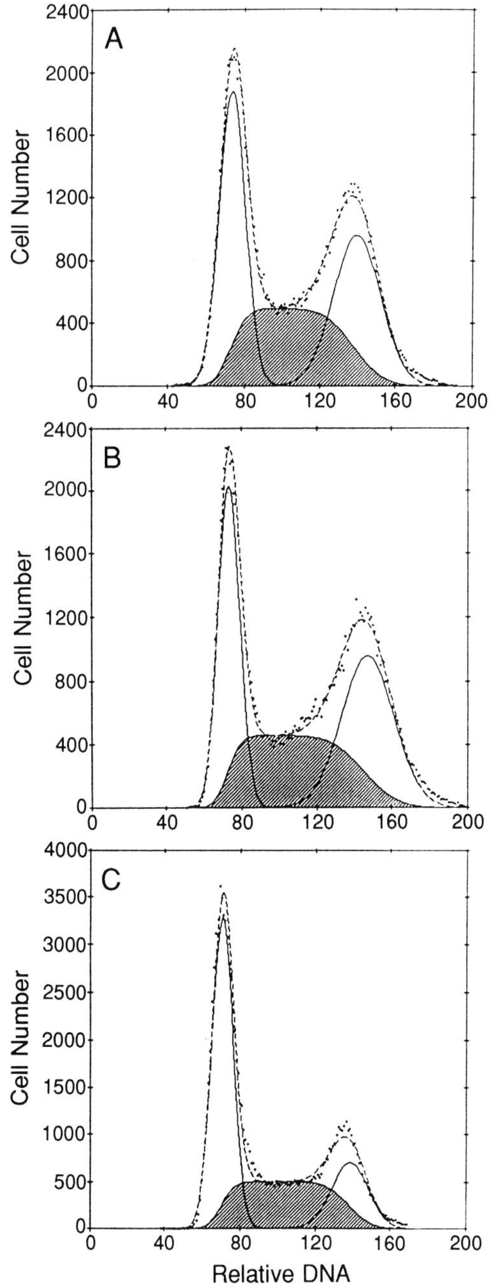

Fig. 1 DNA histograms and cell-cycle analysis for exponentially growing yeast cells stained with (A) acriflavine, (B) mithramycin, and (C) propidium iodide. The dots are the measured frequencies and the solid lines are the fits to the data obtained by using the Dean and Jett model (see text). The DNA distributions of the S phase cells, as predicted by the model, are indicated by the shaded regions.

Table III
Estimated Percentage of Cells in Each Cell-Cycle Phase

DNA stain	G_1	S	G_2 + M	Doublets	CV of G_1 peak
Acriflavine	32.0 ± 2.8	34.8 ± 0.9	33.2 ± 2.0	11.6 ± 0.4	8.8 ± 0.5
Mithramycin	28.5 ± 1.6	35.2 ± 0.6	36.3 ± 1.0	11.8 ± 0.8	8.4 ± 0.7
Propidium iodide	45.1 ± 1.8	35.2 ± 0.1	19.8 ± 1.9	4.2 ± 0.0	7.6 ± 0.1
BrdUrd	48.7	28.3	23.0		

of cells in the G_1 phase than those stained with either acriflavine or mithramycin (Fig. 1, Table III). The cause for this apparent discrepancy is the 10-min zymolyase treatment used in the propidium iodide protocol. While 10 min is not long enough to entirely remove the cell wall (Palmer *et al.*, 1989), it appears to be long enough to separate the G_1 doublets. G_1 doublets are cells which have completed cytokinesis but have not yet completed cell wall separation. Since the mitotic phase ends with cytokinesis, each doublet is comprised of two G_1 phase cells. However, a doublet has the same DNA content as a single G_2+M phase cell and is identified by the flow cytometer as a G_2+M phase cell. Therefore, the propidium iodide-stained population has an average of 8% fewer doublets than the acriflavine or mithramycin-stained populations.

The quality of the DNA histograms measured with the different DNA stains

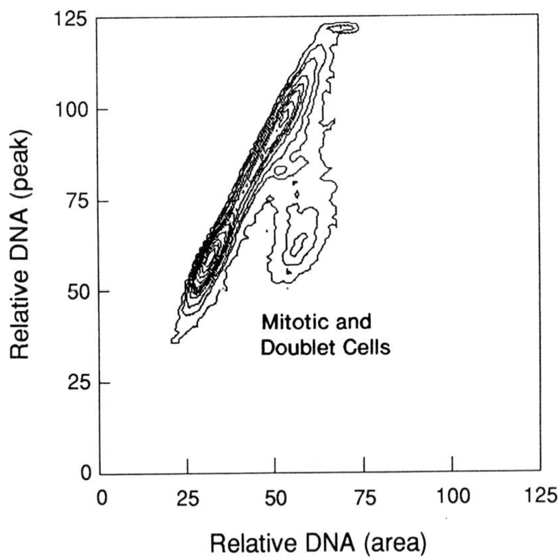

Fig. 2 Cytogram of DNA fluorescence in peak mode versus the DNA fluorescence in area mode used to determine the fraction of mitotic and doublet cells. Cells were stained with propidium iodide.

can be evaluated by comparing the estimated CVs of the Gaussian curves used to fit the G_1 phase distribution (Table III). The G_1 phase distribution CVs from the DNA histograms obtained for mithramycin- and propidium iodide-stained cells were lower than the CV determined for the acriflavine stained cells. However, all of the CVs were within the 1–10% range typically obtained for mammalian cells (Dean, 1986), albeit on the high side of the range.

The duration of each cell-cycle phase was estimated by assuming an exponential age distribution (Table IV). The duration of the mitotic phase was calculated from the fraction of mitotic cells measured for the propidium iodide-stained samples. This subpopulation was identified from the DNA area versus DNA peak cytograms. The doublet phase length is the additional time two newborn cells remain attached by their cell walls after completing mitosis. For both the propidium iodide-stained population and the BrdUrd-pulsed population, which were both treated with zymolyase to dissociate newborn doublets, this time was set to 0. The assumption that no detectable doublets arise from two yeast cells sticking together as they flow past the laser beam has been tested and found to be valid for this strain (manuscript in preparation).

In general, the estimated lengths of the cell-cycle phases are simialar regardless of the DNA staining protocol applied. Estimates for the combined duration of the G_1 and the doublet phases from the different DNA staining procedures differ by 5 min or less from each other. This represents 4% of the total cell-cycle period. Results for the duration of the S phase are also in similar agreement. The largest difference is between the length of G_2+M phase estimated from the mithramycin- and the propidium iodide-stained samples. These times varied by 10 min.

B. BrdUrd-Labeled Cells and Lengths of Cell-Cycle Phases

The percentage of cells in each cell-cycle phase can also be measured from the BrdUrd-labeled cells. A typical BrdUrd/propidium iodide cytogram is displayed in Fig. 3. While the percentage of cells in each of the cell-cycle phases

Table IV
Estimated Duration of Cell-Cycle Phases

	Duration of cell-cycle phase (minutes)				
DNA stain	G_1	S	G_2	M	Doublet
Acriflavine	30	40	31	7.1	12
Mithramycin	27	40	34	7.1	12
Propidium Iodide	44	45	24	7.1	0.0
BrdUrd	49	37	27	7.1	0.0

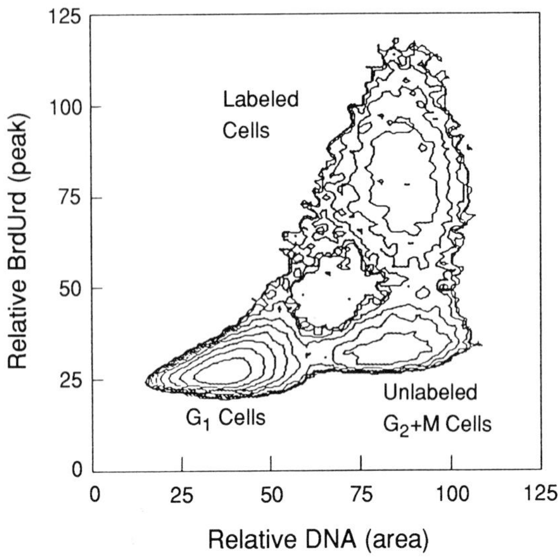

Fig. 3 One of the cytograms of relative BrdUrd versus relative DNA contents used to determine the fractions of labeled cells, unlabeled G_1, and unlabeled $G_2 + M$ cells. This sample was taken 15 min after BrdUrd was added to the culture.

during balanced growth can be estimated from a single cytogram, a more accurate estimate is obtained using a time series of cytograms measured after the application of the BrdUrd pulse (Section VII,C). In the experiment shown, the culture was sampled at 5-min intervals from 10 to 35 min after the pulse to obtain the data needed to calculate the cell-cycle percentages. Duplicate samples removed at each sampling time were stained in parallel. The cell-cycle percentages estimated with Multicycle for the propidium iodide-stained sample and those calculated from the BrdUrd-pulsed cells are very similar (Table III).

The lengths of the cell-cycle phases were also estimated from the cell-cycle fractions for the BrdUrd-labeled population by assuming an exponential age distribution (Table IV). The results obtained with BrdUrd labeling are in good agreement with the values obtained from the DNA histograms discussed earlier (Section VII,A). The largest difference between the two sets of data is in the value predicted for the combined length of the G_1 and doublet phases. Of particular interest is the predicted length of the S phase which has been underestimated by flow cytometric analysis of DNA histograms in the past (Wheals, 1987). From our analysis, the duration of the S phase predicted using Multicycle analysis and BrdUrd labeling are in good agreement. The duration of the G_1 phase represents the average of the amount of time the mother and daughter cells reside in the G_1 phase. It is important to note that the daughter cells would be expected to spend on average much more time in the G_1 phase than do the mother cells.

C. Calculation of Cell-Cycle Parameters from BrdUrd-Pulsed Cells

The percentage of cells in each cell-cycle phase were estimated using a series of samples taken at 5- min intervals once the incorporated BrdUrd pulse became detectable: BrdUrd can be detected 5 min after the pulse is initiated. Each resultant BrdUrd/propidium iodide cytogram was used to determine the percentages of unlabeled G_1 cells, unlabeled G_2+M cells, and BrdUrd-labeled cells; the data are shown in Fig. 4A.

To calculate the percentage of cells in each cell-cycle phase from this series of data, it has been assumed that there is no age dispersion in the G_2+M phase, i.e., all cells spend the same amount of time in the G_2+M phase before cell division and no labeled cell divides until all the unlabeled cells have divided. With this assumption, a cell population balance equation can be formulated which expresses the decrease in the function of the unlabeled G_2+M phase cells as a function of time after the BrdUrd first becomes detectable. For every newborn cell, one unlabeled G_2+M phase cell must divide

$$U_{G_2+M}(0) - U_{G_2+M}(t) = N(t) - \mathrm{N}(0), \tag{1}$$

where U_{G_2+m} is the number of unlabeled G_2+M phase cells, N is the total number of cells in the population, and t is the time after the BrdUrd pulse first becomes detectable. The number of cells at time t is

$$N(t) = N(0)\exp(kt), \tag{2}$$

where k is the specific growth rate of the culture. The fraction of unlabeled G_2+M phase cells is defined by

$$f^U_{G_2+M}(t) = \frac{U_{G_2+M}(t)}{N(0)\exp(kt)}. \tag{3}$$

Division by Eq.(1) by the total number of cells at time t gives

$$f^U_{G_2+M}(0)\exp(-kt) - \mathrm{f}^U_{G_2+M}(t) = 1 - \exp(-kt). \tag{4}$$

Re-arrangement of Eq.(4) gives

$$\ln\{\mathrm{f}^U_{G_2+M}(t) + 1\} = \ln\{\mathrm{f}^U_{G_2+M}(0) + 1\} - kt, \tag{5}$$

where $f^U_{G_2+M}(0)$ is the fraction of unlabeled G_2+M phase cells at time 0, i.e., previous to the time when the BrdUrd pulse first becomes detectable.

By plotting the function of the left side of Eq.(5) versus time (Fig. 4B), the initial fraction of G_2+M phase cells can be calculated from the intercept (Table III). The generation time of the population can be obtained from the slope of this curve. If the culture's growth behavior is not disturbed during the pulse, this generation time should agree with the one measured from the cell number doubling time. For the data displayed in Fig. 4B, the calculated generation time deviated less than 2% from the 2.0-hr cell number doubling time measured by monitoring the increase in cell concentration over time during the exponential growth (data not shown).

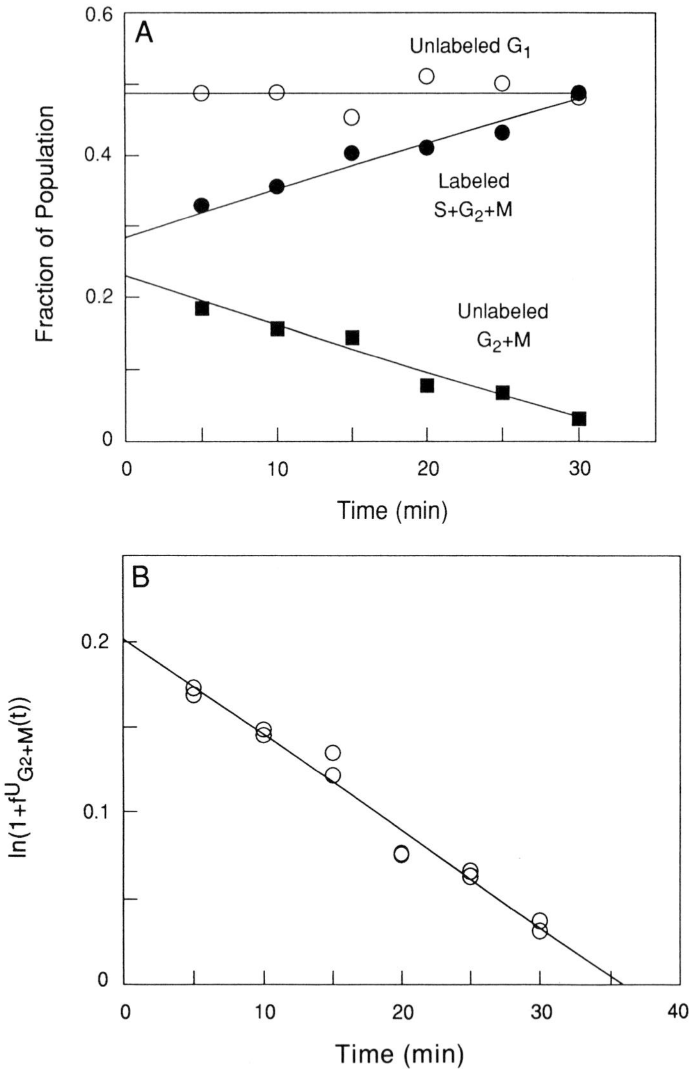

Fig. 4 (A) Predicted and actual changes in the labeled and unlabeled cell fractions with time during a BrdUrd pulse experiment. The data points are indicated by symbols and the predicted values by solid lines. (B) Linearized relationship between unlabeled G_2+M cells and time as suggested by Eq. (5). The correlation coefficient of the linear regression is 0.98. The intercept at Time 0 gives an estimate of the total G_2+M cell-cycle fraction (see text).

The percentage of cells in the G_1 phase can also be obtained from this graph. As can be seen in Fig. 4A, the fraction of unlabeled G_1 phase cells is constant for the first 35 min of the pulse. The time invariant value of the unlabeled G_1 cell fraction is a direct consequence of the (near) absence of age dispersion in the G_2+M phase. That is, in balanced growth, the fraction of unlabeled G_1

phase cells would be expected to be constant until the last unlabeled G_2+M phase cell has divided. Thus, the fraction of G_1 phase cells in the population can be estimated from the mean of the G_1 phase fractions observed in samples collected during the first 35 min of the pulse.

The fraction of S phase cells can be obtained by subtracting the fraction of cells in the G_1 and the G_2+M from unity:

$$f_s(0) = 1 - f_{G_2+M}(0) - f_{G_1}(0). \tag{6}$$

To test the agreement between the model used to calculate the cell-cycle fractions and the experimental data, the values predicted by the model are plotted in Fig. 4A together with the experimentally determined fractions. The fraction of unlabeled G_2+M phase cells was calculated using Eq. (5) with the initial G_2+M fraction calculated from Fig. 4B and specific growth rate determined from the cell number doubling time of the culture. The fraction of G_1 phase cells was calculated from the mean of the G_1 fractions observed in the samples collected during the pulse. The fraction of labeled cells was determined from an equation similar to Eq. (6),

$$f^L(0) = 1 - f^U_{G_2+M}(0) - f^U_{G_1}(0), \tag{7}$$

where f^L is the fraction of labeled cells and $f^U_{G_1}$ the fraction of unlabeled G_1 phase cells. The graph shows that the theoretical and experimental data agree closely.

D. Cell-Cycle-Specific Property Distributions and Single-Cell Growth Rates

The BrdUrd/PI technique has greater utility than simply providing a means for determining cell-cycle fractions. When combined with the simultaneous labeling of other cell properties, this technique can be used to directly analyze cell-cycle regulation and the influence of the operation of the cell cycle on cell properties. Specifically, when combined with the measurement of an additional property, the BrdUrd technique allows for the measurement of property distributions for seven distinct cell-cycle subpopulations: (i) G_1, (ii) S, (iii) G_2+M, (iv) cells entering the S phase, (v) cells exiting the S phase, (vi) dividing cells, and (vii) newborn cells. Furthermore, the single-cell rates of accumulation of the measured property can be determined in the individual cell-cycle phases by applying one-dimensional population balances to these property distributions.

Application of the BrdUrd technique to measure distributions of size—estimated by measurement of the forward angle light scatter (FALS) of the cells— and of total protein at specific points within the cell cycle of *S. cerevisiae* has been described previously (Dien and Srienc, 1991; Srienc and Dien, 1992). Single-cell protein content was measured together with DNA and BrdUrd contents after the cells were stained with AMCA, a protein-specific stain. Measurement of these properties have revealed important information about cell-cycle regulation and the influence of the operaiton of the cell cycle on cell growth in

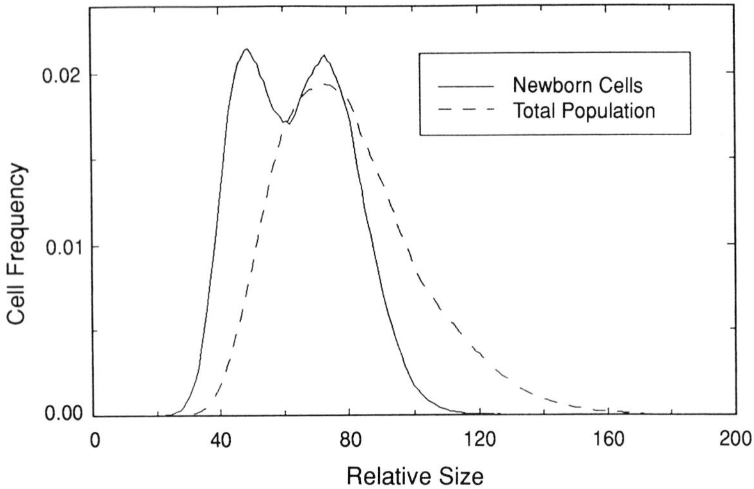

Fig. 5 FALS distributions of the newborn and the total cell population measured for a culture of exponentially growing *S. cerevisiae*. Cells were grown in YPD with added thymidine, pulse chased with BrdUrd, and analyzed with the BrdUrd technique. The newborn cells were identified by following the cohort of BrdUrd-labeled cells through the cell cycle. The bimodal size distribution measured for the newborn cells indicates the large difference in size between the mother and daughter cells at birth.

terms of both size and protein content. Examples of FALS distributions for the newborn and the total cell populations of an exponentially growing yeast culture obtained with the BrdUrd technique are displayed in Fig. 5. The bimodal FALS distribution measured for the newborn population indicates the difference in size between mother and daughter cells at birth which is expected for budding yeasts.

Acknowledgments

We are grateful to S. J. Kromenaker and P. Sweeney for help in preparation of the manuscript. Bruce S. Dien is a recipient of a National Institute of Health biotechnology training grant fellowship. The work has been supported by the National Science Foundation (BCS-9100385).

References

Alberghina, L. and Porro, D. (1993). *Yeast* **9,** 815–823.

Beach, D., Rodgers, L., and Gould, J. (1985). *Curr. Genet.* **10,** 297–311.

Beisker, W., Dolbeare, F., and Gray, J. W. (1987). *Cytometry* **8,** 235–239.

Block, D. E., Eitzman, P. D., Wangensteen, J. D., and Srienc, F. (1990). *Biotechnol. Prog.* **6,** 504–512.

Cross, F., Roberts, J., and Weintraub, H. (1989). *Annu. Rev. Cell Biol.* **5,** 341–395.

Darzynkiewicz, Z., Traganos, F., Kapuscinski, J., Staiano-Coico, L., and Melamed, M. R. (1984). *Cytometry* **5,** 355–363.

Dean, P. N. (1986). *In* "Techniques in Cell Cycle Analysis" (J. W. Gray and Z. Darzynkiewicz, eds.), p. 211. Humana Press, Clifton, NJ.

Dean, P. N., and Jett, J. (1974). *J. Cell Biol.* **60,** 523–527.

Dien, B. S., and Srienc, F. (1991). *Biotechnol. Prog.* **7,** 291–298.

Dolbeare, F., Kuo, W. L., Wolfgang, B., Vanderlaan, M., and Gray, J. W. (1991). *In* "Methods in Cell Biology" (Z. Darzynkiewicz and H. A. Crissman, eds.), Vol. 33, pp. 207–216. Academic Press, San Diego.

Doran, P. M., and Bailey, J. E. (1986). *Biotechnol. Bioeng.* **28,** 73–87.

Dujon, B. (1981). *In* "The Molecular Biology of the Yeast *Saccharomyces*" (J. N. Strathern, E. W. Jones, and J. R. Broach, eds.), pp. 505–635. Cold Spring Harbor Lab., Cold Spring Harbor, NY.

Dvorak, J. A., Whelan, W. L., and McDaniel, J. P., Gibson, C. C., and Kwon-Chung, K. J. (1987). *Infect. Immun.* **55,** 1490–1497.

Fried, J. (1976). *Comp. Biomed. Res.* **9,** 263–276.

Ghiara, J. B., Richardson, H. E., Sugimoto, K., Henze, M., Lew, D. J., Wittenberg, C., and Reed, S. I. (1991). *Cell (Cambridge, Mass.)* **65,** 163–174.

Harlow, E., and Lane, D. (1988). "Antibodies: A Laboratory Manual," p. 633. Cold Spring Harbor Lab., Cold Spring Harbor, NY.

Hutter, K. J., and Eipel, H. E. (1979). *J. Gen. Microbiol.* **113,** 369–375.

Kuchenbecker, C., and Braun, G. (1985). *J. Basic Microbiol.* **25,** 509–512.

Kuo, S. C., and Yamamoto, S. (1975). *In* "Methods in Cell Biology" (D. M. Prescott, ed.), Vol. 11, pp. 169–181. Academic Press, New York.

Munch, T., Sonnleitner, B., and Fiechter, A. (1992). *J. Biotechnol.* **22,** 239–352.

Olaiya, A. F., and Sogin, S. J. (1979). *J. Bacteriol.* **140,** 1043–1049.

Palmer, R., Koval, M., and Koshland, D. (1989). *J. Cell Biol.* **109,** 3355–3366.

Pearse, A. G. E. (1968). "Histochemistry: Theoretical and Applied," Vol. 1. Little, Brown, Boston.

Peterson, M. S. (1991). Ph.D. Thesis, Purdue University, West Lafayette, IN.

Pringle, J. R., Adams, A. E. M., Drubin, D. G., and Haarer, B. K. (1991). *In* "Methods in Enzymology" (C. Guthrie and G. R. Fink, eds.), Vol. **194,** pp. 565–601. Academic Press, San Diego.

Richardson, H. E., Wittenberg, C., Cross, F., and Reed, S. I. (1989). *Cell (Cambridge, Mass.)* **59,** 1127–1133.

Sazer, S., and Sherwood, S. (1990). *J. Cell Sci.* **97,** 509–516.

Scherrer, R., Louden, L., and Gerhardt, P. (1974). *J. Bacteriol.* **118,** 534–540.

Sclafani, R. A., and Fangman, W. L. (1986). *J. Cell Biol.* **114,** 753–767.

Slater, M. L., Sharrow, S. O., and Gart, J. J. (1977). *Proc. Natl. Acad. Sci. U.S.A.* **74,** 3850–3854.

Srienc, F., and Dien, B. S. (1992). *Ann. N.Y. Acad. Sci.* **665,** 59–71.

Vanderlaan, M., Watkins, B., Thomas, C., Dolbeare, F., and Stanker, L. (1986). *Cytometry* **7,** 466–507.

Van Doorn, J., Valkenburg, J. A. C., Scholte, M. E., Oehlen, L. J. W. M., Van Driel, R., Postma, P. W., Nanninga, N., and Van Dam, K. (1988). *J. Bacteriol.* **170,** 4808–4815.

Wehr, C. T., and Parks, L. W. (1969). *J. Bacteriol.* **126,** 1339–1342.

Wheals, A. (1987). *In* "The Yeasts" (A. H. Rose and J. S. Harrison, eds.), Vol. 1, pp. 284–387. Academic Press, San Diego.

CHAPTER 28

Staining and Measurement of DNA in Bacteria

Harald B. Steen, Mette W. Jernaes, Kirsten Skarstad, and Erik Boye

Department of Biophysics
Institute for Cancer Research
Montebello
0310 Oslo
Norway

I. Introduction

Whereas flow cytometry has had a major impact on cell-cycle studies of mammalian cells, similar measurements of bacteria are just in their infancy (Steen, 1990a). This is despite the fact that the cell cycle of bacteria presents a number of essential problems which are difficult to answer without a method which facilitates precise determination of the DNA content of large numbers of individual cells. The cell cycle of bacteria is basically different from that of mammalian cells in that more than one replication cycle can be in progress in the same chromosome at the same time. This makes DNA histograms of bacteria look very different from what we are used to seeing for mammalian cells and complicates cell-cycle analysis significantly. Furthermore, the bacterial cell cycle varies much more with growth conditions than that of mammalian cells.

From an experimental, flow cytometric point of view bacteria differ from mammalian cells in several important respects:

1. The DNA content of bacteria is generally much lower than that of mammalian cells. For example, the DNA content of the *Escherichia coli* chromosome is some 1400× less than that of diploid human cells. This means that measurement of the bacterial DNA content with sufficient precision for applications like determination of cell-cycle distribution, i.e., with coefficients of variation (CV) of the order of a few percent, requires a combination of highly fluorescent staining and a sensitive instrument.

2. On the other hand, bacteria in some situations have a relatively much higher RNA content than typical mammalian cells, notably when they grow under optimal conditions. This means that dyes with some affinity for RNA, like ethidium bromide and propidium iodide, are not suitable, except if RNA has been removed, e.g., by treating the cells with RNase. However, we have not been able to obtain consistent results subsequent to RNase treatment of bacteria. Hence, dyes with higher DNA specificity are desirable, that is, dyes such as DAPI (4′6-diamidino-2-phenylindole), Hoechst 33342 (bis-benzimide) and Hoechst 33258 (bis-benzimide), chromomycin, mithramycin, and 7-AMD (7-aminoactinomycin D). DAPI and the Hoechst dyes require UV excitation, which means that the flow cytometer light source must be either a high-power ion laser or a mercury high-pressure arc lamp. (These lamps have a strong emission line at 366 nm which coincides with the peak absorption of these dyes.) In laser instruments mithramycin and its analogue chromomycin can be excited only marginally by means of the 458-nm line of a tunable high-power argon laser. Furthermore, these dyes are not very bright due to a relatively low-fluorescence quantum yield. The fluorescence yield of 7-AMD is too low to facilitate measurements on bacteria.

3. Bacteria differ from eukaryotic cells in that the chromosome does not contain histones and other proteins which inhibit the binding of many DNA-specific dyes and thereby destroys the stoichiometry of the staining. Presumably the bacteria DNA is more loosely packed so that "chromatin structure" may not be expected to affect the staining of such cells.

4. Many bacterial species, such as *E. coli,* are rods rather than spheres and may therefore create orientation artifacts in some flow cytometers, especially laser-based instruments with near-parallel excitation light.

5. In contrast to mammalian cells which, with few exceptions, are spherical in suspension and with a nucleus which is roughly concentric with the outer cell wall and with a size roughly constant relative to that of the cell, bacteria may vary greatly with respect to the intracellular distribution of their DNA, depending on growth conditions and other factors. Thus, while under certain conditions the DNA appears to be evenly distributed in all of the cytoplasm, it may be concentrated into a minor portion of the cell volume in other cases. Again, this may cause artifacts in some instruments.

6. The volume of bacteria is typically three orders of magnitude smaller than that of mammalian cells. In some instruments this creates problems with the light scattering measurement; the large angle (90°) detection especially does not have sufficient sensitivity.

7. The cell wall of bacteria is quite different from that of mammalian cells. In addition to the cytoplasmic membrane which is similar to that of mammalian cells, bacteria exhibit a complex cell wall consisting primarily of peptidoglycans, lipoproteins, and lipopolysaccharides. The permeability of this envelope is significantly different from that of the plasma membrane. Hence, the knowledge one may have on staining of mammalian cells, and especially vital cells, is not necessarily applicable to bacteria.

8. Vital staining of bacteria is further complicated by the fact that some bacteria have the ability to excrete some dyes very efficiently. Thus, as demonstrated below, even some DNA binding dyes which premeate the cell wall are pumped out so efficiently that hardly any staining occurs.

It may appear that the main reason that flow cytometry has been so little applied to bacterial studies can be found in combination of some of the above problems. In particular the poor overlap between the emission lines of the argon laser, which is the standard excitation light source in most flow cytometers, and the absorption spectra of the most DNA-specific dyes may have caused weak fluorescence yields and discouraged some investigators from taking on a challenge which already has proved most rewarding.

Using an arc lamp-based flow cytometer we have been able to carry out DNA measurements of bacteria with sufficient sensitivity and precision to allow assessment of essential cell-cycle parameters (Boye *et al.*, 1983; Skarstad *et al.*, 1983,1985; Steen, 1990a) and to study the mechanisms of the replication of the chromosome in more detail than was previously possible (Skarstad *et al.*, 1986; Skarstad and Boye, 1988; Løbner-Olesen *et al.*, 1989; Boye and Løbner-Olesen, 1990).

Although bacteria are easy to grow and process for measurement, we have found that the results are generally much more sensitive to growth conditions and other experimental parameters than are corresponding data for mammalian cells. Strict reproducibility of culture conditions is essential to obtain consistent data. The staining also appears more critical than with mammalian cells. In our hands the dye giving the best results is mithramycin. Although DAPI and the Hoechst dyes are known to be highly DNA specific, we have not been able to achieve as good results with these dyes, in terms of sharp peaks and reproducibility, as we do with mithramycin, although the fluorescence signals are of the same magnitude. The fluorescence quantum yield of mithramycin is quite low, i.e., about 0.05 (Langlois and Jensen, 1979), thus putting great demands on instrument sensitivity. We therefore use it in combination with ethidium bromide in order to obtain a higher fluorescence yield. Ethidium bromide, which also binds to RNA and consequently does not produce clean DNA histograms when excited directly, has negligible absorption at the excitation wavelength being used for mithramycin (mainly the strong 436-nm line of the Hg arc lamp) and is therefore excited almost exclusively by resonance energy transfer from adjacent (closer than 5 nm) mithramycin molecules (Langlois and Jensen, 1979). Hence, fluorescence from RNA-bound ethidium bromide appears to be negligible with this staining and excitation wavelength. Because of the higher fluorescence quantum yield of ethidium bromide, excitations transferred to this dye produce more fluorescence than that from mithramycin alone. Thus, the DNA specificity of mithramycin is combined with the higher fluorescence yield of ethidium bromide. The net result is an increase in fluorescence intensity by a factor of about 2. It is interesting that this increase is significantly smaller than the value reported for mammalian cells, i.e., a factor of 3.4 (Zante *et al.*, 1976; Langlois and Jensen, 1979).

Flow cytometry is also an obvious approach to detection and counting of bacteria in various areas, such as clinical bacteriology, food analysis, and environment monitoring including analysis of air, water, and sewage. The most direct and general way to distinguish biological cells from other types of microscopical particles that are often present in such samples is to stain DNA with a fluorescent dye. To simplify sample preparation and avoid the aggregation and denaturation it is preferable in many such applications to eliminate fixation and stain the cells vitally. We describe here a method for vital staining of bacteria which overcomes the above-mentioned problems with the impermeability of the cell wall and the active excretion of dye.

II. Materials

To stain fixed cells, the following solutions are required:

A. Fixing solution: 77% ethanol in H_2O. (If 96% ethanol is used, mix 80.2 parts of 96% ethanol with 19.8 parts of H_2O.)
B. Staining buffer: 10 m*M* Tris, pH 7.4, 10 m*M* $MgCl_2$. (Prepare by adding 1.21 g Trizma base (Sigma Chemical, St. Louis, MO) and 2.03 g $MgCl_2{\cdot}6H_2O$ to 1 liter of distilled H_2O and adjust pH with HCl.)
C. Ethidium bromide stock solution: 1 mg/ml ethidium bromide in staining buffer.
D. Staining solution: 180 μg/ml mithramycin and 40 μg/ml ethidium bromide in staining buffer. (Prepare by dissolving the contents of one 2.5-mg ampoule of Mithracin (Pfizer, New York) into 13.1 ml of the staining buffer and add 1.12 ml of the ethidium bromide stock solution.)

The ethidium bromide stock solution is stable for many months when stored in the dark at 4°C. The staining solution should be stored in the dark at 4°C and be used within a week of being prepared.

Note: Distilled water may contain significant amounts of microscopic particles (presumably mainly silicates) with sizes which produce light scattering signals of the same order of magnitude as do many bacteria. The water may therefore produce a sizable background in the light scattering channel. In order to avoid this artifact, both sheath water and reagent water should be filtered, preferably with a somewhat finer filter than the standard 0.22-μm pore size, which transmits significant numbers of particles larger than 0.3 μm.

III. Fixation and Staining

The bacterial suspension (1–1.5 ml) is washed once in TE (10 m*M* Tris, pH 8.0, 1 m*M* EDTA) by centrifugation in a microtube and resuspended in 100 μl of TE before fixation by the addition of 1 ml 77% ethanol. Vortex mix thoroughly. Fixation is completed within a few minutes. The cells may be stored at 4°C in the fixing solution for a few months and at −20°C for much longer.

Before being stained, the fixed cells are washed once in staining buffer. The cell pellet is resuspended in ice-cold staining buffer to a density which should not exceed 2×10^9 cells/ml. The cell suspension is mixed with an equal volume of staining solution. This procedure ensures a reproducible dye concentration in the samples. The final sample volume is normally 1 ml or less. The stained cells are kept in subdued light on ice until a few minutes before measurement. The cells may be measured after a few minutes of staining. However, to obtain perfectly reproducible results they should be left in the staining solution for at least 1 hr before measurement.

Stained samples may be stored on ice for at least 8 hr before any deterioration can be observed. In our experience the storage times which can be allowed appear vary considerably from one experiment to the next. In some cases samples have been stored for several days without detrimental effects. We suspect that the main reason for the degradation of the samples is the presence of traces of DNase. Hence, the use of autoclaved buffers and clean, sterilized labware is important.

IV. Vital Staining

Vital *E. coli* in exponential phase do not stain under physiological conditions with commonly used DNA binding dyes like ethidium bromide and acridine orange. The same has been found for other gram-negative bacteria that we have studied. The reason for this appears to be a combination of low permeability and the ability of these cells to actively excrete dye molecules that enter the cell (Jernæs and Steen, 1994). When vital *E. coli,* harvested in exponential phase, were incubated with dye at 0°C, they stained readily with ethidium bromide, the degree of staining depending on the buffer in which the staining was performed (Table I). Thus, the cells appear to be permeable to this dye, at least at low temperature. However, when the temperature was raised to about 20°C, the staining dropped almost to zero within a few seconds. The reduction in fluorescence intensity observed upon raising the temperature was a factor of a few hundred. The interpretation is that when brought to room temperature the metabolism of the cells increases so as to activate the pump that brings the dye out of the cells. This interpretation was confirmed by the observation that when the metabolism of the cells was paralyzed by metabolic inhibitors, Na-azide (4 g/liter) and 2-deoxy-D-glucose (5 m*M*), the staining, in the presence of EDTA (10 m*M*), was about the same as that at 0°C and similar also to that obtained for cells permeabilized by 70% ethanol by the method described above (Table I, Fig. 1). On the other hand, in PBS and Tris (100 m*M*) the metabolic inhibitors increased the staining by more than a factor 10, but still only to a level more than one order of magnitude below complete staining. Hence, it appears that even in the presence of the metabolic inhibitors the cells were able to excrete dye, although at a greatly reduced rate. The effect of EDTA seems to be to increase permeability to a level where a fully effective efflux pump is required to keep the dye out, whereas when the pump is inhibited it is not able to keep up with dye influx in this buffer. In the other buffers the permeability is so low that even the inhibited pump is able to keep intracellular dye concentration at a low level, although noticeably above zero.

The practical consequence of these results is that vital staining of *E. coli* can be carried out by the following procedure:

Add to a cell suspension an equal volume of 40 μg/ml ethidium bromide in 20 m*M* EDTA, 8 g/liter Na-azide, and 10 m*M* 2-deoxy-D-glucose, and measure in the flow cytometer at room temperature within 2 hr. Wash the cells once before staining if required to remove excess amounts of debris.

Table I
Median Fluorescence Intensity of Vital *E. coli* Stained with 20 μg/ml Ethidium Bromide in the Buffers noted

	PBS	Tris	EDTA
Room temperature	2 ± 1	5 ± 1	16 ± 3
Ice	597 ± 38	1004 ± 16	988 ± 0
Room temperature w/metabolic inhibition	66 ± 7	83 ± 4	959 ± 61

Note. The fluorescence median for a similar sample fixed in 70% ethanol was 898 ± 0. Error limits represent range of two measurements.

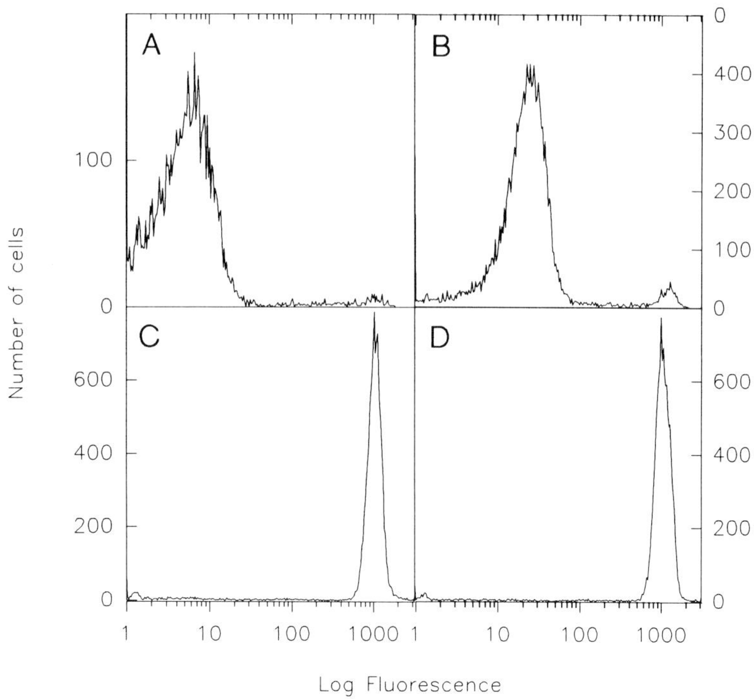

Fig. 1 Fluorescence histograms of vital *E. coli* harvested during exponential growth and stained with 20 μg/ml ethidium bromide in the following solutions: (A and C) phosphate-buffered saline (PBS), pH 7.3; (B) PBS with 10 m*M* EDTA; (D) PBS with 10 m*M* EDTA, 4 g/liter Na-azide, and 5 m*M* 2-deoxy-D-glucose. All samples were made isomolar with NaCl. For (C) the cells were measured within a few seconds after the sample was removed from ice, whereas the other samples were prepared and measured at room temperature.

V. Flow Cytometric Measurement

Due to the small size and low DNA content of bacteria the sensitivity of the flow cytometer with regard to light scattering and fluorescence is critical. We have obtained data with adequate quality using an arc lamp-based instrument (Steen, 1990b). [Our laser-based instrument (Coulter Epics V) has been found insufficient with regard to sensitivity for both fluorescence and light scattering.] The optical configuration of the arc lamp-based instrument is essentially similar to that of the epifluorescence microscope (Steen and Lindmo, 1979). In addition it comprises a dark field configuration which facilitates measurement of light scattering at small and large scattering angles in separate detectors (Steen and Lindmo, 1985). The instruments employs a 100-W high-pressure mercury arc lamp as the excitation light source. For measurement of fixed cells stained with mithramycin and ethidium bromide we use a "B1" filter block which contains a dichroic excitation filter with a transmission band between 390 and 440 nm, a dichroic beam splitter having a characteristic wavelength of 460 nm, a long-pass emission filter with transmission from 470 nm, and an additional short-pass emission filter with transmission below 720 nm. The excitation filter of the B1 block transmits primarily the 405- and 436-nm mercury emission lines. The 436-nm line especially is quite intense and coincides with the absorption peak of DNA-bound mithramycin. The 720-nm short-pass filter is used to eliminate red and infrared background light which would otherwise increase the shot noise on the signal and thereby reduce the effective sensitivity. For the measurement of vital cells stained with ethidium bromide we use a G_1 filter block which has an excitation filter, transmitting the strong Hg line at 546 nm, a long-pass emission filter with transmission above 570 nm, plus the above-mentioned short-pass 720-nm filter. A commercial version of this instrument is currently available from Bio-Rad S.P.D. (Milan, Italy).

Instrument sensitivity is the major concern in measurement of bacteria. It is a function of the background of shot noise on the signal as much as of the signal itself. Thus, the signal to noise ratio is the essential parameter in this regard (Steen, 1992). In order to reduce noise the flow cytometer should be used with the smallest permissible excitation and emission slits. (The latter is equivalent to the "pinhole" in other types of flow cytometers.) To enhance signal the flow velocity may be reduced by lowering the sheath pressure. (In the Argus instrument a pressure around 0.5 kg/cm^2 is suitable.) This can be done without affecting measuring precision since the cell density of samples from bacteria cultures is typically high, i.e., 10^8–10^9 cells/ml, and the sample flow rate can therefore be kept correspondingly low, i.e., 0.1–1 μl/min. An instrument with volumetric sample injection, that is, injection of the sample by means of a syringe pump, is preferable to achieve a stable sample flow at this level.

Simultaneous light scattering detection to gate the fluorescence measurement is always carried out in order to distinguish degraded cells and debris from intact cells. Thus, fluorescence and light scattering are recorded in dual-parameter mode. With samples having a high content of debris light scattering can be a prerequisite for adequate data quality. Dual-parameter low angle and large angle light scattering measurement can be used to distinguish dead cells from vital ones (Steen, 1986).

VI. Standards and Controls

Monodisperse fluorescent particles are always run when the instrument is set up, in order to check the nozzle, ie., to see that the flow is perfectly laminar, and to optimize the instrument with regard to sensitivity and resolution, i.e., narrow histogram peaks. We use 1.5-μm fluorescent particles (Bio-Rad S.P.D., Milan, Italy). The CV of the fluorescence peak should not exceed 2%, provided the particles are sufficiently monodisperse with regard to fluorescence.

DNA histograms of bacteria in many cases have no sharp peaks which can be used to check the preparation of the sample and the instrument function. One or more standard biological samples are therefore most helpful. For this purpose we use *E. coli* cells treated with the antibiotic rifampicin (see method below). This treatment produces cells giving DNA histograms with several prominent and narrow peaks, each of which represents an integral number of chromosomes (Fig. 2), which may thus be used both to calibrate the fluorescence scale versus DNA content and to check the resolution of the measurement. Such histograms usually contain two peaks, representing cells with either two and four chromosomes or four and eight chromosomes, depending on the growth rate of the bacterial culture. A most convenient control sample may be prepared

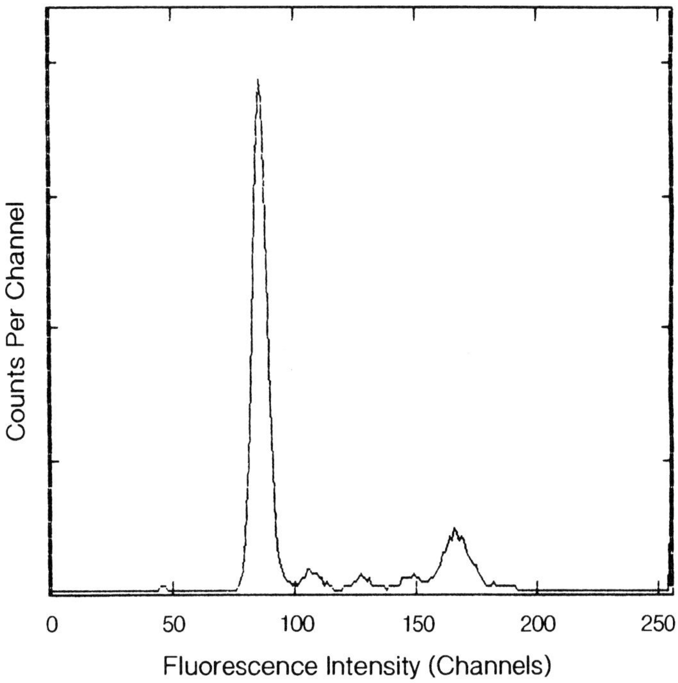

Fig. 2 Fluorescence histogram of *E. coli* K-12 cells, strain CM735, harvested from a culture grown for 3 hr in the presence of the antibiotic rifampicin which was given while the cells were in rapid exponential growth. The cells were fixed and stained with a combination of mithramycin and ethidium bromide according to the present procedure. The two prominent peaks represent cells with four and eight chromosomes, respectively. Peaks due to cells with two, five, six, and seven chromosomes are also evident. The histogram can thus be used to calibrate the fluorescence axis with regard to cellular DNA content. The CV of the main histogram peak is 3.5%.

from *E. coli* bacteria lacking a functional Dam methyltransferase. Such mutant cells lack the ability to coordinate multiple initiations and therefore contain all integral numbers of chromosomes (1,2,3,4, etc. and not just 2 and 4) (Boye *et al.*, 1988; Boye and Løbner-Olesen, 1990).

When examining bacteria for the contents of replication origins by treatment with rifampicin, it is important that cell division is stopped as rapidly as possible, so that the number of origins per cell at the time of rifampicin addition can be measured. If cell division occurs after replication initiation has been stopped, the number of origins per cell will be too low. For *E. coli* this can be achieved by using the cell division inhibitors furazlocillin or cephalexin. At low concentrations, these drugs inhibit septum formation within 1–2 min (Boye and Løbner-

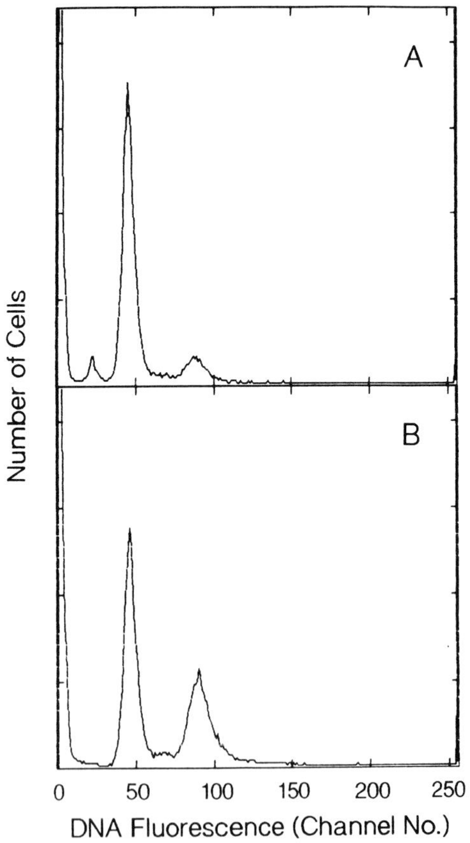

Fig. 3 Fluorescence/DNA histograms of drug-treated bacteria, *Escherichia coli* K-12, strain AB1157, was grown in LB medium with 0.2% glucose at 37°C and treated with either rifampicin (A) or rifampicin plus cephalexin (B) for 3 hr before fixation, staining with mithramycin and ethidium bromide, and flow cytometry. The three peaks in A represent cells with 2, 4, and 8 fully replicated chromosomes.

Olesen, 1991). The effect of cephalexin can be seen easily when comparing cells that have been treated with rifampicin only to those with both drugs in combination (Fig. 3). The number of origins, or the DNA contents after replication run out in the presence of drugs, is clearly higher when cephalexin is also present. This is due to the rapid cessation of cell division after cephalexin addition.

The procedure for preparing this standard sample is as follows: An *E. coli* wild-type strain or a *dam* mutant is grown in LB medium with 0.2% glucose at 37°C to an optical density (600 nm) of 0.1. At this point, when the cells are in exponential growth, rifampicin (15 mg/ml in ethanol) is added to the culture to a final concentration of 150–300 μg/ml. If required, cell division is stopped by the addition of furazlocillin (4 μg/ml) or cephalexin (10 μg/ml). After a further 3 hr of culture, cells are harvested and fixed according to the above procedure. Provided the instrument has sufficient sensitivity and is properly tuned, the CV of the fluorescence peaks should be of the order of 5% or lower.

References

Boye, E., and Løbner-Olesen, A. (1990). *Cell (Cambridge, Mass.)* **62,** 981.

Boye, E., and Løbner-Olesen, A. (1991). *Res. Microbiol.* **142,** 131.

Boye, E., Steen, H. B., and Skarstad, K. (1983). *J. Gen. Microbiol.* **129,** 973.

Boye, E., Løbner-Olesen, A., and Skarstad, K. (1988). *Biochim. Biophys. Acta* **951,** 359.

Jernaes, M. W., and Steen, H. B. (in press). *Cytometry.*

Langlois, R. G., and Jensen, R. H. (1979). *J. Histochem. Cytochem.* **27,** 72.

Løbner, Olesen, A., Skarstad, K., Hansen, F. G., von Mayenburg, K., and Boye, E. (1989). *Cell (Cambridge, Mass.)* **57,** 881.

Skarstad, K., and Boye, E. (1988). *J. Bacteriol.* **170,** 2549.

Skarstad, K., Steen, H. B., and Boye, E. (1983). *J. Bacteriol.* **154,** 656.

Skarstad, K., Steen, H. B., and Boye, E. (1985). *J. Bacteriol.* **163,** 661.

Skarstad, K., Boye, E., and Steen, H. B. (1986). *EMBO J.* **5,** 1711.

Steen, H. B. (1986). *Cytometry* **7,** 445.

Steen, H. B. (1990a). *In* "Flow Cytometry and Sorting" (M. R. Melamed, T. Lindmo, and M. L. Mendelsohn, eds.), pp. 605–622. Wiley-Liss, New York.

Steen, H. B. (1990b). *In* "Flow Cytometry and Sorting" (M. R. Melamed, T. Lindmo, and M. L. Mendelsohn, eds.), pp. 11–25. Wiley-Liss, New York.

Steen, H. B. (1992). *Cytometry* **13,** 822.

Steen, H. B., and Lindmo, T. (1979). *Science* **204,** 403.

Steen, H. B., and Lindmo, T. (1985). *Cytometry* **6,** 281.

Zante, J., Schumann, J., Barlogie, B., Goehde, W., and Buchner, T. (1976). *In* "Pulsecytophotometry" (W. Goehde, J. Schumann, and T. Buchner, eds.), pp. 97–106. European Press, Gent, Belgium.

CHAPTER 29

Detection of Specific Microorganisms in Environmental Samples Using Flow Cytometry

Graham Vesey,* Joe Narai,† Nicholas Ashbolt,‡ Keith Williams,* and Duncan Veal*

* School of Biological Sciences
Macquarie University
Sydney, New South Wales 2109
Australia

† Commonwealth Centre for Laser Applications
Macquarie University
Sydney, New South Wales 2109
Australia

‡ Australian Water Technologies,
Science and Environment
Sydney, New South Wales 2114
Australia

METHODS IN CELL BIOLOGY, VOL. 42

I. Introduction

Flow cytometers are technologically advanced instruments which combine laser interrogation of a fluid stream with sophisticated data handling technology for obtaining information about, and potentially isolating, particles that pass through the laser beam. Traditionally they are among the most expensive of laboratory instruments and require highly skilled personnel to operate them. This combined with the fact that samples must be particulate and of reasonably uniform size has limited their application in biology to well-funded laboratories in biomedical research where they are used to analyze blood cell subpopulations (immunology, AIDS, cancer) or separate chromosomes.

The reduced cost and ease of operation of analytical flow cytometers (which collect data but do not sort particles) means that applications in other areas of biology are now envisaged. The applications of flow cytometry to clinical microbiology have been described by Shapiro (1990). In this chapter we discuss the use of flow cytometry within the environmental microbiology laboratory. In particular we focus on flow cytometric methods for the detection of low numbers of, and even single, specific microorganisms within environmental samples.

Flow cytometric analysis performed in environmental microbiology laboratories is often more stringent than that required for the analysis of mammalian cells and can push sensitivities close to limits of operation. This is because the volume, nucleic acid and protein content of bacteria are approximately 1000× less than that in mammalian cells. Since detection involves identification of light scatter, specific proteins, or DNA, the signals produced by bacteria are generally several orders of magnitude lower than those from eukaryotic cells. However, recently flow cytometers have been used to great effect for microbiological diagnosis and even more recently they have been applied in environmental microbiology.

Developments in both biological techniques and instrumentation, described in this chapter, will considerably increase the range of applications of flow cytometry within environmental microbiology laboratories. Furthermore, these developments result in greatly simplified protocols allowing not only research laboratories but also routine environmental testing laboratories to perform these analyses. We envisage that within the foreseeable future small, robust, relatively cheap and simple to operate flow cytometers will be available for the detection of a vast range of microorganisms in environmental samples.

A. The Instrument

The principles of flow cytometry have been described by Shapiro (1988) and here we provide only the basic features and terminology.

Flow cytometry is used to quantitatively measure physical or chemical characteristics of cells as they are presented, in single file, into a focused light beam. The flow rate of fluid through the focused beam is usually around 10 $msec^{-1}$ which allows approximately 2,000 to 10,000 cells sec^{-1} to be analyzed. The light source can be either a high-pressure mercury lamp or one of an assortment of different lasers. Various parameters of an individual cell, such as fluorescence (FL), forward angle light scatter (FALS), and side (90°) angle light scatter (SALS), can be measured simultaneously. Forward angle light scatter provides information on the size of a cell while side angle light scatter correlates with cell refractibility and, thus, it is thought, provides information on the surface properties and internal structure. Fluorescence can be used to detect autofluorescence in the cell emanating from cellular components such as flavin nucleotides, pyridine, chlorophyll, and other photosynthetic pigments. In natural environments FALS, SALS, and autofluorescence can provide sufficient information to enable clustering of cells into particular types (Yentsch, 1990). However, alone, these parameters normally do not provide sufficient information to identify specific microorganisms in environmental samples. To identify specific microorganisms, it is usually necessary to label cells of interest with fluorescent molecules such as fluorescein isothiocyanate (FITC). Cells can be tagged using specific antibodies, lectins, or nucleic acid probes conjugated to particular fluorochromes. In addition to analyzer flow cytometers described above, some machines have the capability of physically sorting cells using information from the various detectors as discriminators. Sorting enables purification of a particular cell type from a mixture. To sort cells, the fluid stream is broken into droplets by the vibration of a piezoelectric crystal. If a particle is to be collected, the droplet containing the particle is electrically charged and then deflected, using charged plates, from the main stream into a collection vessel.

A major hurdle limiting to the use of flow cytometry in environmental microbiology has been the initial cost of the instrument. A basic analyzer cytometer (such as the Coulter XL) costs around $100,000 (US). The XL is an excellent instrument and will certainly find applications in the area of routine monitoring in environmental microbiology. However, to sort cells doubles the cost of the basic instrument. Instruments with the ability to sort include the Coulter Elite and the Becton–Dickinson FACStar plus or FACS Vantage. Further costs are involved with cell sorting since the instrument takes considerably longer to set up and requires a highly skilled operator. Nevertheless, the ability to sort cells is an important feature for environmental microbiology since it enables the collection of the organisms of interest and confirmation of results. We regard the ability to sort as essential to most research and development applications in environmental microbiology. However, one aim of this research and development should be to develop protocols which can be used reliably in routine situations on analysis only instruments.

B. Environmental Microbiology

Despite the high cost of the instrumentation, flow cytometers are now found in a few environmental microbiology laboratories around the world. This preparedness to invest in sophisticated instrumentation is due, in part, to an increasing appreciation of the potential applications of flow cytometry in environmental microbiology (Edwards *et al.*, 1992; Button and Robertson, 1993) and awareness of the limitations of traditional techniques (Mills and Bell, 1986). The availability of funds to purchase flow cytometers for environmental microbiology also reflects growing awareness and concern for environmental issues. Certainly environmental microbiology is crucial to the major global environmental issues and flow cytometry is now beginning to play its role in the development of this discipline. For example, global warming has been described as a most significant environmental issue. In the oceans two groups of microorganisms are essential for carbon cycling and, therefore, have important impacts on future trends in global warming. These are the phytoplankton, which are responsible for about half of the total primary production on the earth, and the heterotrophic bacteria. Of the phytoplankton, picophytoplankton (including cyanobacteria, prochlorophytes, diatoms, and cryotophytes), due to their autofluorescence, are well suited to flow cytometric analysis, as are their predators (Balfoort *et al.*, 1992; Olson *et al.*, 1993). On the other hand, flow cytometric quantification of heterotrophic aquatic bacteria is in its infancy (Robertson and Button, 1989; Button and Robertson, 1993), although methods based on fluorescent ribosomal probes for aquatic bacteria (Lim *et al.*, 1993) could be adapted for flow cytometry (Amann *et al.*, 1990b). The latter approach is already leading to a better understanding of the identification and biodiversity of the predominant bacteria in environments (Manz *et al.*, 1993; Wagner *et al.*, 1993).

Pollution is another major environmental issue. Flow cytometry has been used to study the hydrocarbon degrading bacteria in Resurrection Bay after the Exon Valdez grounding (Button *et al.*, 1992). In these studies, microbial biomass was computed from the population and mean cell volume (obtained from forward scatter histograms). Significantly, this study concluded that turnover times for toluene were decades longer than expected and that dissolved spill components such as toluene should enter the world ocean pool of hydrocarbons rather than biooxidize in place. There is surprisingly little information on flow cytometry for human or animal health microbiology. The analysis of foods, body fluids, or culture media should, however, pose fewer problems than environmental samples and in fact the challenge to apply flow cytometry to the bioreactor has been taken up by the fermentation industry (Betz *et al.*, 1984; Scheper *et al.*, 1987; Kell *et al.*, 1991). Also, pathogens in food or body fluid will be present at relatively high concentrations (Humphreys *et al.*, 1993; Pinder and McClelland, 1993). Other environmental areas awaiting flow cytometry study include monitoring the fate of probiotics or nuisance species in aquaculture.

C. Environments

1. Water

Water is routinely tested for microorganisms on a large scale by water utilities and government agencies to ensure that the water is safe for consumption, recreation, and shell fish production. The detection of microorganisms in water depends largely on the use of selective and differential media to isolate, culture, and identify specific types of microorganisms. These culture-based methods are limited because they rely on the growth of the microorganisms of interest on laboratory medium. For most microorganisms this culturing takes between a few hours and a few days, thus results are never immediately available. Furthermore, in our experience, only between 0.001 and 10% of bacteria from natural environments are culturable, including introduced pathogens.

Due to the resources required to screen for a wide range of water-borne pathogens (including viruses, bacteria, protozoa, nematodes, and flukes), the fact that many pathogens are currently nonculturable or are very slow growing, and the potential risks associated with culturing pathogens to laboratory workers, microbial water quality is routinely assessed by enumerating the presence of certain indicator bacteria, generally fecal coliforms and enterococci (Ashbolt and Veal, 1994). These bacteria are used to indicate recent fecal contamination and depend on the indicator organism not growing in the environment, being present at greater numbers than any pathogen, and dying in the environment at a rate no faster than that of any pathogen. Unfortunately, it is now well established that dependence on these indicator bacteria can lead to false conclusions as all of these assumptions can fail some of the time; for example, human viruses and pathogenic protozoa of human origin have been reported in waters when indicator bacteria were not detected (Ashbolt and Veal, 1994; Smith and Rose, 1990; Badenoch, 1990). Thus there is renewed interest in detection of the specific pathogens themselves in water.

Work in our laboratories has concentrated on the detection of the enteric protozoan parasites *Giardia* and *Cryptosporidium* in water. These protozoa are among the most frequent causative agents of diarrheal disease in humans (Adam, 1991; Current 1986). No effective chemotherapy is available for *Cryptosporidium* and this organism can be life threatening in immunosuppressed individuals (Current 1986). Although the most common route of infection of *Cryptosporidium* and *Giardia* is direct person to person, data from the United Kingdom and the United States indicate that the zoonotic water-borne route is important (Smith and Rose, 1990; Badenoch, 1990).

Cryptosporidium and *Giardia* form robust oocysts and cysts, respectively, which are shed in the feces. These oocysts and cysts are very infective with as few as 10 cysts or oocysts leading to the development of disease (Rendtorff, 1954; Miller *et al.*, 1986).

One of the features of the oocysts and cysts of *Cryptosporidium* and *Giardia* is that they are robust and survive for extended periods in the environment. Bacterial indicators, such as fecal coliforms are an unsatisfactory standard by which to evaluate the health risk of the protozoan parasites, since bacterial indicators survive for days or at most weeks in aquatic environments, while cysts and oocysts may survive for several months. Establishment of the health risks of these organisms in drinking water using indicator organisms is further complicated because indicator organisms are inactivated by levels of chlorine used routinely in drinking water, whereas cysts and oocysts are not (Korich *et al.*, 1990; Adam, 1991). Several water-borne outbreaks of *Cryptosporidium* and *Giardia* have been reported in situations where the water meets all statutory requirements of microbial quality (Badenoch, 1990; Smith and Rose, 1990). Some of these outbreaks have resulted in hundreds of thousands of individuals becoming infected (e.g., cryptosporidiosis in Milwaukee in 1993).

Flow cytometry has the potential for rapid detection of nonculturable microorganisms in water samples at high levels of sensitivity. Vesey and his co-workers (Vesey *et al.*, 1993a,c,1994) have described the use of a sorting flow cytometer to detect *Cryptosporidium* oocysts and *Giardia* cysts in concentrated water samples. This flow cytometric detection method involves flocculation with calcium carbonate to concentrate cysts and oocysts from large volumes (1–40 liters) of water (Vesey *et al.*, 1993b). The flocculant is dissolved and then stained with FITC-labeled monoclonal antibodies and analyzed by flow cytometry. Presumptive cysts and oocysts are then sorted on to microscope slides and confirmed by epifluorescence microscopy.

The method described by Vesey and co-workers is a significant advance on the American Society for Testing Materials (ASTM) method evaluated by LeChevallier *et al.* (1991). The method is, however, far from perfect. At present, to achieve the level of sensitivity required for detecting *Cryptosporidium* oocysts, the method is restricted to an instrument fitted with cell sorting facility. Nonsorting cytometers, however, are less expensive, much simpler to operate, and better suited to routine analysis within a water microbiology laboratory.

Features such as an automatic sample loader enable samples to be loaded and left to run unattended. Within our laboratory the development of a nonsorting detection method is underway. We believe that in the near future a flow cytometric method which can achieve the sensitivity required for the routine monitoring of *Cryptosporidium* oocysts in water will be developed. Already we have developed a reliable method for the detection of *Giardia* cysts in water that does not depend on cell sorting (Vesey *et al.*, in preparation).

2. Foods and Beverages

In the food and beverage industries, routine microbiological testing is performed on a large scale to ensure freedom from pathogens and acceptable levels of spoilage organisms. Most of this testing is still performed using traditional culture-based techniques. These techniques are laborious and relatively expensive in terms of culture media and reagents, and it is generally several days before results are available. Flow cytometry has considerable potential in the food and beverage industry since it can be used to detect nonculturable organisms, and in the analytical mode it is readily automated and can be used to rapidly enumerate microorganisms in a sample. Immediate microbial assessment has both stock control and production advantages as microbiological problems can be immediately identified and rectified, reducing the amount of lost product and negating the need to store products while awaiting results of microbiological analyses.

Despite the potential applications of flow cytometry in food microbiology there have been few reports detailing its use. Pettipher (1991) has evaluated the ChemFlow flow cytometer system (Chemunex S.A.) for the rapid detection of spoilage yeasts in soft drinks. Minimum sensitivity achieved using this system varied from 50 to 14,000 yeasts/ml depending on the product being tested. Pettipher (1991) concluded that the ChemFlow system looks promising for the rapid detection of relatively low numbers of yeast in soft drinks, but without some concentration or preincubation it was not sufficiently sensitive for use on site by a production laboratory. Using 24-hr preincubation with a ChemFlow Autosystem II sensitivities as low as 1 viable yeast per gram of product have been achieved (Bankes and Richard, 1993). Without preincubation these authors achieved sensitivities $<10^3$ cells/ml of product within 1 hr of sampling.

Infections with wild yeast are a severe problem for breweries, with spoilage of the beer occurring at infection levels as low as 1 wild yeast in 10^3–10^4 culture yeast. Here the problem is to detect the presence of relatively low numbers of wild yeast against a high background of culture yeast. Jespersen *et al.* (1993) have described a flow cytometric procedure based on preincubation on wild yeast-selective medium followed by staining with the Fluorassure Substrate B.W. (Chemunex S. A., Paris). By examining fluorescence intensity, differentiation of the wild yeast from culture yeast was possible. Using this method the authors claimed to be able to detect 1 wild yeast in the presence of 10^6 culture yeast after 48–72 hr of incubation.

Donnelly and Baigent (1986) have described the use of flow cytometry for the detection of *Listeria monocytogenes* in milk using FITC-labeled polyclonal antibodies. This work is of particular significance because food-borne listeriosis can be fatal and because culturing of *L. monocytogenes* is tedious, may take several days of enrichment, and can be unreliable as cultures are frequently overgrown by competing microorganisms. However, Haslett *et al.* (1991) have reported problems with the application of flow cytometry to the detection of this organism since commercially available antibodies cross-react with other *Listeria* species and various other gram-positive organisms.

3. Soils and Sediments

A rapidly growing area of biotechnology is in the development of microbial plant and soil inoculants. Microorganisms are used to break down pollutants in soil, control pests and disease of plants, and promote plant growth through N_2 fixation or by producing plant growth stimulating factors. In developing new products in this area it is important to be able to monitor the inoculated microoganisms, to determine their fate, and to assess the success or failure of inoculation (Ryder *et al.,* 1994). Many microbial inoculants which are being developed for environmental applications contain recombinant DNA. The ability to monitor the survival, growth, and migration of such genetically modified products is an essential prerequisite for their licensing as products (Ford and Olsen, 1988). Methods for monitoring genetically modified microorganisms need to be sensitive and able to detect both culturable and viable but nonculturable cells (Colwell *et al.,* 1985).

The ability to enumerate bacteria introduced into soil using flow cytometry has been investigated by Page and Burns (1991). When compared to plate counts and direct counts, flow cytometry was found to be more accurate at low numbers. These authors concluded that flow cytometry is a more rapid and objective method for bacterial enumeration in soil than direct counting.

An unresolved problem with soils and sediments is the difficulty of separating microorganisms from particulates (Hopkins *et al.,* 1991). Enumeration of particle-bound pathogens is important as they act as reservoirs readily resuspended into the water column by a number of forces including storms, dredging, bioturbation, current, and wave actions, impacting on recreational use and shellfish culture (Grimes *et al.,* 1986; Lewis *et al.,* 1986). Virtually all methods for enumerating bacteria in soils or sediments, including flow cytometry, are dependent on separating the microorganisms from the particulates. Generally the highest yields of desorbed bacteria are achieved by sonicating sediments, often in the presence of a surfactant (McDaniel and Capone, 1985). Once in suspension, microorganisms may be separated from detritus by differential centrifugation, such as sedimention field-flow fractionation which may recover 59–87% of a seeded bacterial population (Sharma *et al.*, 1993). However, in our experience of protozoan parasites differential centrifugation of cysts results in poor quantitation and sonication can lead to the removal of the identifying epitopes required for monoclonal antibody binding (Vesey and Slade, 1990). Hence, we are investigating the chemical encrusting of iron and humic salts that may bind environmental cysts.

D. Current Technologies for Rapid Detection of Specific Microorganisms in Environmental Samples

1. Antibody-Based Methods

Most rapid techniques used to detect specific microorganisms in environmental samples depend on either antibody- or nucleic acid-based methodologies. Most of the antibody-based techniques employ some form of colorimetric detection system such as immunofluorescence, ELISA, or chemiluminescence. These techniques normally involve immobilization of the sample onto a solid substance and then probing with an antibody labeled with the detection reagent. In the case of immunofluorescence the detection reagent is normally FITC and the sample is visualized using epifluorescence microscopy. The entire sample is scanned for fluorescing particles with the morphological characteristics of the target organism. The microscopic examination requires highly skilled operators and is tedious, time consuming, and, with dirty samples, may result in low detection efficiencies (Vesey *et al.*, 1994).

ELISA techniques often employ an antibody bound to a solid substrate to capture the target organism and then a horseradish peroxidase-labeled antibody, which also identifies the target organism, is attached to the captured sample. The horseradish peroxidase is then developed using hydrogen peroxide and tetramethylbenzidine and the sample is examined for color development either visually or with an ELISA reader.

The sensitivity of colorimetric detection systems can be increased by employing chemiluminescence. In this technique light emitting substrates are utilized in conjunction with enzyme-linked antibodies which allow the detection of the target organism to be performed by exposing the sample to photographic film.

The advantage of ELISA and chemiluminescence techniques is that they are simple to perform and enable the analysis of multiple samples simultaneously. However, these techniques often lack the specificity required to detect low numbers of specific microorganisms within environmental samples (Campbell *et al.*, 1993). This is because detection is based on a single paramater, namely the binding of an antibody to the organism of interest. In our experience, antibodies always bind to some interfering particles found in environmental samples, no matter how specific the antibody is and how good the blocking agents used. Flow cytometry can overcome this problem because additional detection parameters, such as light scatter and autofluorescence can be used to be certain that the antibody is binding an authentic target organism.

2. Gene Probes

Methods that employ specific sequences of nucleic acids to probe samples utilize either colorimetric enzyme-labeled probes or radiactively labeled probes. These techniques are potentially extremely sensitive and have been successfully used to detect pathogens in environmental samples (Abbaszadegan *et al.*, 1991). However, radioactively labeled probes are often required to achieve the required sensitivity. The facilities needed to handle radioactive materials are not generally available in environmental microbiology laboratories. As with antibody-based techniques (see above), the detection relies on a single parameter. The use of fluorescently labeled nucleic acid probes with flow cytometry allows additional parameters and, therefore, increased selectivity. The use of these probes is discussed later.

3. Polymerase Chain Reaction

The polymerase chain reaction (PCR) is a technique which can detect single microorganisms within environmental samples with high specificity (Mahbubani *et al.*, 1991; Tsai and Olson, 1992; Brooks *et al.*, 1992). These techniques have been successfully used for the detection of specific microorganisms within a range of samples including the detection of lactic acid bacteria in beer (Dimichele and Lewis, 1993), *Escherichia coli* in sewage and sludge (Tsai *et al.*, 1993) and raw milk (Keasler and Hill, 1992), *L. monocytogenes* in food (Wernars *et al.*, 1991; Furrer *et al.*, 1991; Niederhauser *et al.*, 1993), *Carnobacterium* spp. in meat (Brooks *et al.*, 1992), and *Campylobacter jejuni* in raw milk and dairy products (Wegmuller *et al.*, 1993). These techniques often enable inexpensive and rapid analysis of multiple samples. However, there are problems with the use of these techniques in many environmental samples. Substances present in some samples interfere with the polymerase enzyme and inhibit the reaction (Herman and De-Ridder, 1993; Tsai and Olson, 1992). Procedures for removing the inhibitors such as Sephadex columns have been developed for some types of samples (Abbaszadagen *et al.*, 1991; Tsai and Olson, 1992). Unfortunately, these procedures are often tedious and labor intensive and are not applicable to many microorganisms. The use of magnetic-activated cell sorting to remove inhibitors may prove to be successful.

II. Preparation of Water Samples for Flow Cytometric Analysis

A. Concentration

Most microbial pathogens occur in water at very low concentrations. Thus, the detection of specific pathogens in water by flow cytometry requires some concentration step prior to analysis. The degree of concentration depends on the infective dose of the organism and its frequency of occurrence. For the detection of many organisms, such as *Legionella,* which needs to occur in waters in high numbers to cause a serious health threat (Badenoch, 1987), sample concentration simply involves a 10- to 50-fold concentration using centrifugation. With organisms that have a very low infectious dose, such as *Cryptosporidium,* much higher levels of concentration are required, and typically 10–40 liters of water must be processed. Because it is impractical to routinely centrifuge tens of liters of water, alternative concentration methods have been developed. These include filtration using wound cartridge filters (Rose *et al.*, 1986), tangentinal flow filtration (Issac-Renton *et al.*, 1986), continuous flow centrifugation (Bee *et al.*, 1991), and vortex filtration. However, most of these techniques do not solve the basic problems associated with concentrating microorganisms from environmental samples, i.e., achieving good recovery without damaging the target microorganism and with the minimum of labor and expense.

We now routinely use the following calcium flocculation method for concentrating parasites from samples up to 20 liters in volume. The method is inexpensive and simple and results in high recoveries of *Cryptosporidium* oocysts and *Giardia* cysts (Vesey *et al.*, 1993b).

1. Calcium Flocculation (Vesey *et al.*, 1993b)

1. Collect a 10-liter sample in a flat-bottomed plastic container.
2. Add 100 ml of 1 *M* sodium bicarbonate and 100 ml of 1 *M* calcium chloride.
3. Adjust the pH to a value of 10.0 (± 0.1) by the addition of sodium hydroxide and allow the sample to stand for at least 4 hr.
4. Remove the supernatant by aspiration without disturbing the calcium deposit.
5. Dissolve the calcium carbonate residue by adding sulfamic acid.
6. Decant the sample and centrifuge.
7. Resuspend the pellet in 50 ml of distilled water and centrifuge. Wash a second time in phosphate-buffered saline (PBS), pH 7.4, and centrifuge.
8. Resuspend the pellet in a small volume of PBS and analyze (e.g., by flow cytometry).

Many organisms, including *Cryptosporidium* oocysts, are very sticky and will adhere to any containers used. To improve recoveries all containers should be rinsed with 0.01% Tween 80 and the washings combined with the sample.

Samples over 40 liters in volume are too large to be conveniently flocculated in any number. To overcome this logistical problem we use a membrane filtration method (Ongerth and Stibbs, 1987) to concentrate samples up to 5000 liters in volume on site.

2. Membrane Filtration

1. Place a 293- or 142-mm polycarbonate membrane of appropriate pore size (2 μm for *Cryptosporidium* oocysts, 5 μm for *Giardia* cysts, 0.22 μm for bacteria) in a stainless steel housing and filter the sample. Ensure that the pressure across the filter does not exceed 15 psi or the membrane may be damaged.
2. Once filtration is complete and no water is left in the housing, carefully remove the membrane and place it onto a perspex plate slightly larger in size than the membrane.
3. Hold the perspex plate upright above a narrow plastic box. Use a spray bottle a squirt 0.01% (v/v) Tween 80 onto the membrane. Allow the washings to drip into the plastic box.
4. While squirting Tween 80 at the membrane, rub the entire surface with a squeegee (we use a modified windshield wiper). Start at the top of the membrane and work downward.
5. Continue the process until the membrane appears clean. Use a minimum amount of Tween 80 (less than 100 ml).
6. Transfer the washings to a centrifuge pot and centrifuge at 3000*g* for 10 min.
7. Resuspend in a small volume of PBS and analyze (e.g., by flow cytometry).

The methods described here were developed for concentrating *Cryptosporidium* oocysts and *Giardia* cysts. However, with slight modifications they should be applicable to a range of other microorganisms.

Prior to staining and analysis, all samples should be prefiltered through a small disk of stainless steel with 100-μm pores.

B. Magnetic-Activated Cell Sorting

Concentration by magnetic-activated cell sorting is an area which has received considerable attention within environmental microbiology. In particular, the technique has been used extensively within the food industry to selectively enrich food-borne pathogens (Patel and Blackburn, 1991). The technique has some of the advantages of fluorescence-activated cell sorting, but is much cheaper to set up. Two magnetic separation systems are commercially available. The first, MACS, is produced by Becton–Dickinson. In this system very small magnetic particles (50 nm diameter) are conjugated with a ligand specific to the target organism. The magnetic particle-conjugated ligands are then reacted with the sample containing the target cells. The cell suspension is passed through a column, filled with steel wool, contained within a magnetic field. Magnetically labelled material is retained in the column until the magnetic field is removed and then the labeled fraction is eluted. The second magnetic system is produced by Dynal. The technique employs 2- or 4-μm magnetic beads and a simple sample tube holder with a rare earth magnet in the base. The ligand-conjugated beads are reacted with the sample and then placed in the tube holder. Within seconds the magnetic particles are concentrated in the bottom of the tube allowing the supernatant to be discarded. Thus the sample is concentrated to a small (<1 ml) volume which is an ideal volume for flow cytometric analysis. Furthermore, although the Dynal beads change the optical properties of the target organism, Dynal now markets a reagent for detaching the beads from their target. The MACS system has the advantage that the small magnetic particles used do not aggregate, do not change the optical properties of the target organism, and provide a stronger attachment to the target organism. However, the MACS system is expensive [about $5000 (US)] compared to the Dynal system [<$1000 (US)]. Also, the steel wool columns have a pore size of 50 μm which may limit their application in environmental microbiology.

Although considerable literature is available on the use of magnetic-activated cell sorting within food microbiology, very little is available within water or soil microbiology. A method has been described by Bifulco and Schaerfer (1993) to purify *Giardia* cysts from water which involves the use of a custom-made magnetic capture system. However, losses averaged 18% and very turbid samples could not be analyzed.

Research in our laboratory on the use of magnetic-activated cell sorting for purification of microorganisms from water samples has been hampered by the large numbers of magnetic particles present in environmental samples. These particles are sufficiently magnetic to interfere with the method but not magnetic enough to be easily removed prior to addition of the magnetically conjugated ligands. Further details on magnetic cell sorting are provided in Chapter 23.

III. Staining of Organisms from Water Samples for Flow Cytometric Analysis

Detection of specific single organisms within the diverse array of other particles normally found in environmental samples requires a high level of discrimination. To achieve this, highly specific labels are used which have spectral properties dissimilar to the autofluorescent material found in environmental samples.

Described below are a range of staining methods including gene probes, lectins, and various DNA stains.

At present, the simplest and most reliable method of staining organisms within environmental samples is with antibodies.

A. Antibody-Based Staining

1. Polyclonal and Monoclonal Antibodies

Commerical companies now produce antibodies to a range of environmental microorganisms. Armed with catalogs from these antibody companies, Molecular Probes Inc., (Eugene, OR) "Handbook of Fluorescent and Research Chemicals," and a reasonable quantity of dollars the environmental microbiologist can label a range of microorganisms with almost whatever color takes their fancy.

In our experience monoclonal antibodies are far superior to polyclonal antibodies for staining microorganisms within environmental samples. Considerably more binding of polyclonal antibody to debris occurs than with monoclonal antibodies. Figure 1 illustrates the superior staining qualities of monoclonal antibodies in an environmental water sample seeded with *Cryptosporidium* oocysts and stained with a FITC-labeled monoclonal antibody specific to the oocyst wall. Figure 1B shows the same sample stained with a FITC-labeled polyclonal antibody. Note the higher fluorescence of the background debris in Fig. 1B. Similar results were obtained when comparing indirect immunofluorescence staining of microorganisms within environmental samples and staining

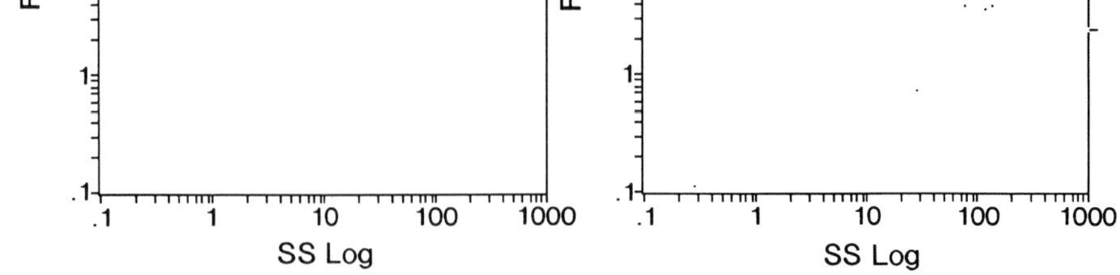

Fig. 1 Results from analyzing a river water sample seeded with *Cryptosporidium* oocysts and stained with monoclonal (A) or polyclonal FITC-labeled antibodies (graph B). The areas labeled "A" show the position of oocysts. Note the higher fluorescence of the debris in B.

with directly conjugated antibodies. Again we have found that directly conjugated antibodies are far superior.

The concentration and the incubation temperature at which antibodies are used are critical to good staining. Concentrations used for environmental samples are often considerably higher than those for other applications. The antibody should be titered out to determine the optimal concentration. These titrations need to be performed using typical environmental samples seeded with realistic numbers of the target organism. Environmental samples from different locations may require different titers of antibody.

2. Blocking Agents

The use of blocking agents to prevent nonspecific binding of antibodies to debris in environmental samples is essential. Even when highly specific, directly conjugated monoclonal antibodies are used, considerable nonspecific binding may occur unless adequate blocking agents are employed. This is demonstrated in Fig. 2 which represents a river water sample seeded with *Cryptosporidium* oocysts and stained with a FITC-conjugated mAb with blocking agents (Fig. 2A) and without blocking agents (Fig. 2B). The choice of blocking agent and concentration to use must be determined empirically for each antibody that is used. Table I gives examples of some blocking agents and the range of concentrations at which we use them. Determining the optimal blocking agent(s) is performed by staining and analyzing seeded environmental samples using

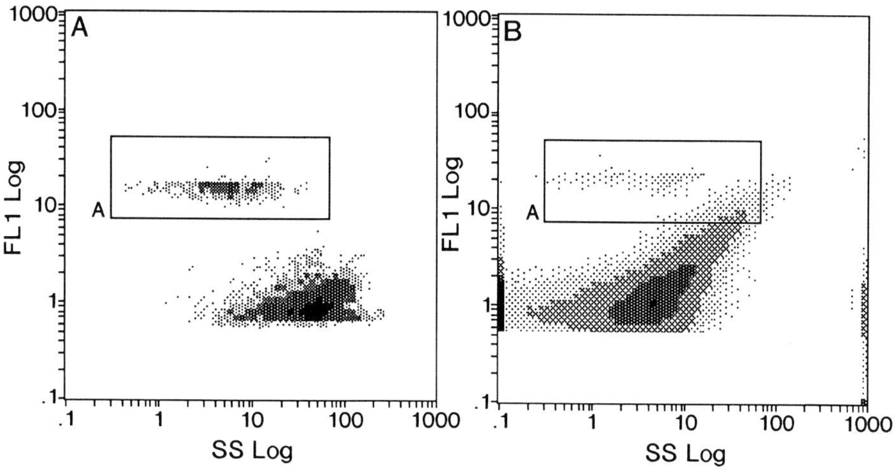

Fig. 2 Comparison of analyses of a river water sample seeded with *Cryptosporidium* oocysts and stained with monoclonal antibodies labeled with FITC with (A) and without (B) blocking agents. The areas labeled "A" show the position of oocysts. Note the higher fluorescence of the debris in B.

Table I
Examples of Blocking Agents and the Concentrations at Which They Are Employed

Blocking agent	Range of concentrations
Bovine serum albumin	0.1–5% (w/v)
Skim milk	0.1–5% (w/v)
Fetal calf serum	0.1–10% (v/v)
Glycine	0.01–1% (w/v)
Tween 20	0.001–1% (v/v)

different combinations of blocking agents. However, results may differ from one type of sample to another. The easiest way to overcome this is to prepare a cocktail sample by combining all types of samples that will be analyzed. The method used to concentrate the sample may also influence the choice of blocking agents.

3. Loss of Epitopes

Commercially available antibodies are not normally developed or prepared for staining environmental samples for flow cytometric analysis. Instead they are developed for clinical samples using epifluorescence microscopy or ELISA. This can create problems when they are used in novel applications. For example, the antigens which the antibodies recognize may be stable and always present on the organism in clinical samples but this may not be true for organisms which have been in the environment for some time. We have found that the carbohydrate antigen on *Cryptosporidium* oocyst walls which is recognized by all commercially available antibodies is slowly shed by oocysts in the aquatic environment (data not presented). Furthermore, this antigen is removed by oxidizing agents, such as chlorine, and is not essential to viability (Moore *et al.*, in preparation). Therefore oocysts which have been treated with high levels of chlorine may still be infectious but they cannot be detected using the commercially available antibodies.

4. Dual Staining

The use of two different antibodies to stain a microorganism increases the sensitivity of a detection method to such a degree that nonsorting detection is often possible (see Section IV). The antibodies must not compete with each other. Staining can be performed with both antibodies simultaneously. If uneven staining is seen then stain with one antibody, fix in 10% formalin, wash in PBS, and stain with the second antibody.

5. Fluorescence Amplification

The sensitivity of a detection method can often be increased by using a mAb to the fluorochrome bound to the antibody. These antibodies, such as an anti-FITC mAb, can be conjugated with a second fluorochrome such as R-phycoerythrin (RPE) or allophycocyanine (APC) (a HeNe laser is required for APC). This technique can help to discriminate the target organism from autofluorescent particles. However, any particle to which the first antibody is attached will be stained and, thus, amplified, with the second fluorochrome.

The use of biotinylated probes with avidin or streptavidin as secondary reagents is, in our experience, not suitable for staining environmental samples. We have experienced considerable increases in background staining when using these techniques. This is probably attributable to the fact that avidin's positively charged residues and its oligosaccharide component can interact nonspecifically with negatively charged cell surfaces and nucleic acids. Similarly, streptavidin binds to integrins and related cell-surface molecules (Alon *et al.,* 1990) causing background problems. However, Molecular Probes, Inc., now markets a new form of avidin, NeutraLite avidin, which appears to be less susceptible to background staining problems.

B. DNA and Vital Stains

Fluorescence methods are available for accurate, reproducible determinations of cellular viability and proliferation. These methods normally rely on a metabolic process, such as enzymatic activity or membrane potential, to identify living cells and employ a membrane impermeable nuclear stain to identify dead cells. Considerable research has been performed into the use of these techniques on bacteria (Kaprelyants and Kell, 1992,1993; Diaper *et al.,* 1993). However, all of these studies involve the analysis of pure cultures of bacteria. There is very little literature available on the detection of specific microorganisms and simultaneous evaluation of their viability within environmental samples. Donnelly and Baigent (1986) combined the DNA stain propidium iodide (PI) with immunofluorescence staining for the detection of *Listeria* in milk. However, the authors did not use PI as an indicator of viability but fixed the cells prior to staining so that PI would enter all cells. The authors then used the PI signal in conjunction with the immunofluorescence FITC signal to detect the *Listeria* cells. Pettipher (1991) reported on the use of the ChemFlow system for detecting yeasts in soft drinks. The ChemFlow system is a flow cytometer and bacterial stain developed by Chemunex (France) specifically for microbiology laboratories. The fluorochrome used is a propriety viability stain. The system is capable of enumerating viable microorganisms but is not able to distinguish between different types of microorganisms.

To be able to detect specific microorganisms and assess their viability simultaneously, separate fluorochromes for viability and detection must be used which do not spectrally overlap. If two different fluorochromes are being used for detection then the choice of fluorochrome for viability is limited. This type of three-color cytometry normally requires a second laser beam operating at a different wavelength. A UV laser enables the use of stains such as 4′,6-diamidino-2-phenylindole (DAPI) used in combination with two detection fluorochromes such as FITC and APC.

We have found that Molecular Probes new DNA stains YOPRO-1 and TOTO-1 are far superior to other DNA stains. These stains are considerably brighter than PI or ethidium bromide. We routinely use the stains at concentrations of 1 in 10,000 in conjunction with immunofluorescent staining.

The tetrazolium salt iodonitrotetrazolium violet (INT) is a useful nonfluorescent viability stain. Respiring cells oxidize the stain with their electron transport system and deposit the red formazan product (Zimmerman *et al.*, 1978; Vesey *et al.*, 1990). These formazan deposits are clearly visible as pink spots within cells when examined with light microscopy. Furthermore, the formazan deposits cause a large increase in SALS during flow cytometric analysis. We routinely use INT at a concentration of 0.02% (w/v) in PBS, pH 9.2.

A fluorescent tetrazolium dye 5-cyano-2,3-ditolyl tetrazolium chloride (CTC) has been reported for the direct visualization of actively respiring bacteria (Rodriguez *et al.*, 1992). The dye is similar in principle to INT except that the CTC-formazan that accumulates intracellularly fluoresces at 602 nm. However, in our experience fluorescence is often considerably less than when other fluorescent viability dyes are used.

The use of fluorescent dyes to determine the viability of *Cryptosporidium* oocysts and *Giardia* cysts is an area which is receiving considerable attention at present. Schupp and Erlandsen (1987) reported on the use of fluorescein diacetate (FDA) and PI to distinguish between live and dead *Giardia* cysts. Live cysts stained with FDA and dead cysts with PI. By staining and flow cytometrically sorting cysts the authors were able to show that FDA-positive cysts caused infection in a mouse model whereas PI-stained cysts did not. However, the authors occasionally observed cysts which did not stain with FDA or PI. Further evaluation of PI as an indicator of cyst viability was performed by Sauch *et al.* (1991). The authors demonstrated that PI staining is not satisfactory for determining the viability of cysts exposed to commonly employed water disinfection methods.

A viability assay for oocysts of *Cryptosporidium* based on the inclusion or exclusion of the dyes DAPI and PI has been reported by Campbell *et al.* (1993). The authors report that oocysts which stain with PI are nonviable and oocysts which stain with only DAPI are viable. The authors found that not all oocysts stained with the two dyes unless a pretreatment involving incubation in acid was performed. Korich *et al.* (1993) investigated the use of a range of methods for determining oocyst viability and in contrast to Campbell *et al.* (1993) the authors reported that the frequency of DAPI-stained nonviable oocysts was 18 times higher than stained viable oocysts. The authors state that the ability of

DAPI to stain nuclei within sporozoites of oocysts was not sufficient to establish oocyst viability and that the most promising technique involved the use of a mAb which is specific for partially opened oocysts. The mAb will only bind to nonviable oocysts but in our opinion it is very unlikely that the mAb will stain all nonviable oocysts.

C. Lectins

The use of fluorescently conjugated lectins for staining environmental samples is a simple and inexpensive method to attach an additional probe to the target organism. A range of fluorescently conjugated lectins is commercially available. On their own lectins are not specific enough to enable detection of individual strains or species of microorganisms. However, by combining lectins with antibodies or gene probes the sensitivity of a detection method can be increased. A lectin does not have to stain the target organism to be useful. When screening a range of fluorescently conjugated lectins against samples seeded with *Giardia* cysts we found that several lectins stained the cysts, several stained nothing, and some stained the debris but not the cysts. Further investigations revealed that the lectins which stained cysts were not specific enough to be of use as an additional probe in a detection method. However, by combining several of the lectins which stained the debris particles we were able to stain virtually all debris particles without staining the cysts. By combining this lectin staining with cyst-specific antibodies labeled with a different fluorochrome to the lectins we were able to significantly increase the sensitivity of our detection method.

D. Fluorescent *in Situ* Hybridization

The use of fluorescent *in situ* hybridization (FISH) is a relatively new and exciting method to label specific microorganisms. The technique relies on identification of a specific sequence of DNA or RNA within the target organism. Thousands of copies of a complementary sequence are then chemically synthesised and labeled with a fluorochrome. The fixation procedure leads to permeation of the wall and membranes of microorganisms allowing the probe to penetrate (Amann *et al.*, 1990a). The probe is added to the sample and incubated at a temperature which allows hybridization to the organism. The probe will only attach to complementary sequences within the target organism. However, these specificities are strongly dependent on the hybridization temperatures, concentrations of monovalent cations, and denaturing agents.

Considerable research has been performed on the use of ribosomal RNA-targeted oligonucleotide probes with flow cytometry by Amann *et al.* (1990a,b). Molecules of rRNA are ideal targets for fluorescently labeled nucleic acid probes. High sensitivity can be achieved due to the target being present in very high numbers (normally more than 10,000 per cell) and a denaturation step is not required as the target is single stranded. Also, in active cells the degree of fluorescence depends on the integrity of the organism and an organism which is nonviable will have little or no fluorescence.

We have successfully used the following method for staining both bacteria and protozoa (Fig. 3) with fluorescently labeled rRNA-targeted probes.

1. Staining with FISH

a. Fixation

1. Add three volumes of 4% (w/v) paraformaldehyde in PBS to one volume of sample.
2. Hold for 2 hr at 4°C and then wash twice in PBS.
3. Add one volume of ice-cold ethanol. Samples can be stored at this stage for several months at −20°C.

b. Hybridization

1. Add to a 1.5-ml Eppendorf tube:

50 μl fluorescent probe (25 ng/μl in distilled water)

500 μl hybridization buffer (0.1% SDS, 0.9 *M* sodium chloride, 20 m*M* Tris/HCl, pH 7.2) pre-warmed to 50°C

50 μl of sample

2. Incubate at 50°C for 2 hr.

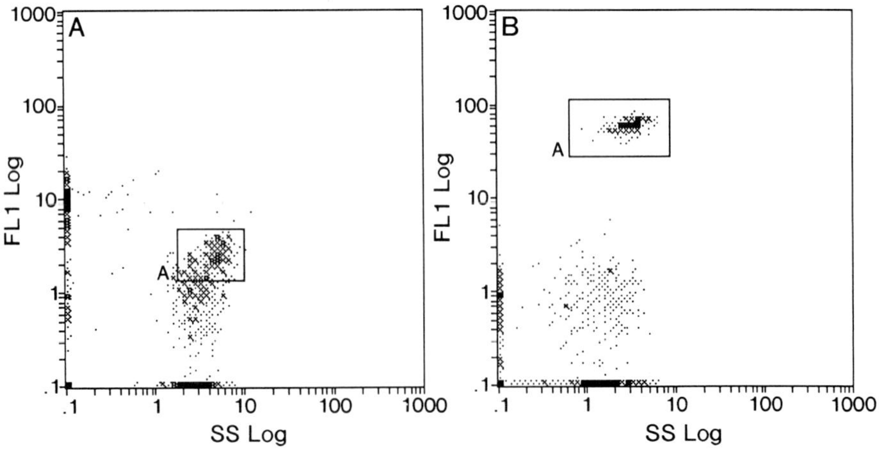

Fig. 3 Results from analyzing *Cryptosporidium* oocysts labeled with a FISH probe specific to bacteria (A) and a probe specific to Eukaryote rRNA (B). The areas labeled "A" show the position of oocysts. Note the high fluorescence of oocysts in B and the high fluorescence of the contaminating bacteria in A (the bacteria are next to the *y*-axis).

3. Wash cells in prewarmed (50°C) hybridization buffer.
4. Wash cells in PBS and analyze immediately.

One of the major attractions of FISH is that probes of desired specificity can be simply and inexpensively produced. Compared to the production of monoclonal antibodies where numerous fusions and screening of hundreds of cell lines is often required before a suitable antibody is found, the technique appears very attractive. FISH is still in its infancy and there are still many practical problems associated with the application of this technique. Often, the amount of fluorescence achieved is considerably less than that with antibodies. We are currently trying to increase the fluorescence signal by increasing the amount of fluorochrome attached to each probe.

Unfortunately, at present, when more probe is attached the fluorescence increases but so does the background. We are confident that this and similar problems will be overcome in the near future and the technique of FISH will become a routine procedure.

IV. Flow Cytometric Analysis of Water Samples

A. Instrumentation

The type of samples which are of interest to environmental microbiologists are very different from the cell suspensions which flow cytometers are designed to analyze. A water or sediment sample may contain up to 10^{11} particles/ml of sufficient size to be detected by a flow cytometer. Against a diverse background composed of bacterial, algal, plant, animal, and mineral particles the analyst may be trying to detect a single microorganism. These types of samples create problems which have not been encountered previously in flow cytometry.

A major problem with flow cytometric analysis of environmental samples is blockages. Blockages within the flow cell and within the sample tubing can result in hours of frustration. Extensive prefiltering of samples is often not possible due to losses of organism of interest. We have overcome blockages on a Coulter Elite cytometer by using a quartz flow cell with a 140-μm orifice and the shortest possible length of sample tubing from the sample tube to the flow cell. Problems were encountered with unstable droplet breakoff due to the large size of the droplets. However, by replacing the pietzoelectric crystal wafer with a cylindrical bimorph and replacing the flow cell body with a prototype flow cell body made from thinner plastic and which has a finer sample insertion rod permanently glued in position, the problem was overcome. The voltage across the deflection plates was increased to 3000 V to enable deflection of the large droplets.

The Becton–Dickinson (BD) Facstar Plus has the potential to solve blockage problems with their optional extra the Macro-Sort. The Macro-Sort enables the use of air flow cells with an orifice size of up to 400 μm. However, we have not yet successfully used a flow cell of greater than 100 μm diameter on the Facstar plus. We have also encountered problems due to the turbid and viscous nature of environmental samples causing disturbances within the sample stream. The Facstar plus analyzes the sample in air after it has left the flow cell. The disturbance of the sample stream caused by the injection of viscous and turbid samples into the stream causes noise in both the light scatter and fluorescence signals.

The two instruments discussed above are both state of the art flow cytometers which can sort. Analysis only instruments, such as the Coulter XL or the BD Facscan, offer considerable advantages, over the sorters, to the environmental microbiologist. They are half the cost of sorters, they are simple to operate, they can be fully automated, they have a large orifice which is seldom blocked, and they require little or no alignment or maintenance. However, because they cannot sort their level of sensitivity may limit their application, as discussed later in this chapter.

B. Discriminators

Once an environmental microbiologist has managed to persuade a flow cytometer to analyze a sample without blocking up, decisions need to be made on which particles the cytometer should collect data from, which particles it should ignore, and what signals should be processed. The cytometer is not capable of collecting data from all particles present in a typical sample since there are too many. Therefore a discriminator is used to reduce the amount of data to be collected to a level within the capabilities of the instrument.

During flow cytometric analysis a particle enters the laser intersection point and measurements of one parameter, the parameter chosen as the discriminator, are made. If the signal for this parameter is higher than the level set for the discriminator then the acquisition of data for all other parameters is performed. When analyzing environmental water samples for the presence of low numbers of fluorescently labeled particles the fluorescent signal is the most appropriate parameter to use as the discriminator. By setting the discriminator at a level of fluorescence slightly lower than that of the particles to be detected, the number of measurements the cytometer needs to make per second (the data rate) is typically less than 500/sec. If FALS or SALS were used as the discriminator for the same sample the data rate would be >10,000 per sec. The reason for this is that environmental samples contain large numbers of particles of which only a small proportion have high fluorescence.

The next question is what type of fluorescence signal should be used as the discriminator: a log, a linear, or a peak signal? The quick answer is if recovery is important then use a peak signal as the discriminator because a peak signal is processed much faster than a log or linear signal. The reason for wanting to use the fastest signal as the discriminator is explained below.

C. Signals and Interference

We have experienced problems with the analysis of very turbid river samples and sediment slurries that contain large quantities of small particulate matter. In these samples a significant reduction in the recovery of the target organism was observed. When fluorescently labeled beads were seeded into a very turbid sample and flow cytometric analysis was performed, a high proportion of the beads produced light scatter signals quite different from those obtained for pure beads (Fig. 4).

Further investigations revealed that when analyzing turbid environmental samples, considerable reductions in recovery can occur due to interference caused by small particles. These losses can be completely overcome by careful control of the length of time a particle remains within the laser intersection point and by using peak signals for the discriminator and for analysis. We have observed that by adjusting the laser spot size and the acquisition start delay value we can still operate the Elite at the sheath pressure which results in optimal sorting. Using this set up we can obtain 100% recovery of organisms seeded into our most turbid samples.

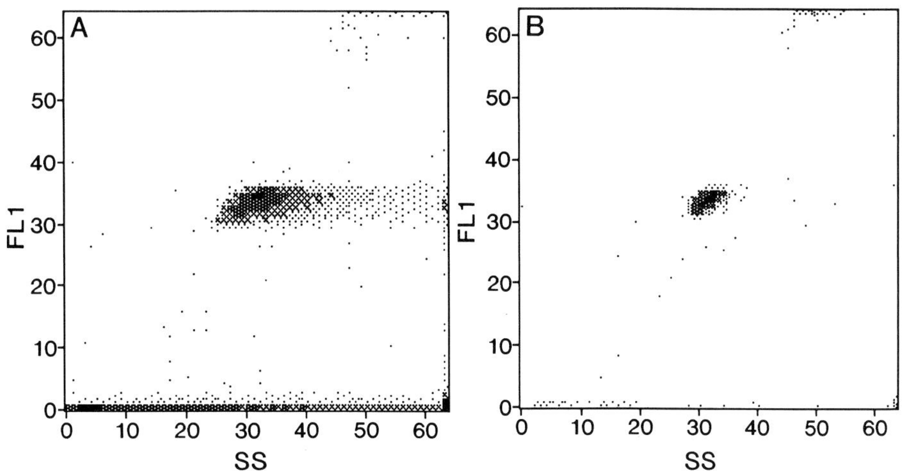

Fig. 4 Analysis of fluorescently labeled beads seeded into a turbid river water sample (A) and the same beads alone (B). Note how the population in graph A has spread due to interference caused by small particulate matter present in the sample.

D. Analysis

The two major concerns while analyzing a sample are that the cytometer is detecting or sorting all the target organisms present in the sample and that there is no contamination of the target organism from a previous sample.

To confirm that the cytometer is detecting or sorting all target organisms present, two quality control procedures should be performed. The first procedure involves the analysis of a sample seeded with a known number of the target organism within each batch of samples analyzed. The second procedure involves seeding samples with a known number of fluorescent beads. If sorting is being performed then the beads should be sorted along with the target organism. If the correct number of beads is not detected or sorted then the sample should be reanalyzed.

When analyzing environmental samples for low numbers or single microorganisms it is essential to ensure that carryover of target organisms from one sample to the next does not occur. The best way to achieve this is to run a sample of 20% sodium hypochlorite. Sodium hypochlorite removes the antibody from the target organism, so even if an organism is carried over into the next sample it will not be fluorescing. A 2-min run of sodium hypochlorite followed by a 30-sec run of Coulter Cleanze is the protocol we use routinely.

E. Detection

In an ideal situation, flow cytometric enumeration of specific microorganisms in environmental samples simply involves recording the number of particles within a predefined area on a scatter plot. The area should be defined by analysis of real samples seeded with the target organism. A scatter plot or scatter plots are then set up and an area is defined which encloses all of the organism of interest. The ability of the software to allow you to define amorphous regions is a considerable aid which actually increases the sensitivity of the detection method.

Defining the correct area in which the target organism will appear is the most important stage in developing a detection method. As an example we examine the development of a method for the detection of an organism which can be stained with a FITC-labeled antibody. The first step is to analyze a seeded environmental sample and to set up a scatter plot of FALS or SALS versus FITC. The choice between FALS or SALS depends on which produces the tightest population as well as separation from debris. Once the graph of choice has been created it is worth running the sample at several different laser power and PMT high-voltage settings to optimize the separation between the organisms and the debris. The autofluorescence of the debris and the fluorescence of the FITC have different fluorescence lifetimes. By adjusting the laser power and the high-voltage settings it is possible to obtain the maximum amount of FITC emissions without increasing the autofluorescence of the debris. To increase the sensitivity of the method it is important to use as much information as possible to define the target organisms. Although we only have one label on the organism, FITC, it is worth looking at fluorescence at different wavelengths. The organism will have an autofluorescent signal different from other particles. It does not have to have a high autofluorescent signal to be useful. For example, the organism may have very low autofluorescence at 600 nm, whereas much of the autofluorescent material may be considerably higher. To be able to use this information we need to perform backgating. This is simply performed by setting up histograms of the fluorescent parameter (e.g., FL3) against FALS or SALS (use the light scatter that is not used in the FL1 histogram) and gating the histogram on the area defined in the first histogram. After analyzing a sample or playing list-mode data, define an area on the new histogram which encloses all of the target organisms. Remove the gate from the second histogram and gate the first histogram on the new area. Some wavelengths will be useful and some will not, but by trial and error and using a range of different wavelengths it is possible to increase the sensitivity of the detection method considerably.

Another technique which we have found useful is to look at the ratio of fluorescence from one wavelength to that of another wavelength. Most flow cytometers enable this to be performed by assigning the chosen fluorescence parameters to the ratio parameter. The ratio can then be used on a two-dimensional histogram. For example, one of the limitations in sensitivity is the presence of very bright autofluorescent mineral and algal particles. Because these particles normally autofluoresce across a very wide spectrum it is possible to discriminate between them and the target organism, which only fluoresces across a narrow spectum, by using a backgating method.

The target organism used to seed must be as similar as possible to the organism which is actually present in environmental samples. For example, if the method which is being developed is for the detection of *Legionella* in environmental samples then preparing the seed by simply using *Legionella* growth from an agar plate is not sufficient. Cells of *Legionella* in the environment are considerably smaller and more rounded than freshly cultured cells. In our experience, these type of differences between freshly cultured bacteria and the same organism occurring in the environment are very common. Differences may also occur in organisms which have been purified from fecal material and the same organism once it has been in the environment for a period of time. For example *Cryptosporidium* oocysts which have been freshly isolated from feces fluoresce around the entire oocyts wall when stained with a specific mAb, whereas oocysts which have been in an aquatic environment for more than a few days show patchy and considerably reduced fluorescence. To overcome these differences when setting up our detection method we seeded oocysts into various types of water and stored them at a range of temperatures for periods up to 3 months

(Vesey *et al.*, 1993a). These samples as well as fresh samples were then used to define the area in which oocysts would appear. We recommend that similar experiments are performed before defining an area in which the organism to be detected will appear.

Now that we have defined an area in which our target organism will always appear, we need to determine the sensitivity of the method. This is performed by seeding environmental samples with known numbers of the test organism. When a negative sample is analyzed no particles should appear in the defined area and a sample seeded with 10 organisms should have 10 particles within the defined area. However, at present it is extremely difficult to obtain this level of sensitivity and often a negative sample will contain some particles within the defined area (Vesey *et al.*, 1991). These particles are typically autofluorescent mineral or algal particles or particles which cross-react with the antibody. If the level of sensitivity required is higher than the number particles present in the negative sample then there is no problem. For example, if a method for detecting *Legionella* in cooling tower water is being developed and a negative sample contains 100 particles in the defined region and the level of sensitivity required is 1000 cells per liter [this is a realistic figure for *Legionella* because below this level there is considered to be little health risk (Badenoch, 1987)] then the method is satisfactory. However, for organisms such as *Cryptosporidium*, for which 1 oocyst in 1 liter may be a potential health risk (Badenoch, 1990), then the sensitivity is not satisfactory. To increase the sensitivity we need to be able to obtain more information from the target organism so that we can distinguish it from the interfering particles. To achieve this level of discrimination, one option is to label the organism with a second fluorochrome. This second label can be achieved with a second antibody which does not compete with the first antibody, a gene probe or a simple stain. In our experience a second antibody is the most successful method. The fluorescent signal from the second probe is used as an additional gate for the first histogram.

Figure 5 demonstrates the use of two probes to achieve good sensitivity. The figure shows results of analyzing concentrated river water samples for the presence of seeded *Dictyostelium* spores. Samples were stained with two different antibodies specific to the spore wall. One antibody was labeled with FITC and the other with RPE. Figures 5A and 5B are the results from analyzing a sample of river water which contained no spores. The area within the RPE graph (Fig. 5B) is the defined area in which spores will appear. This area is gated on a second area within the FITC graph (Fig. 5A). Note that the area on the RPE graph contains no particles. Figures 5C and 5D represent the analysis of a similar sample seeded with a single spore. Note that the area on the RPE graph contains a single particle.

If a second probe is not available the way to increase the sensitivity to the single organism level is to sort all particles in the defined area. By using three-droplet sorting virtually 100% recoveries of the target organism can be achieved.

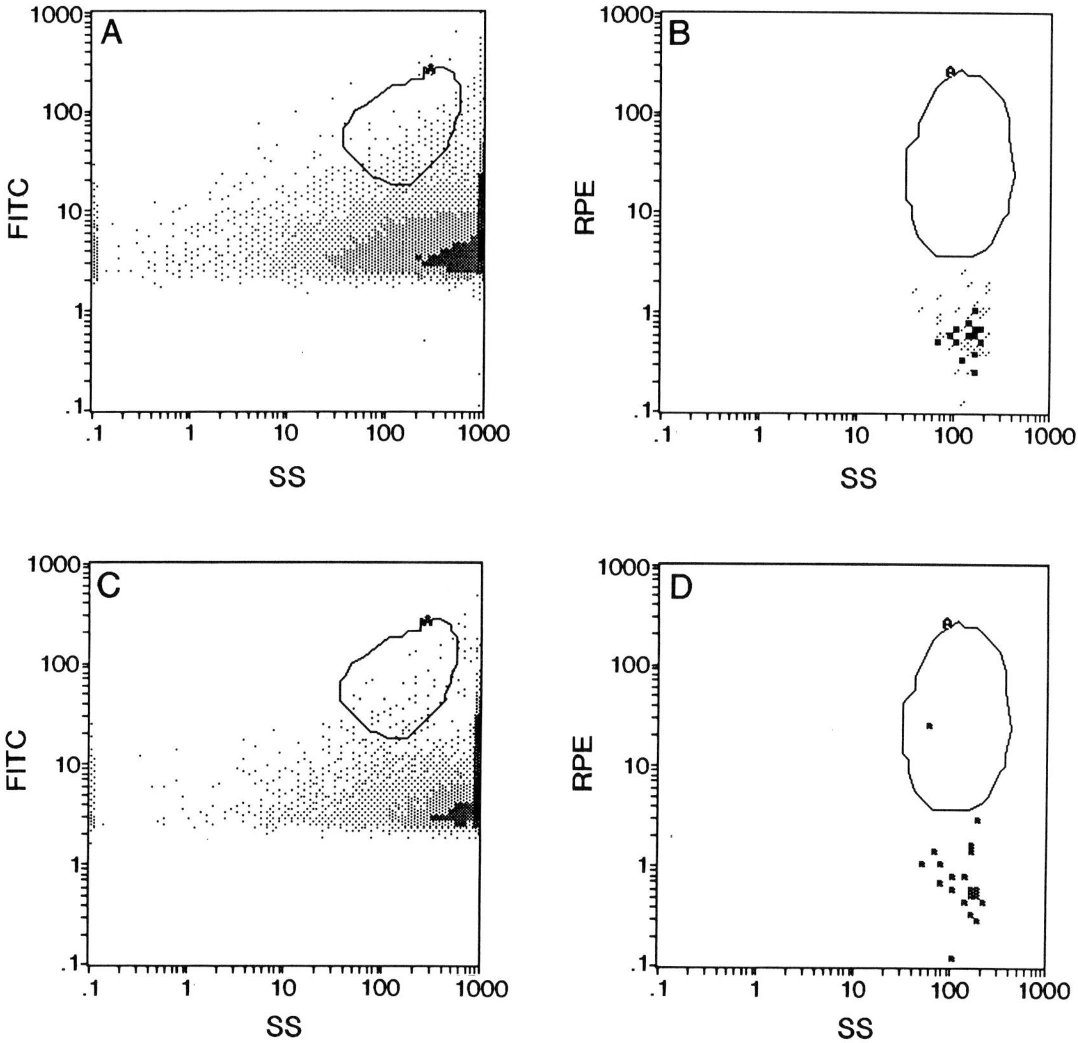

Fig. 5 Analysis of a river water sample for *Dictyostelium* spores. Samples were stained with two different spore coat antibodies labeled with FITC and RPE. (A and B) A negative sample. The area A is used to gate B. All spores will appear in the area in B. Note that there are no particles in this area in B. (C and D) The same analysis performed on a similar sample seeded with a single spore. Note the single particle in the area in D.

Furthermore, the sorted sample contains very little contaminating debris which enables rapid examination using microscopy, gene probes, or PCR.

If an analysis only method is used, i.e., no sorting and confirmation is performed, then we strongly recommend that, initially, some form of confirmation is performed on positive samples.

V. Instrumentation Developments for Environmental Applications

Current cytometers only give peak, integral, or log intensity for each parameter and in our experience, even with fluorescently labeled antibodies in environmental samples, this is not sufficient to discriminate many of the background particles from the target. Several detection techniques currently being developed to increase the amount of information that can be collected from each particle in a flow cytometer are discussed below.

A. Array Detection

Intensified diode arrays and charge coupled device (CCD) arrays have been used extensively for both spectroscopy and imaging for many years (Messenger, 1991). The main reason they have not been suitable for flow cytometry until now is their readout speed. Generally spectroscopic and imaging detectors read out at "video" type rates, up to about 50 Hz. Flow cytometry demands data rates (and therefore readout rates) of up to 10,000 particles per second or more. Recent advances in array detector have seen extremely fast CCD arrays become available; for example, 128-element linear CCD arrays are now available which have readout at rates over 100 kHz. This means that we can collect either 128 spectral channels or a 128-element line image for each particle passing through the cytometer.

B. Spectral Fingerprinting

As it is now possible to read out from an array detector at sufficient speeds we can apply the array detector in a variety of ways to increase the amount of information obtained from each particle passing through the cytometer. If we remove all conventional filters and photomultiplier tubes and replace them with a dispersing prism or grating and a high-speed array detector we have the equivalent of 128 spectral channels! [CCD array detectors have sensitivities approaching those of PMTs (Anon, 1992)]. We are examining the possibility of characterizing the autofluorescence of both the background particles found in the environmental samples and also the various fluorochromes. The concept of spectral fingerprinting may then allow the discrimination of particles solely from autofluorescence spectra or allow enhanced discrimination of fluorescent tags. This is demonstrated in Fig. 6 where the spectral fingerprint of three dyes and the light scatter from milk are compared. The three dyes can be easily separated.

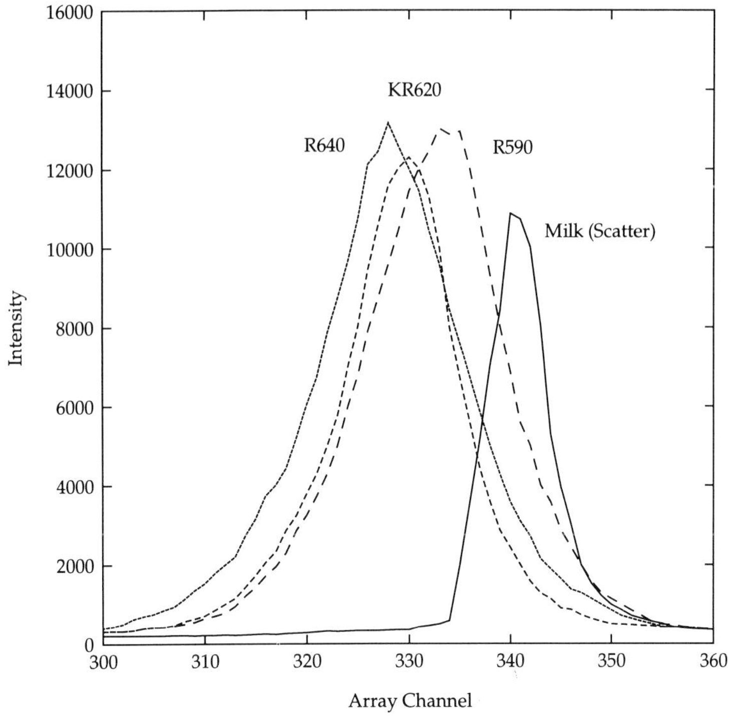

Fig. 6 Spectral fingerprinting of three fluorescent dyes: kiton red 620 (KR620), rhodamine 590 (R590), and rhodamine 640 (R640) and the light scatter from milk. The sample was illuminated with a pulsed, frequency-doubled, diode-pumped laser producing at 532 nm.

C. Pulsed Lasers

Most current laser flow cytometers employ countinuous wave lasers as the excitation source. Generally, these are continuous wave argon lasers, operating in the green, blue, and ultraviolet. The use of pulsed lasers in flow cytometry systems allows the possibility of "in-flow" particle imaging and the use of time-resolved fluorescence techniques. Also, solid-state diode-pumped lasers have the potential for reducing flow cytometry costs as they have high efficiencies, low power consumption, and long lifetimes [estimated 50,000 hr (Byer, 1988; Fan, 1990)]. They also have the potential for the generation of new wavelengths in pulse mode by nonlinear processes, as well as temporally multiplexing different excitation wavelengths to the interaction region. Another advantage of solid-state lasers is that they are very small and robust thus enabling the development of portable machines which would be ideal for environmental monitoring.

Amann, R. I., Krumholz, L., and Stahl, D. A. (1990b). *J. Bacteriol.* **172,** 762–770.

Anon (1992). "CCD Image Sensors Data Book." Dalsa, Inc., Waterloo, Ontario, Canada.

Ashbolt, N. J., and Veal, D. A. (1994). *Todays Life Sci.*, **6**(6), 28–29.

Badenoch, J. (1987). "Second report of the Committee of Inquiry into the Outbreak of Legionnaires' Disease in Stafford in April 1985." London: HMSO.

Badenoch, J. (1990). "*Cryptosporidium* in Water Supplies." London: HMSO.

Balfoort, H. W., Berman, T., Maestrini, S. Y., Wenzel, A., and Zohary, T. (1992). *Hydrobiologia* **238,** 89–97.

Bankes, P., and Richard, F. (1993). *In* "The Society for Applied Bacteriology. 62nd Annual Meeting and Summer Conference. University of Nottingham. 13–16 July 1993." p. xviii.

Bee, C. A., Christy, P. E., and Robinson, B. S. (1991). "Methods for Detection of Giardia and *Cryptosporidium* in Water: A Preliminary Assessment." Report Number 25, Urban Water Research Association of Australia, Melbourne. (Published for the Urban Water Research Association of Australia by the Melbourne and Metropolitan Board of Works.)

Betz, J. W., Aretz, W., and Härtel, W. (1984). *Cytometry* **5,** 145–150.

Bifulco, J. M., and Schaerfer III, F. W. (1993). *Appl. Environ. Microbiol.* **59,** 772–776.

Brooks, J. L., Moore, A. S., Pratchett, R. A., Collins, M. D., and Kroll, R. G. (1992). *J. Appl. Bacteriol.* **72,** 294–301.

Button, D. K., and Robertson, B. R. (1993). *In* "Handbook of Methods in Aquatic Microbial Ecology" (P. F. Kemp, B. F. Sherr, E. B. Sherr, and J. J. Cole, eds.), pp. 163–173. Boca Raton, FL: Lewis.

Button, D. K., Robertson, B. R., McIntosh, D., and Juttner, F. (1992). *Appl. Environ. Microbiol.* **58,** 243–251.

Byer, R. L. (1988). *Science* **239,** 742–747.

Campbell, A. T., Robertson, L. J., and Smith, H. V. (1993). *Appl. Environ. Microbiol.* **58,** 3488–3493.

Colwell, R. R., Brayton, P. R., Grimes, D. J., Roszak, D. B., Huq, S. A., and Palmer, L. M. (1985). *Biotechnology* **3,** 269–277.

Current, W. L. (1986). *Crit. Rev. Environ. Control* **17,** 21–51.

Diaper, J. P., Thither, K., and Edwards, C. (1993). *Appl. Microbiol. Biotechnol.* **38,** 268–272.

Dimichele, L. J., and Lewis, M. J. (1993). *J. Am. Soc. Brewing Chem.* **51,** 63–66.

Donnelly, C. W., and Baigent, G. J. (1986). *App. Environ. Microbiol.* **52,** 689–695.

Edwards, C., Porter, J., Saunders, J. R., Diaper, J., Morgan, J. A. W., and Pickup, R. W. (1992). *SGM Q.* **19**(4), 105–108.

Fan, T. Y. (1990). *Lincoln Lab. J.* **3,** 413–425.

Ford, S., and Olsen, B. H. (1988). *Adv. Microb. Ecol.* **10,** 45–79.

Furrer, B., Candrian, U., Hoefelein, C., and Luethy, J. (1991). *J. Appl. Bacteriol.* **70,** 372–379.

Grant, K. J., Piper, J. A., Ramsay, D. J., and Williams, K. L. (1993). *Appl. Optics* **32,** 416–417.

Grimes, D. J., Atwell, R. W., Brayton, P. R., Palmer, L. M., Rollins, D. M., Roszak, D. B., Singleton, P. L., Tamplin, M. L., and Colwell, R. R. (1986). *Microbiol. Sci.* **3,** 324–329.

Haslett, N. G., Adams, M. R., Cordier, J. I., and Cox, L. J. (1991). *J. Appl. Bacteriol.* **71,** XXIII–XXIV.

Herman, L., and De-Ridder, H. (1993). *Netherlands Milk Dairy J.* **47,** 23–29.

Hopkins, D. W., O'Donnell, A. G., and Macnaughton, S. J., (1991). *Soil Biol. Biochem.* **23,** 227–232.

Humphreys, M. J., Allman, R., and Lloyd, D. (1993). *In* "Proceedings of the Society for Applied Bacteriology 62nd Annual Meeting and Summer Conference, University of Nottingham, 13–16 July, 1993. Society for Applied Bacteriology, Nottingham." p. xix.

Issac-Renton, J. L., Fury, C. P., and Lochen, A. (1986). *Appl. Environ. Microbiol.* **52,** 400–402.

Jespersen, L., Lassen, S., and Jakobsen, M. (1993). *Int. J. Food Microbiol.* **17,** 321–328.

Kaprelyants, A. S., and Kell, D. B. (1992). *J. Appl. Bacteriol.* **72,** 410–422.

Kaprelyants, A. S., and Kell, D. B. (1993). *Appl. Environ. Microbiol.* **59,** 3187–3196.

Keasler, S. P., and Hill, W. E. (1992). *J. Food Protect.* **55,** 382–384.
Kell, D. B., Ryder, H. M., Kaprelyants, A. S., and Westerhoff, H. V. (1991). *Antonie van Leeuwenhoek* **60,** 145–158.
Korich, D. G., Mead, J. R., Madore, M. S., Sinclair, M. A., and Sterling, C. R. (1990). *Appl. Environ. Microbiol.* **57,** 2610–2616.
Korich, D. G., Yozwiak, M. L., Marshal, M. M., Sinclair, N. A., and Sterling, R. S. (1993). Report to the American Water Works Association Research Foundation.
LeChevallier, M. W., Norton, W. D., and Lee, R. G. (1991). *In* "Monitoring Water in the 1990's: Meeting New Challenges" (J. R. Hall and D. Glysson, eds.), pp. 483–498. American Society for Testing and Materials.
Leif, R. C., Vallarino, L., Harlow, P. M., and Cayer, M. L. (1993). *Cytometry,* Suppl. 6, Abstract 404B.
Lewis, G., Loutit, M. W., and Austin, F. J. (1986). *N.Z.J. Mar. Freshwater Res.* **20,** 431–437.
Lim, E. L., Amaral, L. A., Caron, D. A., and DeLong, E. F. (1993). *Appl. Environ. Microbiol.* **59,** 1647–1655.
Mahbubani, M. H., Bej, A. K., Perlin, M., Schaefer, F. W., Jakubowski, W., and Atlas, R. M. (1991). *Appl. Environ. Microbiol.* **57,** 3456–3461.
Manz, W., Szewzyk, U., Ericsson, P., Amann, R., Schleifer, K. H., and Stenstrm, T. A. (1993): *Appl. Environ. Microbiol.* **59,** 2292–2298.
McDaniel, J. A., and Capone, D. G. (1985). *J. Microbiol. Methods* **3,** 291–302.
Messenger, H. W. (1991). *Laser Focus World* **27,** 77–82.
Miller, R. A., Bronsdon, M. A., and Morton, W. R. (1986). *Ann. Meeting Am. Soc. Microbiol.* (Washington). 23–28 March 1986. p. 48.
Mills, A. L., and Bell, P. E. (1986). *In* "Microbial Autoecology: A Method for Environmental Studies" (R. L. Tate, ed.) pp. 27–60. New York: Wiley.
Narai, J., Vesey, G., Champion, A. C., Piper, J., and Williams, K. (1993). *In* "Proceedings of the Australian Flow Cytometry Group Annual Symposium, 30 June, Melbourne, Australia."
Niederhauser, C., Candrian, U., Hoflein, C., Jermini, M., Buhler, H. P., and Luthy, J. (1992). *Appl. Environ. Microbiol.* **58,** 1564–1568.
Olson, R. J., Zettler, E. R., and DuRand, M. D. (1993). *In* "Handbook of Methods in Aquatic Microbial Ecology" (P. F. Kemp, B. F. Sherr, E. B. Sherr, and J. J. Cole, eds.), pp. 163–173. Boca Raton, FL: Lewis.
Ongerth, J. E., and Stibbs, H. H. (1987). *Appl. Environ. Microbiol.* **53,** 672–676.
Page, S., and Burns, R. G. (1991). *Soil Biol. Biochem.* **23,** 1025–1028.
Patchett, R. A., Back, J. P., Pinder, A. C., and Kroll, R. G. (1991). *Food Microbiol. (London)* **8**(2), 119–126.
Patel, P. D., and Blackburn, C. (1991). *In* "Magnetic Separation Techniques Applied to Cellular and Molecular Biology" (J. T. Kemshead, ed.), pp. 93–106. Wordsworths' Conference Publications, Bristol, UK.
Pettipher, G. L. (1991). *Lett. Appl. Microbiol.* **12,** 109–112.
Pinder, A. C., and McClelland, R. G. (1993). *In* "Proceedings of the Society for Applied Bacteriology 62nd Annual Meeting and Summer Conference, University of Nottingham, 13–16 July, 1993. The Society for Applied Bacteriology, London." pp. ix.
Rendtorff, R. C. (1954). *Am. J. Hyg.* **59,** 209–220.
Robertson, B. R., and Button, D. K. (1989). *Cytometry* **10,** 70–76.
Rodriguez, G. C., Phipps, D., Ishoguro, K., and Ridgeway, H. F. (1992). *Appl. Environ. Microbiol.* **58,** 1801–1808.
Rose, J. B., Madore, M. S., Riggs, J. L., and Gerba, C. P. (1986). *In* "Proceedings of the AWWA Water Quality Technology Conference, American Water Works Association, Denver, CO." pp. 417–424.
Ryder, M. H., Pankhurst, C. E., Rovira, A. D., and Correll, R. L. (1994). *In* "Microbial Ecology

of the Rhizosphere" (O'Garra, F., Dowling, D., and Boestein, B. eds.). Weinheim, Germany: VCH.

Sauch, J. F., Flanagan, D., Galvin, M. L., Berman, D., and Jakubowski, W. (1991). *Appl. Environ. Microbiol.* **57,** 3243–3247.

Scheper, T., Hitzmann, B., Rinas, U., and Schugerl, K. (1987). *J. Biotechnol.* **5,** 139–148.

Schupp, D. G., and Erlandsen, L. S. (1987). *App. Environ. Microbiol.* **53,** 704–707.

Shapiro, H. M. (1988). "A Practical Guide to Flow Cytometry." New York: A. R. Liss.

Shapiro, H. M. (1990). *Am. Soc. Microbiol. News* **56,** 584–588.

Sharma, R. V., Edwards, R. T., and Beckett, R. (1993). *Appl. Environ. Microbiol.* **59,** 1864–1875.

Smith, H. V., and Rose, J. B. (1990). *Parasitol. Today* **6,** 8–12.

Tsai, Y. L., and Olson, B. H. (1992). *Appl. Environ. Microbiol.* **58,** 2292–2295.

Tsai, Y. L., and Palmer, C. J., and Sangermano, L. R. (1993). *Appl. Environ. Microbiol.* **59,** 353–357.

Venkateswaran, K., Shimada, A., Maruyama, A., Higashihara, T., Sakou, H., and Maruyama, T. (1993). *Can. J. Microbiol.* **319,** 506–512.

Vesey, G., and Slade, J. S. (1990). *Water Sci. Technol.* **24,** 165–167.

Vesey, G., Nightingale, A., James, D., Hawthorne, D. L., and Colbourne, J. S. (1990). *Lett. Appl. Microbiol.* **10,** 113–116.

Vesey, G., Slade, J. S., and Fricker, C. R. (1991). *Lett. Appl. Microbiol.* **13,** 62–65.

Vesey, G., Slade, J. S., Bryne, M., Shepherd, K., Dennis, P. J., and Fricker, C. R. (1993a). *J. Appl. Bacteriol.* **75,** 87–90.

Vesey, G., Slade, J. S., Byrne, M., Shepherd, K., and Fricker, C. R. (1993b). *J. Appl. Bacteriol.* **75,** 82–86.

Vesey, G., Slade, J. S., Byrne, M., Shepherd, K., Dennis, P. J. L., and Fricker, C. R. (1993c). *In* "Application of New Technology in Food and Beverage Microbiology," Society of Applied Bacteriology Technical Series No. 31. Oxford, UK: Blackwell Scientific.

Vesey, G., Hutton, P. E., Champion, A. C., Ashbolt, N. J., Williams, K. L., Warton, A., and Veal, D. A. (1994). *J. Cytometry,* **16,** 1–6.

Wagner, M., Amann, R., Lemmer, H., and Schleifer, K. H. (1993). *Appl. Environ. Microbiol.* **59,** 1520–1525.

Wegmuller, B., Luthy, J., and Candrian, U. (1993). *Appl. Environ. Microbiol.* **59,** 2161–2165.

Wernars, K., Heuvelman, C. J., Chakraborty, T., and Notermans, S. H. W. (1991). *Listeria J. Appl. Bacteriol.* **70,** 121–126.

Yentsch, C. M. (1990). *In* "Methods in Cell Biology" (Z. Darzynkiewicz and H. A. Crissman, eds.), Vol. 33, pp. 572–612. San Diego: Academic Press.

Zimmerman, R., Iturriaga, R., and Becker-Birch, J. (1978). *Appl. Environ. Microbiol.* **36,** 926–935.

CHAPTER 30

Strategies for Flow Cytometric Analysis of Marine Microalgae and Sponge Cells

Clarice M. Yentsch* and Shirley A. Pomponi†

* Bigelow Laboratory for Ocean Sciences
West Boothbay Harbor, Maine 04575
and Biology Department
Bowdoin College
Brunswick, Maine 04011

† Division of Biomedical Marine Research
Harbor Branch Oceanographic Institution, Inc.
Fort Pierce, Florida 34946

I. Introduction

The objective of this chapter is to detail two strategies important in order to overcome constraints inherent in flow cytometric analysis of marine microalgae and sponge cells. The two strategies are: (1) selective cell enrichment, employing physical separation techniques, and (2) selective signal enhancement, optimizing the optics. Specifically, we detail cell dissociation and density gradient separation of sponge cells, concentration of microalgae from seawater samples, determination of absolute (vs relative) abundance of microalgal subpopulations, and a metabolic index protocol that can be used for both microalgae and sponge cells.

II. Application

Imagine yourself in a submersible 10 miles off the northeast coast of Bermuda. These waters are some of the least productive regions known, yet as you descend through the upper tens of meters it is obvious that these are particle-rich waters. By day or white light illumination, light scatter from the particles is brilliant. By night, a three-dimensional constellation is revealed due to stimulation of bioluminescence by turbulence. Ultraviolet or blue light illumination reveals fluorescence of red from chlorophyll orange from phycoerythrin, and green from a yet unnamed fluorescing substance.

In fact these particles are far smaller than they appear underwater. A majority are less than 1 μm in diameter. In the sea, single particles range from 0.1 to 100 μm in diameter and include bacteria, cyanobacteria, microalgae, and microflagellates, as well as detritus (decaying organic matter) and suspended inorganic sediments. Concentrations of these particles are inversely proportional to size, 10^6 to 10^7 cells per milliliter of the smaller to 10^1 per milliliter of the larger.

This unicell world transforms into a multicell world which is obvious as the submersible reaches bottom. There, sponges, corals, and anemones abound. These organisms depend on the particle "rain" or "snow" for nutrition and in many cases are symbiotic associations of bacteria and/or microalgae on or within the primitive tissues. Both the metazoan hosts and the unicellular symbionts are amenable to flow cytometric analysis. Nutritional requirements, biochemical processes controlling the production of biologically active secondary metabolites, signal transduction events, immunological responses, the nature of host–symbiont relationships, and physiological responses to toxins in the environment are poorly understood for most marine invertebrates. One approach to gaining a fundamental understanding of these phenomena is by flow cytometric analysis of dissociated and/or cultured cells of primitive, yet abundant invertebrates, such as sponges.

The questions posed by biological oceanographers are the same whether targeting the drifting microalgae or the anchored sponges.

In general we ask:

- what is the particle makeup of this environment/organism?
- what are the abundance and heterogeneity of the particles?
- how many of the particles are cells?
- are the cells dead or alive?

Of the living cells, we probe further:

- how do these cells respond to environmental stimuli?
- what are cell function responses?
- what are cell constituent responses?

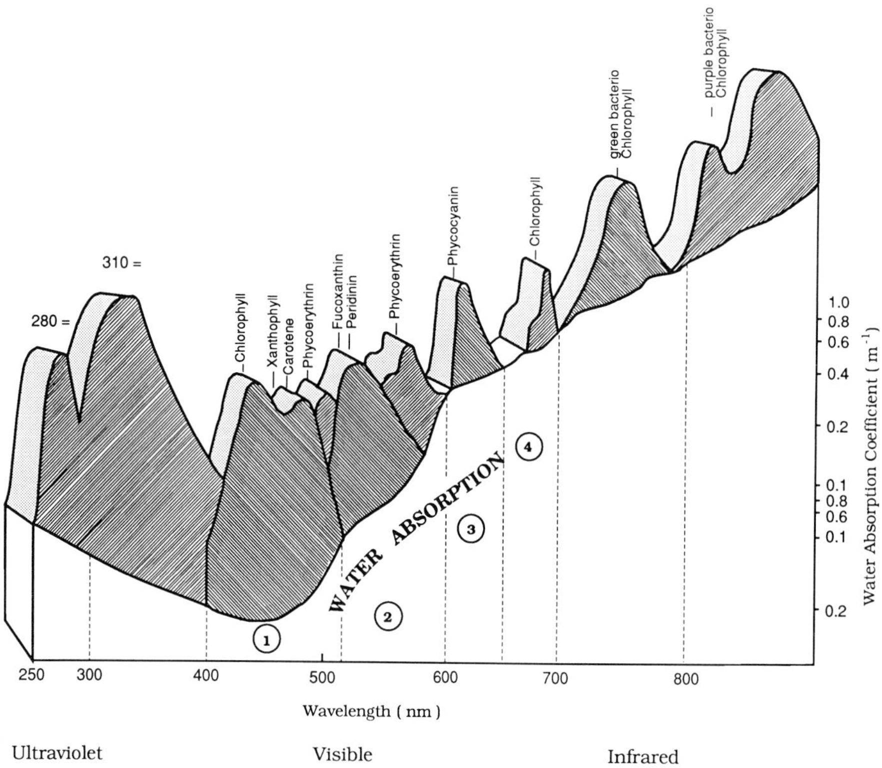

Fig. 1 Absorption characteristics of pigments present in various bacterial and microalgal particles, superimposed on water absorption. Spectral range from the ultraviolet (250 nm) to the infrared (900 nm). Absorbance peaks at 280 and 310 nm represent ultraviolet photoprotective pigments. From Yentsch and Yentsch (1992), with permission.

We mentioned the "light show" and spectral variability from our ocean descent. A spectacular array indeed, yet a limiting constraint for flow cytometric analysis. Each color represents a pigment with autofluorescing properties. These pigments also serve as a basis for discriminating various subpopulations (Fig. 1). In other words, these cells have natural substances which, when excited, emit light. If one chooses to do flow cytometric assays of cell constituents on preserved or dead cells, then one can extract these pigments with methanol. If the objective, however, is to test cell function using viable cells, it is necessary to work around the "interfering" autofluorescing compounds present. Only by judicious selection of stains, wavelengths, and filters does the flow cytometer become powerful as an analytical tool for autofluorescing cells.

III. Materials

1. Calcium- and magnesium-free artificial seawater (CMF)

NaCl	39 g
KCl	1.15 g
Na_2SO_4	1.44 g
$NaHCO_3$	0.25 g
Deionized H_2O	1 liter.

Adjust to proper salinity, e.g., 32–35 parts per thousand for seawater, by titration with NaCl solution.

2. Fluorescein diacetate (FDA) (Sigma Chemical Co., St. Louis, MO, F 7378, Lot No. 42F-5066).
3. Nitex nylon mesh (Research Nets, Inc., PO Box 249, Bothell, WA 98041).
4. Percoll (Sigma Chemical Co., P 1644).

IV. Cell Preparation and Staining

A. Obtaining a Monodisperse Suspension

1. Microalgae

Microalgae are unicells and, in general, do not clump. No enzymatic or chemical techniques are necessary to obtain a monodisperse suspension. Clumps are easily dispersed with gentle agitation.

2. Sponges

Sponges are multicellular animals and must be dissociated into single cells prior to flow analysis. Because of their primitive level of organization, sponges are relatively easy to disperse using a combination of physical and chemical methods (for review of dissociation and enrichment methods, see Yentsch and Pomponi, 1986). We have successfully used the following technique to dissociate cells from a variety of sponge species.

a. Protocol for Sponge Cell Dissociation

1. Cut the sponge into small pieces (2 to 3 cm) in a dish or beaker of 0.45-μm filtered seawater without exposing the sponge to air.
2. Rinse the sponge pieces in filter-sterilized, CMF to remove surface debris, and mince with scissors or scalpels into smaller pieces (2 to 5 mm).
3. Soak in CMF with 10 m*M* EDTA (10 : 1, CMF vol : sponge vol) for 20 to 60 min. Critical aspect: viability of the sponge cells decreases with the amount of time they are exposed to EDTA, so an optimum dissociation time should be established for each species. Cells should not be stored in CMF with EDTA.
4. Gently squeeze the pieces through gauze to release cells.
5. Filter the cell suspension through 70-μm mesh Nitex to remove large cell aggregates and spicule debris and to yield a crude cell suspension.
6. Concentrate the filtrate to 10^6 to 10^8 cells/ml by centrifugation at 300 rpms (Sorvall RT6000B centrifuge) for 5 min.

Many sponges will release an adequate number of cells for flow cytometric evaluation in less than 30 min. Dissociated cells should be stored in seawater or CMF. Immediately before flow cytometric analysis, the cells should be examined microscopically for the presence of aggregates. If aggregates are observed, gently centrifuge the cells (5 min at 300 rpm), and resuspend in CMF-EDTA to disperse the aggregates.

B. Enriching the Sample

1. Microalgae

A discrete volume (20 to 1000 ml) is obtained from a water sampling bottle which has collected water from a specific depth in the water column. Remember that, at most, flow cytometers will analyze 5 ml/hr. Generally, analysis times for natural populations of marine water samples are 5 to 15 min. Thus at most, 1 ml of the sample is consumed. If information is desired on the smaller cells (bacteria, cyanobacteria), the procedure is straightforward because the cells are present at concentrations of 10^5 to 10^6 cells/ml. If the cells are in the large size range, for example, dinoflagellates at 35 μm diameter, their concentration even during bloom conditions will be only 10^6 cells/liter, or 10^3 cells/ml. During nonbloom conditions, concentrations run 10^4 cells/liter or 10 cells/ml. At best one can expect analysis of 10 cells! It is obvious that cell enrichment is necessary.

We have found success with gentle centrifugation in some cases, but not in others. For viable cell function assays, a "reverse filtration" is gentle and effective.

a. Protocol for Reverse Filtration

Figure 2 illustrates the technique for reverse filtration.

1. Accurately fill a 1000-ml graduated cylinder with seawater sample.
2. Insert an inner cylinder of slightly smaller diameter that has been covered with nylon plankton netting of desired mesh size; 20-μm mesh Nitex netting is ideal for enriching 35-μm dinoflagellate cells.
3. Let the inner cylinder gently fall to the bottom through the sample in the graduated cylinder. In effect, the sample within the inner cylinder consists of particles less than 20 μm in diameter, and the sample between the inner and outer cylinders consists of particles greater than 20 μm in diameter.

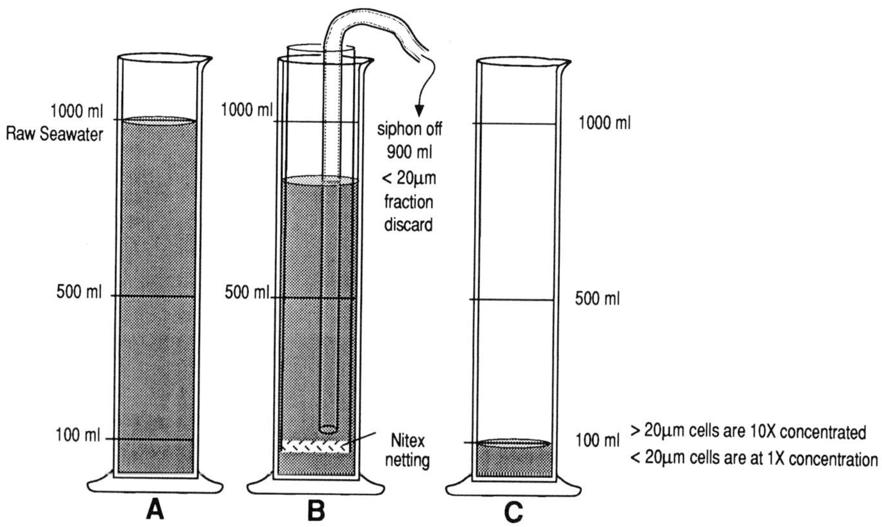

Fig. 2 Schematic representation of simple reverse filtration technique. (A) Fill a 1000-ml graduated cylinder with raw seawater. (B) One end of an inner cylinder of polyvinylchloride (PVC) is fitted with Nitex plankton netting (here 20-μm mesh size has been used) and allowed to gravity settle in the filled graduated cylinder. Once the cylinder has reached bottom, an aspirating tube is used to siphon off the water from inside the inner cylinder. In this way the size fraction less than 20 μm is siphoned off and discarded, and the size fraction greater than 20 μm is concentrated between the inner and outer cylinders. Remove the inner cylinder and siphoning apparatus. (C) Calculate concentration factor. If 1000 ml is taken to 100 ml, the concentration factor is 10× for the particles greater than 20 μm (note that the particles less than 20 μm are not concentrated). If 1000 ml is taken to 10 ml, the concentration factor is 100× for the particles greater than 20 μm. The mesh size of Nitex netting and volumes may be altered as necessary.

4. Suction off the inner volume of water until the desired volume is obtained in the 1000-ml graduated cylinder. For example, for a 10-fold concentration of the sample, the volume will be reduced to 100 ml. To concentrate 100 times, the volume in the cylinder will be reduced to 10 ml. A gentle swish of the netting generally releases any cells which might have adhered to it, but this should be checked with each cell suspension.

2. Sponge Cells

Dissociation with CMF and EDTA generally yields such a heterogeneous suspension of sponge cells that flow cytometric identification of subpopulations is not possible. It is, therefore, necessary to enrich the dissociated cells to obtain a more homogeneous sample. Density gradient centrifugation yields relatively homogeneous fractions of sponge cells that remain viable for establishment of primary cell cultures and are suitable for flow cytometric analysis. In our laboratories, Percoll has consistently yielded clean bands at the interfaces of the gradients.

a. Protocol for Percoll Density Gradient Centrifugation

1. Prepare a discontinuous gradient of Percoll by pipetting 2 ml of each Percoll gradient solution (diluted in CMF) into a 15-ml centrifuge tube (see Fig. 3). Critical aspect of procedure: establishment of an optimal gradient is empirical and will depend on the types of cells desired and the species of sponge. A good starting point is a gradient of 15–30–45–60% Percoll which will yield relatively homogeneous fractions of cells, regardless of species. For creating the gradient and layering the cell suspension, a Pasteur pipette and a Pipet Helper (Brinkmann, Cat. No. 50 09 000-1) work well.

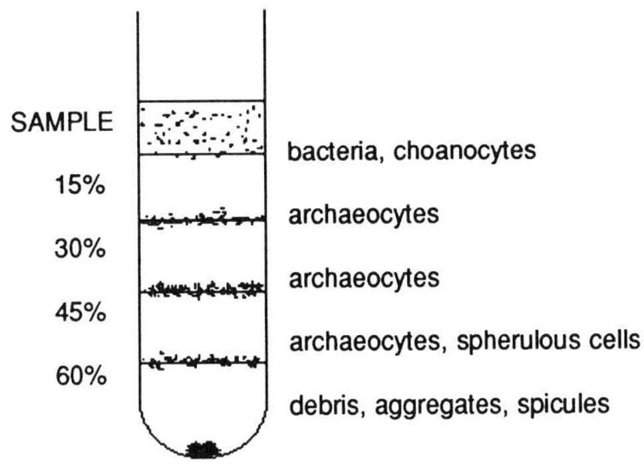

Fig. 3 Distribution of sponge cells at the interfaces of discrete density gradients of 15, 30, 45, and 60% Percoll in CMF, after centrifugation. Bands of cells are removed by careful pipetting. See text for details of procedure.

2. Layer 2 to 4 ml of the crude suspension of cells on top of the Percoll gradient and centrifuge at 600 rpm (Sorvall RT6000B centrifuge) for 10 min.

3. Collect the bands at the Percoll interfaces by aspiration with a Pasteur pipette, and transfer each fraction into a clean tube (note: the same fractions from several tubes may be pooled to collect a sufficient number of cells of each type).

4. Rinse twice by diluting with CMF and centrifuging for 5 min at 300 rpm.

5. Resuspend in 0.45-μm-filtered seawater or CMF to the desired concentration. Examine the cell fractions microscopically, count using a hemocytometer to verify concentration, and assess viability by trypan blue exclusion.

In general, bacteria, microflagellates, and sponge choanocytes are retained at the 0–15% layer (Fig. 3). Depending on the species, archaeocytes may be concentrated at the 15–30, 30–45, or 45–60% layers. Spherulous cells and other cells with inclusions are found in the 45–60% interface. Cell aggregates, debris, and spicules are pelleted. For a discussion of sponge cell types, see Bergquist (1978) and Simpson (1984).

C. Counting Cells and Estimating Volume with the Flow Cytometer

Biological oceanographers have traditionally described a water mass by the numbers and sizes of particles present; therefore, cell concentration is a highly useful parameter.

A few instruments are capable of analyzing discrete sample volumes; thus, the calculation of the total number of cells per milliliter as well as the various subpopulations can be determined easily. Most instruments, however, give only relative data—highly sensitive information about several parameters measured simultaneously—but only an indication of the numbers of cells measured over a given span of time. To extrapolate cell concentration one can employ one of a variety of techniques. One approach is to calculate the flow rate and use time to infer the cell concentration. Another is to add a known concentration of standard beads, for example, (1×10^6)/ml. When 10^5 of the standard beads have been analyzed, the operator knows that 0.1 ml has been analyzed. With functional assays, however, adding standard beads to the sample is not desirable. Thus, the most effective means described to date is to gravimetrically measure the sample in the container before and after sample analysis. To do this, one uses an accurate balance, weighs the sample before and after analysis, and calculates the difference. Assuming that the sample is pure water (1 cc = 1 ml = 1 g), a good approximation of the volume analyzed can be obtained. For further discussion of this problem, see Phinney and Cucci (1989).

D. FDA Metabolic Activity Assay

1. Control: Killed Cells

As a control, a sample of nonviable cells is prepared by adding 0.5% glutaraldehyde and heating in a microwave oven for 10–60 sec or in a constant temperature water bath (approximately 100°C). The sample of killed cells must be cooled to room temperature (approximately 20°C) for 1 hr before stain is added.

2. Preparation and Storage of FDA

FDA is colorless, does not fluoresce, and moves rapidly across cell membranes. Once inside metabolically active cells, esterases cleave off the two acetates, and the resulting fluorescence is from the naked fluorescein molecule.

A stock solution of FDA is made in reagent grade dimethyl sulfoxide (DMSO) at a concentration of 5 mg/ml and stored at 4°C. The stock solution is thawed (DMSO freezes at 18°C) and diluted 400-fold into filtered seawater or cold NaCl (32 parts per thousand, pH 7.8 to 8.0) for staining phytoplankton, or diluted into CMF with 10 m*M* EDTA for sponge cells. The solution should be made fresh each day. Since FDA is only slightly soluble in aqueous solutions, it tends to flocculate at concentrations greater than 1 mg/liter and thus a technique was developed for avoiding this problem (Dorsey *et al.,* 1989). If stock FDA is injected directly into the cold NaCl or CMF/EDTA and mixed quickly, it does not precipitate and although it appears slightly opaque, it stays evenly dispersed. If the stock is added by running it down the tube of NaCl or CMF/EDTA, it forms a precipitate.

The 400-fold dilution made daily is kept on ice in the dark to minimize degradation. Other researchers favor making fresh working solutions every 3 hr. When the solutions are made fresh daily, however, there is no noticeable degradation. If any obvious yellow-green color is noted, the solution should be discarded. Desired final concentrations of FDA are obtained by adding the diluted material directly to measured volumes of celll suspensions or, if necessary, making a second dilution of the stock and adding that directly to the cell suspensions. Most frequently, we use final concentrations of 0.75 μM (1 : 16,000); however, the optimal concentration for sponge cells was 3.0 μM.

3. Staining Protocol: Cell Loading and Fluorescence Development

We found that the optimal protocol for microalgae and sponge cells was to measure fluorescence development for a 10- to 15-min time course immediately after the addition of FDA to the cells. The cells are not washed, but kept in the FDA solution. Both the cells and FDA are kept at room temperature. Bright light should be avoided.

V. Instrumentation

A. Instrument Setup and Optimization

Cells should be examined microscopically prior to flow cytometric analysis to verify the presence of desired particles and uptake of stain. For epifluorescence microscopy, a Zeiss IV FL epifluorescence microscope with an Osram 50-W mercury vapor lamp source is used to detect fluorescein using Zeiss No. 487 716 BP 485/20 blue-green band pass excitation and 520-nm long-pass emission filters.

Flow cytometry was developed to quantify particle subpopulations in a cell suspension. The technique offers the ability to analyze heterogeneous cell populations, whether in a seawater sample or dissociated metazoan, by measuring attributes of individual cells. It allows precise, simultaneous measurements of individual particle volume, fluorescence, and light scatter at rapid rates with a maximum of about 5 ml of sample consumed per hour.

The instrument is configured to maximize the number of attributes. Engage as many of the following as possible:

1. Light scatter—both forward angle and 90°.
2. Fluorescence—for photosynthetic microalgae:

   Chlorophyll (emission at 680 nm)
   Phycoerythrin (emission at 590 nm).

3. Cell volume—if possible on your instrument.

Time-dependent accumulation/development of fluorescein in cells is measured using a flow cytometer with one photomultiplier tube (PMT) optimized for chlorophyll fluorescence, one PMT for phycoerythrin fluorescence, and one PMT for fluorescein fluorescence (Fig. 4). Any flow cytometer is appropriate for these analyses. We have used EPICS V, EPICS Profile, and EPICS Elite (Coulter Corporation, Hialeah, FL), and FACS Analyzer and FACScan (Becton–Dickinson, San Jose, CA).

Cells pass through 488-nm excitation from an argon ion laser beam or a mercury lamp beam (Fig. 4). Fluorescence signals are collected at 90° to the light beam, split by a 488-nm dichroic mirror, and detected by photomultiplier tubes. Scattered light is removed from fluorescence measurements using a 488-nm laser blocking filter. Autofluorescence (from chlorophyll and phycoerythrin pigments) is separated from the yellow-green fluorescence of fluorescein using a 550-nm beam splitter. The final wavelength of collected light at each PMT was established using a 600-nm short-pass and a 590-nm long-pass filter for phycoerythrin autofluorescence and a 635-nm band pass filter for chlorophyll autofluorescence. For analysis of sponge cells (which do not autofluoresce), the filter combinations and PMTs for chlorophyll and phycoerythrin are not used.

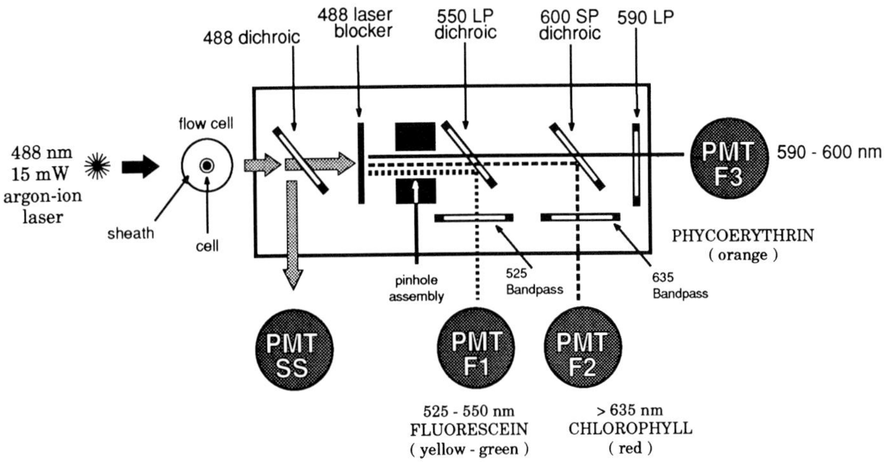

Fig. 4 Optical setup for three-color fluorescence using the EPICS profile. With this setup, phycoerythrin, chlorophyll, and fluorescein fluorescence can be measured simultaneously. Reprinted from Yentsch (1990), with permission.

B. Data Analysis

For heterogeneous samples, log amplification is important. For homogeneous samples, such as lab cultures or "bloom" samples, a linear presentation may be more informative.

For the first run, collect two scattergrams or bivariate plots of: (1) forward angle vs 90° light scatter (Fig. 5b), and (2) chlorophyll vs phycoerythrin fluorescence for microalgal samples (Fig. 5a). These data will give you a general impression of your sample diversity.

For microalgal studies, we gate on chlorophyll fluorescence (and, therefore, only accumulate measurements on cells which have chlorophyll) (Figs. 5c and 5d). We can then determine if these cells are viable and metabolically active. For studies on specific organisms for which we have a cell-surface antibody and which are present in complex mixtures of multiple species in field samples, we gate on the antibody fluorescence and then determine viability and metabolic activity.

Accumulation/development of fluorescein fluorescence is followed intermittently over a 10- to 120-min period (Fig. 6), or continuously for 15 min (Fig. 7). For the intermittent analysis, 5000 cells are measured at each time point. The mean fluorescence per cell is then obtained and plotted against time. For establishment of baseline metabolic activity of unstimulated sponge archaeocytes (Fig. 7), a scatterplot of relative fluorescence vs time is generated. High-voltage and laser power settings should be appropriate for the cell type studied. Standard beads are used to normalize mean fluorescence per cell values.

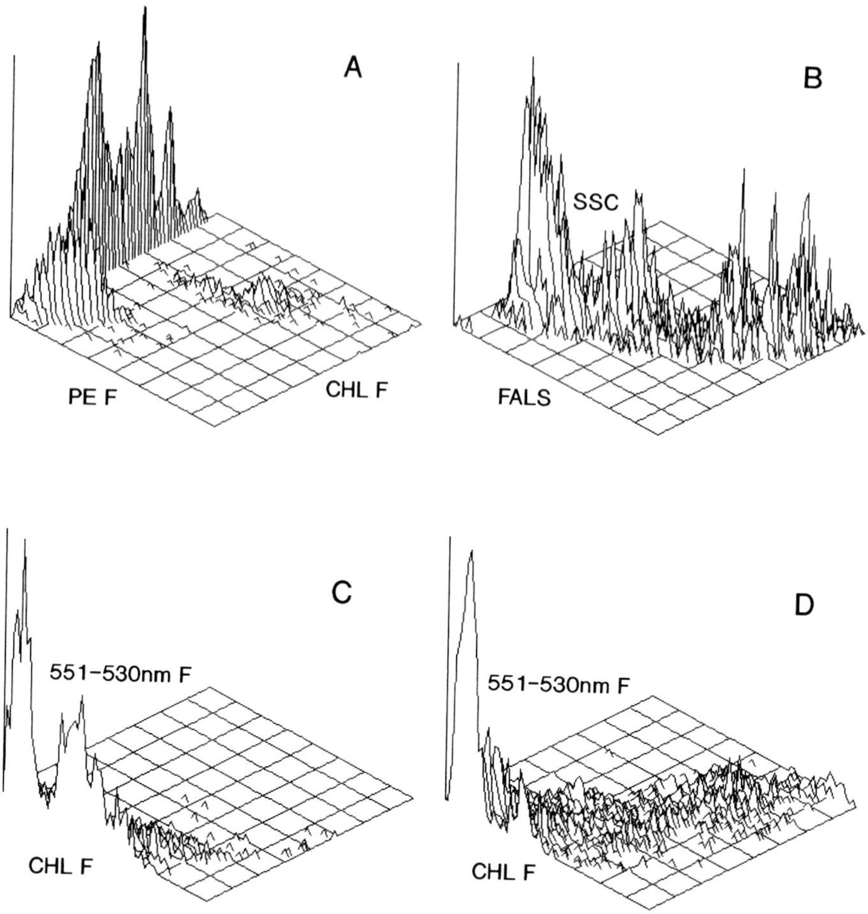

Fig. 5 Three-dimensional plots of microalgal cells: (a) clustered according to pigment groups (and termed "ataxonomy" because this type of sample identification bears little resemblance to classical taxonomy; see Yentsch, 1990); (b) clustered according to forward angle and side scatter characteristics; (c) gated on chlorophyll autofluorescence without FDA and (d) with FDA staining. Data were obtained from a seawater sample taken from the Gulf of Maine, at a depth of 8 m, during a cruise aboard the R/V Cape Hatteras in April, 1991. Data courtesy of T. L. Cucci, Bigelow Laboratory.

C. Calibration of Fluorescence Intensity

A major advantage in using a fluorescein compound is the fact that standard fluorescent beads are available (Coulter fluorescein calibration beads, Coulter EPICS Division, Hialeah, FL, or Flow Cytometry Standards, San Juan, PR). By using these standard beads (calibrated at specific numbers of molecules of

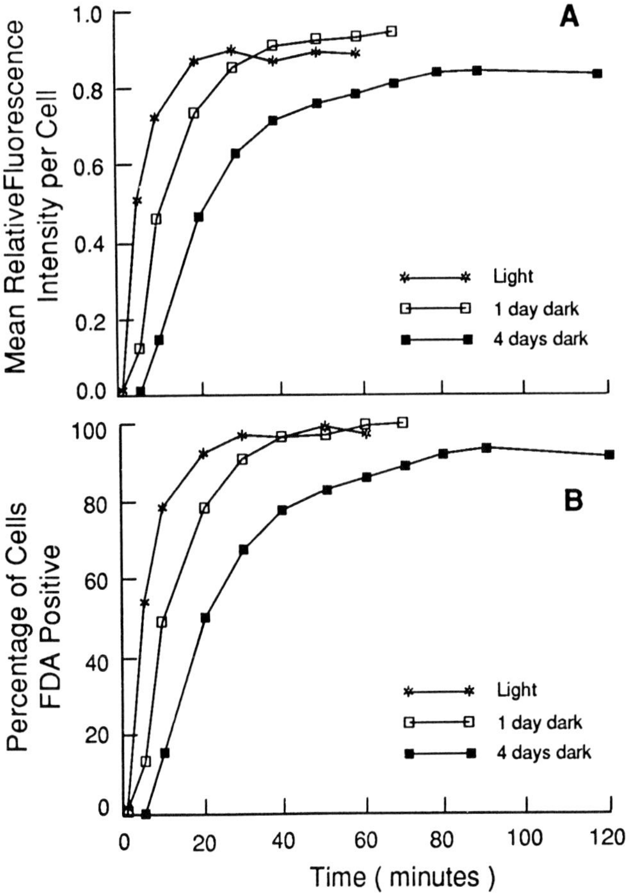

Fig. 6 Cells of a prasinophyte microalga (clone Omega 48-23) are grown in one light regime and then aliquots kept in darkness to demonstrate variations in rates of FDA development. Data are presented against time, in two formats: (A) mean relative fluorescence intensity per cell and (B) percentage of cells which are FDA positive. The patterns are similar. Initial and saturated rates were reduced for cells removed from the light which is necessary for photosynthesis. From Yentsch *et al.* (1989), with permission.

fluorescein equivalents) one can compare different gain settings on one instrument or various instruments to derive equivalent fluorescence intensities for comparison of cell viability among different groups of cells or between laboratory and shipboard instruments. We conduct bead checks at approximately 2-hr intervals throughout the experiments to check for instrument drift.

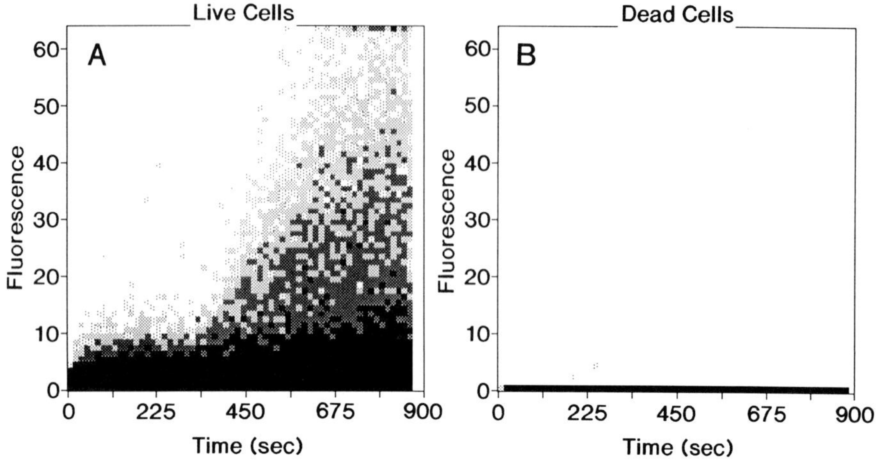

Fig. 7 Scatterplot of linear fluorescence intensity (from 3.0 μM FDA uptake) vs time for (A) freshly dissociated and enriched archaeocytes from sponge *Hymeniacidon heliophila* and (B) cells from the same fraction after heat treatment (microwaving for 30 sec). Live cells (A) showed a significant increase in metabolic activity, as measured by relative fluorescence intensity, during the 15-min analysis. Dead cells (B) did not fluoresce.

VI. Results

A. Microalgae

Figure 6 shows changes in fluorescence per cell with changes in growth irradiance. All cells were grown in the same medium and same conditions. Four days before the FDA metabolic activity assay was run, replicates were covered; thus, these cells were in the dark for 4 days. One day prior to the FDA metabolic activity assay, other batches of replicates were covered; thus, these cells were in the dark for 1 day. Curve shapes, rates, and peak fluorescence levels vary. Curve shapes may have significance in membrane transport of the dye and fluorescein accumulation/development. In general, stability in fluorescence from the cells is reached in 10 to 20 min. However, in cells which have been manipulated to have reduced metabolic activity (e.g., cells kept 4 days in darkness), the time course for stability approaches 60 min. Data expressed as mean relative fluorescence intensity per cell (as Fig. 6A) and as percentage of cells FDA positive (as Fig. 6B) have similar patterns.

B. Sponge Cells

Unstimulated sponge cells do not cycle rapidly, with cell cycling times being on the order of several days. To assess physiological responses of the cells to nutrients, mitogens, and immunologic challenges, we have developed alternative techniques to cell-cycle analysis and measurement of cell division rates. The fact that sponge cells remain viable and are capable of differentiation after dissociation makes them amenable to flow cytometry analysis. The use of FDA as an indicator of metabolic activity offers a solution to the problem of monitoring cellular responses on the scale of minutes or hours rather than days. Unstimulated sponge cells take up FDA and cleave the molecule within 5 min after staining. Further development of this technique for use on cells stimulated with mitogens, nutrients, or foreign cells, for example, are in progress in our laboratory.

VII. Discussion

The FDA assay is a simple, nonisotopic method by which we can assess cell-to-cell variability in metabolic activity in microalgae and sponge cells. It is well suited for flow cytometry. While many reagents can be used for viability studies, we found FDA adequate and inexpensive. Some researchers suggest derivatives of FDA because their esterase-cleaved products are less likely to leach out of the cell. We did comparative experiments with FDA and analogues and found no advantage of the more expensive compounds. In our experiments, retention of fluorescein in cells was sufficient to be measured easily. Development and accumulation studies avoid the cell washing steps of efflux studies.

The methods discussed in this chapter have been successfully used in flow cytometric analyses of microalgae and marine sponges. Preparative techniques to obtain a monodisperse suspension for marine sponges may be applied to other marine invertebrates. Although dissociation of sponge cells may be easier due to their relatively loose organization, tissues and organs of more "advanced" invertebrates should be amenable to dissociation using these same techniques.

Functional flow cytometric assays, including but not limited to FDA, are simple, rapid, and easily adapted from protocols for mammalian cells. The use of functional stains can be a powerful technique to study responses of cells (both unicells and dissociated cells of multicellular marine organisms) at the molecular level to environmental perturbations—whether their environment is the open ocean, a polluted bay, or a culture flask. Application of these functional assays to questions related to nutritional status, growth dynamics, biosynthesis of secondary metabolites, cell communication, and cell cycling should provide us with new perspectives on the biochemical ecology of these marine cells and, ultimately, a better understanding of the marine ecosystem as a whole.

Acknowledgments

We thank the National Science Foundation and the Office of Naval Research for instrumentation and research support. Both the NSF and the ONR have been supporters of the application of new

I. Introduction

Flow cytometry is a technique that involves measurements on populations of single cells. With the exception of special cases such as pollen, microspores, and sperm cells, higher plants are not found in the form of single cells. Instead they are formed of complex three-dimensional tissues, comprising individual cells interconnected by means of their walls. This is a result of the way in which plant cells divide. Cytokinesis is accompanied by the production of a cell plate, which progressively extends from the mother cell wall. After cell division is complete, a continuous wall structure links the daughter cells, and the individual cells do not separate. In order to convert plant tissues into cell suspensions, so that they can be used in flow cytometry, the cell wall must be removed. Wall-less plant cells are termed *protoplasts*.

Protoplasts are prepared through the use of commercially available hydrolases (cellulases, hemicellulases, and pectinases). These enzymes solubilize the cell wall polymers, and this leads to the release of intact protoplasts as a single-cell suspension. After the removal of the cell wall, the protoplasts are usually perfectly spherical, even if derived from nonspherical cells. This makes them well suited for flow cytometric analysis, from an optical point of view. However, protoplasts (and other single plant cells such as pollen) are almost always larger than the animal cells commonly analyzed using flow cytometry; occasionally their sizes exceed the diameter of the flow cell tips normally employed in flow cytometry. For successful flow analysis of plant protoplasts and cells, flow tips larger in diameter (100–200 μm) than those employed for animal cells (60–80 μm) are therefore required.

Flow analysis has been employed for the examination of light scatter and fluorescence signals derived from different types of protoplasts, and these have provided information concerning a variety of different parameters, including cell viability, size, and chlorophyll content; protoplast cell-surface architecture; and interactions between protoplasts and microorganisms (reviewed in Galbraith, 1989,1990). More recently, flow cytometry has been employed for analysis of protoplast membrane fluidity (Gantet *et al.*, 1990) and for the characterization of microprotoplasts (Verhoeven and Sree Ramulu, 1991) and of maize sperm cells (Zhang *et al.*, 1992).

In terms of sorting, it is clear both from theoretical considerations and empirically that the principles governing droplet formation in flow cytometry can be scaled up to accommodate large particles such as protoplasts and pollen (Harkins and Galbraith, 1987; Kachel *et al.*, 1990). This involves several elements, governed for jet-in-air sorters by equations describing the amplification due to surface tension of microscopic undulations in the diameter of the fluid jet induced by the action of a mechanoelectric transducer (Donnelly and Glaberson, 1966). Specifically, production of droplets requires that the wavelength of the imposed undulation be longer than π times the diameter of the fluid jet. This wavelength is a function both of the transducer drive frequency and of the fluid velocity, which itself is a function of the system pressure and the jet diameter. Empirical observations parallel theoretical treatment and can be used to produce tables listing those drive frequencies and system pressures that allow droplet production for flow tips of different diameters (Harkins and Galbraith, 1987). In work of this type, the availability of indestructible particles of defined diameters covering the size ranges of interest is important. Conditions optimal for sorting small standard microspheres are not necessarily optimal for larger particles, due to an increased degree of interference between the particle and the process of droplet formation as the particle diameter approaches that of the fluid stream (Harkins and Galbraith, 1987). Commercial microspheres larger than about 20 μm are not readily available. However, pollen from a variety of different plant species provides standard particles spanning diameters from 20–95 μm that are convenient, cheap, and intrinsically autofluorescent. Pollen precursors such as microspores can also be flow analyzed and sorted based on differences in light scatter and autofluorescence (Pechan *et al.,* 1988), and this led to populations enriched according to embryogenic potential.

Whereas microspores and pollen are mechanically extremely robust, protoplasts are very fragile and, therefore, require careful handling during flow sorting (Galbraith, 1989). Successful recovery of viable protoplasts after flow sorting was first reported using suspension culture cells and leaf mesophyll as source tissues (Galbraith *et al.,* 1984; Harkins and Galbraith, 1984). Following development of methods for fluorescent labeling of different populations of parental protoplasts (Galbraith and Galbraith, 1979; Galbraith and Mauch, 1980), similar conditions have subsequently been used by a number of different groups for the sorting of heterokaryons formed by protoplast fusion followed by regeneration of somatic hybrid and cybrid plants (Afonso *et al.,* 1985; Glimelius *et al.,* 1986; Pauls and Chuong, 1987; Sjödin and Glimelius, 1989; Hammatt *et al.,* 1990; Sundberg and Glimelius, 1991; Walters *et al.,* 1992). A different type of electromechanical cell sorter has also been successfully employed for heterokaryon sorting (Bromova and Knopf, 1991, and references therein). Other applications of flow sorting to plant protoplasts include the purification of protoplasts representative of different leaf tissue types for analysis at the molecular level (Harkins *et al.,* 1990).

As previously indicated, there are no theoretical limitations either on the sizes of the cells that can be analyzed and sorted using flow cytometry or on the sizes of the flow cells that can be employed. However, traditional procedures of flow cytometry and cell sorting and the designs of the instrumentation have been optimized using small biological cells and artificial microspheres (i.e., those falling in the range of 3–20 μm in diameter). When dealing with the flow analysis and sorting of larger particles, a variety of practical considerations enter the picture. This article, which updates methods previously described in this series (Galbraith, 1990), outlines some of the problems and their resolution in the sorting of biological cells as large as 100 μm using commercially available flow cytometric instrumentation. In principle, the procedures are applicable to all biological cells and cell aggregates within this size range, although they are illustrated for plant protoplasts, pollen, and spores of the tree-club moss *Lycopodium*.

II. Application

Protoplasts can be prepared from a wide variety of different plant species. The experimental conditions optimal for release of viable protoplasts vary according to the particular plant tissue under study. Details concerning protoplast preparation are beyond the scope of this article, and the reader is referred elsewhere (see, for example, Lindsey, 1991). This article outlines methods for the analysis and sorting of a variety of different protoplast and cell types. In principle, since all protoplasts comprise spherical structures, the methods described should be universally applicable, although some possible limitations are discussed in Section V.

III. Materials

A. Plant Materials

Tobacco (*Nicotiana tabacum* cv Xanthi) seed was obtained from the USDA Tobacco Research Laboratory (Oxford, NC). Seeds were germinated under sterile conditions, and the resultant plantlets maintained as vegetative plantlets in Magenta boxes on basal MS medium containing 3% (w/v) sucrose, solidified with 0.8% agar. Plants were subcultured every 14–21 days by excision of the portion of the plant containing the apical meristem and one to two expanded leaves. This process was continued for not more than three cycles before the plantlets were discarded. Pollen (*Zea mays, Carya illinoiensis,* paper mulberry and ragweed) and *Lycopodium* spores were obtained from Polysciences (Warrington, PA). A cell suspension culture of *Nicotiana sylvestris,* kindly provided by Dr. Roy Jensen, was maintained in darkness under sterile conditions as 100-ml aliquots in 500-ml Erlenmeyer flasks at 25°C, with constant orbital agitation (100 rpm). The growth medium comprised basal MS medium at pH 5.7 supplemented with 3% (w/v) sucrose and 1 mg/l 2,4-dichlorophenoxyacetic acid. Subculture was done at 5-day intervals by transfer of 50 ml of cells into 100 ml fresh medium.

B. Chemicals

Macerase and cellulysin were obtained from Calbiochem, Inc. (La Jolla, CA), cellulase from Worthington (Freehold, NJ), aniline blue from the Fisher Scientific Co. (Pittsburgh, PA), and fluorescent microspheres from Polysciences (Warrington, PA). MS (Murashige and Skoog) medium was obtained from

Gibco (Grand Island, New York). All remaining chemicals were from the Sigma Chemical Co. (St. Louis, MO).

IV. Procedures

A. Protoplast Preparation

1. Tobacco Leaf Tissues

Protoplasts are prepared from leaves selected from vegetative plants growing vigorously in Magenta boxes in basal MS medium containing 3% (w/v) sucrose, solidified with 0.8% agar. Leaf tissue (0.5 g) is excised under sterile conditions, sliced into 1 × 10-mm segments, and incubated within an 85-mm-diameter sterile plastic petri dish in 10 ml of a filter-sterilized osmoticum containing 10 m*M* $CaCl_2$, buffered with 3 m*M* 2-[*N*-morpholino]ethane sulfonic acid (Mes), pH 5.7; the osmotic species depends on the particular application. For isolation of mesophyll protoplasts either 0.7 *M* mannitol or 0.35 *M* KCl can be employed; for applications involving epidermal protoplasts, an ionic osmoticum is essential since these protoplasts have a lower buoyant density than isotonic mannitol. The cell walls are removed by including in the osmotica 0.1% (w/v) driselase, 0.1% (w/v) macerase, and 0.1% (w/v) cellulysin. Incubation is continued for 12–15 hr at room temperature. The protoplast digest is filtered through sterile cheesecloth into plastic conical centrifuge tubes and is centrifuged at 50*g* for 10 min. The protoplast pellet is gently resuspended in 5 ml of a solution containing 25% (w/w) sucrose dissolved in 3 m*M* Mes and 10 m*M* $CaCl_2$, pH 5.7, and is overlaid with 5 ml of osmoticum. The protoplasts are centrifuged at 50*g* for 2 min. The interface, which contains the viable protoplasts, is collected and diluted with 10 ml KCl-osmoticum, and the protoplasts are pelleted by centrifugation at 50*g* for 5 min. The protoplasts are resuspended in osmoticum to a final concentration of about 10^5/ml. The total protoplast yield is determined through hemocytometry. The proportion of viable protoplasts is found by resuspension of the protoplasts in osmoticum containing 0.1% v/v of a freshly prepared solution of FDA (1 mg/ml in acetone). Viable protoplasts accumulate fluorescein and appear bright green when illuminated with blue light under the fluorescence microscope (using the standard FITC filter set and mercury arc lamp illumination).

2. *Nicotiana sylvestris* Cell Suspension Cultures

Protoplasts are prepared from actively growing cell cultures 3 days after subculture. The cells contained within 150 ml of the cultures are allowed to settle under gravity for about 15 min. The supernatant is removed and the cells (about 14 ml) are resuspended in 5 volumes of an osmoticum (NTTO; Galbraith and Mauch, 1980) containing 0.5% (w/v) cellulase, 0.1% (w/v) cellulysin, 0.05%

(w/v) driselase, and 0.02% (w/v) macerase. After incubation with gentle orbital shaking at 25°C for 20 hr, the protoplasts are filtered through five layers of cheesecloth into 50-ml conical tubes. The protoplasts are sedimented by centrifugation at 50*g* for 5 min, resuspended in 5 ml 25% (w/v) sucrose, and are overlaid with 5 ml of osmoticum. After centrifugation at 50*g*, the intact protoplasts are collected from the interface (approximately 2 ml), diluted with 5 volumes of osmoticum, and collected by centrifugation at 50*g* for 5 min.

B. Fluorescent Standards

Protoplasts are fixed by incubation for 60 min at 20°C in 2% (w/v) paraformaldehyde in osmoticum. The paraformaldehyde is dissolved in 4% (w/v) by heating in water with gradual dropwise addition of 0.1 *M* NaOH until the solution clarifies, prior to being chilled and mixed with an equal volume of 2× concentration osmoticum. Pollen from the listed species (Table I) is autofluorescent, but fluorescence can be enhanced by staining with aniline blue. Pollen (10 mg) is stained by resuspension in 4 ml of 0.1% (w/v) aniline blue dissolved in phosphate-buffered saline, pH 9.0. The mixing of pollen and *Lycopodium* spores with water is facilitated by addition of 0.5% (v/v) Triton X-100.

C. Flow Cytometry

1. EPICS Series Instruments

The EPICS series flow sorters (Coulter Electronics, Hialeah, FL) are operated as detailed for the various applications.

Table I
Biological Particle Size Ranges

Particle type	Diameter (μm)
Broussonetia papyrifera pollen	13.8 ± 1.3; *N* = 20
Ambrosia elateior pollen	20.9 ± 1.2; *N* = 22
Lycopodium spores	28.6 ± 2.5; *N* = 12
Nicotiana tabacum leaf protoplasts (fixcd)	33.7 ± 6.2; *N* = 25
N. tabacum leaf protoplasts	42.3 ± 6.9; *N* = 20
N. tabacum suspension culture protoplasts	41.2 ± 10.4; *N* = 21
Carya illinoensis pollen	51.1 ± 3.3; *N* = 11
Zea mays pollen	95.3 ± 4.1; *N* = 11

Note. Diameters are expressed as means ± standards deviations (adapted from Harkins and Galbraith, 1987).

a. Analysis of Intact Mesophyll Protoplasts

The laser is tuned to 457 nm with a power output of 100 mW. We employ two barrier filters to exclude scattered light; these have half-maximal transmittance at 510 and 515 nm (termed LP510 and LP515). A further barrier filter (LP610) screens the red channel photomultiplier. Most chlorophyll autofluorescence emission occurs above 620 nm; the first two filters are routinely present in the light path but may not be needed for this application. Integral or log integral red fluorescence, one-parameter frequency distributions are typically accumulated to a total count of 20,000.

b. Analysis of Viable Mesophyll Protoplasts

Protoplasts are stained with FDA. The laser is tuned to 457 nm with a power output of 100 mW. Fluorescence emission is detected using the green channel photomultiplier. Barrier filters LP510 and LP515 are used to eliminate scattered light. Light is reflected into the photomultiplier using a dichroic mirror (DC590) that splits at a wavelength of 590 nm (for this application, a fully silvered mirror would also be appropriate; the dichroic is used for convenience in other applications). Two blue glass BG38 filters (Optical Instrument Laboratory, Houston, TX) are used to screen the photomultiplier. Integral or log integral green fluorescence, one-parameter frequency distributions are accumulated to a total count of 20,000.

c. Analysis of Viable Protoplasts from Suspension Cultures

Protoplasts are stained with FDA. The flow cytometer is operated as described in Section 2 with the exception that the BG38 barrier filters can be omitted.

d. Sorting Pollen

Pollen is stained with aniline blue. The laser is tuned to 514 nm with an output of 200 mW. Barrier filters LP530 and LP540 are used to screen the green channel photomultiplier. Pulse width time-of-flight (PW-TOF) analysis is performed based on peak green fluorescence. The TOF module is set to analyze the time that the pulses of fluorescence remain above thresholds set to 50% of the peak value, although this threshold can be lowered if necessary (Galbraith *et al.*, 1988). One-parameter frequency distributions of protoplast size result from this analysis; further one-parameter frequency distributions can be accumulated based on peak or integral fluorescence and on forward angle light scatter.

e. Sorting Fixed Mesophyll Protoplasts

This signal derives mostly from chlorophyll autofluorescence although there is some contribution to autofluorescence from paraformaldehyde fixation. The laser is tuned to 457 nm with an output of 100 mW. Barrier filters LP510 and LP515 and 590 are used to screen the red channel photomultiplier. Log integral red fluorescence signals are accumulated to give one-parameter frequency distributions. Sort windows are positioned according to the location of the peak.

f. Separation of Mesophyll and Epidermal Protoplasts

The protoplasts are prepared in ionic osmoticum. The purified protoplasts are stained with FDA, as described previously. The laser is tuned to 457 nm with a power output of 100 mW. Fluorescence emission is detected using both the red and green channel photomultipliers. Barrier filters LP510 and LP515 are used to eliminate scattered light. Light is reflected into the green photomultiplier using a dichroic mirror (DC590), with two BG38 barrier filters to eliminate entry of chlorophyll autofluorescence. Light transmitted through the dichroic mirror enters the red photomultiplier screened by barrier filter LP610. Some signal subtraction may be necessary to eliminate spectral overlap. Cell size is determined through PW-TOF based either on peak green (FDA) fluorescence (for epidermal protoplasts) or peak red fluorescence (for mesophyll protoplasts). Two-dimensional PW-TOF (FDA) versus integral red fluorescence allows the generation of contour plots permitting the sorting of epidermal from mesophyll protoplasts (Galbraith *et al.*, 1988). Frequency distributions are typically accumulated to a total count of 20,000.

2. Coulter Elite

The Elite flow cytometer/cell sorter (Coulter Electronics, Hialeah, FL) is operated using a flow-in-quartz cell having a 100-μm-diameter orifice and a fixed sample injection assembly (Coulter part 6856762). For the described applications, the primary argon laser provides illumination at 488 nm, and the light paths leading to the four PMTs are as follows: light scattered orthogonally to the flow stream is detected by PMT1 after being reflected by a 488-nm dichroic mirror and passing through a 488-nm band pass filter. Fluorescence emission is split according to wavelength to access the three remaining PMTs. Fluorescence transmitted by the 488-nm dichroic encounters a 488-nm laser blocking filter and then two further dichroic mirrors splitting at 550 and 625 nm. Light reflected by the 550-nm dichroic enters PMT2 after passage through a 525-nm band pass filter. Light transmitted by the 550-nm dichroic, but reflected by the 625-nm dichroic, enters PMT3 after passage through a 575-nm band pass filter. Light transmitted by the 625-nm dichroic enters PMT4 after passing through a 675-nm band pass filter.

a. Analysis of Mesophyll Protoplasts Based on Chlorophyll Autofluorescence

The laser is tuned to a power output of 15 mW. The forward scatter detector is operated at 370 V and a gain of 3.0. PMT1 (90° light scatter) is operated at 350 V and a gain of 2.0. PMT4 is operated at 620 V with a gain of 3.0. The forward scatter discriminator is set at 100; all other discriminators are off.

b. Analysis of Lycopodium Spores

Laser power output is set to 15 mW. For measurement of forward and 90° angle light scatter (linear or log signals), the foward scatter detector is operated at 100 V with an amplification of 7.5, and PMT1 at 140 V with an amplification of 7.5. Autofluorescence is detected using PMT2 at 650 V with an amplification of 7.5. Uniparametric histograms (linear or log integral signals) are accumulated. Multiparametric histograms based on forward and 90° angle light scatter (linear or log signals) can also be accumulated. For all applications, the forward scatter discriminators are set at 100, and the remaining discriminators are turned off.

D. Cell Sorting

1. The Coulter EPICS Series Instruments

The EPICS series flow sorters can be employed for the successful sorting of protoplasts and pollen using the 100-, 155-, or 200-μm flow tips. The system pressures and transducer drive frequencies suitable for the various flow tips are listed in Harkins and Galbraith (1987). For the 200-μm flow tip, we routinely use a system pressure of 6 psi and a drive frequency of 8 kHz. The sheath fluid comprises mannitol (50.5 g/l) and glucose (68./4 g/l) buffered with 3 m*M* Mes, pH 5.7. We employ two 2.5-liter sheath tanks connected in parallel; this provides approximately 2–3 hr of sorting and analysis between refills.

Sort alignment and optimization is achieved as follows:

1. Using the sort test mode, the transducer drive amplitude is adjusted to provide a uniform and stable sorted stream.
2. The number of undulations are counted to provide an estimate of the position of droplet breakoff and hence the charge delay setting.
3. Analysis is initiated using standard particles that approximate the size of the protoplasts that are to be sorted (either paraformaldehyde-fixed protoplasts or aniline blue-stained pollen). Sort windows are set on the appropriate frequency distributions.
4. A sort matrix analysis is performed, to precisely define the charge delay setting that yields a sort efficiency of 100%. This can be conveniently measured using a single-cell deposition device (the Coulter Autoclone) to sort 10 particles/well in a 96-well culture plate; the actual numbers of particles recovered are then determined using an inverted light microscope.
5. Analysis and sorting of the protoplasts are initiated, using the analysis parameters described above and the sort conditions defined by use of the standard particles.

For sterile sorting, we utilize a 0.2-μm in-line filter from Pall (Ultipor Type DFA4001ARP) that can be autoclaved. The sample tube, sample pickup, and sample introduction line are autoclaved. The sheath tanks and sheath lines are thoroughly cleaned with dilute bleach and are rinsed with 70% ethanol. The sheath fluid is sterilized either by autoclaving or by passage through millipore 0.22-μm filters. Prior to sorting, the sample lines are cleared of residual ethanol by passage of sterile sheath fluid. Sorting is performed in 96-well plates prefilled with 50–100 μl of sterile growth medium. Protoplast density affects further development in culture. Either sufficient protoplasts must be sorted (about 1000 per well) or feeder cells (nonmorphogenic *N. tabacum* cell suspensions) must be included in the wells (Galbraith, 1989).

2. The Coulter Elite

Protoplasts and pollen can be sorted using the Elite with the 100-μm quartz flow tip. The system is operated at a pressure of 8 psi and a transducer drive frequency of 11.5 kHz (for pollen) and 8.9–14 kHz (for protoplasts), using the same sheath fluid as described for the EPICS series instruments.

Sort alignment and optimization is as follows:

1. Prior to instrument optimizing for sorting, the transducer is allowed to warm up completely. For the piezoelectric transducer, this involves running the drive at an amplitude of about 70% for at least 1 hr. The deflection plate assembly is then moved as close as possible to the flow cell tip without blocking the laser beam (about 5 mm). The video camera is adjusted to observe the flow stream such that one can see both the ground plate on the left edge of the screen and the laser beam intercept on the right while using only the "pan" function of the camera adjustment. The transducer drive frequency is adjusted at constant amplitude (typically about 20%) to provide as short a droplet breakoff point as possible. The machine is then switched to sort test mode, and the transducer amplitude is adjusted in order to give a stable sorted stream.

2. The deflection plate assembly is lowered to allow observation of two to three free droplets above the ground plane. The undulations in the fluid stream are examined using the video camera. It is important that the last attached droplet be well rounded on the left-hand side and clearly connected by a ligament to the flow stream on the right-hand side. Satellite droplets are usually seen. These can be either "fast" or "slow," depending on whether they merge with the major droplet ahead of or behind the satellite. The shape of the last attached droplet and the behavior of satellites are dependent on the amplitudes and frequencies of transducer activation. Conditions should be established that produce fast satellites, since these carry charge of the same sign as the droplets with which they shall merge and so will not affect their electrostatic deflection.

3. The cursor is moved to mark the second well-defined undulation to the right of the last attached droplet. A second point is marked between the first two free drops above the ground plate. The number of droplets between the two cursors is entered into the delay calculation program. This calculated delay

setting should be very close to the optimal delay setting (as determined through sorting of particles, described below).

4. Frequency distributions are acquired, using standard particles approximating the size of the cells within the samples of interest. Sort windows are set based on these distributions.

5. A sort matrix analysis is performed, in order to empirically define the delay setting which yields a sort efficiency as close as possible to 100%. This is done by programming the sorting of batches of 25 particles onto a standard 3 × 1″ glass microscope slide and counting the actual numbers recovered using microscopy. For each batch of 25, the sort delay setting is adjusted by 1.0-step increments to span a range of ± 5.0 around the calculated delay setting.

6. The phase settings and deflection plate assembly high-voltage amplitude are adjusted to obtain side streams with minimal fanning.

7. Histograms are accumulated for the cells of interest, the appropriate sort windows are defined, and the cells are sorted.

For sterile sorting, the sheath and rinse tanks are cleaned using dilute detergent and are filled with 70% ethanol. The instrument is placed in run mode for 15 min without a sample tube at the sample station; this operates the sample station on backflush. The machine is then switched to shutdown mode several times. The run/shutdown mode sequence is repeated twice, after the sheath tank is refilled first with sterile (autoclaved) water and then with sterile sheath fluid. Finally, the desired cells are sorted into a sterile collection tube.

E. Transformations of Data

1. Protoplast Size Measurement

Previous work has shown that the TOF parameter can be used for the accurate measurement of protoplast size, assuming appropriate calibration (Galbraith *et al.,* 1988). For this calibration, we require particles that approximate the size ranges over which the TOF measurements are to be made; these particles must be fluorescent and must not deform or degrade during passage through the flow cytometer. In general, artificial fluorescent microspheres with diameters similar to those of protoplasts are not readily available. We therefore have employed natural microspheres (pollen) stained with aniline blue for calibrating the TOF parameter (Galbraith *et al.,* 1988). Since pollen is essentially indestructible, it can also be employed for the purpose of optimizing the sort process (See Section VI). The linearity of the relationship between the TOF parameter and actual protoplast diameter can be determined directly through sorting of the protoplasts using a series of nonoverlapping sort windows spaced across the TOF frequency distribution, followed by measurement of protoplast diameters from light micrographs (Galbraith *et al.,* 1988).

2. Protoplast Chlorophyll Measurement

Correlation between the emission of red autofluorescence and protoplast chlorophyll content can be achieved in an manner analogous to that described for TOF/size measurements. Thus, different, nonoverlapping sort windows are placed on the one-dimensional frequency distributions of red autofluorescence emission and defined numbers of protoplasts are sorted. Cellular chlorophyll amounts can then be quantitated through spectrofluorometric analysis (Galbraith *et al.*, 1988).

V. Critical Aspects of the Procedures

A. Sample Preparation

The first critical aspect relates to the biological materials under study. Familiarity with procedures of preparation and culture of protoplasts is essential for their successful use in flow analysis and sorting. Protoplasts are perhaps the most fragile type of eukaryotic cell yet subjected to flow techniques. Correspondingly, the researcher must learn to recognize features of protoplasts that indicate their viability before attempting work in flow. For this, a good light microscope equipped with differential interference contrast and epifluorescence capabilities is essential. Optical features of viable protoplasts include a turgid cytoplasm, a perfectly spherical shape, and the observation of fluorochromasia following FDA staining and of cytoplasmic streaming. Further, the organelles should be evenly dispersed through the cytoplasm and, for protoplasts prepared from cell suspension cultures, the nucleus should be well defined, centrally located, and attached to the peripheral cytoplasm by well-defined cytoplasmic strands.

In manipulation of protoplasts, care should be taken to avoid mechanical damage. Resuspension steps using Pasteur pipettes should involve minimal force. Gross changes in the osmotic potential of the various solutions encountered by the protoplasts should also be minimized, noting that different procedures of protoplast preparation often call for different levels of osmotica. One of the most critical aspects of protoplast preparation concerns the physiological status of the donor plant tissues. Variation in light intensity, day length, plant stage, and leaf number can affect protoplast yield and viability. Plantlets grown as described in sterile culture *in vitro* lack an extensive cuticle, and this appears to facilitate enzymatic degradation of the cell walls. There are reports that suggest the act of excision of leaves from these plants itself adversely affects the ability to produce protoplasts from further leaves excised at later times (Walker-Simmons *et al.*, 1984; Farmer and Ryan, 1992). Consequently, we routinely discard the plantlets after three leaf harvests and typically do not subculture them more than twice prior to starting over with new seed-derived plantlets. Finally, it is a truism to state that no one set of experimental conditions is applicable to all types of protoplasts. Differences in the composition of the various media are encountered in reports concerning the successful culture of different protoplast types. A sensible approach to flow analysis and sorting of different protoplast types is to employ optimal culture media as the sheath fluid.

B. Identification of the Population of Interest

The second critical aspect of the procedures relates to the apparently simple task of defining the objects of interest within the populations that will be subjected to flow analysis and sorting. For pollen, this is not a problem, since they usually comprise the majority of the particles within the suspension. However, protoplasts contain a very large number of internal structures and organelles which when released from broken cells can provide light scatter and autofluorescence signals during flow analysis. Use of techniques for purification of the protoplasts is recommended, such as sucrose gradient centrifugation. Protoplast purification may not always be sufficient to eliminate this problem. For example, protoplasts from mature tobacco leaves contain about 70 chloroplasts (Harkins and Galbraith, 1984). Therefore, a population of 90% intact protoplasts (which would be considered excellent according to most tissue culture standards) would produce flow cytometric frequency distributions comprising 89% free chloroplasts and 11% intact protoplasts. For most applications, therefore, the desired objects (intact protoplasts) comprise a significant minority of the particles that are being analyzed. Triggering on fluorescence rather than light scatter can be employed to eliminate the contribution of nonfluorescent, light-scattering particles. Acquisition of frequency distributions on logarithmic scales can be helpful, since it tends to highlight the presence of discrete populations. Linear frequency distributions can then be accumulated by gating on the log distributions. Finally, use of the fluorescence microscope is strongly recommended in order to determine whether appropriate staining of the protoplasts has been achieved.

C. Selection of Flow Tips

As previously noted, plant protoplasts are generally much larger than animal cells, and the sizes of those commonly employed in flow cytometric applications (30–60 μm) approach the diameters of the standard flow tips (60–75 μm). Evidently, it is important to employ flow tips larger than the cells to be analyzed, although use of flow tips much larger in diameter could lead to errors associated with the hydrodynamic process of centering the particles in the fluid stream. Larger flow tips produce larger rates of volume accumulation of sheath fluid, and this requires larger sheath tanks. The lower transducer drive actuation frequencies required by larger tips leads to the production of larger droplets, and the deflection of these droplets by the conventional defection assemblies and power supplies may not be adequate. This problem can be alleviated by increasing the length of the deflection plate assemblies or by increasing the high voltages applied to these plates.

D. Competing Pigments

A feature of higher plants is the presence of a variety of intracellular pigments, particularly those associated with photosynthesis. The presence of these pigments complicates flow analysis of protoplasts, depending on the degree of spectral overlap between these pigments and the specific fluorochromes to be used in the particular study. The researcher should be aware of these potential problems and be prepared to deal with them through appropriate selection of fluorochromes, fixation procedures, and excitation and emission wavelengths.

E. TOF Limitations

Theoretical considerations indicate that the relationship between protoplast diameter and the PW-TOF parameter will increasingly deviate from the linearity as the size of the protoplast approaches that of the beam. For the EPICS system using standard optics, deviation becomes noticeable at a protoplast diameter of about 15 μm (Galbraith *et al.*, 1988). Below this point, correct sizing can still be achieved, although it is necessary to use a series of appropriately sized standard particles in order to properly calibrate these measurements. We have not empirically established the upper limit to TOF size analysis which will be dictated by the maximal time domain that can be accommodated by the TOF circuitry. However, brief calculations suggest that this will not be a limitation: for the standard EPICS hardware, the upper limit of TOF analysis is 40 μsec. For large particle sorting using the larger flow tips, the system pressure must be lowered in order to obtain a point of droplet breakoff above the position of the deflection assembly (Harkins and Galbraith, 1987). For the 200-μm flow tip, the highest fluid stream velocity is typically 10.8 m/sec. This means that the effective upper limit for TOF measurements set by the electronics is around 400 μm, which is an order of magnitude larger than tobacco leaf protoplasts.

Biological factors might be expected to influence the linearity of the TOF analysis. For example, deformation of cellular shapes at the laser interrogation point due to hydrodynamic forces experienced within the flow cytometer might be expected to introduce errors in sizing; our data indicate that this is not the case for protoplasts (Galbraith *et al.*, 1988). On the other hand, the presence of chloroplasts within mesophyll protoplasts introduces errors in protoplast sizing when TOF measurements are performed based on FDA fluorescence, but not when they are based on chlorophyll autofluorescence (Galbraith *et al.*, 1988). This is probably due to quenching to FDA fluorescence by the highly absorbent pigments within the chloroplasts.

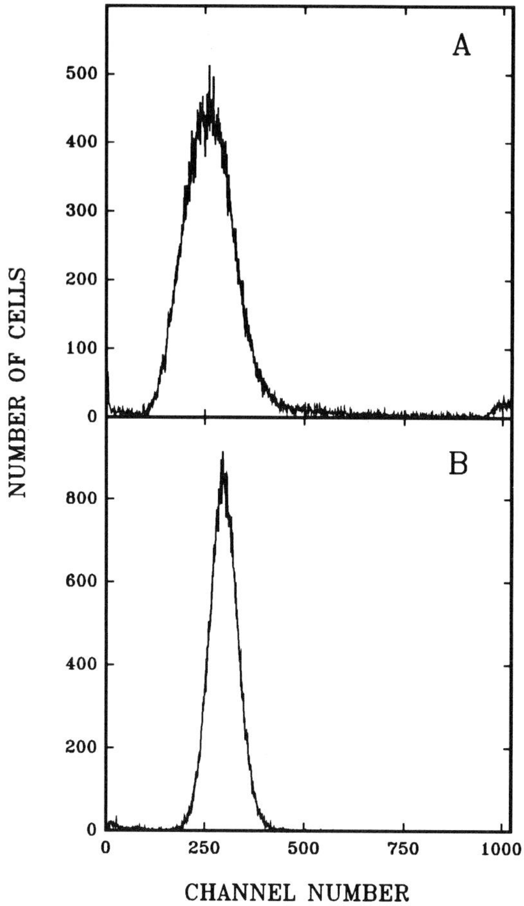

Fig. 1 Flow cytometric analysis of (A) autofluorescence (PMT2; 505–545 nm) and (B) 90° light scatter from *Lycopodium* spores. The one-dimensional frequency distributions were accumulated to a total count of 70,000. The mean and CVs (%) for the distributions were 255 (21.5%) for autofluorescence and 295 (11.1%) for light scatter.

F. Protoplast Sorting

Critical to the successful sorting of protoplasts are the following elements: the generation of high-quality frequency distributions in which the location of the desired protoplast populations are well defined, the optimization of sort parameters in order to provide accurate and high-efficiency sorting of the desired protoplasts, the provision of sterile conditions, and the establishment of conditions for protoplast growth after sorting. Optimization of sort parameters is most easily achieved with large flow cell tips, coupled to the use of appropriate standard particles. These are detailed in further sections. Sterile conditions are achieved through the use of a combination of liquid sterilants and autoclaving. Use of a single-cell deposition device to sort protoplasts into 96-well plates also greatly facilitates maintenance of sterile conditions. Inclusion of certain antibiotics (for example, penicillin derivatives at 50–500 mg/l) can be helpful; these are innocuous to protoplasts. In terms of optimization of culture conditions after sorting, maintenance of a minimal cell density is important. This can be

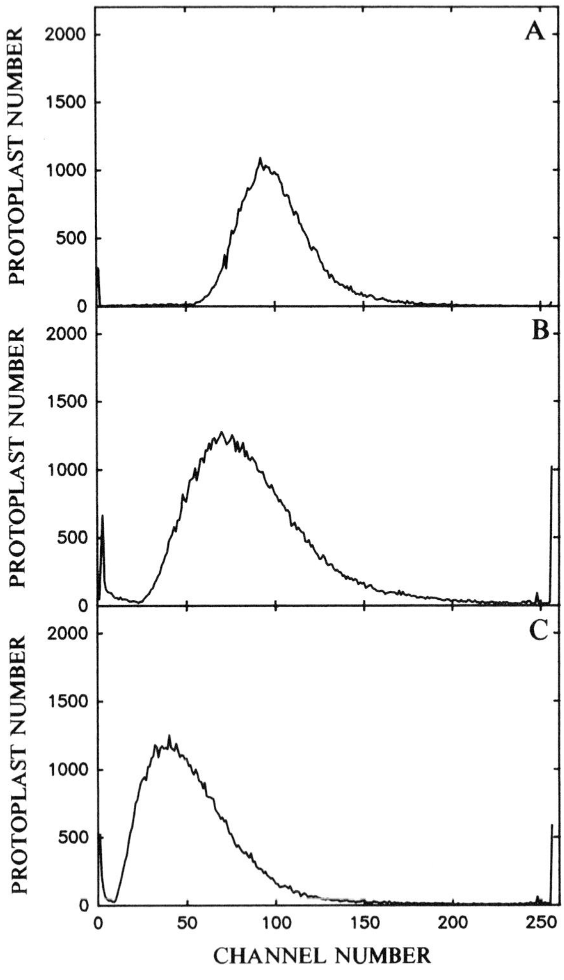

Fig. 2 Flow cytometric analysis of size according to time of flight. One-dimensional frequency distributions were accumulated to a total count of 20,000, corresponding to (A) TOF, (B) TOF^2, and (C) TOF^3 signals derived from chlorophyll autofluorescence of leaf protoplasts. The means and CVs for the distributions are 98 (18.4%), 80 (33.3%), and 48 (47.6%), respectively. From Galbraith (1990).

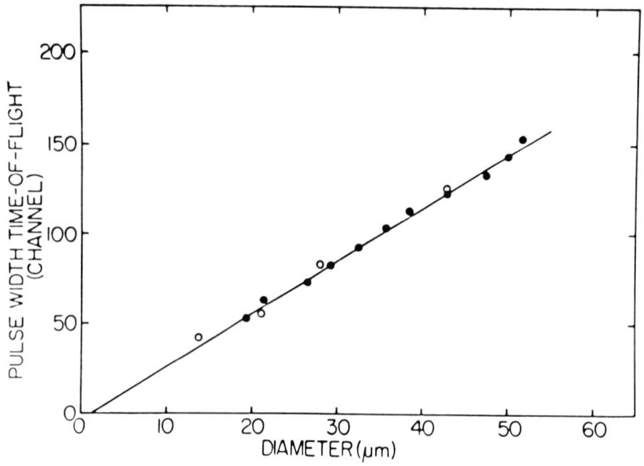

Fig. 3 Correlation between the pulse width time-of-flight parameter and protoplast or pollen/spore diameters, based on chlorophyll or aniline blue-induced fluorescence, respectively. Protoplasts, filled circles; pollen/spores, open circles. From Galbraith *et al.* (1988).

achieved either by sorting large numbers of protoplasts (>1000) within the wells of the culture plates or through the use of conditioned media or feeder cells (Afonso *et al.*, 1985). Feeder cells can be provided as a liquid cell suspension or can be embedded in agarose, with or without a physical barrier. Finally, it should be noted that some types of tissue culture plates are toxic to protoplasts.

VI. Controls and Standards

We routinely employ pollen or *Lycopodium* spores for setting up and standardizing the sort process (Table I). It should be reemphasized that correct selection of sort parameters for fragile plant protoplasts requires the use of particles that approximate the size of the protoplasts, since conditions for optimal sorting of small particles may not necessarily be optimal for large cells (Harkins and Galbraith, 1987).

VII. Instruments

The above procedures have been worked out using the Coulter EPICS and ELITE systems. Large particle sorting has also been achieved using a Becton–Dickinson FACStar Plus equipped with a 200-μm flow tip. The procedures should apply to other flow instruments operating on the same hydrodynamic principles.

VIII. Results

Use of these procedures is illustrated with five examples.

A. Analysis of a Large Particle Standard: *Lycopodium*

Typical flow cytometric analyses of the optical characteristics of *Lycopodium* spores are given in Fig. 1. The Coulter ELITE was used to collect one-dimensional histograms of green autofluorescence and 90° light scatter. Unimodal, near-normal distributions are observed for both parameters.

B. Analysis of the Dimensions of Plant Protoplasts and Cells

Characterization of the TOF distributions of populations of tobacco leaf protoplasts is presented in Fig. 2A. This is also a near-normal distribution with a mode channel of 98. A correlation between the TOF parameter and actual protoplast size can be obtained by sorting protoplasts using a series of defined,

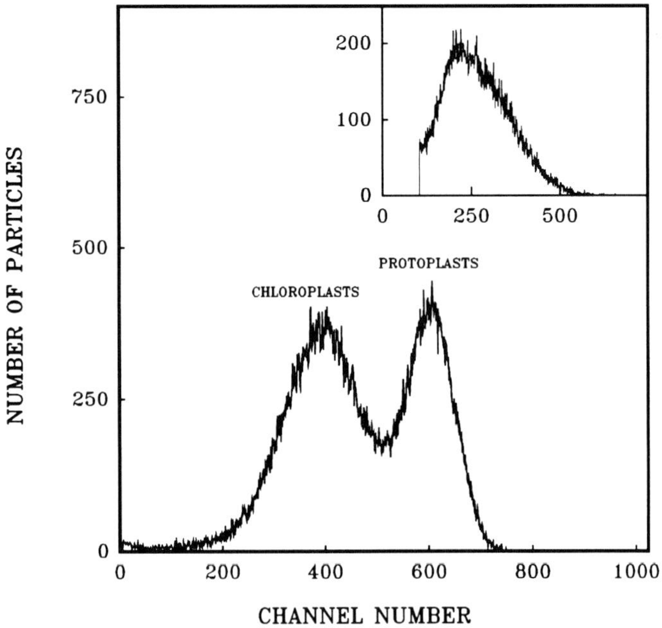

Fig. 4 Uniparametric flow analysis of the emission of chlorophyll autofluorescence from a suspension of tobacco leaf protoplasts, using a logarithmic abscissa. The lower peak corresponds to chloroplasts released from broken protoplasts. Insert: The logarithmic distribution is gated to include for analysis only the intact protoplasts, and the resultant distribution is displayed on a linear scale.

nonoverlapping sort windows. Protoplast sizes are then subsequently measured using light microscopy. Figure 3 illustrates the linear relationship obtained in this manner (Galbraith *et al.*, 1988). Use of standard particles allows the TOF/size relationship to be calibrated rapidly; thus, for the data presented in Fig. 2A, the true size of the protoplasts is 41 μm.

In order to produce real-time accumulations of distributions corresponding to surface area and volume, I have previously described an analog circuit that can be used to square and/or cube the TOF signal (Galbraith, 1990). This circuit operates with an accuracy that is not less than that of the EPICS MDADS. Analysis and processing of real-time distributions of TOF signal produced by protoplasts are illustrated in Figs. 2B and 2C. Since protoplasts are spherically symmetrical, these distributions provide a measure of protoplast surface area and volume. The major advantage of this circuit is its low cost. Future developments in this area are likely to involve manipulation of the digitized pulses, rather than operating in the analog realm, as the digital approach offers more flexibility and accuracy and less performance degradation over time.

C. Analysis of the Chlorophyll Contents of Leaf Protoplasts

Uniparametric analysis of the autofluorescence emission from a tobacco leaf protoplast suspension is presented in Fig. 4. Two peaks of fluorescence are

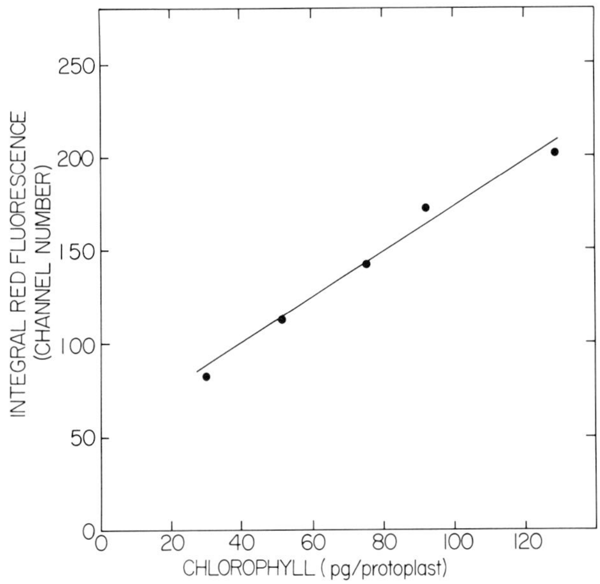

Fig. 5 Correlation between the chlorophyll autofluorescence emission and the protoplast chlorophyll content. From Galbraith *et al.* (1988).

observed on the logarithmic scale. The lower peak corresponds to chloroplasts and the upper to intact protoplasts (Harkins and Galbraith, 1984). This profile can be gated to exclude free chloroplasts and linear autofluorescence distributions obtained from the protoplasts, as shown in the insert to Fig. 4. In order to obtain a correlation between fluorescence emission and chlorophyll content, this linear distribution is divided into a series of nonoverlapping sort windows. In our previous work, we selected five windows that were 15 channels wide and that were spaced by intervals 5 channels wide (Galbraith *et al.*, 1988). Specific numbers of protoplasts were sorted and their chlorophyll content was measured by spectrofluorometry. Protoplast chlorophyll content and protoplast autofluorescence yield were highly correlated (Fig. 5).

D. Combined Chlorophyll/TOF Analysis of Leaf Protoplasts

The tissues of dicotyledonous leaves comprise a variety of different cell types. Predominant are the photosynthetic tissues of the mesophyll, which contain large numbers of mature chloroplasts, and the nonphotosynthetic tissues of the epidermis and perivascular parenchyma. Biparametric flow analysis of protoplasts prepared from leaves according to TOF and chlorophyll content readily permits an identification of two populations of protoplasts (Fig. 6). As

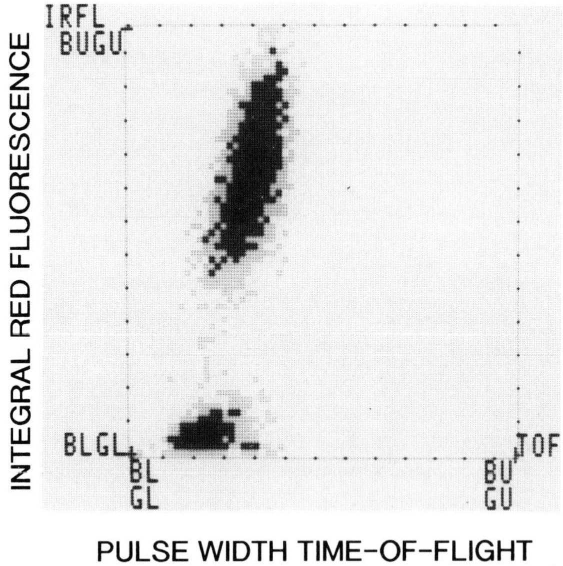

Fig. 6 Contour analysis of two-parameter measurements of the chlorophyll autofluorescence versus pulse width time-of-flight characteristics of leaf protoplasts. FDA fluorochromasia was used as the source of signals for the TOF analysis. The contour levels correspond to 5, 15, and 50% of the peak channel. From Galbraith *et al.* (1988).

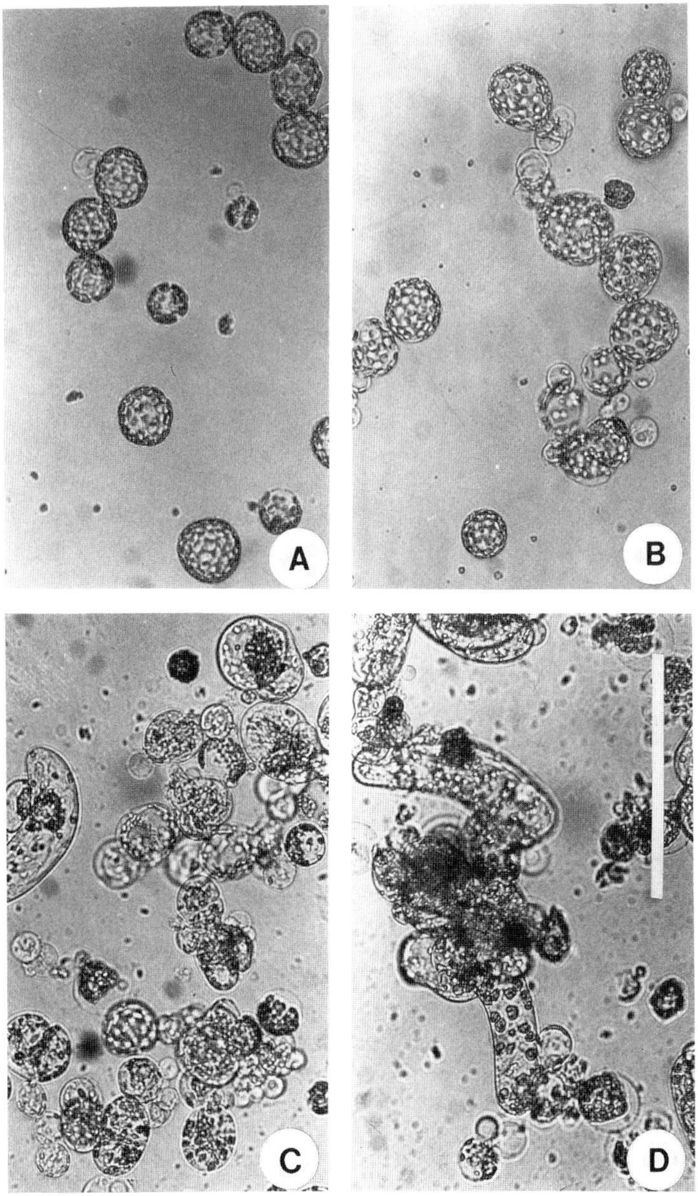

Fig. 7 Growth in culture of sorted *N. tabacum* protoplasts (A) after 1 day, (B) after 2 days, (C) after 5 days, and (D) after 12 days. (200×). Bar represents 150 μ. From Galbraith (1984).

would be expected from the uniparametric analysis (Fig. 5), the mesophyll protoplasts exhibit a broad range of chlorophyll content, whereas those from the epidermis and perivascular parenchyma are essentially devoid of chlorophyll. It should be noted that for accurate sizing of the mesophyll protoplasts, TOF processing of the chlorophyll autofluorescence signal rather than the FDA fluorochromasia signal is required (Galbraith *et al.*, 1988); obviously, it is not possible to analyze nonphotosynthetic protoplasts in this way. Thus for combined, accurate TOF/size analyses of these different protoplast populations, two independent TOF modules would be required.

E. Sorting and Culture of Leaf Protoplasts

Under optimal conditions, protoplasts can be sorted at efficiencies approaching 100% with no loss of viability (Afonso *et al.,* 1985; Harkins and Galbraith, 1984,1987). The integrity of mesophyll protoplasts can be conveniently and continuously assayed by monitoring the numbers of free chloroplasts in the protoplast suspension (Fig. 4). Under optimal conditions, the sorted protoplasts are metabolically active and will eventually resynthesize cell walls and enter into cell division (Fig. 7). We have successfully sorted and cultured protoplasts following induced fusion (Afonso *et al.*, 1985) and following transfection (Harkins *et al.,* 1990). Finally, we have been able to sort and culture protoplasts derived from both the photosynthetic and the nonphotosynthetic tissues of the leaf with equal efficiencies and to demonstrate cell-specific patterns of transgene expression within these sorted protoplasts (Harkins *et al.*, 1990).

Acknowledgments

I thank Georgina Lambert for valuable assistance. This work was supported by the National Science Foundation.

References

Afonso, C. L., Harkins, K. R., Thomas-Compton, M., Krejci, A. E., and Galbraith, D. W. (1985). *Bio Technology* **3,** 811–815.
Bromova, M., and Knopf, U. C. (1991). *Plant Sci.* **74,** 127–133.
Donnelly, R. J., and Glaberson, W. (1986). *Proc. R. Soc. London, Ser. A* **290,** 547–556.
Farmer, E. E., and Ryan, C. A. (1992). *Plant Cell* **4,** 129–134.
Galbraith, D. W. (1984). *In* "Cell Culture and Somatic Cell Genetics of Plants" (I. K. Vasil, ed.), pp. 433–447. Academic Press, New York.
Galbraith, D. W. (1989). *Int. Rev. Cytol.* **116,** 165–227.
Galbraith, D. W. (1990). *In* "Methods in Cell Biology" (Z. Darzynkiewicz and H. Crissman, eds.), Vol. 33, pp. 527–547. Academic Press, San Diego.
Galbraith, D. W., and Galbraith, J. E. C. (1979). *J. Plant Physiol.* **93,** 149–158.
Galbraith, D. W., and Mauch, T. J. (1980). *J. Plant Physiol.* **98,** 129–140.
Galbraith, D. W., Afonso, C. L., and Harkins, K. R. (1984). *Plant Cell Rep.* **3,** 151–155.

Galbraith, D. W., Harkins, K. R., and Jefferson, R. A. (1988). *Cytometry* **9,** 75–83.
Gantet, P., Hubac, C., and Brown, S. C. (1990). *Plant Physiol.* **94,** 729–737.
Glimelius, K., Djupsjöbacka, M., and Fellner-Felldeg, H. (1986). *Plant Sci.* **45,** 133–141.
Hammatt, N., Lister, A., Blackhall, N. W., Gartland, J., Ghose, T. K., Gilmour, J. B., Power, J. B., Davey, M. R., and Cocking, E. C. (1990). *Protoplasma* **154,** 34–44.
Harkins, K. R., and Galbraith, D. W. (1984). *Physiol. Plant* **60,** 43–52.
Harkins, K. R., and Galbraith, D. W. (1987). *Cytometry* **8,** 60–70.
Harkins, K. R., Jefferson, R. A., Kavanagh, T. A., Bevan, M. W., and Galbraith, D. W. (1990). *Proc. Natl. Acad. Sci. U.S.A.* **87,** 816–820.
Kachel, V., Fellner-Feldegg, H., and Menke, E. (1990). *In* "Flow Cytometry and Sorting" (M. R. Melamed, T. Lindmo, and M. L. Mendelsohn, eds.), 2nd ed. pp. 27–44. Wiley-Liss, New York.
Lindsey, K. (1991). "Plant Tissue Culture Manual: Fundamentals and Applications." Kluwer Acad. Publ., Dordrecht, The Netherlands.
Pauls, P. K., and Chuong, P. V. (1987). *Can. J. Bot.* **65,** 834–838.
Pechan, P. M., Keller, W. A., Mandy, F., and Bergeron, M. (1988). *Plant Cell Rep.* **7,** 396–398.
Sjödin, C., and Glimelius, K. (1989). *Theor. Appl. Genet.* **77,** 651–656.
Sundberg, E., and Glimelius, K. (1991). *Plant Sci.* **79,** 205–216.
Verhoeven, H. A., and Sree Ramulu, K. (1991). *Theor. Appl. Genet.* **82,** 346–352.
Walker-Simmons, M., Hollaender-Czytko, H. J., Andersen, J. K., and Ryan, C. A. (1984). *Proc. Natl. Acad. Sci. U.S.A.* **81,** 3737–3741.
Walters, T. W., Mutschler, M. A., and Earle, E. D. (1992). *Plant Cell* **10,** 624–628.
Zhang, G., Campenot, M. K., McGann, L. E., and Cass, D. D. (1992). *Plant Physiol.* **99,** 54–59.

CHAPTER 32

Use of Fluorescence-Activated Cell Sorting for Rapid Isolation of Insect Cells Harboring Recombinant Baculovirus

Maja A. Sommerfelt* and Eric J. Sorscher†,[1]

* National Centre for Research in Virology
University of Bergen
Bergen High Technology Centre
N-5020 Bergen
Norway

† Department of Physiology and Biophysics
University of Alabama at Birmingham
Birmingham, Alabama 35294

[1] To whom correspondence should be addressed.

I. Introduction

The baculovirus expression system provides a very powerful means of obtaining high-level expression of foreign genes in eukaryotic insect cells (Fraser, 1992; Luckow and Summers, 1988; Miller, 1988). The system exploits the Baculoviridae group of insect viruses, most often *Autographa californica* nuclear polyhedrosis virus (AcNPV), because these viruses contain very active late gene promoters. Replacement of a viral gene with a foreign gene under the control of such a late gene promoter results in the production of stable recombinant viruses that can be grown to high titer. Such recombinant viruses, on infection of a suitable insect cell line, usually *Spodoptera frugiperda* (Sf9) or *Trichoplusia ni* (Ti5), result in the synthesis of large amounts of foreign protein (up to 50% total cellular protein). In addition, as this is a eukaryotic expression system, the foreign protein is synthesized with many of the relevant post-translational modifications (Kang, 1988; Summers and Smith, 1987).

The genome of AcNPV is very large (120-130kb) and so recombinant viruses cannot be generated using conventional cloning techniques. Instead the gene of interest is cloned into a suitable transfer vector under the control of a late gene promoter, flanked by viral sequences. Cotransfection of recombinant vector with wild-type AcNPV DNA results in the generation of recombinant virions following homologous recombination. Such recombination events are rare (0.1% progeny virus being recombinant), and so a system is required to distinguish and isolate recombinant virions from the vast excess of wild-type virus. This has proved in many instances to be difficult and time consuming, although methods are currently being developed to facilitate this process.

The transfer vectors that are most commonly used to generate recombinant baculovirus replace the viral-encoded polyhedron gene with a foreign gene of interest. Expression of the polyhedron gene from wild-type AcNPV results in the formation of occlusion bodies (occ) also known as polyhedra within the nucleus of infected cells. Polyhedra can be visualized easily using light microscopy and confer an occ^+ phenotype to the virus-infected cells. Replacement of the polyhedron gene with a heterologous gene results in recombinant virions that lack polyhedra and virus-infected cells acquire an occ^- phenotype. This phenotypic difference between wild-type and recombinant AcNPV has provided the standard means of identifying cells infected with recombinant virions; however, because these viruses are identified by their lytic effect on cells in a plaquc assay, identification of recombinant occ^- plaques among a vast array of occ^+ plaques requires experience and patience. The process of identifying recombinant plaques has been simplified in recent years with the introduction of transfer vectors that mediate coexpression of β-galactosidase with the foreign gene of interest. Recombinant AcNPV therefore produce blue occ^- plaques in the presence of the β-galactosidase substrate 5-bromo-4-chlor-3-indolyl-β-D-galactopyranoside (X-gal) (Vialard *et al.*, 1990), while wild-type virus plaques remain occ^+ and white. Although this greatly simplifies the identification of recombinant virus, several rounds of plaque purification are still required to obtain a pure stock and this may take several weeks or months to accomplish.

We describe a means of enhancing the blue–white selection system by efficiently sorting cells infected with recombinant AcNPV from the wild-type or uninfected cell population using fluorescence-activated cell sorting (FACS). An alkyl-derivatized fluorescein di-β-D-galactopyranoside is used as an alternate substrate for β-galactosidase. The hydrolysis product is hydrophobic and remains intracellular. Recombinant virus can be obtained from the highly fluorescent cellular fraction allowing for the production of a pure recombinant viral culture within 2–3 weeks of insect cell transfection.

II. Application

The method of rapid purification of recombinant baculovirus using fluorescence-activated cell sorting described here relies on the use of transfer vectors that coexpress β-galactosidase. Although this approach utilizes AcNPV, the methodology should be broadly applicable to members of the Baculoviridae family other than AcNPV.

This procedure provides a quick and simple means of separating and thereby enriching the population of cells infected with recombinant AcNPV. Such enrichment of recombinant-infected cells is necessary because the frequency of homologous recombination is so low and the presence of wild-type virus can be overwhelming. The utilization of β-galactosidase coexpression for the identification and isolation of rare cells in a population using FACS analysis has also been useful in other systems (Jasin and Zalamea, 1992).

The method described here has the potential to be used further in the formation of double-recombinant AcNPV. Once one foreign gene is stably expressed from the polyhedron promoter, with β-galactosidase being expressed from the p10 promoter, viral DNA can be isolated from this recombinant and used in cotransfections to replace the β-galactoside gene with another gene of interest. In this instance, the population of low fluorescenating cells will be desired. The use of double recombinants is gaining popularity and technically it is easier to have a single recombinant expressing both genes than doubly infecting cells with two distinct recombinants. At present, the means to generate double recombinants have involved the construction of dual-transfer vectors that have multiple polyhedron promoters but do not involve the coexpression of β-galactosidase (Takehara *et al.*, 1988; French and Roy, 1990).

III. Materials

AcNPV was a gift from R. D. Possee (NERC Institute of Virology and Environmental Microbiology, Oxford, U.K.). Viral DNA can be isolated according to the method of Summers and Smith (1987). Virus is stable when stored at 4°C.

Sf9 cells (ATCC CRL1711) or Sf21 (Vaughn *et al.*, 1977) are grown in Graces insect medium (Gibco Grand Island, N.Y.) supplemented with 10% fetal bovine serum (FBS) at 27°C.

Transfer vector pJVETLZ (gift from Chris Richardson, Biotechnology Research Institute, NRC Canada, Montreal) carries the β-galactosidase gene under the control of the p10 late gene promoter and a Nhe1 cloning site just downstream of the polyhedron promoter. This vector is equivalent to the BlueBac vector produced by InVitrogen (San Diego, CA). Cesium chloride gradient-purified recombinant vector DNA is used for transfection and is stored at −20°C.

Lipofectin (Gibco/BRL, Life Technologies, Gaithersburg, MD) is stored at 4°C.

Sterile double-distilled water is used for all transfections. Imagene alkyl-derivatized fluorescein di-β-D-galactopyranoside substrate (Molecular probes, Eugene, OR) is used at a final concentration of 33 μM and is stored at −20°C.

X-gal (Sigma Chemical Co., St. Louis, MO) is made as a 100× stock solution (20 mg/ml) in DMSO and stored protected from light at 4°C.

Agar overlay is composed of molten 3% Sea Plaque agarose in water (FMC BioProducts, Rockland, ME) mixed with an equal volume of Graces insect medium; X-gal is added to give a final concentration of 0.2 mg/ml. The overlay, once made, must be used immediately and cannot be stored. Sea Plaque agarose (3% in water) is sterilized by autoclaving and is stored at room temperature.

Neutral red (Sigma) is made as a 0.5% solution in water and sterilized by autoclaving. A 1/40 dilution of this in medium is used to stain cells.

IV. Methods

A. Transfection and Isolation of Recombinant AcNPV

1. A T25 flask is seeded with 2×10^6 Sf9 or Sf21 cells in a 5-ml volume. The cells are allowed to settle overnight at 27°C.

2. The cells are washed in serum-free medium and finally overlayed with a volume in 1 ml.

3. Wild-type AcNPV genomic DNA (100 ng) is combined with 1 μg recombinant vector DNA in a total volume of 12 μl in sterile double-distilled water. Lipofectin, 8 μl, is mixed with 4 μl sterile double-distilled water and added to the DNA mixture. After 15 min incubation at room temperature, the DNA/lipofectin mixture is added dropwise to the cells.

4. After overnight incubation (5–24 hr later), 1 ml serum containing medium is added.

5. The cells are harvested for FACS analysis on the 7th day post-transfection.

B. Fluorescence-Activated Cell Sorting and Plaque Purification

1. Medium is removed from the transfected cells and 1 ml fresh serum-free medium containing 3 μl Imagene alkyl-derivatized fluorescein di-β-D-galactopyranoside substrate (final concentration, 33 μM) is applied and incubated for 15 min at room temperature.

2. The cells are then washed from the sides of the flask and subjected to FACS analysis.

3. Prior to FACS analysis the sorter is calibrated with fluorescence standard beads (Becton–Dickinson, San Jose, CA) and sterilized with 20% Clorox in normal saline for 60 min. The sheath fluid in the FACS analysis is serum-free Graces insect medium. Serum-free medium (300 μl) is added to the collecting tube prior to sorting.

4. Cells with the greatest fluorescence intensity (termed highly fluorescent cells) are separated from the cell population and subject to either of two protocols that lead to the rapid purification of recombinant virus.

Protocol A

1. Forty-five cells from the highly fluorescent population are cocultivated with 2×10^6 cells in a T25 tissue culture flask.

2. Seventy-two hours later, the medium is harvested, filtered through a 0.45 μM syringe top filter, and titrated to determine the proportions of recombinant and wild-type virus.

Protocol B

1. Of the highly fluorescent cells, 20, 200, or 2000 are added directly to 2×10^6 uninfected cells in a six-well tray.

2. Four hours later, the medium is removed and the cells overlaid with 4 ml agarose containing X-gal.

3. The agarose is allowed to polymerize at room temperature and then 2 ml medium is added to each well. The cells are incubated at 27°C for 2 to 7 days and observed for the presence of blue plaques.

C. Titration of Infected Cell Supernatants

1. Six-well trays are seeded with 2×10^6 Sf9 or Sf21 cells/well and allowed to settle overnight at 27°C.

2. The medium to be titered is serially diluted (10^{-1}, 10^{-2}, 10^{-3}, etc.) in 0.4-ml volumes, and each 0.4-ml dilution is used to cover one of the wells of a six-well tray.

3. Following 1 hr incubation at 27°C the inoculum is removed and 4 ml/well molten agarose overlay is added gently to the cells.

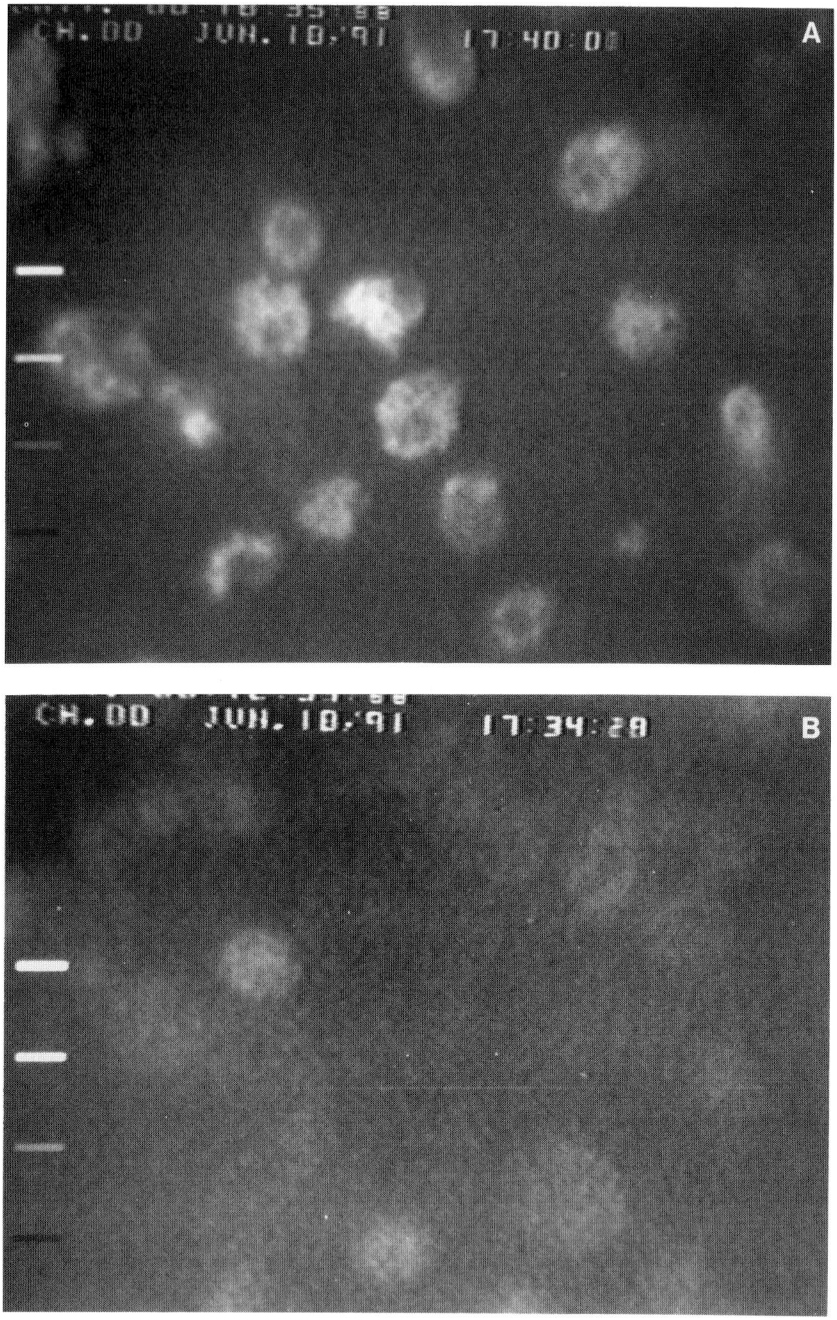

Fig. 1 The use of the fluorescent β-galactosidase substrate to identify cells infected with recombinant AcNPV. (A) Insect cells infected with a recombinant AcNPV expressing β-galactosidase and a nucleotide binding domain of the cystic fibrosis transmembrane conductance regulator (CFTR) (Riordan *et al.*, 1989). The cells were grown on glass coverslips and infected using a multiplicity

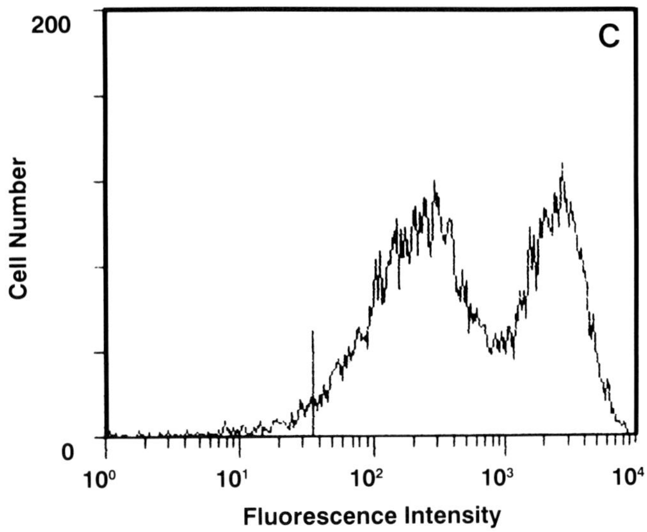

Fig. 1 (*Continued*) of infection of 2. Two days later, the cells were washed and incubated for 15 min in serum-free medium containing 33 μM fluorescent β-galactosidase substrate (Section III). The coverslips were placed in a chamber mounted on a Zeiss IM-35 inverted microscope. Images were relayed by means of a multichannel plate image intensifier (Videoscope International ICS-1380, Washington, DC) to a Newicon video camera (MTI, Michigan City, IN) and were recorded on a high-resolution optical memory disc recorder (Panasonic). (B) Insect cells infected with wild-type AcNPV (not expressing β-galactosidase). Only a weak fluorescent signal can be detected. (C) Fluorescence-activated cell sorting of a cell population comprised of equal numbers of wild-type and recombinant AcNPV-infected insect cells (as shown in A and B). The two cell populations are clearly distinguished and can be separated on the basis of their fluorescence intensity.

4. When the agarose has polymerized, 2 ml medium is added and the cells are incubated at 27°C. Three days later the trays can be examined for the presence of plaques.

5. To accentuate the presence of contaminating wild-type AcNPV-derived plaques, the cells can be stained with neutral red. The 2 ml medium is removed from each well and replaced with 2 ml medium containing neutral red. After 30 min incubation at 27°C, the medium is removed and the cells can be examined for the presence of white wild-type plaques.

D. Plaque Purification

1. An isolated blue plaque is chosen. A Pasteur pipette is used to stab the plaque and remove the agarose plug immediately above the plaque.

2. The agarose plug is placed in 1 ml medium and stored overnight at 4°C to allow elution of virus from the agarose.

3. The medium containing virus is then titrated to confirm homogeneity of recombinant virus plaques and to obtain the virus titer, which is typically 10^2 plaque forming units (pfu) per ml.

V. Critical Aspects of the Procedure

Once the population of cells expressing recombinant AcNPV has been enriched and virus obtained by either protocol A or B, the blue plaques must also be checked visually to confirm that they have an occ$^-$ phenotype.

Care must be taken to ensure that the cells are not killed because the agarose overlay is applied while still too hot (above 40°C). The cells in six-well trays must also not be left without medium for more than a few minutes because they may dry out.

VI. Instruments

A FACStar Plus instrument (Beckton–Dickinson) was used. The parameters for FACS analysis are 535 emission wavelength and an excitation of 488 nm. DF 530/30 filters (Omega Optics, Brattleboro, VT) were used. Fluorescence amplification was logarithmic. A 70-μm nozzle was used and sheath fluid passed through a 0.45-μm filter prior to the addition of cells.

VII. Results and Discussion

The rapid purification of recombinant AcNPV using fluorescence-activated cell sorting has been successfully used to select and purify recombinant baculovirus coexpressing β-galactosidase with a low-density lipoprotein receptor (LDLR), and the entire and also the nucleotide binding domain of the cystic fibrosis transmembrane conductance regulator (CFTR) (Peng *et al.*, 1993).

Figure 1 shows the use of the fluorescent β-galactosidase substrate to demonstrate the presence of cells expressing recombinant AcNPV (Fig. 1A) compared to cells infected with wild-type AcNPV (Fig. 1B). If cells infected with wild-type AcNPV and recombinant AcNPV are combined, they can be easily separated into two distinct cell populations on the basis of their fluorescence intensity following fluorescence-activated cell sorting (Fig. 1C). This shows that the process of fluorescence-activated cell sorting described provides a very efficient means of separating cell populations without the use of fluorescently tagged antibodies commonly used to separate cell populations expressing different cell-surface antigens.

Tables I and II show that cells derived from the highly fluorescent cell population produce exclusively recombinant AcNPV which can be purified from cell

Table I
Enrichment of Recombinant AcNPV by Fluorescence-Activated Cell Sorting Using Protocol A

	Dilution of supernatant		
	Undiluted	10^{-1}	10^{-2}
Highly fluorescent cells			
Blue plaques	29	5	0
White plaques	0	0	0
Less-fluorescent cells			
Blue plaques	0	0	0
White plaques	>100	>100	>100

Titration of cell supernatants harvested from cultures derived from both the highly fluorescent and less-fluorescent cell populations (Section IV, B). The highly fluorescent cell population was enriched for the presence of recombinant virus (in this case expressing the full-length CFTR) and produced only blue plaques. The less-fluorescent cell population produced only white plaques signifying infection with wild-type AcNPV.

Table II
Enrichment of Recombinant AcNPV by Fluorescence-Activated Cell Sorting Using Protocol B

	Number of cells plated		
	2500	200	20
Highly fluorescent cells			
Blue plaques	>100	16	3
White plaques	0	0	0
Less-fluorescent cells			
Blue plaques	0	0	0
White plaques	TNTC	TNTC	TNTC

Note. Cocultivation of highly and less fluorescent cell populations with uninfected insect cells (Section IV,B). Over a period of 7 days, a marked predominance of blue plaques was observed only following cocultivation of the highly fluorescent cell population. Cocultivation using the less-fluorescent cell population resulted in only white (wild-type AcNPV) plaques. TNTC, too numerous to count. The results shown here were obtained using a recombinant AcNPV that coexpressed β-galactosidase with the human LDL receptor.

supernatants following short-term culture (protocol A, Section IV,B) or directly from the cells (protocol B, Section IV,B). The less fluorescent population of cells clearly contains predominantly wild-type AcNPV. Supernatants from both the highly fluorescent and less-fluorescent cell populations arrested SF9 cell growth, but occlusion bodies were observed only in the experiments involving the less-fluorescent, wild-type AcNPV producing cells, further indicating substantial enrichment for recombinant virus in the highly fluorescent cell group.

For either of the two protocols used involving any of the three viral constructs (coding for the expression of the full-length CFTR, the human LDL receptor, or a large segment of the CFTR containing the first nucleotide binding domain), a viral shock was obtained with one or two plaque purifications in 2 to 3 weeks in each case.

VIII. Comparison of Methods

The approach described here for the rapid isolation of recombinant AcNPV is one of many developing methods. A similar approach has been used in which magnetic beads coated with anti-transferrin receptor monoclonal antibody have been used to immunoselect insect cells expressing the transferrin receptor (Domingo and Trowbridge, 1988). Although this approach is also rapid it is limited because the magnetic beads must be coated with an antibody specific for the gene product of interest which must also be localized at the cell surface. The rapid purification of recombinant AcNPV using FACS analysis is independent of foreign gene expression of localization of the gene product within or secreted from cells.

The use of linear baculoviral genomic DNA in cotransfection methods has also been described (Kitts *et al.*, 1990) to facilitate homologous recombination and increase the yield of recombinant versus wild-type AcNPV. This method increases the frequency of homologous recombination approximately 10-fold. Although wild-type AcNPV DNA is now commercially available, care must be taken that there is no circular DNA present as this will more readily give a higher yield of wild-type progeny.

Wild-type AcNPV DNA is also now available where restriction sites have been introduced on either side of an essential gene downstream of the polyhedron expression locus. The wild-type virus is viable; however, digestion of this AcNPV DNA removes the essential gene and provides linear DNA more suitable for homologous recombination. The transfer vector used in this system provides the missing essential gene following homologous recombination in addition to a gene of interest cloned under the control of the polyhedron promoter. This method ensures that only recombinant virions are viable and the system is reported to generate recombinant virions at a frequency approaching 100% (Kitts and Possee, 1993).

Another means of selecting recombinant AcNPV has been described (Patel *et al.*, 1992), where the homologous recombination event occurs within yeast cells. The strain of AcNPV utilized has been engineered so that recombinant AcNPV containing yeast cells can be isolated as surviving colonies on selective growth media. DNA from these yeast colonies can then be used to transfect insect cells and produce recombinant virus without the need of plaque purification. The method is reported to take only 10–12 days to obtain a pure recombinant. However, this procedure has not been widely used and is limited by the present low availability of the engineered wild-type AcNPV used in this system. The advantage is that the selection process occurs rapidly in yeast and there is no need for plaque purification.

Fluorescence-activated cell sorting has provided a very powerful means of sorting cell populations. FACS analysis can enrich cell populations using fluorescently tagged antibodies specific for antigens expressed at the cell surface of a subpopulation of cells. This could provide an alternate means of separating cells expressing recombinant AcNPV, although this method is dependent on antibodies specific for the antigen being expressed. However, there may be some antigens which are expressed at an intracellular site or for which no antibody is currently available. Coexpression of β-galactosidase in such cases provides a more desirable means of cell separation as it is not dependent on availability of antigen-specific antibodies.

Acknowledgments

This work was supported by the Cystic Fibrosis Foundation, NIH 5 PO1 DK38518. The authors are grateful to Dr. Eric Hunter and Dr. Kevin Kirk for support and useful suggestions and to Ms. Bonnie Parrott for administrative assistance. Eric J. Sorscher is a Lucille P. Markey Scholar in the Biomedical Sciences and this work was supported by a grant from the Lucille P. Markey Charitable Trust.

References

Domingo, D. L., and Trowbridge, L. S. (1988). *J. Biol. Chem.* **263,** 133386–133392.
Fraser, M. J. (1992). *Curr. Top. Microbiol. Immunol.* **158,** 131–172.
French, T. J., and Roy, P. (1990). *J. Virol.* **64,** 1530–1536.
Jasin, M., and Zalamea, P. (1992). *Proc. Natl. Acad. Sci. U.S.A.* **89,** 10681–10685.
Kang, C. Y. (1988). *Adv. Virus Res.* **35,** 177–192.
Kitts, P. A., and Possee, R. D. (1993). *BioTechniques* **14,** 810–817.
Kitts, P. A., Ayres, N. D., and Possee, R. D. (1990). *Nucleic Acids Res.* **18,** 5667–5672.
Luckow, V. A., and Summers, M. D. (1988). *Virology* **167,** 56–71.
Miller, L. K. (1988). *Annu. Rev. Microbiol.* **42,** 177–199.
Patel, G., Nasmyth, K., and Jones, N. (1992). *Nucleic Acids Res.* **20,** 97–104.
Peng, S., Sommerfelt, M. A., Berta, G., Berry, A. K., Kirk, K. L., Hunter, E., and Sorscher, E. J. (1993). *BioTechniques* **14,** 274–277.
Riordan, J. R., Rommens, J. M., Kerem, B.-S., Alon, N., Rozmahel, R., Grzelezak, Z., Zielenski,

J., Lok, S., Plavsic, N., Chou, J.-L., Drumm, M. L., Iannuzzi, M. C., Colins, F. S., and Tsui, L.-C. (1989). *Science* **245,** 1066–1073.
Summers, M. D., and Smith, G. E. (1987). *Tex., Agric. Exp. Stn.* [*Bull.*] **B1555,** 10–39.
Takehara, K., Ireland, D., and Bishop, D. H. L. (1988). *J. Gen. Virol.* **69,** 2763–2777.
Vaughn, J. L., Goodwin, R.-H., Tompkins, G. J., and McCrawley, P. (1977). *In Vitro* **13,** 213–217.
Vialard, J., Lalumiere, M., Vernet, T., Briedis, D., Alkhatib, G., Henning, D., Levin, D., and Richardson, C. (1990). *J. Virol.* **64,** 37–50.

CHAPTER 33

Flow Microsphere Immunoassay for the Quantitative and Simultaneous Detection of Multiple Soluble Analytes

Thomas M. McHugh
Department of Laboratory Medicine
University of California, San Francisco Medical Center
San Francisco, California 94143

I. Introduction

The technique known as the flow microsphere immunoassay (FMIA) is a quantitative and highly sensitive assay for the simultaneous quantitation of several analytes in a sample. The technique relies upon the ability of the flow cytometer to accurately detect different classes of microspheres based upon a physical characteristic such as size or color. The different microsphere classes are coated with different capture reagents and the fluorescence associated with each microsphere class is quantitated with the flow cytometer. The use of different microsphere classes, each coated with a different capture reagent, then allows for the rapid and simultaneous detection of multiple analytes. This provides the potential to perform multiple assays in the same reaction mixture reducing cost and hands-on time as well as generating results using the same method between analytes.

The FMIA requires that microspheres are coated with a unique capture reagent. While the FMIA is intriguing in that multiple analytes can be assayed simultaneously using multiple microsphere classes most studies have been published using a single microsphere class. The capture reagent varies dependent upon the assay to be performed. Studies have used a variety of capture reagents including antigens (from infectious agents, cell surfaces, or other soluble proteins) to capture antibodies, antibodies to capture soluble antigens, receptors for immunoglobulins, oligonucleotides to capture products from the polymerase chain reaction, proteins for competitive immunoassays with soluble protein, DNA, etc. Once the microsphere classes are coated with the appropriate capture reagent the different microsphere classes can be mixed. The test sample is then added to the mixed microspheres and the reaction is allowed to proceed. Usually after the reaction of the test sample with the microspheres a fluorescent detection reagent is added to the microspheres and allowed to react. The microsphere classes are then analyzed by using flow cytometry and the different classes separated by size or color. Each microsphere class is then analyzed independently of the other microsphere classes. The fluorescence associated with each microsphere class is quantitated and used to indicate the presence or absence of the test substance. Original work in our laboratory was directed at performing the TORCH panel using the FMIA. The TORCH panel is the detection of serum antibodies to *Toxoplasma gondii,* rubella virus, cytomegalovirus (CMV), and herpes simplex virus (HSV). This panel is often used in newborns in whom symptoms are suggestive of a congenitally acquired infection. Often the sample size is small and interpretation of the results from different assays is difficult (e.g., HSV antibody by indirect immunofluorescence, antibody to *T. gondii* by EIA, antibody to CMV by latex agglutination, and antibody to rubella by EIA). Using FMIA the detection of IgG antibodies to these four infectious agents was possible using a single 10-μl sample of serum. Figure 1 shows a diagram of the FMIA for the TORCH panel. There are four different microsphere classes of different sizes shown in Fig. 1 of 4, 5, 7, and 9-μm in diameter. The 4-μm microspheres are coated with antigens from *T. gondii,* the 5-μm microspheres are coated with antigens from CMV, the 7-μm microspheres are coated with antigens from rubella, and the 9-μm microspheres are coated with antigens from HSV. A serum sample is added and allowed to incubate for 1 hr at 37°C after which excess serum is removed by washing; a fluorescent anti-human

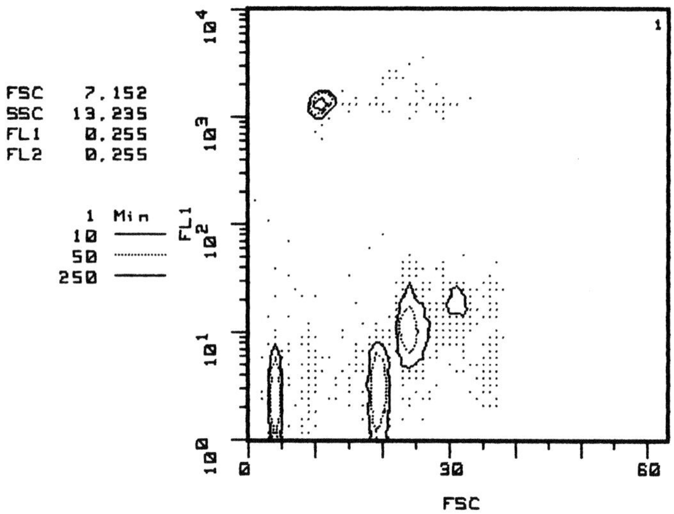

Fig. 1 A two-parameter contour plot of microsphere size as measured by FSC versus FITC fluorescence (FL1). Four major microsphere populations are seen, 4, 5, 7, and 9-μm in diameter (as well as a low proportion of aggregated microspheres seen just to the right of the 9-μm microsphere population). The 5-μm microspheres are coated with CMV and react strongly with the test serum. A low level of reactivity is seen to the 9-μm microspheres which are coated with HSV. The test serum is negative for antibodies to *T. gondii* (4 μm microspheres) and rubella (7 μm microspheres).

immunoglobulin reagent [fluorescein isothiocyanate (FITC) goat F $(ab')_2$ anti-human IgG] is then added and incubates for 0.5 hr at 37°C. The excess fluorescent reagent can be either removed by washing or left in the reaction mixture. The microsphere populations are separated by forward scatter (FSC) versus green FITC fluorescence. The four different sized microspheres can be seen easily and the strong reaction with the 5-μm microspheres indicates positivity for antibody to CMV. The 4- and the 7-μm microspheres are negative (no antibody present to *T. gondii* or rubella) and the 9-μm microspheres are weakly positive (possible low level of antibody to HSV, cross-reactivity between CMV and HSV antigens, or higher background fluorescence due to the size or type of microsphere used). Figure 1 demonstrates a typical use of the FMIA in which microsphere classes are separated by size as measured by forward scatter and a single fluorescent dye is used as the signal. Many variations on this basic technique have been tried and some of these are detailed in the brief review of published applications in Section IV at the end of the chapter.

The first description of the FMIA was by Fulwyler in 1976 in a United Kingdom patent by Coulter Electronics. The possible utility of this idea was published by Horan and Wheeless in 1977 in which they detected rheumatoid factor (human IgM directed against human IgG) using a single-sized microsphere population. The possible use of multiple microsphere classes separated by size is described in this paper. A few papers appeared in the early 1980s (Lisi *et al.* 1982, Horan *et al.* 1979) using this technique and since 1985 a number of new studies have been published (McHugh *et al.* 1986, Saunders *et al.* 1985, Lin *et al.* 1989, Yang *et al.* 1993) focusing on developing assays using the FMIA or using the FMIA as a tool in other investigations.

II. Instrumentation and Materials

A. Flow Cytometer Requirements

In its simplest form the FMIA requires that the flow cytometer be able to detect microspheres of different sizes. This detection can be by either the scattering of incident light (e.g., forward angle light scatter) or electrical impedance sizing. Orthogonal light scatter (SSC) is a relatively poor measure of particle size and is not generally used alone to separate microsphere classes; however, in conjunction with the forward angle light scatter signal, SSC helps to differentiate different microsphere classes. In addtion to the detection of microsphere size, the flow cytometer must be able to detect microsphere-associated fluorescence. Most studies have used microsphere size and one fluorescent channel for the FMIA. As such, the measurement needs of the flow cytometer are simple. This has prompted a number of groups to discuss designing a small and simple instrument which would lend itself well to the use of FMIA. However, the flexibility of most flow cytometers can be advantageous to studies employing the FMIA. The combined use of FSC and SSC allows for the complete discrimination of microspheres which may be very close in size but vary in their ability to scatter light at right angles. In the FMIA if different microsphere classes are used for the simultaneous detection of multiple analytes it is important to reduce or eliminate overlap among the microsphere populations. At times the combined use of two light scatter angles provides for more complete discrimination. The use of more than one fluorescent channel also has benefits for the FMIA. An example of two different fluorescent dyes being used is when microspheres of the same size are coated separately with two different capture reagents and then mixed. Test sample is added followed by two fluorescent detection reagents of different colors [e.g., FITC and phycoerythrin (PE)]. The detection reagents bind to the appropriate microsphere and while the microspheres are the same size they fluoresce in two different wavelengths depending upon the presence of the analyte in question. Another approach to the use of different colors in the FMIA is the use of two different microsphere populations, each of the same size but one with a fluorescent dye imbedded within the polymer. In this case FSC and SSC may be the same for the two microsphere populations but one of the populations is fluorescent. This fluorescence, if selected properly, does not interfere with the detection of analyte using a fluorescence detection reagent (e.g., FITC or PE) provided the emission is at a different wavelength. The ability to separate microspheres by both size and color then greatly expands the number of analytes which can be simultaneously measured in the FMIA.

Along with the detection of size and color of the microspheres the flow cytometer also needs some simple event processing software to count the number of microspheres and determine the relative intensity of the fluorescence of the population. Unlike the software needed for cell analysis in flow cytometers, the FMIA requires little in the way of data analysis. Some gating capabilities are required to select populations discriminated by size or color as well as to exclude debris or aggregates of microspheres. Accurate counting of the number of microspheres in each population is important as well as the expression of the relative fluorescence intensity. The benefits of flow cytometry for the FMIA are (1) the ability to detect different populations within a complex mixture and (2) the high sensitivity achieved in analyte detection based upon the fluorescence and the optical systems used.

In addition to the flow cytometer requirements, the performance of the assay requires some instrumentation. Most assays can be performed using standard laboratory incubation devices (water baths and incubators), pipettors, and centrifuges. The most time consuming and variable portion of the performance of the FMIA is the washing of the reacted microspheres and the delivery of the reacted microspheres to the flow cytometer for analysis. Assays in which the microspheres are washed between steps to remove unbound material may experience considerable loss of microspheres and the process may be time consuming. Most wash steps for the FMIA have used centrifugation to separate the microspheres from the unbound reagent. The pelleted microspheres are then resuspended in a wash solution for multiple washes and finally in the next reagent. Additionally, if very small microspheres are used or microspheres are of low density, high centrifugal forces are required to separate the microspheres from the liquid medium. Centrifugal forces which are too high may cause aggregation of the microspheres. The use of microtiter plate vacuum filtration devices allows for the rapid washing of microspheres. Additionally some of these microplate devices have a delivery system which will semiautomate the delivery of the reacted microspheres to the flow cytometer. The use of the microtiter plate washing and delivery system greatly reduces the hands-on time needed to perform these assays.

B. Microspheres

Most studies have used polystyrene-based microspheres 2–15 μm in diameter. These polystyrene microspheres without surface functional groups are used for the noncovalent attachment of proteins. Polystyrene microspheres which contain surface functional groups (amide, carboxylic acid, etc.) allow for the covalent attachment of proteins (or other molecules). Most applications use noncolored microspheres; however, microspheres are available both with visible dyes (dyes are added either after polymerization by adsorption of the dye to the microsphere surface or during the polymerization process) as well as with fluorescent dyes. Microspheres with visible dyes are most often used in assays such as macroagglutination where the reaction is read visually and separation of different populations is based upon visual detection of the color of the microsphere. However these colored microspheres often have different orthogonal light scatter patterns as compared to the same microsphere without the dye and can be advantageously used in the FMIA. The fluorescent microspheres are often very bright and the dyes used may have very broad emission spectra. The broad emission spectra may create problems when using a fluorescent tag such as FITC or PE due to spectral overlap.

III. Procedures

A. Preparation of Microspheres

Most polystyrene-based microspheres are supplied in solution which contains a small amount of surfactant such as Triton X-100 or Tween 20. The surfactant helps to prevent aggregation of the microspheres but must be removed prior to the coating of the microspheres with the capture reagent. Surfactant (both ionic and nonionic) even in low concentrations can reduce or prevent the attachment of the capture reagent. Microsphere solutions should be stored at 22°C to reduce aggregation although aggregates can usually be dissociated by inversion, vortexing, or bath sonication. If surfactant is present in the microsphere preparation it can be removed most simply by repeated washing of the microspheres. For a more complete removal of surfactant ion exchange can be used; however, in practice it is generally not needed. To remove surfactant:

1. Dilute an aliquote of microspheres 1 : 5 in water or the buffer which is to be used in the coating process and mix well.
2. Centrifuge the microsphere population for 5 min at sufficient *g* force to sediment the microspheres (for microspheres of 2 μm diameter and above, 500*g* is usually sufficient).
3. Aspirate and discard the supernatant and gently resuspend the microsphere pellet in fresh water or buffer.
4. Centrifuge again as in step 2 and resuspend final pellet in coating buffer.
5. For microspheres with a high concentration of surfactant or in applications where attachment of the capture reagent is affected by any remaining surfactant, either the microspheres can be washed in a solvent such as 70% ethanol or the surfactant can be removed by running the microspheres over an ion exchange resin.
6. Often when the surfactant is removed the microspheres can become unstable and form irreversible aggregates. Stock microsphere suspensions should be

stored at 22°C in the solution provided by the manufacturer or in coating buffer with a low concentration of a surfactant such as Triton X-100 (0.01%).

7. If during the washing process the microspheres become unstable and start to aggregate a low concentration of the capture reagent (0.1 μg/ml) in coating buffer can be used in place of water or buffer alone to wash the microspheres. This low concentration of protein will help to stabilize the charges on the microsphere surface reducing aggregation.

8. Once the microspheres are washed determine the concentration by counting in a particle counter calibrated to read the correct size particle and adjust the concentration to 10^7/ml in coating buffer.

Passive (Noncovalent) Coating

Most proteins can be passively coated to polystyrene microspheres. Polystyrene microspheres will attach proteins strongly and often irreversibly. The reaction of proteins with polystyrene is thought to be due to hydrophobic interactions and should proceed at a wide range of pH. However, the pH and the ionic strength of the coating buffer will affect the conformation of, and hence the attachment of, the protein to the microsphere surface. Smaller proteins such as synthetic peptides often do not attach well to polystyrene or their conformation is so altered by the attachment that they no longer function normally or are not recognized by antibodies directed to the protein. In these cases where a very small protein is being used direct coating to the polystyrene is not the method of choice. For these situations use either a covalent coupling technique to a reactive group added to the synthetic peptide (described in Section III,D, Covalent Coating) or an intermediate reagent such as biotin or possibly glutaraldehyde. In addition to small proteins the passive coating of antibodies to polystyrene often is deleterious to antibody function. Conformational changes to antibodies often occur when they are passively coated onto polystyrene and this conformational change reduces the ability of the antibody to capture antigen. As for synthetic peptides, coupling of antibodies to polystyrene probably requires covalent attachment or the use of an intermediate reagent such as avidin : biotin or glutaraldehyde.

B. Passive Coating

1. To 1 ml of washed microspheres at 10^7/ml add an optimal concentration of protein in coating buffer. The optimal concentration of protein must be determined experimentally but the range of concentrations is from 1 to 25 μg protein/ml. The coating buffer is most often 0.1 *M* Na_2CO_3 buffer at pH 9.5; if incomplete coating occurs try either a neutral pH phosphate-buffered saline (PBS) solution or a pH 8.4 Tris buffer as the coating buffer.

2. With constant gentle mixing incubate the microspheres and protein for 3 hr at 37°C followed by 18 hr at 4°C. Depending upon the protein and the

affinity of the bond to polystyrene, the time needed for complete coating may be reduced.

3. Remove excess coating protein and buffer by centrifugation and wash the microspheres with coating buffer.

4. Resuspend the microspheres in PBS + 0.5% bovine serum albumin + 0.1% Tween 20, pH 7.2 (PBS-BSA-Tw) and incubate for 1–4 hr at 37°C. The BSA and Tween 20 serve to block unoccupied sites on the microsphere and reduce nonspecific reactions in the assay. If BSA or Tween 20 are problematic for the given application (e.g., studying albumin) the other irrelevant proteins such as gelatin, casein hydrolysate, or other serum proteins may be used as well as other detergents such as Triton X-100 or NP-40.

5. Centrifuge the microspheres and resuspend in fresh PBS-BSA-Tw at 10^7/ml. The several washing steps may have reduced the level of microspheres so the preparation should be recounted.

C. Use of Intermediate Reagents for Passive Coating of Microspheres

1. Avidin/Streptavidin and Biotin

Polystyrene microspheres can be coated with avidin or streptavidin using the passive coating technique described above. These coated microspheres can then be used to capture reagents which have incorporated biotin. Avidin is a protein isolated from egg whites and streptavidin is a protein isolated from *Streptomyces* spp. each with a mass of 50,000–60,000 Da. Streptavidin has some benefits over avidin in that streptavidin lacks the carbohydrate moieties present in avidin and has a lower isoelectric point (pI of streptavidin = 5.0, pI of avidin = 10.5). The absence of the carbohydrate regions and the lower pI of streptavidin reduce nonspecific binding which is often seen with avidin. The binding of biotin to avidin/streptavidin occurs over a range of pH and in most assay conditions could be considered essentially irreversible (affinity of avidin for biotin is $10^{15}\ M^{-1}$). The avidin/biotin complex is not dissociated over a range of temperatures and is relatively unaffected by the presence of proteolytic enzymes (except where avidin is degraded). Proteins with primary amines can be labeled with biotin which then will link to the avidin or streptavidin coated onto the surface of the microspheres.

1. At pH 9.2 *N*-hydroxysuccinimide-biotin (NHS-biotin) will react with primary amines (lysine epsilon groups) in proteins forming a stable protein–biotin complex; generally a 50 : 1 molar ratio of NHS-biotin to protein is used. Incubate NHS-biotin and protein for 60 min at 22°C.

2. Remove nonreacted NHS-biotin and released NHS (without the biotin group) from the protein by running the protein over a G25 desalting column.

3. Add biotinylated protein to microspheres and incubate for 3 hr at 37°C (while the affinity between avidin and biotin is quite high and therefore the

reaction occurs rapidly, extended incubation times often produce more reproducible conjugates).

4. Wash excess biotinylated protein from the microspheres with PBS-BSA-Tw by centrifugation.

5. Resuspend microspheres in PBS-BSA at 10^7/ml.

2. Glutaraldehyde

Microspheres can be precoated with glutaraldehyde and then the protein of interest. Glutaraldehyde will coat the microsphere surface and then attach the protein through cross-linking of amine groups (excess glutaraldehyde must be removed by washing to reduce protein-to-protein crosslinking). The recommended glutaraldehyde grade is an electron microscope (EM) grade which contains a small proportion of polymerized glutaraldehyde. Lower grades which contain higher proportions of polymerized glutaraldehyde may fluoresce to an extent to make the assay unreadable.

1. Wash microspheres as described and resuspend 10^7 microspheres in 1 ml of 1% glutaraldehyde in PBS.
2. Incubate for 2–8 hr at 22°C and wash microspheres in PBS (no BSA or Tw).
3. Add protein to be coated at 1–25 μg protein/ml of coating buffer (usually PBS) and incubate for 4–24 hr at 22°C with constant gentle mixing.
4. Centrifuge the microspheres and wash with PBS-BSA-Tw.
5. Incubate to block excess sites using PBS-BSA-Tw (some studies recommend the blocking of glutaraldehyde active sites with 0.2 *M* ethanolamine in PBS).
6. Resuspend final microsphere preparation in PBS-BSA-Tw at 10^7/ml.

3. Protein A or G

Protein A (from *Staphylococcus aureus*) and protein G (from group G streptococci) bind IgG through the Fc portion (protein A also binds IgA and IgM). Protein A has a mass of 42,000 to 56,000 Da with a pI of 5.1. There appear to be at least two active binding sites on protein A and G for the Fc portion of immunoglobulin. Protein A and G also attach well to polystyrene and can be coated onto microspheres and, thus, capture antibodies, leaving the antibody binding sites available. The binding between protein A or G and immunoglobulin is reversible (affinity of protein A for IgG is 10^8 M^{-1}) and antibody may be replaced from the microsphere surface by immunoglobulin in the test sample (e.g., test serum).

1. Coat protein A or protein G (there is a recombinant fusion protein which

has the characteristics of both proteins) onto microspheres as described above in Section III,B, Passive Coating.

2. Resuspend coated microspheres in 0.1–5 μg antibody protein/ml of PBS. Incubate for 3 hr at 37°C (and 18 hr at 4°C if needed).
3. Wash the microspheres and block unoccupied sites with PBS-BSA-Tw.
4. Resuspend final microspheres at 10^7/ml in PBS-BSA-Tw.

4. Anti-immunoglobulin

While antibodies often do not function optimally once attached to a solid surface it is often possible to coat microspheres with a polyclonal antibody which then will function well enough to capture a second antibody (often a monoclonal) which then captures the analyte of interest. This second captured antibody is then the capture reagent for the assay (e.g., a goat polyclonal anti-mouse IgG is coated onto microspheres, a mouse monoclonal directed to the soluble antigen of interest is added and captured by the goat anti-mouse, the mouse monoclonal then functions normally in the FMIA to capture the appropriate antigen). The polyclonal antibody coated to the microsphere should be directed against the Fc portion of the second antibody to reduce capturing of the second antibody through its antibody finding sites.

1. Coat microspheres with the first antibody (e.g., goat polyclonal) as described in Section III,B, Passive Coating.
2. Resuspend coated microspheres in 0.1–5 μg second antibody (mouse monoclonal)/ml of PBS. Incubate for 3 hr at 37°C (and 18 hr at 4°C if needed).
3. Wash the microspheres and block unoccupied sites with PBS-Tw and 0.1% of an irrelevant mouse IgG (e.g., mouse serum).
4. Resuspend final microspheres at 10^7/ml in PBS-BSA-Tw.

D. Covalent Coating

The covalent coating of proteins or other macromolecules to the surface of microspheres is most useful when attaching antibodies or small molecules which lose function when passively adsorbed to the microsphere surface. Numerous methods exist for coupling macromolecules to surface groups on microspheres (Lundblad, 1991). The method of choice depends upon the type of molecule which is being coupled and the type of surface active groups available on the microsphere surface. Two methods each of which uses a heterobifunctional crosslinking reagent are outlined below, the first uses a water-soluble carbodiimide which reacts with a carboxylic acid group on the microsphere surface and an amine group on the protein to be attached; the second method uses an NHS-maleimide which reacts with an amine on the microsphere surface and a free sulfhydryl group (cysteines) on the protein to be attached. These covalent coupling methods can also be used to attach intermediate reagents such as avidin or a capture antibody which then captures the reagent used to capture the analyte to be measured.

1. Carbodiimide (EDC) Coupling

This procedure requires a carboxylic acid group on the microsphere and an amine group on the protein to be coated.

1. Wash the microspheres as for passive coating.
2. Incubate microspheres which contain surface carboxylic acid residues in 0.1% (w/v) *N*-ethyl-*N'*-(3-dimethylaminopropyl)carbodiimide (EDC), in water with the pH adjusted to 4.5 with HCl, for 4 hr at 22°C.
3. Wash the microspheres twice in 0.025 *M* 2-(*N*-morpholino)ethanesulfonic acid (MES) buffer and add the protein to be coated onto the microsphere at 10–100 μg/ml in MES buffer; incubate for 24 hr at 22°C with constant gentle mixing.
4. Wash the microspheres and resuspend in 0.1 *M* ethanolamine for 1 hr at 22°C.
5. Wash the microspheres twice and resuspend the final pellet in PBS-BSA-Tw; incubate for 1 hr at 37°C.
6. Centrifuge and adjust the microsphere concentration to 10^7/ml in PBS-BSA-Tw.

2. NHS-Maleimide Coupling

This procedure requires an amine group on the microsphere and a sulfhydryl group on the protein to be coated. The protein should be reduced [using dithiothreitol (DTT) or 2-mercaptoethanol (2-ME)] just prior to the addition to the microspheres.

1. Generate free sulfhydryls on the protein to be coupled (e.g., immunoglobulin) by reduction using 20–200 m*M* of either 2-ME or DTT per 10 mg protein in 1 ml.

2. Incubate for 60 min at 22°C.

3. Separate the reduced protein from the 2-ME or DTT by running the protein over a G25 desalting column (G25 column will exclude all molecules over 5000 Da). Store reduced protein at 1–10 mg/ml. Ethylenediamine tetraacetic acid (EDTA) can be added to the storage buffer to help remove oxidizing metals which may oxidize the sulfhydryls.

4. Wash the microspheres as for passive coating; however, ensure that none of the buffers contain extraneous amines or sulfhydryls.

5. Dissolve the NHS-maleimide in dimethyl formamide (DMF) or dimethyl sulfoxide (DMSO) at 5 mg/ml.

6. Add 50 μl of stock NHS-maleimide to 10^8 microspheres which contain surface amine groups in 0.1 *M* Na_2CO_3, pH 8.5 (higher pH increases the availability and rate of amine coupling of the microspheres).

7. Incubate with gentle rocking for 0.5–2 hr at 22°C.

8. Centrifuge the microspheres to remove unbound NHS-maleimide.

9. Add 0.01–0.1 mg of the reduced protein to 10^7–10^8 microspheres at pH 6.5 and incubate with gentle rocking for 60 min at 22°C.

10. Centrifuge the microspheres to remove unbound protein and wash in PBS.

11. Block any remaining sites with an irrelevant protein (sulfhydryl containing proteins may be the most optimal) and adjust the final concentration to 10^7 microspheres/ml.

E. Storage of Coated Microspheres

Microspheres passively coated with protein are usually stable when stored concentrated (10^7 or higher microspheres/ml) in PBS-BSA at 4°C for up to 4 weeks. For longer storage the microspheres can be placed in PBS-BSA containing 40% glycerol and stored at −20°C. The glycerol concentration prevents the freezing of the buffer and the microspheres often remain stable for as long as 1 year. After storage at −20°C, in glycerol containing buffer, the microspheres must be washed two to three times in PBS to remove the glycerol before the microspheres are used in an assay. The stability of microspheres coated with compounds other than proteins has not been described in most studies and is no doubt dependent upon the capture reagent used, the medium, and the storage temperature. Microspheres which are covalently coupled with protein appear to be stable when stored for extended periods in PBS-BSA at 4°C. Long-term storage in PBS-BSA + 40% glycerol seems to preserve the coating. It may be prudent to add protease inhibitors to the storage buffer.

F. Sample Protocols for Assay Performance

1. Homogeneous Competitive FMIA for the Detection of Antigen X

1. Reduce monoclonal antibody to antigen X and couple using NHS-maleimide to amine containing microspheres.
2. Add 10^6 microspheres in 0.1 ml to 0.1 ml test sample which contains unknown concentration of antigen X.
3. Add 0.1 ml of known concentration of antigen X coupled with FITC.
4. Incubate at 37°C for 2 hr.

5. Analyze microspheres by using flow cytometry; select singlet populations of microspheres by gating using FSC and SSC signals and display FITC fluorescence.
6. Fluorescence is inversely proportional to the concentration of antigen X in the test sample.
7. Using known concentrations of antigen X construct a standard curve from which the concentration of antigen X in the unknown samples is determined.

2. Heterogeneous Noncompetitive FMIA for the Detection of Antibody to Antigen X

1. Coat polystyrene microspheres passively with antigen X.
2. Add 10^6 microspheres in 0.1 ml to 0.1 ml test sample which may contain antibody to antigen X.
3. Incubate for 2 hr at 37°C.
4. Wash twice with PBS-BSA-Tw.
5. Resuspend microsphere pellet in 0.1 ml FITC-conjugated goat F $(ab')_2$ anti-human IgG.
6. Incubate for 0.5 hr at 37°C.
7. Wash as before or analyze directly by using flow cytometry; select singlet populations of microspheres by gating using FSC and SSC signals and display FITC fluorescence.
8. Fluorescence is proportional to the concentration and affinity of the antibodies in the test sample directed to antigen X.

G. Assay Development and Data Interpretation

The number of analytes to be assayed and the anticipated concentration will dictate the number of different microspheres to use an other assay parameters. The decision as to how to distinguish the microsphere classes must be determined. Usually the size of the microsphere is the distinguishing feature. A difference of 1 μm in diameter is generally sufficient to accurately differentiate microsphere populations by flow cytometry. The variation of the size of the microspheres is usually expressed as the standard deviation or coefficient of variation (CV). CVs are generally <5% for most polystyrene microspheres. If colored or fluorescent microspheres are to be used, the color or fluorescence must be matched to the particular assay. For fluorescent microspheres ensure that the emission from the dye does not overlap with the fluorescent detection dye.

The easiest and often the most effective coating process is the passive coating of polystyrene microspheres. This procedure should probably be tested and only abandoned if the capture reagent loses function, will not coat, or is too unstable after coating. If passive coating does not produce the desired results a covalent method should be evaluated. The covalent coupling method depends upon the available groups on the capture reagent as well as the type of coupling chemistry desired.

The assay must be configured as homogeneous (all reactants incubate together without separation) or as heterogeneous (reactants incubate in distinct phases). In addition the assay must be configured either as a competitive assay in which a known concentration of analyte is added and competes with analyte in the test sample or as a noncompetitive assay in which the analyte in the test sample reacts without exogenous analyte. The most commonly developed immunoassays are noncompetitive heterogeneous assays. The selection of homogeneous versus heterogeneous and competitive versus noncompetitive is governed by the analyte to be measured and the type of assay the investigator requires. The availability of required assay components also helps dictate the assay type to be developed. The detection of soluble analyte using antibody-coated microspheres often can be done in a competitive homogeneous assay; however, competitive assays often have a narrow dynamic range as compared to noncompetitive assays. The incubation time and temperatures needed are dependent upon the type of assay performed. Some antibodies react very well at 4°C while others react better at 37°C and some other nonantibody binding processes may proceed rapidly at a number of temperatures. For instance the reaction of biotin and avidin can proceed rapidly in a wide temperature and pH range. Since most flow cytometric analysis is based upon the detection of the fluorescent signal associated with a given size particles (or FSC pulse) only fluorescence-associated with the particle is measured. This indicates that in either a homogeneous or a heterogeneous assay the final reactant which is usually the fluorescently tagged reagent does not have to be removed by washing as the unbound fluorescent reagent will not be measured.

An important consideration is that of what constitutes a positive signal. It is important to start with test samples which are known to lack the analyte of interest as well as test samples which contain the analyte of interest in varying concentrations. Once a clear differentiation between a negative and a strong positive is demonstrated then coating conditions (capture reagent concentration), incubation times and temperatures, and detection reagent concentration can be altered to maximize analytic sensitivity of the FMIA. In assays for serum antibodies or antigens using FMIA, 50–100 test sera which lack the analyte of interest are tested. The distribution of results from these negative samples is used to establish a cutoff for negative versus positive. If the distribution is Gaussian then mean signal + 3 standard deviations (SD) is often selected as the cutoff. However if the data are not normally distributed then a nonparametric method (e.g., a percentile ranking test using any signal over the 2.5th percentile as positive) is used to determine the cutoff. Once a cutoff is established samples which contain a low concentration of the analyte in question and samples which contain analytes which may cross-react are tested to determine the validity of the selected cutoff. Rather than using a cutoff some studies have used the nonparametric Kolmogorov–Smirnov test to compare the fluorescence distribution of the test sample to that of a known negative sample. If the fluorescence distribution of the test sample is significantly different from the known negative then the test sample is considered positive. This approach is especially useful when there is not strong fluorescence separation between a negative and a positive signal.

When measuring analytes such as soluble antigens a standard curve is generally used to determine analyte concentration. Microspheres are coated with antibody to the analyte to be measured and known concentrations of the analyte are added and followed by a fluorescent anti-analyte reagent. The fluorescent signal is plotted against the known analyte concentrations and the concentration of analyte in unknown samples is determined from the standard curve. The standard curve method is widely used in other immunoassays and is relevant to the use of FMIA. The quantitation of soluble antigen using antibody-coated microspheres is possible since the antibody coated to the microspheres reacts with the analyte in all test samples with the same antibody affinity. The factors influencing the validity of the reading are test sample-specific interfering substances such as the presence of other analytes which may cross-react or other sample components (e.g., rheumatoid factor binding to the IgG antibody on the microsphere) which will adversely affect the signal. The quantitation of antibody in test samples presents difficulties due to the fact that different test samples have antibody, directed toward the antigen of interest, in different affinities, directed at different antigenic epitopes on the antigen, etc. As such the concentration of the antibody is not the only factor affecting the fluorescent signal. For these reasons most specific antibody immunoassays do not attempt to report the antibody concentration in gravimetric units (e.g., μg/ml). Instead the results are usually expressed as positive or negative or in some cases as a semi-quantitative result referenced to an internal standard. In this case a known sample is assayed in each run and the test samples are expressed as the ratio of the internal control, either a fluorescence ratio (internal standard produces a fluorescent signal in channel 100, negative samples produced fluorescent signals in channels <20, and the test sample produces a fluorescent signal in channel 75; test sample result is 0.75 units with >0.20 units being positive). Alternatively if the internal control antibody (e.g., mouse monoclonal antibody) is serially diluted and the standard curve generated shows fluorescence versus mouse monoclonal antibody equivalents in weight per volume (e.g., μg/ml) then test samples can be assayed and read from the standard curve and reported in equivalent units. In using an internal standard it is important to always use the same internal standard since the antibody affinity and the epitope reactivity both affect the reaction of the antibody to the antigen. While the semi-quantitation can be used to express antibody levels it does not take into account the differences in antibody affinity and epitope distribution between the test samples. Another approach is to report the antibody level as a titer. This requires serial dilutions of the test sample and then comparing the fluorescent signal from each dilution to a cutoff. The reciprocal of the highest dilution producing a signal over the cutoff is used as the titer.

A number of parameters of the FMIA must be selected such as the fluorescent dye, the number of microspheres per reaction mixture, and how to wash the microspheres. The choice of fluorescent reagent used in the FMIA depends upon the assay configuration, reagent availability, and excitation/emission possibilities in the flow cytometer. Most published studies have used FITC primarily due to the widespread availability of the reagent in a variety of conjugates, the ease of excitation at 488 nm with argon ion lasers, and the availability of appropriate optical filters to detect emission at 525 nm. Other dyes, such as PE, have been used but little work has been published on the optimum dyes for FMIA. Most studies have used 105 to 106 microspheres per reaction mixture and analyzed 104 microspheres in the flow cytometer. The number of microspheres in the reaction mixture is dependent upon the assay parameters; however, in practice the FMIA can be performed using low numbers of microspheres. In heterogeneous assays with wash steps the number of microspheres added must be high as there is loss of microspheres during the washing steps. However, in assays without wash steps the number of microspheres needed is quite small. Since the reaction on the microsphere surface is relatively uniform between microspheres the number of microspheres which need to be analyzed by the flow cytometer is small. Enough microspheres must be collected to allow gating of single populations or to separate distinct microsphere classes and to produce an accurate fluorescent peak for measurement. In practice the analysis of 500 microspheres produces results that are not statistically different from results generated from the analysis of 10,000 microspheres. If analyzing 500 microspheres, 5,000 microspheres is usually a sufficient starting number to account for some loss during the assay and for allowing repeat analysis.

IV. Review of Published Applications

A number of papers have been published describing FMIA. Most of these assays have used antigen : antibody reactions. A small number of studies have used fluorochromes other than FITC such as PE and methods other than antigen : antibody binding such as the antibiotic binding to DNA coating microspheres of Saunders *et al.* (1990) or the PCR-based assay of Yang *et al.* (1993). A brief summary of the major applications to date is listed in Table I. The analyte which was measured, the type of microsphere, and the coupling method used, along with a comment about the study, are listed.

Table I

Analyte measured	Microsphere type/size	Results/comments	Reference
		Detection of specific antibody	
IgG (rheumatoid factor, RF)	19.5 μm polystyrene, passive coating with human IgG	Different IgG subclasses evaluated as antigen for RF, indirect assay, FITC	Horan *et al.* (1979)
Eimeria tenella	3.0 μm carboxylic acid polystyrene, covalent coating with sporozoite antigens using EDC	Detection of antibody levels in chickens immunized with *E. tenella* antigen, indirect assay, FITC	Zemcik (1986)
Cytomegalovirus and herpes simplex virus (CMV and HSV)	5 and 7 μm polystyrene, passive coating with viral antigens	Simultaneous detection of antibodies in human serum, compared to standard methods, indirect assay, PE	McHugh *et al.* (1988a)
Various antigens	3 μm polyacrylamide and 10 μm polystyrene, passive coating with human light chains or cell-surface proteins	Screening of monoclonal antibodies for reactivity to a variety of antigens, indirect assay, FITC	Wilson and Wotherspoon (1988)
Candida albicans	5, 7, and 9.5 μm polystyrene passive coating with three different preparations	Simultaneous detection of antibodies from patients with systemic candidiasis, compared to EIA, indirect assay, FITC	McHugh *et al.* (1989)
Human immunodeficiency virus (HIV)	5, 7, 10, and 15 μm polystyrene passive coating with recombinant HIV proteins	Simultaneous detection of human antibodies to p24, p31, gp41, and gp120, indirect assay, FITC	Scillian *et al.* (1989)
OKT3 (CD3)	9.5 μm polystyrene, covalent coating with mouse antiCD3	Detection of human anti-CD3 in patients with organ transplant, compared to EIA, indirect assay, FITC	Lim *et al.* (1989)

(*continued*)

Table I (*continued*)

Analyte measured	Microsphere type/size	Results/comments	Reference
Gliadin	5.2 μm carboxylic acid polystyrene, covalent coating with EDC and α-gliadin	Detection of anti-gliadin in pediatric patients with coeliac disease, indirect assay to detect IgA, FITC	Presani (1989)
IgA	3 μm polystyrene, passive coating with human IgA	Detection of anti-IgA in IgA-deficient patients, indirect assay, FITC	Syrjala *et al.* (1991)
Pyruvate dehydrogenase (PDH)	5 μm polystyrene, passive coating with bovine PDH	Detection of anti-PDH in patients with primary biliary cirrhosis, indirect assay, FITC	Elkhalifa *et al.* (1992)
Helicobacter pylori	0.9 μm carboxylic acid polystyrene, passive coating with two different bacterial antigen preparations	Detection of antibody in patients undergoing endoscopy for upper gastrointestinal symptoms, compared to Western blot and EIA, indirect assay, FITC	Best *et al.* (1992)
Mouse monoclonal antibody KS1/4	8- to 10-μm microspheres coated with goat anti-mouse captured the mouse monoclonal KS1/4 which is used to treat lung and colon adenocarcinoma	Detection of human antibodies to KS 1/4, compared to an EIA, indirect assay, FITC	Labus and Petersen (1992)
Detection of specific antigen or other analyte			
Human IgG	1–5 μm and 40–50 μm polyacrylamide and 30–40 μm dextran, covalent coating with EDC and anti-human IgG	Sensitivity measured by Scatchard analysis with ^{125}I-labeled IgG to determine sensitivity between different-sized microspheres, indirect assay with no wash step, FITC	Lisi *et al.* (1982)

Horseradish peroxidase (HRP)	10 μm polystyrene, passive coating with anti-HRP; 0.1 μm fluorescent green polystyrene (used as the signal) passive coating with anti-HRP or HRP	Competitive and sandwich assay, sensitivity to 10^{-14} *M*, one-step, no wash assay, 0.1 μm microspheres emit at 510 nm	Saunders *et al.* (1985)
Human IgG immune complexes	5 μm polystyrene, passive coating with human C1q	Compared to standard fluorescent assay, indirect assay, FITC	McHugh *et al.* (1986)
Human IgG immune complexes containing HIV antigen	5 μm polystyrene, passive coating with human C1q	Detection of HIV immune complexes in patients using mouse monoclonal antibodies to a variety of antigens, indirect assay, PE	McHugh *et al.* (1988b)
Carcinoembryonic antigen (CEA)	7 and 10 μm polystyrene, passive coating with two monoclonal antibodies to CEA each with a different affinity	Broadened dynamic range with two particles, one step, no wash assay, PE	Lindmo *et al.* (1990)
Actinomycin-D	10 μm polystyrene, passive coating with calf thymus DNA	One-step competitive assay using mithramycin as the fluorescent antibiotic competing with actinomycin-D for binding sites on the DNA	Saunders *et al.* (1990)
Mouse immunoglobulin, human chorionic gonadotropin, or progesterone	100- to 150-μm agarose beads, covalent coupling with specific antibody using cyanogen bromide	Three different assays compared to EIA, indirect assay, FITC	Kim *et al.* (1992)
Hepatitis B virus DNA by polymerase chain reaction (PCR)	2.8 μm paramagnetic polystyrene, coated with streptavidin to capture a biotinylated oligonucleotide probe	Detection of hepatitis B virus DNA using the PCR, fluorescence and flow cytometry, FITC	Yang *et al.* (1993)

V. Summary

The FMIA is an assay which has been shown to work in numerous applications. The assay compares well to standard assays and a number of studies demonstrate an increase in sensitivity as compared to standard immunoassays such as EIA and Western blotting. Relatively few studies have taken advantage of the size or color discriminating capability of the flow cytometer for the FMIA. The technique offers a versatile tool to the investigator using flow cytometry. The ability to simultaneously detect multiple analytes with high sensitivity are attributes making the FMIA an attractive assay method. Improvements in washing and delivery systems are making the FMIA easier to use and are providing for greater reproducibility of the results.

VI. Suppliers

Bangs Laboratories, Inc.
979 Keystone Way
Carmel, IN 46032
(317) 844–7176

Supplier of a variety of microspheres in varying sizes and surface chemistries, large variety of different types of microspheres including colored and fluorescent particles.

Cytek Development, Inc.
46560 Fremont Blvd., Unit 116
Fremont, CA 94538
(510) 657–0102

Supplier of a microsphere sample preparation system, a vacuum microplate device for washing, incubating and delivering microspheres to the flow cytometer.

Interfacial Dynamics, Inc.
17300 SW Upper Boones Ferry Rd., Suite 120
Portland, OR 97224
(503) 684–8008

Supplier of a variety of microspheres in varying sizes and surface chemistries, large variety of different types of microspheres including colored and fluorescent particles.

Polysciences
400 Valley Road

Warrington, PA 18976
(215) 343–6484

Supplier of a variety of microspheres in varying sizes and surface chemistries.

Seradyn
1200 Madison Avenue
Indianapolis, IN 46206
(317) 266–2915

Supplier of a variety of microspheres in varying sizes and chemistries.

References

Best, L. M., Veldhuyzen van Zanten, S. J. O., Bezanson, G. S., Haldane, D. J. M., and Malatjalian, D. A. (1992). *J. Clin. Microbiol.* **30,** 2311–2317.

Elkhalifa, M. Y., Kiechle, F. L., Gordon, S. C., Chen, J., and Poulik, M. D. (1992). *Am. J. Clin. Pathol.* **97,** 202–208.

Fulwyler, M. J. (1976). UK patent #1561042.

Horan, P. K., and Wheeless, L. L. (1977). *Science* **198,** 149–157.

Horan, P. K., Schenck, E. A., Abraham, G. N., and Kloszewski, M. D. (1979). "Immunoassays in the Clinical Laboratory." Liss, New York.

Kim, K., Han, M. Y., Yoon, D. Y., Cho, B., Choi, M. J., Choe, I. S., and Chung, T.-W. (1992). *Immunol. Lett.* **31,** 267–272.

Labus, J. M., and Petersen, B. H. (1992). *Cytometry* **13,** 275–281.

Lim, V. L., Gumbert, M., and Garovoy, M. R. (1989). *J. Immunol. Methods* **121,** 197–201.

Lindmo, T., Bormer, O., Ugelstad, J., and Nustad, K. (1990). *J. Immunol. Methods* **126,** 183–189.

Lisi, P. J., Huang, C. W., Hoffman, R. A., and Teipel, J. W. (1982). *Clin. Chim. Acta* **120,** 171–179.

Lundblad, R. L. (1991). "Chemical Reagents for Protein Modification," 2nd ed. CRC Press, Boca Raton, FL.

McHugh, T. M., Stites, D. P., Casavant, C. H., and Fulwyler, M. J. (1986). *J. Immunol. Methods* **95,** 57–61.

McHugh, T. M., Miner, R. C., Logan, L. H., and Stites, D. P. (1988a). *J. Clin. Microbiol.* **26,** 1957–1961.

McHugh, T. M., Stites, D. P., Busch, M. P., Krowka, J. F., Stricker, R. B., and Hollander, H. (1988b). *J. Infect. Dis.* **158,** 1088–1091.

McHugh, T. M., Wang, J. Y., Chong, H. O., Blackwood, L. L., and Stites, D. P. (1989). *J. Immunol. Methods* **116,** 213–219.

Presani, G., Perticarari, S., and Mangiarotti, M. A. (1989). *J. Immunol. Methods* **119,** 197–202.

Saunders, G. C., Jett, J. H., and Martin, J. C. (1985). *Clin. Chem. (Winston-Salem, N.C.)* **31,** 2020–2023.

Saunders, G. C., Martin, J. C., Jett, J. H., and Perkins, A. (1990). *Cytometry* **11,** 311–313.

Scillian, J. J., McHugh, T. M., Busch, M. P., Tam, M., Fulwyler, M. J., Chien, D. Y., and Vyas, G. N. (1989). *Blood* **73,** 2041–2048.

Syrjala, M. T., Tolo, H., Koistinen, J., and Krusius, T. (1991). *J. Immunol. Methods* **139,** 265–270.

Wilson, M. R., and Wotherspoon, J. S. (1988). *J. Immunol. Methods* **107,** 225–230.

Wilson, M. R., Mulligan, S. P., and Raison, R. L. (1988). *J. Immunol. Methods* **107,** 231–237.

Yang, G., Ulrich, P. P., Aiyer, R. A., Rawal, B. D., and Vyas, G. N. (1993). *Blood* **81,** 1083–1088.

Zemcik, (1986). *J. Immunol. Methods* **91,** 265–269.

CHAPTER 34

Calibration of Flow Cytometer Detector Systems

Ralph E. Durand
Medical Biophysics Department
British Columbia Cancer Research Centre
Vancouver, British Columbia V5Z 1L3

I. Introduction

Despite the fact that flow cytometers are designed to quantify the fluorescence of individual cells with high precision, it is rare to see the fluorescence intensity of a sample presented other than on the relative scale of "channel number." This is largely due to the number of factors that influence that channel number to which a given measurement will ultimately be assigned. Principal among these are instrument alignment, laser power, optical filters, photomultiplier tube (PMT) sensitivity, and display mode (linear or logarithmic) and resolution. A further uncertainty is the linearity of the preamplifier, amplifier, and analog-to-digital converter (ADC) of the cytometer.

The potential uncertainties just listed notwithstanding, the preferred method to generate highly quantitative data is actually quite simple: the fluorescence intensity of an unknown sample can be expressed relative to that of a known standard (preferably, with *simultaneous* measurement if the standard and unknown can be distinguished on the basis of other parameters). This technique is particularly suitable if the fluorescence intensity of the unknown is very similar to that of the standard, thus minimizing concerns about instrument linearity. Unfortunately, however, many flow cytometrists eventually face the "real world" problem of quantifying fluorescence when a standard of comparable intensity is not available. A mathematical approach to this problem, which in essence provides experimentally determined "correction factors" for the electronic components of the flow cytometer, is the subject of this chapter.

For the purposes of our discussion, two parameters must first be established. First, our discussion does not directly address instrument linearity. We assume *absolute* amplifier and ADC linearity; the methodology described by Bagwell *et al.* (1989) provides a quick and convenient way to check this and, if necessary, to compensate for at least some types of nonlinearity. Second, one must question when the sample comparison to an external standard is "adequate." We previously stated that this is appropriate and preferable if the fluorescent intensities of the standard and unknown are similar. "Similar" in this sense is, however, actually quite restrictive: given typical measurement CVs of a few percent, distributions centered at channels 10 and 200 (for 256 channels full scale) probably represent the maximum range feasible. This represents a dynamic range of only 20! Logarithmic amplification ostensibly increases that range somewhat, but at the expense of decreased overall resolution.

The starting point for our normalization technique is the well-known fact that "off-scale" signals can be easily brought back on scale by adjusting the instrument gain or PMT voltage. What is perhaps not generally appreciated, however, is the predictability and reproducibility of the electronic components used in flow cytometers. In essence, this reproducibility allows the net amplification introduced by any combination of amplifier gains and PMT voltages to be "normalized out," providing the means to quantitatively express sample intensity on a scale which can extend over several decades (far greater than any log amplifier currently available). In view of the operational convenience of acquiring data with logarithmic amplification, we will also briefly describe a method for conversion of logarithmic data to linear format. Using this procedure, data acquired using either log or linear amplification can then be reduced to the *same* scale of relative fluorescence intensity, giving a dynamic range approaching at least 10^6.

II. Materials

The amplifier and PMT calibration procedure subsequently described is not specific to any particular instrument type, nor are any specialized materials necessarily required. A highly useful addition, however, is a set of neutral density (ND) filters (e.g., 0.3, 0.6, 1.0, 2.0, and 3.0) to decrease signal intensity by known increments. Oriel Corp. (Stamford, CT) sells filters of various sizes and provides wavelength-specific calibration curves for each filter.

A. Amplifier Calibration

To check the absolute (as opposed to nominal, that is, that listed by the manufacturer) gain of the amplifiers of Coulter or Becton–Dickinson cytometers, availability of a precision signal generator is desirable. Use of the "built-in" signal generator on either instrument is convenient; alternately, one can also use appropriate fluorescent particles (microspheres, or homogeneous cells).

B. Photomultiplier Tube Calibration

All PMTs have a response that is highly dependent on the wavelength of the detected light. Calibration for a particular wavelength is possible with, for example, a light emitting diode (LED). However, since LED and fluorochrome emissions seldom overlap precisely, we have found it to be preferable to use appropriately stained cells or fluorescent microspheres, with the same optical filters that would be used under typical working conditions. A separate "calibration" is necessary for each fluorochrome-filter combination.

III. Procedures

Two separate steps are required to complete the basic calibration procedure for Coulter and Becton–Dickinson instruments, and a third step provides calibration of the log amplifiers. Although it is not essential that the following sequence be followed, it is suited to laboratories that do not have sophisticated electronic test instruments (oscilloscopes and signal generators) available.

A. Amplifier Calibration

Signal amplification is handled somewhat differently in Ortho, Coulter, and Becton–Dickinson instruments. Ortho instruments rely only on PMT voltage; in the EPICS and FACS series, amplifiers nominally adjustable over at least a 32-fold range are provided. However, individual amplifiers (even those for the different channels of a single instrument) may have actual gains that differ from the nominal gain by as much as 20% (Durand, 1981). Depending on the ultimate accuracy desired in determining the relative fluorescence of experimental samples, the individual investigator must determine whether precise calibration of each amplifier is of benefit.

As indicated in Section II, availability of an electronic signal generator results in maximal precision, but it is not essential. Regardless of the signal source, the recommended procedure is as follows:

1. Choose a signal that, when the instrument is operated at maximum gain, produces a mean (mean = $\Sigma n_i x_i / \Sigma n_i$, where n_i is the number of cells in channel x_i) or modal (peak) signal near the maximal channel number of the frequency histogram. Since all events recorded contribute to determination of the mean, better reproducibility can be achieved with less data accumulated when using means rather than modes.

2. For the same signal, record the mean or mode signal position for each reproducible amplifier setting (for amplifiers with continuously adjustable gains, the minimum and maximum of the range are likely to be the only easily reproduced settings).

3. The actual amplification at each setting in steps 1 and 2 can be calculated based on the ratio of distribution means or modes [e.g., means of 161, 81.4, 40.6, 21.2, and 10.0 for amplifier gains of 16, 8, 4, 2, and 1, translate to actual gains (G) of 161/10 or 16.1, 8.14, 4.06, 2.12, and 1]. Repetitive measurements with several (slightly) different initial signal levels are recommended; the reproducibility of each calculated gain should be within 1%.

B. Photomultiplier Calibration

Since each PMT is critically sensitive to the wavelength of the light detected, it is recommended that this calibration be undertaken with each fluorochrome and filter combination that will utlimately be employed. Also, band pass filters are recommended to increase further the accuracy of this procedure. The use of (wavelength-calibrated) ND filters greatly simplifies the procedure.

1. Select a brightly stained sample, and determine its mean or modal fluorescence (as previously described), using the minimum PMT voltage that produces an easily analyzed frequency distribution with a near-minimum intensity. If a complete set of ND filters is available, any convenient amplification setting can be used; in the absence of ND filters, start with the maximum amplifier gain available.

2. Increase the PMT voltage by a convenient increment (e.g., 50 V), and determine the new mean or mode. This will be the second point on a (log–log) plot of signal mean (or mode) versus PMT voltage.

3. Increase the PMT voltage by another increment, and again determine and plot the mean. In the event that a voltage increment produces an overflow (intensities greater than full scale), three options are available:

a. Insert an appropriate ND filter in front of the PMT. The calculated mean is then the product of the observed mean and the attenuation factor of the ND filter at the wavelength being used.

b. Decrease the amplifier gain. The calculated mean is then the product of the observed mean and the ratio of the initial/present amplification (both are accurately known from Section III,A).

c. In the event that further reductions of amplifier gain are impossible, or that additional ND filters are unavailable, it is possible to substitute a less intensely stained sample at this point. Choose a sample that will give reasonable histograms at two or more amplifier/voltage settings used with the initial sample; this allows normalization of new measurements to those taken previously and thus expands the calibration range.

4. The appropriate combinations of step 3 and option a, b, or c should be repeated until data have been acquired over the usable voltage range of the PMT (generally 300–1500 V). The log–log plot of signal intensity (calculated means or modes) versus PMT voltage should be linear over most of the range of voltages chosen, indicating that PMT output is a power function of the tube voltage (the power is indicated by the slope, *k*, of the line). Typical data for one of the PMTs of a FACS 440 are shown in Fig. 1. Note that the power function is maintained over at least six decades (i.e., over the range indicated by the filled circles).

C. Derivation of "Calibration Factors"

The procedure described in Section III,A is a method of determining the actual (linear) amplifier gain (G), and in Section III,B, the PMT gain coefficient

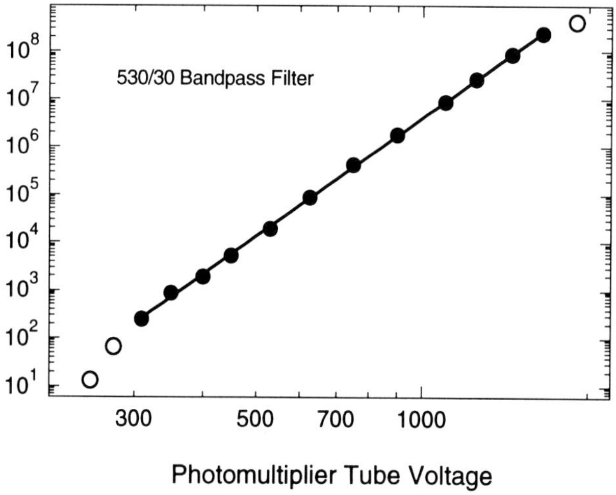

Fig. 1 Typical response of a PMT as a function of operating voltage. Results were obtained using fluorescent microspheres and ND filters as described in the text, with the calculated mean being the product of observed mean channel number and attenuation factor for the ND filter combinations used at each experimental point. The response is adequately described by a power function over the range indicated by the closed circles.

(k) was determined. It then follows that the relative intensity RI of signals accumulated in channel I is given by

$$\mathrm{RI} = (A\mathrm{I}/G)(V_r/V)^k, \tag{1}$$

where V_r is a PMT "reference" voltage, and A is a normalization constant. We have found it convenient to choose a PMT reference voltage near that normally used for the PMT (e.g., 600 V); this is, however, a completely arbitrary choice, as is the value assigned to A. It is often convenient to choose a value of A that results in unstained cells having a relative intensity of 1.0. Alternatively, if direct intercomparison of results from several instruments is desired, a separate set of calibration constants can be derived for each instrument, so that the fluorescence intensity of a (commercially available) fluorescent microsphere preparation has a defined value.

It should also be noted that this expression for RI can be applied channel by channel as written in the equation or can be used to normalize previously determined mean or modal fluorescences by simply substituting those values for I in Eq. (1).

D. Conversion of Logarithmic Data

The advantages of visualizing data having a large dynamic range by using logarithmic amplification have been reviewed by Muirhead *et al.* (1983). Mathematical description of such data is, however, accomplished most easily by performing a logarithmic-to-linear transform, and this can be done in a manner consistent with the normalization conditions derived here (or, as shown in greater detail by Schmid *et al.*, 1988). A simple procedure for this conversion follows:

1. Select a PMT voltage and log amplifier gain that provide a near-maximum mean or modal signal. Record the (log) channel number of the distribution mean (or mode), then determine the RI of the sample using suitable voltage and linear amplifier settings (and the constants previously determined).
2. Repeat step 1 (with log data acquired at the same PMT voltage and amplifier gain as for the initial determination) for several ND filter combinations that reduce the signal intensity (or, alternatively, with cells or microspheres of different intensities).

Data collected in this manner can be plotted directly as in Fig. 2, or if linear conversion without normalization is adequate, the (log) distribution mean or mode can be plotted as a function of relative signal intensity (calculated according to signal attenuation by ND filters). Data obtained using a logarithmic amplifier of a FACS 440 show a dynamic range as large as five decades (Fig. 2). Since RI and channel number (I) are obviously related for any curve of Fig. 2 by the expression $\mathrm{RI} = a\, e^{b\mathrm{I}}$, it follows that the logarithmic data can be converted channel by channel to relative (linear) intensities, and the (weighted) mean fluorescence or other statistical parameters can then be easily calculated.

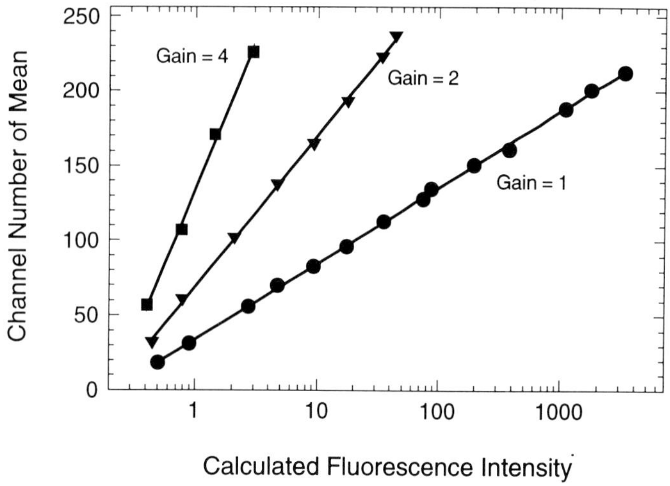

Fig. 2 Typical response of a FACS 440 log amplifier, as a function of the relative fluorescence intensity calculated with linear amplification and the procedures described in the text. Data were accumulated using fluorescent microspheres, with signal intensity varied by interposing appropriate combinations of ND filters. The instrument "multiplier gain" was held at 0.5, with "base gains" as shown, and PMT voltage at 600 V. Note that all curves extrapolate to a common intercept and that the relative slopes of the curves were essentially equal to the nominal gains shown. Gain 1 (●); gain 2 (▲); gain 4 (■).

Data acquired at different PMT voltages produce parallel curves for any amplifier gain and are displaced by the same factor $(V_r/V)^k$ previously determined for PMT dependence on operating voltage (at a particular wavelength). Similarly, the slopes of the curves in Fig. 2 differ by exactly the same amplifier gain factor (G) previously determined for linear amplification. Thus, it follows that the general expression

$$\mathrm{RI} = a(V_r/V)^k e^{bI/g}, \tag{2}$$

will convert log data acquired in channel I at any amplifier (relative) gain (g) and PMT voltage, V (note that g is the gain relative to that used to define the "standard" curve, with slope b, at voltage V_r). After performing this channel-by-channel conversion, the mean of the linearized distribution can be calculated easily.

IV. Concluding Comments

The PMT calibration procedure requires ≤30 minutes for each filter–fluorochrome combination; we have found the procedures to be sufficiently useful that we now express all our quantitative measurements only in terms of calculated intensities. To facilitate these calculations further, it is useful to monitor PMT voltages and amplifier settings electronically in a manner that permits automated recording and use of these values by software packages.

It should be obvious that the procedures just described allow standardization of fluorescence measurements, but do not substitute for the usual techniques of instrument alignment or for the necessity of internal controls in high-precision studies. These procedures do, however, permit quantization of fluorescence intensity (relative to an arbitrary standard) over the entire usable range of the detectors, without ADC or amplifier limitations.

Acknowledgments

Supported by the Medical Research Council of Canada and the National Cancer Institute of Canada.

References

Bagwell, C. B., Baker, D., Whetstone, S., Munson, M., Hitchcox, S., Ault, K. A., and Lovett, E. J. (1989). *Cytometry* **10,** 689–694.
Durand, R. E. (1981). *Cytometry* **2,** 192–193.
Muirhead, K. A., Schmitt, T. C., and Muirhead, A. R. (1983). *Cytometry* **3,** 251–256.
Schmid, I., Schmid, P., and Giorgi, J. V. (1988). *Cytometry* **9,** 533–538.

CHAPTER 35

Standardization for Flow Cytometry

A. Schwartz[*] and Emma Fernández-Repollet[†]

[*] Flow Cytometry Standards Corporation
San Juan, Puerto Rico 00919

[†] Department of Pharmacology
School of Medicine
University of Puerto Rico
San Juan, Puerto Rico 00936

I. Introduction

Flow cytometry provides a unique tool for the quantitative analysis of individual particles or cells. To perform this in a reliable manner, appropriate standards are necessary. The emphasis in standardizing flow cytometry has been limited to obtaining the percentage of cells in a population that binds a specific antibody. This has been beneficial for determining both the percentage and absolute number of CD4 lymphocytes in peripheral blood in conjunction with the diagnosis of the onset of AIDS and its therapy. Many factors, however, affect this measurement, ranging from the acquisition and preparation of the sample to the gating and analysis, to ensure that all populations of the lymphocytes have been accounted for. The effects of these factors on flow cytometric analysis, as well as commercially available standards have been extensively discussed in the recent literature (Horan *et al.*, 1990; National Committee for Clinical Laboratory Standards, 1992; Muirhead, 1992).

This chapter focuses on the comprehensive perspective of standardization, specifically identifying cell population patterns, quantitating fluorescence intensities, and determining the antibody binding capacity. Attempts at standardizing these parameters have involved descriptions of weak or strong light scatter to describe size and degree of granularity, as well as references to dim or strong fluorescence intensities. Unfortunately, without standardizing the performance of the flow cytometers, inter-instrument comparisons of data on these basis are, at best, rather inconclusive and the innate potential of these instruments has not been utilized.

The need for developing effective standardization procedures is magnified by the dramatic increase in commercial and custom-built flow cytometers which collect four, five, or more parameters. The sensitivity and accuracy in comparing cell population patterns over these parameters require that proper standards are available to normalize the performance of the instruments. In turn, the standards need to have characteristics which closely resemble the samples being analyzed, including apparent size, granularity, fluorescence spectra, and intensity (Schwartz, 1988). It is our intent to review the status of flow cytometry standardization and suggest several new standardization procedures.

II. Considerations for Qualitative Standardization

To avoid confusion in our discussion of standardization, "channel"-related terms need to be defined.

Fluorescence channels: the optical and electronic section of the flow cytometer which processes signals from specific portions of the spectra, e.g., FL1 and FL2, or FITC channel and PE channel.

Histogram channels: the linear scaler divisions into which the number of events are stored in list-mode data. These are geometric factors of 2, e.g., 64, 256, and 1024. Histogram channels are often displayed as a linear output of the data.

Relative channels: the continuum of linear channels obtained by factoring in gain settings to extend the range covered by linear amplifiers.

Relative linear channels: the conversion of histogram channels into a log scale, usually to the base 10, to correspond to the number of decades covered by the log amplifier. This log scale of relative linear channels may also be found on the graphic output displays of data.

A. Theoretical Considerations

1. Sample Space

Comparison of cell population patterns may be aided by considering that the sample will have as many dimensions in space as the number of parameters being measured by the flow cytometer. The sample space contains the characteristics of a cell population such as forward (FALS) and right (RALS) angle light scatter, as well as its fluorescent properties. Therefore, when using two fluorescent probes, e.g., fluorescein isothiocynate (FITC) and phycoerythrin (PE), along with the FALS, RALS, the cell population patterns can be described in four-dimensional space. Adding a third fluorescent probe would increase the sample space to five dimensions. Attempting to visualize five-dimensional sample space is a demanding task even for the most abstract thinkers. Although some software is available where three dimensions are presented on the sides of a cube, the usual presentation consists of a series of two-dimensional squares. These two dimensional presentations are usually referred to as dot plots or bivalent histograms. Improvements in software presentation, in which coloration of a particular population is consistent throughout its presentation across the parameters, have greatly helped this visualization task (Loken *et al.*, 1991; Bagwell, 1989).

2. Window of Analysis

The concept of a window of analysis is presented to help simplify standarization of the multidimensional flow cytometry data. A window of analysis is the range which is covered by the 256 or 1024 histogram channels of the various parameters of the flow cytometer. It can also be defined as the area of the sample space that can be viewed at particular instrument settings. Consider the bivarient or even single variant histograms of a window of analysis when using a log amplifier. The position of the "window" is dependent on the instrument setting (i.e., PMT or gain settings). Remember, that changing a PMT or amplifier gain does not change the sample properties, but merely how the instrument perceives the sample. It just repositions the window relative to the sample emissions. For example, when using a log amplifier, increasing the PMT voltage appears to move the sample distribution to the right. Some may mistakenly believe that this represents an increase in the fluorescence intensity of the sample; however, the sample is unchanged, and the window of analysis is simply being shifted to the left as illustrated in Figs. 1a and 1b. Note that the shape and relative positions of the distributions do not change when using a log amplifier.

The range of sample space covered by the window of analysis when using a linear amplifier is very limited, relative to a log amplifier. The use of relative channels, i.e., (channel number) × (maximum gain)/actual gain, can extend the

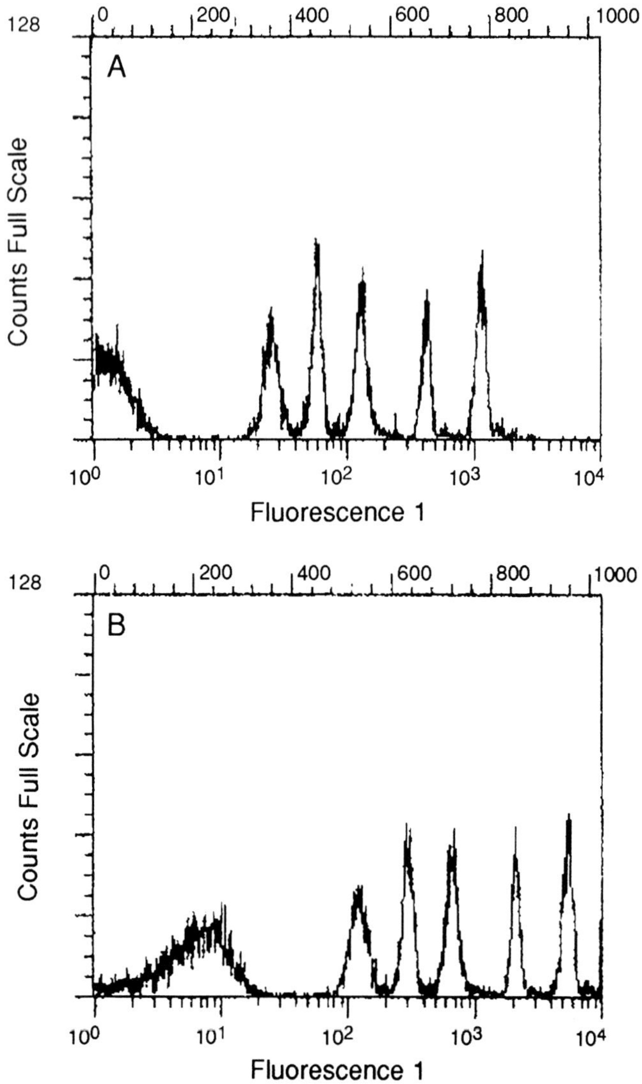

Fig. 1 FL1 window of analysis using a log amplifier showing six microbead populations where (a) the PMT is set at 600 V and (b) the PMT is set at 700 V. Note that all microbead populations remain in their relative positions and the window of analysis shifts to the left with the increase of PMT voltage.

range of sample space into a linear continuum. With a linear amplifier, a population distribution appears to widen, the further to the right it is positioned in the window of analysis, even though the sample does not change. With an increase of PMT voltage or amplifier gain, the zero point remains fixed and the sample space is stretched to the right with the higher intensities going off scale, as shown in Figs 2a and 2b.

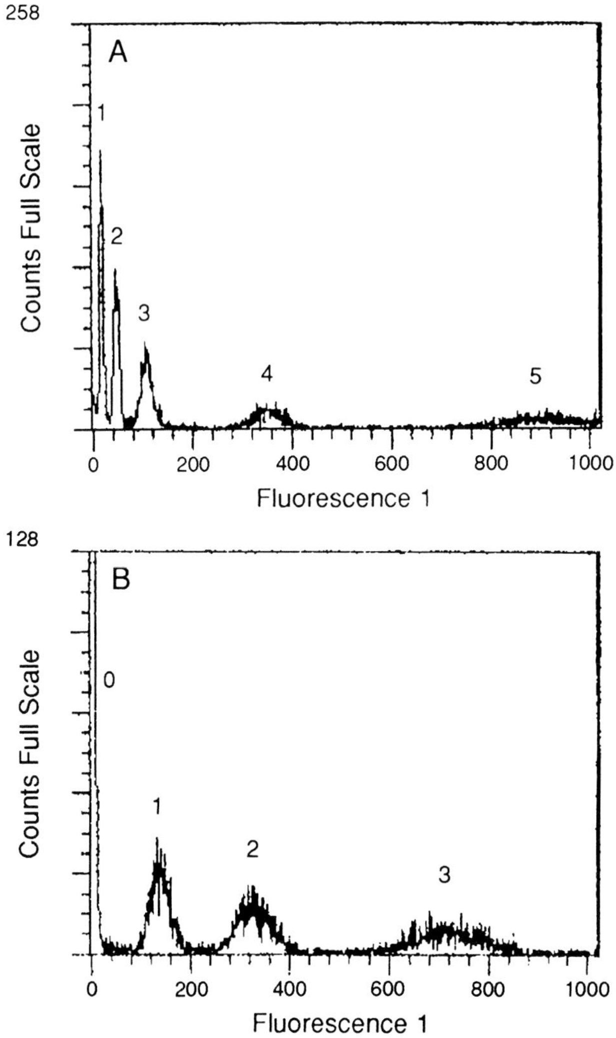

Fig. 2 FL1 window of analysis using a linear amplifier (a) at a gain setting of 1.0 showing five fluorescent microbead populations (1–5) and (b) at a gain setting of 9.99 showing four microbead populations (0–3). Note that the (0) microbead is up against the *y*-axis and cannot be seen in (a); however, it is shown in (b) because the amplifier "stretched out" the sample space near the origin at high gain settings.

3. Defining the Window of Analysis

A window of analysis can be defined by two parameters: (1) a reference point within the window and (2) the window's dimensions. The reference point in the window may be defined by a target channel. The dimensions of the window are related to the slope of the calibration lines of the respective parameters (e.g., FALS, RALS, and FL1).

The reference material, which matches the properties of the samples, is placed in the target channel by adjusting the instrument settings, usually PMT voltages. In the case of fluorescence measurements, the spectra of the reference material must match the spectra of the sample for the relationship to hold among different instruments.

The dimensions of the window of analysis are related to the coefficients of response, i.e., the slopes of the calibration lines, of the specific parameters defining the window. As illustrated in Fig. 3, the range of sample space which the histogram channels display is determined by the slope of the calibration line or the projection of the calibration line onto the axis. It should be noticed that a bivarient histogram has two independent dimensions for the window of analysis and the slope of each paramater needs to be determined independently.

B. Practical Considerations

1. Qualitative Standardization of Light Scatter Data

The size of a cell or particle is related to the FALS signal, whereas its granularity is related to the RALS signal. Since the major components of peripheral blood, i.e., lymphocytes, monocytes, and granulocytes, each have different size and granularity, they can easily be distinguished using FALS and RALS on most commercial flow cytometers. However, it is difficult to obtain complete separation of these populations using only light scatter data. For example, large lymphocytes and degranulated neutrophils may occur in the monocyte region, whereas small lymphocytes may occur where erythrocytes and debris appear.

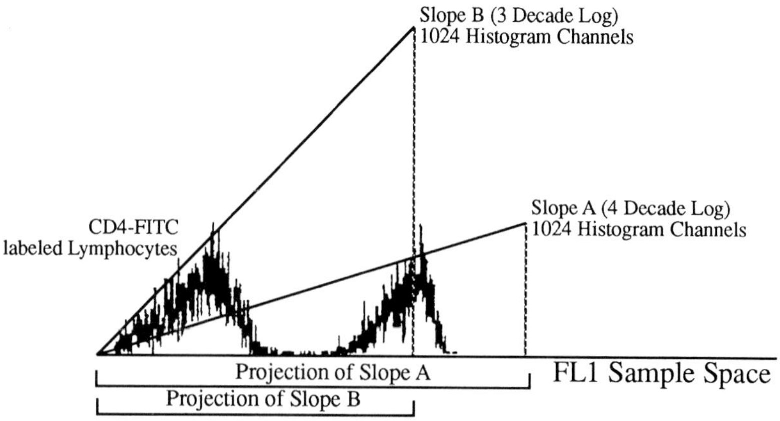

Fig. 3 An illustration of the ranges in samples space covered by two windows of analysis which have different slopes, A and B, due to having a different number of log decades, four and three, respectively, for their amplifiers. The slopes are represented by equal lengths of the 1024 histogram channels.

Such overlaps can cause significant errors in quantitating the percentage of subpopulations of cells labeled with a specific antibody. Corrective statistics regarding purity of the light scatter gate should be included in the report.

a. Reference Point

Comparisons of the cluster patterns of leukocytes require a common reference point in the light scatter window of analysis. A population of highly uniform microbeads may serve this purpose since their light scatter properties are not significantly affected by the aqueous solution in which they are suspended. Since polymeric microbeads have a different refractive index than biological cells (Lakowicz, 1983), they will have different light scatter properties than cells, especially RALS, even if they have the same physical size, and thus will usually appear as a distinct population providing a reference point.

b. Effect of Instrument Differences

Light scatter results are very sensitive to the configuration of the instrument optics. Significant differences in the position of cell populations occur in the light scatter window of analysis among different makes and models. Figures 4a and 4b depict the appearance of microbeads mixed with whole normal peripheral blood lysed with the same lysing solution, but analyzed on two different instruments. Note the significant difference of the forward angle light scatter profile of each of the populations on the different instruments.

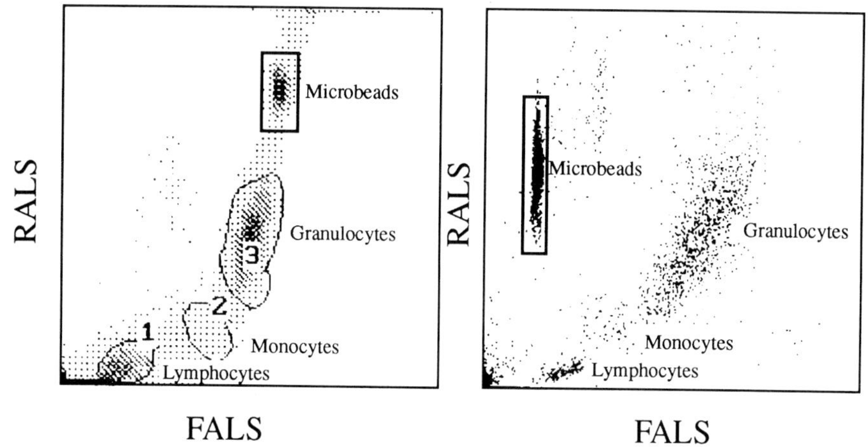

Fig. 4 Light scatter windows of analysis showing the relative positions of polymeric microbeads and leukocytes in whole lysed blood in (a) the Profile II (Coulter Electronics Inc.) and (b) the FACScan (Becton–Dickinson Immunocytometry Systems, Inc.) flow cytometers. Note, that all leukocyte populations look smaller, i.e., lower FALS, on the Profile II than the FACScan, relative to the microbeads.

c. Effect of Lysing Solution

The degree of swelling induced by different lysing solutions on a given cell population is also a factor which affects the relative positions of different cell populations in the light scatter window of analysis. The effect of different lysing solutions, i.e., FACS lyse (BDIS) and Coulter lyse (Coulter Electronics) on the light scatter profile of normal whole peripheral human blood was compared using the same flow cytometer (Schwartz and Fernández-Repollet, 1991). Figures 5a and 5b illustrate the variability in light scatter properties, due primarily to the swelling response of the cells to a particular lysing solution. This finding is in agreement with that of Carter and collaborators (Carter *et al.*, 1992).

It is evident from the above discussion that the goal of obtaining a common window of analysis for light scatter data is extremely difficult to achieve, especially with the current status of having various optical configurations in the different commercial flow cytometers and a variety of lysing solutions in use by the different laboratories. Qualitative standardization of light scatter where cluster patterns fall in specific positions in a common window of analysis will require an organized effort within the flow cytometry community, combined with focused cooperation among the manufacturers.

2. Qualitative Standardization of Fluorescence Intensity Data

Currently, the majority of the fluorescence data analysis obtained from a flow cytometer involves the use of region makers to determine the percentage of positive labeled cells in a population. As analysis and interpretation of flow

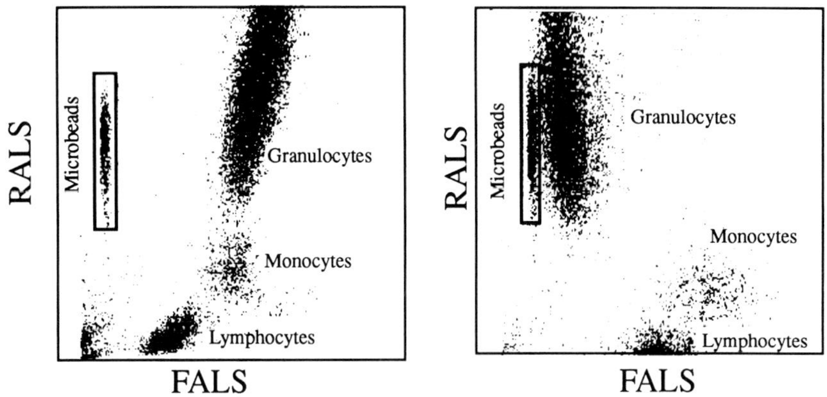

Fig. 5 Light scatter windows of analysis (FACScan, BDIS) showing the relative positions of polymeric microbeads and leukocytes in whole blood when lysed with (a) BDIS lysing solution and (b) Coulter lysing solution. Note that the granulocytes appear "shrunken" and the lymphocytes and monocytes appear "swollen" when lysed with Coulter lysing solution as compared to those lysed with BDIS lysing solution, when using the microbeads as a reference point.

cytometry data progress, more emphasis is being placed on cluster pattern recognition. This has proven to be especially important in the diagnosis, therapy monitoring, and prognosis of bone marrow-related disorders (Terstappen *et al.*, 1988), as well as activation of subpopulations in peripheral blood (Terstappen *et al.*, 1990).

a. Fluorescence Window of Analysis

The key to standardizing qualitative fluorescence intensity data is obtaining common windows of analysis for the fluorescence sample space. As previously mentioned, this can be accomplished by having a common reference point, e.g., target channels, and having the response coefficients (i.e., the slopes of the calibration lines) be the same for all instruments. Fortunately, the differences in design of flow cytometers with respect to their fluorescence measurements is in some ways less critical than that of the light scatter measurements (Shapiro, 1988). Moreover, the barrier filters, PMTs, and amplifiers of commercial instruments are rather similar, often coming from the same sources. These factors, together with the availability of proper standards to establish normalized target channels, permit qualitative standardization of fluorescence intensity data and direct comparison of cell cluster position data among instruments.

b. Fluorescence Target Channels

When considering individual instruments and if the comparison of data among instruments is disregarded, any highly uniform fluorescent material (e.g., fixed chicken red blood cells or fluorescent microbeads of nonspecified spectra) may be employed as the reference material. However, when fluorescence data need to be compared among different instruments, a common window of analysis is essential and, consequently, the fluorescent reference material must have excitation and emission spectra which match those of the labeled samples. With matching spectra, the relative positions of the reference material and labeled cells are fixed in sample space. Therefore, a particle labeled with both fluorescein and phycoerythrin would be a suitable reference material to set target channels for samples labeled with the same two fluorochromes, as shown in Fig. 6.

c. Noise Level

As with other kinds of laboratory instruments, the noise level of a flow cytometer is the limit of detection of a real sample signal. The fluorescence threshold is equivalent to the noise level and nonfluorescent particles may be used for its determination. When the flow cytometer is set to be triggered by forward angle light scatter, it will also collect data in all other channels when a particle passes through the laser beam. If the particle is nonfluorescent, then the readings in the fluorescence channels represent only the noise of that particular channel. Noise may arise from both optical (fluorescent or scattered light from the laser, mirrors, filters, lenses, etc.) and electronic sources (PMTs,

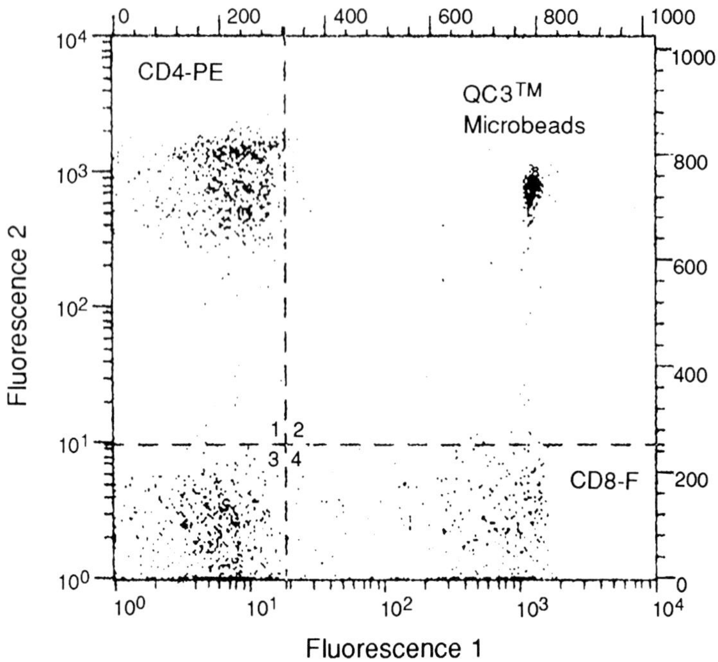

Fig. 6 Illustration of the relationships between unstained cells, FITC- and PE-labeled cells, and FITC/PE-labeled microbeads after positioning the labeled microbeads in their respective target channels in the window of analysis. Instruments using this method of setup will have a common window of analysis with cells appearing in comparable positions.

amplifiers, power supply, etc.). Specific sources of noise may be isolated by feeding electronic signals into the circuits, but the aggregate noise signal can only be determined by analyzing nonfluorescent particles which combines all noise sources, including scatter, that would be present when analyzing actual samples.

In qualitative terms, knowing the position of the noise level with respect to the unlabeled cells being assayed is extremely important since noise can mask the detection of low-level antigen expression. As long as the fluorescence level of the unlabeled cells is higher than the noise level, then the noise cannot interfere with the assay. On the contrary, when the noise level of the instrument is equal or greater than the autofluorescence of the cells, then the autofluorescence of the cells will be masked by the noise and they will appear in the same location as the nonfluorescent particles. Thus, the noise level needs to be determined in a quantitative manner and compared to the known values for that particular cell, as described below.

It is best to use unlabeled cells for this "sensitivity check" rather than isotypic controls since the lowest fluorescence intensity that a cell can exhibit will be evident before exposure to any fluorescent antibody. Furthermore, if it is shown that the instrument noise level is below the fluorescence of the unlabeled cell, then it follows that the instrument can detect the real autofluorescence signal of the cell, and, in turn, an approximation of the degree of nonspecific binding from the isotypic controls can be made.

To ensure that the noise level evaluation is meaningful, it is essential that the nonfluorescent particles used in the test are known to have a lower fluorescence than the unlabeled cells in all the fluorescent channels being used. Polymeric microbeads are ideal particles for this application; however, many such "unlabeled" microbeads contain impurities or are made of materials which are autofluorescent, and thus are not suitable for this application. Therefore, the fluorescent levels of the unlabeled microbeads must be "certified" to be lower than the autofluorescence of the specific unstained cells being analyzed.

The noise level of a flow cytometer has an additional significance with respect to the compensation circuits. These circuits are designed to subtract a percentage of the signal in the secondary channels as determined from the intensity in the primary channel. For example, fluorescein signals are subtracted from the FL2 channel as per the expression FL2-%FL1. A current practice in setting the compensation circuits is to place the unstained cell population in the lower left-hand corner of the window of analysis and use cells labeled with antibodies conjugated with specific fluorochromes, e.g., FITC and PE, to adjust these populations such that they are orthogonal to each other with respect to the unstained cell population. However, often populations which are dimmer than the population used to adjust the circuit are found to be overcompensated, whereas populations which are brighter are found to be undercompensated. This indicates that the instrument is not correctly compensated because the noise level is not being properly considered. This is represented graphically in Fig. 7. In this representation, the line of compensation described by expressions like FL2-%FL1 can be described by a line that pivots around the point of zero fluorescence, which can be approximated by the noise level.

When the circuit is properly adjusted, the compensation line will be parallel to the axes of the window of analysis. This yields a correct compensation across the intensity range. Since cells are autofluorescent, the position of the cells will be offset, respectively, when the instrument is adjusted such that the noise level is on scale. This setup has the additional advantage of knowing when overcompensation is present since the populations are not "crammed" against the axes.

This noise check has an additional application in determining whether the offset voltages of the amplifiers are properly adjusted. For this test, the compensation circuit is turned off to ensure that the unstained cells have a higher fluorescence than the nonfluorescent particles. Then, the compensation circuits are turn on and the particles and cells are reanalyzed. If the relationship remains such that the cells have a higher fluorescence than the particles, the pivot point is in the correct position and the instrument is performing properly. However, if the particles now appear more fluorescent than the cells, then the pivot point

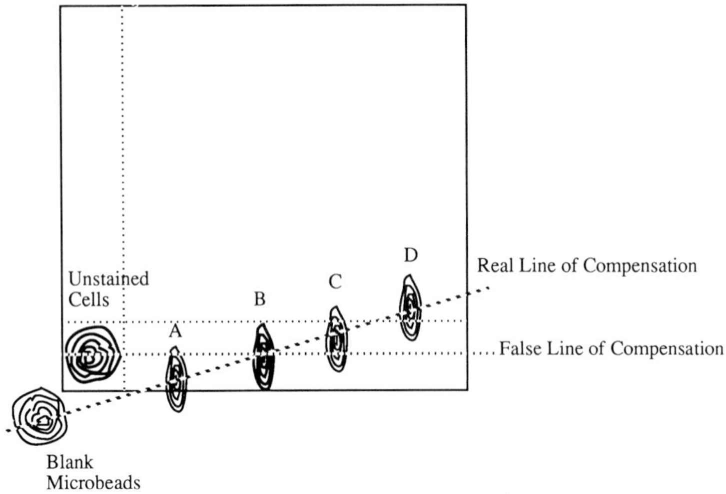

Fig. 7 Illustration of how compensation is set incorrectly by placing the unstained cell population in the corner of the window of analysis. Note that although the compensation appears to be correct for population (B), relative to the unstained cells, the other populations are not compensated correctly. If the blank microbeads were place on scale and the compensation readjusted, then all populations would be parallel to the axis and uniformly compensated across the intensity range.

is not in the correct position, indicating that the offset voltages of the amplifiers require adjustment.

In summary, detection of the noise level can be used to determine the limit of detection with respect to an assay, as well as to set correct compensation.

d. Instrument Setup and the Use of Appropriate Fluorescent Conjugates

One of the major notions working against standardization for flow cytometry is the perception that individual instrument setups are needed for each type of assay. This is fostered by the common practice of placing the unlabeled or isotype controls in the corner of the window of analysis allowing the autofluorescence and/or nonspecific binding of isotype controls to dictate the instrument settings. This approach to instrument setup is responsible for the loss of important information, namely, determination of the level of autofluorescence and the level of nonspecific antibody binding. However, if the instrument is using a log amplifier and it is a setup such that the noise level is just on scale, then this setup will be applicable for any assay.

It is thought by some that the ability to detect a fluorescent signal (i.e., sensitivity) can be enhanced by raising the PMT voltages. However, with log amplifiers, the noise level is, for the most part, as stated before, also fixed in sample space and only the window of analysis is shifted by PMT adjustment. In other words, since fluorescence levels cannot be measured below the noise level, changing the PMT will not improve the measurements.

Objections are raised that if the noise level is on scale, then the cells with high antigen expression will be off scale. This problem can be solved easily by the proper selection of the fluorochrome conjugated to a particular marker antibody. For example, CD8-PE-labeled lymphocytes do fall into the fourth decade of many instruments due to the large number of CD8 binding sites and the intense brightness of PE (i.e., high extinction coefficient). In contrast, CD4-FITC-labeled cells barely fall into the third decade because there are fewer CD4 binding sites and FITC has a lower extinction coefficient. These positions are "normalized" by labeling with CD4-PE and CD8-FITC (Stewart, 1990).

By using such labeling strategies, a single common window of analysis may be obtained for all instruments where the noise level, as well as the labeled populations, remains on scale.

e. Practical Applications of the Window of Analysis

In general terms, the hardware of most commercial flow cytometers is very similar. For example, most instruments use PMTs and linear and log amplifiers to detect and process signals from samples. The raw data (list-mode files) are stored and analyzed as histogram channels. However, differences in data begin to surface depending on the resolution of the scalar histogram channels used, i.e., 64, 256, 1024, or 4096, and how many decades the log amplifier covers.

Significant efforts from the flow cytometer manufacturers and third-party sources have led to the development of software that analyzes and presents data in a more meaningful manner. Unfortunately, these efforts have not been conducted in a coordinated manner among the manufacturers, resulting in a significant problem with respect to standardizing the presentation of flow cytometry analysis.

Data processed from software programs may be displayed as histogram channels (HC), usually 64, 256, or 1024 channels on a linear scale, or they may be presented as relative linear channels (RLC), which are displayed on a log scale if the log amplifier covers 3, 3.5, or 4 decades, which is specific to the make and model of the flow cytometer. It becomes clear that comparison of data using unspecified channels numbers, even if all instruments performed exactly the same, would be extremely difficult when using different software.

The development of software is a major task and is usually approached by a particular philosophy which in turn has particular tradeoffs. For example, some manufacturers want the program to be as flexible as possible, which requires greater understanding by the user of what the program does and how the results are presented. For instance, the user can have various options as to the output scale of histograms, e.g., 256 or 1024 histogram channels (linear scale) or 10,000 relative linear channels (log scale), as is the case with much of Becton–Dickinson Immunocytometry System acquisition and analysis software. Such flexibility makes standardization within even the same instrument model more difficult and complicated.

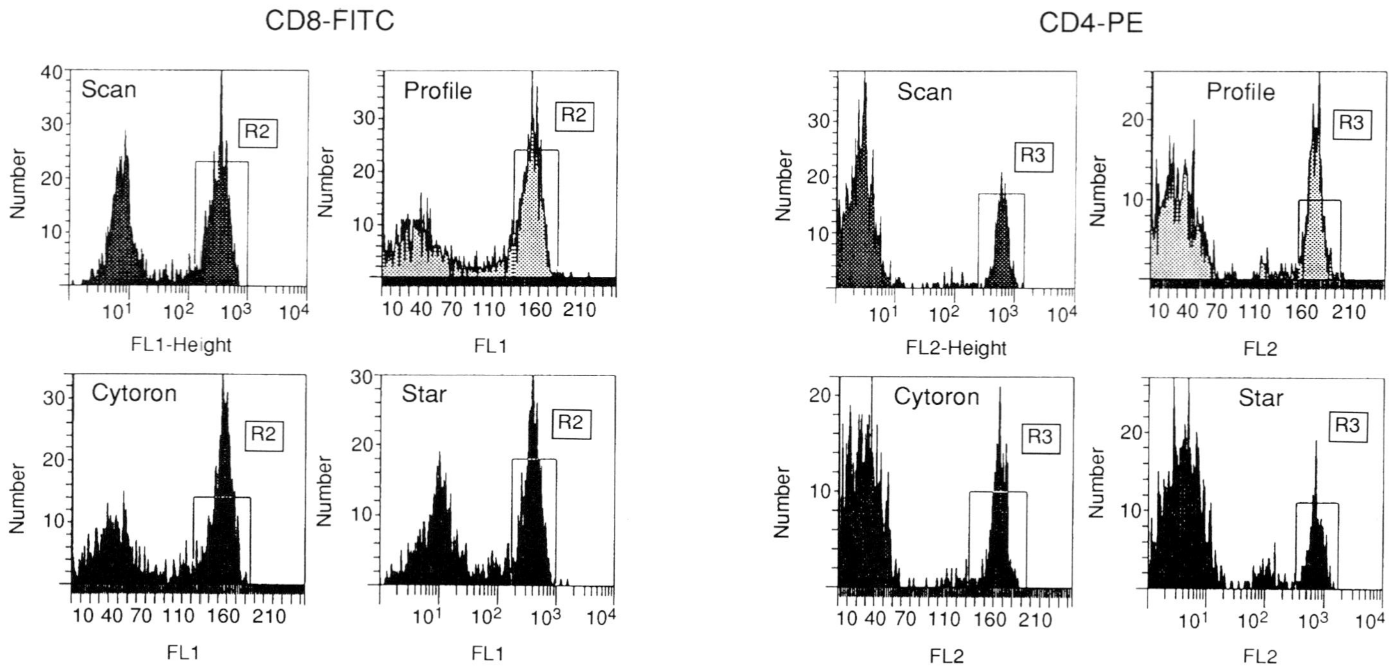

CD8-FITC
Scan
R2
Number
FL1-Height
Profile
FL1
Cytoron
Star
CD4-PE
R3
FL2-Height
FL2

Another approach to software is to have all output scales the same, e.g., 10,000 RLC, but this is still not free of confusion because the output scales may start at different points. For example, Becton–Dickinson Immunocytometry System, Inc., and Ortho Diagnostic Systems, Inc., start their relative linear scales at channel 1 (i.e., the log to the base 10 of channel 0), whereas Coulter Electronics, Inc., starts their RLC scales at 0.1. Knowledge of these differences and their effects on flow cytometry data analysis is essential when data comparison is required. It is evident that significant progress in standardization of flow cytometry interpretation can be made if just this one area is addressed.

To examine how the concept of a common window of analysis can be applied in comparing flow cytometry data obtained with different instruments and acquisition software, the following experiment was conducted by G. Stelzer from Cytometry Associates.

Four different models of flow cytometers manufactured by three different companies were used in this study. Attempts were made to limit sources of variables other than the instrumentation, i.e., the same reference standard, labeled cell samples, and list-mode analysis software were used with each instrument. Normal human peripheral whole blood was prepared by labeling with CD8-FITC and CD4-PE monoclonal antibodies, followed by lysing and washing. With the compensation circuits off, QC3 reference microbeads (Flow Cytometry Standards Corp.) labeled with FITC and PE were placed in initial target channels (normalized to the acquisition output scale of each instrument). The labeled cells were then run on each instrument and the compensation circuits adjusted. All the list-mode files from each instrument were analyzed with the same analysis program (WinList from Verity Software, Inc.). Figure 8 shows where each of the labeled cell populations appears in the common window of analysis. As can be seen, the positions of both the labeled and unlabeled populations in the one-dimensional window of analysis are quite comparable even though the output scales and amplifiers differ among the instruments.

III. Considerations for Quantitative Standardization

A. Quantitative Standardization of Light Scatter Data

Of the two measurements derived from light scatter signals, determination of cell size can be accomplished in a straightforward manner in the FALS channel with the proper use of matched standards. A set of particles of different

---

Fig. 8 Histograms showing immunophenotyping data obtained from four different flow cytometers (Scan, BDIS FACScan; Profile, Coulter Profile II; Cytoron, Ortho Cytoron Absolute; and Star, BDIS FACStar plus). Note that each of the cell populations labeled with CD8-FITC and CD4-PE, respectively, fall in the same positions when a common window of analysis is used for the different instruments.

sizes, which are made from the same material, may serve as light scatter sizing standards. Size-calibrated populations of microbeads synthesized from the same polymer which have the same refractive index and high uniformity may serve as such standards. These standards should be used to calibrate and determine the linearity of the response of the FALS channel. It may be noted that polymeric microbeads, e.g., 5 μm in diameter, appear to have the same FALS signal as cells which are larger, e.g., lymphocytes which are 8 μm in diameter. This is because the refractive index of cells is different from that of the microbeads (Shapiro, 1988). Correction factors are required to obtain accurate size measurements with FALS. Such corrections may be achieved by shifting the calibration plot while maintaining the slope generated by a cell population of known size (e.g., determined by scanning electron microscopy), until it falls on the proper position of the calibration plot.

Quantitative standardization and interpretation of RALS data are very difficult and complex subjects which are beyond the scope of this discussion. At present, quantitative granularity standards for such calibrations are unavailable. It is hoped that future efforts will provide such standardization tools.

B. Quantitative Fluorescence Intensity

Quantitative fluorescence measurements have the ability to go beyond determining the percentage of positively labeled cells in a particular population. In fact, such measurements are based on relative fluorescence intensities to determine labeled and unlabeled cells. The linear signal to channel response of flow cytometers in the fluorescence channels provides the means to make direct intensity measurements. This requires a unit of fluorescence intensity which is related to the particular fluorochrome and independent of the instrument.

1. MESF Definition and Applications

Expressing fluorescence intensity directly in terms of fluorochrome molecules would seem to be a logical approach, but it has a number of pitfalls. They include reduction of intensity as a result of quenching, e.g., energy transfer between close molecules, as well as changes in extinction coefficient and quantum efficiency due to binding and microenvironment effects. These factors can exert even a greater influence when determining the fluorescence intensity of particles, such as biological cells or microbeads.

A more consistent determination of the fluorescence intensity of particles can be made by comparing their intensity to that of the soluble fluorochrome under the same environment conditions which can be readily controlled and reproduced in the laboratory. These comparisons will have a constant quantitative relationship as long as both the excitation and emission spectra of the solution and labeled particle are the same. Therefore, a set of particles labeled with a specific fluorochrome whose spectra match a solution of the fluorochrome can serve as calibrators for cells labeled with the same fluorochrome when suspended in the same medium.

A convenient way of expressing a unit of fluorescence intensity is in molecules of equivalent soluble fluorochrome (MESF). When applying this unit, the specific fluorochrome must be indicated since, for example, 50,000 MESF of fluorescein is not equivalent to 50,000 MESF of phycoerythrin.

2. Calibration of Fluorescence Intensity

Calibrating the fluorescence channels of a flow cytometer involves determining the instrument response to specific fluorescent signals across the entire fluorescence range. Therefore, when analyzing FITC- or PE-labeled cells, the calibrations must be expressed in units, e.g., MESF units, of the respective fluorochromes.

Because of the high uniformity, sets of fluorescent microbeads labeled with these fluorochromes and having preassigned MESF values can serve as a convenient calibrator for flow cytometers, as well as fluorescent microscopes. By plotting the peak channels of these microbeads against their MESF values (or performing a linear regression) a calibration line is obtained. With such calibration lines, the intensity of labeled cell populations can be determined by finding the corresponding MESF values for the peak channel of the cell distribution, as shown in Fig. 9. Many laboratories have incorporated this methodology in their daily calibration procedure to assure reproducibility in instrument performance (Fay *et al.*, 1991; Schols *et al.*, 1990; Bohmer *et al.*, 1992).

3. Evaluation of Instrument Performance

There is far more to evaluating the performance of a flow cytometer than merely minimizing the coefficient of variation (CV) for a particular population of microbeads. This will only assess the alignment and focus of the optical components. A more complete instrument evaluation will require information regarding the linearity, resolution, and noise level. This cannot be determined with a single population of microbeads but requires a series of quantitative microbead standards.

After plotting a calibration line of MESF values vs peak channels of the quantitative microbead standards, the linearity of the fluorescence response is determined by calculating the coefficient of determination (r^2). Care should be taken in interpreting this value since the coefficient of determination is very insensitive to log data and most fluorescence data are obtained using a log amplifier. For example, a r^2 value of 0.95 may be quite acceptable with linear data, whereas, it takes a value of 0.99 to be acceptable for log data. A better measure of linearity for fluorescence log data is the average residual percent (AvRes%) which is the root mean square difference of where the points fall relative to the regression line. One set of suggested criteria for AvRes% of flow

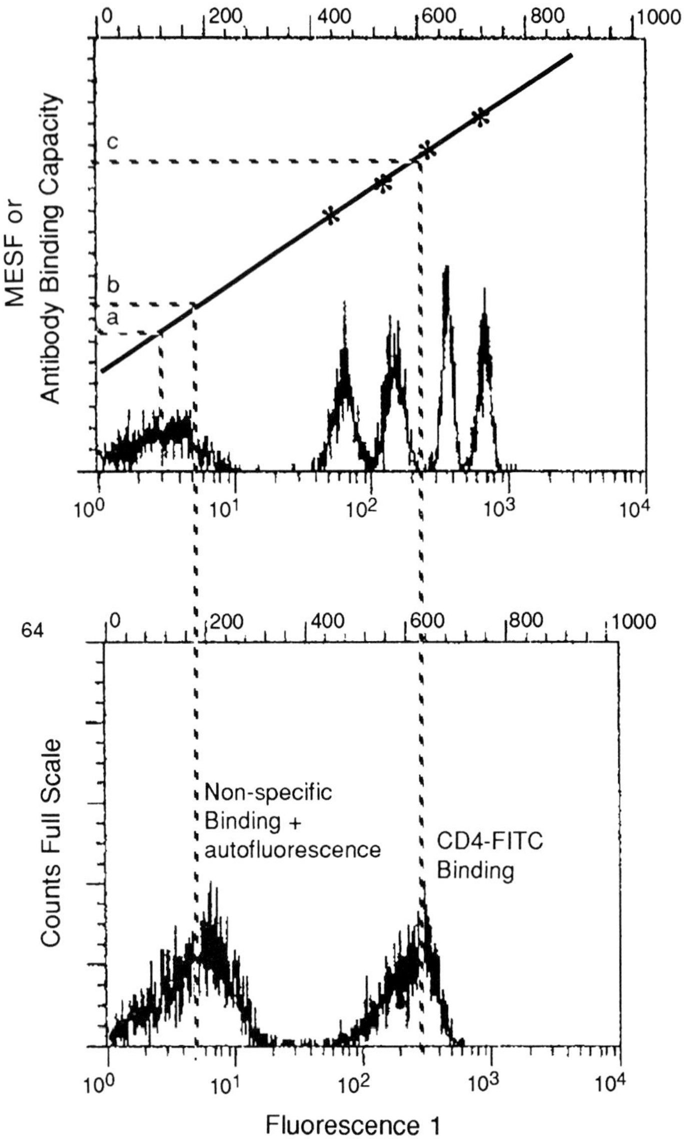

Fig. 9 Illustration of the relationship between the quantitative microbead standards and the intensities of cells as determined from the calibration line. Note that the MESF or ABC value of the noise level (a) of the instrument can be determined from the position of the blank microbead on the calibration line. In addition, both the level of autofluorescence/nonspecific binding (b) and specific binding (c) can be determined from the calibration line.

cytometers currently under investigation is the following: acceptable <1%, marginal 1–3%, and unacceptable >3%. These criteria appear to have held up in a number of instrument performance surveys (Ehman, 1992; Vogt *et al.*, 1991).

Resolution is a term used to describe the smallest difference detectable between two signals of the same properties, e.g., two cells labeled with FITC. It is not accurate to describe resolution between an autofluorescent cell and a cell labeled with FITC. Resolution must be evaluated in several portions of the window of analysis when using a log amplifier. In the lower decade, a difference of two histogram channels may only represent 50 MESF units, whereas, in the fourth decade, two histogram channels may represent 5000 MESF units. A qualitative assessment of resolution may be made by examining the separation among the populations of the quantitative microbead standards.

The quantitative noise level of the fluorescence channels is also assessed with the calibration line by determining the peak channel of a nonfluorescent reference material and finding its corresponding MESF value (Fig. 9). This value allows direct comparison of instrument noise levels. In regard to the assay, the noise level should be lower than the autofluorescence MESF values of the unstained cells.

The calibration line also provides important information with respect to the window of analysis, e.g., the response coefficients (slopes) of the calibration lines are directly related to the size and shape of the window of analysis (Fig. 3). By comparing the slopes of specific fluorescence channels among instruments, size comparisons of their respective windows of analysis can be made directly.

Therefore, a comprehensive evaluation of instrument performance not only requires a low CV on alignment microbeads to ensure the instrument is in alignment, but also an evaluation of the linearity, resolution, and noise level of each fluorescence channel.

C. Quantitative Antibody Binding Capacity

The ultimate goal for flow cytometry is the determination of the number of antibodies binding to specific cell populations. The key to measuring the number of binding antigens on a cell using fluorescent antibodies resides in quantitating the fluorescence intensity of the particular cell and translating that intensity into the number of bound antibodies. Quantitation of fluorescence intensity is an indirect way of making this measurement since the conjugated antibodies have an average fluorescence. If this average antibody fluorescence could be measured in MESF units, using the calibration plots previously described, then the number of antibodies could be determined by dividing the total MESF intensity of the cell population of the average MESF intensity per antibody.

Note that the MESF intensity, or "effective *F/P*," of an antibody is not the same as the *F/P* ratio which is the actual number of fluorochromes bound to an antibody. The *F/P* ratio does not directly translate into the fluorescent intensity associated with the cell because of environmental conditions such as dye ionizability, pH, and quenching. The effective *F/P* ratio is determined by measuring the fluorescence of the antibody, whereas the *F/P* ratio is measured by absorbance (Schwartz, 1988; Hoffman *et al.*, 1992).

1. Indirect Quantitation of Binding Antibodies

An easy method to determine the effective *F/P* ratio of antibodies, or the average fluorescence intensity per antibody molecule, consists of saturating a microbead population which has a calibrated number of binding sites and determine its MESF intensity on a flow cytometer precalibrated with the MESF calibration standards. This methodology has been used successfully with reticulocytes (Schimenti *et al.*, 1992), epidermal growth factor receptor expression (Lopez *et al.*, 1992), nucleoside transporter sites (Jamieson *et al.*, 1993), and the development of anti-OKT3 antibodies (Lim *et al.*, 1989).

2. Direct Quantitation of Binding Antibodies

A new methodology has been developed which allows the direct determination of antibody binding capacity by flow cytometry. Rather than calibrating and evaluating the instrument in terms of fluorescent units, this can be accomplished in terms of the specific antibody being measured. Calibration would be expressed in terms of binding antibodies per histogram channel. The noise level would represent the lowest detectable number of the specific antibody. This would mean that the instrument would have to be calibrated for each antibody used; however, such a calibration would take into consideration all the correction factors related to fluorescence measurements, e.g., quenching, changes in extinction coefficient, and quantum efficiency.

This methodology requires a series of particle populations which are able to bind calibrated numbers of antibodies and which would have to be saturated with the antibody of interest. The peak channels of each particle population would be plotted against their respective binding capacities. Different antibodies would generate different calibration lines depending on a number of factors, e.g., the effective *F/P* ratio and quenching. Finally, the number of antibodies binding to a specific cell population can be directly read from the calibration line for that antibody (Figure 9). This method has the advantage of being independent not only of the instrument, but also of the conjugated fluorochrome.

IV. Conclusions

The ability of flow cytometry instrumentation has far surpassed the simple determination of percentage of positive cells. These instruments have the capabilities of determining fluorescence intensities in quantitative units and even measuring the number of antibodies binding to specific subpopulations of cells. However, proper standardization and calibration are required for full realization of these potential applications. Properly designed standards can provide comparable data on instrument performance, as well as assay analysis which is instrument independent and comparable over time and among all instrument makes and models. This will certainly propel clinical immunophenotyping into exciting new areas.

Acknowledgments

The authors thank Francis Mandy, Kathy Muirhead, Alan Landay, Robert Vogt, and Howard Shapiro for many valuable comments. Mrs. Marinelly Velilla provided excellent technical assistance. The development of the quantitative antibody binding capacity standards was supported in part by the National Cancer Institute of the NIH (SBIR Grant No. R44-CA 48570).

References

Bagwell, C. B. (1989). "ModFit Program." Verity Software House, Inc., Topsham, Maine.

Böhmer, R. H., Trinkle, L. S., and Staneck, J. L. (1992). *Cytometry* **13,** 525–531.

Carter, P. H., Resto-Ruiz, S., Washington, G. C., Ethridge, S., Palini, A., Voght, R., Waxdal, M., Fleisher, T., Noguchi, P. D., and Marti, G. E. (1992). *Cytometry* **3,** 68–74.

Ehman, D. (1992). *Flow Cytometry Stand. Forum* **4**(1), 1–5.

Fay, S. P., Posner, R. G., Swann, W. N., and Sklar, L. A. (1991). *Biochemistry* **30,** 5066–5075.

Hoffman, R. A., Recktenwald, D. J., and Vogt, R. F. (1992). *In* "Clinical Flow Cytometry: Principles and Applications" (K. D. Bauer, R. E. Duque, and T. V. Shankey, eds.), pp. 469–477. Williams & Wilkins, Baltimore, MD.

Horan, P. K., Muirhead, K. A., and Slezak, S. E. (1990). *In* "Flow Cytometry and Sorting" (M. R. Melamed, T. Lindmo, and M. L. Mendelsohn, eds.), pp. 397–414. Wiley-Liss, New York.

Jamieson, G. P., Brocklebank, A. M., Snook, M. B., Sawyer, W. H., Buolamwini, J. K., Paterson, A. R. P., and Wiley, J. S. (1993). *Cytometry* **14,** 32–38.

Lakowicz, J. R. (1983). "Principles of Fluorescence Microscopy." Plenum, New York.

Lim, V. L., Gumbert, M., and Garovoy, M. R. (1989). *J. Immunol. Methods* **121,** 197–201.

Loken, M. R., Civin, C. I., Shah, V. O., Fackler, M. J., Segers-Nolten, I., and Terstappen, L. W. M. M. (1991). *In* "Flow Cytometry in Hematology" (O. D. Laerum and R. Bjerknes, eds.), pp. 31–42. Academic Press, San Diego.

Lopez, J. M., Chew, S. J., Thompson, H. W., Malter, J. S., Insler, M. S., and Beuerman, R. W. (1992). *Invest. Ophthalmol. Visual Sci.* **33,** 2053–2062.

Muirhead, K. A. (1992). *In* "Clinical Flow Cytometry: Principles and Applications" (K. D. Bauer, R. E. Duque, and T. V. Shankey, eds.), pp. 177–199. Williams and Wilkins, Baltimore, MD.

National Committee for Clinical Laboratory Standards (1992). "Clinical Applications of Flow Cytometry: Quality Assurance and Immunophenotyping of Peripheral Blood Lymphocytes: Proposed Guideline," NCCLS Doc. H42-T, No. 12, p. 6. Vol. 12 No. 6, Villanova, PA.

Schimenti, K. J., Lacerna, K., and Wambler, A. (1992). *Cytometry* **13,** 853–862.

Schols, D., Pauwels, R., Desmyter, J., and De Clercq, E. (1990). *Cytometry* **11,** 736–743.

Schwartz, A. (1988). "Monograph: Fluorescent Microbead Standards." Flow Cytometry Standards Corporation, Research Triangle Park, North Carolina.

Schwartz, A., and Fernández-Repollet, E. (1991). *Flow Cytometry Stand. Forum* **3**(2) 7–8.

Shapiro, H. (1988). "Practical Flow Cytometry." Liss, New York.

Stewart, C. C. (1990). *In* "Methods in Cell Biology" (Z. Darzynkiewicz and H. A. Crissman, eds.), Vol. 33, pp. 427–450. Academic Press, San Diego.

Terstappen, L. W. M. M., Shah, V. O., Civin, C. I., Hurwitz, C. A., and Loken, M. R. (1988). *Cytometry* **9,** 477–484.

Terstappen, L. W. M. M., Hollander, Z., Meiners, H., and Loken, M. R. (1990). *J. Leukocyte Biol.* **48,** 138–148.

Vogt, R. F., Cross, G. D., Phillips, D. L., Henderson, O., and Hannon, W. H. (1991). *Cytometry* **12,** 525–536

CHAPTER 36

Phase-Sensitive Detection Methods for Resolving Fluorescence Emission Signals and Directly Quantifying Lifetime*

John A. Steinkamp
Los Alamos National Laboratory
University of California
Los Alamos, New Mexico 87545

I. Introduction

A. Background

Clinical tests and biological experiments performed by flow cytometry (FCM) often require the labeling of cells and subcellular components, e.g., chromosomes, with multiple fluorochromes for correlated analysis of macromolecules, such as DNA, RNA, protein, enzymes, lipids, and cell-surface receptors. A major limitation of these procedures can be the availability of fluorochromes having common excitation regions, i.e., requiring only one excitation source,

* This chapter was written under the auspices of the United States Department of Energy.

and emission spectra that are sufficiently separated to permit measurement by conventional multicolor detection methods that employ dichroic and band-pass filters (Steinkamp *et al.*, 1987). If the fluorochromes have partially separated emission spectra, differential amplifiers can be employed to resolve signals by electronic compensation (Loken *et al.*, 1977). Spectroscopic methods also have been developed to record fluorescence emission spectra on each cell in real time and to resolve multiple emission spectra signals using computational methods (Buican, 1990). If fluorochromes cannot be resolved by these means, but have separated excitation spectra, multiple excitation sources can be employed to sequentially excite cells and spatially resolve the spectrally overlapping fluorescence signals (Steinkamp *et al.*, 1992).

In addition to utilizing the spectral properties of fluorochromes to differentiate and measure cellular features, the excitated state lifetimes also can provide a means to discriminate among the different fluorochromes. A new FCM approach, based on phase-resolved fluorescence spectroscopy methods (Veselova *et al.*, 1970; Lakowicz and Cherek, 1981), that provides unique capabilities for separating signals from multiple overlapping emissions in fluorochrome-labeled cells as they pass across a modulated excitation source, has been developed (Steinkamp and Crissman, 1993a). The measurement of fluorescence lifetime also is of importance because it provides additional information about fluorochrome/cell interactions. An important advantage of lifetime measurements is

Table I
Fluorescence Lifetimes and Corresponding Phase Shifts of Emissions from Fluorochromes Used to Label Cells for Analysis by Flow Cytometry

Fluorescent dye/compound	Excitation wavelength (nm)	Fluorescence lifetime (nsec)	Phase shift[a] at 10 MHz (degrees)
Hoechst 33258	UV	3.5	12.4
DAPI (DNA)	UV	4.0	14.0
Mithramycin	420	3.0	10.7
Propidium iodide	515	1.2	4.3
Propidium iodide (cells)	515	13.0	39.2
Ethidium bromide	515	1.8	6.5
Ethidium bromide (cells)	515	19.0	50.0
Ethidium bromide (DNA)	515	22.5	54.6
Acridine orange (cells)	480	3.0 (Gn), 13.0 (Rd)	10.7, 39.2
Fluorescein	480	4.7	16.4
FITC	480	3.6	12.5
Texas red-avidin	530	4.6	16.1
Phycoerythrin-avidin	530	3.5	12.6

[a] Phase shift equals arctan $\omega\tau$, where $\omega = 2\pi f$ the angular frequency and τ = fluorescence lifetime.

that lifetimes in some cases can be considered as absolute quantities. However, the lifetime of dye molecules bound to cellular macromolecules can be influenced by physical and chemical factors near the binding site, such as solvent polarity, cations, pH, energy transfer, excited-state reactions, and quenching. Thus, lifetime measurements can be used to probe the cellular environment, possibly including structure changes such as those that occur in DNA and chromatin during the cell cycle. Fluorescence lifetime measurements in flow, as expressed by the phase shift from fluorochrome-labeled particles, also has been reported (Pinsky *et al.*, 1993). Table I lists the lifetimes of some typical fluorochromes that are used to quantify cellular DNA, total protein, and antibody labeling to cellular antigens.

B. Theory

Fluorochrome-labeled cells are analyzed as they intersect an intensity-modulated laser excitation beam consisting of a DC and a high-frequency (sinusoidal) excitation component (see Fig. 1). Fluorescence is measured orthogonal to the laser beam–cell stream intersection point using a single detector consisting of a collection lens, a long-pass barrier filter, and a photomultiplier tube. The fluorescence signals, which are shifted in phase (ϕ) relative to the excitation frequency and amplitude demodulated, are processed by phase-sensitive signal detection electronics to resolve signals from heterogeneous fluorescence emissions resulting from differences in their lifetimes and quantify fluorescence lifetimes directly by the two-phase ratio method.

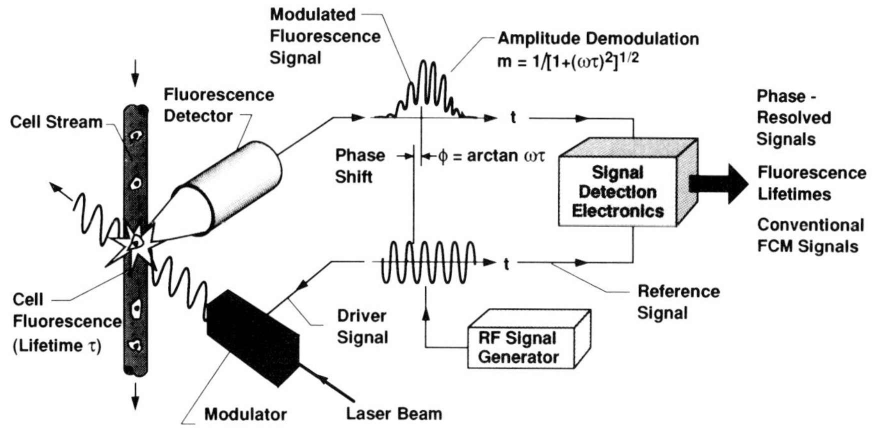

Fig. 1 Conceptual diagram of the phase-sensitive flow cytometer illustrating the laser excitation source, modulator, RF signal generator, laser beam-cell stream intersection point in flow chamber (flow chamber not shown), fluorescence detector, and signal detection/processing electronics. The fluorescence signal phase shift (ϕ) with respect to the reference signal is equal to the arctan $\omega\tau$, where ω is the angular frequency (2πf) and τ is the fluorescence lifetime.

The time-dependent fluorescence emission signal [$v(t)$] is a high-frequency, Gaussian-shaped, modulated pulse that results from the passage of the cell across the focused laser beam and it can be expressed as

$$v(t) = V[1 + m\cos(\omega t - \phi)] \cdot e^{-a^2(t-t_0)^2} \tag{1}$$

where V is the signal intensity, ω is the angular excitation frequency, ϕ and m are the respective signal phase shift and demodulation terms associated with a single fluorescence decay time (τ), t is time, and a is a term related to the velocity of a cell crossing the laser beam at time t_0 (see Fig. 2A). This equation

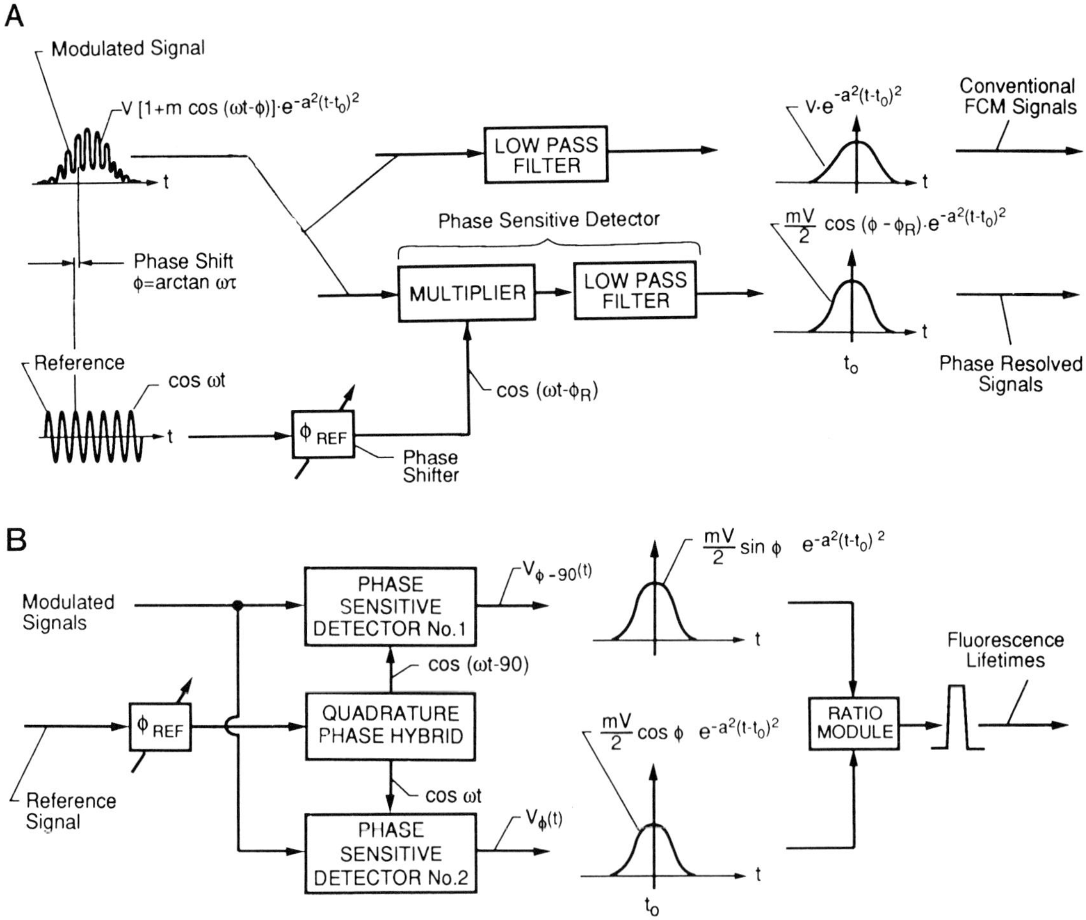

Fig. 2 Block diagram of the signal detection electronics for obtaining conventional FCM signals by low-pass filtering and phase-resolved signals by phase-sensitive detection (homodyne method) (A) and the two-phase ratio method for quantifying fluorescence lifetimes (B).

is derived for cells that are excited by an excitation source with a 100% depth of modulation. A more general expression will take into account the laser excitation depth of modulation factor (m_{ex}) which will reduce the high-frequency signal amplitude term. The CW-excited DC signal component is extracted using a low-pass filter to give conventional fluorescence intensity information. In addition, other optical parameters, e.g., light scatter, are obtainable by low-pass filtering. The high-frequency fluorescence signal component, which is shifted in phase (ϕ) by an amount,

$$\phi = \arctan \omega\tau, \tag{2}$$

relative to the excitation frequency and demodulated by a factor (m), where

$$m = 1/(1 + (\omega\tau)^2)^{1/2}, \tag{3}$$

is processed by a phase-sensitive detector (PSD) consisting of a multiplier and a low-pass filter. A phase shifter is used to shift the phase (ϕ_R) of the reference signal input to the multiplier with respect to modulated fluorescence signal. The PSD output is a Gaussian-shaped signal that is proportional to fluorescence intensity, the demodulation factor, and the cos (ϕ-ϕ_R) expressed as

$$v_o(t) = \frac{1}{2} mV \cos(\phi - \phi_R) \cdot e^{-a^2(t-t_0)^2}. \tag{4}$$

The output intensity can thus be made to vary from both positive and negative values depending upon the sign of the ϕ-ϕ_R phase shift term.

The principle of phase suppression, as applied to flow, for separating two fluorescence signals having different emission decay times, i.e., phase shifts, by phase-sensitive detection, is illustrated below. The output of the PSD is expressed (by superposition) as

$$\begin{aligned} v_o(t) = {} & \frac{1}{2} m_1 V_1 \cos(\phi_1 - \phi_R) \cdot e^{-a^2(t-t_0)^2} \\ & + \frac{1}{2} m_2 V_2 \cos(\phi_2 - \phi_R) \cdot e^{-a^2(t-t_0)^2}, \end{aligned} \tag{5}$$

where V_1 and V_2 are the signal intensities, m_1 and m_2 are the demodulation factors, and ϕ_1 and ϕ_2 are the phase shifts that result when a cell stained with two fluorochromes, each having a different lifetime τ_1 and τ_2, is excited with a modulated excitation source. To resolve either of the two signals the reference phase is shifted by an amount $\pi/2 + \phi_1$ or $-\pi/2 + \phi_2$ degrees. This results in one signal being passed and the other being nulled. For example, if the reference phase is adjusted to equal $-\pi/2 + \phi_2$ degrees, the detector output is expressed as

$$v_o(t) = \frac{1}{2} m_1 V_1 \sin(\phi_2 - \phi_1) \cdot e^{-a^2(t-t_0)^2}. \tag{6}$$

Similarly, if the reference phase is adjusted to equal $\pi/2 + \phi_1$ degrees, the output is expressed as

$$v_o(t) = \frac{1}{2} m_2 V_2 \sin(\phi_2 - \phi_1) \cdot e^{-a^2(t-t_0)^2}. \tag{7}$$

Both signals are resolved, but with a loss in amplitude [sin $(\phi_2 - \phi_1)$].

Fluorescence lifetime is quantified directly by the two-phase method (Steinkamp and Crissman, 1993b) (see Fig. 2B). A quadrature phase hybrid circuit is used to form two reference signals that are 90° out of phase with each other. These signals are input as references to two phase-sensitive detectors, the filtered outputs of which are expressed as

$$v_{\phi-90}(t) = \frac{1}{2} mV \sin\phi \cdot e^{-a^2(t-t_0)^2} \tag{8}$$

and

$$v_{\phi}(t) = \frac{1}{2} mV \cos\phi \cdot e^{-a^2(t-t_0)^2} \tag{9}$$

where ϕ is the signal phase shift (see Eq. 2). The $v_{\phi-90}(t)/v_{\phi}(t)$ ratio expression results in the tan ϕ, which is directly proportional to the fluorescence decay time expressed as

$$\tau = \frac{1}{\omega} \tan\phi = \frac{1}{\omega} [V(\phi - 90)/V(\phi)]. \tag{10}$$

II. Application

This technology is so new that many of its potential applications have not yet been fully explored and developed. Like conventional flow cytometers, the phase-sensitive cytometer can analyze fluorochrome-labeled cells or subcellular components for clinical and biomedical research applications. However, because this cytometer can electronically separate the signals from different fluorochrome emissions based on their lifetimes and also make conventional measurements, it has a wide range of technical applications. Of particular importance are fluorescent probes in which the lifetime changes between the bound and unbound (free) state, e.g., calcium indicators. Also of interest are fluorescent probes, such as Hoechst 33258, in which the lifetime changes as a function of energy transfer/quenching by the surrounding agents, such as DNA-incorporation BrdU. Phase-sensitive detection methods also have the potential to reduce background interferences, e.g., cellular autofluorescence and Raman and Rayleigh scatter, that cause decreased measurement sensitivity and precision in the analysis of cells and subcellular components, e.g., chromosomes, by FCM. In addition, fluorescence lifetime can be measured and used as a spectroscopic probe to study the interaction of fluorochromes with their targets (as in structural biology studies), each other, and the surrounding environment. The present phase-sensitive cytometer can resolve two single-component fluorochrome mixtures and determine the lifetime of single-component fluorochromes using a single-frequency excitation beam. The procedure requires that the precise phase shift for the suppression of each component be determined. Furthermore, only one component can be suppressed at a time, so that in a solution containing more than two emitting species, the individual components cannot be resolved. Alternative methods for the resolution of multicomponent emissions will require that multifrequency excitation be employed (Jameson *et al.*, 1984).

III. Cell Preparation and Staining

Chinese hamster cells (line CHO) maintained in exponential growth in suspension culture were harvested, fixed in 70% ethanol (1 hr), and centrifuged, and the ethanol was removed by aspiration prior to staining. Fixed cells were stained for DNA alone in a PBS solution containing 1.5 μg/ml PI (Polysciences, Warrington, PA) and 50 μg/ml RNase (Worthington, Freehold, NJ, Code R) for 30 min at 37°C; for total protein alone in a PBS solution containing 0.5 μg/ml fluorescein isothiocyanate (FITC), isomer 1 (BBL, Cockeysville, MD), and 50 μg/ml RNase for 30 min at 37°C; and for combined DNA and protein in a PBS solution containing 1.5 μg/ml PI, 0.5 μg/ml FITC, and 50 μg/ml RNase for 30 min at 37°C (Crissman and Steinkamp, 1982). The cell density was maintained at 2.5×10^6 cells/ml in all staining solutions. Stained samples were analyzed by phase-sensitive FCM at room temperature within 1–2 hr after staining.

IV. Critical Aspects of the Procedure

This new technology is readily adaptable to commercial flow instruments by adding an optical modulator, a frequency generator (low phase noise), a high-speed fluorescence detector/preamplifier, and phase-sensitive detection electronics. We selected an electro-optic modulator (Enscoe and Kocka, 1984), driven by a sine wave generator, to modulate the laser beam, because of its simplicity in both excitation and signal demodulation/detection. Acousto-optic modulators also have been used in flow cytometers for quantifying fluorescence lifetimes (Pinsky *et al.*, 1993). Although the fluorescence measurements in the 1- to 10-MHz range are essentially the same as those obtained by conventional

FCM (Steinkamp and Crissman, 1993b), improved measurement precision and sensitivity can be obtained by placing a high-frequency band-pass filter/amplifier between the fluorescence detector and the PSD multiplier input to reduce wide-band noise interference. A mode-locked laser for excitation (harmonic), coupled with frequency heterodyning for synchronous detection of phase-resolved fluorescence emission signals at lower frequencies, can be used to extend the frequency range for quantifying multiple decay times (Alcala *et al.*, 1985).

The resolution of fluorescence signals from cells stained in combination with PI and FITC was achieved by phase-sensitive detection of their respective decay times observed as phase shifts with respect to a reference signal. Although the emissions of these fluorochromes are separable by two-color fluorescence measurement methods, we processed the total fluorescence emission signal electronically and resolved the individual signal components based only on their decay times (phase shifts). By separating the fluorescence emission signal components electronically rather than by optical filters, the entire emission spectrum of each component was used and thus the light loss due to optical filters was eliminated. However, each signal was reduced in amplitude by the sin $(\phi_2 - \phi_1)$ factor.

The longest lifetimes that can be used in the separation of fluorescence emission signals by phase-sensitive detection depends upon the lowest usable excitation frequency. In our system, we have experimentally determined this to be about 0.5 MHz, which corresponds to 318 nsec, i.e., 45° phase shift. The shortest lifetime depends on the maximum usable excitation frequency, which is about 35MHz (reference phase shifter 3 dB half-power point) and corresponds to 0.5–1.0 nsec. The phase-resolved measurement resolution ranges from 0.1° at 0.5 MHz to 6.28° at 35 MHz as determined by the 0.5-nsec minimum switchable increment of the reference phase shifter. We can resolve 0.5- to 1.0-nsec signal-related phase shift differences using simulated test signals and 1- to 1.5-nsec decay time differences between the autofluorescence background and FITC signals have been detected from FITC-labeled autofluorescent macrophages (unpublished data). A Bishop Instruments 0–5 (± 0.002)-nsec delay line recently has been installed as the reference phase shifter (see Fig. 2B) to improve phase measurement resolution.

Direct measurement of fluorescence lifetime on a cell-by-cell basis was demonstrated using the two-phase ratio method. The longest and shortest lifetimes that can be measured depend upon the usable frequency range of the dual-channel two-phase PSD. In our system, the limiting component is the 2-32-MHz quarature phase hybrid module. The maximum lifetime measurement range was determined under ideal conditions to be 0.05 to 450 nsec using 32- and 2-MHz simulated signals, respectively. However, due to staining and biological variability and signal strength to background noise, this translates into a 0.5- to 350-nsec range.

V. Standards

The use of biological cells (stained) and fluorosphere standards is important in the calibration of the two-phase ratio detector for quantifying fluorescence lifetimes. The average lifetimes by phase shift or amplitude demodulation methods of stained cells or fluorospheres in suspension can be measured using a fluorescence lifetime spectrophotometer or by phase-sensitive FCM. The average fluorescence lifetime of cells in flow also can be determined by simply shifting the phase of the reference (ϕ_R) first to null the output of the PSD using nonfluorescent particles (barrier filter removed, zero lifetime reference point) and then to null the PSD output using fluorochrome-labeled cells/particles (barrier filter in place). The lifetime is calculated using Eq. (2), where ϕ is the phase shift difference between the fluorescent and nonfluorescent particle signal null points recorded at the same photomultiplier gain setting. The nonfluorescent particles are analyzed at the same PMT gains by removing the barrier filter and replacing it with neutral density filters to maintain the orthogonal light scatter and fluorescence signal intensity at the same values.

VI. Instrumentation

A Spectra Physics Model 2025-05 argon laser operating at 488 nm wavelength/600 mW (TEM_{oo}) is used as the excitation source and the modulator is an electro-optic Conoptics Model 380 DC-50-MHz bandwidth unit (see Fig. 1). The modulator is located on a precision slide for movement into and out of the laser excitation beam. Micrometer adjustments (*XYZ*, pitch, and yaw) are used to align the modulator coaxially on the laser beam axis. A Hewlett–Packard Model 3335A signal synthesizer (200 Hz–81 MHz) is used as the frequency source for the Conoptics Model 50 drive electronics and to supply the reference phase shifter with a sine wave signal. The laser beam is focused by a pair of crossed cylindrical lenses into an elliptical shape onto the cell stream in a "Biosense" flow chamber (Coulter Corp.). With 600 mW applied to modulator input port, the average (DC) modulator output power is 250 mW. The output varies from 465 mW maximum to 20 mW minimum (1 MHz) and from 340 mW maximum to 160 mW minimum (50 MHz) using a sinusoidal modulation frequency.

The fluorescence detector consists of a f/0.95 CCTV lens to collect and collimate the modulated fluorescence emission to a second CCTV lens which focuses the emitted light onto a 100-μm-diameter pinhole spatial filter located in front of a photomultiplier tube (PMT). A Corning 3-69 colored class filter located between the two lenses functions as a long-pass barrier filter to block scattered laser light and pass the total emitted fluorescence. A Burle Industries Model 4526 dormer window 10-stage PMT, with the anode connected directly to a Comlinear Model 401 150-MHz operational amplifier configured in the transimpedance mode (7.5 kΩ feedback resistor), serves as the photodetector in the fluorescence detector (Steinkamp *et al.*, 1987). Bertan Model 353 power supplies provide high voltage to the PMT detectors. The fluorescence and reference signals are routed to the signal detection/processing electronics as described below.

The conventional DC-excited fluorescence signals are obtained by using the low-pass section of a Krohn-Hite Model 3201 electronic filter set at 160 kHz (see Fig. 2A), which are amplified/integrated. The phase-sensitive detection electronics consist of a Mini-Circuits Model ZRPD-1 phase detector, i.e., multiplier (double-balanced mixer), and one-half of a Krohn-Hite Model 3202 dual-electronic filter set at 160 kHz. Allen Avionics Models V127050 (0–127 nsec) and VAR011 (0–11 nsec) switchable delay lines having 1.0 and 0.5 nsec time resolution, i.e., 3.6° and 1.8° phase resolution at 10 MHz, respectively, are used to shift the phase of the reference signal with respect to the input fluorescence signals. The PSD negative output signals are inverted and then amplified/integrated.

The two-phase ratio detector (phase comparator) for making fluorescence decay time measurements is shown in Fig. 2B. A Model JH-6-4 Anzac 2- to 32-MHz quadrature phase hybrid module supplies two reference signals that are 90° out of phase with each other to PSD circuits for generating outputs $v_{\phi-90}$ (t) and V_{ϕ} (t), the ratio of which is directly proportional to the decay lifetime. A Mini-Circuits Model ZFRSC-2050 two-way resistive signal divider (not shown) is used to divide the fluorescence detector output signal to the PSD circuits consisting of double-balance mixers, low-pass filters, and inverting amplifiers as described above. The ratio module contains gated peak sense and hold circuits to acquire the amplified numerator [$v_{\phi\text{-}90}$ (t)] and the denominator [v_{ϕ} (t)] signals followed by a high-speed analog divider to generate the tan ϕ signal. The above signals are recorded as list-mode data for display as frequency distribution histograms using a computer-based data acquisition system.

VII. Results and Discussion

Prior to demonstrating the ability of phase-sensitive detection to resolve fluorescence emission signals based on differences in their phase shifts and to quantify lifetime directly, we determined the maximum measurement precision and sensitivity using fluorospheres and strained cells. Results from measurement precision tests on Coulter Corp. "DNA Check" alignment fluorospores and PI-stained CHO cells showed coefficients of variations [(CVs), standard deviation divided by the mean] of 1.3% on fluorospheres and 3.4% on G_1 phase cells (DNA content) (Steinkamp and Crissman, 1993b) In sensitivity tests using a mixture of Flow Cytometry Standards, Inc., calibrated fluorospheres labeled with known amounts of soluble fluorescein, the fluorescence signal detection threshold limit was 300–500 fluorescein molecules equivalence for fluorospheres excited at 1 to 30 MHz, respectively, compared to 250 molecules equivalence using conventional laser excitation.

The resolution of signals from fluorescence emissions having different lifetimes by phase-sensitive detection is illustrated in Fig. 3. In this experiment, CHO cells were stained with PI alone, FITC alone, and PI/FITC in combination and measured using a 10-MHz modulation frequency. PI-stained cells were first analyzed separately to determine the maximum PSD output as a function of the reference signal phase shift ($\phi_R = \phi_{PI} = -43.2°$) with respect to the fluorescence signals. The output signals were then recorded in the computer and the histograms were displayed (see Fig. 3A). The relative DNA content distribution recorded on PI-stained CHO cells compared favorably with those obtained by conventional FCM. Peak 1 of the DNA distribution represents 2C diploid DNA content cells in the G_1/G_0 cell-cycle growth phase, prior to DNA replication, and peak 2, the 4C DNA content cells in G_2 and M phase following DNA replication. Cells contained between the two peaks are in S phase, i.e., synthesizing DNA. The peak 2 to peak 1 modal fluorescence intensity ratio is 1.95 and the peak 1 CV is 5.3%. Next, FITC-stained cells were analyzed by phase-sensitive detection. ϕ_R was adjusted (−12.6°) to maximize the PSD output signals and the corresponding histogram was displayed (see Fig. 3B). The total protein distribution is broad, unimodal, and consistent with cells in exponentional growth.

To resolve PI fluorescence emission signals from the FITC signals based on differences in their lifetimes, ϕ_R was adjusted ($-\pi/2 + \phi_{FITC} = -100.8°$) to null (zero) the PSD output while FITC-stained control cells were being analyzed (data not shown). PI-stained cells were then measured at the same reference phase setting and the DNA content histogram was displayed (see Fig. 3C). The G_0/G_1 signal amplitude (peak 1) was reduced in value (53%) compared to the histogram recorded in Fig. 3A. The reduction in signal amplitude is due to the sin (ϕ_2-ϕ_1) relationship (value 0.51) expressed in Eqs.(6) and (7), where ϕ_1 and ϕ_2 are the −12.6° and −43.2° FITC and PI signal phase shifts, respectively. Similarly, PI-stained control cells were analyzed to determine the reference signal phase shift needed ($\pi/2 + \phi_{PI} = 50.4°$) to null the PSD output and thus resolve FITC fluorescence emission signals from the PI signals. FITC-stained cells were then analyzed at the same phase shift setting, the PSD output signals were recorded, and the histogram was displayed (see Fig. 3D).

Cells stained in combination with PI and FITC were next analyzed to illustate the resolution of fluorescence signals from cells labeled with of two fluorochromes having different lifetimes (see Fig. 3E). The reference signal phase was first adjusted to −100.8° to resolve PI signals and the histogram was displayed. The reference phase was then adjusted to 50.4° to resolve FITC signals and the PSD outputs were recorded. This data graphically illustrate the capability of resolving fluorescence signals based on differences in their lifetimes.

Fluorescence lifetime measurement is illustrated in Fig. 4 using DNA Check

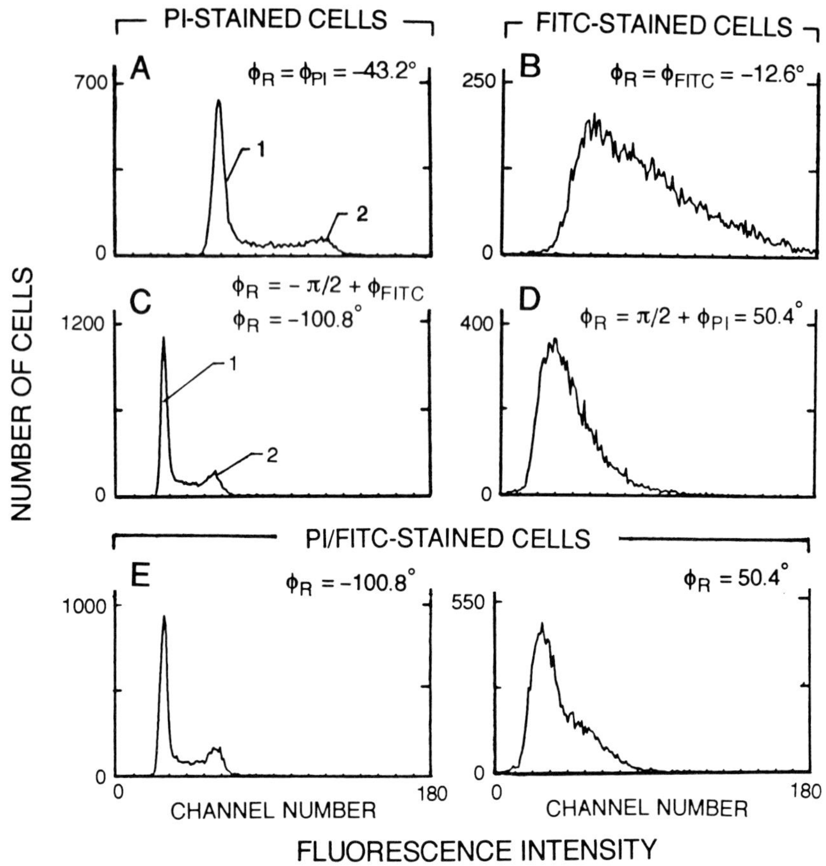

Fig. 3 Frequency distribution histograms of the PSD output signals from CHO cell samples stained separately with PI and FITC and in combination and analyzed using a 10-MHz modulation frequency. (A and B) The phase of the reference signal was adjusted to maximize the PSD output signals (250 mV peak signal input to PSD) for PI-stained cells ($\phi_R = \phi_{PI} = -43.2°$) and for FITC-stained cells ($\phi_R = \phi_{FITC} = -12.6°$), respectively, and the corresponding histograms were recorded. (C) FITC-stained control cells were first analyzed and ϕ_R was adjusted ($\phi_R = -\pi/2 + \phi_{FITC} = -100.8°$) to null the PSD output signals. PI-stained cells were then analyzed at the same phase shift and PMT/amplifier gains and the histogram was recorded. Similarly (D) PI-stained cells were analyzed and ϕ_R was adjusted ($\phi_R = \pi/2 + \phi_{PI} = 50.4°$) to null the PSD output signals. FITC-stained cells were then analyzed at the same phase shift and gain settings and the corresponding histograms were recorded. (E) Cells stained in combination with PI and FITC were analyzed by first recording the PSD signal output histograms with $\phi_R = -100.8°$ (DNA content signals resolved) and $\phi_R = 50.4°$ (protein signals resolved). In (A–E), the PMT and amplifier gains were the same.

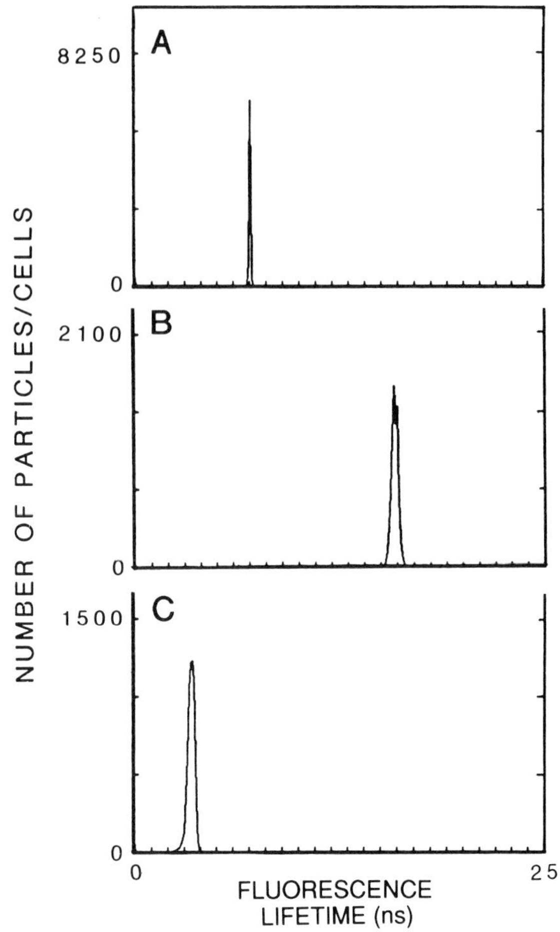

Fig. 4 Fluorescence lifetime frequency distribution histograms recorded on DNA Check alignment fluorospheres (A) and on PI-stained (B) and FITC-stained (C) CHO cells analyzed using a 10-MHz modulation frequency and a ratio module gain of 96×. The signal output of the fluorescence detector from "DNA check" alignment fluorospheres was first adjusted to 500 mV peak using the PMT high voltage (gain). The dual-channel phase-sensitive electronics were initialized using 90° light scatter signals (3–69 barrier filter removed from fluorescence detector) from 7.0-μm-diameter nonfluorescent spheres to align the $V_{\phi\text{-}90}$ (t) and V_{ϕ} (t) PSD outputs for zero (null) and maximum values, respectively, by adjusting the reference signal phase ϕ_R (see Fig. 2B). Neutral density filters (placed in front of the PMT) were used to attenuate the light scatter signals to 500 mV peak levels at the same PMT gain. The alignment fluorospheres were then analyzed with the barrier filter in place and the histogram was recorded. PI- and FITC-stained cells were analyzed following the same procedure at the same ratio module gain (96×), but at different PMT gains.

alignment fluorospheres and PI- and FITC-stained cells. The lifetime histograms were recorded at different PMT gains, but at the same ratio module gain. The alignment fluorospheres, which had previously been measured by phase shift and amplitude demodulation to be approximately 7 nsec, were measured first and the ratio module gain was adjusted to center the histogram in channel 7 (see Fig. 4A). PI- and FITC-stained cells were then measured at the same ratio module gain setting and the corresponding histograms were recorded (see Figs. 4B and 4C). These results vividly illustrate the capability of quantifying fluorescent decay times in flow by the two-phase ratio measurement method and they are in agreement with the lifetime values obtained by independent measurements (see Table I).

Acknowledgments

This work was performed at the Los Alamos National Laboratory, Los Alamos, NM, under the joint sponsorship of the United States Department of Energy; the Los Alamos National Flow Cytometry Research Resource, National Institutes of Health Grant No. P41-RR01315; and the National Institutes of Health Grant No. R01-RR07855.

References

Alcala, J. R., Gratton, D., and Jameson, D. M. (1985). *Anal. Instrum.* **14,** 225–250.

Buican, T. N. (1990). *Proc. SPIE—Int. Soc. Opt. Eng.* **1205,** 126–133.

Crissman, H. A., and Steinkamp, J. A. (1982). *Cytometry* **3,** 84–90.

Enscoe, R. F., and Kocka, R. J. (1984). *Lasers Appl.* June, pp. 91–95.

Jameson, D. M., Gratton, E., and Hall, R. D. (1984). *Appl. Spectrosc. Rev.* **20,** 55–106.

Lakowicz, J. R., and Cherek, H. (1981). *J. Biol. Chem.* **256,** 6348–6353.

Loken, M. R., Parks, D. R., and Herzenberg, L. A. (1977). *J. Histochem. Cytochem.* **25,** 899–907.

Pinsky, B. G., Ladasky, J. J., Lakowicz, J. R., Berndt, K., and Hoffman, R. A. (1993). *Cytometry* **14,** 123–135.

Steinkamp, J. A., and Crissman, H. A. (1993a). *Cytometry* **14,** 210–216.

Steinkamp, J. A., and Crissman, H. A. (1993b). *Proc. SPIE—Int. Soc. Opt. Eng.* **1885,** 278–289.

Steinkamp, J. A., Habbersett, R. C., and Stewart, C. C. (1987). *Cytometry* **8,** 353–365.

Steinkamp, J. A., Habbersett, R. C., and Hiebert, R. D. (1992). *Rev. Sci. Instrum.* **62,** 2751–2764.

Veselova, T. V., Cherkasov, A. S., and Shirokov, V. I. (1970). *Opt. Spectrosc.* (*Engl Transl.*) **29,** 617–618.

CHAPTER 37

Spectra of Fluorescent Dyes Used in Flow Cytometry

Richard P. Haugland
Molecular Probes, Inc.
Eugene, Oregon 97402

I. Introduction

Fortunately for applications in flow cytometry, most biological cells have low intrinsic fluorescence. This both permits and necessitates the use of extrinsic probes such as conjugates of antibodies with fluorescent dyes, nuclear stains, and fluorogenic substrates to obtain an optical signal that can be related to some cellular component or property. This chapter reviews the factors that go into the selection of dyes for use in flow cytometry, identifies the principal dyes being used, and compares their spectral properties. Also, it briefly reviews some of the recent developments in the field related to dye development and applications of new probes in flow cytometry.

Relatively few dyes and fluorescent probes are used extensively in flow cytometry. Much of this limitation results from the dependence of most flow cytometers on the argon laser as the primary or only excitation source. This laser has its principal output at 488 and 514 nm with weaker lines—which are rarely utilized—at 457, 472, and 476 nm. Higher powered argon lasers may have weak output in the ultraviolet between about 350 and 360 nm. Other laser

METHODS IN CELL BIOLOGY, VOL. 42

excitation sources such as the krypton laser (primary output at 568 and 647 nm), red HeNe laser (633 nm), orange HeNe laser (594 nm), green HeNe laser (543 nm), HeCd laser (325 and 441 nm), or tunable dye laser ($>\sim$550 nm) have been used much less frequently in flow cytometry, although incorporation of the red HeNe laser as a second excitation source is becoming more common. Consequently, most of the dyes that are used in flow cytometry need to have appreciable absorption at 488 nm (or at 514 nm).

Three primary types of fluorescent probes have been used in flow cytometry: (1) fluorescent immunoconjugates (and recently probes for fluorescence *in situ* hybridization); (2) nucleic acid stains, and (3) physiological probes. The latter class includes probes for measuring ions, membrane potential, enzymatic activity, viability, organelles, phagocytosis, cell development, and other properties of living cells. This area represents the frontier of new probe development for flow cytometry. Each class of probes will be discussed separately.

II. Spectral Properties of Fluorescent Dyes

A primary attribute of flow cytometry is its ability to make rapid multiparameter measurements on single cells by simultaneously measuring two, three, or more colors of fluorescence and, at the same time, to detect scattering of the excitation light. To accomplish this requires the use of fluorescent dyes that have emissions with low spectral overlap. A fundamental knowledge of the basic properties and spectra of dyes is essential for determining whether the dye can be excited by the excitation source, for selecting appropriate optical filters, and for reducing spectral overlap and the required compensation when one seeks to combine dyes for multiparameter measurements.

The primary factors that go into determining the fluorescence emission from a dye include the following:

1. Dye absorbance at the excitation wavelength. As indicated above, in flow cytometry this is usually absorption at the principal outputs of the argon laser: 488 and 514 nm. The great utility of fluorescein in flow cytometry results from a combination of factors: the close match of its absorption peak with the 488-nm excitation line of the Ar laser, its extinction coefficient—which is high compared to that of most dyes—and its good quantum yield. Longer wavelength absorbing dyes such as tetramethylrhodamine and Texas Red have extinction coefficients at 488 nm that are less than 10% that of fluorescein.

The effect of dye absorbance at the excitation wavelength is a critical limitation of flow cytometry when one is restricted to using a single argon laser, particularly for multicolor applications. The initial event in fluorescence is always absorption of light by the dye, which for a given concentration of a dye is proportional to the extinction coefficient. Most dyes have a relatively symmetrical absorption spectrum that consists of an envelope of overlapping vibronic transitions. Typically the extinction coefficient decreases by about 50% at ±10–20 nm from the peak and by about 90% at ± 20–40 nm from the peak. Thus, when using a laser excitation source, shifting the dye's absorption maximum away from the excitation peak usually results in a considerable loss in ability to excite the fluorphore.

2. Dye concentration. In an ideal situation, the fluorescence signal will be proportional to the concentration of the dye. Although there are many cases and experimental conditions where this is not operative, it remains approximately correct. Consequently the intensity of cell staining is usually related to the number of dyes that are associated with the cell, whether they are bound through an antibody linkage, bound to nucleic acids, or, as in the case of an ion indicator, relatively free in the cytoplasm. Because nucleic acids can simultaneously bind a large number of dyes relative to the number of sites to which fluorescent antibodies can bind, even weakly absorbing or weakly fluorescent dyes can give a high *total* signal when bound to nucleic acids. On the other hand, low dye concentration makes it difficult to determine relatively low abundance receptors, unless a strongly absorbing dye is used.

3. Dye quantum yield. The fluorescence quantum yield of a dye is a measure of the efficiency of its emission relative to its absorption. The higher the quantum yield, the higher the potential fluorescence. Although extinction coefficients of dyes are relatively independent of the environment, the quantum yields are often extremely dependent on the local environment in which the fluorophore is bound. For instance, pure fluorescein has a reported quantum yield of 0.92 in alkaline solutions (Weber and Teale, 1957), but the first molecule of fluorescein isothiocyanate (FITC) that conjugates to an antibody typically has a quantum yield of ~0.3–0.5 and subsequent molecules of FITC that react have an even lower fluorescence yield. Eventually one gets to a point where addition of more FITC dyes to an antibody results in a *decrease* in the total fluorescence yield of the conjugate. In contrast, thiazole orange, which is widely used in flow cytometry for reticulocyte analysis, has a quantum yield that *increases* at least 3000-fold on binding to DNA versus when it is free in aqueous solution (Lee *et al.*, 1986). The property of fluorescence enhancement on binding for the nucleic acid stains is extremely useful in that the fluorescence that results can be separated from that of the free dye in the medium.

4. Spectral distribution. Organic dyes have fluorescence emissions at wavelengths that are characteristic of the dye. Sometimes the wavelengths are also affected by binding of the dye to a cellular component. Although dye emission occurs invariably at a wavelength that is longer than that of the exciting light, both the peak intensity and the spectral width depend greatly on which dye is used. The tables in this chapter list the peak emission that is typical for the dye in its applications. However, it is usually a combination of *both* the peak emission wavelength and the width of the emission spectrum that determines whether two or more dyes can be combined for multicolor applications. For

that reason, full emission spectra are included in this chapter for several of the more important dyes used in flow cytometry. Unlike absorption measurements, fluorescent dyes do not have any easily quantifiable measure of absolute intensity. Consequently, most of the spectra are normalized to the same peak intensity. As mentioned above, it is essential to take into account differences in dye absorption strength at the excitation wavelength, differences in dye concentration from the experimental situation, and differences in dye quantum yields when determining whether two or more dyes can be combined. In general the determining factor for whether two emissions can be resolved is related to the mathematical product of the extinction coefficient at the excitation wavelength (often 488 nm) times a dye's quantum yield times the effective dye concentration in the cell.

Related to the problem of spectral distribution is the ability to resolve two or more overlapping spectra. The typical broadness of emission for all organic dyes means that the emission from the highest energy emitting dye will always overlap the emission from the second (and third) dye to some degree. To accurately measure the emission from the second dye requires a means (usually electronic) to compensate for the part of the long wavelength emission from the first dye that overlaps with the emission that is detected from the second dye (Horan *et al.*, 1990a,b). For quantitative measurements one obviously wants to minimize this compensation by having a reasonable spectral separation of the emission spectra of the two dyes so that the second (and third) dyes have significant signals relative to the signal from the shorter wavelength dye. In most cases the signals from the two dyes are separately isolated and analyzed by collecting the emission through appropriately selected optical filters. Unfortunately, collecting only part of an emission spectrum through a filter rather

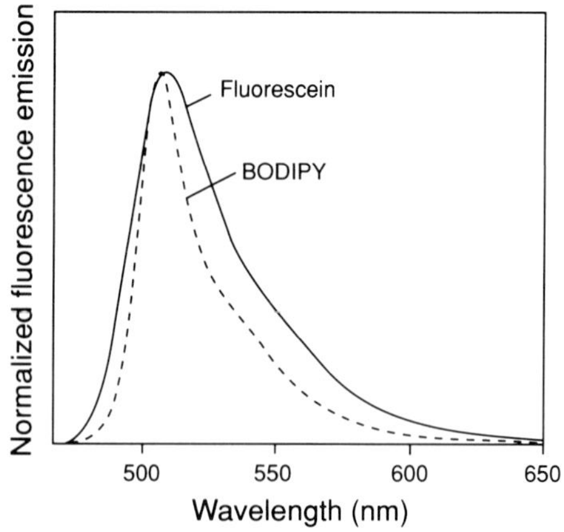

Fig. 1 Normalized fluorescence emission spectra of BODIFY FL and fluorescein.

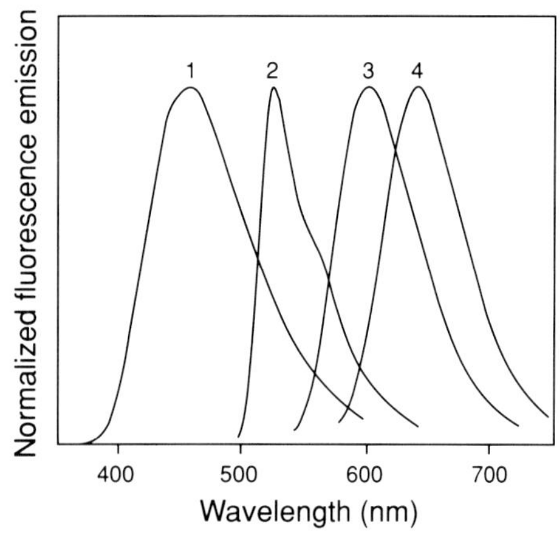

Fig. 2 Normalized fluorescence emission spectra of (1) Hoechst 33258, (2) thiazole orange, (3) ethidium bromide, and (4) 7-aminoactinomycin D bound to calf thymus DNA.

than the entire spectrum results in part of the emission intensity being discarded. Also, the more restricted the spectral bands that are passed through the filter, the less signal one collects. Because of the broadness of most emission spectra, to obtain adequate spectral resolution usually requires that the emission peaks of each of the dyes be separated by a minimum of about 50–60 nm to obtain a good signal for both dyes that is relatively free of spectral overlap. However, if the second dye's mathematical product of absorbance, quantum yield, and effective cellular concentration is particularly low relative to that of the first dye, then a greater spectral separation will be required. The spectral separation required can be reduced somewhat by the use of dyes such as BODIPY-FL that have relatively narrow emission peaks (Fig. 1). Some dyes, such as ethidium bromide and Hoechst 33258 (Fig. 2) have extremely broad emission spectra that overlap that of almost all other dyes that are excited by the Ar laser.

III. Immunofluorescence and Related Applications

Certainly one of the major applications of flow cytometry is its use for detecting immunofluorescence, often in combination with staining for nucleic acids or one of the physiological parameters. An extensive assortment of antibodies is available for use in flow cytometry, particularly in classification of markers on various blood cells. Recently, direct and indirect probes for fluorescence *in situ* hybridization (FISH) have also been used in flow cytometry; their use is likely to increase (Wallner *et al.*, 1993). Despite the plethora of antibodies available, a very narrow selection of dyes predominates in these applications. Properties of the most important of these are listed in Table I.

Table I
Spectral Properties of Dyes of Current Practical Importance in Flow Cytometry

Dye	λ_{abs} (nm)	ε_{max} ($cm^{-1}\ M^{-1}$)	λ_{em} (nm)
Cascade Blue	376, 399	28,000	419
AMCA	356	17,000	442
Fluorescein	492	75,000	515
BODIPY FL	503	90,000	511
Tetramethylrhodamine	549	100,000	574
Cy-3	540	130,000	567
Rhodamine B	556	123,000	577
Texas Red	596	85,000	615
Cy-5	648	200,000	655
R-Phycoerythrin	496, 566	1,960,000	575
Texas Red–R-phycoerythrin	496, 566, 596	1,960,000	613
Cy-5–phycoerythrin	496, 566, 648	1,960,000	670
Allophycocyanin	651	104,000	660
PerCP	470	35,000	680

Note. λ_{abs}, absorption maximum; ε_{max}, maximum molar extinction coefficient; λ_{em}, fluorescence emission maximum.

By far the most common dyes for preparing fluorescent immunoconjugates for flow cytometry are derivatives of fluorescein. The various chemically reactive forms of fluorescein that are used such as FITC, the dichlorotriazine (DTAF) (Blakeslee and Baines, 1976) and the succinimidyl ester of carboxyfluorescein (CFSE) all have approximately the same absorption and emission spectra and similar quantum yields and are all ideally matched to the argon laser. The only major alternatives to fluorescein are BODIPY-FL (Fig. 1), carboxyrhodamine 110, and carboxyrhodol (Haugland, 1990). Although these dyes have certain advantages for imaging applications in which the lower photostability of fluorescein and sensitivity to pH can be a problem, they are probably not significant improvements compared to fluorescein conjugates for flow cytometry.

The choice of the "second dye"—which is typically another dye that can be detected at a longer wavelength emission in combination with fluorescein—illustrates the problems of trying to obtain simultaneous detection of multiple fluorophores. Although Texas Red has excellent spectral resolution from fluorescein with an emission maximum at about 615 nm versus about 515 nm for fluorescein (Fig. 3), excitation of Texas Red conjugates at 488 nm results in relatively low fluorescence compared to when this dye is excited near its absorbance peak at 596 nm. This factor, combined with the intrinsically lower fluorescence yield of conjugates of Texas Red versus fluorescein, results in direct Texas Red conjugates of antibodies being not very useful when excited with the Ar laser at either 488 or 514 nm (Fig. 4). Although rarely used, Texas

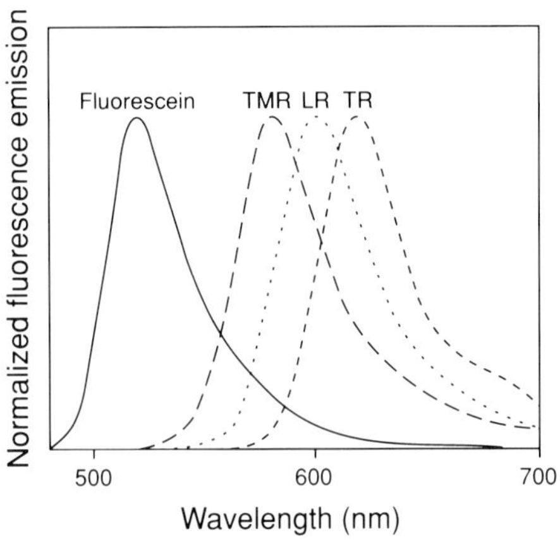

Fig. 3 Normalized fluorescence spectra of protein conjugates of fluorescein, tetramethylrhodamine (TMR), Lissamine rhodamine B (LR), and Texas Red (TR). Because fluorescein and Texas Red exhibit very little spectral overlap, these two dyes are suitable for multicolor applications.

Red conjugates can be used in combination with fluorescein conjugates if the fluorescein-labeled antibody is on a relatively low-abundance epitope and the Texas Red-labeled antibody is used to label a target that is at least 10 times as abundant. This is an example of using the concentration of the dye to overcome the relatively low extinction coefficient at 488 nm. The other potential alternative dyes to Texas Red that are widely used in microscopy—in which the excitation source is commonly the broad band of the mercury arc lamp—such as tetramethylrhodamine (used as its isothiocyanate TRITC or succinimidyl ester

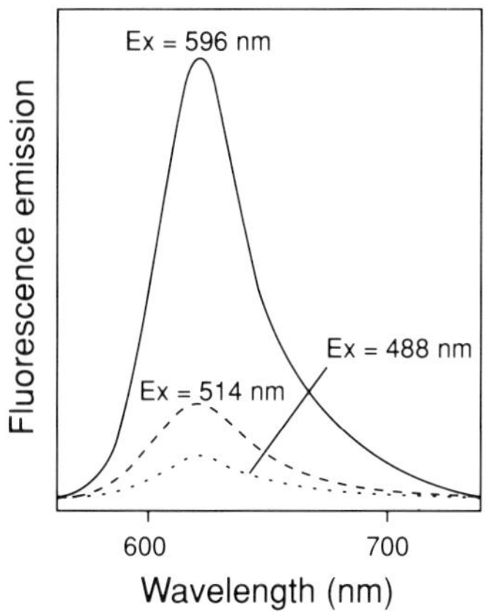

Fig. 4 Relative fluorescence emission from Texas Red excited at 596, 514, and 488 nm. Spectra are of Texas Red bovine serum albumin conjugates.

CTMR-SE) or rhodamine B (conjugated through its isothiocyanate or the sulfonyl chloride, Lissamine Rhodamine B sulfonyl chloride) have greater spectral overlap with fluorescein than does Texas Red (Fig. 3). In addition, the quantum yields of their protein conjugates are lower than those of Texas Red and their absorbance at 488 nm is also quite weak.

As demonstrated empirically (Haugland, 1990), it is not likely that any single chemical dye that could be used as a second dye in combination with fluorescein will have the required properties of a high absorbance at 488 nm, a >100 nm or so Stokes shift for the emission that results in maximal emission beyond about 588 nm, and a high quantum yield. The intensity of fluorescein emission at 588 nm is about 7% of its peak intensity at about 515 nm. At least all single dyes tested so far as possible "second dyes" for combination with fluorescein have suffered from a low absorbance at 488 nm and a low fluorescence yield at >560 nm, or both. The solution that has been almost universally adopted for obtaining a second label for immunofluorescence in flow cytometry has been to use a phycobiliprotein—usually R-phycoerythrin—conjugate because this dye can be excited at 488 nm and has emission at about 580–590 nm (Oi *et al.*, 1982). Although conjugates of phycobiliproteins are relatively difficult to prepare (Kronick, 1986), have limited stability, are sensitive to light, and have a large size, the conjugates of these algal pigments have near-optimal spectral properties for this use (Figs. 5a and 5b). The proteins consist of multiple covalently linked bilin fluorophores, each with strong absorbance and extremely high quantum yields. Significantly, there are two different fluorophores in the phycoerythrins with one having absorption near the 488-nm peak of the Ar laser. Upon absorption into this band, the excitation energy is quantitatively transferred to the longer wavelength absorbing fluorophore in the complex, which is the emitting species. This energy transfer mechanism results in the protein having both strong absorbance at 488 nm and intense emission at about 580–590 nm. The higher content of the 488-nm absorbing pigment in R-phycoerythrin has usually made this protein the dye of choice for protein labeling.

With the difficulties in finding single dyes that can be excited at 488 nm and have significant Stokes shifts, it is obvious that attempting to find simple chemical dyes for use as "third" or potentially "fourth" dyes, all of which can be excited at 488 nm, will also be difficult. Although the cost advantages of having a single excitation source will be evident to those who do flow cytometry, for several years solving this problem necessitated experimentation with longer wavelength excitation sources such as the krypton laser, which has a line at 568 nm (and a stronger line at 647 nm), dye lasers with broad excitation capabilities beyond about 550 nm, and the red HeNe laser. Allophycocyanin, another pigment from algae, is minimally excited by the Ar laser, but it is readily excited by the HeNe laser at 633 nm (Fig. 5a) (Oi *et al.*, 1982). Because it contains fewer fluorophores and its quantum yield is intrinsically lower, allophycocyanin's absorbance and fluorescence are much less than those of the other phycobilipro-

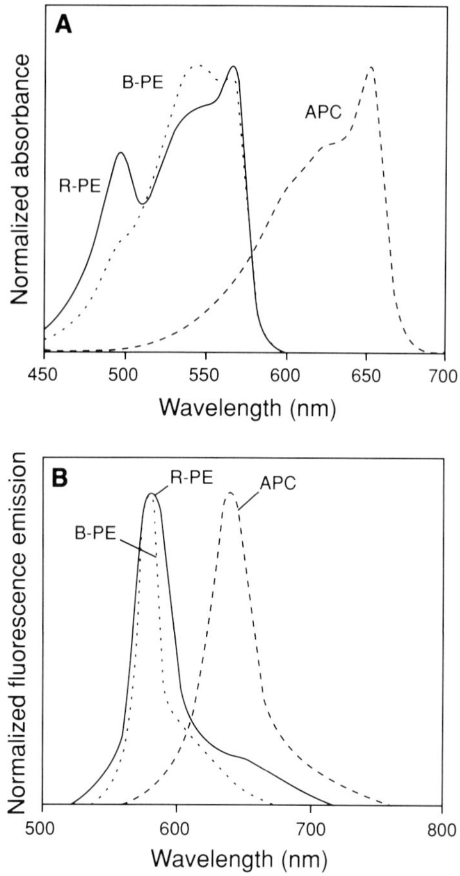

Fig. 5 (A) Absorption spectra of the phycobiliproteins B-PE, R-PE, and APC. (B) Normalized fluorescence emission spectra for B-PE, R-PE, and APC.

teins. Furthermore it is unstable to dissociation into subunits unless it is chemically crosslinked (Yeh *et al.*, 1987). More recently other new dyes such as Cy-5 (Fig. 6) (Southwick *et al.*, 1990) and longer wavelength BODIPY dyes (Haugland, 1992) have been described that can be excited by the red HeNe laser. In addition, some reactive phthallocyanine dyes such as LaJolla Blue that can be excited by laser diodes at 680 nm or longer are under development (Devlin *et al.*, 1993).

Within the past few years several companies have experimented with conjugates of Texas Red (Festin *et al.*, 1990) or of Cy-5 with phycoerythrin. The intent has been to establish an artificial energy transfer situation in which the energy that is initially absorbed at 488 nm is transferred in a multistep process to the Texas Red or the Cy-5 and the emission that results is characteristic of

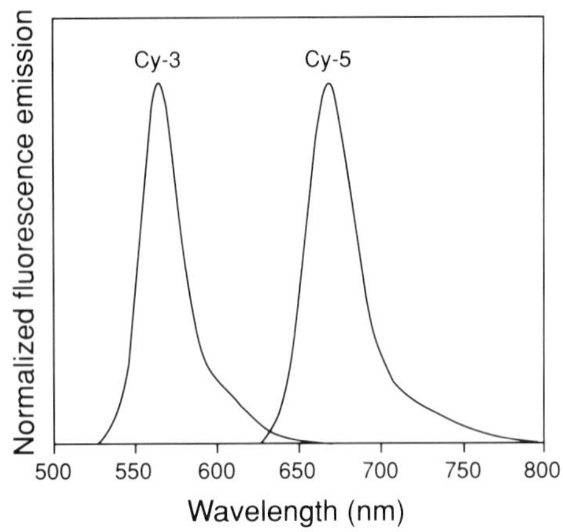

Fig. 6 Normalized fluorescence spectra of the water-soluble cyanine dyes Cy-3 and Cy-5.

the longest wavelength dye, in this case Texas Red or Cy-5. To a certain extent this approach has been successful, but in all of the commercial conjugates now available some residual emission is present from the unquenched phycoerythrin and, therefore, their use as a third color requires careful compensation that may depend on the batch (and supplier) of reagent being used. Allophycocyanin has also been linked to phycoerythrin to create a two-protein energy transfer pair with emission properties characteristic of allophycocyanin (Glazer and Stryer, 1983).

The final solution now being used to obtain long wavelength emission using 488-nm excitation is to conjugate the antibody to a protein from a dinoflagellate, peridinin chlorophyll protein (PerCP) (Recktenwald, 1989,1990). This multipigment protein complex, which has a molecular mass of about 35,000 Da and a quantum yield approaching 1.0, can be excited at 488 nm and has *no* emission below about 650 nm (Fig. 7), meaning that there is minimal compensation required when it is combined with phycoerythrin. Unfortunately, the fluorescence tends to disappear when the complex is excited by the Ar laser used at high power.

Some argon lasers have relatively weak output in the ultraviolet between 350 and 360 nm. There are only two dyes that have blue fluorescence whose protein conjugates have been used to any extent for flow cytometry: AMCA (7-amino-4-methylcoumarin-3-acetic acid) (Khalfan *et al.*, 1986) and Cascade Blue (Whitaker *et al.*, 1991). Of these dyes, Cascade Blue has a better combination of absorbance, quantum yield, and spectral separation from fluorescein than does AMCA (Fig. 8). Fluorescein has a molar extinction coefficient at 355 nm of about 1000 meaning that it is not well excited in the UV. Consequently, when Cascade Blue or AMCA is used as a third dye, it is usually necessary to also use 488-nm excitation to excite the fluorescein. On the other hand, R-

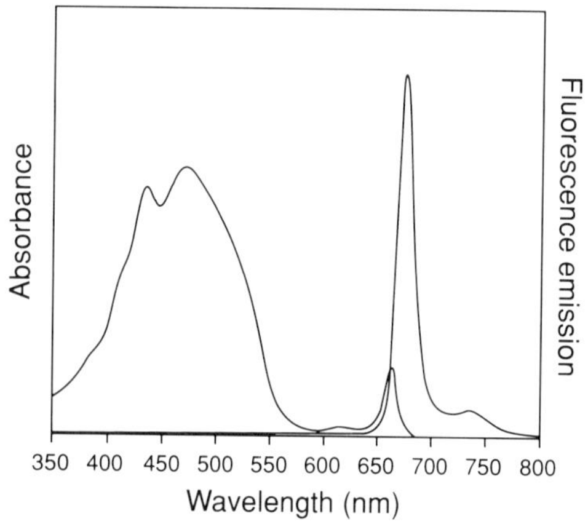

Fig. 7 Absorption and fluorescence spectra of peridinin chlorophyll protein (PerCP).

phycoerythrin has an extinction coefficient of about 192,000 at 355 nm and it can also be excited in the UV.

Although one can anticipate that other dyes will be developed for use in immunofluorescence in flow cytometry, it appears that the current dyes have been optimized for three- or four-color analysis with excitation by the Ar laser. It also is apparent that more widespread utilization of alternative excitation sources such as the HeNe, HeCd, or other lasers yet to be developed would have a greater impact on the selection of fluorophores that could be used in flow cytometry than will efforts at synthesis. However, there are other possible

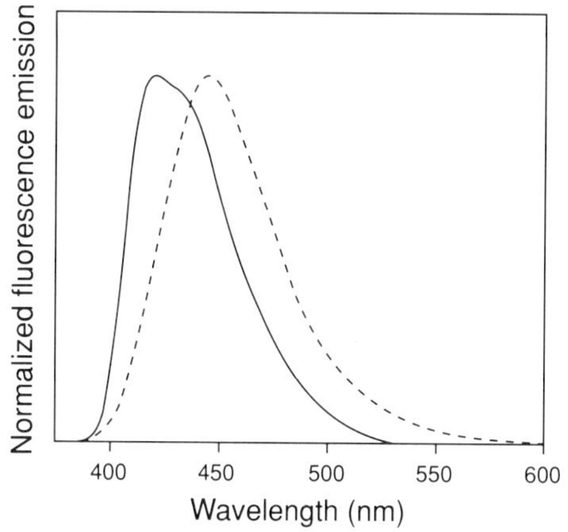

Fig. 8 Normalized fluorescence emission spectra of Cascade Blue (solid line) and AMCA (dashed line).

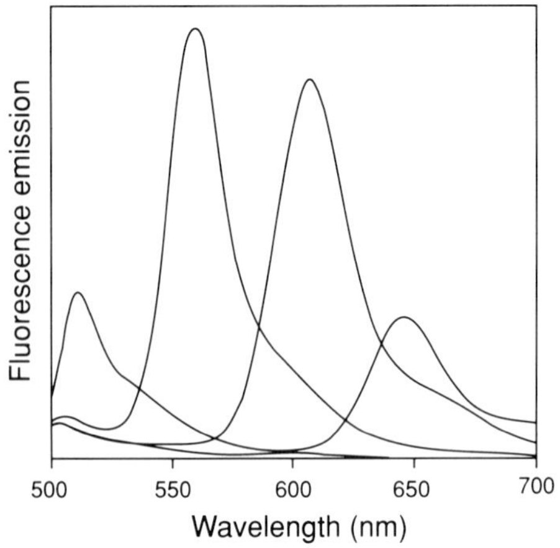

Fig. 9 Fluorescence emission spectra of four types of 0.3-μm latex microsphere, each of which is formulated to produce a distinctive emission spectral distribution relative to a common excitaion wavelength (488 nm).

alternatives using more complex dyes. Scientists at Molecular Probes are experimenting with using fluorescent latex particles that have high fluorescence *and* considerably enhanced Stokes shifts as a way to overcome the problems of synthesizing an organic dye or of stability of any protein dye (Fig. 9). Fluorescent latex has been used previously for detection of low-abundance receptors (Cupp *et al.*, 1984). Liposomes containing multiple fluorescent dyes have also been used as a means for enhancing the flow cytometric detection of low-abundance targets (Truneh and Machy, 1987).

IV. Nucleic Acid Stains

An assortment of stains for nucleic acids is available. These differ greatly in their spectral properties, modes of nucleic acid binding, and extent of use in cytometry. A previous volume of this book contains a very useful table that summarizes the binding selectivity and spectral ranges for 29 nucleic acid stains and gives several leading references (Kapuscinski and Darzynkiewicz, 1990). Several chapters in this current volume also discuss applications of nucleic acid stains in flow cytometry.

Nucleic acids are abundant in cells and staining of fixed and permeabilized cells by any of the probes is usually facile. Because the number of dye binding sites is high, the fluorescence from nucleic acid staining can be strong. Consequently, dyes such as ethidium bromide and propidium iodide that have low absorbance at 488 nm (Table II) still give bright staining. The Stokes shifts for these two dyes are quite large so that their emission can be readily separated

Table II
Spectral Properties of Fluorescent Nucleic Acid Stains Used in Flow Cytometry

Dye	λ_{abs} (nm)	ε_{max} ($cm^{-1} M^{-1}$)	$\varepsilon_{488\,nm}$ ($cm^{-1} M^{-1}$)	λ_{em} (nm)
DAPI	359	24,000	—	461
Hoechst 33342	346	43,000	—	460
Hoechst 33258	346	47,000	—	460
POPO-1	434	92,000	—	456
YOYO-1	491	99,000	97,000	509
TOTO-1	514	117,000	103,000	533
Thiazole orange	509	54,000	33,000	525
Acridine orange (DNA)	502	53,000	34,000	526
Acridine orange (RNA)	460	ND	ND	650
Ethidium bromide	526	3,200	2,700	604
Propidium iodide	536	6,400	4,000	620
7-Aminoactinomycin D	555	27,000	13,200	655
TOTO-3	642	154,000	8,000	660

Note. λ_{abs}, λ_{em}, absorption and fluorescence emission maxima (respectively). ε_{max}, maximum molar extinction coefficient. $\varepsilon_{488\,nm}$, molar extinction coefficient at 488 nm. All parameters relate to complexes of the dye with DNA unless noted otherwise. ND, not determined.

from that of fluorescein (Fig. 2). It is usually preferable to use the nucleic acid stain as the longest wavelength dye, even if it has relatively weak absorbance at 488 nm. LDS 751 (Terstappen *et al.*, 1988) and 7-aminoactinomycin D (Toba *et al.*, 1992) are examples of nucleic acid stains that can be excited by the argon laser and detected as a "third color" along with fluorescein and phycoerythrin.

For detection of nucleic acids in live cells, however, penetration of the dye through the membrane has been a problem, particularly for dyes that can be excited by the argon laser. A few long wavelength absorbing dyes, such as thiazole orange (Lee *et al.*, 1986) and 7-aminoactinomycin D (Toba *et al.*, 1992) have been used to stain certain live cells. The UV absorbing DNA-selective stains Hoechst 33342 and DAPI are readily permeant to live cells. Scientists at Molecular Probes have recently developed a series of probes termed SYTO dyes that can be excited by the argon laser that rapidly stain nucleic acids in all types of living cells, including bacteria (Millard *et al.*, 1993).

Almost all of the dyes are known to stain DNA; however, most of the dyes also stain RNA. Under carefully controlled conditions (Kapuscinski *et al.*, 1982), the "metachromatic shift" of acridine orange can be used to differentiate DNA and RNA (Fig. 10). The mode of dye binding to DNA or RNA, binding to single-stranded or double-stranded nucleic acids, and selectivity for AT-rich or GC-rich regions depends to a large degree on the dye's chemical structure. In some cases this selectivity has not been conclusively established or is still subject to controversy. Most of the dyes have their highest fluorescence on double-stranded DNA. AT-selective DNA stains include DAPI, Hoechst 33258, Hoechst 33342, and quinacrine. GC-selective dyes for DNA include the antibiot-

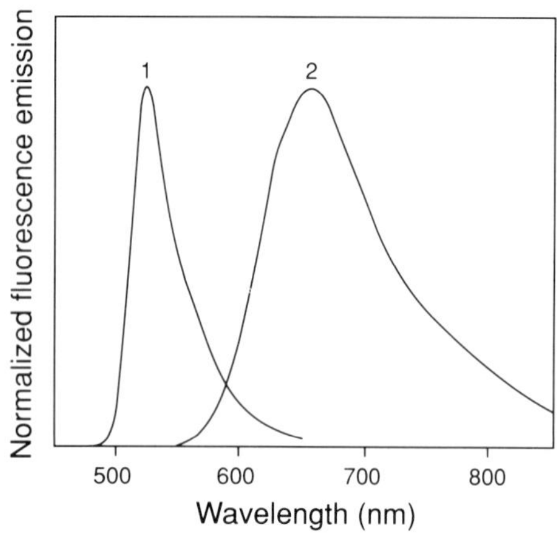

Fig. 10 Normalized emission spectra of acridine orange bound to DNA (1) and RNA (2). Spectrum (2) is adapted from published data of Kapuscinski *et al.* (1982).

ics bleomycin, chromomycin A_3, mithramycin, and olivomycin as well as rhodamine 800. Pyronin Y is one of the few dyes that is reported to have high selectivity for staining double-stranded RNA (Darzynkiewicz *et al.*, 1987; Kapuscinski and Darzynkiewicz, 1987). Thiazole orange is widely used to stain RNA in reticulocytes, which are devoid of DNA, but the dye also stains DNA (Lee *et al.*, 1986).

The ability to detect and quantitate nucleic acids depends on the same factors as described for the dyes used to prepare immunofluorescent conjugates. These are primarily related to the dye's absorbance and quantum yield (Table II). In most cases the fluorescence quantum yield of the dye increases substantially on binding to nucleic acids, probably as the result of both dye rigidification and protection from fluorescence quenching by water and other diffusible quenchers. For quantitative measurements of nucleic acid content in cells the specificity of staining nucleic acids versus other structures in the cells is particularly important. The cationic nature of most of the nucleic acid stains frequently results in nonspecific uptake of the dye into mitochondria or lysosomes or binding to polyanionic surfaces, which can obscure quantitative nucleic acid staining. This occurs particularly in live cells in which the accumulation may be based on difference in membrane potential or pH of the compartment. RNase treatment is commonly used to eliminate staining of RNA in fixed cell preparations, but this cannot be done in live cells. Problems with quantitation can result from restricted permeability of the dye toward condensed nucleic acids, failure to saturate dye binding sites, or dye–dye interactions that result in the fluorescence intensity not being linearly proportional to the nucleic acid content.

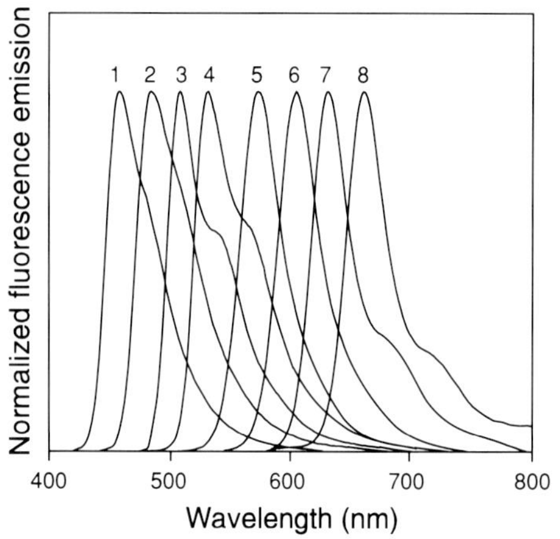

Fig. 11 Normalized emission spectra of cyanine dimer nucleic acid stains bound to DNA. Emission peaks for each dye are labeled as follows: (1) POPO-1, (2) BOBO-1, (3) YOYO-1, (4) TOTO-1, (5) POPO-3, (6) BOBO-3, (7) YOYO-3, and (8) TOTO-3.

Most of the dyes that have been used to stain nucleic acids have been available for several years, and therefore their applications in flow cytometry are relatively well developed and their spectral properties well understood. Properties of the most common dyes are listed in Table II. Scientists at Molecular Probes have recently prepared high-affinity dimeric derivatives related to thiazole orange. Two of these—YOYO-1 and TOTO-1—are particularly well excited by the argon laser (Fig. 11). Most of these dyes are essentially nonfluorescent until they bind to nucleic acids. Also, they have much stronger absorption than ethidium bromide (Table II). The high affinity and fluorescence intensity of their nucleic acid complexes has permitted the detection of single *molecules* of nucleic acid fragments in a flow cytometer (Goodwin *et al.,* 1993) and by imaging (Castro *et al.,* 1993). There is a brief report on their use in staining cells (Hirons and Crissman, 1993). Analogues such as TO-PRO-1 and YO-PRO-1 (Haugland, 1992) may be more useful for staining cells because they bind to nucleic acids much faster than TOTO-1 and YOYO-1 and they are smaller, making them more likely to bind to condensed nucleic acids. The AT versus GC selectivity of both thiazole orange and these dyes has not been conclusively established.

V. Probes for Live Cell Physiology

A growing area of research in flow cytometry is its use in the study of properties of living cells and the use of these observations in the same way as cell-surface markers are used to discriminate different types or states of cells.

Flow cytometry provides a useful means for screening large numbers of single cells; however, it is limited in that it cannot provide information with the same spatial resolution as an optical imaging experiment. For many laboratories the selection of useful probes is limited by the constraints of the visible argon laser. However, there are many properties of live cells that have been or could be studied other than just cell-surface markers: enzymatic activity, ions, membrane potential, cell viability (and cytotoxicity), phagocytosis, secretion, glutathione levels, proliferation, drug and hormone receptors, and cell tracing.

The most common physiologically relevant measurement of this type has been estimation of the flux of Ca^{2+} ions following cell stimulation. Indo-1 (Grynkiewicz *et al.*, 1985) has been the dye of choice for flow cytometry (Davies *et al.*, 1990) because it can be excited by the UV lines of the argon laser and its Ca^{2+}-dependent emissions (Fig. 12) can be separately collected (usually at about 405 and 480 nm). The ratio of these intensities can be related to the intracellular Ca^{2+} concentration. Dyes such as Fura-2—which is widely used in imaging of Ca^{2+} transients—are not very useful in flow cytometry because they do not undergo this emission shift. Of course, many laboratories do not have the UV argon laser that is required to use indo-1 as a Ca^{2+} indicator. Fluo-3 is a calcium indicator whose fluorescence can be excited by the argon laser at 488 nm, but its fluorescence intensity simply increases on Ca^{2+} mobilization with no emission shift (Minta *et al.*, 1989). Thus the intensity depends on the degree of cell loading and other factors and the response of $[Ca^{2+}]_i$ to a stimulant is difficult to calibrate. Fluo-3 (and Calcium Green) (Fig. 13) are still useful indicators for measuring the kinetics of a response in a relatively uniform population of cells by doing a "time-of-flight" experiment in which the stimulus is added immediately before the flow is started and the Ca^{2+}-dependent signal

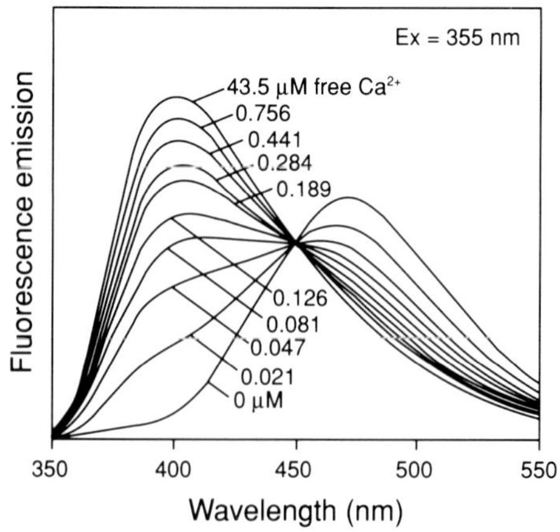

Fig. 12 Fluorescence emission spectra of indo-1 in aqueous solutions with defined free Ca^{2+} concentrations as indicated.

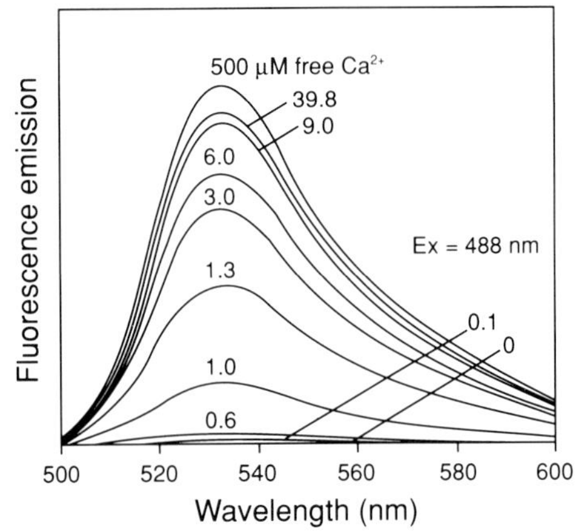

Fig. 13 Fluorescence emission spectra of Calcium Green-2 in aqueous solutions with defined free Ca^{2+} concentrations as indicated.

is dependent on the flow rate (Vandenberghe and Ceuppens, 1990). Although the ideal probe would be an indicator whose Ca^{2+}-dependent emission wavelengths shift on Ca^{2+} binding, a partial alternative has been to simultaneously load two Ca^{2+}-responsive indicators—fluo-3 and Fura Red—into cells and to use their combined response (Fig. 14) to estimate the intracellular Ca^{2+} concentration (Lipp and Nigli, 1993). Fluo-Rhod-2, an emission shifting Ca^{2+} indicator that can be excited at 488 nm, was recently described; however, it has not yet been tested in cells (Smith *et al.*, 1993).

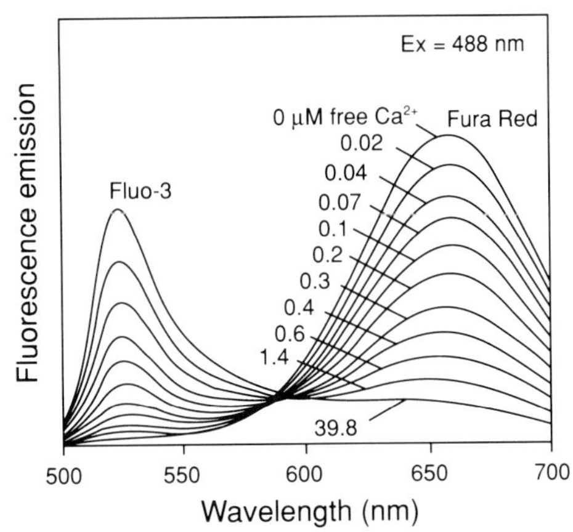

Fig. 14 Fluorescence emission spectra of fluo-3 and Fura Red, both excited at 488 nm, in aqueous solutions with defined free Ca^{2+} concentrations as indicated. The concentration of Fura Red in the sample is about 10 times that of fluo-3.

A number of laboratories have measured cytoplasmic pH, sometimes in combination with Ca^{2+} or cell-surface markers. Valet has used dicyanohydroquinone for this purpose, but it must be excited in the UV (Rothe and Valet, 1988). The pH indicator carboxy SNARF-1 is easily excited at 488 nm and undergoes an emission shift to shorter wavelengths on acidification (Fig. 15). This dye has frequently been used for flow cytometric measurements of intracellular pH (Van Graft *et al.*, 1993); Ziegelstein *et al.*, 1992; van Erp *et al.*, 1991); however, its pK_a (about 7.5) may be too high for some applications. BCECF (pK_a about 6.98), which is widely used for imaging measurements of intracellular pH, has also been used in flow cytometry, despite its lack of an emission shift with pH (Davies *et al.*, 1990).

The original fluorescent indicator for sodium, SBFI (Minta and Tsien, 1989), cannot be excited at long wavelengths. A new sodium-selective indicator developed by Molecular Probes, Sodium Green, can be excited at 488 nm and has a response to Na^+ (Fig. 16) that mimics the response of Calcium Green-2 to Ca^{2+} (Fig. 13).

Although a number of dyes have been tested for flow cytometric estimation of plasma membrane potential, most of the common dyes such as the carbocyanines have the problem of not being able to distinguish between plasma and mitochondrial membrane potential. Consequently, anionic dyes such as the bis-oxonol, $DiBaC_4(3)$, are probably the most suited for measurements of plasma membrane potentials (Wilson and Chused, 1985).

Detection of enzymatic activity in cells remains an almost untapped area for discriminating types of cells and their physiological states. Part of this problem has been that the common fluorogenic substrates form fluorescent products

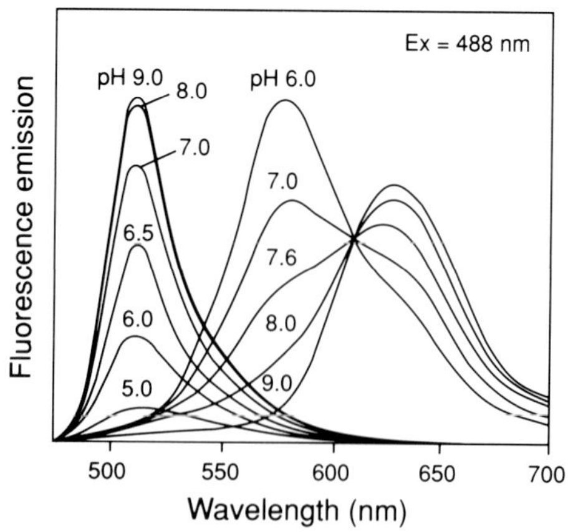

Fig. 15 Fluorescence emission spectra of fluorescein and carboxy SNARF-1 excited at 488 nm, as a function of pH. The peak wavelength for fluorescein emission (514 nm), is pH independent, whereas for carboxy SNARF-1 the emission maximum shifts from 585 to 635 nm with increasing pH.

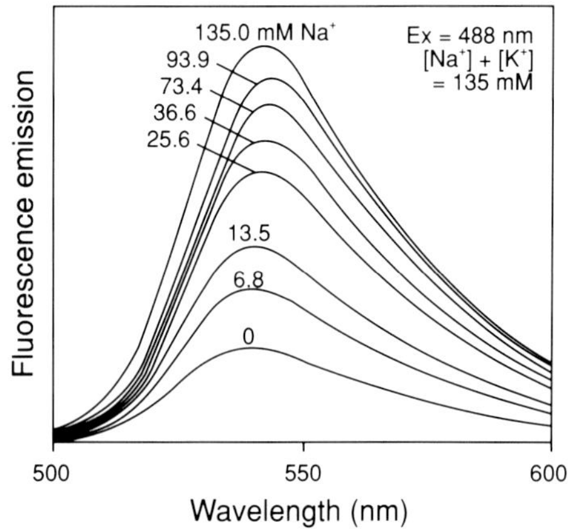

Fig. 16 Fluorescence emission response of Sodium Green in aqueous solutions with defined free Na^+ concentrations as indicated. The total $[Na^+ + K^+]$ concentration of each solution is maintained at 135 m*M* by making compensating changes in the K^+ concentration.

that rapidly leak from cells. Scientists at Molecular Probes have used formation of lipophilic fluorescent products from the fluorogenic substrate, reaction of the fluorescent product with intracellular glutathione, precipitation of a fluorescent product, or targeting of the product to organelles to yield fluorescent products that stay associated with the same cell in which the product is formed by enzymatic action (Haugland, 1992). Permeability of the substrate through the membrane can still be a problem. Also, there is difficulty in making quantitative measurements of enzymatic activity inside cells under potentially nonsaturating concentrations of substrate.

Probably the most extensively used fluorogenic substrates are those that can monitor oxidative bursts in neutrophils. 2′,7′-Dichlorodihydrofluorescein diacetate (H_2DCF-DA) is a colorless dye that penetrates membranes and is cleaved to nonfluorescent dichlorodihydrofluorescein by cytoplasmic nonspecific esterases. Superoxide release following stimulation results in formation of the highly fluorescent 2′,7′-dichlorofluorescein (Bass *et al.,* 1983). By chemically coupling this substrate to an immune complex the kinetics of Fc receptor-mediated internalization and substrate oxidation in the phagovacuole can be followed (Ryan *et al.,* 1990). A similar probe is dihydrorhodamine 123, which is oxidized to rhodamine 123 during respiratory bursts; the product accumulates in the mitochondria (Rothe *et al.,* 1988). Dihydroethidium is a chemically reduced form of ethidium bromide. Upon intracellular oxidation the product accumulates in cells by staining nucleic acids (Rothe and Valet, 1990).

Nolan and Herzenberg have been leading developers of the use of flow cytometry to detect genetic transformation through the incorporation of the *lacZ* gene into eukaryotic cells (Roederer *et al.,* 1991). This gene codes for the marker enzyme β-galactosidase. They have used the colorless nonfluorescent substrate, fluorescein digalactoside (FDG), to detect *lacZ*-positive cells. However, hydrolysis of FDG yields fluorescein, which rapidly leaks from cells and makes it necessary to do the flow cytometric analysis at 4°C. C_{12}FDG, a lipophilic analogue of FDG (Zhang *et al.*, 1991), has recently been used for flow cytometric analysis and sorting of low-abundance transfected human sperm (Jasin and Zalamea, 1992). Also, 5-chloromethylfluorescein digalactoside (CMFDG) has properties similar to those of FDG, but its reactivity with intracellular glutathione permits the analysis for β-galactosidase activity to be done at 37°C (Haugland, 1992).

Recently Valet and collaborators have prepared several fluorogenic peptidase substrates based on rhodamine 110, a dye whose spectral properties are similar to those of fluorescein. They have shown that the fluorescent hydrolysis product stays associated with live cells for a period sufficient for the analysis (Banati *et al.*, 1993).

Phospholipase A activity is another marker of stimulus–response coupling. Using a synthetic bis-BODIPY phospholipid the activation of this enzyme can be measured by flow cytometry in live cells (Meshulam *et al.*, 1992).

A common application of fluorescent probes has been to determine the membrane integrity of live cells and the associated phenomenon of cell lysis. Usually this has been accomplished using a red fluorescent membrane-impermeant nucleic acid stain such as propidium iodide or ethidium homodimer to detect cells in which the membrane integrity has been lost ("dead cells") and a green fluorescent dye such as BCECF AM or calcein AM that is reasonable well retained in live cells, but which leaks out of cells on loss of membrane integrity (Radcliff *et al.*, 1991). This approach has been used to measure the interaction between effector lymphocytes and their target cells by flow cytometry (Callewaert *et al.*, 1991).

Cell tracers are fluorescent probes that stay associated with the cell through cell division and are apparently not toxic to cells. They can be used to mark a population for future study, to follow cell division, for "homing" of lymphocytes, and for other applications. In general two types of markers have been used: lipophilic tracers and cytoplasmic tracers. The lipophilic tracers are exemplified by the lipophilic carbocyanines DiI and DiO (Gant *et al.*, 1992) and also by several "PKH" dyes developed by Dr. Paul K. Horan and collaborators at Zynaxis Cell Sciences that have been used for *in vivo* tracking of hematopoietic cells (Horan and Slezak, 1989). General viability probes such as BCECF AM and calcein AM are not sufficiently retained in cells to survive through cell division. However, certain chemically reactive fluoresceins such as chloromethylfluorescein diacetate (CMFDA) react with intracellular glutathione and other thiols and stay in the cell through several cell divisions (Zhang *et al.*, 1992a). Labeling of one type of cell with CMFDA and a second type with a similar rhodamine-based reactive tracer permits analysis and sorting of cells following fusion (Zhang *et al.*, 1992b).

Besides being useful to trap fluorescent products as described above, glutathione is an important intracellular scavenger for reactive molecules and is also essential for maintaining the oxidation—reduction state inside cells. Measurement of glutathione levels by flow cytometry, however, is complicated by its very high concentration—often 1 m*M* or higher—inside cells. Glutathione transferase, which chemically couples the dye to glutathione, also consists of a mixture of isoenzymes that have widely differing substrate specificity. None of the probes that has been used to measure glutathione is completely satisfactory because they cannot be used under conditions of concentration and reaction kinetics where their rate of reaction totally reflects the glutathione concentration. The dyes may also react with other intracellular thiols. Monobromobimane and monochlorobimane are the most widely used probes for this purpose, with both becoming fluorescent on reaction with intracellular thiols, including glutathione (Ublacker *et al.,* 1991). However, bimanes must be excited in the UV and so use of the 488-nm excitable CMFDA for this application is now being explored (Barhoumi *et al.,* 1993).

There also remains a significant potential to study cell-surface receptors for drugs, peptides, growth factors, and hormones on live cells by flow cytometry by using fluorescent analogues of the natural or synthetic products. However, relatively few laboratories are doing so at this time. Among the receptors in live cells that have been labeled with fluorescent probes are those for chemotactic peptides (Finney and Sklar, 1983), transferrin (Torres *et al.,* 1990), low-density lipoproteins (Hassall, 1992), epidermal growth factor (Chatelier *et al.,* 1986), glucose transporter (Rauchmann *et al.,* 1992), and methotrexate (Assaraf *et al.,* 1989).

VI. Summary

Flow cytometry uses a relatively small set of dyes for immunochemistry and nucleic acid detection, most of which have been known and used reliably in flow cytometry for several years. These can usually be combined to make simultaneous two-color measurements of multiple cell-surface antigens and nucleic acid content. Because of the overlap of dye spectra and difficulties in finding dyes with substantial Stokes shifts that can be excited by the argon laser, simultaneous three-color or more detection can be more difficult. A basic knowledge of the factors that go into producing the fluorescent signal, including the spectra of dyes and their overlap, is necessary in planning multicolor experiments.

By contrast, there have been a number of new fluorescent probes developed for detecting ions, membrane potential, metabolism, organelles, and other properties of living cells as well as for determining cell viability, proliferation, and cell tracking. So far most of these physiological probes are being used only for fundamental research rather than for cell classification. However, as research activity expands in this area, its diagnostic potential is likely to be increasingly appreciated.

References

Assaraf, Y. G., Molina, A., and Schimke, R. T. (1989). *Anal. Biochem.* **178,** 287–293.

Banati, R. B., Rothe, G., Valet, G., and Kreutzberg, G. W. (1993). *Glia* **7,** 183–191.

Barhoumi, R., Bowen, J. A., Stein, L. S., Echols, J., and Burghardt, R. C. (1993). *Cytometry* **14,** 747–756.

Bass, D. A., Parce, J. W., Dechatelet, L. R., Szejda, P., Seeds, M. C., and Thomas, M. (1983). *J. Immunol.* **130,** 1910–1917.

Blakeslee, D., and Baines, M. G. (1976). *J. Immunol. Methods* **13,** 305–320.

Callewaert, D. M., Radcliff, G., Waite, R., LeFevre, J., and Poulik, M. D. (1991). *Cytometry* **12,** 666–676.

Castro, A., Fairfield, F. R., and Shera, E. B. (1993). *Anal. Chem.* **65,** 849–852.

Chatelier, R. C., Ashcroft, R. C., Lloyd, C. J., Nice, E. C., Whitehead, R. H., Sawyer, W. H., and Burgess, A. W. (1986). *EMBO J.* **5,** 1181–1186.

Cupp, J. E., Leary, J. F., Cernichiari, E., Woods, J. C. S., and Doherty, R. A. (1984). *Cytometry* **5,** 138–144.

Darzynkiewicz, Z., Kapuscinski, J., Traganos, F., and Crissman, H. A. (1987). *Cytometry* **8,** 138–145.

Davies, T. A., Weil, G. J., and Simons, E. R. (1990). *J. Biol. Chem.* **265,** 1522–1526.

Devlin, R., Studholme, R. M., Dandliker, W. B., Fahy, E., Blumeyer, K., and Ghosh, S. S. (1993). *Pap., 25th Annu. Oak Ridge Conf.: Return to the Future,* Oak Ridge, TN.

Festin, R., Bjorkland, A., and Totterman, T. H. (1990). *J. Immunol. Methods* **126,** 69–78.

Finney, D. A., and Sklar, L. A. (1983). *Cytometry* **4,** 54–60.

Gant, V. A., Shakoor, Z., and Hamblin, A. S. (1992). *J. Immunmol. Methods* **156,** 179–189.

Glazer, A. N., and Stryer, L. (1983). *Biophys. J.* **43,** 383–386.

Goodwin, P. M., Johnson, M. E., Martin, J. C., Ambrose, W. R., Marrone, B. L., Jett, J. H., and Keller, R. A. (1993). *Nucleic Acids Res.,* **21,** 803–806.

Grynkiewicz, G., Poenie, M., and Tsien, R. Y. (1985). *J. Biol. Chem.* **260,** 3440–3450.

Hassall, D. G. (1992). *Cytometry* **13,** 381–388.

Haugland, R. P. (1990). *In* "Optical Microscopy for Biology" (B. Herman and K. Jacobson, eds.), pp. 143–157. Wiley, New York.

Haugland, R. P. (1992). "Handbook of Fluorescent Probes and Research Chemicals." Molecular Probes, Eugene, OR.

Hirons, G. T., and Crissman, H. A. (1993). *Cytometry, Suppl.* **6,** 87.

Horan, P. K., and Slezak, S. E. (1989). *Nature (London)* **340,** 167–168.

Horan, P. K., Muirhead, K. A., and Slezak, S. E. (1990a). *In* "Flow Cytometry and Sorting" (M. R. Melamed, T. Lindmo, and M. L. Mendelsohn, eds.), 2nd ed., pp. 397–414. Wiley-Liss, New York.

Horan, P. K., Melnicoff, M. J., Jensen, B. D., and Slezak, S. E. (1990b). *In* "Methods in Cell Biology" (Z. Darzynkiewicz and H. A. Crissman, eds.), Vol. 33, pp. 469–490. Academic Press, San Diego.

Jasin, M., and Zalamea, P. (1992). *Proc. Natl. Acad. Sci. U.S.A.* **89,** 10681–10685.

Kapuscinski, J., and Darzynkiewicz, Z. (1987). *Cytometry* **8,** 129–137.

Kapuscinski, J., and Darzynkiewicz, Z. (1980). *In* "Methods in Cell Biology" (Z. Darzynkiewicz and H. A. Crissman, eds.), Vol. 33, pp. 655–669. Academic Press, San Diego.

Kapuscinski, J., Darzynkiewicz, Z., and Melamed, M. R. (1982). *Cytometry* **2,** 201–211.

Khalfan, H., Abuknesha, P., Rand-Weaver, M., Price, R. G., and Robinson, D. (1986). *Histochem. J.* **18,** 497–499.

Kronick, M. N. (1986). *J. Immunol. Methods* **92,** 1–13.

Lee, L. G., Chen, C.-H., and Chiu, L. A. (1986). *Cytometry* **7,** 508–517.

Lipp, P., and Niggli, E. (1993). *Cell Calcium* **14,** 359–372.

Meshulam, T., Herscovitz, H., Casavant, D., Bernardo, J., Roman, R., Haugland, R. P., Strohmeier, G. S., Diamond, R. D., and Simons, E. R. (1992). *J. Biol. Chem.* **267,** 1465–1470.

Millard, P. J., Wells, S., Yue, S., and Roth, B. (1993). *Abstr., 93rd Gen. Meet. Am. Soc. Microbiol.,* p. 391.

Minta, A., and Tsien, R. Y. (1989). *J. Biol. Chem.* **264,** 19449–19457.

Minta, A., Kao, J., and Tsien, R. Y. (1989). *J. Biol. Chem.* **264,** 8171–8178.

Oi, V. T., Glaser, A. N., and Styer, L. (1982). *J. Cell Biol.* **93,** 981–984.

Radcliff, G., Waite, R., Lefevre, J., Poulik, M. D., and Callewaert, D. M. (1991). *J. Immunol. Methods* **139,** 281–292.

Rauchman, M. I., Wasserman, J. C., Cohen, D. M., Perkins, D. L., Hebert, S. C., Milford, E., and Gullans, S. R. (1992). *Biochim. Biophys. Acta* **1111,** 231–238.

Recktenwald, D. J. (1989). U.S. Pat. 4,876,190.

Recktenwald, D. J. (1990). *Proc. SPIE—Int. Soc. Opt. Eng.* **1206,** 106–111.

Roederer, M., Fiering, S., and Herzenberg, L. A. (1991). *Methods: Comp. Methods Enzymol.* **2,** 248–261.

Rothe, G., and Valet, G. (1988). *Cytometry* **9,** 316–324.

Rothe, G., and Valet, G. (1990). *J. Leukocyte Biol.* **47,** 440–448.

Rothe, G., Oser, A., and Valet, G. (1988). *Naturwissenschaften* **75,** 354–355.

Ryan, T. C., Weil, G. J., Newburger, P. E., Haugland, R., and Simons, E. R. (1990). *J. Immunol. Methods* **130,** 223–233.

Smith, G. A., Metcalfe, J. C., and Clarke, S. D. (1993). *J. Chem. Soc., Perkin Trans.* **2,** pp. 1195–1204.

Southwick, P. L., Ernst, L. A., Tauriello, E. W., Parker, S. R., Mujumdar, R. B., Mujumdar, S. R., Clever, H. A., and Waggoner, A. S. (1990). *Cytometry* **11,** 418–430.

Terstappen, L. W. M. M., Shah, V. O., Conrad, M. P., Recktenwald, D., and Loken, M. R. (1988). *Cytometry* **9,** 477–484.

Toba, K., Winton, E. F., and Bray, R. A. (1992). *Cytometry* **13,** 60–67.

Torres, J. M., Esteban, C., Aguilar, J., Mishai, Z., and Uriel, J. (1990). *J. Immunol. Methods* **134,** 163–170.

Truneh, A., and Machy, P. (1987). *Cytometry* **8,** 562–567.

Ublacker, G. A., Johnson, J. A., Siegel, F. L., and Mucahy, R. T. (1991). *Cancer Res.* **51,** 1783–1788.

Vandenberghe, P. A., and Ceuppens, J. L. (1990). *J. Immunol. Methods* **127,** 197–205.

van Erp, P. E. J., Jansen, M. J. J. M., de Jongh, G. J., Boezeman, J. B. M., and Schalkwijk, J. (1991). *Cytometry* **12,** 127–132.

Van Graft, M., Kraan, Y. M., Seggers, I. M. J., Radošević, K., De Grooth, B. G., and Greve, J. (1993). *Cytometry* **14,** 257–264.

Wallner, G., Amann, R., and Beisker, W. (1993). *Cytometry* **14,** 136–143.

Weber, G., and Teale, F. W. J. (1957). *Trans. Faraday Soc.* **6,** 465–471.

Whitaker, J. E., Haugland, R. P., Moore, P. L., Hewitt, P. C., Reese, M., and Haugland, R. P. (1991). *Anal. Biochem.* **198,** 119–130.

Wilson, H. A., and Chused, T. M. (1985). *J. Cell. Physiol.* **125,** 72–81.

Yeh, S. W., Ong, L. J., Clark, J. H., and Glazer, A. N. (1987). *Cytometry* **8,** 91–95.

Zhang, Y-Z., Naleway, J. J., Larison, K. D., Huang, Z., and Haugland, R. P. (1991). *FASEB J.* **5,** 3108–3113.

Zhang, Y.-Z., Kang, H. C., Kuhn, M., Roth, B., and Haugland, R. P. (1992a). *Mol. Biol. Cell* **3,** 90a.

Zhang, Y.-Z., Olson, N., Mao, F., Roth, B., and Haugland, R. P. (1992b). *FASEB J.* **6,** A1835.

Ziegelstein, R. C., Cheng, L., and Capogrossi, M. C. (1992). *Science* **258,** 656–659.

INDEX

B

D

M

T

U

V

W

Y

Z

VOLUMES IN SERIES

Founding Series Editor
DAVID M. PRESCOTT

Advisory Board Chairman
KEITH R. PORTER

Volume 21B (1980)
Normal Human Tissue and Cell Culture, Part B: Endocrine, Urogenital, and Gastrointestinal Systems
Edited by Curtis C. Harris, Benjamin F. Trump, and Gary D. Stoner

Volume 22 (1981)
Three-Dimensional Ultrastructure in Biology
Edited by James N. Turner

Volume 23 (1981)
Basic Mechanisms of Cellular Secretion
Edited by Arthur R. Hand and Constance Oliver

Volume 24 (1982)
The Cytoskeleton, Part A: Cytoskeletal Proteins, Isolation and Characterization
Edited by Leslie Wilson

Volume 25 (1982)
The Cytoskeleton, Part B: Biological Systems and *in Vitro* Models
Edited by Leslie Wilson

Volume 26 (1982)
Prenatal Diagnosis: Cell Biological Approaches
Edited by Samuel A. Latt and Gretchen J. Darlington

Series Editor
LESLIE WILSON

Volume 27 (1986)
Echinoderm Gametes and Embryos
Edited by Thomas E. Schroeder

Volume 28 (1987)
***Dictyostelium discoideum:* Molecular Approaches to Cell Biology**
Edited by James A. Spudich

Volume 29 (1989)
Fluorescence Microscopy of Living Cells in Culture, Part A: Fluorescent Analogs, Labeling Cells, and Basic Microscopy
Edited by Yu-Li Wang and D. Lansing Taylor

Volume 30 (1989)
Fluorescence Microscopy of Living Cells in Culture, Part B: Quantitative Fluorescence Microscopy—Imaging and Spectroscopy
Edited by D. Lansing Taylor and Yu-Li Wang

Volume 31 (1989)
Vesicular Transport, Part A
Edited by Alan M. Tartakoff

Volume 32 (1989)
Vesicular Transport, Part B
Edited by Alan M. Tartakoff

Volume 33 (1990)
Flow Cytometry
Edited by Zbigniew Darzynkiewicz and Harry A. Crissman

Volume 34 (1991)
Vectorial Transport of Proteins into and across Membranes
Edited by Alan M. Tartakoff

Selected from Volumes 31, 32, and 34 (1991)
Laboratory Methods for Vesicular and Vectorial Transport
Edited by Alan M. Tartakoff

Volume 35 (1991)
Functional Organization of the Nucleus: A Laboratory Guide
Edited by Barbara A. Hamkalo and Sarah C. R. Elgin

Volume 36 (1991)
***Xenopus laevis:* Practical Uses in Cell and Molecular Biology**
Edited by Brian K. Kay and H. Benjamin Peng

Series Editors
LESLIE WILSON AND PAUL MATSUDAIRA

Volume 37 (1993)
Antibodies in Cell Biology
Edited by David J. Asai

Volume 38 (1993)
Cell Biological Applications of Confocal Microscopy
Edited by Brian Matsumoto

Volume 39 (1993)
Motility Assays for Motor Proteins
Edited by Jonathan M. Scholey

Volume 40 (1994)
A Practical Guide to the Study of Calcium in Living Cells
Edited by Richard Nuccitelli

Volume 41 (1994)
Flow Cytometry, Second Edition, Part A
Edited by Zbigniew Darzynkiewicz, J. Paul Robinson, and Harry A. Crissman

Volume 42 (1994)
Flow Cytometry, Second Edition, Part B
Edited by Zbigniew Darzynkiewicz, J. Paul Robinson, and Harry A. Crissman

Volume 43 (1994)
Protein Expression in Animal Cells
Edited by Michael G. Roth

Volume 44 (1994)
***Drosophila melanogaster:* Practical Uses in Cell and Molecular Biology**
Edited by Lawrence S. B. Goldstein and Eric A. Fyrberg

Volume 45 (1994) (in preparation)
Microbes as Tools for Cell Biology
Edited by David G. Russell

Volume 46 (1995) (in preparation)
Cell Death
Edited by Lawrence M. Schwartz and Barbara A. Osborne